Handbook of GaN Semiconductor Materials and Devices

SERIES IN OPTICS AND OPTOELECTRONICS

Series Editors: **E Roy Pike**, Kings College, London, UK
Robert G W Brown, University of California, Irvine, USA

Handbook of GaN Semiconductor Materials and Devices

Edited by

Wengang (Wayne) Bi
State Key Laboratory of Reliability and Intelligence of Electrical Equipment
Hebei University of Technology
and
School of Electronic and Information Engineering
College of Information and Electrical Engineering
Hebei University of Technology
Tianjin, People's Republic of China

Hao-Chung (Henry) Kuo
Department of Photonics
and
Institute of Electro-Optical Engineering
and
Institute of Photonic System
College of Photonics
National Chiao Tung University
Hsinchu, Taiwan

Pei-Cheng Ku
Department of Electrical Engineering
and Computer Science
Department of Applied Physics
Department of Macromolecular Science and Engineering
University of Michigan
Ann Arbor, Michigan

Bo Shen
Research Center for Wide Gap Semiconductors
School of Physics
Peking University
Beijing, People's Republic of China

CRC Press
Taylor & Francis Group
Boca Raton London New York

CRC Press is an imprint of the
Taylor & Francis Group, an **informa** business

CRC Press
Taylor & Francis Group
6000 Broken Sound Parkway NW, Suite 300
Boca Raton, FL 33487-2742

First issued in paperback 2019

© 2018 by Taylor & Francis Group, LLC
CRC Press is an imprint of Taylor & Francis Group, an Informa business

No claim to original U.S. Government works

ISBN-13: 978-1-4987-4713-4 (hbk)
ISBN-13: 978-0-367-87531-2 (pbk)

Library of Congress Cataloging-in-Publication Data

Names: Bi, Wengang, editor.
Title: Handbook of GaN semiconductor materials and devices / Wengang (Wayne) Bi, Hao-chung (Henry) Kuo, Pei-Cheng Ku, Bo Shen, editors.
Description: Boca Raton : Taylor & Francis, CRC Press, 2017. | Series: Series in optics and optoelectronics | Includes bibliographical references.
Identifiers: LCCN 2017011253 | ISBN 9781498747134 (hardback : alk. paper)
Subjects: LCSH: Semiconductors--Handbooks, manuals, etc. | Gallium nitride--Handbooks, manuals, etc.
Classification: LCC TK7871.15.G33 H36 2017 | DDC 621.3815/2--dc23
LC record available at https://lccn.loc.gov/2017011253

Visit the Taylor & Francis Web site at
http://www.taylorandfrancis.com

and the CRC Press Web site at
http://www.crcpress.com

Contents

SECTION I Fundamentals

SECTION II Growth and Processing

SECTION III   Power Electronics

SECTION IV   Light Emitters

SECTION V   Emerging Applications

Series Preface

This international series covers all aspects of theoretical and applied optics and optoelectronics. Active since 1986, eminent authors have long been choosing to publish with this series, and it is now established as a premier forum for high-impact monographs and textbooks. The editors are proud of the breadth and depth showcased by published works, with levels ranging from advanced undergraduate and graduate student texts to professional references. Topics addressed are both cutting edge and fundamental, basic science and applications-oriented, on subject matter that includes lasers, photonic devices, nonlinear optics, interferometry, waves, crystals, optical materials, biomedical optics, optical tweezers, optical metrology, solid-state lighting, nanophotonics, and silicon photonics. Readers of the series are students, scientists, and engineers working in optics, optoelectronics, and related fields in the industry.

Proposals for new volumes in the series may be directed to Lu Han, senior publishing editor at CRC Press, Taylor & Francis Group (lu.han@taylorandfrancis.com).

Foreword

III-nitride semiconductors and devices have emerged as a rising star in numerous electronic and optoelectronic applications. The breakthrough in epitaxial growth and p-type doping back in the early 1990s transformed the once unremarkable material into one of the most important for the worldwide semiconductor industry. The unique properties of GaN, AlN, InN, and related compounds presented us with both opportunities and challenges. Their large energy bandgaps enabled highly energy efficient ultraviolet and visible wavelength lasers, and pyro- and piezoelectric properties have been utilized to generate two-dimensional electron gas in high-power, high-frequency, high-electron mobility transistors. Light-emitting diodes (LEDs), in particular, promise a revolution in lighting efficiency unprecedented in the past hundred years. The ubiquity of LEDs in flat-panel displays and increasing popularity in general lighting, the advent of visible and ultraviolet lasers for optical data storage, and the superior power-handling performance of GaN-based transistors are just a few examples of how III-nitride materials and devices have been utilized in many fields.

Thanks to research and development efforts in both industry and academia in the past two decades, I believe we are just seeing the beginning of the GaN revolution. There are still many challenges to overcome before we can fully uncover the potential of III-nitride materials. These include fundamental understandings of the droop mechanism, continuing advancement of epitaxial technologies for both short-wavelength and long-wavelength applications, as well as the development of better substrates for improved materials quality and device performance. However, holistic approaches toward system-level solutions are needed as well. For example, the use of III-nitride lasers instead of LEDs for lighting can be just as efficient but without the issue of droop. It was the persistent efforts by only a handful of researchers in the early days that brought about the revolution we enjoy today. It is equally important that we keep the enthusiasm going forward, which is why I think this book serves an important role for this endeavor.

This book is devoted to the latest development of III-nitride materials and technologies. It compiles the latest results in materials growth, understanding of electronic and optical properties, and device physics and architectures from experts in this field. These developments are important as they not only continue to improve the performance of existing devices but establish a solid foundation to new capabilities and new applications of III-nitride materials. This book strikes a good balance between fundamental materials science and technology developments with diverse topics ranging from epitaxial growth of high-aluminum and high-indium composition III-nitride thin-film and nanostructural materials to development of native GaN substrates, from fundamental understandings of electronic and optical properties to processing techniques, which are of central importance to improve device performance, and from a continued push to new emitter wavelengths to new frontiers in solar cells, intersubband (ISB) optoelectronics, lighting communications, and quantum light emitters. These topics and their promising results reassert my belief that the potential of nitride materials and technologies are only limited by our creativity.

I believe this book will be a great resource to researchers and engineers, in both academia and industry, to learn and be inspired by the latest developments and potentials of III-nitride materials and devices. I thank the editors and CRC Press/Taylor & Francis Group for their efforts in putting everything together. I am happy to contribute this Foreword, and I sincerely wish the success of this book.

Shuji Nakamura
University of California, Santa Barbara
2014 Nobel Laureate in Physics

Preface

Gallium nitride (GaN) and related alloys or GaN materials in short possess remarkable electronic and optoelectronic properties, unparalleled in other semiconductors, including an extremely tunable direct bandgap from 0.7 to 6.2 eV, a large exciton binding energy, a high-electron mobility ($>10^4$ cm^2/V s in InN), and a large saturation velocity (10^8 cm/s in InN). The research and development of the GaN-based devices in the past 25 years have created an enormous impact on our daily life, leading to a revolution in energy-efficient lighting and electronic displays, resilient power electronic systems for vehicle electrification, and ultracompact semiconductor lasers in the ultraviolet, blue, and green wavelengths for data storage as well as many scientific and industrial applications. However, it has also become clear to the research community that what has been unearthed is only the tip of the iceberg. Many potential applications of GaN have yet to be realized. These include—but are not limited to—deep ultraviolet, deep visible, and near-infrared light-emitting diodes (LEDs) and lasers; next-generation power transistors; sensors that operate in harsh environments; and sources for quantum photonics. Grand challenges must be overcome to achieve breakthroughs in these emerging applications and require continued push on several research frontiers, including epitaxial growths, materials science, manufacturing technologies, characterization techniques, device architecture, and modules as well as system optimizations. This book is devoted to gathering the latest developments in these areas, presented by renowned researchers from both academia and industry.

The book is organized into sections with a total of 22 chapters. Thanks to our contributors, the breadth and depth of this book are incredible. Section I (Chapters 1 through 4) focuses on the fundamental properties of GaN materials, encompassing their structural, electronic, and optical properties. Section II (Chapters 5 through 8) is devoted to the latest developments in GaN and related material growth and device fabrication technologies. Section III (Chapters 9 through 12) covers research frontiers in the area of GaN power electronics. Section IV (Chapters 13 through 18) presents the state-of-the-art developments of GaN-based light emitters including both LEDs and lasers. Section V (Chapters 19 through 22) includes a diverse range of emerging applications of GaN-based materials and devices. We think readers from both academia and industry will find this book beneficial. We also put our best efforts to make each chapter independent on its own so that readers who are specifically interested in certain topics can easily navigate through selected chapters across different parts with ease.

A more detailed overview of this book is following. Updates on the fundamental properties of GaN materials will be presented in Section I. In Chapter 1, Shen et al. give a comprehensive overview of the basic physical properties of GaN materials and discuss various growth and characterization techniques. In Chapter 2, Ponce discusses in more detail on the microstructures and polarization properties of GaN materials, including the evolution of microstructures, characteristics of dislocations, and polarization effects. In Chapter 3, Paskov and Monemer give an in-depth overview for the optical properties of GaN and related materials, including both band-to-band absorption and emission as well as emission properties from structural defects. In Chapter 4, Wu will review electronic band structures and carrier transport in GaN materials and how they are influenced by strain, alloy, and quantization.

Section II presents the latest developments in GaN and related material growth and processing technologies, which have been the key to the successful commercialization of GaN devices. In Chapter 5, Paskova et al. review recent advances in substrate technologies and discuss the relationship between growth techniques and substrates. In Chapter 6, Dr. Koleske gives a comprehensive overview on metal–organic vapor-phase epitaxy of GaN materials with an emphasis on the chemistry, growth mechanisms, and material properties that transformed a once backwater research topic into the ultimate resolution of solid-state lighting industry we have witnessed in recent years. In Chapter 7, Mi et al. discuss recent advances made in molecular beam epitaxy of GaN nanostructures, which have helped pushing GaN materials toward both deep visible and deep ultraviolet wavelengths. In Chapter 8, Jiang describes the state-of-the-art GaN device fabrication technologies with a focus on LEDs.

Sections III through V focus on devices and applications of GaN materials, including power electronics (Section III), light emitters (Section IV), and emerging applications (Section V). Section III covers a wide range of topics in GaN power devices, including material properties, device physics, and device architecture as well as reliability. In Chapter 9, Huang et al. review the fundamental properties of GaN-based heterostructures for electronic devices, including both depletion and enhancement mode transistors. In Chapter 10, Chowdhury presents the latest developments of GaN power devices, including vertical transistors. In Chapter 11, Chen and Yang discuss the status and challenges of GaN transistors on silicon. In Chapter 12, Zanoni et al. present an overview of the characterization methodologies and the latest results on the understanding of parasitic effects and reliability of GaN power devices.

Light emitters, including both LEDs and lasers, are perhaps the most researched applications for GaN materials. As Nakamura, the Nobel laureate in Physics in 2014 for his seminal contributions in GaN LEDs, points out in the Foreword, the breakthroughs in epitaxial growth and p-type doping back in the early 1990s led to the revolution in GaN-based LEDs and lasers. In Section IV, we gather six chapters to present the latest developments in this area. In Chapter 13, Zhang et al. summarize recent progress in continuing to push for a higher efficiency for GaN LEDs and introduce several novel concepts recently developed in this area. In Chapter 14, Wang et al. report the state-of-the-art white light emitters based on blue LEDs. In Chapter 15, Dr. J. Wang et al. summarize the progress in the growth and fabrication of indium-rich InGaN green LEDs with discussions on various substrate orientations. At the shorter end of the spectrum, Kim et al. present the challenges facing UV LEDs and recent efforts to overcome them in Chapter 16. Chapters 17 and 18 include the latest developments in GaN-based lasers. In Chapter 17, Bhattacharya et al. pave a new ground with low-threshold red-emitting GaN lasers by using the quantum dot to replace a more conventional quantum-well active region. In Chapter 18, Lu et al. present the latest developments in GaN surface-emitting lasers and their unique challenges as well as properties compared to their III–V counterparts.

GaN materials possess unique optical and electronic properties. Section V covers a series of emerging yet exciting opportunities for GaN-related materials and devices. In Chapter 19, Lin et al. discuss the potential application of GaN in photovoltaic cells by exploiting the small bandgap of InN and bandgap tunability of InGaN. In Chapter 20, Monroy et al. describe optoelectronic devices based on intersubband transitions in GaN. Unique applications toward ultrafast near-infrared devices and room-temperature terahertz lasers are discussed. In Chapters 21 and 22, the applications of GaN light emitters for novel forms of data networks are presented. In Chapter 21, Lin et al. discuss free-space communication using GaN transmitters. In Chapter 22, Ku et al. present scalable GaN quantum light emitters exploiting the large exciton binding energy in GaN.

We take this opportunity to thank all the authors, professors, researchers, and students for their contributions. Without them, this book would not be possible. We are grateful for their efforts in delivering the manuscripts in a very tight schedule. We hope this book will not only be useful for the GaN research community, but most importantly will benefit students who are just entering this field. It is by no means an encyclopedia, but the diverse range of topics will likely stimulate the next-generation engineers and scientists to discover something new. Seasoned engineers should also find new insight and opportunity presented by the latest developments of GaN devices.

We also thank Prof. Shuji Nakamura, a pioneer in this field, for delivering an insightful and inspiring foreword. If the readers have any suggestion or criticism, please feel free to let us know. *Enjoy Reading!*

Editors

Wengang (Wayne) Bi is distinguished chair professor and associate dean in the College of Information and Electrical Engineering, chief scientist in the State Key Laboratory of Reliability and Intelligence of Electrical Equipment, Hebei University of Technology, Tianjin, People's Republic of China. He earned his PhD in applied physics in the department of electrical and computer engineering at the University of California-San Diego, California. He has worked in industry for over two decades, including at Hewlett-Packard Laboratory (Palo Alto, California), Agilent Technologies Laboratory (Palo Alto, California), and Philips Lumileds (San Jose, California), developing cutting-edge optoelectronic and photonic materials and device structures by molecular beam epitaxy and metal–organic vapor-phase epitaxy with their applications to optoelectronic and electronic devices. Previously, he was chief engineer and senior vice president of Nanjing Technology (Hangzhou, Zhejiang, People's Republic of China)/NNCrystal (Fayetteville, Arkansas). Dr. Bi has authored or coauthored over 60 refereed journal publications, has presented numerous conference talks, and is the inventor of 20 patents. He is an elected fellow of the Optical Society of America.

Hao-Chung (Henry) Kuo is distinguished professor and associate director of the Photonics Center at National Chiao-Tung University, Hsin-Tsu, Taiwan, where he has supervised more than 40 PhD and Master's-level scientists and engineers. He earned his doctorate in the department of electrical and computer engineering at the University of Illinois at Champaign Urbana, Champaign, Illinois. He has worked in industry for many years, including at Bell Labs (Murray Hill, New Jersey), Lucent Technologies (Murray Hill, New Jersey), Agilent Technologies (San Jose, California), and LuxNet Corporation (Fremont, California). He has more than 20 years of experience in the field of III–V optical devices and materials, solid-state lighting process development, and fabrication and measurement of quantum devices. He is an associate editor of the *IEEE Journal of Selected Topics in Quantum Electronics* and *Journal of Lightwave Technology*. He has published more than 300 papers in peer-reviewed journals and is an elected fellow of the International Society for Optics and Photonics (SPIE), the Optical Society of America, and the Institution of Engineering and Technology.

Pei-Cheng Ku is an associate professor in the department of electrical engineering and computer science at the University of Michigan, Ann Arbor, Michigan. He earned his PhD in electrical engineering and computer science from the University of California-Berkeley, California, and worked at Intel Corporation, Santa Clara, California, prior to joining the University of Michigan faculty. He co-founded Arborlight (Ann Arbor, Michigan) in 2011, a solid-state lighting technology company.

Bo Shen is a Cheung Kong professor at Peking University, Beijing, People's Republic of China. He received his PhD from Tohoku University, Sendai, Japan, and obtained the Outstanding Youth Foundation of the National Science Foundation of China in 2003. His main research field is III-nitride wide bandgap semiconductor physics, materials, and devices. He has published more than 160 contributed papers and owned 18 invention patents in this field.

Contributors

Akhil Ajay
CEA-Grenoble
INAC-PHELIQS
Grenoble, France

Pallab Bhattacharya
Department of Electrical Engineering and
    Computer Science
University of Michigan
Ann Arbor, Michigan

Davide Bisi
Department of Information Engineering
University of Padova
Padova, Italy

David A. Browne
CEA-Grenoble
INAC-PHELIQS
Grenoble, France

Kevin J. Chen
Department of Electronic and Computer
    Engineering
The Hong Kong University of Science and
    Technology
Clear Water Bay, Hong Kong

ZhiZhong Chen
Research Center for Wide Gap Semiconductors
School of Physics
Peking University
Beijing, People's Republic of China

Yu-Chieh Chi
Graduate Institute of Photonics and Optoelectronics
National Taiwan University
Taipei, Taiwan

Yu-Hsun Chou
Department of Photonics
National Chiao Tung University
Hsinchu, Taiwan

Faqrul A. Chowdhury
Department of Electrical and Computer
    Engineering
McGill University
Montreal, Quebec, Canada

Srabanti Chowdhury
Department of Electrical and Computer
    Engineering
University of California, Davis
Davis, California

John Dallesasse
Department of Electrical and Computer
    Engineering
University of Illinois at Urbana-Champaign
Champaign, Illinois

Hilmi Volkan Demir
LUMINOUS Centre of Excellence for
    Semiconductor Lighting and Displays
School of Electrical and Electronic Engineering
Nanyang Technological University
Singapore, Singapore

Steve Denbaars
Department of Materials
and
Department of Electrical and Computer
    Engineering
University of California, Santa Barbara
Santa Barbara, California

Hui Deng
Department of Physics
University of Michigan
Ann Arbor, Michigan

Bingfeng Fan
Institute of Advanced Technology
Sun Yat-sen University
Guangzhou, People's Republic of China

and

Foshan Institute of Sun Yat-sen University
Foshan, People's Republic of China

Shizhao Fan
Department of Electrical and Computer
    Engineering
Micro and Nanotechnology Laboratory
University of Illinois at Urbana-Champaign
Champaign, Illinois

YuXia Feng
Research Center for Wide Gap Semiconductors
School of Physics
Peking University
Beijing, People's Republic of China

Thomas Frost
Department of Electrical Engineering and
    Computer Science
University of Michigan
Ann Arbor, Michigan

WeiKun Ge
Research Center for Wide Gap Semiconductors
School of Physics
Peking University
Beijing, People's Republic of China

Shin-Yi Ho
Graduate Institute of Photonics and Optoelectronics
National Taiwan University
Taipei, Taiwan

Kuo-Bin Hong
Department of Photonics
National Chiao Tung University
Hsinchu, Taiwan

Dan-Hua Hsieh
Department of Photonics
National Chiao Tung University
Hsinchu, Taiwan

Lung-Hsing Hsu
Institute of Photonic System
College of Photonics
National Chiao Tung University
Tainan City, Taiwan

Shen-Che Huang
Department of Photonics
National Chiao Tung University
Hsinchu, Taiwan

Jian-Jang Huang
Graduate Institute of Photonics and
    Optoelectronics
National Taiwan University
Taipei, Taiwan

Fengyi Jiang
National Institute of LED on Si Substrate
Nanchang University
Nanchang, People's Republic of China

Dong Yeong Kim
Department of Materials Science and
    Engineering
Pohang University of Science and Technology
Pohang, South Korea

Jong Kyu Kim
Department of Materials Science and
    Engineering
Pohang University of Science and Technology
Pohang, South Korea

Daniel D. Koleske
Advanced Materials Sciences
Sandia National Laboratories
Albuquerque, New Mexico

Jonas Lähnemann
CEA-Grenoble
INAC-PHELIQS
Grenoble, France

Wei-Chih Lai
Institute of Photonic System
College of Photonics
National Chiao Tung University
Tainan City, Taiwan

Chun-Hsun Lee
Graduate Institute of Photonics and
    Optoelectronics
National Taiwan University
Taipei, Taiwan

Jong Won Lee
Department of Materials Science and
    Engineering
Pohang University of Science and
    Technology
Pohang, South Korea

Caroline B. Lim
CEA-Grenoble
INAC-PHELIQS
Grenoble, France

Chien-Chung Lin
Institute of Photonic System
College of Photonics
National Chiao Tung University
Tainan City, Taiwan

Gong-Ru Lin
Graduate Institute of Photonics and
    Optoelectronics
and
Department of Electrical Engineering
National Taiwan University
Taipei, Taiwan

Lianghong Liu
Department of Electrical and Computer
    Engineering
North Carolina State University
Raleigh, North Carolina

Zhe Liu
Semiconductor Lighting R&D Center
Institute of Semiconductors
Chinese Academy of Sciences
Beijing, People's Republic of China

Tien-Chang Lu
Department of Photonics
National Chiao Tung University
Hsinchu, Taiwan

Gaudenzio Meneghesso
Department of Information Engineering
University of Padova
Padova, Italy

Matteo Meneghini
Department of Information Engineering
University of Padova
Padova, Italy

Zetian Mi
Department of Electrical and Computer
    Engineering
McGill University
Montreal, Quebec, Canada

and

Department of Electrical Engineering and
    Computer Science
Center for Photonics and Multiscale
    Nanomaterials
University of Michigan
Ann Arbor, Michigan

Bo Monemar
Department of Physics, Chemistry and Biology
Linköping University
Linköping, Sweden

and

Global Innovation Research Organization
Tokyo University of Agriculture and
    Technology
Tokyo, Japan

and

The Solid State Lighting Center
Solid State Physics Division
Lund University
Lund, Sweden

Eva Monroy
CEA-Grenoble
INAC-PHELIQS
Grenoble, France

John F. Muth
Department of Electrical and Computer
    Engineering
North Carolina State University
Raleigh, North Carolina

Sujie Nakamura
Department of Materials
and
Department of Electrical and Computer
    Engineering
University of California, Santa Barbara
Santa Barbara, California

Jun Hyuk Park
Department of Materials Science and
    Engineering
Pohang University of Science and Technology
Pohang, South Korea

Plamen P. Paskov
Department of Physics, Chemistry and Biology
Linköping University
Linköping, Sweden

Tania Paskova
Department of Electrical and Computer Engineering
North Carolina State University
Raleigh, North Carolina

Fernando A. Ponce
Department of Physics
Arizona State University
Tempe, Arizona

ZhiXin Qin
Research Center for Wide Gap Semiconductors
School of Physics
Peking University
Beijing, People's Republic of China

Isabella Rossetto
Department of Information Engineering
University of Padova
Padova, Italy

Jinn-Kong Sheu
Institute of Photonic System
College of Photonics
National Chiao Tung University
Tainan City, Taiwan

Michael Slomski
Department of Electrical and Computer
    Engineering
North Carolina State University
Raleigh, North Carolina

Guan-Lin Su
Department of Electrical and Computer
    Engineering
University of Illinois at Urbana-Champaign
Champaign, Illinois

Xiao Wei Sun
LUMINOUS Centre of Excellence for
    Semiconductor Lighting and Displays
School of Electrical and Electronic
    Engineering
Nanyang Technological University
Singapore, Singapore

and

Department of Electrical and Electronic
    Engineering
College of Engineering
South University of Science and Technology
Guangdong, People's Republic of China

Ning Tang
Research Center for Wide Gap Semiconductors
School of Physics
Peking University
Beijing, People's Republic of China

Chu-Hsiang Teng
Department of Electrical Engineering and
    Computer Science
University of Michigan
Ann Arbor, Michigan

Yu-Ling Tsai
Institute of Photonic System
College of Photonics
National Chiao Tung University
Tainan City, Taiwan

An-Jye Tzou
Department of Photonics
and
Institute of Electro-Optical Engineering
National Chiao Tung University
Hsinchu, Taiwan

Gang Wang
State Key Laboratory of Optoelectronic Materials
    and Technologies
and
School of Electronics and Information Technology
Sun Yat-sen University
Guangzhou, People's Republic of China

Junxi Wang
Semiconductor Lighting R&D Center
Institute of Semiconductors
Chinese Academy of Sciences
Beijing, People's Republic of China

Renjie Wang
Department of Electrical and Computer
    Engineering
McGill University
Montreal, Quebec, Canada

WeiYing Wang
Research Center for Wide Gap Semiconductors
School of Physics
Peking University
Beijing, People's Republic of China

XinQiang Wang
Research Center for Wide Gap Semiconductors
School of Physics
Peking University
Beijing, People's Republic of China

JieJun Wu
Research Center for Wide Gap
    Semiconductors
School of Physics
Peking University
Beijing, People's Republic of China

Yuh-Renn Wu
Graduate Institute of Photonics and
    Optoelectronics
and
Department of Electrical Engineering
National Taiwan University
Taipei, Taiwan

FuJun Xu
Research Center for Wide Gap
    Semiconductors
School of Physics
Peking University
Beijing, People's Republic of China

Shu Yang
Department of Power Electronics
College of Electrical Engineering
Zhejiang University
Hangzhou, People's Republic of China

XueLin Yang
Research Center for Wide Gap
    Semiconductors
School of Physics
Peking University
Beijing, People's Republic of China

TongJun Yu
Research Center for Wide Gap
    Semiconductors
School of Physics
Peking University
Beijing, People's Republic of China

Enrico Zanoni
Department of Information Engineering
University of Padova
Padova, Italy

Ning Zhang
Semiconductor Lighting R&D Center
Institute of Semiconductors
Chinese Academy of Sciences
Beijing, People's Republic of China

Yonghui Zhang
Key Laboratory of Electronic Materials and
    Devices of Tianjin
School of Electronics and Information
    Engineering
Hebei University of Technology
Tianjin, People's Republic of China

Zi-Hui Zhang
Key Laboratory of Electronic Materials and
    Devices of Tianjin
School of Electronics and Information
    Engineering
Hebei University of Technology
Tianjin, People's Republic of China

Songrui Zhao
Department of Electrical and Computer
    Engineering
McGill University
Montreal, Quebec, Canada

Yi Zhuo
School of Electronics and Information
    Technology
Sun Yat-sen University
Guangzhou, People's Republic of China

I

Fundamentals

1

III-Nitride Materials and Characterization

Bo Shen, Ning Tang,
XinQiang Wang,
ZhiZhong Chen,
FuJun Xu,
XueLin Yang,
TongJun Yu,
JieJun Wu,
ZhiXin Qin,
WeiYing Wang,
YuXia Feng,
and WeiKun Ge

1.1 Introduction

Semiconductor materials and devices play a major role in science and technology of modern society. Historically, there would neither be microelectronic technique and industry without the development of silicon material and an integrated circuit, nor optical communication, mobile communication, and digital high-speed information network technology without the development of compound semiconductor materials such as GaAs and InP. Today, driven by the demand from highly advanced science and technology, the semiconductor material system has evolved into a new era of the third generation, represented by III-nitride wide gap semiconductors which have become one of the predominant areas of modern semiconductor science and technology.

The III-nitrides (or GaN-based) semiconductors consist of three direct bandgap InN, GaN, and AlN materials and their composition tunable alloys. Their bandgap varies from 0.7 eV of InN to 6.2 eV of AlN, and is tunable continuously, covering a huge wavelength ranging from infrared to ultraviolet, hence becoming such a semiconductor system with the widest tunable range in wavelength, and meanwhile keeping the high quality of physical and chemical properties. For the development and applications of important artificial optoelectronic functional materials, the epitaxial growth of the III-nitride semiconductors and their low-dimensional quantum structures is now the key issue.

Since the great breakthrough of the epitaxial growth and doping techniques in the early 1990s of the last century, taking GaN as the core material, the research and industrial development of III-nitride semiconductors have achieved significant advancement in terms of blue and white LEDs based on InGaN/GaN quantum wells (QWs), and high electron mobility transistors (HEMTs) devices based on AlGaN/GaN heterostructures, leading to a rapid development of semiconductor lighting and high-power microwave technology and industry. Entering the new century, the accelerating advances in science and technology push forward the development of III-nitride semiconductor materials and devices toward the direction of high power, low energy consumption, multiwavelength band, ultrafast response, miniaturization, and highs integration degree. The high-tech industry in the areas of information, energy, and transportation, and so on, makes new requests to III-nitride wide gap semiconductor materials and devices, and those demands become a new and strong driving force for a more rapid development of the latter.

This chapter will introduce the basic physical properties and main characterization techniques of III-nitride semiconductor materials, aiming at promoting the relevant research as well as the industrialization of III-nitride semiconductors.

1.2 Fundamental Properties of III-Nitride Semiconductors

1.2.1 Crystal Structure and Symmetry

Group III nitrides can be of various crystalline structures: wurtzite, zinc blende, and rock salt. Under ambient conditions, the thermodynamically stable structure is wurtzite for bulk AlN, GaN, and InN. The zinc blende structure for GaN and InN can be stabilized by epitaxial growth of thin films on {0 1 1} crystal planes of cubic substrates such as Si, SiC, MgO, and GaAs.

The rock salt or NaCl structure (with space group Fm3m in the Hermann–Mauguin notation and O_i^5 in the Schoenflies notation) can be induced in AlN, GaN, and InN under very high pressures. The reason for that is that the reduction of the lattice dimensions causes the interionic Coulomb interaction to favor the ionicity over the covalent nature.

The space grouping for the zinc blende structure is $\overline{F43m}$ in the Hermann–Mauguin notation and T_d^2 in the Schoenflies notation. The zinc blende structure has a cubic unit cell, containing four group III elements and four nitrogen elements. The wurtzite structure has a hexagonal unit cell, and thus, two lattice constants, c and a, containing six atoms of each type. In both cases, each group III atom is coordinated by four nitrogen atoms. Conversely, each nitrogen atom is coordinated by four group III atoms. The main difference between these two structures lies in the stacking sequence of closest packed diatomic planes (Figure 1.1).

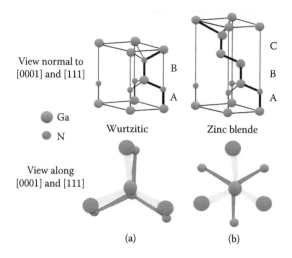

FIGURE 1.1   A stick-and-ball stacking model of crystals with (a, both top and bottom) 2H wurtzitic and (b, both top and bottom) 3C zinc blende polytypes. The bonds in an A-plane (11$\overline{2}$0) are indicated with heavier lines to accentuate the stacking sequence. The figures on top depict the three-dimensional view. The figures at the bottom indicate the projections on the (0001) and (111) planes for wurtzitic and cubic phases, respectively.

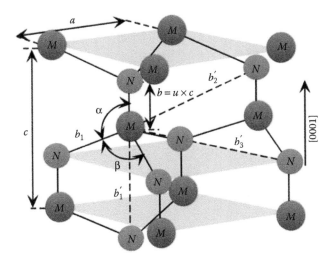

FIGURE 1.2   Schematic representation of a wurtzitic metal nitride structure with lattice constants a in the basal plane, c in the basal direction, and u parameter that is expressed as the bond length or the nearest neighbor distance b divided by c (0.375 in ideal crystal); a and b (109.47 in ideal crystal) are the bond angles, and b_1, b_2, and b_3, represent the three types of second nearest neighbor distances.

TABLE 1.1   Lattice Constants, Internal Displacement Parameters, and TECs of the Three Wurtzite III-Nitride Binary Compounds

Parameters (T = 300 K)	GaN	AlN	InN
$a(\text{Å})^1$	3.189	3.112	3.545
$c(\text{Å})^1$	5.185	4.982	5.703
u^2	0.376c	0.382c	0.377c
TEC(10^{-6}K^{-1})^3	a5.59/c3.17	a4.15/c5.27	–

In the hexagonal structure, the lattice parameter in (0001) plane—the edge length a, is distinct from the one along [0001] direction—prism height c, indicating that nitrides are anisotropic crystals. Indeed, such anisotropy gives rise to some unique properties and will be referred to from time to time in this book. The internal displacement parameter u is defined as the anion-cation bond length along c-axis, in the unit of c, as shown in Figure 1.2. The values of a, c, and u for bulk GaN, AlN, and InN at room temperature are listed in Table 1.1.

A close look at the unit cell reveals that the wurtzite structure is noncentrosymmetric, as it is asymmetrical along c-axis. It is referred to as polarity of the layer, defined by the direction of the III-N bonds parallel to c-axis.

In addition to the polar c-plane, there are two other important types of crystal planes: the nonpolar and the semipolar ones. The nonpolar ones, such as $\{11\bar{2}0\}$ a- and $\{1\bar{1}00\}$ m-planes, are perpendicular to the polar c-plane. The angle between two neighboring crystal directions from the same set, say between $[10\bar{1}0]$ and $[1\bar{1}00]$, is 60°; and the angle between the two adjacent directions, say between $[10\bar{1}0]$ and $[11\bar{2}0]$, is 30°. The semipolar ones are the planes with nonzero h or k and nonzero l, for example, (10$\bar{1}$3) and (11$\bar{2}$2) Figure 1.3.

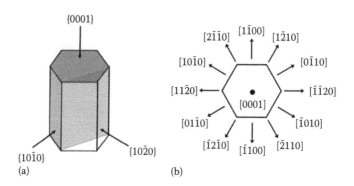

FIGURE 1.3 (a) Diagram of the crystal planes of III-nitrides and (b) top view of the wurtzite (0001) plane.

1.2.2 Energy-Band Structures

In recent years, group III nitrides and their alloys have gained recognition as being among the most important materials systems for optoelectronic and electronic applications. Most of the intensive research studies have been focused on Ga-rich $In_{1-x}Ga_xN$ and $Al_xGa_{1-x}N$ alloys whose energy gaps cover the short wavelength visible and near ultraviolet parts of the electromagnetic spectrum. A major breakthrough in 2002, stemming from the much improved quality of InN films grown using molecular beam epitaxy, resulted in the bandgap of InN being revised from 1.9 eV to a much narrower value of 0.7 eV. This finding triggered a worldwide research thrust into the area of narrow-band-gap group-III nitrides. It extends the fundamental bandgap of the group III-nitride alloy system over a wider spectral region, ranging from near infrared at 1.9 μm (0.7 eV) for InN to ultraviolet at 0.36 μm (3.4 eV) for GaN, or even deep ultraviolet at 0.2 μm (6.2 eV) for AlN.

GaN, InN, and AlN, usually crystallize in wurtzite lattice. If the crystal structure of a nitride semiconductor is not specified in the following text, the wurtzite phase is implied, whereas the zinc blende phase is always explicitly stated.

The temperature dependence of the direct bandgap is well described by Varshni's equation,

$$E_g(T) = E_g(T=0) - \frac{\alpha T^2}{T+\beta}$$

The parameters α and β are listed in Table 1.2.

Since nowadays the epitaxially grown heterostructures routinely combine layers of lattice-mismatched constituents, the material properties under strain must also be clarified. That is conventionally done within the deformation potential theory. The wurtzite materials generally require as many as two potentials for the shift of the energy gap, along with six valence-band deformation potentials. For completeness, six elastic constants for the wurtzite forms of nitride materials are also listed in Table 1.2.

The composition dependences of the energy gaps for the ternary alloys AlGaN, GaInN, and AlInN are assumed to follow a simple quadratic form

$$E_g(A_{1-X}B_X) = (1-X)E_g(A) + XE_g(B) - X(1-X)C$$

where the so-called "bowing parameter" C accounts for the deviation from a linear interpolation between the two binaries A and B. The bowing parameter is always positive for these materials, which reflects a reduction of the alloy energy gaps.

Wurtzite AlN has the distinction of being the only Al-containing III–V semiconductor compound with a direct energy gap. Furthermore, it is the largest-gap material that is still commonly considered to be a semiconductor. The crystal-field splitting in AlN is believed to be negative (different from GaN and InN, as shown in Figure 1.4), which implies that the topmost valence band is crystal hole-like. This makes the polarization of the emitted deep ultraviolet light special, which will be discussed in detail in Section 1.4.2.

TABLE 1.2 Recommended Band Structure Parameters for Wurtzite Nitride Binaries

Parameters	GaN	AlN	InN
E_g (eV)	3.4	6.2	0.7
α (meV/K)	0.909	1.799	0.245
β (K)	830	1462	624
Δ_{cr} (eV)	0.010	−0.169	0.040
Δ_{so} (eV)	0.017	0.019	0.005
$m_e\!//$	0.20	0.32	0.07
$m_e^{\perp}$	0.20	0.30	0.07
A_1	−7.21	−3.86	−8.21
A_2	−0.44	−0.25	−0.68
A_3	6.68	3.58	7.57
A_4	−3.46	−1.32	−5.32
A_5	−3.40	−1.47	−5.11
A_6	−4.90	−1.64	−5.96
A_7 (eVÅ)	0.0937	0	0
a_1 (eV)	−4.9	−3.4	−3.5
a_2 (eV)	−11.3	−11.8	−3.5
D_1 (eV)	−3.7	−17.1	−3.7
D_2 (eV)	4.5	7.9	4.5
D_3 (eV)	8.2	8.8	8.2
D_4 (eV)	−4.1	−3.9	−4.1
D_5 (eV)	−4.0	−3.4	−4.0
D_6 (eV)	−5.5	−3.4	−5.5
c_{11} (GPa)	390	396	223
c_{12} (GPa)	145	137	115
c_{13} (GPa)	106	108	92
c_{33} (GPa)	398	373	224
c_{44} (GPa)	105	116	48

Source: Vurgaftman, I. and Meyer, J. R., *J. Appl. Phys.*, 94, 3675, 2003.

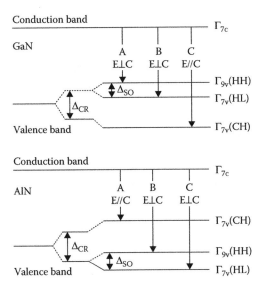

FIGURE 1.4 The energy-band structure of GaN and AlN.

1.2.3 Polarization Properties

A crystal like GaN-based materials with wurtzite structure is pyroelectric, and its polarization is oriented parallel to the direction along the c-axis. Based on a classical charge model, the electric polarization could be summarized as a permanent electric dipole moment. The cation–anion bond vector $\vec{l_i}$ and charges into a deviation of internal parameter u produces polarization, including spontaneous polarization ($\mathbf{P}_{SP}$) and piezoelectric polarization ($\mathbf{P}_{PE}$). Therefore, the polarization is usually presented as a sum

$$\mathbf{P} = \mathbf{P}_{SP} + \mathbf{P}_{PE}$$

Based on a classical charge model, $\mathbf{P}_{SP}$ could be summarized as a permanent electric dipole moment

$$\mathbf{P}_{SP} = -\frac{q}{\Omega} \sum_i Z^* \vec{l_i}$$

where
   q is the electron charge
   Ω is the unit cell volume of the tetrahedral unit
   Z^* is the Born effective charge of the dipole
   $\vec{l_i}$ is the ith cation–anion bond vector

In the regime of linear response, the components of $\mathbf{P}_{PE}$ vector are given by

$$\mathbf{P}_{PE} = \frac{q}{\Omega} Z^* \mathbf{u} + \kappa \mathbf{e}$$

where e_{ij} defines the components of the piezoelectric tensor. The symmetry of the wurtzite crystal determines the nonzero elements of e_{ij}, among which e_{31} and e_{33} are usually used in the polarization along (0001) axis.

Ab Initio theory is a typical approach in calculating electronic polarization. Table 1.3 summarizes the results of the spontaneous polarization ($\mathbf{P}_{SP}$) charge and elastic constants in group-III nitride semiconductors based on the density functional theory (DFT). The $\mathbf{P}_{SP}$ and piezoelectric constants are calculated by using Berry-phase approach [2], local-density approximation (LDA) [3–7], generalized gradient approximation (GGA) [5,6], and Heyd–Scuseria–Ernzerhof (HSE) screened-exchange hybrid functional [7].

Polarization charge is an exhibition of spontaneous polarization and piezoelectric polarization of nitride semiconductors. Usually, noncentrosymmetric compound crystals exhibit two different sequences of atomic layering in the two opposing directions parallel to certain crystallographic axes. Consequently, crystallographic polarity along these axes can be observed. In GaN-based heterostructures with the most common growth direction normal to the {0001} plane, the atoms are arranged in bilayers, which consist of two closely spaced hexagonal layers, one formed by cations and the other formed by anions, thus leading to polar faces. In the case of GaN, a basal surface should be either Ga- or N-faced. Ga-faced means Ga atoms are placed on the top position of the {0001} bilayer, corresponding to the [0001] polarity as shown in Figure 1.5a, and polarization charges occur on the surface and at the interface as illustrated in Figure 1.5b.

For ternary alloy of group-III nitrides ABN, polarization can be expressed as

$$\mathbf{P}_{ABN} = \mathbf{P}_{ABN}^{pz} + \mathbf{P}_{ABN}^{SP}$$

At the interfaces of a $A_x B_{1-x} N/GaN$ heterostructure, the polarization can change abruptly, which causes a fixed polarization charge σ:

$$\sigma_{ABN/GaN} = \mathbf{P}_{GaN} - \mathbf{P}_{ABN} = (\mathbf{P}_{GaN}^{pz} + \mathbf{P}_{GaN}^{SP}) - (\mathbf{P}_{ABN}^{pz} + \mathbf{P}_{ABN}^{SP})$$

TABLE 1.3 Spontaneous Polarization Charge and Elastic Constants of Nitride Semiconductors

Parameter	Groups	Methods	AlN	GaN	InN
P_{SP} (C/m²)	Bernardini et al. [2]		−0.081	−0.029	−0.031
	Wei et al. [3]	LDA	−0.094	−0.032	−0.042
	Bechstedt et al. [4]	LDA	−0.120	−0.074	−0.050
	Bernardini et al. [5]	LDA	−0.099	−0.032	−0.041
	Bernardini et al. [5]	GGA	−0.090	−0.034	−0.042
	Zoroddu et al. [6]	LDA	−0.100	−0.032	−0.041
	Zoroddu et al. [6]	GGA	−0.900	−0.034	−0.042
	Caro et al. [7]	LDA	−0.096	−0.029	−0.041
	Caro et al. [7]	HSE	−0.091	−0.040	−0.049
e_{31} (C/m²)	Bernardini et al. [2]		−0.60	−0.49	−0.57
	Bernardini et al. [5]	GGA	−0.53	−0.34	−0.14
	Zoroddu et al. [6]	LDA	−0.64	−0.44	−0.52
	Zoroddu et al. [6]	GGA	−0.53	−0.34	−0.41
	Caro et al. [7]	LDA	−0.69	−0.49	−0.63
	Caro et al. [7]	HSE	−0.63	−0.44	−0.58
e_{33} (C/m²)	Bernardini et al. [2]		1.46	0.73	0.97
	Bernardini et al. [5]	GGA	1.50	0.67	0.81
	Zoroddu et al. [6]	LDA	1.80	0.86	1.09
	Zoroddu et al. [6]	GGA	1.50	0.67	0.81
	Caro et al. [7]	LDA	1.59	0.83	1.09
	Caro et al. [7]	HSE	1.46	0.74	1.07

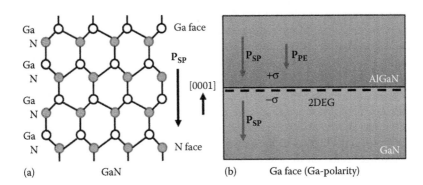

FIGURE 1.5 (a) The [0001] polarity of GaN and (b) polarization charges occur on the surface and at the interface in the heterostructure.

That induces a two-dimensional electron gas (2DEGs) whose density can be calculated by the following equation [8]:

$$n_s(x) = \frac{\sigma_{ABN/GaN}(x)}{e} - \frac{\varepsilon_0 \varepsilon_{ABN}(x)}{d_{ABN} e^2}\left[e\phi_{ABN}(x) + E_F - \Delta E^C_{ABN/GaN}(x) \right]$$

where

ε_0 is the dielectric constant of the vacuum
$\varepsilon_{ABN}(x)$ is the relative dielectric constant
d_{ABN} is the thicknesses of the barrier layer

The Fermi level with respect to the GaN conduction-band-edge energy is

$$E_F(x) = E_0(x) + \frac{\pi \hbar^2}{m^*} n_s(x)$$

and the ground subband level of the sheet carrier is

$$E_0(x) = \left(\frac{9\pi \hbar e^2}{8\varepsilon_0 \sqrt{8m^*}} \frac{n_s(x)}{\varepsilon_{GaN}} \right)^{2/3}$$

with $\hbar$ as the reduced Plank's constant, m^* the effective electron mass, $e\phi_{ABN}(x)$ the Schottky barrier height, and $\Delta E_{ABN/GaN}^C(x)$ the conduction band offset.

With the parameters of dielectric constants, the Schottky barrier for metal-contact, band gaps, band offsets, an effective electron mass of the sheet carrier concentrations and carrier distribution profiles can be verified by solving a one-dimensional Schrödinger-Poisson equation.

Spontaneous polarization and piezoelectric polarization have temperature dependences and give influence on the properties of 2DEG in heterostructures. In $Al_{0.25}Ga_{0.75}N/AlN/GaN$ heterostructure, $\mathbf{P}_{SP}$ is found to decrease when the temperature rises from 300 to 600 K, and $\mathbf{P}_{PE}$ varies oppositely. The change of $\mathbf{P}_{SP}/e$ in an order of mid-$10^{11}cm^{-2}$ is several times larger than that of $\mathbf{P}_{PE}/e$ and is similar to n_s in trend, demonstrating a dominant role of $\mathbf{P}_{SP}$ in temperature dependent properties of 2DEG in AlGaN/GaN heterostructures [8].

2DEGs confined in AlGaN/GaN [9–14], InGaN/GaN [15–17], and AlInN/GaN [18–21] heterostructures were widely investigated in both experimental and calculation. Figure 1.6a, b show the polarization induced bound interface charges as a function of the alloy composition of the barrier for Ga-face GaN/AlGaN/GaN heterostructure. The bound interface charge is calculated for barriers grown on relaxed GaN. And sheet carrier concentrations increase with alloy composition of the barrier because both spontaneous and piezoelectric polarizations are larger in AlGaN with higher Al-content. Comparison of experimental (Figure 1.6b) and theoretical data of 2DEG calculated by taking into account the depletion caused by the metal Schottky contact, shows as nearly consistent.

The polarization induced other effect is the quantum-confined Stark effect (QCSE) in optoelectronic quantum structures. The polarization charges result in a strong built-in electric field in the (QWs, which is usually on the order of MV/cm [22,23] in nitrides. This built-in electric field gives rise to separation of electrons and holes in the QWs, leading to an obvious decrease in the radiative recombination rate.

Solving Schrödinger equation to determine the wave functions and the energy levels in the low-density regime [24,25], the radiative lifetime of the optical transition can be approximated [26]. Some reported results showed a nearly exponential increase of radiative lifetime with sizes for QWs above a certain width (height) [27–29]. From experiment and calculation results of Berkowicz et al. in Figure 1.7, the radiative lifetime rises remarkably from ~10 ns to ~10 μs with the QWs width (height) enlarged from ~1.0 to ~6.2 nm [27].

Other effects arising from the QCSE are the occurrence of a Stokes shift and a decrease of the exciton binding energy, respectively. Berkowicz et al. [27], as shown in Figure 1.8, and Lefebvre et al. [30] have unambiguously shown that for a given indium concentration, emissions from InGaN/GaN QWs exhibit a Stokes shift increasing with well thickness. The reason for such an effect has to be found in the progressive decrease in the oscillator strength of the fundamental transition energy with increasing thickness/height due to the reduced overlap of electron and hole envelope wave functions in the triangular part of the confining potentials. A decrease of the exciton binding energy with increasing well width is a further signature of the QCSE. Above a certain width, the impact of the built-in electric field dominates over quantum confinement effects, which leads to a progressive separation of bound electron-hole pairs by the electric field screening the Coulomb interaction.

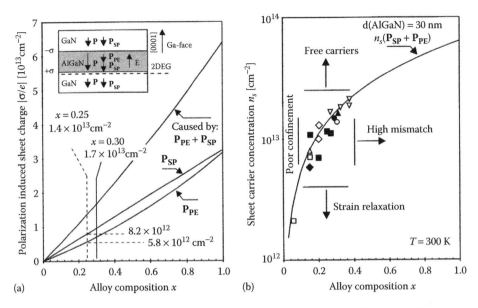

FIGURE 1.6 (a) The polarization induced bound interface charges as a function of alloy composition of the barrier for Ga-face GaN/AlGaN/GaN heterostructure. The bound interface charge is calculated for barriers grown on relaxed GaN and (b) calculated and measured sheet carrier concentrations with alloy composition of the barrier with Al composition. (From Ambacher, O. et al., *J. Appl. Phys.*, 85, 3222, 1999. With permission.)

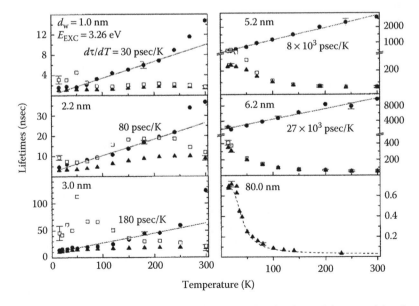

FIGURE 1.7 The measured effective lifetime τ_{eff} (▲) and the deduced radiative lifetime τ_R (●) and nonradiative lifetimes τ_{NR} (□) as a function of sample temperature. Solid lines represent the fitted linear temperature dependence of τ_R. (From Berkowicz, E. et al., *Phys. Rev. B.*, 61, 10994, 2000. With permission.)

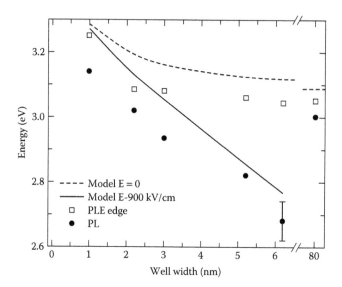

FIGURE 1.8   PL energy and photoluminescence excitation (PLE) absorption edge as a function of QW width. The solid (dashed) line represents the calculated band gap with (without) a piezoelectric field. (From Berkowicz, E. et al., *Phys. Rev. B.*, 61, 10994, 2000. With permission.)

1.2.4 Other Basic Physical Properties

Owing to their advantages of long spin relaxation time and high Curie temperature, III-nitride semiconductor materials are the favorite materials in developing spintronic devices. The magnetotransport and the circular photogalvanic effect (CPGE) are the useful methods in the investigation of the spin properties of III-nitride semiconductor materials.

Beating patterns in the oscillatory magnetoresistance in $Al_xGa_{1-x}N/GaN$ heterostructures, which are attributed to the zero-field spin splitting of the 2DEG, have been investigated by means of SdH measurements. The zero-field spin splitting energy is determined to be 2.5 meV, and the spin-orbit coupling (SOC) parameter is determined to be 2.2×10^{-12} eVm [31]. Despite the energy bandgap in GaN being very large, the zero-field spin splitting is still observed, which is due to a strong polarization-induced electric field and high 2DEG density in the heterostructures.

The zero-field spin splitting in $Al_xGa_{1-x}N/GaN$ heterostructures with various Al compositions has been investigated at low temperatures and high magnetic fields. It is found that the SOC parameter can be tuned by the polarization-induced electric field [32]. It is observed that the SdH beating nodes shift to lower magnetic fields after illumination. Therefore, the spin splitting energy decreases after the illumination. The illumination also decreases the electric field at the $Al_xGa_{1-x}N/GaN$ heterointerface. Based on the fact that the electric field changes the spin splitting energy, it is suggested that the zero-field spin splitting of the 2DEG in $Al_xGa_{1-x}N/GaN$ heterostructures mainly arises from the Rashba effect [33]. A large Rashba SOC was realized in a quasi-1D quantum point contacts fabricated on $Al_xGa_{1-x}N/GaN$ heterostructures. The enhanced effective Landé factor g* is anisotropic and increases as the lateral confinement becomes stronger. Both the Rashba SOC and the exchange interaction in the 1D system contribute to this enhancement [34]. The $Al_xGa_{1-x}N/GaN$ heterostructure is one of the promising materials for the spin-polarized field effect transistor.

The spin splitting of the 2DEG in $Al_xGa_{1-x}N/GaN$ heterostructures has been investigated systemically using CPGE, anomalous CPGE (ACPGE), and photo-induced anomalous Hall effect (PIAHE). The CPGE of the 2DEG in $Al_xGa_{1-x}N/GaN$ heterostructures has been observed at room temperature [35]. The differentiation of structure inversion symmetry (SIA) and bulk inversion symmetry (BIA) spin splitting of the 2DEG in $Al_xGa_{1-x}N/GaN$ heterostructures has been investigated, and the SIA/BIA ratio is estimated to be about 13.2/1, indicating that the SIA is the dominant mechanism to induce the k-linear spin splitting of the subbands in the triangular QW at $Al_xGa_{1-x}N/GaN$ heterointerfaces [36]. A swirly current named ACPGE has been observed first by moving the incident light spot [37]. PIAHE has been researched in $Al_xGa_{1-x}N/GaN$ heterostructure. The anomalous Hall conductivity is determined to be $\sigma_{AH} = 2.59 \times 10^{-9} \Omega^{-1}$ [38]. The contributions of the ACPGE and PIAHE were differentiated, and the ratio of the spin diffusion coefficient to photo-induced anomalous spin Hall mobility $D_s/\mu_s = 0.08$ V was extracted at room temperature [39].

1.3 Epitaxial Growth of III-Nitride Semiconductors

1.3.1 Metalorganic Chemical Vapor Deposition

The Metalorganic chemical vapor deposition (MOCVD) process was developed by Manasevit (1968) [40], who first demonstrated its potential in depositing single crystal GaAs in an open tube cold-wall reactor. Subsequently, this process has been adapted to grow nearly all III–V and II–VI semiconductors, including antimonides, arsenides, phosphides, sulfides, selenides, tellurides, and nitrides, and also ternary and quaternary alloys. Besides, specially designed reactors have made it possible to grow very sharp interfaces and multilayered structures with precise control and excellent uniformity of low-dimensional quantum structures necessary for devices such as laser diodes, LEDs, and HEMTs. The ease of scaling the MOCVD process also makes it very suitable for commercial production, as well as laboratory research of compound semiconductor devices. For III-nitrides semiconductors, the approach developed by Maruska and Tietjen is a precursor to the modern MOCVD growth reactor for GaN, which can also be used to deposit AlN [41].

As a nonequilibrium growth technique, MOCVD relies on vapor phase transport process operating at atmospheric or low pressure (e.g., 100 mba). The group-III metalorganic precursors and group-V hydrides for III-nitrides can be transported into the reactor for growing GaN-based materials, and the precursors react with NH_3 at the surface of a single-crystal substrate which is heated up to a high temperature. Figure 1.9 shows a schematic configuration of MOCVD for III-nitrides, which includes several important parts, such as gas delivery and blending system, reactor, and exhausting system, and so on. The group-III metalorganic precursors, such as TriMethylaluminum (TMAl), TriMethylindium (TMIn), and TriMethylGallium (TMGa) are stored in stainless steel bubblers through which a carrier gas (hydrogen or nitrogen) flows. These metalorganic group III sources are either solids such as TMIn at room temperature, or liquids, such as TMAl and TMGa. By controlling the bubble temperature, the vapor pressure of the source materials can be precisely adjusted, and then

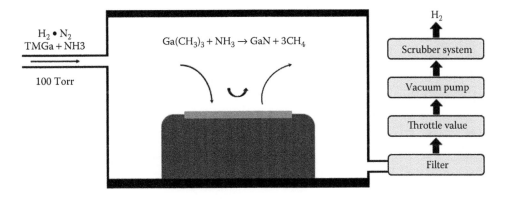

FIGURE 1.9 A schematic configuration of MOCVD for III-nitrides.

the carrier will be saturated by the vapor from the source and is then transported into the heated substrates via stainless steel pipe lines with inner-face electrochemical polishing. The group-V hydride is commonly gaseous ammonia (NH_3) for nitride growth that is contained in high-pressure cylinders. In addition to the binary semiconductor, such as GaN and AlN, ternary or quaternary alloys can also be grown by introducing more sources simultaneously, and the change in composition and growth rate can be realized by ramping/switching the source flows, as well as precisely controlling the mass flow rate and dilution of the various mixtures of the gas stream. Furthermore, the grown layers can also be doped by introducing small concentrations of the appropriate sources, and the doping source can be metalorganic precursors, such as cyclopentadienylmagnesium (Cp_2Mg), or hydrides, such as silane (SiH_4). The substrates are usually placed on a graphite pallet called susceptor that can be heated by a radio frequency (RF) coil or resistance heater. MOCVD processes are pyrolytic in nature, as a consequence, cold-wall reactors are used almost exclusively, which is also an important feature of the MOCVD process to minimize the wall deposits and reduce reaction depletion effects that always occur in a hot-wall case.

The basic MOCVD reaction for the GaN deposition process is:

$$Ga(CH_3)_3 + NH_3 \rightarrow GaN + 3CH_4$$

where the reactants and products are all gaseous except the solid-state GaN. Generally, the growth processes in MOCVD are highly complex, being closely related with the reactor configuration. Hence, the reactor is the key for MOCVD design, as well as the basis to understand the fundamentals of the growth process. There are several problems that must be resolved for the reactor designing for GaN growth, such as high growth temperature, prereaction, and film uniformity, and so on. Generally, a high temperature is necessary for GaN growth, due to the very high bond strength of the N-H bond in ammonia. However, due to the strong thermodynamic tendency of ammonia to react with the group III metalorganic source, any prior interaction, especially for the case of Al and ammonia, will lead to nonvolatile adducts in the gas phase, destroying the growth process. So, the common feature of all the reactors is that the reactants should be controlled as much as possible to only act upon each other on the surface of the substrate.

A typical design is the atmospheric-pressure (AP) MOCVD reactor that is widely adapted by major research groups in Japan, as it is very beneficial to afford the high partial pressure of ammonia, or nitrogen involved precursors. A modified AP-MOCVD reactor structure was put forward by Nakamura et al. in 1991 as shown in Figure 1.10a [42], which has two different flows. One flow carries a reactant

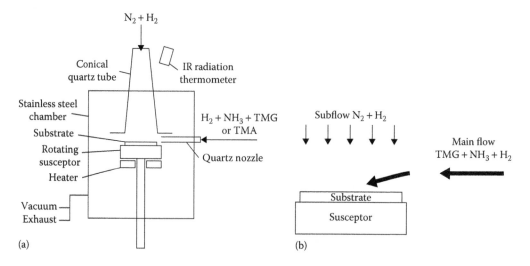

FIGURE 1.10 (a) Two-flow MOCVD design and (b) schematic principle gas flow of two-flow MOCVD.

gas parallel to the substrate, and the other adds an inactive gas perpendicular to the substrate for the purpose of changing the direction of the reactant gas flow as clearly shown in Figure 1.10b. A rotating susceptor is adopted to improve the uniformity of the deposited films. By adopting this reactor structure, the breakthrough in bright blue LEDs was achieved by Nakamura et al. (1994) [43].

Another typical design is the close coupled showerhead (CCS) low-pressure MOCVD reactor developed by Thomas Swan Corporation as shown in Figure 1.11, which features the very short distance (10–20 mm) [44,45] between the substrate and a showerhead used to inject group-III metalorganic precursors and ammonia. This short distance is beneficial to suppress the formation of whirlpool above the susceptor from the point of fluid mechanics. However, the short distance will result in undesired deposition due to high temperature of the showerhead caused by thermal radiation, which must be avoided by water cooling. Figure 1.12 further illustrates the details of the showerhead structure. It is composed of many stainless steel tubes with inner diameter 0.6 mm as the injecting orifice, where there are two sets

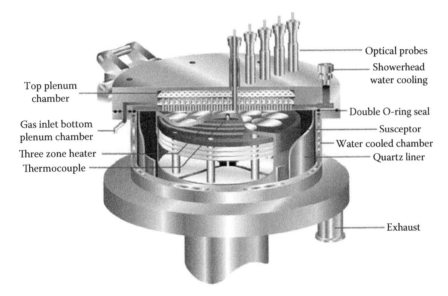

FIGURE 1.11 Schematic close coupled showerhead low-pressure MOCVD reactor. (From Aixtron, Herzogenrath, Germany.)

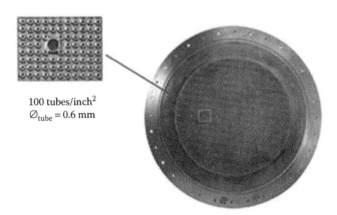

100 tubes/inch^2
$\varnothing_{\text{tube}} = 0.6$ mm

FIGURE 1.12 Structure of the showerhead for the CCS-MOCVD. (From Aixtron, Herzogenrath, Germany.)

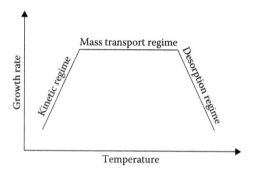

FIGURE 1.13 Three possible growth regimes in MOCVD.

of tubes in staggered form to avoid the prereaction. One set is for group-III metalorganic precursors, and the other is for ammonia. The distance between the adjacent tubes must be carefully designed to be 100 tubes/inch^2 to ensure the metalorganic precursors and ammonia being fully blended before reaching the substrate, and to guarantee the uniform concentration distribution of the reactant gas above the susceptor to form uniform-thickness boundary layer that is vital for the wafer uniformity. When operating under low pressure, laminar flow will be maintained without laminar flow. For this reactor, the total flow rate (Q), pressure (P), distance between the susceptor and showerhead (H), and the speed of rotation (ω) are the main factors influencing the growth rate, uniformity, and material utilization. The growth rate is roughly proportional to $H^{-1/2}$, $Q^{1/2}$ and $(\omega p)^{1/2}$ according to the flow field characteristics and the so-called "boundary layer" theory. A small H will lead to a fast reaching of the precursors and ammonia to the substrate, resulting in high material utilization and heterostructures with sharp interfaces.

Generally, there are three possible growth regimes (process windows) in MOCVD: kinetic limited, mass transport limited and desorption limited, which are influenced greatly by temperature, as schematically shown in Figure 1.13. The mass transport limited process is the generally preferred window, where the growth rate is dominated by diffusion of the reagent to the surface.

For III-nitride semiconductors, the breakthrough of high-crystalline GaN growth on sapphire by so-called "Two-Step" method using buffer layer in MOCVD process opens the door for great applications in optoelectronics, such as in blue LED. The MOCVD growth for GaN follows a sequence of substrate cleaning, low temperature (480°C–600°C) AlN or GaN buffer-layer (nucleation layer) growth followed by high-temperature annealing and finally growing of the GaN epilayer. The details can be found elsewhere [46–49]. As will be detailed below, a low-temperature AlN or GaN buffer layer improves the GaN-layer quality substantially. A short-time nitridation for sapphire usually is adopted to promote the growth of the buffer layer. The schematic reflectivity features during the "Two-Step" process is shown in Figure 1.14, depicting the evolution of salient structural features of GaN [49]. The reflectivity and corresponding morphology evolution at different stages are further illustrated in Figure 1.15 [50], including nucleation layer before (a) and after recrystallization (b), at the beginning of the epilayer growth (c) and the final surface with atomic steps after coalescence (d).

In recent years, some growth techniques are further developed for III-nitrides, such as AlN epitaxy for ultraviolet light-emitting diodes (UV LEDs). Due to the poor mobility of Al adatom and severe gas-phase reaction, it is hard to grow AlN and high-Al-content AlGaN to a large extent. Sensor Electronic Technology (SET), Inc. introduced a new growth technique called migration-enhanced MOCVD (MEMOCVD®) [51–52]. This technique deposits ternary $Al_xGa_{1-x}N$ by repeating unit cells grown using sequential metal–organic precursor pulses of Al, Ga, and ammonia (NH_3). The durations and waveforms of precursor pulses are optimized, and the pulses might overlap, providing a continuum of growth techniques ranging from pulsed atomic layer epitaxy (PALE) to conventional MOCVD. Therefore, it can realize a high growth rate for thick films and a slower growth rate for active layers with relatively low

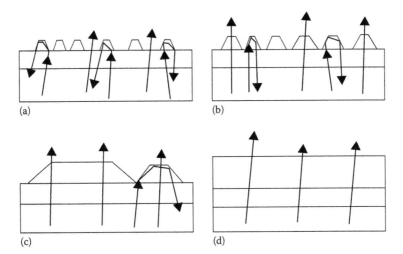

FIGURE 1.14 Schematic reflectivity features during a "Two-Step" process for GaN: (a) Island growth, (b) lateral growth, (c) coalescence, and (d) Quasi-2D growth.

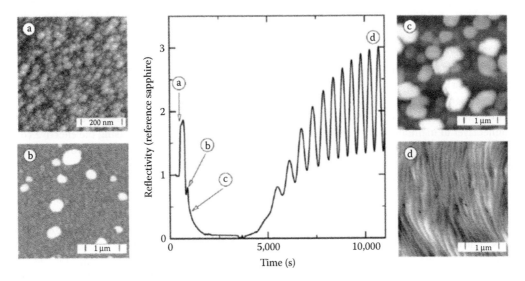

FIGURE 1.15 Typical reflectance transient taken from a GaN growth run. The insets show the morphology at different stages of growth measured by atomic force microscopy.

growth temperature. In the meantime, improved crystal quality can also be achieved. Using this new technique, SET Inc. fabricated UV LEDs with peak emission from 235 to 338 nm. The UV LEDs exhibit high external quantum efficiency (EQE) and narrow EL line, evidencing the superiority of this epitaxial technique. In 2012, SET Inc. obtained an EQE of 10.4% at 20 mA CW current with output power up to 9.3 mW at 278 nm for AlGaN-based deep UV LEDs grown on sapphire substrates [52]. There is no doubt that MEMOCVD® has significantly promoted the development of UV LED.

On the other hand, epitaxial growth of GaN-based heterostructures for electronic application has also made great progress by means of MOCVD technique. The first observation of enhanced electron mobility in AlGaN/GaN heterostructures prepared by MOCVD was reported in 1991 [53]. These structures were deposited on basal plane sapphire using low-pressure MOCVD system. In 1993, M. Asif Khan et al. fabricated a high mobility transistor based on an n-GaN/AlGaN heterostructure grown by MOCVD

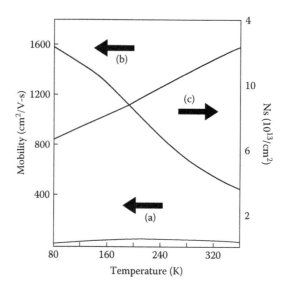

FIGURE 1.16   Mobility and sheet carrier concentration versus temperature. (a) Mobility for 3,000 A GaN layer. (b) Mobility for single AlGaN/GaN heterojunction. (c) Sheet carrier concentration for single AlGaN/GaN heterojunction. (From M. Asif Khan et al., *Appl. Phys. Lett.*, 58, 2408, 1991. With permission.)

for the first time. For a device with a 4 μm gate length (10 μm channel opening, that is, source-drain separation), a transconductance of 28 mS/mm at 300 K and 46 mS/mm at 77 K was obtained at +0.5 V gate bias (Figure 1.16) [54].

A very thin AlN interfacial layer (~1 nm) was inserted between the AlGaN barrier layer and the GaN channel layer to maintain high mobility at high sheet charge densities by increasing the effective conduction band offset and decreasing the alloy scattering [55]. Furthermore, in order to improve the buffer isolation and reduce the buffer leakage current, the GaN buffer is usually doped with Fe, or C atoms to achieve a high resistance [56,57].

Recently, AlGaN/GaN heterostructures grown on Si substrate have attracted much attention due to its lower cost, larger wafer size, and high thermal conductivity [58,59]. However, it is difficult to grow GaN-based epilayers on Si substrate due to large lattice and thermal mismatch. Although several stress-control approaches have been reported to handle this issue [60,61], there still remains a high density of threading dislocations in AlGaN/GaN heterostructures grown on Si substrate. Therefore, it is still highly desirable to develop a simple and cost-effective GaN-on-Si technology. Using large lattice-mismatch induced stress control technology, the dislocation density in GaN layer reported in reference [62] was significantly reduced due to the dislocation annihilation in the low Al content AlGaN intermediate layer, being assisted by compressive stress originated from the large lattice mismatch between the AlGaN layer and AlN buffer. Based on this high-quality GaN buffer, high-quality AlGaN/GaN heterostructures with 2DEG mobility as high as 2,040 cm²/(V·s) were achieved.

1.3.2 Molecular Beam Epitaxy

Molecular beam epitaxy (MBE) growth is operated in an ultrahigh vacuum (UHV) environment. Constituent elements of a semiconductor in the form of "molecular beams" are deposited onto a heated substrate to grow thin epitaxial layers. In the case of III-nitrides growth, the metal beam fluxes are provided by conventional Knudsen cells while several different methods are used to provide nitrogen. For Gas Source MBE (GSMBE), ammonia is used as the nitrogen source in spite of its high thermal stability. Since nitrogen incorporation is difficult at low temperature when ammonia is used, plasma cells are widely used to provide the nitrogen species, which are often called plasma-assisted MBE. The generated nitrogen species is chemically so active that the growth temperature of nitrides is low.

Generally, the basic MBE system includes several vacuum chambers/systems: a growth chamber/ multiple growth chambers, a transfer chamber, and a load lock system. The load lock chamber is used to bring samples into and out of the vacuum environment while the vacuum integrity of the other chambers is maintained. The transfer chamber is used for transferring samples through the load lock system between the growth chamber and the load lock chamber, and for the preparation and storage of samples. In the growth chamber, the Ga and Al, or In molecular/atom beams and nitrogen species arrive onto the substrate to deposit the nitride films, alloys, and quantum structures. Other sources such as Mg and Si are usually used for p- and n-type doping of the GaN. A great advantage of MBE is the high capability for *in situ* monitoring due to the UHV working environment. Besides the beam flux monitor and reflection high-energy electron diffraction (RHEED), spectroscopic ellipsometry (SE), and coaxial impact collision ion scattering spectroscopy (CAICISS), angle-resolved photoemission spectroscopy (ARPES) and STM/AFM can also be used for in-suite monitoring and measuring. Figure 1.17 shows a schema of a common PAMBE growth chamber used for III-nitrides growth.

The InN as an important member of III-nitrides has attracted great attention in the past decades. It is, however, difficult to obtain high-quality InN due to the low dissociation temperature of InN and the extremely high equilibrium vapor pressure of nitrogen, which hindered the understanding of fundamental physical properties of InN and their applications. In that sense, the metalorganic vapor phase epitaxy (MOVPE) has an inherent disadvantage because it must satisfy the conditions for NH_3 pyrolysis and prevention of InN dissociation, which impose conflicted temperature requirements. In PAMBE, the excited nitrogen radicals can be generated separately in a plasma source, enabling us to select growth temperature without considering the requirements of the NH_3 pyrolysis. Due to this inherent advantage, the quality of InN grown by PAMBE has improved very quickly over a relatively short period [63–65].

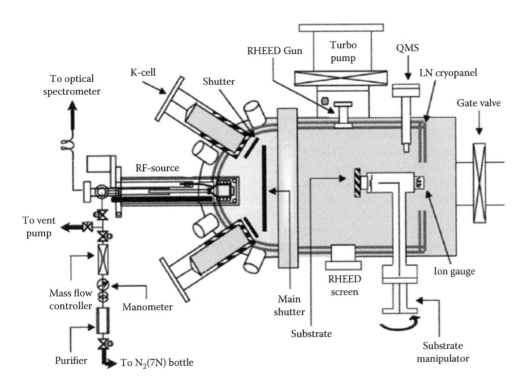

FIGURE 1.17 Schema of PAMBE growth chamber used for III-nitrides growth.

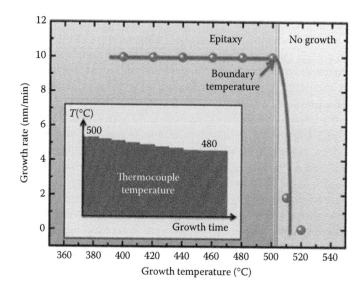

FIGURE 1.18 Growth rate of InN as a function of growth temperature. Schema of boundary-temperature-controlled epitaxy is shown in the inset, where the thermocouple temperature is decreased step by step from 500°C to 480°C. (From X. Wang et al., *Appl. Phys. Exp.*, 5, 015502, 2012. With permission.)

It is known that one of the key problems for InN epitaxy is its low maximum growth temperature. It is found that the maximum growth temperature for In-polarity InN is about 500°C (Figure 1.18). The growth is usually kept at the boundary temperature since higher temperature leads to better crystalline quality. However, the growth front temperature is slightly increased with increasing layer thickness, and thus InN is often dissociated when the thermocouple temperature remains constant, particularly when the layer becomes thick. X.Q. Wang et al. [66] put forward the boundary temperature controlled epitaxy, that is, keeping the growth temperature just below the boundary between crystallization and dissociation. High-quality InN layers on the sapphire substrate were thus obtained by using this method. The mobility was improved with increasing layer thickness, and a recorded value of 3,280 cm²/V⁻¹s⁻¹ was obtained with the electron concentration of 1.47×10^{17} cm⁻³.

With the significant improvement in the quality of InN films grown by molecular beam epitaxy, a narrow band gap of around 0.7 eV [67] has widely been recognized after several years of discussion. This new finding expanded the field of applications of group III nitrides, with the alloy systems spanning from 6.2 eV of AlN to 0.7 eV of InN. Unfortunately, the epitaxy of high-quality $In_xGa_{1-x}N$ films over the entire composition range is notoriously difficult due to the large gap between the growth temperature and lattice constant for InN and GaN. It is known that one big problem for the epitaxy of high-quality $In_xGa_{1-x}N$ alloys is the low growth temperature, which is mainly limited by InN since the maximum growth temperature of InN is much lower than the usual growth temperature of GaN. Figure 1.19 shows the In composition in $In_xGa_{1-x}N$ alloys grown at various temperatures and different ratios between the incident In and Ga beam flux (Φ_{In}/Φ_{Ga}). As shown in Figure 1.19, the In composition increases with decreasing growth temperature under the same ratio of Φ_{In}/Φ_{Ga} and the In composition increases with increasing Φ_{In}/Φ_{Ga}. As is known, the bond energies between In-N and Ga-N are 1.93 and 2.20 eV [68], respectively, indicating that a low growth temperature is beneficial for In incorporation while the supplied N adatoms would bond with Ga adatoms with priority.

A growth temperature controlled epitaxy (GTCE) [69] has been set up to modulate the In composition in the $In_xGa_{1-x}N$ alloys, where the samples with different In compositions were grown at their

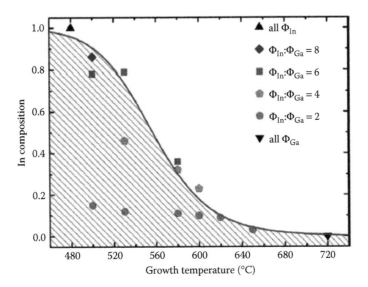

FIGURE 1.19 Dependence of the In composition of $In_xGa_{1-x}N$ films on the growth temperature and In/Ga flux ration. The solid line is a fitting curve for the maximal In composition at different temperatures.

highest temperature. Figure 1.20 shows the PL spectra at room temperature, where all samples show strong emission. However, the linewidth of the PL peak is clearly broadened with increasing In composition and reaches a maximum of about 280 meV at around $x = 0.5$ in comparison to 100 meV at a low In composition, and then it narrows down. The broader FWHMs of the PL peaks for $In_xGa_{1-x}N$ films with mid-In-composition may be due to the fact that the alloy disorder is more serious within this region.

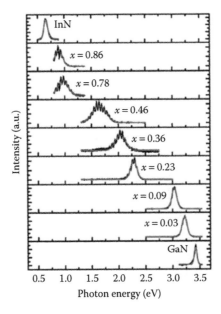

FIGURE 1.20 The PL spectra of the $In_xGa_{1-x}N$ films at room temperature. The fringes are Fabry–Pérot oscillations. The solid lines are Gauss fitting results.

1.3.3 Hydride Vapor Phase Epitaxy

Hydride vapor phase epitaxy (HVPE) is a very old growth technology, which is widely used for film growth in silicon and GaAs industry [70]. When the HVPE method is applied to III-nitrides, it is quickly found that this method is a highly practical method to obtain GaN thick layers, due to a relatively high growth rate (usually greater than 100 μm/h) and a possibility to crystallize high-purity material. Thus, thick GaN deposited on a foreign substrate (such as sapphire) by HVPE can serve as quasi-bulk substrates after removing the foreign substrates. It leads to a quick development of free-standing GaN substrate in recently 10 years.

Typical HVPE systems fall into two basic types, being vertical [71] and horizontal [72], respectively, just as shown in Figure 1.21. The chemical reaction equations in different zones are shown below:

$$\text{In the source zone [73]: } Ga(l) + HCl(g) = GaCl(g) + 1/2H_2(g) \tag{1.1}$$

$$GaCl(g) + HCl(g) = GaCl_3(g) + 3/2H_2(g) \tag{1.2}$$

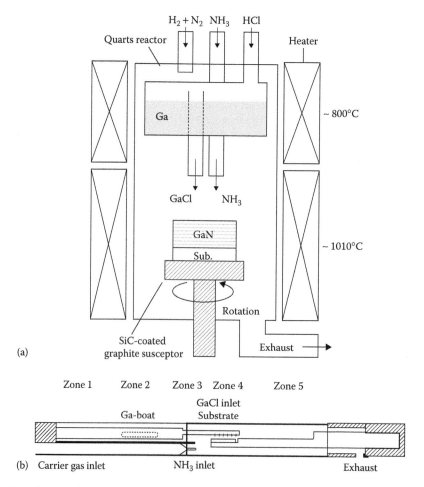

FIGURE 1.21 Schematic drawings of HVPE system: (a) vertical type (From Fujito, K. et al., *J. Cryst. Growth*, 311, 3011–3014, 2009. With permission.) and (b) horizontal type (From Richter, E. et al., *J. Cryst. Growth*, 277, 6–12, 2005. With permission.)

In the deposition zone [74]: $GaCl(g) + NH_3(g) = GaN(s) + HCl(g) + H_2(g)$ (1.3)

$$3GaCl(g) + 2NH_3(g) = 2GaN(s) + GaCl_3(g) + 3H_2(g) \qquad (1.4)$$

Two key processes for HVPE-growth GaN substrate are (1) crack-free GaN thick layer (>100 μm) with high crystal quality on foreign substrates, and (2) removing GaN from foreign substrates.

Epitaxial lateral overgrowth (ELOG) is a traditional method to effectively reduce the dislocation density and release the growth and thermal stress. The standard processes include depositing SiO_2 mask film, etching the window area, and laterally overgrowing the mask. Crack-free thick HVPE GaN layers with 300–500 μm were demonstrated by using ELOG on 2-in sapphire [75] and GaAs substrates. Alternatives of W [76], SiN [77] and WSiN [78] masks instead of SiO_2, have also been tried. A successful developing method based on ELOG is inverse-pyramidal pits (DEEP) for further dislocation elimination, provided by Sumitomo Electric [79]. Dislocations were found to concentrate on the center of each large pit, leading to the low, dislocation density of about 2×10^5 cm^{-2} in the other areas.

The other techniques by inserting a voids [80] or nanorods [81] layer, were have also been used to increase the crack-free thickness, and be also helpful for removing the GaN thick layer from the substrate. A simple method without mask and etching processes was developed by us, named as pulsed gas flow modulation (PFM) method [82]. Crack-free thickness was increased up to 300 μm by PFM method. The improved method named as slowly periodic adjust growth (SPAG), is schematically shown in Figure 1.22 [83]. Growth conditions are slowly and periodically adjusted between two states of stress released and low defect. In the stress released state, a high growth rate and rough surface was obtained, which is helpful to release the stress in GaN film. However, in the other state, it was dominated by the process of lateral growth, and thus, leading to a smooth surface with high crystal quality [83]. With these two growth states repeat alternately, the stain release and surface rehabilitation work periodically, so that stress could hardly be accumulated and high crystal quality and crack-free GaN epilayers of 370 μm are achieved with thickness over 350 μm. The atomic force microscopy (AFM) image of the sample surface shows a surface roughness of 0.61 nm (10 × 10 μm) in Figure 1.22c.

Although crack-free GaN with high quality by HVPE represents a good intermediate step toward FS-GaN substrate, it is still a great challenge in removing the foreign substrate without breakage. One removal way is a chemical etching in case of using GaAs or $LiGaO_2$ as a substrate. Sumitomo Electric Industries Ltd successfully manufactured a 6-inch free-standing GaN wafer and removed it from the GaAs substrate [84]. But it is not easy to grow GaN on GaAs substrate, and few companies grow GaN layer in such a way. A more popular technique was called the laser lift-off (LLO) process using pulsed UV lasers [85,86]. GaN near the interface absorbs the laser energy to decompose, and the GaN layer is

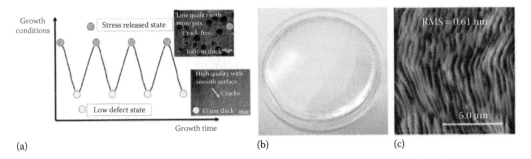

(a)   (b)   (c)

FIGURE 1.22 (a) Schematic drawings of slowly periodic adjust growth (SPAG) method, in which growth conditions are slowly adjusted periodically between stress released state and low-defect state. (b) The photo of a 370 μm thick GaN without crack grown by SPAG method. (c) The AFM image of relatively thick GaN layer over 10 × 10 μm scanned area.

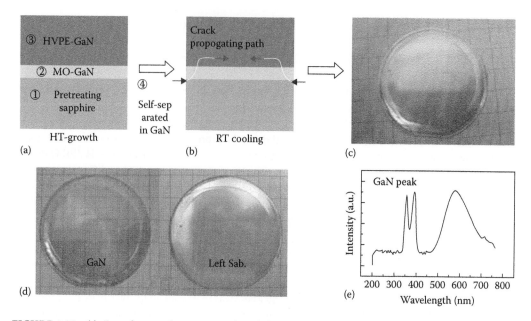

FIGURE 1.23 (a) Growth procedure in an order of the numbers for the sapphire pre-h-treatment method; (b) Delamination of self-separation caused by the large thermal mismatch stress and the edge shear stress; (c) Self-separated GaN layer on sapphire substrate; (d) 2-inch free-standing GaN substrate and sapphire substrate with non-mirror surface on top after GaN layer removal, and (e) CL spectrum from the sapphire surface after separation, strong GaN peak indicating GaN film remained as the proof of part of separation path inside the GaN layer. (From Tadashige, S. et al., *Jpn. J. Appl. Phys.*, 52, 08JA08, 2013. With permission.)

then separated from the substrate. This method succeeded in the manufacturing process of Lumilog Ltd [87]. The other approach was developed with a special buffer or a weak interface region to help separation. A good example is the void-assisted separation acted by sputtered TiN, which allows crack-free 3-inch FS-GaN substrate to be fabricated in Hitachi Cable [88]. Other void-assisted separation processes, such as Ni mask with nanoimprint lithography method [89] and growth of a void or pit inducing GaN buffer layer [82,90], were also introduced successfully to fabricate 2-inch and nearly 4-inch FS-GaN wafers.

Our lab develops a self-separation method to remove the foreign substrates. Sapphire is prethermally treated in a high-temperature range of 1,200°C–1,500°C before the deposition. It is better for the GaN thick layer to peel from sapphire when cooling down to room temperature. The detailed processes and the obtained 2-in FS-GaN substrates are shown in Figure 1.23. The separation was observed to take place in the GaN layer, not along the interface (indicated by CL spectrum). It indicates that the shear stress exists in the interface near the edge, which being the driving force to self-separation. High crystal quality was proved by the full width at half maximum (FWHM) of XRC, which are 108 and 81 arcsec for (002) and (102) reflection, respectively. The dislocation density was estimated by both etch pit density and CL method to be about 2–4×10^6 cm^{-2}.

1.4 Characterization of III-Nitride Semiconductors

1.4.1 Characterization of Microstructure and Morphology

The III-nitride semiconductors are a kind of smart materials which preserve high optical and electrical properties in spite of extremely high density of defects in their crystalline structures. In most III-nitride devices, the defects are inert in the carrier transportation and recombination because the potential barriers and localization centers prevent the carriers from moving toward these defects. However, the crystal quality is always the key issue for high-performance devices including III-nitrides. The lattice structure and defect formation, development and multiplication, and strain status should be aware for optimization of the growth condition and devices epi-structures. The diffraction from the lattice atoms by X-ray or electron de Broglie wave can characterize the crystalline structure including the information for lattice and defects of III-nitrides. On the other hand, through the tiny sharp tip interaction with the surface atoms of the samples, the morphology can be measured for characterizing the solid structures.

1.4.1.1 X-ray Diffraction Analysis

X-ray diffraction (XRD) is one of the most popular methods for III-nitrides to demonstrate the crystalline quality, polarity, components, multiple quantum wells (MQWs), alloy disorder, and so on. When the X-rays penetrate in the solid and the path difference between the rays from the neighboring crystal planes is an integer number of wavelength, there will be constructive interference, and the diffracted intensity will be a maximum. It is the Bragg's law, which can be expressed by the formula:

$$n\lambda = 2d\sin\theta \tag{1.5}$$

where:
 λ is the wavelength of the X-ray
 d is the spacing of the crystal plane

However, not all reflections satisfying the Bragg's condition will be detected. According to the dynamical calculation [91], some reflections are absent because the atomic arrangements of the particular material within the unit cell give rise to destructive interference. In order to find the Bragg's reflection, the sample and detector should be rotated to an appropriate position, which is shown in Figure 1.24. With reference to an epitaxial wafer, θ rotation is used to find a Bragg's reflections, φ rotation is usually used to find the correct azimuth angle and χ rotation is used for a fine adjustment to optimize a peak alignment.

2θ is the detected beam angle and ω represents the incident beam angle. In 2θ/ω scan, the step size for the 2θ scan has twice the value of the change in ω, which is used to measure the crystal plane spacing by Equation 1.5. The plane spacing can be used to calculate the alloy content [92] and strain [93] using the Vigard's law and combination with other experimental results. Figure 1.25a shows the 2θ/ω XRD scan curves of *a*-plane GaN grown on *r*-plane sapphire [94]. The *in*-plane orientation in

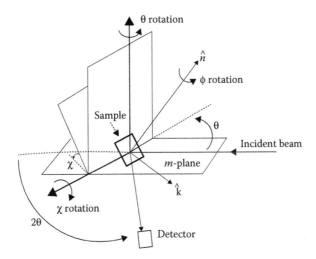

FIGURE 1.24 Schematic diagram of triple axis XRD diffractometer.

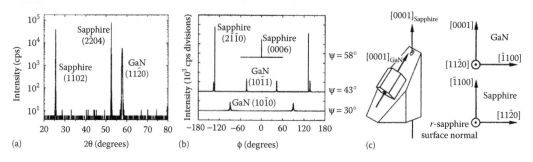

FIGURE 1.25 (a) Scan of GaN on *r*-sapphire which identifies the growth surface as $(11\bar{2}0)$ *a*-plane GaN. (b) Compilation of off-axis scans used to determine the *in*-plane epitaxial relationship between GaN and *r*-sapphire. The angle of inclination c used to access the off-axis reflections is noted for each scan. (c) Schematic illustration of the epitaxial relationship between GaN and *r*-plane sapphire. (From Zheng, X. H. et al., *J. Cryst. Growth*, 255, 63–67, 2003. With permission.)

a-plane GaN can be determined by χ and ϕ rotations, as shown in Figure 1.25b. The correlation between the ϕ positions of these peaks determined the following epitaxial relationship: $[0001]_{GaN}//[1\bar{1}01]_{sapphire}$, $[1\bar{1}00]_{GaN}//[11\bar{2}0]_{sapphire}$, as shown in Figure 1.25c.

The ω scan is a special method for scanning through a reciprocal lattice spot or a group of reciprocal lattice features. The rocking curve is collected by an open detector meaning that the 2θ angular acceptance is large. By using the triple-axis diffractometer, the rocking curve widths of (0002), (0004), and (0006) reflections are influenced by the tilt of the GaN crystal columns. If one assumes that both screw dislocations ($\mathbf{b} = [0001]$) are related to an edge type misfit dislocation in the interface with the same Burgers vector that leads to a corresponding tilt of the single crystallites, the density of these screw dislocations N_S can be obtained using the following equation [95]:

$$N_S = \frac{\alpha_\Omega^2}{4.35 b_c^2} \tag{1.6}$$

where:

α_Ω is the tilt angle which can be derived by the rocking curve widths of GaN (0002), (0004), and (0006)

$b_c = 0.5185$ nm

The density of the edge threading dislocation can also be obtained by:

$$N_E = \frac{\alpha_\Phi^2}{4.35 b_E^2} \tag{1.7}$$

where:

α_Φ is the twist angle which can be measured by performing Φ-scans on asymmetric reflections

$b_E = 0.3189$ nm

Simpler empirical formulas are also used to calculate the dislocation densities [96]:

$$N_S = \frac{\beta_{(0002)}^2}{9 b_c^2}, N_E = \frac{\beta_{(10-12)}^2}{9 b_E^2} \tag{1.8}$$

where $\beta(0002)$, $\beta(10-12)$ are the full width at half maximum (FWHM) measured by XRD rocking curve.

Lateral epitaxial overgrowth (LEO) is a very important technique to improve the crystal quality. The tilt of the overgrown wings to the window region should be measured and controlled.

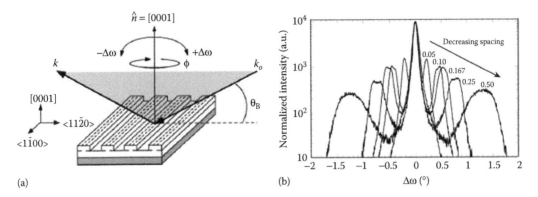

FIGURE 1.26 (a) Schema of diffraction geometry employed to measure the tilt of crystal planes in stripe wings relative to those in windows and (b) X-ray rocking curves of (a) series A stripes for FFs from 0.05 to 0.50. (From Fini, P. et al., *J. Cryst. Growth*, 209, 581–590, 2000. With permission.)

The measurement is performed in Figure 1.26 [97]. The diffraction plane is perpendicular to the stripe to measure the wings' tilt. The shoulder peaks in the rocking curve appear when the wing crystal satisfies for the Bragg's law. When the fill factor (FF) of GaN windows is increased, the wings peak in rocking curve become broad and far away from the main peak of windows, as shown in Figure 1.26b. From these XRD results, local V/III ratio and interfacial force are attributed to the lateral overgrowth of GaN on SiO_2 mask.

A 2-axis reciprocal space map (RSM) can be obtained when $2\theta/\omega$ scans are repeated for a sequence of offset values. When plotted in reciprocal space units the 2-axis maps can represent the reciprocal lattice spots. Prismatic stacking faults (PSFs) in *a*-plane films can be detected by collecting RSMs in the asymmetric diffraction geometry, in which the scan progresses in a plane perpendicular to the [0001] direction, and energy streaking due to the PSFs would be observed in the RSM [98]. According to the invisibility criterion, $g \cdot R = n$, where $R = 1/2$, the PSFs should be invisible for the entire $h0$-$h0$ series of reflections, but visible for the $11\bar{2}0$ reflection, when the sample is rotated perpendicularly to the [0001] direction. As shown in Figure 1.27, significant streaking in the lateral direction can be seen even in the $20\bar{2}0$ and $30\bar{3}0$ reflections, where the PSFs should not affect the data. Figure 1.27 also shows an RSM of the $11\bar{2}0$ reflection taken with the sample rotated *in*-plane perpendicular to the [0001] direction, indicating that comparatively similar broadening scattering from PSFs is not considered to be significant for the present sample set. However, it can be seen that broadening in the direction is expected due to mosaic tilt decreases as 2θ increases.

1.4.1.2 Scanning and Transmission Electron Microscopy

Scanning electron microscope (SEM) is to scan the high energy electron beam focused on the specimen surface, and obtain an image by collecting the signals at each position. The signals produced by a SEM include secondary electron, light, back-scattered electrons (BSEs), characteristic X-rays, and so on. Second electron imaging is the most common function for a SEM equipment, which produces very high resolution (~1 nm) images of the sample's morphology. The magnification can reach up to 500,000 times. In the earlier reference reported by Amano and Akasaki [99], the GaN nucleation and coalescence on AlN buffer are clearly observed by SEM. The AlN buffer is a milestone in the research field for III-nitride. Nowadays, SEM has become a popular method to study the earlier stage of III-nitride epitaxy, which can provide a lot of information on nucleation, island growth, coalescence, and so on. Besides GaN epitaxy, GaN nanostructures can also be observed and analyzed by SEM, such as VLS synthetized GaN nanowire [100] and photon-assisted etching GaN whisker [101], as shown in Figure 1.28.

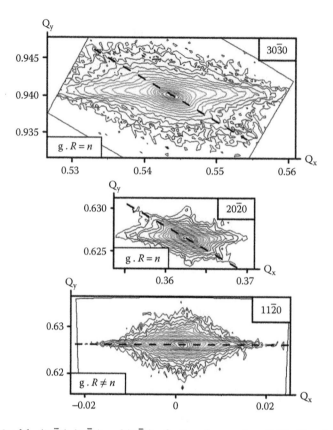

FIGURE 1.27   RSMs of the $(20\bar{2}0)$, $(30\bar{3}0)$, and $(11\bar{2}0)$ reflections from *a*-plane GaN with a single ScN interlayer.

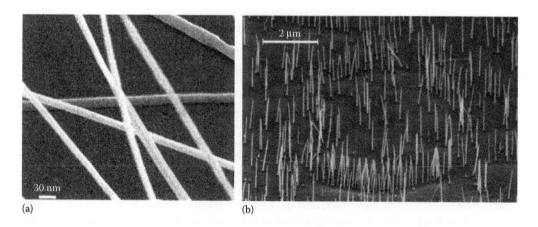

FIGURE 1.28   SEM images of (a) the straight and smooth GaN nanowires. (From Ambacher, O. et al., *J. Appl. Phys.*, 87, 334, 2000. With permission.) (b) Whiskers produced by selective etching of dislocations in a HVPE GaN film. (From Smorchkova, I. P. et al., *J. Appl. Phys.*, 86, 4520, 1999. With permission.)

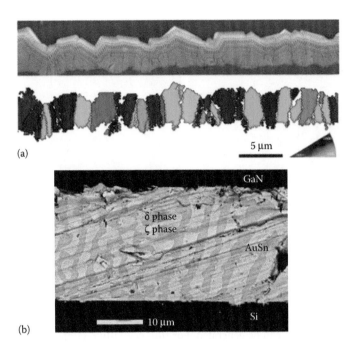

FIGURE 1.29 (a) SEM and EBSD images of laterally unconstrained growth of GaN from textured AlN seed. (From Leung, B. et al., *Adv. Mater.*, 25, 1285–1289, 2013. With permission.) (b) EBSD image of GaN/Au$_{0.8}$Sn$_{0.2}$/Si, which is bonded at 350°C, 10 min, and 0.276 MPa. (From Tian, P. et al., *Mater. Sci. Eng. B.*, 175, 213–216, 2010. With permission.)

Because BSEs are reflected from the sample by elastic scattering, the BSE images can provide information about the distribution of different elements in the sample. The crystalline orientation and phase of microdomains can be shown in electron back-scattering diffraction (EBSD). Figure 1.29 shows the EBSD images of GaN islands from textured AlN seed and AuSn phases after high-temperature and high-pressure treatment. In Figure 1.29a the position and specific *in*-plane orientations of each grain are found, and these initial conditions are input into the simulation of growth obeying ES selection rules. The agreement of the simulated grain morphology based on EBSD orientation map provides a quantitative support to the claim that evolutionary selection is responsible for the realization of single-crystalline GaN on SiO$_2$ [102]. Figure 1.29b shows two phases, δ and ζ phases in the AuSn eutectic bonding layer, corresponding to the bright and gray contrasts. It is found that phase penetration across AuSn layer enhanced the thermal and electrical conductivity, which could reduce the junction temperature and spread the current uniformly [103].

Transmission electron microscopy (TEM) is a microscopy technique analogy to optical microscopy. But the magnification is about a thousand times than that for an optical one. The highest resolution of TEM is achieved as 0.5 Å [104]. When a beam of electrons is transmitted through an ultrathin specimen, it will interact with the specimen. An image is then formed by the interaction of the electrons transmitting through the specimen. The most common mode of operation for a TEM is the bright field imaging mode. Typical TEM micrographs taken from two cross-sectional specimens, with the direction of incident electron beam parallel to [11-20]$_{GaN}$//[1-100]$_{Al2O3}$ or [1-100]$_{GaN}$//[11-20]$_{Al2O3}$ zone axes, respectively, are shown in Figure 1.30 [105]. A high density of dislocations parallel to the growth direction (i.e., the *c*-axis) can be seen. They are also distributed uniformly throughout the film except for a region within about 50 nm from the interface. From numerous tilting experiments, the planes on which the curved dislocations lay were determined to be close to [1100] GaN. Using the conventional **g·b** technique (where **g** is the reflection vector used for imaging a dislocation with Burgers vector **b**), the direction of **b** was readily determined.

 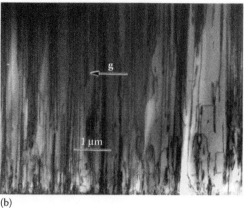

(a)   (b)

FIGURE 1.30 Typical low magnification bright-field micrographs showing the whole thickness of the film: (a) $[11\bar{2}0]_{GaN}$ cross-sectional TEM specimen ($\mathbf{g} = 2\bar{2}02$), (b) $[1\text{-}100]_{GaN}$ cross-sectional TEM specimen with $\mathbf{g} = 11\bar{2}0$. (From Ning, X. J. et al., *J. Mater. Res.*, 11, 580, 1996. With permission.)

Selected area diffraction (SAD) is a strong tool to analyze the crystalline structure in a micro–area by adjusting the field slit at the first imaging plane. The back focal plane is placed on the imaging apparatus, and then a diffraction pattern can be generated. For thin crystalline samples, this produces an image that consists of a pattern of dots in the case of a single crystal or a series of rings in the case of a polycrystalline or amorphous solid material. Figure 1.31 is a typical SAD pattern from the interface region of a cross-sectional TEM specimen with the incident electron beam parallel to the $[11\text{-}20]_{GaN}$ direction [105]. This Figure is a superposition of the $[1\text{-}120]_{GaN}$ and $[1\text{-}100]_{Al2O3}$ zone axes diffraction patterns of the GaN film and sapphire substrate, respectively. The composite diffraction pattern of Figure 1.31 indicates that the GaN film is single crystalline with an epitaxial orientation relationship (OR) with respect to the sapphire substrate given by $(0001)_{GaN}//(0001)_{Al2O3}$, or $(11\text{-}20)_{GaN}//(1\text{-}100)_{Al2O3}$.

The formation, propagation, and multiplication of threading dislocations in GaN films can be observed by TEM. Figure 1.32a shows the substrate scheme for GaN laterally overgrowth. SiO$_2$ mask was deposited

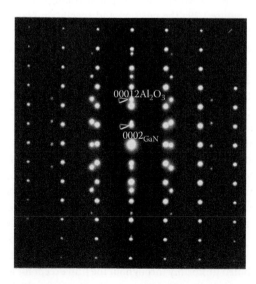

FIGURE 1.31 Selected area electron diffraction pattern of the interface region from a $(11\bar{2}0)_{GaN}$ cross-sectional TEM specimen. The pattern is a superimposition of diffraction patterns of GaN and sapphire along the $[11\bar{2}0]_{GaN}$ and $[1\bar{1}00]_{Al2O3}$ zone axes, respectively.

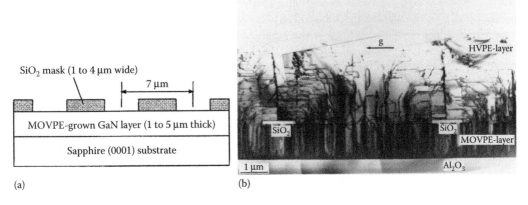

(a) (b)

FIGURE 1.32 (a) Schematic diagram of the substrate structure used for HVPE growth, and (b) cross-sectional TEM [11-20] image taken from the area near the interface between HVPE and MOVPE-grown GaN layers with **g** = 1–100. (From Ning, X. J. et al., *J. Mater. Res.*, 11, 580, 1996. With permission.)

and then fabricated into mask/window stripes with a period of about 7 μm aligned along the <11-20> of the GaN film [105]. Figure 1.32b is a cross-sectional TEM [11–20] images near the interface of GaN/ sapphire. The image was taken under two-beam diffraction conditions with **g** = 1–100 (**g** is the reflection vector used for imaging). In TEM, HVPE-grown GaN, MOVPE- grown GaN, SiO_2 and sapphire are observed in the windows region, with all the dislocations in the MOVPE-GaN propagated into the HVPE-GaN without any new defects generated. The SiO_2 mask blocks the dislocations from extending to the HVPE-GaN. More interestingly, the dislocation is inclined to the lateral direction and piled up along the [0001] direction in the wing region. These dislocation propagations lead to the formation of a dislocation loop above the SiO_2 mask and the reduction of vertical aligned threading dislocation density.

In order to investigate the GaN/sapphire interface structure in greater detail, a high-resolution electron microscopy (HREM) study of the interface region was carried out, as shown in Figure 1.33 [106]. The lattice images can be seen from both the GaN epilayer and sapphire substrate. The closest separation between the bright spots along the *c*-axis of GaN is $1/2c_{GaN}$, while it is $1/3\ c_{Al2O3}$ in sapphire. In the horizontal direction, the closest separation between the bright spots in GaN and sapphire are $d_{GaN}(1\text{-}100) = \sqrt{3}a_{GaN}/2$ and $d_{Al2O3}(11\text{-}20) = a_{Al2O3}/2$, respectively. It is clearly seen that the $(1120)_{Al2O3}$ planes of sapphire are parallel to the $(1100)_{GaN}$ planes of the GaN film. According to Figure 1.33, on the average, every eight

FIGURE 1.33 HREM micrograph from the film/substrate interface region. The incident beam is along the $[1120]_{GaN}$ zone axis. (From Sakai, A. et al., *Appl. Phys. Lett.*, 71, 2259, 1997. With permission.)

(1120)$_{Al2O3}$ planes of sapphire match seven (1100)$_{GaN}$ planes of GaN. Sets of Moiré fringes are observed in the TEM image on the tilted samples around <11-20> $_{Al2O3}$ [107]. At the GaN/Al2O3 interface, surface steps with a height equal to 1/6 c_{Al2O3} of the sapphire substrate are clearly seen (one is arrowed in Figure 1.33). The few (0001)$_{GaN}$ planes of the GaN film closest to the interface are elastically bent to accommodate the stepped surface of the substrate. The prismatic partial dislocation of the type $\mathbf{b}_p = 1/2<0001>$ bounded at the basal faults can also be observed in TEM.

1.4.1.3 Atomic Force Microscopy

Atomic force microscopy (AFM) was invented by Binnig, Quate, and Gerber in 1986 [108]. As one of the important members of scanning probe microscopy (SPM), AFM can do nondestructive testing, reaching a lateral resolution of 1 nm and a vertical resolution of 0.1 nm. The topography image of the sample can be obtained by measurement of the force on a sharp tip (insulating or not) created by the proximity to the surface of the sample, being as small as 10^{-18} N. This force is kept small and at a constant level with a feedback mechanism. When the tip is moved sideways, it will follow the surface contours such as trace B in Figure 1.34 [109].

As shown in Figure 1.35, on the basis of the different intervals of the interaction force between the probe and sample surface, AFM testing can be divided into three kinds of modes: the contact mode,

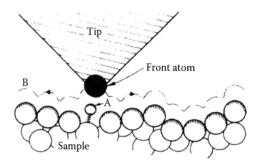

FIGURE 1.34 Description of the principle operation of an AFM. (From Keller, S. et al., *J. Appl. Phys.*, 86, 5850–5857, 1999. With permission.)

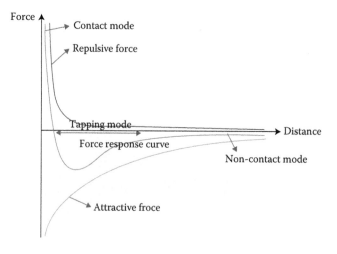

FIGURE 1.35 Three kinds of AFM work modes in different interaction intervals between the probe and the sample surface.

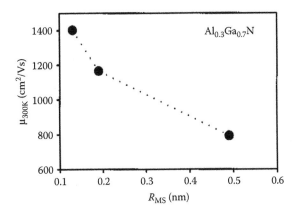

FIGURE 1.36 Dependence of the electron mobility on the surface roughness prior to $Al_xGa_{1-x}N$ deposition for heterostructures with $x = 0.3$. The RMS values were calculated using an AFM software.

which is particularly useful for the relatively rough surface, the tapping mode, which has a higher lateral resolution than the contact made and does less damage to a soft surface, and finally the noncontact mode.

Other kinds of force besides inter-atomic forces, such as electrostatic force, magnetic force, adhesive force, and friction force can also have an influence on the force on the tip created from the sample surface. As such, various material properties can be characterized by the application of the electrostatic force microscopy (EFM), kelvin force microscopy (KFM), scanning capacitance microscopy (SCM), magnetic force microscopy (MFM), chemical force microscopy (CFM), and lateral force microscopy (LFM).

In addition, there are few limits on the structure and properties of materials to the AFM testing as it depends only upon the interaction force between the probe and sample surface. On the other hand, the AFM also has significant potential to micromachining. Therefore, the AFM became more and more popular with the rapid development of semiconductor and microelectronic fields in recent years.

As we all know, the surface roughness of III-nitrides epilayer is an important index for the performance of devices. According to a previous study, for high mobility AlGaN/GaN heterostructures, the increase in surface roughness from the root mean square (RMS) 0.13 to –0.49 nm leads the electron mobility to drop dramatically from 1,400 to 800cm²/V s, as shown in Figure 1.36. The AFM can thus play a great role in assessing the surface roughness, and hence, advance the research and development of devices.

1.4.2 Characterization of Optical Properties

1.4.2.1 Photoluminescence and Cathodoluminescence Spectra

Photoluminescence (PL) is a process of spontaneous light emission from a solid after absorbing photons, in which electrons in excited states relax to ground state through radiative transitions. The energy and intensity of the emission are determined by energy-band structure, carrier thermal distributions in the band and the levels of impurity states and defect states, so that a PL spectrum brings much principal information about the materials. Now, PL has not only been a powerful tool for studying band structure and emission mechanisms but also widely applied in the characterization of semiconductor materials and devices.

PL spectra can be recorded with an experimental arrangement such as the one shown in Figure 1.37. The sample is mounted on the cold finger of a variable-temperature (10–300 K) closed-cycle He compressor. Samples are excited by laser or bright lamp, and the emitted lights are collected by a convex lens and focused onto the entrance slit of the spectrometer. Through a spectrometer and a sensitive detector, such

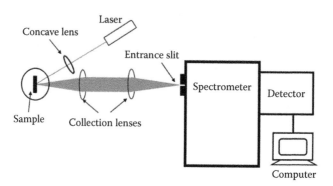

FIGURE 1.37   Set up of PL measurement.

as photomultiplier tube (PMT) and charge coupled device (CCD), the intensities at each wavelength are measured and recorded by a computer.

With advantages of high spectral resolution, easy adjustment of temperature and intensity of the excitation light, and the possibility of excitation with a short wavelength, the PL technique has many applications in characterizing GaN materials. The most common and simplest use of PL is for measuring the fundamental band gap in GaN and its alloys with InN and AlN. In 2002, the PL of InN showed an energy gap between 0.7 and 0.8 eV, which is much lower than the commonly accepted value at that time [110].

The PL is also used to probe the optical signatures associated with impurities in doped semiconductors. Due to high spectral resolution, the emission mechanisms can be studied in detail based on the PL spectra. In turn, according to the wavelength and intensity of the light emission associated with impurities in the PL spectra, we can sometimes identify the intentional or unintentional impurities. There is often a 2.8 eV emission from Mg-doped GaN, which even dominates PL spectrum at room temperature when the Mg concentration is very high. Kaufmann et al. reported measurements of the intensity density dependence of PL from Mg-doped GaN grown by metalorganic chemical vapor deposition. The dependence of luminescence intensity and peak position on excitation density is shown in the Figure 1.38. There are three peaks, among which the one of 3.4 eV is related to the recombination of free exciton, and the one of 3.2 eV is due to the conduction band-to-impurity transition involving shallow Mg impurities [111]. In the case of 2.8 eV, the peak energy shifts to higher energy with increasing excitation density, which is a clear indication for recombination of electrons and holes trapped at spatially separated donors and acceptors [112]. For the acceptor, the isolated Mg_{Ga} is a natural choice. The deep donor is believed to be a nearest-neighbor associate of Mg_{Ga} and a nitrogen vacancy, formed by self-compensation.

Cathodoluminescence (CL) is a phenomenon of photon emission when electrons are impacting on a luminescent material. Similar to PL, CL is a spontaneous light emission from radiative recombination of nonequilibrium carriers and usually used to study the emission mechanisms and characterize the properties of materials with impingement of a high energy electron beam. CL measurements are performed in an electron microscope equipped with a CL detector. As sketched in Figure 1.39, a sample is impinged by a focused electron beam and the emitted light is collected by an optical system, such as a parabolic mirror. Through the detection of monochromatic light passing through a spectrometer by PMT or CCD, the CL spectrum can be obtained. By moving the electron beam on the surface of a sample and measuring the light emitted at each point, the optical activity of the specimen can be mapped.

CL measurement is a very powerful technique often used to assess the optical properties of semiconductors. The primary advantage is the ability to obtain high spatial resolution.

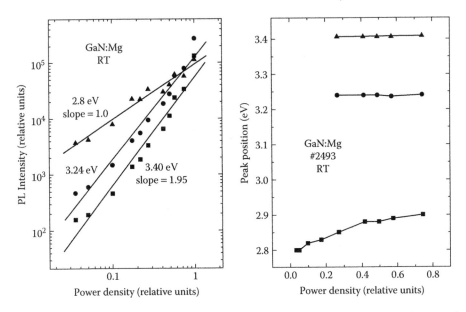

FIGURE 1.38 Luminescence intensity and peak energies of the 2.8, 3.2, and 3.4 eV band as a function of excitation density. (From Takeuchi, T. et al., *Appl. Phys. Lett.*, 73, 1691, 1998; Wu, J. J., *Appl. Phys.*, 106, 011101, 2009. With permission.)

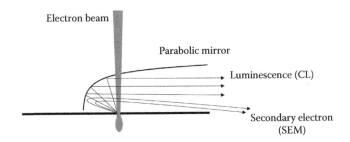

FIGURE 1.39 A schematic of the setup of a CL measurement.

Secondary electron images and optical properties on specific areas can be obtained in a measurement. In this way, the optical properties of a specimen can be correlated to surface morphology, which is useful for examining semiconductors, especially low-dimensional structures, such as QWs and quantum dots [113,114]. In 1996, Ponce et al. found in their room temperature study of CL (Figure 1.40) that yellow band emissions occurs at the outer edges of the crystallites and hypothesized that the sources of yellow emission are either dislocations at low angle grain boundaries in the material, or point defects which nucleate at the dislocations [115].

Analyzing electron energy dependence of CL signal by changing accelerating voltage provides the possibility to probe sample under the surface at different depths. In 1997, Chichibu et al. [116] demonstrated an exciton localization into InN-rich quantum-disks originating from compositional undulation in InGaN SQWs with spatially resolved CL spectrum mapping, as shown in Figure 1.41. In 2001, Pereira et al. studied the compositional profile of InGaN epilayer using depth-resolved CL [117]. Figure 1.42 shows in detail the $In_xGa_{1-x}N$ related emission dependence on the electron beam energy. They calculated the content of In using the peak energy of CL spectra and the electron energy deposition profile for different accelerating voltages using Monte Carlo simulations and found that the In mole fraction decrease linearly from the near surface. The result was confirmed by the Rutherford backscattering spectrometry.

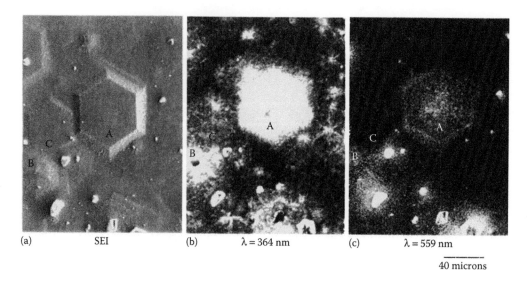

(a) SEI (b) $\lambda = 364$ nm (c) $\lambda = 559$ nm

40 microns

FIGURE 1.40 Undoped film with large hexagonal crystallites (a) secondary electron image, (b) CL at 364 nm, and (c) CL at 559 nm. (From Ponce, F. A. et al., *Appl. Phys. Lett.*, 68, 57–59, 1996. With permission.)

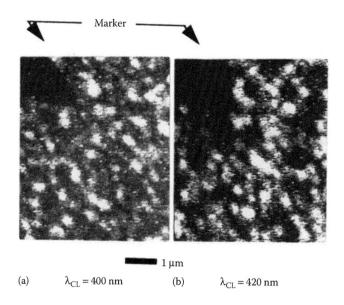

Marker

1 μm

(a) $\lambda_{CL} = 400$ nm (b) $\lambda_{CL} = 420$ nm

FIGURE 1.41 Monochromatic scanning CL images of GaN-capped $In_{0.2}Ga_{0.8}N$ SQW taken at wavelengths of (a) 400 nm (3.100 eV) and (b) 420 nm (2.952eV). (From Chichibu, S. et al., *Appl. Phys. Lett.*, 71, 2346–2348, 1997. With permission.)

Another important application of CL is to evaluate the density and spatial distribution of dislocations. The dislocations act as nonradiative recombination centers in GaN and are shown as dark dots in CL images. Figure 1.43 is a room temperature (RT) CL image at 360 nm for an epitaxial lateral overgrowth (ELOG) GaN sample [118]. In the three distinctive regions, the densities of dark spots within the window region and along the coalescence fronts are about 2 and 3×10^8 cm^{-2}, respectively, and consistent with the densities observed from wet chemical etching.

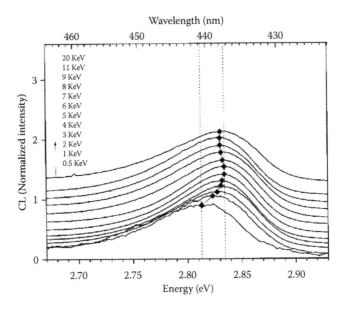

FIGURE 1.42 InGaN related CL emission acquired at electron energies ranging from 0.5 to 20 keV. (From Pereira, S. et al., *Phys. Rev. B.*, 64, 205311, 2001. With permission.)

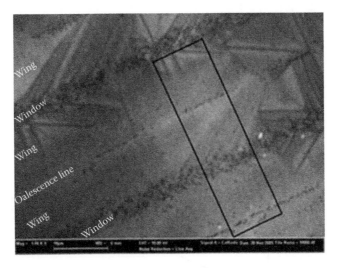

FIGURE 1.43 RT CL image taken at 360 nm for an ELOG GaN sample. (From Bretagnon, T. et al., *Phys. Rev. B.*, 73, 113304, 2006. With permission.)

Similar to PL spectrum, the CL spectrum is often used to study the band gaps and impurities in the GaN-based material. Sometimes, the application of PL is limited when large bandgap energies of group-III nitride for $Al_xGa_{1-x}N$ example with high Al composition, the excitation energies are beyond the photo energy of the commonly used He-Cd lasers. CL can work for wider band gap materials with excitation of a focused electron beam impingement. Rare-earth-doped semiconductors have been of considerable interest for their unique optical and electrical properties. In 1999, Lozykowski et al. demonstrated the observation of CL from GaN implanted with Dy, Er, and Tm. The rare-earth ions such as

Dy, Er, and Tm ions implanted into GaN at atmospheric pressure is found to be activated as luminescent centers emitting in the near-UV, visible, and infrared regions [119].

1.4.2.2 Absorption and Transmission Spectra

Ternary nitride alloys, being direct band-gap materials, have a wide range of potential applications in optoelectronic devices such as blue LEDs, blue lasers, UV LEDs, UV lasers, and UV photodetectors. Therefore, it is significant to know the fundamental optical properties of GaN and related compounds. Optical transmission (or absorption) is an effective and accurate method to determine the optical constants of GaN-based materials.

The transmission spectrum at room temperature for a 0.4-μm-thick GaN layer is shown in Figure 1.44 [120]. The inset in that Figure shows an expanded scale for the transmission above the band gap. As shown in Figure 1.45, the corresponding absorption coefficient is obtained by computing the optical transmission data of GaN. The fine structure of the valence bands can be determined by absorption spectrum measured

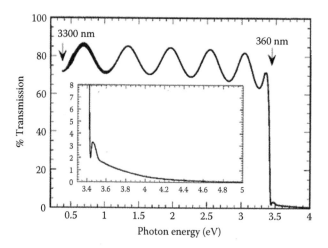

FIGURE 1.44   Transmission spectrum for GaN.

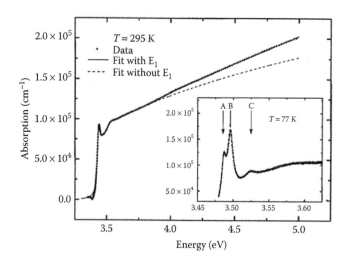

FIGURE 1.45   Absorption spectrum for GaN.

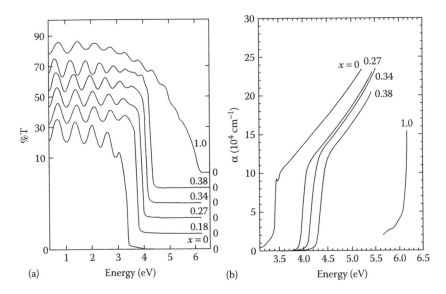

FIGURE 1.46 (a) Transmission spectrum and (b) absorption spectrum for AlGaN.

at low temperature. The inset is an expanded view of the excitonic structure at 77 K and the A, B, and C excitons are clearly resolved to show the excitonic absorption due to the three valence bands.

Figure 1.46a, b shows the transmission and absorption spectra for a sequence of thin $Al_xGa_{1-x}N$ samples [121]. In the binary compound, the GaN exciton is clearly visible, and an absorption tail due to the buffer layer can be observed just below the excitonic band edge. In the alloy films, $x = 0.27, 0.34$, and 0.38, the exciton is suppressed by the alloy broadening.

Figure 1.47 shows the square of the absorption coefficient as a function of photon energy. S.T. Liu et al. used the absorption edges to evaluate the band gap energies of the $In_xGa_{1-x}N$ alloys [122], and then a bowing parameter of $b = 1.9 \pm 0.1$ eV was obtained by using the following equation [123]:

$$E_g(x) = 0.64 \cdot x + 3.43 \cdot (1-x) - b \cdot x \cdot (1-x) \tag{1.9}$$

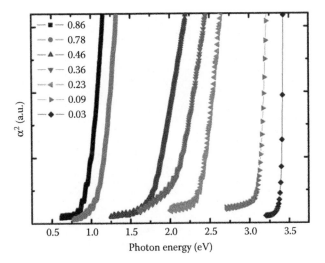

FIGURE 1.47 The room-temperature absorption coefficient squared as a function of photon energy.

1.4.2.3 Time-Resolved Luminescence Spectrum

Time-resolved optical spectroscopy system can be used to study the time evolution of the optical emissions or the dynamical processes of photoexcited carriers, which is beyond the reach of continuous-wave (CW) spectroscopy techniques. A CW spectroscopy study provides time-integrated (or average) information, while time-resolved spectroscopy studies reveal both time-resolved and time-integrated information. Understanding the dynamical processes of optical transitions is fundamental and technologically important for III-nitride semiconductor materials.

Time-resolved luminescence usually employs pulse excitation source, and the sample then emits luminescence signal that varies with time. According to the characteristics of signals, there are several kinds of techniques for time-resolved luminescence, such as streak camera, single-photon counting, pump-probe, and up-conversion. Up to now the only detector that can provide 1ps time resolution is the streak-camera system. Figure 1.48 shows a streak camera image of one AlGaN sample at 90 K, where the brightness of the image is proportional to the intensity of the incident optical pulses, the position in the horizontal direction corresponds to the decay time, and the vertical direction means different wavelengths or emission energies.

The luminescence strength in Figure 1.48 can be expressed as a function of time and wavelength:

$$I = I(\lambda, t) \tag{1.10}$$

If we integrate $I(\lambda, t)$ at some center wavelength λ_i, the decay curve $I_{\lambda_i}(t)$ at different wavelength can be obtained:

$$I_{\lambda_i}(t) = \int_{\lambda_i - \frac{\Delta\lambda}{2}}^{\lambda_i + \frac{\Delta\lambda}{2}} I(\lambda, t) d\lambda \tag{1.11}$$

Figure 1.49 shows a typical transient decay curve of PL emission from an AlGaN sample measured at its peak energy. Obviously, the PL emission curve exhibits two different stages, the rising part reflects the relaxation information of photo carriers, while the decay part can be used to investigate the carrier recombination dynamics. A lifetime τ can be obtained by fitting the decay curve, and it is a measure of various recombination processes in the material that include radiative and nonradiative channels.

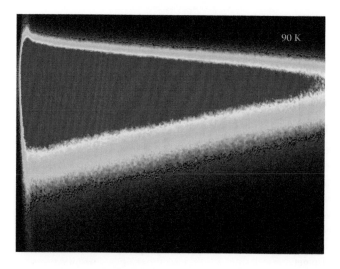

FIGURE 1.48 Streak-camera image recorded for AlGaN sample at 90 K.

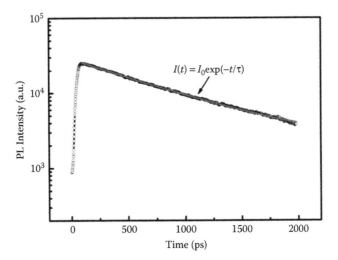

FIGURE 1.49 Transient decay curve of AlGaN sample.

The total recombination rate R is inversely proportional to the observed lifetime τ, being a summation of the radiative and nonradiative recombination rates. It can be expressed by the following equation:

$$R = R_{rr} + R_{nr} = \frac{1}{\tau_{rr}} + \frac{1}{\tau_{nr}} = \frac{1}{\tau} \tag{1.12}$$

where R_{rr} and R_{nr} represent radiative and nonradiative recombination rates, while τ_{rr} and τ_{nr} denote their corresponding lifetimes, respectively. Normally, the radiative and nonradiative lifetimes are inferred from the observable total lifetime τ, and the internal quantum efficiency (IQE) η, can be estimated from the following relation:

$$\eta = \frac{I(T)}{I(0K)} = \frac{R_{rr}}{R} = \frac{1}{1 + (\tau_{rr}/\tau_{nr})} \tag{1.13}$$

If we integrate $I(\lambda, T)$ at some specific delay time t_i, the luminescence spectrum $I_{t_i}(\lambda)$ at different delay time can be obtained

$$I_{t_i}(\lambda) = \int_{t_i - \frac{\Delta t}{2}}^{t_i + \frac{\Delta t}{2}} I(\lambda, t) dt \tag{1.14}$$

Figure 1.50 displays the time-resolved luminescence spectra at different delay times of AlGaN sample at 90 K. As time goes by, we could see the entire process changing, including the shift of the peak energy, the decay of the luminescence strength, as well as the evolvement of full width at half maximum; the energy relaxation information can thus be obtained from above results.

1.4.2.4 Pressure-Dependent and Polarized Luminescence Spectra

Hydrostatic pressure and carefully controlled uniaxial stress have long been used to study the properties of semiconductors. Hydrostatic pressure results in crystal volume compression without changing the symmetry; it is an isotropous pressure. The effect of hydrostatic pressure on the energy band and the optical properties of semiconductor materials is achieved by the variation of lattice constants. With the increasing of hydrostatic pressure and the compression of the lattice of constants, the overlap of wave function of electrons in the conduction band and the holes in the valence band will be changed, and the relative position of band edge in k space, as well as the energy difference between different energy valleys will also get changed. As to the apparatuses for hydrostatic pressure, the diamond anvil cells are usually used to apply the external pressure.

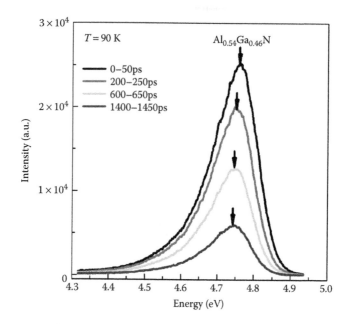

FIGURE 1.50   An example of time-resolved spectra recorded at different delay times.

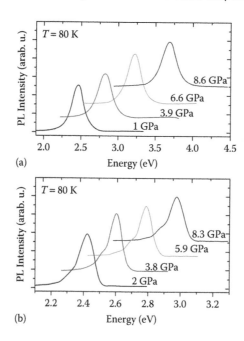

FIGURE 1.51   PL spectra measured at different pressure values (T = 80 K) for InGaN/GaN MQW structures. Sample A (a) and sample B (b).

Figure 1.51 shows the measured PL spectra for two InGaN/GaN MQW samples at different pressures. The values of dE_{pl}/dp can be determined form a fit to the dependence of E_{pl} versus pressure [124]. Hydrostatic pressure techniques are not only useful in determining the characteristics of the fundamental band gap and its evolution with pressure, but can also be used as a tool for studying the properties of structural defects and impurities. In the case of a impurity mediated radiative-recombination, its evolution of PL energy with pressure would be different from that of the band edge luminescence. While for low dimensional semiconductor structures with the built-in electric field, such as nitrides (due to piezoelectric and spontaneous polarization), the confinement and excitons binding energies usually depend weakly on applied external pressure, but the contribution of the built-in electric field could be large.

In addition to the hydrostatic pressure, the uniaxial pressure is also significant in the study of energy-band structure and optical properties. But different from the hydrostatic pressure, the uniaxial pressure changes the symmetry of the crystal structure, resulting in a variation of relevant properties, and it can especially lift up the degeneracy of valence band.

For polarized luminescence spectra, it can be used to investigate the optical polarization properties of materials or the light output characters of devices. Taking AlGaN for example, in the unstrained bulk GaN, the top valence subband is of Γ_{9v} symmetry, whereas in bulk AlN, due to the negative value of the crystal field splitting energy, the sequence of subbands is inverted with the topmost subband having Γ_{7v} symmetry [125]. This inversion results in the polarization of emitted light will switch from TE (polarized perpendicularly to the c axis) to TM (polarized along the axis) for AlGaN alloys at a particular composition. Consequently, the light extraction from surface-emitting devices based on [0001] oriented AlGaN structures becomes extremely difficult, while light extraction is a significant factor for the performance improvement of the device, such as LED. It has been demonstrated that the Al composition, strain, quantum confinement or polarization electric field plays key roles in determination of the optical polarization properties, all these factors change the valence band structure. In order to reflect the change of the band structure, the optical polarization experiments are always performed.

1.4.3 Characterization of Electrical Properties

1.4.3.1 Van der Pauw Hall Measurements

The Van der Pauw Hall measurements are frequently employed to measure the electrical properties of III-nitride, such as resistance, doping type, carrier concentration and mobility of the majority carrier.

With the rapid development of HEMT devices, Van der Pauw Hall measurements are widely used to determine the mobility and density of two-dimensional electron or hole gas (2DEG/2DHG) as well as the resistance of AlGaN/GaN or InAlN/GaN heterostructures. Haoran Li and co-workers at the University of California, Santa Barbara, obtained a 2DEG sheet density of $1.2 \times 10^{13} cm^{-2}$ and 2DEG mobility of 1,400 $cm^2V^{-1}S^{-1}$ by growing a 2.4 nm GaN cap layer on AlGaN/GaN heterostructures [126]. Researchers in Nagoya Institute of Technology optimized the electrical properties of AlGaN/GaN structure by inserting a 1 nm AlN space layer and obtained the mobility of 1,451 $cm^2V^{-1}S^{-1}$ and the carrier density of $1.1 \times 10^{13} cm^{-2}$ [127]. Researchers in the Peking University group reported high mobility AlGaN/AlN/GaN heterostructures grown on 4-inch Si substrate with electron mobility of 2,040 $cm^2V^{-1}S^{-1}$ at sheet charge density of $8.4 \times 10^{12} cm^{-2}$ [128]. They also reported high mobility InAlN/GaN heterostructures with a 2DEG mobility of 2,220 $cm^2V^{-1}S^{-1}$ at sheet density of $1.25 \times 10^{13} cm^{-2}$ [129].

1.4.3.2 Low-Temperature Magnet-Transport Measurements

The Shubnikov–de Haas (SdH) oscillations arising from the formation of Landau levels are visible at low temperatures and high magnetic fields. The oscillatory part $\Delta\rho_{xx}$ of the magnetoresistivity is expressed as [130]

$$\frac{\Delta\rho_{xx}}{2\rho_0} = 2\frac{\chi}{\sinh(\chi)}\exp\left(\frac{-\pi}{\omega_c\tau_q}\right)\cos\left(\frac{2\pi\varepsilon}{\hbar\omega_c}-\pi\right) \tag{1.15}$$

where:

$$\chi = \frac{2\pi^2 k_B T}{\hbar\omega_c}$$

$$\omega_c = \frac{eB}{m^*}$$

$\varepsilon = \dfrac{\pi \hbar^2 n}{m^*}$ is the Fermi energy

τ_q is the quantum scattering time

B is the magnetic field

k_B is the Boltzmann constant

T is the absolute temperature

$\hbar$ is the reduced Plank constant

n is the sheet electron concentration of the 2DEG

The SdH frequency f only depends on the 2DEG concentration $f = hn/2e$, where h is the Planck's constant. From the frequencies, the 2DEG concentrations can be obtained. Furthermore, the Fermi energy level can be obtained from $E_F - E_i = n_i \pi \hbar^2/m^*$. The effective mass of a 2DEG is determined from the temperature dependence of the SdH oscillation amplitude at a fixed magnetic field. Approximating $\sinh(\chi)$ by $\exp(\chi)/2$, we can express the amplitude A of the SdH oscillation at a given magnetic field as

$$\ln\left(\frac{A}{T}\right) \approx C - \frac{2\pi^2 k_B m^*}{e \hbar B} T \tag{1.16}$$

where C is a temperature-independent term. A plot of $\ln(A/T)$ versus T yields a straight line with a slope of $\left(-2\pi^2 k_B m^*/e\hbar B\right)$ from which m^* can be determined. The quantum scattering time can be obtained from a plot of Y versus 1/B

$$Y \equiv \ln\left[AB \sinh\left(\frac{2\pi^2 k_B T}{\hbar \omega_C}\right)\right] = C' - \frac{\pi m^*}{e \tau_q} \frac{1}{B} \tag{1.17}$$

The double periodicity of the SdH oscillations was observed in modulation-doped $Al_{0.22}Ga_{0.78}N/GaN$ heterostructures. It was found that the occupation of the first two subbands by the 2DEG in the triangular QW at the heterointerface took place. The energy separation of the first and the second subbands in the QW was determined to be 75 meV just before the second subband was occupied [131]. It is found that the mobility of the 2DEG in the second subband is much higher than that in the first one. This is explained by the fact that the interface roughness scattering and alloy disorder scattering have much stronger influence on the transport properties of the 2DEG in the first subband than that in the second subband in $Al_xGa_{1-x}N/GaN$ heterostructures [132].

Effective-mass values of the 2DEG in the triangular QW at the heterointerfaces are obtained by analyzing the temperature-dependent SdH oscillations. It is found that the values have strong dependence on the magnetic field and 2DEG density. Such behavior is thought to be due to the conduction band nonparabolicity in GaN [133].

The double periodic SdH oscillations modulated by magneto-intersubband scattering (MIS) oscillations have been observed due to the intersubband scattering of the 2DEG at the lowest two subbands in the triangular QW at the heterointerface. It is found that the MIS oscillations become slightly weaker with increasing temperature. The modulation between SdH and MIS oscillations is strong between 10 and 17 K [134].

The nonequilibrium status of the 2DEG in the triangular QW at the $Al_xGa_{1-x}N/GaN$ heterointerfaces has been investigated. The GaN layer is thought to be the primary contributor of the excited electrons by illumination. The illumination decreases the electric field and weakens the quantum confinement of the triangular QW at $Al_xGa_{1-x}N/GaN$ heterointerfaces. It is also found that the energy separation between the subbands decreases after the illumination. The mobility of the 2DEG in the weakened triangular QW increases and the SdH oscillation amplitudes are enhanced when there is no additional subband occupation [135].

Magnetotransport properties of the 2DEG in lattice-matched $In_{0.18}Al_{0.82}N/GaN$ heterostructures have been studied. The clear SdH oscillations indicate the high crystal quality of the heterostructures. The double subband occupancy of the 2DEG in the triangular QW at the heterointerface is observed.

The 2DEG density is determined to be 2.09×10^{13} cm^{-2} and the energy separation between the first and the second subbands is 191 meV. Both of them are significantly higher than those in Al$_x$Ga$_{1-x}$NGaN heterostructures owing to stronger spontaneous polarization effect. The evident difference of the quantum scattering time constants of the 2DEG in the two subbands indicates that interface roughness scattering plays an important role in the transport properties of the 2DEG in In$_x$Al$_{1-x}$N/GaN heterostructures [136].

1.4.3.3 High Field Transport Measurements

With the development of semiconductor technology, GaN microwave power device has entered the submicron size range, the electric field in the active channel is far more than the typical values where the Ohm's law playing a role; hence, the device working performance depends not only on the material low-electric mobility properties but also the hot electron transport characteristics under high electric field. Therefore, in order to better understand and improve the device performance, the transport characteristics of GaN-based materials under high electric field is of crucial importance.

Compared with GaAs semiconductor material, GaN-based materials have a wider bandgap, higher polar optical phonon energy (LO), better thermal conductivity and low dielectric constant. These features ensure the high field transport properties of the GaN-based semiconductors superior to that of GaAs-base semiconductors. The most important high-field transport features include:

1. Peak electron drift velocity in theory forecasts in GaN epitaxial material up to 3×10^7 cm/s, far higher than that of GaAs (about 2×10^7 cm/s), this is due to the larger LO phonon energy (about 92 meV, in comparison to 36 meV in GaAs) which ensures the electrons can effectively be accelerated before reaching the LO phonon energy [137,138];
2. The corresponding electric field value for electronic peak drift velocity in GaN is more than 200 kV/cm, much higher than the threshold electric field of GaAs (about 3.8 kV/cm) [139], this is a consequence of the wide bandgap and high intervalley energy in GaN;

It is difficult to make a direct measurement of the velocity of electrons as a function of the electric field. Femtosecond time of flight method was used to study the velocity [140]; this approach requires that the sample has a very low background concentration and require an ultrashort optical pulse and a homogeneous substrate, making the experimental condition very demanding. While a more direct measurement method for conductance method [140] measures the high-field I-V characteristics under the pulse voltage to calculate the drift velocity. Samples are usually made H-shape geometry to form a limiting-current region, at the same time the pulse signal is used to avoid the generation of the Joule heating effect. Based on this method, N. Ma and B. Shen et al. investigated the influence of the channel geometry on the high-field electron transport properties in n-type GaN and found that electrons in narrow channels drift much faster than those in wide channels. They believed that the boundary-enhanced momentum relaxation of LO phonons can increase electron drift velocity in GaN through enhancing the energy relaxation process while weakening the momentum relaxation [141]. Very recently, the same research group also investigated the effect of light illumination on electron velocity of AlGaN/GaN heterostructures under high electric field and suggested an alternative way of improving the electron energy relaxation rate, and hence the electron velocity in GaN-based heterostructures [142] (Figure 1.52).

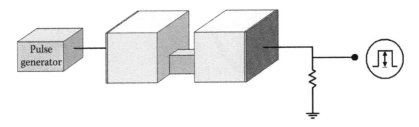

FIGURE 1.52 Testing measurement structure for conductivity.

1.4.3.4 C-V and Frequency-Dependent C-V Measurements

Regular C-V measurements: C-V characterization method has been widely adopted in silicon-based devices. The basic intention of high frequency C-V measurement is to determine impurity distribution. The basic principle is the same for GaN-based and silicon-based devices.

When a reverse voltage is applied to a p-n junction or a Schottky junction, there will be a space charge region depleted of carriers. By changing the reverse voltage by dV, the depletion layer width indicated by W will change with it and results in the change of space charge by dQ. The capacitance indicated by C can be expressed as:

$$C = \frac{dQ}{dV} = \frac{A\varepsilon_0\varepsilon_s}{W} \tag{1.18}$$

where:

A is the junction area
ε_0 is the vacuum permittivity
ε_s is the relative permittivity of the semiconductor

After a complex deduction, the carrier density can be expressed as:

$$N = \frac{C^3}{qA^2\varepsilon_0\varepsilon_s} \left(\frac{dC}{dV}\right)^{-1} \tag{1.19}$$

According to Equations 1.18 and 1.19, impurity distribution can be extracted from the measured C-V curve.

For AlGaN/GaN Schottky structures, due to the high-density 2DEG at the AlGaN/GaN interface, the C-V curve usually displays a platform as shown in Figure 1.53. In addition, a shift or a hysteresis of the C-V curve may indicate some trapping mechanism according to specific experiment conditions.

Frequency-dependent C-V measurements: Interface traps at oxide–semiconductor interface in metal–oxide–semiconductor (MOS) structures have been extensively studied using frequency-dependent measurements. Adapting the model to GaN-based HFETs with traps located at the heterojunction (referred to as interface traps) is straightforward, and thus we will introduce this model in AlGaN/GaN heterostructure as an example. The equivalent circuit is shown in Figure 1.54, where C_b is the barrier

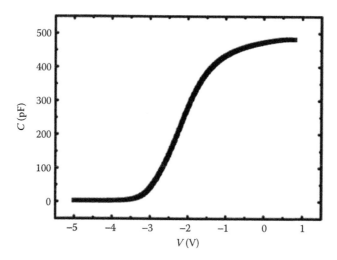

FIGURE 1.53   C-V curve of AlGaN/GaN heterostructure.

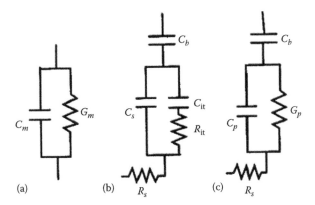

FIGURE 1.54 Models of the HFET structure used to extract trap parameters from the experimental measurements. To make parameter extraction from the measurement circuit (a) more straightforward, the assumed model with the trap states (b) is converted to a simplified circuit (c). (From Miller, E. J. et al., *J. Appl. Phys.*, 87, 8070, 2000. With permission.)

capacitance, C_s is the capacitance of the spacer layer near the heterojunction, R_s is the series resistance of the drain contact, and C_{it} and R_{it} are the interface trap capacitance and associated loss term, respectively, for the traps [143]. Assuming the trap states comprise a continuum of levels, C_p is given by:

$$C_p = C_s + \frac{qD_{it}}{\omega\tau\tan(\omega\tau)} \tag{1.20}$$

By fitting the measurement results with Equation 1.20, the trap density D_{it} and trap state time constant can be extracted.

For the case of evaluating the interface states between the insulating dielectric and AlGaN barrier, the model should be revised. Figure 1.55 is the result from the group at the Hong Kong University of Science and Technology [144], which evaluates the states at Al_2O_3/GaN-cap interface in the Al_2O_3/GaN/AlGaN/GaN MIS-structures using frequency and temperature dependent C-V measurements.

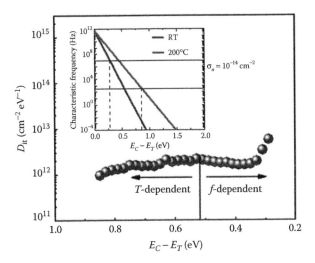

FIGURE 1.55 D_{it}-E_T mapping in an Al_2O_3(NIL)/GaN/AlGaN/GaN MIS diode. (From Yang S. et al., *Proceedings of the IEDM*, pp. 6.3.1–6.3.4, 2013. With permission.)

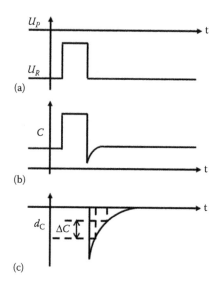

FIGURE 1.56   (a) Applied voltage; (b) Capacitance signal; (c) Capacitance transient.

1.4.3.5 Deep Level Transient Spectroscopy and Other Capacitance Spectroscopy

Deep-level transient spectroscopy (DLTS) is a high-frequency capacitance transient thermal scanning method and is useful for observing a wide variety of traps in semiconductors [145]. The presence of each trap is indicated by a positive or negative peak on a flat baseline plotted as a function of temperature. The heights of these peaks are proportional to their respective trap concentrations, the sign of each peak indicates whether it is due to a majority- or minority-carrier trap and the positions of the peaks are simply and uniquely determined by the integrator gate settings and the thermal emission properties of the respective traps. By a proper choice of experimental parameters, it is possible to measure the thermal emission rate, activation energy, concentration profile, and capture rate of each trap.

The physics of DLTS lies in the capacitance transients. After a trap pulse is applied on the tested junction, a capacitance recovery transient follows. For example, take a Schottky junction with electrons as majority carrier. Figure 1.56 is a schematic illustration of the time dependences of the various experimental parameters associated with a majority-carrier trap pulse sequence.

DLTS method has been widely employed to identify the deep levels in GaN-based HEMTs. Despite the advantages of GaN HEMTs in high-power and high-frequency device applications, the reliability of HEMTs is limited by various trapping phenomena such as "current collapse." The analysis of the properties of the deep levels located in a HEMT structure is important to address the current collapse issues. However, due to the small gate capacitance of the HEMTs, conventional capacitance-mode deep-level transient spectroscopy (C-DLTS) is difficult to be applied to actual transistors. A more accurate description of the properties of the traps related to GaN-based HEMTs reliability can be obtained by investigating the current transients (in the ON-state) induced by exposure to a trapping voltage in the OFF-state, through current mode DLTS (I-DLTS) [146]. An extensive investigation of the DLTS related methods of the deep levels responsible for current collapse in AlGaN/GaN HEMTs can be found in a review paper [146].

References

1. I. Vurgaftman and J. R. Meyer, *J. Appl. Phys.* 94, 3675 (2003).
2. F. Bernardini, V. Fiorentini, and D. Vanderbilt, *Phys. Rev. B* 56, R10024 (1997).
3. S.-H. Wei and A. Zunger, *Appl. Phys. Lett.* 69, 2719 (1996).

4. F. Bechstedt, U. Grossner, and J. Furthmüller, *Phys. Rev. B* 62, 8003 (2000).
5. F. Bernardini, V. Fiorentini, and D. Vanderbilt, *Phys. Rev. B* 63, 193201 (2001).
6. A. Zoroddu, F. Bernardini, P. Ruggerone, and V. Fiorentini, *Phys. Rev. B* 64, 045208 (2001).
7. M. A. Caro, S. Schulz, and E. P. O'Reilly, *Phys. Rev. B* 88, 214103 (2013).
8. Y. Xiang, X. J. Chen, C. Ji, X. L. Yang, F. J. Xu, Z. J. Yang, X. N. Kang, B. Shen, G. Y. Zhang, and T. J. Yu, *Appl. Phys. Lett.* 108, 063503 (2016).
9. E. T. Yu, G. J. Sullivan, P. M. Asbeck, C. D. Wang, D. Qiao, and S. S. Lau, *Appl. Phys. Lett.* 71, 2794 (1997).
10. O. Ambacher, J. Smart, J. R. Shealy et al. *J. Appl. Phys.* 85, 3222 (1999).
11. O. Ambacher, B. Foutz, J. Smart et al. *J. Appl. Phys.* 87, 334 (2000).
12. I. P. Smorchkova, C. R. Elsass, J. P. Ibbetson, R. Vetury, B. Heying, P. Fini, E. Haus, S. P. DenBaars, J. S. Speck, and U. K. Mishra, *J. Appl. Phys.* 86, 4520 (1999).
13. L. Hsu and W. Walukiewicz, *J. Appl. Phys.* 89, 1783 (2001).
14. J. A. Garrido, J. L. Saánchez-Rojas, A. Jimeénez, E. Munoz, F. Omnes, and P. Gibart, *Appl. Phys. Lett.* 75, 2407 (1999).
15. J. Wagner, A. Ramakrishnan, H. Obloh, and M. Maier, *Appl. Phys. Lett.* 74, 3863 (1999).
16. A. Ramakrishnan, J. Wagner, M. Kunzer, H. Obloh, and K. Köhler, *Appl. Phys. Lett.* 76, 79 (2000).
17. R. Neuberger, G. Müller, O. Ambacher, and M. Stutzmann, *Phys. Stat. Sol. (a)* 183, R10 (2001).
18. K. Jeganathan and M. Shimizu, *AIP Advan.* 4, 097113 (2014).
19. S.W. Kaun, E. Ahmadi, B. Mazumder, F. Wu, E. C. H. Kyle, P. G. Burke, U. K. Mishra, and J. S. Speck, *Semicond. Sci. Technol.* 29, 045011 (2014).
20. S. Gökden, R. Tülek, A. Teke, J. H. Leach, Q. Fan, J. Xie, Ü. Özgür, H. Morkoç, S. B. Lisesivdin, and E. Özbay, *Semicond. Sci. Technol.* 25, 045024 (2010).
21. M. Mikulics, R. Stoklas, A. Dadgar, D. Gregušová, J. Novák, D. Grützmacher, A. Krost, and P. Kordoš, *Appl. Phys. Lett.* 97, 173505 (2010).
22. T. Takeuchi, C. Wetzel, S. Yamaguchi, H. Sakai, H. Amano, I. Akasaki, Y. Kaneko, S. Nakagawa, Y. Yamaoka, and N. Yamada, *Appl. Phys. Lett.* 73, 1691 (1998).
23. L. Zhou, D. A. Cullen, D. J. Smith, M. R. McCartney, A. Mouti, M. Gonschorek, E. Feltin, J. F. Carlin, and N. Grandjean, *Appl. Phys. Lett.* 94, 121909 (2009).
24. G. Bastard, *Wave Mechanics Applied to Semiconductor Heterostructures, Monographies de Physique.* les éditions de physique, Les Ulis, (1988).
25. P. Harrison, *Quantum Wells, Wires and Dots: Theoretical and Computational Physics of Semiconductor Nanostructures*, 2nd edn. John Wiley & Sons Ltd: Chichester, New York, (2005).
26. J. Feldmann, G. Peter, E. O. Göbel, P. Dawson, K. Moore, C. Foxon, and R. J. Elliott, *Phys. Rev. Lett.* 59, 2337 (1987).
27. E. Berkowicz, D. Gershoni, G. Bahir, E. Lakin, D. Shilo, E. Zolotoyabko, A. C. Abare, S. P. Denbaars, and L. A. Coldren, *Phys. Rev. B* 61, 10994 (2000).
28. V. A. Fonoberov and A. A. Balandin, *J. Appl. Phys.* 94, 7178 (2003).
29. T. Bretagnon, P. Lefebvre, P. Valvin et al. *Phys. Rev. B* 73, 113304 (2006).
30. P. Lefebvre, A. Morel, M. Gallart, T. Taliercio, J. Allègre, B. Gil, H. Mathieu, D. Damilano, N. Grandjean, and J. Massies, *Appl. Phys. Lett.* 78, 1252 (2001).
30. N. Tang, B. Shen, M. J. Wang, K. Han, Z. J. Yang, K. Xu, G. Y. Zhang, T. Lin, B. Zhu, W. Z. Zhou, and J. H. Chu, *Appl. Phys. Lett.* 88, 172112 (2006).
31. N. Tang, B. Shen, K. Han, F. C. Lu, F. J. Xu, Z. X. Xu, and G. Y. Zhang, *Appl. Phys. Lett.* 93, 172113 (2008).
32. N. Tang, B. Shen, X. W. He et al. *Phys. Rev. B* 76, 155303 (2007).
33. F. C. Lu, N. Tang, S. Y. Huang, M. Larsson, I. Maximov, M. Graczyk, J. X. Duan, S. D. Liu, W. K. Ge, F. J. Xu, and B. Shen, *Nano Lett.* 13, 4654 (2013).
34. Y. Q. Tang, B. Shen, X. W. He et al. *Appl. Phys. Lett.* 91, 071920 (2007).
35. X. W. He, B. Shen, Y. Q. Tang, N. Tang, C. M. Yin, F. J. Xu, Z. J. Yang, G. Y. Zhang, Y. H. Chen, C. G. Tang, and Z. G. Wang, *Appl. Phys. Lett.* 91, 071912 (2007).

36. X. W. He, B. Shen, Y. H. Chen et al. *Phys. Rev. Lett.* 101, 147402, (2008).
37. C. M. Yin, N. Tang, S. Zhang et al. *Appl. Phys. Lett.* 98, 122104 (2011).
38. F. H. Mei, S. Zhang, N. Tang, J. X. Duan, F. J. Xu, Y. H. Chen, W. K. Ge, and B. Shen, *Sci. Rep.* 4, 4030 (2014).
39. H. M. Manasevit, *Appl. Phys. Lett.* 12, 156 (1968).
40. H. P. Maruska and J. J. Tietjen, *Appl. Phys. Lett.* 15, 327 (1969).
41. S. Nakamura, Y. Harada, and M. Seno, *Appl. Phys. Lett.* 58, 2021 (1991).
42. S. Nakamura, T. Mukai, and M. Senoh, *Appl. Phys. Lett.* 64, 1687 (1994).
43. D. W. Weyburne and B. S. Ahem, *J. Cryst. Growth* 170, 77 (1997).
44. X. Zhang, I. Moerman, C. Sys, P. Demeester, and J. A. Crawley, *J. Cryst. Growth* 170, 83 (1997).
45. K. Hiramatsu, S. Itoh, H. Amano, I. Akasaki, N. Kuwano, T. Shiraishi, and K. Oh, *J. Cryst. Growth* 115, 628 (1991).
46. S. Nakamura, *Jpn. J. Appl. Phys* 30, L1705 (1991).
47. S. D. Hersee, J. C. Ramer, and K. J. Malloy, *MRS Bulletin* 22, 45 (1997).
48. Hiramatsu, S. Itoh, H. Amano, and I. Akasaki, *J Cryst. Growth* 115, 628 (1991).
49. S. Nakamura, *Jpn. J. Appl. Phys PART 1-Regular papers short notice & review papers* 30, 1620 (1991).
50. S. Figge, T. BoKttcher, S. Einfeldt, and D. Hommel, *J Cryst. Growth* 221, 262 (2000).
51. M. S. Shur and R. Gaska, *IEEE Transcat. on Electr. Dev.* 57, 12 (2010).
52. M. Shatalov, W. H Sun, A. Lunev et al. *Appl. Phys. Express* 5, 082101 (2012).
53. M. Asif Khan, J. M. Van Hove, J. N. Kuznia et al. *Appl. Phys. Lett.* 58, 2408 (1991).
54. M. Asif Khan, J. M. Van Hove, J. N. Kuznia et al. *Appl. Phys. Lett.* 63, 1214 (1993).
55. L. Shen, S. Heikman, B. Moran et al. *IEEE Electr. Device Lett.* 22, 457 (2001).
56. C. Poblenz, P. Waltereit, S. Rajan et al. *J. Vac. Sci. Technol. B*, 22, 1145 (2004).
57. S. Heikman, S. Keller, S. DenBaars et al. *Appl. Phys. Lett.* 81, 439 (2002).
58. H. Marchand, L. Zhao, N. Zhang et al. *J. Appl. Phys.* 89, 7846 (2001).
59. A. Dadgar, A. Strittmatter, J. Bläsing et al. *Phys. Stat. Sol. C* 0, 1583 (2003).
60. E. Feltin, B. Beaumont, M. Laugt et al. *Appl. Phys. Lett.* 79, 3230 (2001).
61. K.Cheng, M. Leys, S. Degroote et al. *J. Electron. Mater.* 35, 592 (2006).
62. J. P. Cheng, X. L. Yang, B. Shen et al. *Appl. Phys. Lett.* 106, 142106 (2015).
63. M. Higashiwaki and T. Matsui, *Jpn. J. Appl. Phys.* 41, L540–L542 (2002).
64. Y. Nanishi, Y. Saito, and T. Yamaguchi, *Jpn. J. Appl. Phys.* 42, 2549–2559 (2003).
65. K. Xu, N. Hashimoto, B. Cao, T. Hata, W. Terashima, M. Yoshitani, Y. Ishitani, and A. Yoshikawa, *Phys. Status Solidi (c)*, 0, 2790–2793 (2003).
66. X. Wang, S. Liu, N. Ma, L. Feng, G. Chen, F. Xu, N. Tang, S. Huang, K. J. Chen, S. Zhou, and B. Shen, *Appl. Phys. Exp.* 5, 015502 (2012).
67. V. Y. Davydov, A. A. Klochikhin, V. V. Emtsev et al. *Phys. Status Solidi B* 234, 787 (2002).
68. J. H. Edgar, *Group-III Nitrides*. INSPEC: London, UK, (1994).
69. S. T. Liu, X. Q. Wang, G. Chen, Y. W. Zhang, L. Feng, C. C. Huang, F. J. Xu, N. Tang, L. W. Sang, M. Sumiya, and B. Shen, *J. Appl. Phys.* 110, 113514 (2011).
70. S. Lourdudoss, N. Gopalakrishnan, R. Holz, M. Deschler and R. Beccard, *Value-Addition Metallurgy*, Vol. 177. TMS: Warrendale, PA, (1998).
71. K. Fujito, S. Kubo, H. Nagaoka, T. Mochizuki, H. Namita, and S. Nagao, *J. Cryst. Growth* 311, 3011–3014 (2009).
72. E. Richter, C. Hennig, M. Weyers, F. Habel, J.–D. Tsay, W.–Y. Liu, P. Bruckner, F. Scholz, Y. Makarov, A. Segal, and J. Kaeppeler, *J. Cryst. Growth* 277, 6–12 (2005).
73. V.S. Ban, *J. Electrochem. Soc.* 119 (6), 761 (1972).
74. R. Cadoret, *J. Cryst. Growth* 205, 123–125 (1999).
75. C. Hennig, E. Richter, U. Zeimer, M. Weyers, and G. Tränkle, *Phys. Stat. Sol.* 3 (6), 1466–1470 (2006).

76. H. Sone, S. Nambu, Y. Kawaguchi, M. Yamaguchi, H. Miyake, K. Hiramatsu, and N. Sawaki, *Jpn. J. Appl. Phys.* 38 (4A), L356 (1999).
77. R. F. Davis, A. M. Roskowski, E. A. Preble, J. S. Speck, B. Heying, Jr J. A. Freitas, and W. E. Carlos, *Proc. IEEE* 90 (6), 993–1005 (2002).
78. C. Hennig, E. Richter, M. Weyers, and G. Tränkle, *J. Cryst. Growth* 310 (5), 911–915 (2008).
79. K. Motoki, T. Okahisa, S. Nakahata et al. *J. Cryst. Growth* 237, 912 (2002).
80. T. Yoshida, Y. Oshima, T. Eri, K. Ikeda, S. Yamamoto, K. Watanabe, and T. Mishima, *J. Cryst. Growth* 310 (1), 5–7 (2008).
81. C. L. Chao, C. H. Chiu, Y. J. Lee, H. C. Kuo, P. C. Liu, J. D. Tsay, and S. J. Cheng, *Appl. Phys. Lett.* 95 (5), 051905–051905 (2009).
82. W. K. Luo, J. J. Wu, J. Goldsmith, Y. H. Du, T. J. Yu, Z. J. Yang, and G. Y. Zhang, *J. Cryst. Growth* 340, 18 (2012).
83. X. Zhang, P. D. Dapkus, and D. H. Rich, *Appl. Phys. Lett.* 77 (10), 1496–1498 (2000).
84. K. Motoki, *SEI Tech. Rev.* 70, 28 (2010).
85. C. R. Miskys, M. K. Kelly, O. Ambacher et al. *Phys. Stat. Sol. (c)*, 6, 1627–1650 (2003).
86. X. J. Su, K. Xu, Y. Xu, G. Q. Ren, J. C. Zhang, J. F. Wang, and H. Yang, *J. Phys. D: Appl. Phys.* 46, 205103 (2013).
87. A. Koukitu, S. I. Hama, T. Taki, and H. Seki, *Jpn. J. Appl. Phys.* 37, 762 (1998).
88. T. Yoshida, Y. Oshima, T. Eri et al. *J. Cryst. Growth* 310, 5 (2008).
89. V. Nikolaev, A. Golovatenko, M. Mynbaeva, I. Nikitina, N. Seredova, A. Pechnikov, V. Bougrov, and M. Odnobludov, *Phys. Status Solidi C* 11, 502 (2014).
90. S. Tadashige, O. Shinya, G. Takenari, Y. Takafumi, S. Ritsu, S. Akira, and G. Hideki, *Jpn. J. Appl. Phys.* 52, 08JA08 (2013).
91. P. F. Fewster, *X-ray Scattering from Semiconductors.* Imperical College Press: London, UK, (2003).
92. Z. X. Qin, Z. Z. Chen, G. Y. Zhang et al. Estimation of InN phase inclusion in InGaN films grown by MOVPE, *Appl. Phys. A* 74 (5), 655–658 (2002).
93. J. Z. Li, Y. B.Tao, Z. Z.Chen et al. Quasi-homoepitaxial GaN-based blue light emitting diode on thick GaN template, *Chin. Phys. B* 23 (1), 016101 (2014).
94. M. D. Craven, S. H. Lim, F. Wu, J. S. Speck, and S. P. DenBaars, *Appl. Phys. Lett.* 81, 469 (2002).
95. T. Metzger, R. Höpler, E. Born, O. Ambacher, M. Stutzmann, R. Stömmer, M. Schuster, H. Göbel, S. Christiansen, M. Albrecht, and H. P. Strunk, *Philos. Mag. A* 77 (4), 1013–1025 (1998).
96. X. H. Zheng, H. Chen, Z. B. Yan, Y. J. Han, H. B. Yu, D. S. Li, Q. Huang, and J. M. Zhou, *J. Cryst. Growth* 255, 63–67 (2003).
97. P. Fini, H. Marchand, J. P. Ibbetson, S. P. DenBaars, U. K. Mishra, and J. S. Speck, *J. Cryst. Growth* 209, 581–590 (2000).
98. M. A. Moram, C. F. Johnston, J. L. Hollander, M. J. Kappers, and C. J. Humphreys, *J. Appl. Phys.* 105, 113501 (2009).
99. I. Akasaki, H. Amano, Y. Koide, K. Hiramatsu and N. Sawaki, *J. Cryst. Growth* 98, 209–219 (1989).
100. X. Chen, J. Li, Y. Cao, Y. Lan, H. Li, M. He, C. Wang, Z. Zhang, and Z. Qiao, *Adv. Mater.* 12 (19), 1432–1434 (2000).
101. C. Youtsey, L. T. Romano, R. J. Molnar, and I. Adesida, *Appl. Phys. Lett.* 74, 3537 (1999).
102. B. Leung, J. Song, Y. Zhang, and J. Han, *Adv. Mater.* 25, 1285–1289 (2013).
103. P. Tian, Z. Chen, Y. Sun, Y. Qi, H. Zhang, J. Deng, F. Yu, T. Yu, X. Kang, Z. Qin, G. Zhang, *Mater. Sci. Eng.B* 175, 213–216 (2010).
104. E. Rolf, M. D. Rossell, C. Kisielowski, and U. Dahmen. *Phys. Rev. Lett.* 102 (9), 096101 (2009).
105. X. J. Ning, F. R. Chien, P. Pirouz, J. W. Yang, and M. Asif Khan. *J. Mater. Res.* 11 (3), 580 (1996).
106. A. Sakai, H. Sunakawa, and A. Usui. *Appl. Phys. Lett.* 71, 2259 (1997).
107. Z. Z. Chen, B. Shen, Z. X. Qin, J. M. Zhu, R. Zhang, Y. D. Zheng, and G. Y. Zhang, *Physica B* 324, 59–62 (2002).
108. G. Binnig, C. F. Quate, and C. Gerber. *Phys. Rev. Lett.* 56 (9), 930–933 (1986).

109. S. Keller, G. Parish, P. T. Fini, S. Heikman, C.-H. Chen, N. Zhang, S. P. DenBaars, U. K. Mishra, and Y.-F. Wu. *J. Appl. Phys.* 86 (10), 5850–5857 (1999).
110. J. Wu, W. Walukiewicz, K. M. Yu, J. W. Ager Iii, E. E. Haller, H. Lu, and Y. Nanishi, *Appl. Phys. Lett.* 80 (21), 3967–3969 (2002).
111. M. Smith, G. D. Chen, J. Y. Lin, H. X. Jiang, A. Salvador, B. N. Sverdlov, and B. Goldenberg, *Appl. Phys. Lett.* 68 (14), 1883–1885 (1996).
112. U. Kaufmann, M. Kunzer, M. Maier, H. Obloh, A. Ramakrishnan, B. Santic, and P. Schlotter, *Appl. Phys. Lett.* 72 (11), 1326–1328 (1998).
113. P. Waltereit, O. Brandt, A. Trampert, H. T. Grahn, J. Menniger, M. Ramsteiner, and K. H. Ploog, *Nature* 406, 865–868 (2000).
114. K. Okamoto, I. Niki, A. Shvartser, Y. Narukawa, T. Mukai, and A. Scherer, *Nat. Mater.* 3 (9), 601–605 (2004).
115. F. A. Ponce, D. P. Bour, W. Götz, and P. J. Wright, *Appl. Phys. Lett.* 68 (1), 57–59 (1996).
116. S. Chichibu, K. Wada, and S. Nakamura, *Appl. Phys. Lett.* 71 (16), 2346–2348 (1997).
117. S. Pereira, M. R. Correia, E. Pereira, K. P. O'donnell, C. Trager-Cowan, F. Sweeney, and E. Alves, *Phys. Rev. B* 64 (20), 205311 (2001).
118. J. Chen, J. F. Wang, H. Z. J. J. Wang, J. J. Zhu, S. M. Zhang, D. G. Zhao, and K. H. Ploog, *Semi. Sci. Tech.* 21 (9), 1229 (2006).
119. H. J. Lozykowski, W. M. Jadwisienczak, and I. Brown, *Appl. Phys. Lett.* 74 (8), 1129–1131 (1999).
120. J. F. Muth, J. H. Lee, I. K. Shmagin et al. *Appl. Phys. Lett.* 71 (18), 2572–2574 (1997).
121. J. F. Muth, J. D. Brown, M. A. L. Johnson, Z. Yu, R. M. Kolbas, J. W. Cook, and J. F. Schetzina, *MRS Proceedings.* Vol. 537, pp. G5–2. Cambridge University Press: Cambridge, UK, (1998).
122. S. T. Liu, X. Q. Wang, G. Chen, Y. W. Zhang, L. Feng, C. C. Huang, and B. J. Shen, *Appl. Phys.* 110 (11), 113514 (2011).
123. J. J. Wu, *Appl. Phys.* 106 (1), 011101 (2009).
124. T. Suski, S. P. Łepkowski, G. Staszczak, R. Czernecki, P. Perlin, and W. J. Bardyszewski, *J. Appl. Phys.* 112 (5), 053509 (2012).
125. R. G. Banal, M. Funato, and Y. Kawakami, *Phys. Rev. B* 79 (12), 121308 (2009).
126. H. Li, S. Keller, S. P. DenBaars, and U. K. Mishra, *Jpn. J. Appl. Phys.* 53, 095504 (2014).
127. S. Tan, T. Suzue, S. L. Selvaraj, and T. Egawa, *Jpn. J. Appl. Phys.* 48, 111002 (2009).
128. J. Cheng, X. Yang, L. Sang et al. *Appl. Phys. Lett.* 106, 142106 (2015).
129. L. Sang, X. Yang, J. Cheng et al. *Appl. Phys. Lett.* 107, 052102 (2015).
130. P. T. Coleridge, R. Stoner, and R. Fletcher, *Phys. Rev. B* 39, 1120 (1989).
131. Z. W. Zheng, B. Shen, R. Zhang et al. *Phys. Rev. B* 62, R7739 (2000).
132. Z. W. Zheng, B. Shen, Y. S. Gui et al. *Appl. Phys. Lett.* 82, 1872 (2003).
133. N. Tang, B. Shen, M. J. Wang, Z. J. Yang, K. Xu, G. Y. Zhang, T. Lin, B. Zhu, W. Z. Zhou, and J. H. Chu, *Appl. Phys. Lett.* 88, 172115 (2006).
134. N. Tang, B. Shen, Z. W. Zheng et al. *J. Appl. Phys.* 94, 5420 (2003).
135. N. Tang, B. Shen, K. Han et al. *Appl. Phys. A* 96, 953 (2009).
136. Z. L. Miao, N. Tang, F. J. Xu et al. *J. Appl. Phys.* 109, 016102 (2009).
137. M. Littlejohn, J. Hauser, and T. Glisson, *Appl. Phys. Lett.* 26, 625 (1975).
138. J. G. Ruch and G. S. Kino, *Appl. Phys. Lett.* 10, 40 (1966).
139. N. Mansour, K. Kim, and M. Littlejohn, *J. Appl. Phy.* 77, 2834 (1995).
140. W. Shockley, Hot electrons in germanium and Ohm's law. *Bell. Syst. Tech. J* 30, 990 (1951).
141. N. Ma, B. Shen, L.W. Lu et al. *Appl. Phys. Lett.* 100, 052109 (2012).
142. L. Guo, X. Yang, B. Shen et al. *Appl. Phys. Lett.* 105, 242104 (2014).
143. E. J. Miller, X. Z. Dang, H. H. Wieder, P. M. Asbeck, and E. T. Yu, *J. Appl. Phys.* 87, 8070 (2000).
144. S. Yang, Z. Tang, K. Y. Wong, et al. *Proceedings of the IEDM*, pp. 6.3.1–6.3.4 (2013).
145. D. V. Lang, *J. Appl. Phys.* 45, 3023 (1974).
146. D. Bisi, M. Meneghini, C. de Santi et al. *IEEE Trans. Electron Dev.* 60, 3166 (2013).

2

Microstructure and Polarization Properties of III-Nitride Semiconductors

Fernando A. Ponce

2.1 Crystal Growth and Microstructure of III-Nitride Thin Films

Much research in the past five decades has been done on the group III nitrides (III-N) comprising the Al-Ga-In-N alloys, which have a characteristic wide bandgap necessary for many applications, such as short-wavelength visible-light emission [1–3]. The success of the nitride semiconductor technology rests significantly on the control of its microstructure [4,5]. The early development of these materials dates back from the growth of AlN in 1907 [6] to the successful epitaxial growth of GaN in 1969 [7]. Some of the key steps in the development of the III-N semiconductors, which culminated with the demonstration of blue and ultraviolet diode lasers in 1996, are listed in Table 2.1.

TABLE 2.1　Development of Group III Nitrides. Key Steps Leading to the First Demonstration of Blue and Ultraviolet Laser Diodes in 1996

Year	Event	Authors	Reference
1969	GaN by hydride vapor phase epitaxy	Maruska and Tietjen	[7]
1971	Metal-insulator-semiconductor LEDs	Pankove et al.	[8]
	GaN by MOCVD	Manasevit et al.	[9]
	Ultraviolet stimulated emission at 2 K	Dingle et al.	[10]
1974	GaN by sublimation	Matsumoto and Aoki	[11]
	GaN by MBE	Akasaki and Hayashi	[12]
1975	AlN by reactive evaporation	Yoshida et al.	[13]
1982	Synthesis at high pressures	Karpinski et al.	[14]
1983	AlN intermediate layer (MBE)	Yoshida et al.	[15]
1986	Specular films using AlN buffer layers	Amano et al.	[16]
1989	p-type doping with Mg and LEEBI	Amano et al.	[17]
	GaN p-n junction LED	Amano et al.	[17]
	InGaN epitaxy (XRRC = 100 arcmin)	Nagamoto et al.	[18]
1991	GaN buffer layer by MOCVD	Nakamura	[19]
1992	Mg activation by thermal annealing	Nakamura et al.	[20]
	High-brightness AlGaN ultraviolet/blue LED	Akasaki et al.	[21]
	InGaN epitaxy (XRRC = 5 arcmin)	Nakamura et al.	[22]
1993	InGaN MQW structure	Nakamura et al.	[23]
	InGaN/AlGaN DH blue LEDs (1 candela)	Nakamura et al.	[24]
1994	InGaN/AlGaN DH blue-green LEDs (2 candela)	Nakamura et al.	[25]
1995	InGaN QW blue, green, and yellow LEDs	Nakamura et al.	[26]
	InGaN SQW green LEDs (10 candela)	Nakamura et al.	[27]
	Blue laser diode, pulsed operation	Nakamura et al.	[28]
1996	Ultraviolet laser diode	Akasaki et al.	[29]
	Blue laser diode, pulsed operation	Itaya et al.	[30]
	Blue laser diode, c.w. operation	Nakamura et al.	[31]

Source: Ponce, F. A. and Bour, D. P., *Nature*, 386, 351–559, 1997.

The atomic arrangement of these materials is tetrahedrally coordinated, in a similar fashion as in other semiconductor materials such as silicon, GaAs, and ZnO. This is due to the *s* and *p* orbitals in the outer electron shells that combine and produce four sp³ hybrid orbitals. Thus, each atomic site is located at the center and the four nearest neighbors at the corners of a regular tetrahedron. Slight deviations from the positions in a perfect tetrahedron, due to the ionic nature of the chemical bond, result in asymmetric coulomb charge distribution, also referred as spontaneous polarization. An important characteristic of these materials is the large range of energy gaps. This is shown in Figure 2.1, where the bandgap ranges from 0.7 eV (InN), 3.4 eV (GaN), to 6.2 eV (AlN). Another characteristic is that the lattice parameter changes rapidly with chemical composition, with the basal-plane lattice parameter varying from 0.311 nm (AlN), 0.319 nm (GaN), to 0.354 nm (InN).

2.2 Evolution of Microstructure in III-N Epitaxial Films

The III-nitride semiconductors, with the hexagonal wurtzite crystal structure, have a tendency to grow along the [0001] direction, and in the simplest configuration, the substrates should accommodate the hexagonal symmetry of the basal planes. In the absence of bulk GaN crystals, the initial studies were hindered by the ability to epitaxially grow on foreign substrates. The substrates were required to be

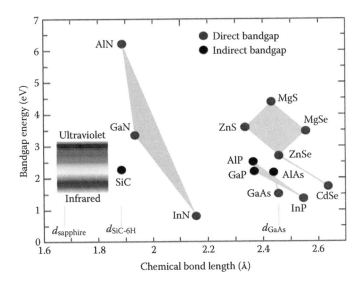

FIGURE 2.1 Energy gap and chemical bond length for semiconductors used in optoelectronic applications. (Reprinted with permission from Macmillan Publishers Ltd. *Nature*, Ponce, F. A. and Bour, D. P. 1997, copyright 1997.)

stable at the high-growth temperatures, which are needed to thermally dissociate the stable nitrogen-containing molecules. The substrates were also required to withstand the highly corrosive environments created by dissociated nitrogen-containing radicals. Traditional thinking leads to the consideration that substrates should have appropriate lattice parameters and thermal expansion coefficients. SiC provides those characteristics.

2.2.1 Growth of GaN on SiC

The lattice of silicon carbide is very similar to AlN and GaN. It is tetrahedrally coordinated, with each basal plane arranged in the same fashion. The stacking sequence of the basal planes leads to several polymorphs, yielding cubic, hexagonal, and rhombohedral configurations [4]. The most commonly commercially available form is 6H-SiC, where the period normal to the basal plane consists of six basal planes. Figure 2.2 is a cross-section lattice image of the AlN/SiC interface grown by metalorganic

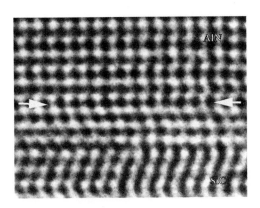

FIGURE 2.2 Atomic arrangement at the AlN/SiC interface. (Reprinted with permission from Ponce, F. A. et al., *Appl. Phys. Lett.*, 67, 410–412, 1995. Copyright 1995, American Institute of Physics.)

chemical vapor deposition (MOCVD) [32]. The heterointerface tends to be atomically flat and can be identified from the atomic arrangement, which has a zig-zag pattern in the 6H lattice, to the vertical arrangement in the wurtzite lattice. The atomic structure at the interface can be determined by analysis of the intensity of the lattice image and taking into consideration the need for charge neutrality at the interface [33].

2.2.2 Growth of GaN on Sapphire

While at first glance sapphire would not be a good candidate as a substrate for growth of GaN thin films, it has become most widely used in commercial applications. Sapphire has a hexagonal/orthorhombic structure composed of layers with hexagonal arrays of aluminum atoms separated by hexagonal planes of oxygen [4]. Sapphire has a large lattice mismatch (14.8% with GaN and 25.4% with InN) and a significant difference in thermal expansion characteristics, both of which should provide difficulties for growth of epitaxial thin films.

A cross-section lattice image of the AlGaN/AlN/Al$_2$O$_3$ interface region in Figure 2.3 shows that, in spite of the differences in the lattice structure, high-crystalline quality is possible [34]. The lattice image shows a flat AlN/Al$_2$O$_3$ interface, and the absence of threading dislocations that could result from the large lattice mismatch.

The lattice image shows a sharp transition between the aluminum oxide lattice of sapphire to an aluminum nitride lattice. Thus, the cation (aluminum) remains the same, while the anion changes (from oxygen to nitrogen). However, the lattice arrangement is different; aluminum is six-fold coordinated in sapphire and four-fold coordinated in the wurtzite structure, leading to some interesting possible configurations at the interface, similar to those that had been previously observed at the silicon/sapphire interface [35].

The first attempts to grow high-quality GaN on sapphire encountered a tendency toward three-dimensional island growth. This was first overcome in 1986 using low-temperature AlN buffer layers, [15,16] leading to the ability to control the morphology and the n-type doping of GaN thin films. The use of low-temperature GaN buffer layers was introduced in 1991 [19]. These developments led to continuous improvements in the quality of thin epitaxial films, the development of p-GaN, [17,20] the growth of InGaN quantum wells, [18,22–27] and the rapid development of high-efficiency light-emitting devices that culminated with the blue diode laser [28,29]. As it is well known, some of these achievements were recognized with the Nobel Prize in Physics in 2014 [36].

FIGURE 2.3 Atomic arrangement at the AlN/Sapphire interface. (Reprinted with permission from Ponce, F. A. et al., *Appl. Phys. Lett.*, 65, 2302–2304, 1994. Copyright 1994, American Institute of Physics.)

2.2.3 Growth of GaN on Silicon

Growth of III–V compounds on silicon substrates is desirable for the integration of optoelectronics and silicon-based electronic devices [37]. The tendency to form an amorphous silicon nitride layer on the silicon surface makes it difficult to produce high-quality GaN layers on silicon substrates. Significant advances were realized early in the millennium by A. Krost and A. Dadgar [38,39]. By first depositing an aluminum layer, followed by nitrogen to create a wetting AlN buffer layer, smooth flat layers with semiconductor properties were produced. Figure 2.4 shows the structure used to optimize the crystal quality of the film [40]. The low-temperature AlN nucleation layer "A" is important in order to avoid island growth and to avoid the formation of an amorphous silicon nitride layer over the silicon substrate. The threading dislocation density in the early stages of growth is significantly reduced in two steps. The introduction of a low-temperature AlN layer "B" has the effect of terminating some of the threading edge dislocations by pairing them into loops. In a similar way, the subsequent introduction of silicon delta doping has the effect of pinning the screw dislocations and bending them to form loops.

Silicon (111) substrates were first used because they match the crystal symmetry of the (0001) wurtzite structure. The atomic arrangement associated with the epitaxial structure is observed by transmission electron microscopy (TEM) in Figure 2.5, which is a cross-section image of the (0001)AlN/(111)Si interface.

The (0001) AlN basal planes and the (111) silicon planes appear horizontally. Another set of {111} planes is observed in this <110> projection of silicon. The coherency at the interface is evident, exhibiting continuity of the oblique set of {111} Si planes into the vertical {1$\bar{1}$00} AlN planes. The lattice mismatch between AlN and Si is resolved by the presence of a periodic array of misfit dislocations, as indicated in the image. It is also observed that a step on the silicon surface does not break the coherency at the interface [41].

The nature of the misfit dislocation at the (0001) AlN/(111) Si interface is observed in Figure 2.6. The transition between the silicon lattice and the AlN wurtzite structure has been determined from the atomic positions in the lattice image using charge balance and neutrality considerations [41].

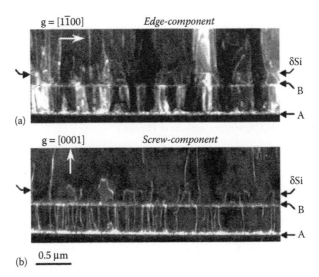

FIGURE 2.4 Cross-section TEM images showing (a) edge component and (b) screw component of threading dislocations in epitaxial GaN film on (111) silicon. A is the AlN buffer layer and B is the low temperature AlN layer. Delta doping of silicon is indicated. (Reprinted with permission from Contreras, O. et al., *Appl. Phys. Lett.*, 81, 4712–4714, 2002. Copyright 2002, American Institute of Physics.)

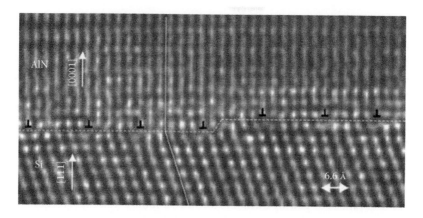

FIGURE 2.5 Cross-section TEM image of the [0001] AlN/[111] Si interface. It shows an abrupt transition from the cubic silicon to the hexagonal AlN lattices (horizontal dashed line). A periodic misfit dislocation "⊥" array is observed. The coherence between the $\{1\bar{1}00\}_{AlN}$ and the $\{111\}_{silicon}$ planes is indicated by the inclined-vertical dashed line. (Reprinted with permission from Liu, R. et al., *Appl. Phys. Lett.*, 83, 860–862, 2003. Copyright 2003, American Institute of Physics.)

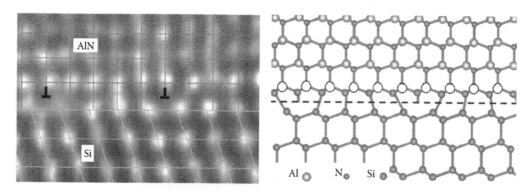

FIGURE 2.6 Atomic arrangement at the (0001) AlN/(111) Si interface. The lattice image on the left is matched with the schematic atomic array on the right. The atomic ordering at the interface plane on the right is determined by using charge balance and neutrality considerations. (Reprinted with permission from Liu, R. et al., *Appl. Phys. Lett.*, 83, 860–862, 2003. Copyright 2003, American Institute of Physics.)

Silicon (110) substrates provide another option for epitaxial growth. The (110) silicon surface has a two-fold symmetry that is different to the six-fold symmetry of (0001) AlN. The top image in Figure 2.7 shows a projection where we observed edge-on the $\{1\bar{1}00\}_{AlN}$ and the $\{001\}_{silicon}$ planes. The interplanar separation between these planes is similar, with only a 0.8% mismatch, providing relatively good coherency. The atomic arrangement at the interface is shown in Figure 2.8. On the other hand, along the orthogonal projection in the lower image in Figure 2.7, the $\{11\bar{2}0\}_{AlN}$ planes are parallel to the $\{1\bar{1}0\}_{silicon}$ planes, with the lattice mismatch of about 19%. Figure 2.7 shows cross-section TEM images corresponding to two orthogonal projections [42].

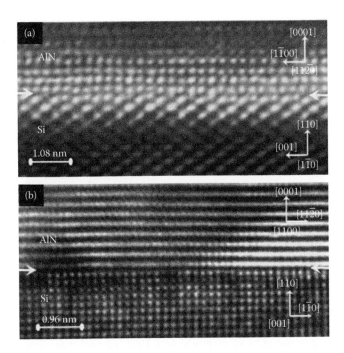

FIGURE 2.7 Cross-section TEM images of the [0001] AlN on [1̄10] Si interface along two orthogonal projections, showing an abrupt crystal interface. (a) Good lattice match is observed in this projection between the [1100]$_{AlN}$ and the [001]$_{silicon}$ planes. (b) Large lattice mismatch between the [112̄0]$_{AlN}$ and the [110]$_{silicon}$ occurs along this projection. (Reprinted with permission from Contreras, O. E. et al., *Appl. Phys. Exp.*, 1, 061104, 2008. Copyright 2012, The Japan Society of Applied Physics.)

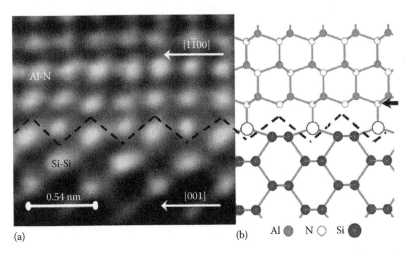

FIGURE 2.8 Lattice image of the [0001] AlN on (1̄10) Si interface viewed along the [112̄0]$_{AlN}$ //[110]$_{silicon}$ projection. (a) The chemical interface is indicated by the dashed line, which follows the contour of the silicon trenches. (b) Schematic diagram of the bonding configurations. The larger open circles can be filled with an Al, N, or Si atom in order to retain charge neutrality at the interface. (Reprinted with permission from Contreras, O. E. et al., *Appl. Phys. Exp.*, 1, 061104, 2008. Copyright 2012, The Japan Society of Applied Physics.)

2.3 Microstructure of GaN on Sapphire

As already mentioned, sapphire substrates are by far the most widely used in commercially available devices. The resulting nitride semiconductor quality is determined by the use of low-temperature nucleation buffer layers that produce two-dimensional growth. A study of the effect of GaN buffer layers indicated that a thickness of about 20 nm has the effect of optimizing the carrier mobility, minimizing background carrier concentration, and achieving a high value of peak width for X-ray rocking curves [19]. It was a surprise to learn that the first high-efficiency blue light-emitting devices from Nichia had a large density of threading dislocations [43]. In retrospect, it made sense that a wide rocking curve should be associated with a high-dislocation density.

2.3.1 Threading Dislocations

A cross-section TEM image of a Nichia blue light-emitting diode (LED) (ca. 1994) is shown in Figure 2.9 [43]. The device structure is grown on (0001) sapphire, using a GaN buffer layer, followed by ~4 μm thick film of silicon-doped GaN. The p-n junction includes an InGaN thin layer. The interesting aspect is the large density of threading dislocations of the order of 10^{10} cm^{-2}.

Analysis of the distribution of threading dislocations indicates a columnar structure, depicted in Figure 2.9 (right) and characterized by the misorientation of columns that involve tilt and rotation of the *c*-axis. The dislocations originate at the buffer layer and propagate through the thin film structure. The density of dislocations was quite surprising at first. Data from other semiconductor thin films for LED applications indicated an inverse correlation between the light-emitting efficiency and dislocation density, as depicted in Figure 2.10 [43].

2.3.2 Characteristics of Dislocations

The threading dislocations in GaN thin films result from a competition from small crystallites that form at the buffer layer. Crystallites that are oriented with the *c*-axis closer to the substrate normal

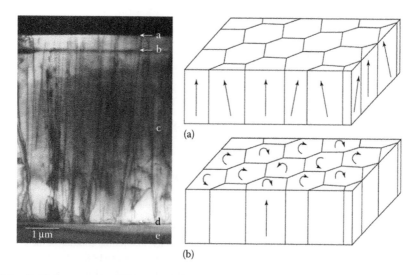

FIGURE 2.9 (Left) Cross-section TEM image of a Nichia blue LED. (Right) Schematic diagram of columnar microstructure responsible for the distribution of threading dislocation following columns misoriented by (a) tilt and (b) twist. (Reprinted with permission from Lester, S.D. et al., *Appl. Phys. Lett.*, 66, 1249–1251, 1995. Copyright 1995, American Institute of Physics.)

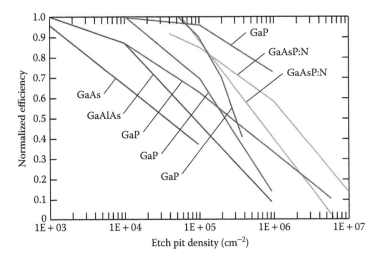

FIGURE 2.10 Dependence of LED efficiency on dislocation density for several III–V semiconductors. High efficiency GaN epilayers have dislocation densities of 10^{-2} cm^{-2}. (Reprinted with permission from Lester, S. D. et al., *Appl. Phys. Lett.*, 66, 1249–1251, 1995. Copyright 1995, American Institute of Physics.)

grow preferentially, but there will be some remaining misorientation between neighboring crystallites, resulting in a columnar structure with columnar diameters of the order of ~200 nm. The misorientation distribution in Figure 2.9 (right) is measured by X-ray diffraction, with tilt being of the order of 5 arcmin, and a twist of the order of 8 arcmin [44].

Dislocations are characterized by a displacement of the lattice, which is known as Burgers vector, **b**. The displacement associated with the dislocation can be measured in various ways. Diffraction contrast analysis (known as **g.b** analysis, where **g** is the diffraction vector) can give the direction of **b**. A screw dislocation has **b** parallel to the dislocation line, whereas for an edge dislocation, it is perpendicular.

Determining the sense and magnitude of **b** is more difficult, and has been achieved using large-angle convergent electron diffraction imaging (LACBED) [45]. Threading screw dislocations are associated with a displacement equal to a full *c*-lattice-parameter, **b** = **c**; while threading edge dislocations are associated with a displacement equal to an *a*-lattice-parameter, **b** = **a**. Mixed dislocations are also common with **b** = **a** + **c**. These are large values when compared to the Ga-N bond length, and can lead to instability of the dislocation core structure, with the existence of coreless dislocations, that is, dislocations where the strain energy at the core is too large, resulting in an open core (Figure 2.11).

2.3.3 Stacking Faults in GaN

Although rarely observed in (0001) GaN epitaxial films grown on sapphire, stacking fault do appear in films grown along directions other than the basal plane, such as in films on off-axis SiC and nonpolar or semipolar GaN [46–48]. The presence of stacking faults has been associated with specific optical transitions [46]. A direct spatial correlation in the *a*-plane $\langle 11\bar{2}0 \rangle$ GaN epitaxy between stacking faults and luminescence peaks (in the range of 3.29–3.1 eV) was obtained by performing cathodoluminescence (CL) imaging in a particular region of a plan-view TEM sample [47]. The CL spectrum in Figure 2.12 shows four resolved peaks, in addition to the main GaN emission peak at 3.474 eV. After CL analysis, the same region was examined in the TEM to identify the defects responsible for the optical emissions.

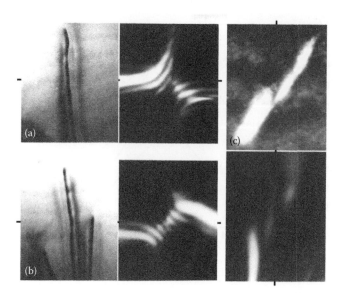

FIGURE 2.11 Determination of the magnitude and sense of the Burgers vector of threading dislocations using the LACBED technique. (a) and (b) correspond to a screw dislocation with **b** = [0001]. (c) Corresponds to an edge dislocation with b = 1/3 11$\bar{2}$0. (Reprinted with permission from Ponce, F. A. et al., *Appl. Phys. Lett.*, 69, 770–772, 1996. Copyright 1996, American Institute of Physics.)

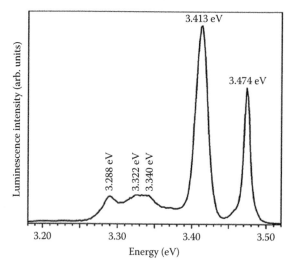

FIGURE 2.12 CL spectrum of *a*-plane GaN epilayer taken at 5 K. The 3.474 eV peak is the donor-bound exciton (D^oX). The other peaks are due to the presence of stacking faults and partial dislocations in the film. (Reprinted with permission from Liu, R. et al., *Appl. Phys. Lett.*, 86, 021908, 2005. Copyright 2005, American Institute of Physics.)

Figure 2.13 shows the spatial correlation between the TEM images and the CL emissions. Figure 2.13a is a plane-view diffraction contrast TEM image, where the *c*-plane appears in the horizontal direction. Six regions are identified, which are then correlated with the CL images in Figure 2.13b–e.

Higher-resolution TEM and CL images have been used to demonstrate the direct correlation between crystal defects and the optical emissions, which are as follows: *c*-plane stacking faults emit at 3.47 eV, *a*-plane stacking faults emit at 3.29 eV, and partial dislocations emit at 3.32–3.34 eV. Similar results have been obtained for lateral epitaxial growth of GaN [48].

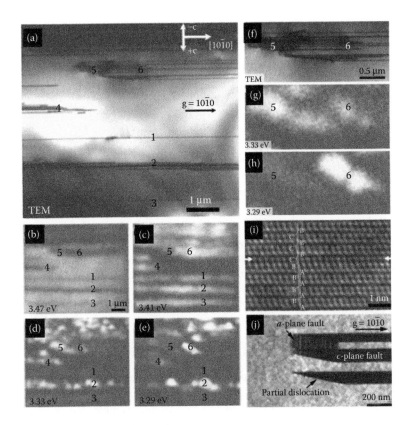

FIGURE 2.13 (a) Plan-view TEM image of nonpolar *a*-plane GaN thin film. The crystal orientation and the diffraction contrast condition are shown. (b–e) CL images taken from the same region in (a) for the emission peaks at 3.29, 3.33, 3.41, and 3.47 eV. (f–h) Higher magnification TEM images and matching CL images. (i) Lattice image of a *c*-plane stacking fault. (j) Basal plane stacking faults intersecting a prismatic *a*-plane staking fault. (Reprinted with permission from Liu, R. et al., *Appl. Phys. Lett.*, 86, 021908, 2005. Copyright 2005, American Institute of Physics.)

2.4 Polarity and Polarization Effects in III-N Semiconductors

2.4.1 Spontaneous and Piezoelectric Polarization

The III-nitride compounds with the hexagonal wurtzite structure exhibit polarization effects due to the alignment of the lattice along the *c*-direction. The *c*-axis direction is defined as the vector from the cation (Ga) to the anion (N) across the basal plane, that is, along the long vertical bond in the schematic diagram in Figure 2.14. Each Ga-N pair contributes a dipole moment. In the absence of external fields, the total *spontaneous* polarization is the sum of the dipole moments. Strain, such as due to lattice mismatch, introduces *piezoelectric* polarization. Several techniques exist to determine the polarity of a crystal, such as selective etching and X-ray diffraction [49].

The polarity of GaN thin films can be experimentally determined using convergent beam electron diffraction (CBED) in the transmission electron microscope. Experimental measurements indicate that control of the electrical properties of III-nitride thin films is facilitated by growth along the +*c* direction [49].

The Coulomb charge distribution due to spontaneous and piezoelectric polarization has a significant effect on the electronic properties of III-nitrides. This is especially expected at defects and at interfaces, where significant displacements occur due to threading and misfit dislocations, and strain due to not-relaxed lattice mismatch.

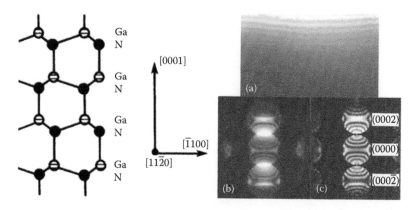

FIGURE 2.14   Left: Atomic arrangement in GaN along a [1120] projection. The basal planes are horizontal in the diagram. Along the [0001] direction, nitrogen atoms are located directly above the gallium atoms. The c-axis is defined as the direction from Ga to N across basal planes. Right: Determination of polarity of a GaN layer grown on sapphire by MOCVD. (a) Cross-section TEM image. (b) Convergent beam electron diffraction pattern of GaN in (a). (c) Simulation of (b) to index the polar direction. (Reprinted with permission from Ponce, F. A. et al., *Appl. Phys. Lett.*, 69, 337–339, 1996. Copyright 1996, American Institute of Physics.)

2.4.2 Electrostatic Potential Profiles with Sub-Nanometer Resolution

In electronic devices, electron transport depends on the carrier density and the internal electrostatic potential distribution. With device dimensions reaching the atomic scale, it becomes necessary to develop probing techniques that allow measurements with a sub-nanometer resolution of the variations of the electrostatic potential at defects such as dislocations and at interfaces in heterostructures. Broken bonds and impurity segregation at dislocations produce net Coulomb charges that result in localized electric fields. Band offsets at the interface between dissimilar materials represent abrupt electrostatic potential variations that signify changes in the carrier kinetic energy. In noncentrosymmetric materials, such as those semiconductors with the wurtzite structure, variations in spontaneous polarization produce electric dipoles at the interfaces. Strain due to lattice mismatch results in piezoelectric fields in epitaxial layers. These variations in the electrostatic potential and the presence of electrostatic charges in semiconductors have strong effects on electron transport and carrier recombination.

Electron holography in the TEM has the capability to profile the electrostatic potential and charge distribution in solids with spatial resolutions at the atomic level [50]. This technique has been used to visualize the potential energy profile across p-n junctions in silicon devices used in microelectronics [51]. Electron holography has been particularly useful to measure the potential and charge density variations in the nitride semiconductors, which are widely used in high-efficiency, long-lasting LEDs [3]. While the light-emitting efficiency in the violet to the blue region is near the physical limit, the green to red portion is still well below the expected values. This is mainly due to the large variations in the lattice parameters in the InN-GaN-AlN ternary system, which results in a significant misfit strain in heterostructures. The strain produces strong piezoelectric fields, an understanding of which has become one of the main challenges for achieving high-efficiency LEDs across the whole range of the visible spectrum. Electron holography was first used by Cherns et al. in 1999 to measure the piezoelectric fields in InGaN quantum wells [52].

2.4.3 Electron Holography in the Transmission Electron Microscope

Holograms are commonly used today to generate three-dimensional images. They are typically recorded on a photographic plate by recording the interference of a coherent visible light beam being reflected by the object with the portion of the same beam that travels directly to the plate. The interference pattern contains amplitude and phase information that can be reconstructed by coherent illumination of the hologram, resulting in a three-dimensional image. This technique was first proposed by Dennis Gabor in 1947 with the purpose of improving the resolution of transmission electron microscopes [53]. The use of electron holography has increased significantly in the past two decades [54]. It is particularly useful for the direct measurement of the electrostatic fields and charges in nitride semiconductor materials. In conventional TEM techniques, like those used for diffraction contrast analysis, only the intensity of the electron beam is recorded, while the phase information is lost. Electron holography is an interferometric technique that provides information about the spatial distribution of the phase shift $\Delta\theta$. In order to obtain the phase information, a highly coherent electron beam is obtained by using a field emission gun. A sample is prepared to electron transparency, but not too thin to avoid surface related effects such as strain relaxation and Coulomb charge screening. The phase of the electron beam is modulated by spatial variations of the electrostatic potential of the specimen. Figure 2.15 shows the configuration used for electron holography. The sample is introduced into the electron beam path covering about half of the beam diameter. The other half of the beam travels through a vacuum, and it is used as a reference beam. The electrons that traverse the sample experience a change in velocity, in part due to the variation of the reference potential, and gain in phase compared to the electrons traveling through a vacuum. The phase change is directly proportional to the electrostatic potential as we will see later. The phase information can be retrieved from the wave interference pattern [55,56].

Electron holography is an electron-interference technique that makes the recovery of the phase information possible. A conducting filament that acts as an optical biprism is used to produce an interference pattern. The biprism is aligned manually so that its image matches the boundary of the specimen at the image plane. When a positive bias is applied to the biprism, the beams traveling through vacuum (reference) and the sample (object) overlap at the image plane, and their interference pattern is recorded. The geometric configuration around the biprism is depicted in Figure 2.16. The biprism filament is along the y-direction, and the beam propagates along the z-direction. The biprism is at a given potential, while the two plates are grounded. In some microscopes, these two plates are the edges of the selected area aperture.

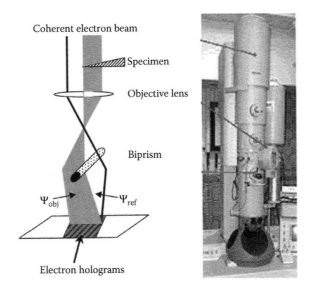

FIGURE 2.15 Schematic diagram of the electron beam path for electron holography in a transmission electron microscope. (Reprinted with permission from Cai, J. and Ponce, F. A., *J. Appl. Phys.*, 91, 9856–9862, 2002. Copyright 2002, American Institute of Physics.)

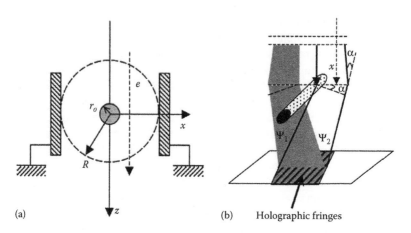

(a)   (b)   Holographic fringes

FIGURE 2.16 Schematic diagram of formation of the holographic image. (a) Configuration around the biprism. (b) Interference between two plane waves. (Reprinted with permission from Cai, J. and Ponce, F. A., *J. Appl. Phys.*, 91, 9856–9862, 2002. Copyright 2002, American Institute of Physics.)

As already indicated, the specimen under investigation is positioned to cover only half of the object plane, in that way one beam passes through the specimen (referred as the object wave $\psi_{obj} = A_o e^{i\theta_o}$) and the other beam passes through a vacuum (referred as the vacuum wave $\psi_{vac} = A_v e^{i\theta_v}$). A biprism, which resides in the plane of the selected area aperture, deflects the two electron beams in order to cause their interference. Applying a positive bias to the filament of the biprism, the object wave and vacuum wave converge and yield an interference pattern in the overlapping region. The intensity distribution can be represented by:

$$I = \left| A_v e^{i(\theta_v - \Delta)} + A_o e^{i(\theta_o + \Delta)} \right|^2$$

$$= A_v^2 + A_o^2 + 2 A_v A_o \cos(\theta_o - \theta_v - 2\Delta)$$

where the first two terms represent the background intensity, and the third term represents the superimposed interference fringes. A complete derivation of this relation is given in Reference 56. The extra phase shift Δ is due to the biprism deflection, which depends on the applied field and the geometric configuration of the biprism, but is not a property of the specimen. The information that we want to obtain is the phase shift $\theta_o - \theta_v$, which is due to the modulation of the crystal potential V. The crystal electrostatic potential V is typically in the order of 10 eV, which is much less than the electron beam energy, so that the difference of momentum between object and vacuum waves is proportional to V. Considering relativistic effects, the relationship between phase shift $\theta_o - \theta_v$ and crystal potential V is:

$$\theta_o - \theta_v = \int \frac{p_{obj} - p_{vac}}{\hbar} \, dz = \frac{eV(E - E_0)}{E(E + 2E_0)} \frac{2\pi t}{\lambda}$$

where:
  E is the kinetic energy of the incident electron beam
  E_0 is the rest energy of the electron
  t is the specimen thickness
  λ is the wavelength of the electron beam

With a fixed accelerating voltage in the TEM, this expression can also be written as

$$\theta_o - \theta_v = C_E V t$$

where C_E is a constant. Therefore, the crystal electrostatic potential profile can be obtained from the phase shift information that is extracted from the electron hologram.

The phase reconstruction process starts by taking the Fourier transform of the intensity equation shown earlier [55]. Three delta functions appear involving convolutions represented by the symbol $\otimes$.

$$FT(I) = \delta(q) \otimes FT(A_v{}^2 + A_o{}^2) + \delta(q - q_\Delta)$$

$$\otimes FT[A_v A_o e^{i(\theta_v - \theta_o)}] + \delta(q + q_\Delta)$$

$$\otimes FT[A_v A_o e^{-i(\theta_v - \theta_o)}]$$

The background intensity is transferred into an autocorrelation function and appears to be in the center, while the Fourier transform of the co-sinusoidal interference fringes provides two sidebands, at a distance from the center, q_Δ. The next step is to shift one of the sidebands to the center, and take the inverse Fourier transform. A complex object image with a modulus $A_v A_o$ and phase $\theta_v - \theta_o$ is obtained. In addition, if we remove the specimen and take a hologram, a complex image with a modulus $A_v{}^2$ and zero phase shift can be obtained. The damping of amplitude of the electron beam across the specimen is due to the inelastic scattering process in the material and can be expressed as

$$\frac{A_0}{A_v} = e^{-t/2\xi}$$

where ξ is the mean free path for inelastic scattering of electrons in the crystal. Therefore, the thickness profile of a specimen can be obtained from the amplitude image, while the electrostatic potential profile of the specimen can be obtained from the phase image. Furthermore, in a one-dimensional potential profile, the internal fields are proportional to the slope of the electrostatic potential. The corresponding electrostatic charges are found using Poisson's equation and are related to the curvature of the potential profile.

2.5 Coulomb Charges at Threading Dislocations

Dislocations play an important role in GaN. Thin epitaxial films with optimized electronic properties have threading dislocation densities ranging from 10^8 to 10^{10} cm^{-2} [43]. They play an important role in accommodating epitaxial growth under adverse conditions associated with lattice mismatch and thermal mismatch with the substrate and at AlGaInN compound heterojunctions. Threading dislocations have been found to be of the edge, screw, and mixed types [45]. The Burgers vectors of the dislocations are large with respect to the chemical bond length, which often leads to the formation of coreless dislocations [57]. Dislocations are frequently associated with nonradiative recombination centers [58]. The electrostatic charges at dislocations lines were first studied using electron holography by Cherns et al [59].

The charge depends on the dislocation type. Figure 2.3 consists of two TEM images taken under different diffraction conditions, with four dislocations that are visible labeled with capital letters. In Figure 2.17a, the diffraction vector is $\mathbf{g} = [0002]$, and two dislocations, A and C, are visible. Figure 2.17b has $\mathbf{g} = \langle 11\bar{2}0 \rangle$ with three dislocations, A, B, and C visible. This means that dislocation C is of the screw type (corresponding to a lattice displacement parallel to the dislocation line), dislocations B and D are of the edge type (displacement perpendicular to the dislocation line), and dislocation A is of the mixed type (total displacement is the sum of a screw- and an edge-types) [60].

The region in Figure 2.17 was analyzed by electron holography as described in the previous section. Figure 2.18a shows the deconvoluted profile of the phase and thickness data. Using the equation $\Delta\theta = C_E V t$, previously derived, the potential profile $V(x)$ along the horizontal and through the dislocations was obtained and is shown in Figure 2.18b.

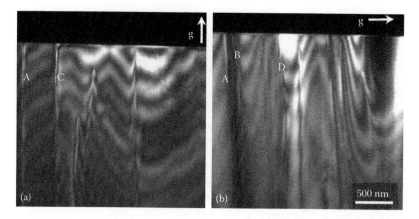

FIGURE 2.17 Two-beam dark-field images to determine the nature of threading dislocations in GaN epitaxial layers, taken under diffraction conditions (a) $\mathbf{g} = [0002]$, (b) $\mathbf{g} = \langle 11\bar{2}0 \rangle$. They show that dislocation A is of the mixed type, B and D are of the edge type, and C is screw type. (From Cai, J. and Ponce, F.A.: *Phys. Stat. Sol. A*. 2002. 192. 407–411. Copyright Wiley-VCH Verlag GmbH & Co. KGaA. Reproduced with permission.)

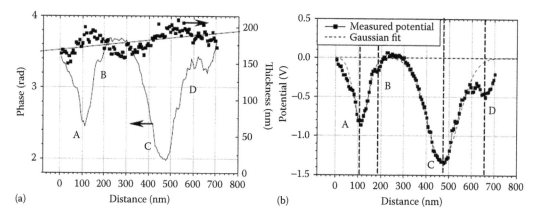

FIGURE 2.18 Electron holography results on same region as in Figure 2.3. (a) Phase and thickness profiles. (b) Electrostatic potential profiles. All types of dislocations in n-type GaN are negatively charged. (From Cai, J. and Ponce, F.A.: *Phys. Stat. Sol. A*. 2002. 192. 407–411. Copyright Wiley-VCH Verlag GmbH & Co. KGaA. Reproduced with permission.)

The potential profiles are fit with Gaussian curves to obtain the individual dislocation profiles. The potential and charge profiles across dislocation C in Figure 2.19a can be selected for analysis. The charge profile shows a negative charge density of 10^{17} cm^{-3} trapped at the core of the dislocation. Positive charges are distributed in the vicinity of the core, screening the negative charges. The data support the linearly charged line model, [61] where deep levels are present at dislocation cores due to either lower atomic coordination or distortion of the chemical bonding configuration. Majority carriers are trapped at these deep levels, turning them into negatively charged lines and causing band bending in the vicinity of the dislocations.

A statistical average from several holograms indicates that the line-charge densities for screw, edge, and mixed dislocations are 1.0, 0.3, and 0.6 (± 0.2) e/c, respectively. The analyses indicate that screw dislocations have the highest line charge density, the largest core radius, and depletion range.

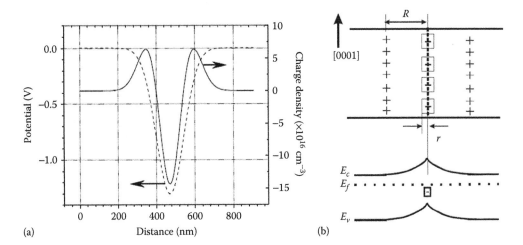

(a)   (b)

FIGURE 2.19 (a) Potential and charge distribution across dislocation C. (b) Charge distribution in the vicinity of the dislocation. E_c, and E_v are the conduction and valence band energies, and E_f is the Fermi level. (From Cai, J. and Ponce, F.A.: *Phys. Stat. Sol. A.* 2002. 192. 407–411. Copyright Wiley-VCH Verlag GmbH & Co. KGaA. Reproduced with permission.)

2.6 Polarization Fields in InGaN/GaN Heterostructures

The hexagonal wurtzite structure does not have a center of symmetry, and as such, it exhibits noticeable spontaneous and piezoelectric polarization effects [62]. Polarization fields up to a few MeV/cm in the nitride semiconductors were first determined by applying a reverse bias to the p-n junction, [63] and later by electron holography [50]. The strong internal fields can have beneficial and detrimental effects in the performance of LEDs. Reduced radiative recombination efficiency can result from the spatial separation of electrons and holes. Spatially separated by the Stark effect, their transition matrix elements and the radiative recombination rates can be significantly decreased. In addition, the internal fields facilitate the transport of carriers across the active region and reduce the carrier capture cross section by the quantum wells.

Another detrimental effect is instability of the emission wavelength for devices device operating under high-injection conditions, where screening of the polarization field results in a blue-shift of the emission wavelength. The magnitude of this shift could be up to 100 nm for large polarization fields, such as for InGaN/GaN green LEDs. Therefore, it is important to find ways to prevent polarization effects in the active region. Nonpolar and semipolar growth has been proposed, and their fabrication has been successfully demonstrated. However, *in*-plane polarization can still be a problem for the nonpolar growth quantum wells, as we will see later.

On the other hand, strong polarization fields could play a positive role in the conductivity along p- and n-type layers in LEDs. In n-type AlGaN/GaN heterostructures, the positive polarization sheet charges at the interface promote the formation of a two-dimensional electron gas (2DEG) with a sheet carrier concentration as high as 10^{13} cm^{-2}. Similarly, in p-type AlGaN/GaN heterostructures, the negative polarization sheet charges at the interface can lead to the formation of a two-dimensional hole gas (2DHG). The lateral conductivity of LEDs can be enhanced by substituting the conventional n-GaN or p-GaN layer with the appropriate AlGaN/GaN heterostructures. Improved current spreading can be expected along with increased light emission efficiencies.

In the following section, the fundamental aspects of the polarization fields in nitrides are discussed, followed by examples of the measurement of polarization effects by electron holography, such as mapping of internal electrostatic potential in the InGaN/GaN quantum wells, and the observation of 2DEG and 2DHG in n-type and p-type AlGaN/GaN heterostructures, respectively.

2.6.1 Polarization Fields and Crystal Orientation

The wurtzite structure belongs to the space group C_{6v}, and it has the highest symmetry compatible with the existence of spontaneous polarization. The magnitude of its spontaneous polarization has been calculated *ab-initio* [62] and depends on the cation–anion bond-length along the [0001] threefold axis. It can be understood as an intrinsic property determined only by the crystal symmetry and the ionicity of the elements. On the other hand, the piezoelectric polarization is due to the internal strain fields, and the relative orientation of the piezoelectric fields depends on the orientation of the growth plane [64]. The total magnitude of the polarization (spontaneous plus piezoelectric, with *in*-plane plus normal components) for arbitrary crystal orientations should be taken into account for pseudomorphic growth conditions [65]. The calculated values of the polarization as a function of crystal orientation are shown in Figure 2.20. There, the polar angle is defined as the angle between the *c*-direction and the normal to the growth plane.

For those specific cases of nonpolar (*a*-plane and *m*-plane) and polar (*c*-plane) growth, the piezoelectric polarization is along the *c*-direction. Thus, the sum of spontaneous and piezoelectric polarization leads to a net dipole along the *c*-direction for the polar and nonpolar growth cases. The perpendicular and parallel components do vanish but at different polar angles; thus the total polarization vector will not be zero at any angle [65]. For the *c*-plane growth, the bound charges act as an infinite parallel capacitor. For the *a*-plane and *m*-plane growth, the bound charge distribution is a linear array at the ends of the film that are perpendicular to the *c*-axis, resembling two wires separated by the corresponding lateral dimensions of the crystal. Due to the intrinsic atomic arrangement and tendency to form facets, it is difficult to achieve perfect planarity for nonpolar growth. The high density of misfit dislocations and stacking faults is necessary in order to relax the strain fields that develop due to the large lattice mismatch typical of the nitride semiconductors. As a result, the characteristic lateral dimensions should significantly affect the *in*-plane electrostatic fields and potentials.

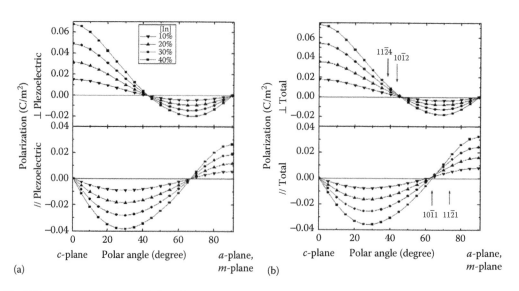

FIGURE 2.20 Perpendicular and *in*-plane components of the (a) piezoelectric and (b) total polarization vectors in the InGaN/GaN system as a function of polar angle, calculated for indium compositions ranging from 10% to 40%. Some low-index crystallographic planes are indicated. (From Wei, Q. Y. et al.: *Phys. Stat. Sol. A.* 2010. 207. 2226–2232. Copyright Wiley-VCH Verlag GmbH & Co. KGaA. Reproduced with permission.)

2.6.2 Electrostatic Fields and Charges in LEDs

Various aspects of the polarization fields in InGaN/GaN quantum wells have been reported [66–69]. In particular, there is much interest in the polarization field and charge distribution in InGaN/GaN multiquantum wells used in the active region of green LEDs. An important feature of electron holography is that it can provide a mapping of the electrostatic potential distribution at an atomic scale, while other techniques usually give average values. An example of a green LED structure is shown in Figure 2.7, where the layer structure was grown by metal organic chemical vapor deposition on a *c*-plane sapphire substrate, the active region containing five InGaN/GaN quantum well periods, emitting light at ~2.3 eV [69]. Cross-section samples were prepared for TEM using standard mechanical polishing and argon-ion milling techniques. Figure 2.21a is a high-resolution image of the active region. The InGaN quantum well (QW) and GaN barriers are uniform with a thickness of 4.6 and 9.3 nm, respectively. The interfaces are flat and atomically abrupt.

Electron holograms of the active region were obtained with the sample aligned off-axis in order to minimize diffraction conditions while keeping close to an orientation where the electron beam was parallel to the interfaces. The holograms were digitally recorded using a charge-coupled device camera. The phase and amplitude values were deconvoluted from the electron hologram; the results are shown in Figure 2.21b,c. The resulting spatial profiles of the electrostatic potential and charge distribution are plotted in Figure 2.22. The electrostatic potential shift of ~3.2 eV due to the p-n junction is clearly observed in Figure 2.22a. Also observed is the potential profile due to the quantum wells. It is noticed that the field and the polarization charge density in the quantum wells vary with distance to the p-n junction. The sheet polarization charge densities of the first three InGaN quantum wells have a similar value of ~0.018 C/m^2, corresponding to a 20% indium composition, for a nonscreening case in Figure 2.20, and corresponding to piezoelectric fields of ~1.95 MV/cm. However, the sheet polarization charge densities in the fourth and fifth periods are estimated to be 0.007 and 0.0005 C/m^2, respectively. This corresponds to a significant screening of the fourth and fifth QWs.

The observation of inhomogeneous polarization fields within the multi-quantum well is consistent with a depth-resolved study using CL in which a low-energy peak appears when the accelerating voltage is

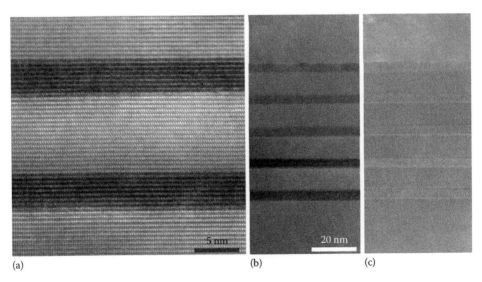

(a) (b) (c)

FIGURE 2.21 Active region of a green LED. (a) High-resolution TEM of the quantum well region. (b) Phase contrast image of the hologram containing the five quantum wells in the active region, and (c) its corresponding amplitude contrast image. (Reprinted with permission from Wu, Z. H., *Appl. Phys. Lett.*, 91, 041915, 2007. Copyright 2007, American Institute of Physics.)

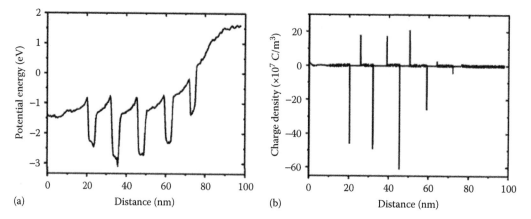

FIGURE 2.22 Active region of a green LED. (a) Electrostatic potential profile showing the p-n junction and quantum well region. (b) Charge density distribution due to polarization effects. (Reprinted with permission from Wu, Z. H., *Appl. Phys. Lett.*, 91, 041915, 2007. Copyright 2007, American Institute of Physics.)

increased [69]. The decay of piezoelectric fields in the fourth and fifth QWs could be due to strain relaxation [70]. It is possible that the first three periods remain fully strained, while the fourth and fifth periods undergo plastic relaxation. Another possibility, believed to be most likely, since this phenomenon has been observed only in the vicinity of p-n junctions, involves the migration of hydrogen ions from the p-type region. The p-type region has an abundance of hydrogen compared to the n-type region [71]. Hydrogen ions would passivate the negative polarization charges in the n-type region near the junction, and free electrons would move on and passivate the positive polarization charges, thus achieving charge neutrality.

2.7 Polarization Effects in AlN/GaN Heterostructures

2.7.1 Two-Dimensional Electron Gas (2DEG) at n-Type AlN/GaN Interfaces

Strong polarization fields can result in two-dimensional carrier confinement, such as in 2DEG AlGaN/AlN/GaN heterostructures. The high-electron conductivity of these thin film structures can be used to provide low resistant current paths to n- and p-type layers of the LED [72–75]. Figure 2.23 shows part of

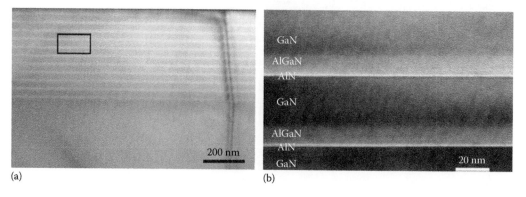

FIGURE 2.23 AlGaN/AlN/GaN n-type superlattice with a 2DEG at each GaN/AlN interface. (a) Bright field TEM image showing the eleven periods of the superlattice. (b) Higher magnification TEM image of small region marked in (a) corresponding to two periods. (Reprinted with permission from Wu, Z. H., *Appl. Phys. Lett.*, 91, 142121, 2007. Copyright 2007, American Institute of Physics.)

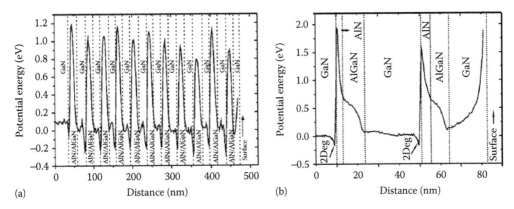

FIGURE 2.24 AlGaN/AlN/GaN n-type superlattice. (a) Bright field TEM image. (b) Potential energy profile of the superlattice showing the features of the 2DEG in the top two periods. (Reprinted with permission from Wu, Z. H., *Appl. Phys. Lett.*, 91, 142121, 2007. Copyright 2007, American Institute of Physics.)

an n-type AlGaN/AlN/GaN superlattice, exhibiting a 2DEG at the AlN/GaN interface. The AlGaN/AlN/GaN superlattice was epitaxially grown by MOCVD on an n-doped (10^{18} cm^{-3}), 3-μm thick, GaN layer on a *c*-plane sapphire substrate. The superlattice consists of 11 periods; each period starts with a 0.5 nm AlN layer, followed by a 12 nm AlGaN barrier layer with an Al concentration that is nominally graded linearly from 30% to 0%, and ends with a 28 nm GaN layer. The AlN layer is used to avoid interface alloy scattering effects, thus improving the lateral mobility of the 2DEG. Using a very thin AlN layer prevents significant reduction of the vertical conductivity. The AlGaN layer has an n-type doping level of 2×10^{19} cm^{-3}, while the n-type GaN layer is doped at 2×10^{18} cm^{-3}. In order to avoid impurity scattering, doping of the AlGaN layer was delayed by 4 seconds during growth, that is, the first nanometer of AlGaN is undoped, and the doping of the GaN layer was stopped 10 nm before starting the growth of the next AlN/AlGaN barrier. The structural characteristics of the superlattice are observed in the TEM image in Figure 2.23a, with a higher magnification image of the two top periods of the superlattice in Figure 2.23b. The AlN layer appears with the brightest contrast, followed by the graded AlGaN layer with fading brightness and the GaN layer with relatively darker contrast.

The electrostatic potential distribution obtained from the reconstructed phase and amplitude images is shown in Figure 2.24a, and a higher resolution profile in Figure 2.24b. An energy dip close to the GaN/AlN interface represents the accumulation of the 2DEG. The curvature of the potential around the 2DEG region is negative as expected. These results are consistent with calculated energy-band diagrams [73].

2.7.2 Two-Dimensional Hole Gas (2DHG) at p-Type AlN/GaN Interfaces

For p-type AlGaN/GaN heterostructures, the band diagram is different from the n-type case due to the alignment of the Fermi level. In order to achieve the accumulation of a 2DHG, the growth sequence of GaN/AlN/AlGaN should be adjusted, and the device structure should be optimized. We have studied a p-type AlGaN/GaN multi-heterostructure that was designed for the purpose of high conductivity with the accumulation of a 2DHG [74,75].

The AlGaN/AlN/GaN heterostructure in Figure 2.25 was epitaxially grown by metal-organic chemical vapor deposition on a nominally undoped, 3-μm thick, GaN layer on a *c*-plane sapphire substrate. The 2DHG heterostructure consists of six periods; each period starts with a 0.5 nm AlN layer, followed by a 5 nm nominally undoped GaN layer, and ending with a 12 nm AlGaN barrier layer with an aluminum concentration that is nominally graded linearly from 0% to 10%. The thin AlN layer is used to avoid interface alloy scattering effects and to improve the lateral mobility of the 2DHG. In order to optimize the vertical conductivity, a very thin AlN layer is used. The initial portion of the AlGaN layer

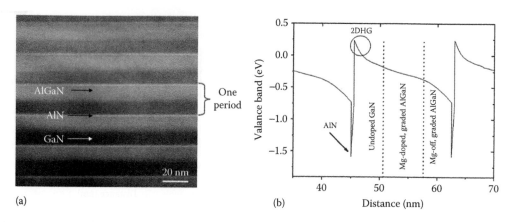

(a)   (b)

FIGURE 2.25 AlGaN/AlN/GaN superlattice structure. (a) TEM image of the superlattice. Brighter regions indicate high aluminum content. (b) Valence band profile of one period of the superlattice. The 2DHG accumulation region is shown by a circle. (Reprinted from Wei, Q. Y., *Appl. Phys. Express*, 2, 121001, 2009. Copyright 2009, The Japan Society of Applied Physics.)

has an Mg-doping level of ~1×10^{19} cm^{-3}, but the Mg flux is turned off for the last few nanometers. The nominally undoped GaN shows, in secondary ion mass spectroscopy, a Mg background concentration in the range of 1–5×10^{18} cm^{-3}, due to memory effects.

Figure 2.26 shows the electrostatic potential profile for three superlattice periods. The locations of the interface are determined by a careful match between diffraction contrast and the phase image. The formation of 2DHG is revealed by a positive curvature in the GaN immediately to the right of the AlN layer in the potential energy profile. The 2DHG at the interface cancels the spontaneous polarization charges, and no discontinuity in the potential profile is observed. But the depletion of holes produces a positive curvature in the profile, in contrast to a negative curvature in the 2DEG case.

Generally, in semiconductor structures of this type, we find four major types of charges that are responsible for the variations of the electrostatic potential. (1) Fixed charges like the spontaneous and piezoelectric polarization, (2) ionized impurities, (3) interface fixed charges (band offset), and (4) free carriers (in our case holes) that redistribute under the effect of the other charges. For the 2DHG case, the major contribution to the potential variation is from the holes confined in a planar region a few

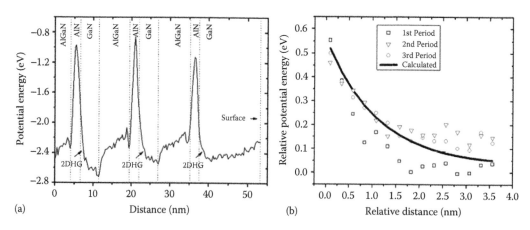

(a)   (b)

FIGURE 2.26 Potential profile of an AlGaN/AlN/GaN superlattice doped p-type, exhibiting a 2DHG. No discontinuity is observed at the AlN/GaN interface. (Reprinted from Wei, Q. Y., *Appl. Phys. Express*, 2, 121001, 2009. Copyright 2009, The Japan Society of Applied Physics.)

nanometers thick. In our case, the experimental and calculated potential distributions show a noticeable curvature in the vicinity (~1 nm) of the interface but become relatively flat away from the interface. This indicates that the 2DHG accumulation happens only over a few atomic monolayers of the interface [75].

2.7.3 Free-Carrier Accumulation at Cubic AlGaN/GaN Interfaces

As previously stated, the nitride semiconducting materials used in optoelectronics typically have the wurtzite structure with the space group $P6_3mc$, which exhibits large spontaneous and piezoelectric polarization fields, and hence, a significant quantum-confined Stark effect that limits the free-carrier recombination efficiency. On the other hand, the cubic phase of GaN with space group $F\bar{4}3m$ exhibits no spontaneous polarization and has been a subject of interest for the past two decades [76–78]. Furthermore, all piezoelectric polarization components (along with the growth direction and in the growth plane) should vanish for films grown along a <100> direction, [79] which is a different case from the wurtzite GaN where the total polarization is never zero for any orientation [65]. The absence of polarization effects is one of the advantages of cubic group III nitrides over the hexagonal group III nitrides.

There is limited experimental understanding of the electronic band structure of cubic nitride heterostructures. Accumulation of carriers at the AlGaN/GaN interface has been observed, but its origin is not well understood. This section deals with a study of the crystal structure and electronic properties of such an interface [80].

A cubic AlGaN/GaN heterostructure was grown by plasma-assisted molecular beam epitaxy on a 3C-SiC (001) substrate, with an aluminum nominal composition of 30%, with GaN and AlGaN layer thickness of 600 nm and 30 nm, respectively. The background carrier concentration for the cubic GaN layer was measured to be ~1 × 10^{17} cm^{-3} and for the AlGaN epilayers ~2 × 10^{18} cm^{-3}, respectively. Secondary ion mass spectroscopy measurements [81] indicated that oxygen might be the reason for this background carrier concentration.

Figure 2.27 is a cross-section TEM image of the cubic $Al_{0.3}Ga_{0.7}N$/GaN heterostructure. The crystal defects are mainly microtwins along {111} planes, which are common in cubic thin films [82,83].

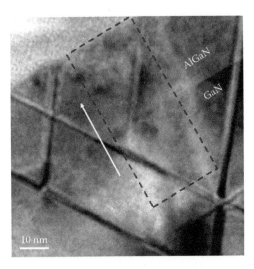

FIGURE 2.27 Cross-section TEM image of the AlGaN/GaN region, taken along a 110 projection and showing microtwins typical of cubic GaN epitaxy. The growth direction is indicated by an arrow. Electron holography data taken from the rectangular region in the box are used to determine the electrostatic potential energy profile. (Reprinted with permission from Wei, Q. Y., *Appl. Phys. Lett.*, 100, 142108, 2012. Copyright 2012, American Institute of Physics.)

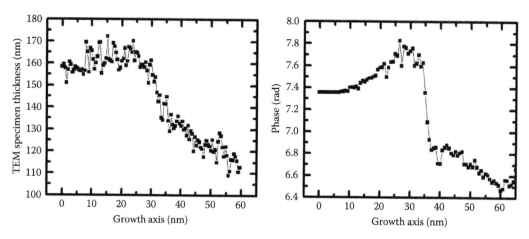

FIGURE 2.28 Thickness (left) and phase (right) profiles obtained from a hologram corresponding to the region depicted in Figure 2.13. (Reprinted with permission from Wei, Q. Y., *Appl. Phys. Lett.*, 100, 142108, 2012. Copyright 2012, American Institute of Physics.)

Electron holography was performed on the region marked by the rectangle, and the integrated thickness and phase profiles are shown in Figure 2.28.

The TEM specimen has a uniform thickness in the GaN region, but the AlGaN layer becomes thinner toward the surface. This is due to preferred oxidation and removal of the Al-containing region during ion milling. Thus, a polynomial fitting is applied to obtain the thickness profile in the AlGaN layer. The electrostatic potential profile is then obtained by dividing the phase shift by the thickness profiles; the result is shown in Figure 2.29.

In the GaN layer, the potential has a negative curvature, which indicates carrier depletion and free electrons accumulation at the GaN/AlGaN interface. The potential becomes flat in GaN at ~20 nm away from the interface, suggesting that the free electrons are confined within that range, and hence have

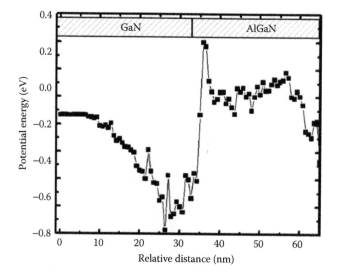

FIGURE 2.29 Potential energy profile obtained by dividing the phase by the amplitude from the electron hologram. (Reprinted with permission from Wei, Q. Y., *Appl. Phys. Lett.*, 100, 142108, 2012. Copyright 2012, American Institute of Physics.)

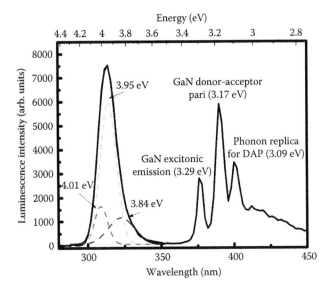

FIGURE 2.30 Cathodoluminescence spectrum of the cubic AlGaN/GaN heterostructure taken at a temperature of 4 K. The spectrum shows a broad AlGaN emission at 3.95 eV with a FWHM of 176 meV and three separate GaN near-band-edge peaks. (Reprinted with permission from Wei, Q. Y., *Appl. Phys. Lett.*, 100, 142108, 2012. Copyright 2012, American Institute of Physics.)

a quasi-two-dimensional characteristic. The 2DEG is characterized by the negative curvature in the potential energy profile with a density of $5.1 \times 10^{11}/\text{cm}^2$, obtained from the electrostatic potential energy profile using Poisson's equation. A positive curvature in the AlGaN potential energy profile is also observed, indicating a significantly ionized donor density. The latter is important in the formation of the 2DEG since it satisfies the necessary charge neutrality condition for the existence of the Coulomb sheet charges.

Further information about the electron band profile can be obtained from the optical emission of this heterostructure. This is provided by the CL spectrum in Figure 2.30.

The spectrum was taken at liquid helium temperature (4 K). Three separated lines at 3.29, 3.17, and 3.09 eV correspond to GaN band-edge emissions. The 3.29 eV peak is assigned to an excitonic transition in cubic GaN, while the 3.17 and 3.09 eV peaks are assigned to a donor–acceptor recombination and its phonon replica [84,85], in good agreement with other reports [86–88]. The AlGaN peak is at 3.95 eV with a full width at half maximum (FWHM) of 176 meV, and can be fitted by three Gaussian curves. These peaks are not as well resolved as the GaN band-edge peaks (with ~40 meV FWHM), probably due to alloy fluctuations and residual oxygen contamination. The peak at 3.95 eV gives a range for the bandgap difference between cubic AlGaN and GaN, from 0.66 to 0.78 eV consistent with the nominal composition of the AlGaN film.

Our experimental results suggest a conduction-to-valence band offset ratio at least as large as 5:1 for cubic AlGaN/GaN, while the first-principle calculations predict a value of 3:1 for cubic AlN/GaN [88]. With our measured energy band values, the electronic band diagram can be calculated by solving the Poisson and Schrodinger equations self-consistently, where charge neutrality from donors and free electrons is satisfied. The calculated potential energy profile compared with the experimental data is shown in Figure 2.31.

In the calculation, an effective n-doping level of $2 \times 10^{18}/\text{cm}^3$ in the AlGaN depletion region is used, and a zero polarization field for cubic nitride is assumed. The first three electron states are also shown in the figure, which indicates that the carriers' distribution is limited in a characteristic length less than 20 nm, in a quasi-2-dimensional nature. The consistency of experimental and theoretical data

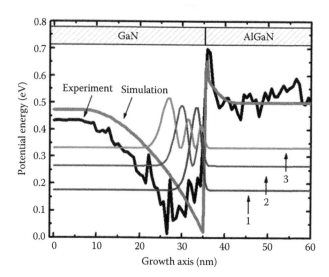

FIGURE 2.31 Calculated conduction band profile compared with the electrostatic potential obtained from electron holography data. The first three electronic states are also shown. (Reprinted with permission from Wei, Q. Y., *Appl. Phys. Lett.*, 100, 142108, 2012. Copyright 2012, American Institute of Physics.)

demonstrates that the free electron accumulation is induced by the donor states in the AlGaN layer, and promoted by the large conduction band-offset of the cubic AlGaN/GaN.

2.7.4 AlN/GaN Quantum Disks in GaN Nanowires

Semiconductor nanostructures are currently being developed for optoelectronics and photovoltaic applications. One approach is the growth of nanowires as an alternative to thin films. Nanowires can be grown with a relatively lower density of defects, provide a larger surface-to-volume ratio, and reduce the strain effects due to the lattice and thermal expansion mismatch. The dimensions of the nanowires (diameter and length) depend on several parameters such as growth temperature, choice of substrates, and conditions for nucleation and growth [89]. The lower dimensionality leads to strain reduction and defect-free interfaces, resulting in improved optical and electronic properties [90–92]. Strain relaxation may lead to the reduction of piezoelectric polarization fields in AlN/GaN and InGaN/GaN quantum well structures. These effects are at the picometer (subnanometer) scale. In this section, the properties of AlN/GaN quantum disks in GaN nanowires are studied using electron holography to determine the variation of the electrostatic potential with subnanometer resolution [93].

A nanowire structure consisting of ~30 wires/μm^2 was grown by plasma-assisted molecular beam epitaxy on (111) silicon substrate. The nanowires grow with nitrogen polarity, along the [0001] crystal direction, with an average diameter of ~100 nm, and a hexagonal cross section with {1$\bar{1}$00} facets. The nanowire structure consists of a silicon-doped GaN wire with a length of 500 nm, followed by twenty periods of AlN/GaN quantum disk superlattice with AlN and GaN layers of 3.5 nm in thickness. It ends with a 450 nm thick silicon-doped GaN overlayer. The GaN in the superlattice is doped with silicon at a concentration of 10^{19} cm^{-3} [94]. Figure 2.32 is a bright-field TEM image of one such nanowire, taken along a 11$\bar{2}$0 zone-axis. The diameter increases monotonically from 85 to 116 nm after the insertion of the AlN layer; this is due to the lateral growth rate for AlN.

The light-emitting properties of the nanowires were studied by CL spectroscopy in a scanning electron microscope. The spectra of a nanowire in Figure 2.33 were taken along its length at 4 and 300 K. The inset shows a secondary electron image of the nanowire and a corresponding monochromatic luminescence image at the peak light emission at 3.55 eV of the AlN/GaN quantum disks.

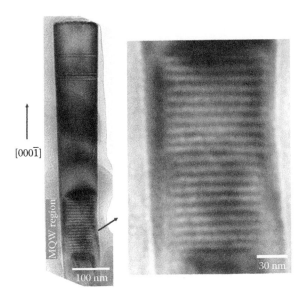

FIGURE 2.32 TEM micrograph of a nanowire with AlN/GaN superlattice. (Reprinted from Fischer, A. M., *Appl. Phys. Express*, 5, 025001, 2012. Copyright 2012, The Japan Society of Applied Physics.)

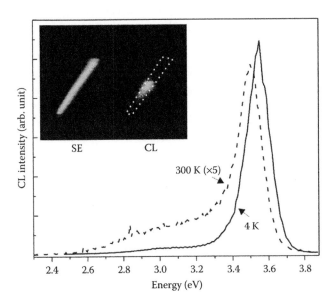

FIGURE 2.33 Cathodoluminescence spectra of a nanowire, taken at 4 and 300 K. The peak emission at 3.55 eV corresponds to the AlN/GaN quantum disks. The inset shows a secondary electron (SE) image of the nanowire and a monochromatic (CL) image taken at the peak emission of the quantum disks. (Reprinted from Fischer, A. M., *Appl. Phys. Express*, 5, 025001, 2012. Copyright 2012, The Japan Society of Applied Physics.)

The variation of the electrostatic potential along the quantum disk superlattice is determined using electron holography in a TEM. A hologram was taken along a $\bar{1}100$ projection perpendicular to the length of the nanowire. The variation of the holographic phase intensity is displayed in Figure 2.34.

The electrostatic potential energy is found by dividing the phase data in Figure 2.20 by the thickness variation. The region inside the rectangle is chosen using geometrical considerations related to

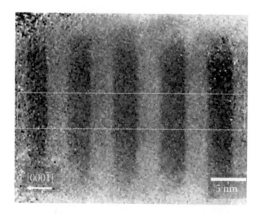

FIGURE 2.34 Phase component of an electron hologram of the quantum disk region. The area inside the rectangle has a constant thickness. The average image intensity is proportional to the electrostatic potential and it is plotted in Figure 2.21. (Reprinted from Fischer, A. M., *Appl. Phys. Express*, 5, 025001, 2012. Copyright 2012, The Japan Society of Applied Physics)

the faceted nature of the nanowire; one of which is that the nanowire thickness along this projection (~85 nm) does not change in the middle section and can be accurately determined from the TEM image. The phase is averaged in the direction perpendicular to the length of the nanowire in order to reduce the noise. Figure 2.35 shows the potential energy profile across five GaN quantum disks. The change in potential energy between AlN and GaN in the three central disks is ~1.5 eV and closely matches the reported theoretical values corresponding to the conduction band offset between GaN and AlN [95].

The interpolation of data points in the AlN and GaN layers in Figure 2.35 results in internal fields with an average value of ~1.8 MV/cm. The internal fields in the GaN and AlN layers have the same value but opposite signs. The variation in the potential profile of the disks may be associated with the surface effects and electron beam damage. The internal fields are related to spontaneous and piezoelectric polarization and have been proposed to be ~6.65 MV/cm for strained AlN/GaN superlattice films [62,96]. The reduced values observed in the nanowires is attributed to strain relaxation from the free surfaces that significantly suppresses the piezoelectric polarization fields, and screening by the silicon doping, which is relatively high in the quantum disks.

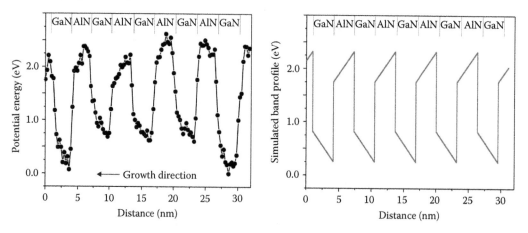

FIGURE 2.35 Potential energy profile of the quantum disk region extracted from Figure 2.20, showing an average internal field of ~1.8 MV/cm and an AlN/GaN band offset of ~1.5 eV. The band profile shown on the right was calculated using the information obtained from the electron holography data. (Reprinted from Fischer, A. M., *Appl. Phys. Express*, 5, 025001, 2012, Copyright 2012, The Japan Society of Applied Physics.)

2.8 Challenges and Perspectives

The chemical bond between elements in column III of the periodic table and nitrogen results in the hexagonal wurtzite lattice. This crystal structure exhibits strong dipole moments and a microstructure that plays a strong role in the III-N semiconductor properties. Thus, understanding the relationship between the atomic arrangement and the electronic properties at the nanometer scale becomes essential in the development of advanced optoelectronic devices.

The development of light-emitting devices has reached high acclaim as it is continuously approaching the physical limits for conversion of electricity into visible light. High-conversion efficiency is allowing the replacement of traditional interior and exterior illumination sources with LEDs based on the III-N semiconductors. But challenges still exist, such as in extending the technologies into monochromatic light emitters, where efficient green LEDs are still lacking. High-power LEDs encounter limitations with the *droop problem* where the efficiency drops with carrier injection. Several proposals exist about its origins, one of them being the internal electrostatic fields in the materials.

InGaN materials with high-indium content are necessary for the development of green and longer wavelength optoelectronic devices and of high-efficiency solar cells that can operate at high temperatures in high concentrator photovoltaics. Improvements in growth techniques show high promise for success in this area. These involve layer by layer growth approach for control of the surface growth morphology and the understanding of critical thickness for uniform control of misfit strain relaxation [96].

AlGaN films with high electric and optical properties are actively being pursued. As an example, vertical cavity surface-emitting lasers (VCSELs) have numerous applications including optical-fiber data transmission, optically pumped solid-state lasers, and chip-scale atomic clocks. For this purpose, high reflectivity distributed Bragg reflectors are necessary. They are extremely difficult in the III-N materials due to the large lattice mismatch between GaN and AlN, and the relatively low refractive index contrast of this materials system. The understanding of the microstructure role on the electronic properties becomes highly important [97].

In summary, the field has grown significantly in the last two decades, and a full review of the field results quite difficult in a limited amount of time and space. This chapter represents the limited view of the author, based on his own experience.

References

1. I. Akasaki and H. Amano. Breakthroughs in improving crystal quality of GaN and invention of the p-n junction blue-light-emitting diode. *Jpn. J. Appl. Phys.* **45**, 9001–9010 (2006).
2. S. Nakamura and M. R. Krames. History of gallium-nitride-based light-emitting diodes for illumination. *Proc. IEEE* **101**, 2211–2220 (2013).
3. F. A. Ponce and D. P. Bour. Nitride-based semiconductors for blue and green light-emitting devices. *Nature* **386**, 351–559 (1997).
4. F. A. Ponce. Microstructure of epitaxial III-V nitride thin films. In *GaN and Related Materials*, Pearton, S. J., (Ed.). Gordon and Breach Publishers, Philadelphia, PA, pp. 141–170, (1996).
5. F. A. Ponce. Crystal defects and device performance in LEDs and LDs. In *Introduction to Nitride Semiconductor Blue Lasers and Light Emitting Diodes*, Nakamura, S. and Chichibu, S. F., (eds.). Taylor & Francis Group, Abingdon, UK, pp. 105–152, (2000).
6. F. Fichter. Über aluminiumnitrid. *Z. Anorg. Chem.* **54**, 322–327 (1907).
7. H. P. Maruska and J. J. Tietjen. The preparation and properties of vapor-deposited single-crystalline GaN. *Appl. Phys. Lett.* **15**, 327–320 (1969).
8. J. L. Pankove, E. A. Miller, and J. E. Berkeyheiser. GaN electroluminescent diodes. *RCA Rev.* **32**, 383–392 (1971).
9. H. M. Manasevit, F. M. Erdmann, and W. I. Simpson. The use of metalorganics in the preparation of semiconductor materials. *J. Electrochem. Soc.* **118**, 1864–1868 (1971).

10. R. Dingle, K. L. Shaklee, R. F. Leheny, and R. B. Zetterstrom. Stimulated emission and laser action in gallium nitride. *Appl. Phys. Lett.* **19**, 5–7 (1971).

11. T. Matsumoto and M. Aoki. Temperature dependence of photoluminescence from GaN. *Jpn. J. Appl. Phys.* **13**, 1804–1807 (1974).

12. I. Akasaki and I. Hayashi. Research on blue emitting devices. *Ind. Sci. Technol.* **17**, 48–52 (1976) (in Japanese).

13. S. Yoshida, S. Misawa, and A. Itoh. Epitaxial growth of aluminum nitride films on sapphire by reactive evaporation. *Appl. Phys. Lett.* **26**, 461–462 (1975).

14. J. Karpinski, S. Porowski, and S. Miotkowska. High pressure vapor growth of GaN. *J. Cryst. Growth* **56**, 77–82 (1982).

15. S. Yoshida, S. Misawa, and S. Gonda. Improvements on the electrical and luminescent properties of reactive molecular beam epitaxially grown GaN films by using AlN-coated sapphire substrates. *Appl. Phys. Lett.* **42**, 427–429 (1983).

16. H. Amano, N. Sawaki, I. Akasaki, and Y. Toyoda. Metalorganic vapor phase epitaxial growth of a high quality GaN film using an AlN buffer layer. *Appl. Phys. Lett.* **48**, 353–355 (1986).

17. H. Amano, M. Kito, K. Hiramatsu, and I. Akasaki. P-type conduction in Mg-doped GaN treated with low-energy electron beam irradiation (LEEBI). *Jpn. J. Appl. Phys.* **28**, L2112–L2114 (1989).

18. T. Nagamoto, T. Kuboyama, H. Minamino, and O. Omoto. Properties of $Ga_{1-x}In_xN$ films prepared by MOVPE. *Jpn. J. Appl Phys.* **28**, L1334–L1336 (1989)

19. S. Nakamura. GaN growth using GaN buffer layer. *Jpn. J. Appl. Phys.* **30**, L1705–L1707 (1991).

20. S. Nakamura, T. Mukai, M. Senoh, and N. Iwasa. Thermal annealing effects on p-type Mg-doped GaN films. *Jpn. J. Appl. Phys.* **31**, L139–L142 (1992).

21. I. Akasaki, H. Amano, K. Itoh, N. Koide, and K. Manabe. GaN-based ultraviolet/blue light emitting devices. *Inst. Phys. Conf. Ser.* **129**, 851–856 (1992).

22. S. Nakamura and T. Mukai. High-quality InGaN films Grown on GaN films. *Jpn. J. Appl. Phys.* **31**, L1457–L1459 (1992).

23. S. Nakamura, T. Mukai, M. Senoh, S. Nagahama, and N. Iwasa. $In_xGa_{1-x}N/In_yGa_{1-y}N$ superlattices grown on GaN films. *J. Appl. Phys.* **74**, 3911–3915 (1993).

24. S. Nakamura, T. Mukai, and M. Senoh. Candela-class high-brightness InGaN/AlGaN double-heterostructure blue-light-emitting diodes. *Appl. Phys. Lett.* **64**, 1687–1689 (1994).

25. S. Nakamura, T. Mukai, and M. Senoh. High-brightness InGaN/AlGaN double-heterostructure blue-green light-emitting diodes. *J. Appl. Phys.* **76**, 8189–8191 (1994).

26. S. Nakamura, M. Senoh, N. Iwasa, and S. Nagahama. High-brightness InGaN blue, green and yellow light-emitting diodes with quantum well structures. *Jpn. J. Appl. Phys.* **34**, L797–L799 (1995).

27. S. Nakamura, M. Senoh, N. Iwasa, S. Nagahama, T. Yamada, and T. Mukai. Superbright green INGaN SQW structure LEDs. *Jpn. J. Appl. Phys.* **34**, L1332–L1335 (1995).

28. S. Nakamura, M. Senoh, S. I. Nagahama, N. Iwasa, T. Yamada, T. Matsushita, H. Kiyoku, and Y. Sugimoto. InGaN-based multi-quantum-well-structure laser diodes. *Jpn. J. Appl. Phys.* **35**, L74–L76 (1996).

29. I. Akasaki, S. Sota, H. Sakai, T. Tanaka, M. Koike, and H. Amano. Shortest wavelength semiconductor laser diode. *Electron. Lett.* **32**, 1105–1106 (1996).

30. K. Itaya et al. Room temperature pulsed operation of nitride based multi-quantum-well laser diodes with cleaved facets on conventional c-face sapphire substrates. *Jpn. J. Appl. Phys.* **35**, L1315–L1317 (1996).

31. S. Nakamura. Characteristics of InGaN multi-quantum-well-structure laser diodes. *Mater. Res. Soc. Proc.* **449**, 1135–1142 (1996).

32. F. A. Ponce, J. S. Major Jr., W. E. Plano, and D. F. Welch. Microstructure of GaN epitaxy on SiC using AlN buffer layers. *Appl. Phys. Lett.* **67**, 410–412 (1995).

33. F. A. Ponce, C. G. Van de Walle, and J. E. Northrup. Atomic arrangement at the AlN/SiC interface. *Phys. Rev. B* **53**, 7473–7478 (1996).

34. F. A. Ponce, J. S. Major Jr., W. E. Plano, and D. F. Welch. Crystalline structure of AlGaN epitaxy on sapphire using AlN buffer layers. *Appl. Phys. Lett.* **65**, 2302–2304 (1994).

35. F. A. Ponce. Fault–free silicon at the silicon-sapphire interface. *Appl. Phys. Lett.* **41**, 371–373 (1982).

36. *2014 Nobel Prize in Physics*, awarded to I. Akasaki, H. Amano, and S. Nakamura.

37. D. K. Biegelsen, F. A. Ponce, A. J. Smith, and J. C. Tramontana. Initial stages of epitaxial growth of GaAs on (100) silicon. *J. Appl. Phys.* **51**, 1856–1859 (1987).

38. A. Krost and A. Dadgar. GaN-based optoelectronics on silicon substrates. *Mat. Sci. Eng. B* **93**, 77–84 (2002).

39. A. Dadgar, J. Blasing, A. Diez, A. Alam, M. Heuken, and A. Krost. Metalorganic chemical vapor phase epitaxy of crack-free GaN on Si (111) exceeding 1 mu m in thickness. *Japn. J. Appl. Phys.* **39**, L1183–L1185 (2002).

40. O. Contreras, F. A. Ponce, J. Christen, A. Dadgar, and A. Krost. Dislocation annihilation by silicon delta-doping in GaN epitaxy on silicon. *Appl. Phys. Lett.* **81**, 4712–4714 (2002).

41. R. Liu, F. A. Ponce, A. Dadgar, and A. Krost. Atomic arrangement at the AlN/Si (111) interface. *Appl. Phys. Lett.* **83**, 860–862 (2003).

42. O. E. Contreras, F. Ruiz-Zepeda, A. Dadgar, A. Krost, and F. A. Ponce. Atomic arrangement at the AlN/Si(110) interface. *Appl. Phys. Exp.* **1**, 061104 (2008).

43. S. D. Lester, F. A. Ponce, M. G. Craford, and D. A. Steigerwald. High dislocation densities in high efficiency GaN-based LED's. *Appl. Phys. Lett.* **66**, 1249–1251 (1995).

44. F. A. Ponce. Defects and interfaces in GaN epitaxy. *MRS Bull.* **22**, 51–57 (1997).

45. F. A. Ponce, D. Cherns, W. T. Young, and J. W. Steeds. Characterization of dislocations in GaN by transmission electron diffraction and microscopy techniques. *Appl. Phys. Lett.* **69**, 770–772 (1996).

46. J. Bai, M. Dudley, L. Chen, B. J. Skromme, B. Wagner, R. F. Davis, U. Chowdhury, and R. D. Dupuis. Structural defects and luminescence features in heteroepitaxial GaN grown on on-axis and mis-oriented substrates. *J. Appl. Phys.* **97**, 116101 (2005).

47. R. Liu, A. Bell, F. A. Ponce, C. Q. Chen, J. W. Yang, and M. A. Khan. Luminescence from stacking faults in gallium nitride. *Appl. Phys. Lett.* **86**, 021908 (2005).

48. J. Mei, S. Srinivasan, R. Liu, F. A. Ponce, Y. Narukawa, and T. Mukai. Prismatic stacking faults in epitaxially laterally overgrown GaN. *Appl. Phys. Lett.* **88**, 141912 (2006).

49. F. A. Ponce, D. P. Bour, W. T. Young, M. Saunders, and J. W. Steeds. Determination of lattice polarity for growth of GaN bulk single crystals and epitaxial layers. *Appl. Phys. Lett.* **69**, 337–339 (1996).

50. H. Lichte. Electron holography approaching atomic resolution. *Ultramicroscopy* **20**, 293–304 (1986).

51. M. R. McCartney, D. J. Smith, R. Hull, J. C. Bean, E. Voelkl, and B. Frost. Direct observation of potential distribution across Si-Si p-n-junction using off-axis electron holography. *Appl. Phys. Lett.* **65**, 2603–2605 (1994).

52. D. Cherns, J. Barnard, and F. A. Ponce. Measurement of the piezoelectric field across strained InGaN/GaN layers by electron holography. *Solid State Comm.* **111**, 281–285 (1999).

53. D. Gabor. A new microscopic principle. *Nature* **161**, 777–778 (1948); and Microscopy of reconstructed waves, *Proc. R. Soc. A* **197**, 454 (1948).

54. M. R. McCartney and D. J. Smith. Electron holography: phase imaging with nanometer resolution. *Annu. Rev. Mater. Res.* **37**, 719–767 (2007).

55. J. Cai and F. A. Ponce. Study of charge distribution across interfaces in GaN/InGaN/GaN single quantum wells using electron holography. *J. Appl. Phys.* **91**, 9856–9862 (2002).

56. M. R. McCartney and M. Gaidardziska-Josifovska. Absolute measurement of normalized thickness from off-axis electron holography. *Ultramicroscopy* **53**, 283–289 (1994).

57. D. Cherns, W. T. Young, J. W. Steeds, F. A. Ponce, and S. Nakamura. Determination of the atomic structure of inversion domain boundaries in alpha-GaN by transmission electron microscopy. *Phil. Mag. A* **77**, 273–286 (1998).

58. F. A. Ponce, D. P. Bour, W. Goetz, and P. J. Wright. Spatial distribution of the luminescence in GaN thin films. *Appl. Phys. Lett.* **68**, 57–59 (1996).

59. D. Cherns and C. G. Jiao. Electron holography studies of the charge on dislocations in GaN. *Phys. Rev. Lett.* **87**, 205504 (2001).

60. J. Cai and F.A. Ponce. Determination by electron holography of the electronic charge distribution at threading dislocations in epitaxial GaN. *Phys. Stat. Sol. A* **192**, 407–411 (2002).

61. D. C. Look and J. R. Sizelove. Dislocation scattering in GaN. *Phys. Rev. Let.* **82**, 1237–1240 (1999).

62. F. Bernardini, V. Fiorentini, and D. Vanderbilt. Spontaneous polarization and piezoelectric constants of III-V nitrides. *Phys. Rev. B* **56**, 10024–10027 (1997).

63. T. Takeuchi, C. Wetzel, S. Yamaguchi, H. Sakai, H. Amano, I. Akasaki, Y. Kaneko, S. Nakagawa, Y. Yamaoka, and N. Yamada. Determination of piezoelectric fields in strained GaInN quantum wells using the quantum-confined Stark effect. *Appl. Phys. Lett.* **73**, 1691–1693 (1998).

64. A. E. Romanov, T. J. Baker, S. Nakamura, and J. Speck. Strain-induced polarization in wurtzite III-nitride semipolar layers. *J. Appl. Phys.* **100**, 023522 (2006).

65. Q. Y. Wei, T. Li, Z. H. Wu, and F. A. Ponce. In-plane polarization of GaN-based heterostructures with arbitrary crystal orientation. *Phys. Stat. Sol. A* **207**, 2226–2232 (2010).

66. M. R. McCartney, F. A. Ponce, J. Cai, and D.P. Bour. Mapping of electrostatic potential across a AlGaN/InGaN/AlGaN diode by electron holography. *Appl. Phys. Lett.* **76**, 3055–3057 (2000).

67. J. Cai, F. A. Ponce, S. Tanaka, H. Omiya, and Y. Nakagawa. Mapping the internal potential across GaN/AlGaN heterostructures by electron holography. *Phys. Stat. Sol. A* **188**, 833–837 (2001).

68. M. Stevens, A. Bell, M. R. McCartney, F. A. Ponce, H. Marui, and S. Tanaka. Effect of layer thickness on the electrostatic potential in InGaN quantum wells. *Appl. Phys. Lett.* **85**, 4651–4653 (2004).

69. Z. H. Wu, A. M. Fischer, F. A. Ponce, W. Lee, J. H. Ryou, D. Yoo, and R. D. Dupuis. Effect of internal electrostatic potential on light emission in a green LED with multiple InGaN quantum wells. *Appl. Phys. Lett.* **91**, 041915 (2007).

70. J. Bai, T. Wangl, and S. Sakai. Study of the strain relaxation in InGaN/GaN multiple quantum well structures. *J. Appl. Phys.* **90**, 1740–1744 (2001).

71. J. Neugebauer and C. G. Van de Walle. Role of hydrogen in doping of GaN. *Appl. Phys. Lett.* **68**, 1829–1831 (1996).

72. Z. H. Wu, M. Stevens, F. A. Ponce, W. Lee, J. H. Ryou, D. Yoo, and R. D. Dupuis. Mapping the electrostatic potential across AlGaN/AlN/GaN heterostructures using electron holography. *Appl. Phys. Lett.* **90**, 032101 (2007).

73. Z. H. Wu, F. A. Ponce, J. Hertkorn, and F. Scholz. Determination of the electronic band structure for a graded modulation-doped AlGaN/AlN/GaN superlattice. *Appl. Phys. Lett.* **91**, 142121 (2007).

74. J. Hertkorn, S. B. Thapa, T. Wunderer, F. Scholz, Z. H. Wu, Q. Y. Wei, F. A. Ponce, M. A. Moram, C. J. Humphreys, C. Vierheilig, and U. T. Schwarz. Highly conductive modulation doped composition graded p-AlGaN/AlN/GaN multiheterostructures grown by metalorganic vapor phase epitaxy. *J. Appl. Phys.* **106**, 013720 (2009).

75. Q. Y. Wei, Z. H. Wu, K. W. Sun, F. A. Ponce, J. Hertkorn, and F. Scholz. Evidence of two-dimensional hole gas in p-type AlGaN/AlN/GaN heterostructures. *Appl. Phys. Express* **2**, 121001 (2009).

76. H. Okumura, S. Misawa, and S. Yoshida. Epitaxial growth of cubic and hexagonal GaN on GaAs by gas-source molecular beam epitaxy. *Appl. Phys. Lett.* **59**, 1058–1060 (1991).

77. S. Miyoshi, K. Onabe, N. Ohkouchi, H. Yaguchi, R. Ito, S. Fukatsu, and Y. Shiraki. MOVPE growth of cubic GaN on GaAs using dimethylhydrazine. *J. Crystal Growth* **124**, 439–442 (1992).

78. D. Schikora, M. Hankeln, D. J. As, K. Lischka, T. Litz, A. Waag, T. Buhrow and F. Henneberger. Epitaxial growth and optical properties of cubic GaN films. *Phys. Rev. B* **54**, R8381–R8384 (1996).

79. D. Sun and E. Towe. Strain-generated internal fields in pseudomorphic (In-Ga)As/GaAs quantum well structures on (111) GaAs substrates. *Jpn. J. Appl. Phys.* **33**, 702–708 (1994).

80. Q. Y. Wei, T. Li, J. Y. Huang, F. A. Ponce, E. Tschumak, A. Zado, and D. J. As. Free carrier accumulation at cubic AlGaN/GaN heterojunctions. *Appl. Phys. Lett.* **100**, 142108 (2012).

81. A. Zado, J. Gerlach and D. J. As. Low interface trapped charge density in MBE in situ grown Si_3N_4 cubic GaN MIS structures. *Semicond. Sci. Technol.* **27**, 035020 (2012).

82. F. A. Ponce and J. Aranovich. Imaging of the silicon on sapphire interface by high-resolution transmission electron microscopy. *Appl. Phys. Lett.* **38**, 439–441 (1981).

83. F. A. Ponce, W. Stutius, and J. G. Werthen. Lattice structure at ZnSe-GaAs heterojunction interfaces prepared by OMCVD. *Thin Solid Films* **104**, 133 (1983).

84. D. J. As, F. Schmilgus, C. Wang, B. Schöttker, D. Schikora, and K. Lischka. The near band edge photoluminescence of cubic GaN epilayers. *Appl. Phys. Lett.* **70**, 1311–1313 (1997).

85. J. Menniger, U. Jahn, O. Brandt, H. Yang, and K. Ploog. Identification of optical transitions in cubic and hexagonal GaN by spatially resolved cathodoluminescence. *Phys. Rev. B* **53**, 1881–1885 (1996).

86. G. Ramirez-Flores, H. Navarro-Contreras, A. Lastras-Martinez, R. C. Powell, and J. E. Greene. Temperature-dependent optical band-gap of the metastable zincblende structure beta-GaN. *Phys. Rev. B* **50**, 8433–8438 (1994).

87. C. Mietze, M. Landmann, E. Rauls, H. Machhadani, S. Sakr, M. Tchernycheva, F. H. Julien, W. G. Schmidt, K. Lischka, and D. J. As. Band offsets in cubic GaN/AlN superlattices. *Phys. Rev. B* **83**, 195301 (2011).

88. K.-Y. Hsu, C.-Y. Wang, and C.-P. Liu. The growth of GaN Nanorods with different temperature by MBE. *J. Electrochem. Soc.* **157**, K109–K112 (2010).

89. E. Calleja, M. Sánchez-García, F. Sánchez, F. Calle, F. Naranjo, E. Muñoz, U. Jahn, and K. Ploog. Luminescence properties and defects in GaN nanocolumns grown by MBE. *Phys. Rev. B* **62**, 16826–16834 (2000).

90. A. Kikuchi, K. Yamano, M. Tada, and K. Kishino. Stimulated emission from GaN nanocolumns. *Phys. Status Solidi B* **241**, 2754–2758 (2004).

91. E. Ertekin, P. A. Greaney, D. C. Chrzan, and T. D. Sands. Equilibrium limits of coherency in strained nanowire heterostructures. *J. Appl. Phys.* **97**, 114325 (2005).

92. A. M. Fischer, K. W. Sun, F. A. Ponce, R. Songmuang, and E. Monroy. Correlated structural, electronic and optical properties of AlN/GaN multiple quantum disks in GaN nanowires. *Appl. Phys. Express* **5**, 025001 (2012).

93. R. Songmuang, O. Landré, and B. Daudin. From nucleation to growth of catalyst-free GaN nanowires on thin AlN buffer layer. *Appl. Phys. Lett.* **91**, 251902 (2007).

94. P. K. Kandaswamy et al. GaN/AlN short period superlattices for intersubband optoelectronics. *J. Appl. Phys.* **104**, 093501 (2008).

95. O. Ambacher et al. Pyroelectric properties of Al(In)GaN/GaN hetero- and quantum well structures. *J. Phys. Condens. Matter* **14**, 3399–3434 (2002).

96. A. M. Fischer, Y. O. Wei, F. A. Ponce, M. Moseley, B. Gunning, and W. A. Doolittle. Highly luminescent, high-indium-content InGaN film with uniform composition and full misfit-strain relaxation. *Appl. Phys. Lett.* **103**, 131101 (2013).

97. Y.-S. Liu, S. Haq, T. T. Kao, K. Mehta, S. C. Shen, T. Detchprohm, P. D. Yoder, R. D. Dupuis, H. Xie, and F. A. Ponce. Electrically conducting n-type AlGaN/GaN distributed Bragg reflectors grown by MOCVD. *J. Crystal Growth* **443**, 81–84 (2016).

3

Optical Properties of III-Nitride Semiconductors

Plamen P. Paskov
and Bo Monemar

3.1 Introduction

The optical properties of the group-III-nitride materials are obviously of direct relevance for optoelectronic applications, but experiments measuring optical properties also give information on a range of electronic properties. There is already a wealth of data in the literature on the optical properties of III-nitrides [1–4], and here we will concentrate on some of the most recent additions to the scientific knowledge. The focus, looking at the present situation concerning technical applications of these materials, has been on GaN, InGaN, and AlGaN in recent decades. AlGaN materials are important for ultraviolet (UV) emitters and high electron mobility transistor (HEMT) structures and AlGaN optical properties have accordingly been studied over the entire Al composition range. InGaN materials (with In content <50%) have also been studied extensively, and the light-emitting diode (LED) applications based on InGaN/GaN quantum structures have already been awarded a Nobel Prize in 2014. However, the applications of InN are lagging behind. The development of growth procedures for InN and In-rich InGaN has been difficult, and their optical properties were consequently much less studied in the past.

The optical properties of bulk GaN have been studied by several authors by spectrometric ellipsometry, yielding the entire dielectric function over a wide range from deep UV (above bandgap region) to infrared (lattice vibrations). Comparison with band structure calculations has produced an upgraded picture of the near-bandgap region for both valence and conduction bands, although the uncertainties in the theoretical calculations of the band structure still cause problems in the interpretations. Luminescence data has produced accurate knowledge about the excitonic- and impurity-related transitions near the bandgap energy, of relevance for optical emitting devices. Such studies are also very useful to establish optical fingerprints of impurities in the material, helpful for characterization in connection with the growth of device structures. Different growth techniques often give different defects in the material, with corresponding optical spectra.

With the availability of high-quality bulk AlN material, the knowledge about the optical transitions in the near-bandgap region of AlN has been significantly updated. The area of point-defect identification in AlN and AlGaN via the specific optical spectra contains many ongoing activities that are still under development.

The recent results on the optical properties of InN have led to an improved accuracy in the basic data on this material, and the correlation of doping with optical properties has advanced recently. Doping control of InN is now in sight, and the optical data are very helpful in characterizing these properties. InGaN is the core material for visible LEDs and as such has attracted much attention during the last decades. Most studies are on quantum well structures, but there is also basic work on bulk InGaN properties.

In this chapter, we have excluded the discussion of optical phenomena in quantum structures, since these are treated in separate chapters in connection with devices.

3.2 Optical Properties of GaN

3.2.1 Band-to-Band Optical Transitions and Free Excitons

GaN (as the other group-III nitrides) is a direct-bandgap semiconductor with the conduction band minimum and valence band maximum both occurring at the center of the Brillouin zone (Γ point). The Bloch wave function of the bottom conduction band is essentially determined by s atomic functions and has s-like orbital character, while the Bloch wave function of the top valence band is built from p atomic functions and has p-like orbital character. In wurtzite GaN (C^4_{6v} space group symmetry) the crystal field and the spin-orbit interaction split the top valence band into three bands, commonly labeled A, B, and C in order of increasing energy of the holes. In terms of the group theory, the symmetry of the conduction band is Γ_7 and the symmetry of the A, B, and C valence bands is Γ_9, Γ_7, and Γ_7, respectively. This implies that the band-to-band transitions between the conduction band and the A valence band are dipole-allowed only for $\mathbf{E}\perp\mathbf{c}$ polarization ($\mathbf{E}$ is the electric field vector of absorbed/emitted light and $\mathbf{c}$ is the hexagonal axis of the crystal), while the transitions between the conduction band and the B and C valence bands are dipole allowed for both polarizations, $\mathbf{E}\perp\mathbf{c}$ and $\mathbf{E}\|\mathbf{c}$. The electron-hole Coulomb interaction gives rise to three free excitons which correspond to the A, B, and C valence bands. These excitons have almost the same binding energy (25–26 meV) as determined in earlier photoluminescence (PL) and optical reflectance (OR) measurements [5,6]. Generally, three parameters (Δ_1, Δ_2, and Δ_3) are required to describe the energy separation between the valence bands (consequently between the ground states of the three free excitons) and the oscillator strengths for all optical transitions [7–9]. In the quasi-cubic approximation [10] (i.e., for an isotropic spin-orbit interaction) the three band parameters are related to the crystal-field splitting (Δ_{cf}) and the spin-orbit splitting (Δ_{so}) by $\Delta_1 = \Delta_{cf}$ and $\Delta_2 = \Delta_3 = \Delta_{so}/3$ [8]. The most consistent values for the crystal-field and the spin-orbit splitting in GaN are $\Delta_{cf} = 10$ meV and $\Delta_{so} = 17$ meV, experimentally determined from detailed analysis of the strain dependence of the free exciton energies [7,11].

The basic optical properties of GaN in the near-bandgap region are known with some precision since more than a decade. Accurate free exciton energies, the exciton fine structure due to the spin-exchange interaction, as well as the exciton-polariton branches arising from the exciton-photon coupling were derived from temperature-dependent and polarization-resolved PL measurements in strain-free GaN samples [12,13]. These samples are cut from boules grown by hydride vapor phase epitaxy (HVPE) and represent the high-purity GaN material so far; the residual impurity concentration is $<10^{16}$ cm^{-3} (mainly Si and O due to the contamination from the quartzware in the growth

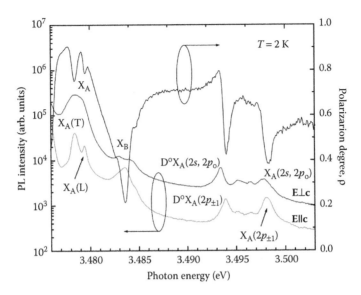

FIGURE 3.1 Low-temperature PL spectra of a bulk GaN samples in the free-exciton region for E⊥c and E∥c polarizations. The polarization degree of the emission defined as $\rho = (I_\perp/I_{II})/(I_\perp + I_{II})$, where $I_\perp(I_{II})$ is the PL intensity for E⊥c and E∥c polarizations, is also shown. (From Monemar, B. et al.: Recombination of free and bound excitons in GaN. *Phys. Status Solidi (b)*. 2008. 1723. 245. Copyright Wiley-VCH Verlag GmbH & Co. KGaA. Reproduced with permission.)

reactor). Figure 3.1 shows the low-temperature PL spectra for E⊥c and E∥c polarizations in the free exciton region [13]. The E⊥c polarized spectrum is dominated by the emission from the Γ_5 state of the A exciton (X_A) broadened due to nonthermalized polaritons and the two polariton branches of the Γ_5 state of the B exciton. For E∥c polarization the Γ_1 state of the B exciton (X_B) is clearly resolved. The doublet in the region of the X_A exciton is related to a mixed-longitudinal-transverse state arising from a slight deviation from k⊥c experimental geometry (k is the light wave vector). At higher energies, the excited states of the X_A, as well as the excited states of the neutral donor bound exciton ($D°X_A$) are observed in both polarizations (for more details see Reference [13]). In such pure samples, the free exciton energies can be followed up to room temperature, and consequently, the bandgap can be precisely determined. The bandgap in unstrained GaN is found to vary from 3.503±0.001 eV at 2 K to 3.437±0.005 eV at 290 K [13]. For the B bandgap, the corresponding values are 3.507±0.001 eV at 2 K and 3.442±0.005 eV at 290 K. We note that due to the very small energy separation between the A and B valence bands (≈4 meV) polarization-resolved measurements are required in order to distinguish between the A and B band-related optical transitions, especially at elevated temperatures. As for the C bandgap, only the room temperature value of 3.460±0.005 eV can be suggested from our studies [13]. These data are more adequate than the previous ones acquired for GaN grown on sapphire and frequently cited [14]. The biaxial strain existing in heteroepitaxial layers, results in a shift of all optical transitions to the higher energies in the case of compressive strain and the lower energies in the case of tensile strain. Moreover, the strain modifies the valence-band splitting (consequently the exciton energies) and the optical transition oscillator strengths [15].

Absorption data from transmission experiments are less straightforward to measure, due to the strong oscillator strengths of the excitons, requiring a sample thickness <1 μm. Reliable data for the above bandgap absorption coefficient at room temperature in undoped and *n*-doped GaN are available from spectroscopic ellipsometry (SE) measurements (Figure 3.2) [16]. Some data at much higher temperature have also been presented [17]. The proper modeling of the measured imaginary part of the dielectric function (ε_2) shows the relevance of the Coulomb enhancement in band-to-band (BB) transitions and the exciton-phonon complexes (EPC) in near-bandgap absorption. The experimental data in

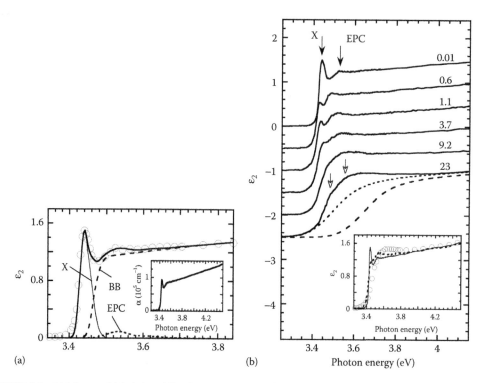

FIGURE 3.2 (a) Measured (circles) and fitted imaginary part of the dielectric function (ε_2) in undoped GaN. The inset presents the absorption coefficient determined from measured dielectric function. (b) Measured ε_2 in sample with different free electron concentrations (the numbers indicate the electron concentration in units of 10^{18} cm^{-3}). The spectra are vertically shifted for clarity. In the inset, ε_2 spectra of samples with electron concentration of 3.7×10^{18} cm^{-3} (dotted) and 2.3×10^{19} cm^{-3} are compared to the spectrum in undoped GaN (solid line). (Reprinted with permission from Shokhovets, S. et al., *Phys. Rev. B.*, 79, 045201, 2009. Copyright 2009 by the American Physical Society.)

Figure 3.2 also implies a value of the Mott density (defined as the electron concentration at which the exciton peak (X) moves into the continuum) of about 2×10^{18} cm^{-3} [16].

While the near-bandgap structures have been explored in detail, the band structure over the entire Brillouin zone is less accurately known. Details of the band structure are important for the modeling of certain electronic processes in devices. As an example, for InGaN-based LEDs in the visible spectral range, there has been a recent debate on the possible importance of intrinsic Auger processes for nonradiative recombination in these LEDs at high injection currents [18]. In the modeling of this process, the energy position of the satellite valleys of the conduction band (the lowest one believed to be at the L point of the Brillouin zone) is crucial in order to evaluate the hot-electron transport. In principle, the high-energy critical points can be studied by vacuum UV (VUV) SE on samples with different crystal orientations, where the ordinary and extraordinary dielectric function can be derived [19,20]. Unfortunately, the SE provides a very rough estimate for the energy separation between Γ and L minima of the conduction band even if the transitions related to the excitation of the Ga 3d core levels are examined [19]. Theoretically, the splitting has been predicted to be about 2.2 eV [21,22], while indirect experimental methods, such as pump-probe transient spectroscopy [23] and photoexcited field emission spectroscopy [24], have given a value of 1.1–1.2 eV. Recently, by near-bandgap photoemission spectroscopy study of *p*-doped GaN activated to negative electron affinity, a value of 0.9 eV has been obtained [25]. The large discrepancy between the experimental and theoretical values for the Γ–L conduction band energy separation as well as the expected complex kinetic energy distribution of hot electrons have provoked the claim for a direct evidence of Auger process in LEDs [26] to be questioned [22].

3.2.2 Bound Excitons in GaN

The optical signatures of the exciton bound to shallow donors (SDs) and acceptors were reviewed for GaN some years ago [13]. All these data were given for strain-free bulk samples, that is, the energies do not suffer from strain shifts, and are accurate within the spectral resolution (<0.2 meV).

There are two dominant SDs in GaN—substitutional oxygen (O) on the nitrogen site (O_N) and substitutional silicon (Si) on the gallium site (Si_{Ga}). In low-temperature PL spectra, the recombination of the excitons bound to these two neutral donors gives rise to two sharp lines below the A free exciton line [13]. The energy separation from the A exciton, 7.0 meV for the O-related line and 6.2 meV for the Si-related line, represents the binding energy of the corresponding donor bound exciton (DBE). At energies above the two main DBE lines, a number of weak emission lines related to the excited state of the DBEs were also resolved. In the lower energy region (3.44–3.45 eV) the PL spectra showed a rich structure of the so-called two-electron transitions (TETs) which occur from the DBE recombination with simultaneous excitation of the donor to its excited states [13,27]. The temperature-dependent and polarization-dependent PL measurements [27] spectra, as well as the detailed analysis of the donor and DBE excited states [28] allows an accurate determination of the binding energy of the Si and the O donors, 30.2±0.3 meV and 33.5±0.4 meV, respectively. The phonon coupling of the DBEs and the polarization selection rules for the phonon-assisted optical transitions were explored both theoretically and experimentally [29]. The dynamics of the DBE recombination was also studied in detail [13,30]. The PL transients of DBEs generally consist of at least two different parts, originating from different processes (near-surface and bulk-related, respectively). There are different processes associated with the excited DBE states and the very commonly used two-state model for the exciton transfer between the free-exciton and DBE state is, in general, inadequate [30]. Also, the exciton transfer processes of DBEs among the neutral donors cannot be neglected, as evidenced by the often observed much shorter lifetimes for DBEs in samples with donor concentrations $\geq 10^{17}$ cm^{-3} [31]. Nonradiative recombination processes in the near-surface region of GaN have recently been analyzed in detail, suggesting that care should be taken to remove the oxide layers as well as subsurface damage and H-related defects, in order to increase the radiative efficiency (by two orders of magnitude) in a PL experiment [32].

Recently, a new donor in GaN, substitutional germanium (Ge) on the gallium site (Ge_{Ga}), has been studied in some detail. The interest of such investigations is stimulated by the observation that much fewer defect problems seem to occur with the Ge doping as compared with the Si doping for GaN grown on sapphire [33,34]. From earlier works, it was known that the binding energy for the Ge donor in GaN is 31 meV (from theoretical calculations [35]) and 30 meV (from optical studies [36]), similar to that of the Si donor. However, the recent PL studies in Ge-doped GaN have not revealed any DBE emission lines because highly doped samples were examined so far [37,38]. An electron carrier concentration above 10^{20} cm^{-3} was achieved by Ge doping in GaN grown on sapphire by metalorganic chemical vapor deposition (MOCVD) [37,38]. The PL spectra of such degenerate material (with the Fermi level in the conduction band) clearly demonstrate a broad band-to-band emission with a high-energy cut-off at the Fermi level and a tapering off at lower energies (Figure 3.3) [38]. The PL line shape can be very well approximated by the model for a recombination of free electrons in the conduction band with localized holes [39]. Attaining a high electron concentration with Ge doping have allowed to properly evaluate the interplay between the effects of the Coulomb interaction screening, the bandgap renormalization (BGR) and the Burstein–Moss shift (BMS) on the bandgap edge optical spectra in n-type GaN. Figure 3.4 shows the free-carrier dependence of the exciton energy (extracted from the fitting of ε_2 measured by SE) together with theoretical modeling which accounts for all above effects [38]. (Note that the Mott density in Reference [38] was defined as the electron concentration at which the exciton binding energy is approaching zero and was estimated to be 1.5×10^{19} cm^{-3}).

Since the beginning of the research on GaN, doping with several acceptors (Mg, Zn, Cd) have been tried in order to achieve p-type conductivity. All these acceptors (substitutional on Ga site) have relatively large binding energies (200–500 meV) which cause a problem with obtaining a high hole

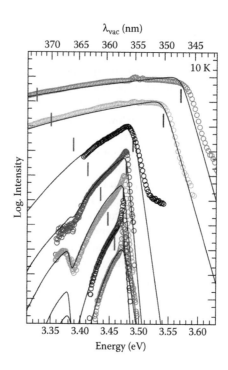

FIGURE 3.3 Low-temperature PL spectra of Ge-doped GaN layers (open symbols) and modeled line shape (continuous lines). The vertical lines show the energy positions of renormalized band gap with and without including the Burstein–Moss shift. (Reprinted with permission from Feneberg, M. et al., *Phys. Rev. B.*, 90, 075203, 2014. Copyright 2014 by the American Physical Society.)

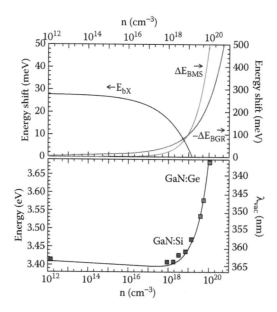

FIGURE 3.4 (Top) Separately calculated contributions of the Coulomb interaction screening (E_{bX}), the bandgap renormalization (BGR) and the Burstein–Moss shift (BMS). (Bottom) The exciton energy as a function of the free electron concentration. The squares are the values extracted from the measured imaginary part of the dielectric function while the line represents the theoretical modeling. (Reprinted with permission from Feneberg, M. et al., *Phys. Rev. B.*, 90, 075203, 2014. Copyright 2014 by the American Physical Society.)

concentration. So far only the Mg acceptor (with a binding energy of $\approx$225 meV) is proved to work fine, and nowadays it is used for all GaN-based *p-n* devices.

The optical signatures of the Mg acceptor (Mg_{Ga}) were briefly reviewed in Reference [13] and then studied in more details in Mg-doped homoepitaxial samples [40–43] where the spectral linewidth of the acceptor bound excitons (ABE) is much smaller than that in the samples grown on sapphire substrates [44]. We want to point out that some researchers apparently are still not convinced that the main donor-acceptor pair (DAP) peaking at 3.27 eV is related to the Mg acceptor [45]. There are several strong indications for that in the literature, however. Here we relate to the ion implantation studies reported in Reference [46]. In Figure 3.5 are shown the DAP signatures of Mg, Zn, and Cd acceptors, as convincing evidence for the attribution of the 3.27 eV DAP to the Mg acceptor [46]. In the same paper, the ABE peak for the Mg-doped sample was reported at 3.466 eV, as also stated in our more recent work [40]. The assignment of the above ABE and DAP signatures to the Mg fulfills the reasonable requirement that these spectra should be strong (or dominant) in all Mg-doped samples up to the high-doping region (10^{19} cm^{-3}). It is well known that the Mg doping in heteroepitaxially grown GaN leads to a formation of pyramidal inversion domains, and thus deteriorate the structural quality of the material [47,48]. For Mg-doped GaN layers grown on bulk GaN substrates, no such defects were found for Mg concentrations <10^{20} cm^{-3} [49]. The main structural defects in such material are numerous small (5–10 nm) basal plane stacking faults (BSFs) which have a shape of two-dimensional islands and do not have any associated threading partials. These defects do however influence the optical properties. It was earlier discovered that there are two separate optical signatures related to the Mg acceptor [40]. The main acceptor bound exciton line at 3.466 eV (labeled ABE1) is accompanied by a lower energy line at about 3.454 eV (ABE2). The broader ABE2 line was recently interpreted as due to a perturbed Mg acceptor, involving interaction with nearby BSFs [43]. The detailed structure of this second acceptor (including the interaction with the BSFs) needs further studies. The two ABE emissions have different transient properties which are consistent with an efficient transfer process of excitons from ABE1 to ABE2 [43]. The second acceptor state related to the Mg acceptor is also evident in the DAP spectra. It was shown that there is a broad spectrum in the background of the Mg-related 3.27 eV DAP, particularly clear in the time-resolved spectrum (see Figure 3.4 in Reference 43).

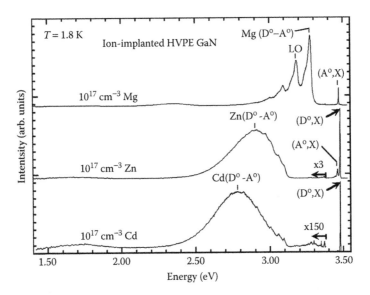

FIGURE 3.5 Low-temperature PL spectra of Mg-, Zn-, and Cd-implanted GaN grown by HVPE, after annealing. (Reprinted from Chen, L. and Skromme, B. J., *Mat. Res. Soc. Symp. Proc.*, 743, L11.35.1, 2003. Copyright 2003, Cambridge University Press.)

Some recent efforts on the theory side on the electronic structure and related properties of acceptors in GaN seem to deviate from the recent experimental results. First-principles calculations on the energy levels of the Mg acceptor in GaN [50] come to the conclusion that the Mg acceptor is somewhat deeper (close to 300 meV) than the one experimentally observed (225 meV) and has a very strong phonon coupling (Huang–Rhys factor >>1) while experimentally a value of about 0.5 is found. The reasons for this discrepancy has so far not been examined, in general, an accurate treatment of acceptors in compounds involving *d*-electrons is difficult. Other authors misinterpret the shallow 3.466 eV ABE PL line as evidence for a nitrogen vacancy shallow donor level [51].

The Zn-related ABEs have also been observed. The low-temperature PL spectrum of bulk GaN grown by the Na-Ga melt method has shown three closely spaced lines around 3.455 eV [52]. The temperature dependence of the spectrum has revealed that these lines are derived from a split ABE ground state. This fact confirms that the ABE ground state in wurtzite semiconductors is not a single state (as obtained if only the top valence band is considered) and a proper treatment including all three top valence bands is needed [53].

3.2.3 Below-Bandgap Emissions Related to Intrinsic and Extrinsic Point Defects

Below the DAP emission at 3.27 eV and its LO phonon replicas, a number of emission bands related to intrinsic or extrinsic defect levels (so-called deep levels) are observed in the PL spectra of GaN. The spectral position of the emission maximum is commonly used for their labeling, for example, red luminescence (RL) at 1.8 eV, yellow luminescence (YL) at 2.2 eV, green luminescence (GL) at 2.4 eV, and blue luminescence (BL) at 2.8–2.9 eV. Low-temperature PL spectra for an undoped bulk GaN grown by HVPE and carbon-doped homoepitaxial GaN layer grown by MOCVD presented in Figure 3.6 show typical below-bandgap emission in GaN.

The presence and the quantum efficiency of a particular emission depend on the method used for growth of GaN material, the growth conditions (temperature, V/III ratio), as well as on the concentration of the impurities (intentional or unintentional). Nevertheless, there are some common characteristics for these emission bands: (i) at low temperatures the recombination occurs between electrons at SDs and holes at deep acceptors (DAs) (DAP emission), while at elevated temperatures the donors are ionized and

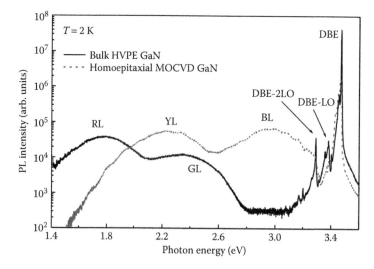

FIGURE 3.6 Low-temperature PL spectra of an undoped bulk GaN layer grown by HVPE and a C-doped homo-epitaxial GaN layer grown by MOCVD. The secondary ion mass spectroscopy (SIMS) on these samples reveals Si and O concentration of 1×10^{16} cm^{-3} and H concentration is of 1×10^{18} cm^{-3}. The C concentration is 1×10^{16} cm^{-3} in the HVPE sample and 5×10^{17} cm^{-3} in the MOCVD layer.

the recombination is between conduction band electrons and DA holes (so-called *e*-A emission); (ii) with increasing temperature the emissions shift toward higher energies, decrease in intensity and eventually quench with a thermal activation energy approximately equal to the energy of the corresponding acceptor level measured from the top of valence band; (iii) at low-excitation intensity an acceptor concentration less than 10^{15} cm^{-3} is enough to give rise to a deep level emission; (iv) with increasing excitation intensity the maximum of the emission bands shift toward higher energies, and the intensity saturate; (v) the emission line shapes are quite broad as a result of the strong electron-phonon coupling and the spectral maximum of the emission band does not correspond to the energy position of the defect in the gap; (vi) the PL transients are nonexponential with an effective PL decay time of 1–100 μs, much longer than the typical one for the exciton transitions. When more than one type of deep-level emissions are present, the behavior of PL intensities and the spectral positions with both the temperature and the excitation intensity becomes more complex due to carrier re-distribution among different recombination channels.

The intrinsic and extrinsic point defects which give rise of various deep level emissions in GaN have been studied theoretically in the past [54]. On the other hand, a comprehensive analysis of the experimental results can be found in Reference [45]. These review papers, however, present the *status quo* in the scientific knowledge as it was more than 10 years ago. In the following, some new theoretical and experimental findings are discussed.

The RL is typically observed in undoped GaN grown by HVPE [45,55,56] (see also Figure 3.6). Among all deep level emissions, the RL exhibits the longest decay time and then saturates at very low-excitation energies [45]. The zero-phonon line (ZPL) of the RL is found to be at 2.36 eV which implies that the DA involved in the emission is located at 1.13 eV above the valence band [57] matching the earlier calculated energy for the (–/2–) transition level of the gallium-vacancy-oxygen-donor complex (V_{Ga}-O_N) [54,58]. New hybrid functional calculations, however, placed this level at 2.2 eV above the valence band, and thus shifted the related emission in the infrared [50]. Low-energy transition level of V_{Ga}-O_N or complexes with hydrogen (H) such as V_{Ga}-O_N-H, V_{Ga}-H and V_{Ga}-2H can be tentatively associated with the RL [50]. An emission at 1.7–1.8 eV has also been observed in Mg-doped GaN [45,59]. In this case, however, the suggested interpretation for the emission is a deep donor (DD) to shallow acceptor (SA) transitions, where the DD is a nitrogen-vacancy-magnesium-acceptor complex (V_N-Mg_{Ga}) and the SA is Mg_{Ga} [59,60].

The YL is by far the most studied mid-bandgap emission in GaN. The YL presents in the PL spectra of undoped or *n*-doped material independent of the growth method (HVPE, MOCVD or molecular beam epitaxy [MBE]) [45]. The earliest interpretation of the YL was a transition between the SD and the carbon (C) related DA, based on the experimentally found enhancement of the YL intensity with the C doping [61]. Later on, a correlation between the YL intensity and the concentration of V_{Ga}, as measured by positron annihilation spectroscopy (PAS), was demonstrated [62]. Since the isolated V_{Ga} was found to be unstable at temperatures above 500°C, while the V_{Ga}-O_N complex was stable up to 1,200°C (the usual growth temperature for HVPE and MOCVD) [63,64], the DA involved in the YL was identified as the V_{Ga}-O_N complex [45,62]. This assignment was in line with the first principles calculations which gave the (–/2–) transition level of the V_{Ga}-O_N acceptor at 1.1 eV above the valence band [54,58]. Then a consensus about the origin of the YL was established around the III-nitride research community [45]. As already mentioned, recent theoretical calculations moved the V_{Ga}-O_N (–/2–) transition level deeper in the gap and some researchers concluded that the V_{Ga}-O_N complex is not responsible for the YL [65,66]. This conclusion, however, contradicts the experimental findings that the YL is suppressed in Mg-doped GaN where the V_{Ga} concentration is negligible [67]. Subsequently, other V_{Ga} complexes such as V_{Ga}-O_N-2H and V_{Ga}-3H were suggested to contribute to the YL [50]. On the other hand, the involvement of C in the YL has recently been reinforced. Hybrid functional calculations have predicted that the substitutional C on nitrogen site (C_N) is a deep acceptor with a (0/–) transition level at 0.9 eV above the valence band [68], instead of a SA as previously assumed [54]. Thus, the optical transitions from the SD to the C_N acceptor are expected to occur at 2.15 eV, that is, at the energy position of the YL. Later on, applying the same theoretical approach, the C_N acceptor-related emission was estimated to be at 1.98 eV, and the YL was attributed to the C_N-O_N complex with a (+/0) transition level at 0.75 eV [65,66]. The stability of the C_N-O_N complex, however, was questioned [69].

The latest theoretical calculations have shown that the isolated C_N acceptor rather than the C_N-O_N complex gives rise to the YL [70]. Having in mind the correlation of the YL intensity with both the V_{Ga} concentration [62,71] and the C doping [72,73] one can conclude that there are at least two DAs involved in the YL.

The GL is most often observed in undoped GaN grown by HVPE [45]. Usually, it emerges at high-excitation intensities, when the YL saturates. The GL intensity increases as a square of the excitation intensities, which implies that the defect responsible for this emission captures two holes before any radiative recombination occurs [45]. Based on the earlier interpretation of the YL (related to the V_{Ga}-O_N) and the experimentally evidenced transition from YL to GL in the PL spectra, the GL has been attributed to the (0/–) transition level of the V_{Ga}-O_N [45]. In view of the new theoretical predictions about the C_N acceptor levels, this interpretation was changed, and the (+/0) transition level of the C_N was suggested as the DA involved in the GL [66]. Recently, from a detailed analysis of the temperature dependence of the GL an excited state of the C_N^+ (C_N occupied with two holes) has been proposed to be involved in the emission [74]. However, since the V_{Ga} complexes cannot be excluded as a source of the YL, most probably there are two sources for the GL as well. The GL has also been observed in high-resistivity GaN grown under Ga-rich conditions by MBE [45] and in Mg-doped GaN also grown by MBE [75]. In this case, the emission band is relatively narrow for a deep level emission (~230 meV) and has a very high Huang–Rhys factor (~25) and small characteristic phonon energy (~23 meV) [75], which is typical for a DD defect [76]. Then this GL (also called GL2) is attributed to internal transitions within the V_N donor [75].

The BL appears in the PL spectra of high-resistivity GaN (C-doped or Fe-doped), n-type and high-resistivity Zn-doped GaN, and p-type Mg-doped GaN [45]. Although in all cases the emission is the 2.8–3.0 region, the defects involved in the emission are different. In C-doped GaN the (+/0) transition level of the C_N acceptor (at 0.35 eV above the valence band) has been assigned as the DA involved in the BL [69]. It was suggested that the C_N acceptor leads to the YL or BL depending on the position of the Fermi level and the excitation intensity. In lightly C-doped material (n-type) the C_N^- is the most stable charge state and the YL dominates. In heavily C-doped GaN optical excitation with a high intensity stabilizes C_N^0 and then the BL emerges. A correlation between the C doping and the BL intensity has been experimentally evidenced in the past [77,78] (see also Figure 3.6). Note that different theoretical results for the C_N transition levels lead to different interpretations of the related emissions: in Reference [69] the two C_N charge states are suggested to produce YL and BL, while in Reference [66] the corresponding emissions are YL and GL. Recently, the BL in high-resistivity GaN was related to C_N or C_N-O_N complexes with hydrogen (H) interstitial (C_N-H_i or C_N-O_N-H_i) [79]. Surprisingly, the same interpretation was proposed for the BL in the Fe-doped GaN [79]. In Zn-doped GaN the BL with a maximum at 2.9 eV and ZPL at 3.05 eV was attributed to the optical transitions between the SD and the Zn_{Ga} acceptor (at 0.45 eV above the valence band) [80]. In heavily Mg-doped GaN, the BL at 2.8 eV was earlier interpreted as a DD to SA transition, where the DD is the V_N-Mg_{Ga} complex, and the SA is the Mg_{Ga} [81]. Since the BL in Mg-doped GaN was known to be enhanced after annealing [43], while the concentration of the V_N-Mg_{Ga} complex significantly decreases at $T > 500°C$ [82] such an interpretation is improbable. An alternative explanation of the BL, in this case, would be the optical transitions between an H-related DD and the Mg_{Ga} acceptor [83] or between the SD and Mg-related DA [43].

3.2.4 Luminescence Related to Extended Structural Defects

The extended structure defects in GaN, such as dislocations, inversion domain boundaries, stacking faults (SFs), usually known as nonradiative defects, can capture electrons or/and holes (or excitons) and then produce characteristic emission lines. Up to eight emission lines in the region of 3.1–3.45 eV was observed in low-temperature PL spectra of GaN layers grown by different epitaxial techniques and was related to structural defects, but their exact identification was not clarified [45].

So far the most studied extended defect-related emissions are those associated with the SFs. The SFs are two-dimensional extended defects which occur when the normal stacking sequences of the crystal are changed. In wurtzite GaN the stacking sequences along (0001) direction can be distorted by introducing one, two or three zinc blende (ZB) bi-layers (molecular monolayers). Then, three types of basal plane SFs

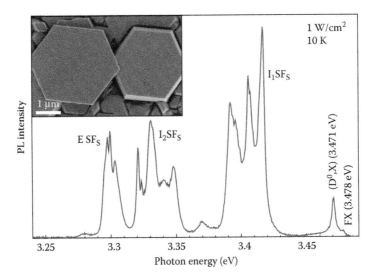

FIGURE 3.7 Low-temperature micro-PL spectrum of a GaN microcrystal showing a rich structure of sharp lines related to different types of BSFs. The inset shows a top-view scanning electron micrograph of the studied microcrystal. (Reprinted with permission from Lähnemann, J. et al., *Phys. Rev. B.*, 86, 081302(R), 2012. Copyright 2012 by the American Physical Society.)

(BSFs) are distinguished—intrinsic I_1-type, intrinsic I_2-type, and extrinsic E-type, with a formation energy increasing in the same order [84]. The SFs are also possible on planes other than the (0001) plane. The faults formed on a prismatic (11-20) plane, called prismatic SFs (PSFs), always connect two I_1-type BSFs with a stair-rod dislocation. The first identification of the optical signatures of the SFs in GaN dates about 20 years ago [85], but extensive studies have started years later with the development of the GaN growth along nonpolar directions [86–88]. It was found that in heteroepitaxial nonpolar and semi-polar GaN layers SFs of high density are formed, and the near-bandgap PL spectra are often dominated by the SF-related emissions. Comprehensive reviews of the optical studies of the SFs in GaN can be found in References [89] and [90]. Figure 3.7 depicts a typical low-temperature PL spectrum of SFs emissions where three group of lines are discriminated around 3.41 eV, 3.32 eV, and 3.29 eV and attributed to I_1-type, I_2-type, and E-type BSFs emission, respectively [91]. In other studies, the lines at 3.30 eV were interpreted as related to the PSFs [92] or to impurity decorated stair-rod partial dislocations terminating the I_1-type BSFs [87]. In a simple modeling of the BSFs, a thin ZB quantum well (QW) embedded into the wurtzite matrix is considered and then the emission occurs from the recombination of electrons and holes confined in the well [85,93]. The thickness of the QW depends on the interface position between ZB and wurtzite phases and is $(1.0\pm0.5)c$, $(1.5\pm0.5)c$, $(2.0\pm0.5)c$ for I_1-type, I_2-type, and E-type BSFs, respectively, where c is the lattice constant along the (0001) direction in wurtzite GaN [93]. This model approximately reproduces the 3.41 eV emissions from the I_1-type BSFs but fails to explain the multiline structure in the PL spectra as well as the energy separation between the lines from different types of BSFs. The model was further developed by including the discontinuity of the spontaneous polarization at the ZB-wurtzite interfaces and the resulting quantum-confined Stark effect [86,91], considering the possibility of type-I or type-II band alignment in the ZB QW [91], and also the interaction between adjacent BSFs [86,88,94]. It was further suggested that donors [95] or acceptors [96] located in a proximity of the BSFs are involved in the emissions. Both the complex physics of the SFs and their experimentally observed rich spectra leave the door open for further investigations in this field.

Dislocations in semiconductors are mostly associated with nonradiative recombination, but it turns out that at low temperatures also radiative processes occur. One such example has been studied with *a*-type screw dislocations in GaN [97], where a bound exciton line is observed at 3.346 eV and related to the dislocations via luminescence topographs. The authors demonstrate via calculations of the electronic structure of this defect that it can bind both electrons and holes in a local potential. Otherwise, such dislocations are known to be serious nonradiative defects at room temperature and higher, participating in the droop process in InGaN-based LED structures [98].

3.3 Optical Properties of AlN and AlGaN Alloys

3.3.1 Free and Bound Excitons in AlN

Like in GaN the top valence band in AlN is split by the crystal field, and the spin-orbit interaction into three bands also labeled A, B, and C in order of increasing energy of the holes. The main difference is that in AlN the crystal-field splitting is negative. *Ab initio* calculations predict quite scattered values for Δ_{cf}, from −59 to −276 meV (see Table III in Reference [99]), while the recent experimentally derived values in bulk AlN are more consistent, Δ_{cf} = −230 meV [100], −225 [101], −220 meV [99]. On the other hand, the calculated and the experimental values for the spin-orbit splitting range between 11 and 22 meV [99]. Comparing all available data we tends to conclude that the most reliable values are Δ_{cf} = −220±2 meV and Δ_{so} = 15±2 meV [99]. The large negative Δ_{cf} has significant consequences on the valence band structure, as well as on near-bandgap optical transitions in AlN (Figure 3.8a). First, the top valence band (A band) is the crystal-field-split-off band (with Γ_7 symmetry) which has a p_z-like orbital character. Second, the optical transitions between the A valence band and the conduction band (Γ_7 symmetry) occur predominantly for E||c polarization. Generally, the Γ_7–Γ_7 transitions are allowed in both E⊥c and E||c polarizations, however, due to the large Δ_{cf} the oscillator strength for E⊥c polarization becomes vanishing small. The estimates within the quasi-cubic model show that for strain-free AlN the oscillator strength ratio between E⊥c and E||c polarizations for the band-to-band transitions involving the A valence band is about 1:1000 [100,102,103]. (Note that in the case of GaN this ratio (now involving the C valence band) is about 1:10 [7]). For the same reason (the large Δ_{cf}) the transitions between the C valence band (Γ_7 symmetry) and the conduction band are mainly E⊥c polarized. As for the transitions related to the B valence band ($\perp_9$ symmetry) they are allowed only for E⊥c polarization.

The exciton binding energy is generally extracted from the energy separation between the ground state (1s) and the excited state (2s) emission lines of the A exciton. However, the simple hydrogenic model for the excitons does not work well for AlN because of the anisotropy in both the dielectric constant and the effective masses of electrons and holes (these values are not exactly known) [104,105]. Nevertheless, the theoretically predicted [105] and the experimentally deduced [103,106–109] binding energy for the A exciton are quite consistent giving a value of 52±1 meV. (For the exciton fine-structure

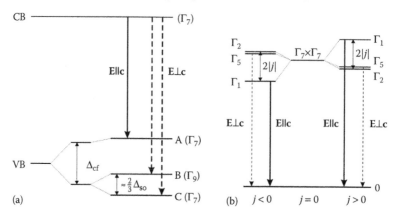

FIGURE 3.8 (a) Valence band structure and the band-to-band optical transitions in AlN. (b) The splitting of the A exciton and the dipole allowed exciton transition for spin-exchange constant $j > 0$ and $j < 0$. (The figure is not drawn to scale.) The Γ_2 and Γ_5 states are virtually degenerate. Note that the exciton oscillator strength for E⊥c polarization is more than two orders of magnitude smaller than that for E||c polarizations.

splitting due to the spin-exchange interaction see the discussion below). There is no available data for the binding energies of B and C excitons, but one can assume values similar to that for the A exciton (as in the case of GaN).

The unusual valence band structure of AlN (in the sense of the optical polarization selection rules) is very challenging for optical investigations. First, polarization resolved measurements are crucial for an accurate determination of the A exciton energy, and correspondingly the bandgap in AlN. This can be done correctly only for the $\mathbf{k} \perp \mathbf{c}$ experimental geometry, that is, samples with a large and clean surface parallel to the *c*-axis are needed (e.g., bulk samples). Second, the large energy separation between the A valence band and the other two valence bands makes the revealing of the B and C excitons practically impossible in PL measurements, because the high-energy bands cannot be populated (even at room temperature). Third, the PL experiments are hampered by the high-background impurity concentration in the present AlN material leading to a number of bound exciton emission lines which have to be distinguished from the A free exciton emission.

The inverse valence band ordering in AlN was experimentally proved by OR and PL measurements more than 10 years ago [100–102]. More accurate data for the A, B, and C exciton energies at low and room temperatures were recently obtained from SE measurements on a bulk AlN sample with (1-100) oriented surface grown by physical vapor transport (PVT) [99]. Figure 3.9 shows the experimental imaginary part of the dielectric function (ε_2) for $\mathbf{E} \perp \mathbf{c}$ and $\mathbf{E} \| \mathbf{c}$ polarizations at $T = 10$ K together with a line-shape modeling that includes the contributions of the free excitons, the exciton continuum and the EPC. The extracted energies are 6.032 eV, 6.255 eV, and 6.264 eV for A, B, and C excitons respectively. The corresponding values at room temperature ($T = 295$ K) are lower by 71 ± 1 meV. Note that the A exciton is seen only for $\mathbf{E} \| \mathbf{c}$ polarization, while the B and C excitons are seen only for $\mathbf{E} \perp \mathbf{c}$ polarization. In PL measurements (where the spectral resolution is usually 4–5 times better than that in SE) more complicated spectra were acquired. Despite that similar strain-free bulk samples or homoepitaxial layers were studied by different groups the energy position of the A exciton was found to span between 6.029 eV

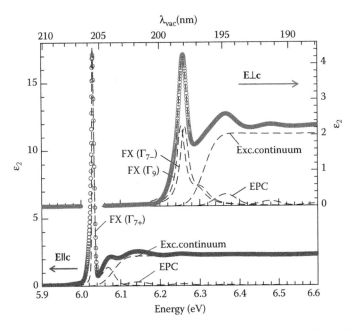

FIGURE 3.9 Experimental and modeled imaginary part of the dielectric function (ε_2) in bulk AlN for $\mathbf{E} \perp \mathbf{c}$ and $\mathbf{E} \| \mathbf{c}$ polarizations at $T = 10$ K. The decomposition of the spectra allows to assign the contributions of the free excitons, the EPC, and the exciton continuum. (Reprinted with permission from Feneberg, M. et al., *Phys. Rev. B.*, 87, 235209, 2013. Copyright 2013 by the American Physical Society.)

and 6.043 eV depending on the experimental geometry used and in some cases double or triple emission lines were observed [99, 107,108,110,111]. Such peculiarities can only be explained by the fine structure of the A exciton.

The 1s ground state of the A exciton in AlN is four-fold degenerate and consists of a two-fold degenerate spin-singlet state (Γ_5 symmetry), a pure spin-triplet state (Γ_2 symmetry) and a mixed singlet-triplet state (Γ_1 symmetry). The electron-hole spin-exchange interaction (characterized by the spin-exchange interaction constant, j) splits the spin-singlet states from the spin-triplet states. The exciton states of Γ_5 symmetry are dipole allowed for **E⊥c** polarization, while the Γ_1 exciton state is dipole allowed for **E∥c** polarization (Figure 3.8b). The spin-triplet state Γ_2 is optically forbidden. The spin-exchange interaction constant in AlN is expected to be quite large following the trend of j with the exciton binding energy (or the exciton Bohr radius) [112,113]. Moreover, due to the large Δ_{cf} the A exciton is almost completely decoupled from the other two valence bands. As a result, the Γ_2 and Γ_5 states are nearly degenerate independent of the j value and the splitting between Γ_5 state and Γ_1 state is $\approx 2j$ (Figure 3.8b).

Two groups have attempted to study the fine structure of the A exciton, both have found a large j, however with an opposite sign [99,107,110,111,114]. In Figure 3.10 low-temperature PL spectra taken in the three common polarization geometries for a bulk AlN sample are shown [111]. The emission lines at 6.032 eV (which coincides with the maximum in the Γ_2) and at 6.040 eV were identified as the Γ_1 and Γ_5 exciton states. Based on both the energy separation and the ordering of the states (the Γ_5 state is at higher energy than the Γ_1 state) the authors concluded that $j = -4$ meV. Different PL spectra were acquired in a homoepitaxial AlN layer (Figure 3.11) [107]. The broad emission at 6.043 eV, not seen in **E⊥c** polarization, was identified as the Γ_1 exciton state, while the sharp emission line at 6.029 eV was assigned the Γ_5 exciton state [107,114]. Then, the spin-exchange interaction constant was found to be $j = 6.8$ meV. Unfortunately, in both cases, there is some uncertainty in the interpretation of the PL spectra which is mainly related to the overlook of the exciton-photon coupling and the oscillator strengths expected for the different exciton states for different polarizations. The free exciton origin of the $X_A(\Gamma_5)$ line in PL spectra on Figure 3.11 was confirmed by temperature-dependent [114] and time-resolved PL measurements [108] but the concerns about its interpretation still remain because the free exciton emission lines are usually much broader than the DBE lines. The detailed theoretical analysis of the A exciton fine structure in AlN along with the experimental finding favors a positive sign j [115]. Moreover, $j > 0$ is

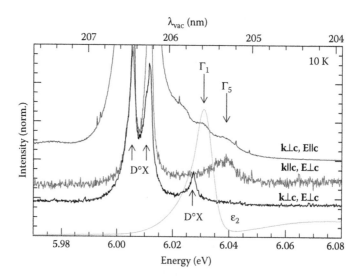

FIGURE 3.10 Imaginary part of the dielectric function (ε_2) for **E∥c** polarization and PL spectra for different polarization geometries in bulk AlN measured at 10 K. (Reprinted with permission from Feneberg, M. et al., *Appl. Phys. Lett.*, 102, 052112, 2013. Copyright 2013, American Institute of Physics.)

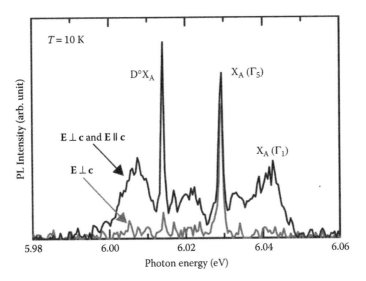

FIGURE 3.11 Low-temperature polarized PL spectra of a homoepitaxial AlN layer (Reprinted from Funato, M. et al., *Appl. Phys. Express*, 5, 082001, 2012. Copyright 2012, The Japan Society of Applied Physics.)

found in all III-V and II-VI semiconductors. One should note, however, that AlN is the only one semiconductor with a negative crystal-field splitting. The conclusion is that more studies are needed in order to clarify the exciton fine structure in AlN.

Due to the uncertainty in the energy of the lowest A free exciton state the bandgap value of AlN is somewhat doubtful. Nevertheless, based on all experimental finding in bulk and homoepitaxial samples, one can claim a bandgap of 6.084±0.003 eV at 2 K and 6.013±0.006 eV at room temperature for pure strain-free AlN.

In high-resolution low-temperature PL spectra of bulk AlN and homoepitaxial AlN layers up to four sharp lines below the free-exciton emission are commonly resolved and assigned as recombination of DBEs related to four different donors [108,110,116] (see Figure 3.12). The localization energies (the binding energies) of these DBEs are 8±1, 13±1, 22±1, and 28±1 meV. The corresponding donors still remain unidentified. Only the DBE with the binding energy of 28±1 meV has been identified as related to the

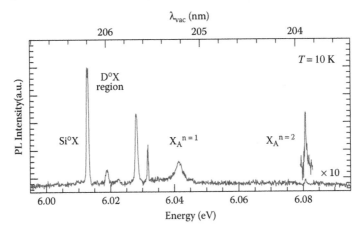

FIGURE 3.12 Low-temperature PL spectrum of a nominally undoped AlN homoepitaxial layer showing four DBE exciton lines. (From Neuschl, B. et al.: Optical identification of silicon as a shallow donor in MOVPE grown homoepitaxial AlN. *Phys. Status Solidi* (*b*). 249, 511, 2012. Copyright Wiley-VCH Verlag GmbH & Co. KGaA. Reproduced with permission.)

Si_{Al} donor in doping experiments [116] and the Si_{Al} binding energy of 63.5±1.5 meV is estimated from the TETs spectra [117]. Recently, more DBE lines were found in m-plane homoepitaxial AlN layers and a DBE with a binding energy of 26 meV (measured from the Γ_1 exciton state) was identified as related to the O_N donor [118]. This assignment, however, is quite doubtful because of the lack of a doping-dependent study and the inappropriate estimation of the DBE binding energies. The proper localization energies (the binding energies of a DBE) should be estimated from the ground 1s state of the A exciton (the Γ_2 state, which in AlN is degenerated with the Γ_5 state). One should also note that the DBE symmetry is reduced from C_{6v} to C_{3v} and the spin-exchange splitting and the polarization selection rules are different for the A free exciton and the DBEs arising from it.

The problem with the identification of the donors related to the DBEs is because all dopants commonly used for n-type conductivity in GaN (O, Si, Ge, Sn) form the so-called "DX centers" in AlN [119]. The DX centers appear when a neutral substitutional donor captures another electron and undergoes a large lattice relaxation, forming a lower-lying negatively charged state (DX⁻) which acts as a deep acceptor. The DX behavior of the O_N in $Al_xGa_{1-x}N$ has been experimentally evidenced by the pressure-dependent Raman scattering two decades ago [120]. It was found that O_N becomes a DX center for Al composition $x > 0.4$, while hybrid functional calculations got $x > 0.61$ [119]. The DX onset for of Ge_{Ga} and Si_{Ga} donors in $Al_xGa_{1-x}N$ was predicted to occur at $x > 0.52$ and $x > 0.94$, respectively [119]. The fact that the Si_{Ga} remains an effective-mass-like donor up to a high-Al composition results in a quite specific situation in AlN, as revealed by temperature-dependent electron paramagnetic resonance (EPR) measurements [121,122]. In these studies, the stable Si_{Ga} DX⁻ state in AlN is found to be at ~240 meV below the conduction band minimum. In addition, an existence of a metastable DX⁻ state at energy ~11 meV below the donor neutral charge state (at ~65 meV) is predicted. (Note that for DX centers in wurtzite crystals the bond rupture can take place either along a bond parallel to the c-axis or along one of the three other equivalent bonds [123].) Then, in non-equilibrium optical measurements (as PL) electrons from the metastable DX⁻ state can easily be excited to neutralize the positively charged donor state, and the exciton bound to a neutral donor can be seen in the spectra [116,117]. On the other hand, to get n-type conductivity electrons from the stable DX⁻ state have to be promoted to the conduction band. This can explain the donor activation energy of 250–280 meV usually obtained in transport measurements [124,125]. For all other donor dopants (O_N, Ge_{Ga}, Sn_{Ga}) the DX⁻ state is lying deeper in the gap and no effective-mass-like behavior is expected in AlN; thus, no DBE lines should appear in PL spectra. Then, the sharp PL lines observed above the Si DBE [108,110,116,117] can be speculatively attributed to the excited DBE states. Recently, it was theoretically predicted that the substitutional sulfur (S) on nitrogen site (S_N) is a very attractive donor dopant for AlN because S_N does not form DX center [126]. However, experimental evidence for this prediction is still missing.

Owing to the large bandgap the acceptors in AlN are expected to be quite deep. In analogy to GaN, an obvious acceptor dopant in AlN is Mg. A PL study of the Mg-doped layers grown by MOCVD on sapphire substrate revealed an ABE with a binding energy of ~40 meV and an acceptor binding energy of ~510 meV [127]. Although these values are not confirmed by other studies the Mg acceptor binding energy is fairly close to that obtained from electrical measurements (~630 meV) [124]. The large Mg acceptor binding energy implies a very low free hole concentration, only ~10^{10} cm⁻³ has been achieved at room temperature [124]. Recently, a surprising improvement of this status was reported for AlN nanowires (NWs) grown by MBE, where axial p-n junctions were incorporated [128]. Under suitable growth conditions, a rather high-Mg concentration could be achieved in the structure, leading to an estimated hole concentration of the order 10^{16} cm⁻³. It was suggested that hopping conduction in the Mg impurity band at high-Mg concentration might be responsible for the very low-activation energy (~23 meV) of the Mg acceptor in this case [129]. These findings, however, are quite ambiguous and need to be confirmed. Other acceptor dopant (Zn, Be) have also been tried in order to achieve p-conductivity in AlN. In a PL study of Zn-doped heteroepitaxial AlN layers an acceptor binding energy of 740 meV was deduced [130] which is larger than that for the Mg acceptor in contrast to the theoretical predictions [131]. For the Be dopant, an acceptor binding energy of 330 meV and an ABE binding energy of 33 meV have been estimated [132].

At high-excitation intensities (~MW/cm²) all DBE-related emissions are saturated, and emission bands related to the bi-excitons (M band), the exciton-exciton scattering (P band) and the electron-hole plasma (EHP band) are observed [110,133,134]. The bi-exciton energy in AlN was estimate as 19 meV [133], 27 meV [134], and 28.5 meV [110]. The P band was found to shift to lower energies with increasing excitation intensity while the energy position of the EHP band (5.83 eV at low temperature) remained constant. At room temperature, the P band merges with the band-to-band recombination and the EHP band becomes the dominant recombination process [110].

As in GaN the biaxial in-plane strain in heteroepitaxial AlN layers causes a shift of all optical transitions from their strain-free energy positions. In the case of a compressive strain (AlN grown on sapphire substrate) the shift is toward higher energies, while for a tensile strain (AlN grown on SiC or Si substrate) the shift is toward lower energies. According to the Bir and Pikus theory [135], the amount of the shift is proportional to the strain via a set of six deformation potentials. In AlN the values of the deformation potentials have been determined from the analysis of the experimental strain dependence of the exciton energies [103,104,136] and by OR measurements under uniaxial stress [114]. Under a higher compressive biaxial strain the valence band ordering is expected to be changed from (Γ_7, Γ_9, Γ_7) to (Γ_9, Γ_7, Γ_7). The critical strain value for this switching was estimated to be 0.7% out-of-plane tensile strain (corresponding to –1.2% in-plane compressive strain) [103]. Note that in Reference [103] the crystal-field splitting of –152.4 meV is assumed, for the more correct value Δ_{cf} = –220 meV [99] the critical out-of-plane (in-plane) strain is –1% (1.7%).

3.3.2 Near-Bandgap Optical Transitions in AlGaN

As in all semiconductor ternary alloys, the bandgap in $Al_xGa_{1-x}N$ is expected to follow the Vegard law, that is, $E_g(Al_xGa_{1-x}N) = xE_g(AlN) + (1-x)E_g(GaN) - bx(1-x)$, where b is the so-called bandgap bowing parameter. Values for b reported in literature vary from 0.7 to 1.3 eV [14]. Two issues should be considered when extracting the bandgap from optical measurements in alloys: (i) in all heteroepitaxial layers the strain should be taken into account for a proper analysis of data; (ii) due to random alloy fluctuations the excitons are localized and the PL peaks do not represent the exact energy of the free excitons particularly at low temperatures (so-called Stokes shift). Thus, the OR and SE measurements provide more accurate data. In the case of $Al_xGa_{1-x}N$ there is another problem, namely the change of the valence band ordering. As already discussed the top valence band in GaN has Γ_9 symmetry, and the lowest free exciton emission is E⊥c polarized, while the top valence band in AlN has Γ_7 symmetry and the lowest free exciton emission it is E∥c polarized. Experimentally, the polarization switching was found to occur at Al content of x = 0.1–0.25 [137–139]. The scattering of data can be attributed to the accuracy of the polarized PL measurements and different strain present in the layers. On the other hand the theoretical examination for bulk strain-free $Al_xGa_{1-x}N$ have predicted x = 0.145 [103] and x = 0.05 [139] for the Γ_9-Γ_7 valence bands crossing. (Note again the lower value for the crystal-field splitting, Δ_{cf} = –152.4 meV, used in Reference [103]) In a refined analysis where the nonlinear composition variation of Δ_{cf} was also included, x = 0.09 was deduced, probably the most accurate value so far [139]. Returning to the bandgap in $Al_xGa_{1-x}N$, the proper way to describe the variation in the Al content is to use two Vegard-like equations (with the same bowing), one for the Γ_9 valence band and another one for the Γ_7 valence band. The bowing parameter for the Γ_9 band was recently determined as $b_{\Gamma9}$ = 0.85–0.9 eV [139,140].

As already mentioned at low temperatures the random alloy fluctuations cause a Stokes shift of the exciton emission peak. The difference between the PL peak energy and the energy of the free exciton is called the localization energy (E_{loc}). With increasing temperature, the excitons start to delocalize and the PL peak moves to the higher energies. At some elevated temperatures, all excitons are delocalized and the PL peak corresponds to the free exciton energy. This behavior is illustrated in Figure 3.13 [139], where the PL spectra for an $Al_xGa_{1-x}N$ (x = 0.142) sample at different temperatures are presented. The characteristic *S*-type shape of the temperature dependence of the emission peak energy is clearly evident. At temperatures above 100 K, the emission energy follows the temperature dependence of the

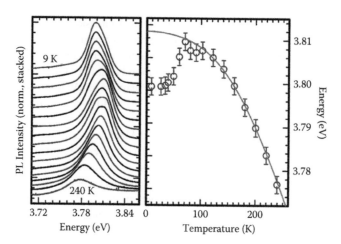

FIGURE 3.13 (Left) PL spectra for an Al_xGa_{1-x} N ($x = 0.142$) layer at different temperatures. (The temperature increases from top to bottom.) (Right) Energy position of the PL band as a function of the temperature. The solid line is the fit using the model in Reference [141]. (Reprinted with permission from Neuschl, B., *J. Appl. Phys.*, 116, 113506, 2014. Copyright 2014, American Institute of Physics.)

bandgap approximated by the Pässler model [141]. It is also seen that the linewidth (or full width at half maximum (FWHM)) of the emission is quite large even in the low-temperature PL spectra. The broadening is also due to the alloy fluctuations. Both the FWHM and the E_{loc} principally depend on the slope of the bandgap composition dependence (i.e., $dE_g(x)/dx$) and the exciton volume, and in $Al_xGa_{1-x}N$ both have maximum values at $x \approx 0.75$ [142,143].

Owing to the alloy broadening the DBEs and ABEs cannot be resolved separately from the localized excitons even in the low-temperature spectra in undoped $Al_xGa_{1-x}N$ materials. The Si and Mg doping further increases the linewidth and quite broad emission bands are usually observed [144,145].

3.3.3 Below-Bandgap Absorption and Emission AlN and AlGaN

The most prominent below bandgap emissions in AlN are observed at 2.8 eV [146,147], 3.1 eV [148,149], 3.7 eV [150–152], and 4.5 eV [148,150,153]. The emissions are quite broad and in different studied samples their energy position varies depending on the growth technique used, as well as on the concentration of the impurities (i.e., on the position of the Fermi level). For bulk AlN, regarded as the best substrate for deep UV LEDs, a number of optical absorption studies have also been reported. Absorption bands at 2.8 eV [153], 3.4 eV [154], 4.0 eV [153], and 4.7 eV [155] were observed. As main defects responsible for these emission and absorption bands, the vacancies of the Al (V_{Al}) and nitrogen (V_N) and their complexes with impurities have been considered. The V_{Al} and V_N have been predicted by the theory to have low formation energy in n-type and p-type AlN, respectively [156–158]. The V_N has been identified by EPR in electron-irradiated PVT grown AlN [159]. In the same type of material V_{Al} of a concentration in the range of 10^{18} cm^{-3} have been measured by PAS and almost all of them are found to be complexed with O_N in as-grown samples [154]. Note that the O is the most abundant impurity in AlN and is present in concentrations above 5×10^{17} cm^{-3} independent of the growth method [147,148,155,160].

Recent hybrid functional calculations for the transition states of the vacancies and vacancy-oxygen defects in AlN and the related optical transitions have provided a satisfactory explanation of the most of emission and absorption bands [158]. The V_{Al}, double and single V_{Al}-O_N complexes have been suggested as DAs involved in the 2.8 eV, 3.1 eV, and 3.7 eV emissions, respectively. The calculations for the V_{Al} and double V_{Al}-O_N acceptor have also predicted absorption peaks at 3.43 eV and 3.97 eV in agreement with the experimental finding [153,154]. Being a deep acceptor C_N was also considered as a source of below-bandgap emission and absorption. In fact, both 2.8 eV and 3.7 eV emissions, as well as the absorption at 4.7 eV have been found to increase with C doping concentration [155,160,161]. From a theoretical point of view, the optical properties related to the C_N in AlN are similar to those in GaN [69]. The excitation from the (0/−) transition level (at 1.9 eV above the valence band) into the conduction band gives rise to an absorption peaking at 4.7–4.8 eV [69,155]. The corresponding emission ($C_N^0 \rightarrow C_N^-$) occurs at 3.6–3.7 eV. At high-C concentration the ground state of the C_N is its neutral charge state and the absorption and the emission associated with the (+/0) transition level are predicted to be at 5.66 eV and 4.5 eV, respectively [69]. The accounting for the (+/0) transition level can explain the shift of the 3.7 eV emission toward high energies with increasing C doping [155,160]. The 2.8 eV emission has been interpreted as a DAP recombination between the V_N and C_N (V_N donors are expected to form as compensating defects at high-C doping) [161,162]. It was also suggested that the C and Si co-doping suppress the 2.8 eV emission and 4.7 eV absorption due to the formation of C_N-Si_{Al} complexes [162].

Studies that address the below-bandgap emissions in $Al_xGa_{1-x}N$ alloys are rare so far. As expected the energy positions of the emissions related to the cation vacancies (V_{Ga}, V_{Al}) and their complexes were found to follow the trend of the bandgap with Al composition [163]. The same was observed in Mg-doped $Al_xGa_{1-x}N$, where the Mg-related emission at 2.8 eV in GaN was transformed to a 4.7 eV emission in AlN [164].

3.4 Optical Properties of InN and InGaN Alloys

3.4.1 Optical Properties of InN in the Near-Bandgap Region

The properties of InN have by tradition been studied on thin film samples grown on foreign substrates, in particular, sapphire (GaN/sapphire templates). These samples typically have a large defect density, with threading dislocation densities as large as 10^{10} cm^{-2}. The reason for this situation is the lack of bulk InN substrates, meaning that studies of low-defect density bulk InN properties could not be performed experimentally. This situation is behind many properties ascribed to the InN material, such as degenerate *n*-type conductivity and Fermi level pinning to the conduction band [165], and problems to produce *p*-type material [166].

The bandgap of InN was for a long time assumed to be about 1.9 eV [167]. However, about 15 years ago the absorption and PL measurements on MBE grown single-crystalline InN epitaxial layers have evidenced a bandgap of ~0.7 eV [168]. The experimentally obtained smaller bandgap has later been confirmed by theoretical calculations of the InN band structure [21,169,170]. The splitting of the top valence band due to the crystal field and the spin-orbit interaction is quite small, just a few meV (as in the case of GaN) [21], and could so far not be resolved experimentally in optical data.

Since the dopant/defect density in the heteroepitaxial InN samples studied are well above 10^{17} cm^{-3}, the free or bound excitons could not be resolved (the free electron concentration at such a doping density is above the Mott density in InN, ~2×10^{17} cm^{-3} [171]). The near-bandgap emission spectra in degenerate *n*-type InN was explained as a recombination of the free electrons with localized holes [172] or as governed by Mahan excitons [173]. Nominally undoped InN material such as NWs grown by MBE under suitable conditions shows a Fermi level position near to the middle of the bandgap, (i.e., no electron accumulation in the conduction band occurs), and thus has the closest to intrinsic properties reported so far [174]. Exciton PL was claimed to be observed for the first time in such nominally undoped InN NWs, with a linewidth down to 9 meV [175] (see Figure 3.14). Future studies need to confirm whether these data at low temperature relate to free excitons (as claimed in Reference [175]) or to bound excitons related to impurities. The PL data are consistent with a binding energy of just a few meV for the excitons (a thermal activation energy about 3 meV was measured). The NW data confirm that the commonly observed degenerate *n*-type property, typical for planar samples grown on sapphire, is related to the high-defect density (the defects are believed to create donor-like states in the conduction band), or to the surface segregation of donor species [175].

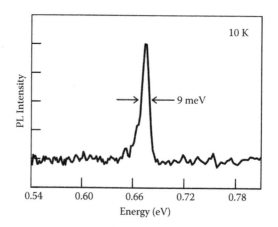

FIGURE 3.14 Low-temperature spectrum of intrinsic InN nanowires taken at an excitation power of 0.5 μW. (Reprinted with permission from Zhao, S., *Nano Lett.*, 12, 2877, 2012. Copyright 2012, American Chemical Society.)

For InN layers grown on sapphire a *p*-type material can be obtained by Mg doping in the 10^{18}–10^{19} cm^{-3} doping window [176,177], but for a Mg concentration of above 10^{20} cm^{-3} the material converts to *n*-type due to a simultaneous creation of a high concentration of structural defects like SFs [178]. The situation appears to be different for the case of InN NWs, which can be grown virtually defect (dislocation) free even on foreign substrates [174,175]. Such material can be doped with Mg in a controlled manner without severe compensation [179]. In Figure 3.15 are shown the PL spectra in Mg-doped NWs at different excitation powers [180]. The lower energy PL peak is Mg-related and is situated about 60 meV below the near-bandgap emission, in agreement with earlier studies on thin films [181].

At this stage, detailed studies of the optical properties of InN are quite difficult because defect free macroscopic crystals are still lacking. A development of growth techniques for bulk InN crystals is highly desired.

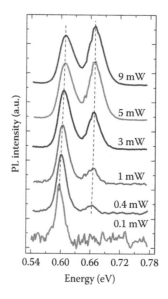

FIGURE 3.15 Low-temperature PL spectra of low Mg-doped InN nanowires measured at different excitation powers. At the lowest excitation power (0.1 mW) the FWHM is 17 meV. (Reprinted with permission from Zhao, S., *Appl. Phys. Lett.*, 103, 203113, 2013. Copyright 2013, American Institute of Physics.)

3.4.2 Absorption and Emission in InGaN Alloys

$In_xGa_{1-x}N$ alloys are the basic materials for visible light emission in nitride-based LEDs. $In_xGa_{1-x}N$ has not been studied much as bulk material, due to difficulties to grow thick layers by MOCVD or MBE. Recently some success has been demonstrated by HVPE growth on GaN substrates, where $In_xGa_{1-x}N$ layers of a thickness of >10 μm have been demonstrated [182]. Near-bandgap emissions of MOCVD grown $In_xGa_{1-x}N$ layers on sapphire have been studied since decades, mainly with the ambition to determine important parameters for the $In_xGa_{1-x}N$ alloy system, such as the bandgap variation according to the Vegard law $E_g(In_xGa_{1-x}N) = xE_g(InN) + (1-x)E_g(GaN) - bx(1-x)$. The studies usually combine optical emission data with a determination of the absorption edge (preferably via SE). Such a study including both N-face and metal-face samples was performed in In-rich alloys and the bandgap values corrected for free carrier density were obtained [183]. Later the data was additionally corrected for the strain and the study was extended to Ga-rich alloys [184]. The In dependence of the transition energies for the high-energy critical points was also examined. The experimental data is shown in Figure 3.16. The best fit for the bandgap energy with the end-point values $E_g(GaN) = 3.435$ eV and $E_g(InN) = 0.675$ eV results in a bowing parameter $b = 1.65 \pm 0.07$ eV in strain-free $In_xGa_{1-x}N$ at room temperature. The bowing parameter for the high-energy critical points was found to smaller (~1 eV) [184]. The bandgap bowing parameter varies between large limits in early works mainly due to the small range of In composition considered or/and to using $E_g(InN) = 1.9$ eV [14]. It has been predicted by the theory that the bowing parameter in $In_xGa_{1-x}N$ depends on In content, and thus, the Vegard law cannot satisfactorily describe the bandgap variation in the entire composition range [185].

Regarding the near-bandgap emission in $In_xGa_{1-x}N$ alloys, all comments about the exciton localization we put above for $Al_xGa_{1-x}N$ are valid for $In_xGa_{1-x}N$ as well. However, for the same In and Al content, the E_{loc} and the emission broadening are larger in $In_xGa_{1-x}N$ alloys. The strong exciton localization in $In_xGa_{1-x}N$ is believed to be a result of the capture of the holes by localized valence states associated

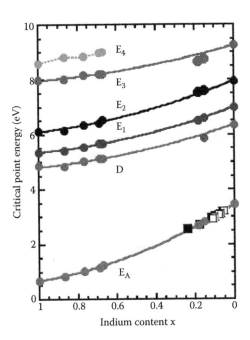

FIGURE 3.16 Bandgap and high-energy inter-band transition energies in $In_xGa_{1-x}N$ as a function of In content. The full symbols present the energies determined from the imaginary part of dielectric function measured by SE [183,184], while the open symbols are literature data obtained from optical absorption measurements. The solid lines represent the best fits for determining the bowing parameters.

with atomic condensates of In-N [186]. In fact, the exciton/carrier localization has been considered as the main reason for the high-internal quantum efficiency InGaN/GaN LEDs emitting in blue and UV spectral region.

3.5 Conclusions

The optical properties of the III-nitrides have been studied over about half a century, along with the development of growth techniques. GaN properties are now quite well covered in the literature, and for AlN, the situation is rapidly improving, since doping is getting under control. Recent results on InN promises more detailed knowledge, but a bulk material of more than nanometer size is indeed needed in the future. The alloys InGaN, AlGaN, and AlInN are very important for devices, and here, the influence of several effects like composition fluctuations and strain from substrates or other layers still restricts the accuracy of data in many cases. The increasing relevance of quantum size structures for optical III-nitride devices has created a new subfield that is not discussed in any detail in this chapter.

Acknowledgments

We would like to thank Prof. R. Goldhahn, Prof. Z. Mi, Prof. B. Skromme, Dr. M. Feneberg, Dr. M. Funato, Dr. J. Lähnemann, Dr. B. Neuschl, and Dr. S. Shokhovets for providing electronic data for the figures. P. P. Paskov acknowledges the partial support from the Swedish Energy Agency.

References

1. B. Monemar, J. P. Bergman, and I. A. Buyanova, Optical characterization of GaN and related materials, in S. J. Pearton (ed.) *GaN and Related Materials*. CRC Press: Boca Raton, FL, p. 85, (1997).
2. B. Monemar, Optical properties of GaN, in J. I. Pankove and T. D. Moustakas (eds.) *Gallium Nitride (GaN) I*. Academic Press: San Diego, CA, p. 305, (1998).
3. H. Morkoç, *Handbook of Nitride Semiconductors and Devices*, Vol. 2. Wiley-VCH: Wienheim, Germany, (2008).
4. B. Monemar and P. P. Paskov, Luminescence studies of impurities and defects in III-nitride semiconductors, in L. Bergman and J. L. McHale (eds.) *Handbook of Luminescent Semiconductor Materials*. CRC Press, Boca Raton, FL, p. 169, (2012).
5. K. Kornitzer, T. Ebner, K. Thonke et al., Photoluminescence and reflectance spectroscopy of excitonic transitions in high-quality homoepitaxial GaN film, *Phys. Rev. B* **60**, 1471 (1999).
6. A. V. Rodina, M. Dietrich, A. Göldner et al., Free excitons in wurtzite GaN, *Phys. Rev. B* **64**, 115204 (2001).
7. B. Gil, O. Briot, and R. L. Aulombard, Valence-band physics and the optical properties of GaN epilayers grown onto sapphire with wurtzite symmetry, *Phys. Rev. B* **52**, R17028 (1995).
8. S. L. Chuang and C. S. Chang, k p method for strained wurtzite semiconductors, *Phys. Rev. B* **54**, 2491 (1996).
9. B. Gil and O. Briot, Internal structure and oscillator strengths of excitons in strained α-GaN, *Phys. Rev. B* **55**, 2530 (1997).
10. J. J. Hopfield, Fine structure in the optical absorption edge of anisotropic crystals, *J. Phys. Chem. Solids* **15**, 97 (1960).
11. A. A. Yamaguchi, Y. Mochizuki, H. Sunakawa et al., Determination of valence band splitting parameters in GaN, *J. Appl. Phys.* **83**, 4542 (1998).
12. P. P. Paskov, T. Paskova, P. O. Holtz et al., Polarized photoluminescence study of free and bound excitons in free-standing GaN, *Phys. Rev. B* **70**, 035210 (2004).
13. B. Monemar, P. P. Paskov, J. P. Bergman et al., Recombination of free and bound excitons in GaN, *Phys. Status Solidi (b)* **245**, 1723 (2008).

14. I. Vurgaftman and J. R. Meyer, Band parameters for nitrogen-containing semiconductors, *J. Appl. Phys.* **94**, 3675 (2003).
15. B. Gil, Stress effects on optical properties, in J. I. Pankove and T. D. Moustakas (eds.) *Gallium Nitride (GaN) II*. Academic Press: San Diego, CA, p. 209, (1999).
16. S. Shokhovets, K. Köhler, O. Ambacher et al., Observation of Fermi-edge excitons and exciton-phonon complexes in the optical response of heavily doped *n*-type wurtzite GaN, *Phys. Rev. B* **79**, 045201 (2009).
17. S. Shokhovets, F. Bärwolf, G. Gobsch et al., Excitons and exciton-phonon coupling in the optical response of GaN, *Phys. Status Solidi (c)* **11**, 297 (2014).
18. J. Piprek, Efficiency droop in nitride-based light-emitting diodes, *Phys. Status Solidi (a)* **207**, 2217 (2010).
19. M. Rakel, C. Cobet, N. Esser et al., GaN and InN conduction-band states studied by ellipsometry, *Phys. Rev. B* **77**, 115120 (2008).
20. C. Cobet, R. Goldhahn, W. Richter et al., Identification of van Hove singularities in the GaN dielectric function: A comparison of the cubic and hexagonal phase, *Phys. Status Solidi (b)* **246**, 1440 (2009).
21. L. C. de Carvalho, A. Schleife, and F. Bechstedt, Influence of exchange and correlation on structural and electronic properties of AlN, GaN, and InN polytypes, *Phys. Rev. B* **84**, 195105 (2011).
22. F. Bertazzi, M. Goano, X. Zhou et al., Looking for Auger signatures in III-nitride light emitters: A full-band Monte Carlo perspective, *Appl. Phys. Lett.* **106**, 0611122 (2015).
23. S. Wu, P. Geiser, J. Jin et al., Time-resolved intervalley transitions in GaN single crystals, *J. Appl. Phys.* **101**, 043701 (2007).
24. M. Semenenko, O. Yilmazoglu, H. L. Hartnagel et al., Determination of satellite valley position in GaN emitter from photoexcited field emission investigations, *J. Appl. Phys.* **109**, 023703 (2011).
25. M. Piccardo, L. Martinelli, J. Iveland et al., Determination of the first satellite valley energy in the conduction band of wurtzite GaN by near-band-gap photoemission spectroscopy, *Phys. Rev. B* **89**, 235124 (2014).
26. J. Iveland, L. Martinelli, J. Peretti et al., Direct measurement of Auger electrons emitted from a semiconductor light-emitting diode under electrical injection: Identification of the dominant mechanism for efficiency droop, *Phys. Rev. Lett.* **110**, 177406 (2013).
27. P. P. Paskov, B. Monemar, A. Toropov et al., Two-electron transition spectroscopy of shallow donors in bulk GaN, *Phys. Status Solidi (c)* **4**, 2601 (2007).
28. B. Gil, P. Bigenwald, M. Leroux et al., Internal structure of the neutral donor-bound exciton complex in zinc-blende and wurtzite semiconductors, *Phys. Rev. B* **75**, 085204 (2007).
29. A. Toropov, Yu. E. Kitaev, T. V. Shubina et al., Polarization-resolved phonon-assisted optical transitions of bound excitons in wurtzite GaN, *Phys. Rev. B* **77**, 195201 (2008).
30. B. Monemar, P. P. Paskov, J. P. Bergman et al., Transient photoluminescence of shallow donor bound excitons in GaN, *Phys. Rev. B* **82**, 235202 (2010).
31. G. Pozina, S. Khromov, C. Hemmingsson et al., Effect of silicon and oxygen doping on donor bound excitons in bulk GaN, *Phys. Rev. B* **84**, 165213 (2011).
32. A. N. Hattori, K. Hattori, Y. Moriwaki et al., Enhancement of photoluminescence efficiency from GaN(0001) by surface treatments, *Jpn. J. Appl. Phys.* **53**, 021001 (20014).
33. S. Fritze, A. Dadgar, H. Witte et al., High Si and Ge *n*-type doping of GaN doping—Limits and impact on stress, *Appl. Phys. Lett.* **100**, 122104 (2012).
34. C. Nenstiel, M. Bügler, G. Callsen et al., Germanium—The superor dopant in *n*-type GaN, *Phys. Status Solidi RRL* **9**, 716 (2015).
35. H. Wang and A. B. Chen, Calculation of shallow donor levels in GaN, *J. Appl. Phys.* **87**, 7859 (2000).
36. A. Shikanai, H. Fukahori, Y. Kawakami et al., Optical properties of Si-, Ge- and Sn-doped GaN, *Phys. Status Solidi (b)* **235**, 26 (2003).
37. R. Kriste, M. P. Hoffmann, E. Sachet et al., Ge doped GaN with controllable high carrier concentration for plasmonic applications, *Appl. Phys. Lett.* **103**, 242107 (2013).

38. M. Feneberg, S. Osterburg, K. Lange et al., Band gap renormalization and Burstein-Moss effect in silicon- and germanium-doped wurzite GaN up to 10^{20} cm^{-3}, *Phys. Rev. B* **90**, 075203 (2014).

39. B. Arnaudov, T. Paskova, E. M. Goldys et al., Modeling of the free-electron recombination band in emission spectra of highly conducting *n*-GaN, *Phys. Rev. B* **64**, 045213 (2001).

40. B. Monemar, P. P. Paskov, G. Pozina et al., Evidence for two Mg related acceptors in GaN, *Phys. Rev. Lett.* **102**, 235501 (2009).

41. B. Monemar, P. P. Paskov, G. Pozina et al., Mg-related acceptors in GaN, *Phys. Status Solidi (c)* **7**, 1850 (2010).

42. B. Monemar, S. Khromov, G. Pozina et al., Luminescence of acceptors in Mg-doped GaN, *Jpn. J. Appl. Phys.* **52**, 08JJ03 (2013).

43. B. Monemar, P. P. Paskov, G. Pozina et al., Properties of the main Mg-related acceptors in GaN from optical and structural studies, *J. Appl. Phys.* **115**, 053507 (2014).

44. G. Callsen, M. R. Wagner, T. Kure et al., Optical signature of Mg-doped GaN: Transfer processes, *Phys. Rev. B* **86**, 075207 (2012).

45. M. A. Reshchikov and H. Morkoc, Luminescence properties of defect in GaN, *J. Appl. Phys.* **97**, 061301(2005).

46. L. Chen and B. J. Skromme, Spectroscopic Characterization of Ion-Implanted GaN, *Mat. Res. Soc. Symp. Proc.* **743**, L11.35.1 (2003).

47. P. Vennegues, M. Benaissa, B. Beaumont et al., Pyramidal defects in metalorganic vapor phase epitaxial Mg doped GaN, *Appl. Phys. Lett.* **77**, 880 (2000).

48. L. T. Romano, J. E. Northrup, A. J. Ptak et al., Faceted inversion domain boundary in GaN films doped with Mg, *Appl. Phys. Lett.* **77**, 2479 (2000).

49. S. Khromov, C. G. Hemmingsson, H. Amano et al., Luminescence related to high density of Mg-induced stacking faults in homoepitaxially grown GaN, *Phys. Rev. B* **84**, 075324 (2011).

50. J. L. Lyons, A. Alkauskas, A. Janotti et al., First-principles theory of acceptors in nitride semiconductors, *Phys. Status Solidi (b)* **252**, 900 (2015).

51. J. Buckeridge, C. R. A. Catlow, D. O. Scanlon et al., Determination of the nitrogen vacancy as a shallow compensating center determination of the nitrogen vacancy as a shallow compensating center in GaN doped with divalent metals, *Phys. Rev. Lett.* **114**, 016405 (2015).

52. B. J. Skromme, K. Palle, C. D. Poweleit et al., Optical characterization of bulk GaN grown by a Na-Ga melt technique, *J. Cryst. Growth* **246**, 299 (2002).

53. B. Gil, P. Bigenwald, P. P. Paskov et al., Internal structure of acceptor-bound excitons in wide-band-gap wurtzite semiconductors, *Phys. Rev. B* **81**, 085211 (2010).

54. C. G. Van de Walle and J. Neugebauer, First-principles calculations for defects and impurities: Applications to III-nitrides, *J. Appl. Phys.* **95**, 23851 (2004).

55. L. Wang, E. Richter, and M. Weyers, Red luminescence from freestanding GaN grown on LiAlO$_2$ substrate by hydride vapor phase epitaxy, *Phys. Status Solidi (a)* **204**, 846 (2007).

56. P. P. Paskov, B. Monemar, T. Paskova et al., Optical characterization of bulk GaN substrates with c-, a-, and m-plane surfaces, *Phys. Status Solidi (c)* **6**, S763 (2009).

57. M. A. Reshchikov, A. Usikov, H. Helava et al., Fine structure of the red luminescence band in undoped GaN, *Appl. Phys. Lett.* **104**, 032103 (2014).

58. J. Neugebauer and C. G. Van de Walle, Gallium vacancies and the yellow luminescence in GaN, *Appl. Phys. Lett.* **69**, 503 (1996).

59. M. W. Bayerl, M. S. Brandt, O. Ambacher et al., Optically detected magnetic resonance of the red and near-infrared luminescence in Mg-doped GaN, *Phys. Rev. B* **63**, 125203 (2001).

60. Q. Yan, A. Janotti, M. Scheffler et al., Role of nitrogen vacancies in the luminescence of Mg-doped GaN, *Appl. Phys. Lett.* **100**, 142110 (2012).

61. T. Ogino and M. Aoki, Mechanism of yellow luminescence in GaN, *Jpn. J. Appl. Phys* **19**, 2395 (1980).

62. K. Saarinen, T. Laine, S. Kuisma et al., Observation of native Ga vacancies in GaN by positron annihilation, *Phys. Rev. Lett.* **79**, 3030 (1997).

63. K. Saarinen, T. Suski, I. Grzegory et al., Thermal stability of isolated and complexed Ga vacancies in GaN bulk crystals, *Phys. Rev. B* **64**, 233201 (2001).

64. F. Tuomisto, K. Saarinen, and T. Paskova, Thermal stability of in-grown vacancy defects in GaN grown by hydride vapor phase epitaxy, *J. Appl. Phys.* **99**, 066105 (2006).

65. D. O. Demchenko, I. C. Diallo, and M. A. Reshchikov, Yellow luminescence of gallium nitride generated by carbon defect complexes, *Phys. Rev. Lett.* **110**, 087404 (2013).

66. M. A. Reshchikov, D. O. Demchenko, A. Usikov et al., Carbon defects as sources of the green and yellow luminescence bands in undoped GaN, *Phys. Rev. B* **90**, 235203 (2014).

67. J. Oila, V. Ranki, J. Kivioja et al., Influence of dopants and substrate material on the formation of Ga vacancies in epitaxial GaN layers, *Phys. Rev. B* **63** 045205 (2001).

68. J. L. Lyons, A. Janotti, and C. G. Van de Walle, Carbon impurities and the yellow luminescence in GaN, *Appl. Phys. Lett.* **97**, 152108 (2010).

69. J. L. Lyons, A. Janotti, and C. G. Van de Walle, Effects of carbon on the electrical and optical properties of InN, GaN, and AlN, *Phys. Rev. B* **89**, 035204 (2014).

70. S. G. Christenson, W. Xie, Y. Y. Sun et al., Carbon as a source for yellow luminescence in GaN: Isolated C_N defect or its complexes, *J. Appl. Phys.* **118**,135708 (2015).

71. F. J. Hu, B. Shen, L. Lu et al., Different origins of the yellow luminescence in as-grown high-resistance GaN and unintentional-doped GaN films, *J. Appl. Phys.* **107**, 023528 (2010).

72. R. Armitage, W. Hong, Q. Yang et al., Contributions from gallium vacancies and carbon-related defects to the "yellow luminescence" in GaN, *Appl. Phys. Rev.* **82**, 3457 (2003).

73. U. Honda, Y. Yamada, Y. Tokuda et al., Deep levels in *n*-GaN Doped with carbon studied by deep level and minority carrier transient spectroscopies, *Jpn. J. Appl. Phys.* **51**, 04DF04 (2012).

74. M. A. Reshchikov, J. D. McNamara, A. Usikov et al., Optically generated giant traps in high-purity GaN, *Phys. Rev. B* **93**, 081202(R) (2016).

75. M. A. Reshchikov, D. O. Demchenko, J. D. McNamara et al., Green luminescence in Mg-doped GaN, *Phys. Rev. B* **90**, 035207 (2014).

76. A. Alkauskas, J. L. Lyons, D. Steiauf et al., First-principles calculations of luminescence spectrum line shapes for defects in semiconductors: The example of GaN and ZnO, *Phys. Rev. Lett.* **109**, 267401 (2012).

77. C. H. Seager, D. R. Tallant, J. Yu et al., Luminescence in GaN co-doped with carbon and silicon, *J. Luminescence* **106**, 115 (2004).

78. R. Armitage, Q. Yang, and E. R. Weber, Analysis of the carbon-related "blue" luminescence in GaN, *J. Appl. Phys.* **97**, 073524 (2005).

79. D. O. Demchenko, I. C. Diallo and M. A. Reshchikov, Hydrogen-carbon complexes and the blue luminescence band in GaN, *J. Appl. Phys.* **119**, 035702 (2016).

80. D. O. Demchenko and M. A. Reshchikov, Blue luminescence and Zn acceptor in GaN, *Phys. Rev. B* **88**, 115204 (2013).

81. U. Kaufmann, M. Kunzer, H. Obloh et al., Origin of defect-related photoluminescence bands in doped and nominally undoped GaN, *Phys. Rev. B* **59**, 5561 (1999).

82. S. Hautakangas, J. Oila, M. Alatalo et al., Vacancy defects as compensating centers in Mg-doped GaN, *Phys. Rev. Lett.* **90**, 137402 (2003).

83. Y. Kamiura, M. Kaneshiro, J. Tamura et al., Enhancement of blue emission from Mg-doped GaN using remote plasma containing atomic hydrogen, *Jpn. J. Appl. Phys.* **44**, L926 (2005).

84. D. N. Zakharov, Z. Liliental-Weber, B. Wagner et al., Structural TEM study of nonpolar a-plane gallium nitride grown on (112-0) 4H-SiC by organometallic vapor phase epitaxy, *Phys. Rev. B* **71**, 235334 (2005).

85. W. Rieger, R. Dimitrov, D. Brunner et al., Defect-related optical transitions in GaN, *Phys. Rev. B* **54**, 17596 (1996).

86. Y. J. Sun, O. Brandt, U Jahn et al., Impact of nucleation conditions on the structural and optical properties of M-plane GaN (11-00) grown on g-LiAlO$_2$, *J. Appl. Phys.* **92**, 5714 (2002).

87. R. Liu, A. Bell, F. A. Ponce et al., Luminescence from stacking faults in gallium nitride, *Appl. Phys. Lett.* **86**, 021908 (2005).

88. P. P. Paskov, R. Schifano, B. Monemar et al., Emission properties of a-plane GaN grown by metal-organic chemical vapor deposition, *J. Appl. Phys.* **98**, 093519 (2005).

89. P. P. Paskov and B. Monemar, Luminescence of GaN layers grown in nonpolar directions, in T. Paskova (ed.) *Nitrides with Nonpolar Surfaces*. Wiley-VCH: Wienheim, Germany, (2008).

90. J. Lähnemann, U. Jahn, O. Brandt et al., Luminescence associated with stacking faults in GaN, *J. Phys. D* **47**, 423001 (2014).

91. J. Lähnemann, O. Brandt, U. Jahn et al., Direct experimental determination of the spontaneous polarization of GaN, *Phys. Rev. B* **86**, 081302(R) (2012).

92. J. Mei, S. Srinivasan, R. Liu et al., Prismatic stacking faults in epitaxially laterally overgrown GaN, *Appl. Phys. Lett.* **88**, 141912 (2006).

93. Y. T. Rebane, Y. G. Shreter, and M. Albrecht, Stacking faults as quantum wells for excitons in wurtzite GaN, *Phys. Status Solidi (a)* **164**, 141 (1997).

94. P. Corfdir and P. Lefebvre, Importance of excitonic effects and the question of internal electric fields in stacking faults and crystal phase quantum discs: The model-case of GaN, *J. Appl. Phys.* **112**, 053512 (2012).

95. P. Corfdir, P. Lefebvre, J. Ristic et al., Electron localization by a donor in the vicinity of a basal stacking fault in GaN, *Phys. Rev. B* **80**, 153309 (2009).

96. I. Tischer, M. Feneberg, M. Schirra et al., I_2 basal plane stacking fault in GaN: Origin of the 3.32 eV luminescence band, *Phys. Rev. B* **83**, 035314 (2011).

97. M. Albrecht, L. Lymperakis and J. Neugebauer et al., Origin of the unusually strong luminescence of a-type screw dislocations in GaN, *Phys. Rev. B* **90**, 241201(R) (2014).

98. G. Pozina, R. Ciechonski, Z. Bi et al., Dislocation related droop in InGaN/GaN light emitting diodes investigated via cathodoluminescence, *Appl. Phys. Lett.* **107**, 251106 (2015).

99. M. Feneberg, M. F. Romero, M. Röppischer et al., Anisotropic absorption and emission of bulk (1-100) AlN, *Phys. Rev. B* **87**, 235209 (2013).

100. L. Chen, B. J. Skromme, R. F. Dalmau et al., Band-edge exciton states in AlN single crystals and epitaxial layers, *Appl. Phys. Lett.* **85**, 4334 (2004).

101. E. Silveira, J. A. Freitas, Jr., O. J. Glembocki et al., Excitonic structure of bulk AlN from optical reflectivity and cathodoluminescence measurements, *Phys. Rev. B* **71**, 041201(R) (2005).

102. J. Li, K. B. Nam, M. L. Nakarmi et al., Band structure and fundamental optical transitions in wurtzite AlN, *Appl. Phys. Lett.* **83**, 5163 (2003).

103. H. Ikeda, T. Okamura, K. Matsukawa et al., Impact of strain on free-exciton resonance energies in wurtzite AlN, *J. Appl. Phys.* **102**, 123707 (2007); **103**, 089901 (2008).

104. B. Gil, Hydrostatic deformation potentials and the question of exciton binding energies and splittings in aluminium nitride, *Phys. Rev. B* **81**, 205201 (2010).

105. B. Gil, B. Guizal, D. Felbacq et al., Quantitative interpretation of the excitonic splittings in aluminum nitride, *Eur. Phys. J. Appl. Phys.* **53**, 20303 (2011).

106. T. Onuma, T. Shibata, K. Kosaka et al., Free and bound exciton fine structures in AlN epilayers grown by low-pressure metalorganic vapor phase epitaxy, *J. Appl. Phys.* **105**, 023529 (2009).

107. M. Funato, K. Matsuda, R. G. Banal et al., Homoepitaxy and photoluminescence properties of (0001) AlN, *Appl. Phys. Express* **5**, 082001 (2012).

108. S. F. Chichibu, K. Hazu, Y. Ishikawa et al., Excitonic emission dynamics in homoepitaxial AlN films studied using polarized and spatio-time-resolved cathodoluminescence measurements, *Appl. Phys. Lett.* **103**, 142103 (2013).

109. C. Reich, M. Feneberg, V. Kueller et al., Excitonic recombination in epitaxial lateral overgrown AlN on sapphire, *Appl. Phys. Lett.* **103**, 212108 (2013).

110. M. Feneberg, R. A. R. Leute, B. Neuschl et al., High-excitation and high-resolution photoluminescence spectra of bulk AlN, *Phys. Rev. B* **82**, 075208 (2010).

111. M. Feneberg, M. F. Romero, B. Neuschl et al., Negative spin-exchange splitting in the exciton fine structure of AlN, *Appl. Phys. Lett.* **102**, 052112 (2013).

112. P. G. Röhner, Calculation of the exchange energy for excitons in two-body model, *Phys. Rev. B* **3**, 433 (1971).

113. H. Fu, L. W. Wang, A. Zunger, Excitonic exchange splitting in bulk semiconductors, *Phys. Rev. B* **59**, 5568 (1999).

114. R. Ishii, M. Funato, Y. Kawakami, Huge electron-hole exchange interaction in aluminum nitride, *Phys. Rev. B* **87**, 161204(R) (2013).

115. R. Ishii, M. Funato, and Y. Kawakami, Effects of strong electron–hole exchange and exciton–phonon interactions on the exciton binding energy of aluminum nitride, *Jpn. J. Appl. Phys.* **53**, 091001 (2014).

116. B. Neuschl, K. Thronke, M. Feneberg et al., Optical identification of silicon as a shallow donor in MOVPE grown homoepitaxial AlN, *Phys. Status Solidi (b)* **249**, 511 (2012).

117. B. Neuschl, K. Thronke, M. Feneberg et al., Direct determination of the silicon donor ionization energy in homoepitaxial AlN from photoluminescence two-electron transitions, *Appl. Phys. Lett.* **103**, 122105 (2013).

118. Z. Bryan, I. Bryan, M. Bobea et al., Exciton transitions and oxygen as a donor in m-plane AlN homoepitaxial films, *J. Appl. Phys.* **115**, 133503 (2014).

119. L. Gordon, J. L. Lyons, A. Janotti et al., Hybrid functional calculations of DX centers in AlN and GaN, *Phys. Rev. B* **89**, 085204 (2014).

120. C. Wetzel, T. Suski, J. W. Ager III et al., Pressure induced deep gap state of oxygen in GaN, *Phys. Rev. Lett.* **78**, 3923 (1997).

121. N. T. Son, M. Bickermann, and E. Janzen, Shallow donor and DX states of Si in AlN, *Appl. Phys. Lett.* **98**, 092104 (2011).

122. X. T. Trinh, D. Nilsson, I. G. Ivanov et al., Stable and metastable Si negative-U centers in AlGaN and AlN, *Appl. Phys. Lett.* **105**, 162106 (2014).

123. C. H. Park and D. J. Chadi, Stability of deep donor and acceptor centers in GaN, AlN, and BN, *Phys. Rev. B* **55**, 12995 (1997).

124. Y. Taniyasu, M. Kasu, and T. Makimoto, An aluminium nitride light-emitting diode with a wavelength of 210 nanometres, *Nature* **44**, 325 (2006).

125. R. Collazo, S. Mita, J. Xie et al., Progress on *n*-type doping of AlGaN alloys on AlN single crystal substrates for UV optoelectronic applications, *Phys. Status Solidi (c)* **8**, 2031 (2011).

126. L. Gordon, J. B. Varley, J. L. Lyons et al., Sulfur doping of AlN and AlGaN for improved *n*-type conductivity, *Phys. Status Solidi (RRL)* **9**, 462 (2015).

127. N. Nepal, M. L. Nakarmi, K. B. Nam et al., Acceptor-bound exciton transition in Mg-doped AlN epilayer, *Appl. Phys. Lett.* **85**, 2271 (2004).

128. S. Zhao, M. Djavid, and Z. Mi, Surface emitting, high efficiency near-vacuum ultraviolet light source with aluminum nitride nanowires monolithically grown on silicon, *Nano Lett.* **15**, 7006 (2015).

129. A. T Connie, S. Zhao, S. M. Sadaf et al., Optical and electrical properties of Mg-doped AlN nanowires grown by molecular beam epitaxy, *Appl. Phys. Lett.* **106**, 213105 (2015).

130. N. Nepal, M. L. Nakarmi, H. U. Jang et al., Growth and photoluminescence studies of Zn-doped AlN epilayers, *Appl. Phys. Lett.* **89**, 192111 (2006).

131. F. Mireles and S. E. Ulloa, Acceptor binding energies in GaN and AlN, *Phys. Rev. B* **58**, 3879 (1998).

132. A. Sedhain, T. M. Al Tahtamouni, J. Li et al., Beryllium acceptor binding energy in AlN, *Appl. Phys. Lett.* **93**, 141104 (2008).

133. Y. Yamada, K. Choi, S. Shin et al., Photoluminescence from highly excited AlN epitaxial layers, *Appl. Phys. Lett.* **92**, 131912 (2008).

134. R. A. R Leute, M. Feneberg, R. Sauer et al., Photoluminescence of highly excited AlN: Biexcitons and exciton-exciton scattering, *Appl. Phys. Lett.* **95**, 031903 (2009).
135. G. E. Pikus and L. G. Bir, *Symmetry and Strained-Induced Effects in Semiconductors.* Wiley: New York, (1974).
136. G. Rossbacsh, M. Feneberg, M. Röppischer et al., Influence of exciton-phonon coupling and strain on the anisotropic optical response of wurtzite AlN around the band edge, *Phys. Rev. B* **83**, 1952002 (2011).
137. K. B. Nam, J. Li, M. L. Nakarmi et al., Unique optical properties of AlGaN alloys and related ultraviolet emitters, *Appl. Phys. Lett.* **84**, 5264 (2004).
138. C. Netzel, A. Knauer, and M. Weyers, Impact of light polarization on photoluminescence intensity and quantum efficiency in AlGaN and AlInGaN layers, *Appl. Phys. Lett.* **101**, 242102 (2012).
139. B. Neuschl, J. Helbing, M. Knab et al, Composition dependent valence band order in c-oriented wurtzite AlGaN layers, *J. Appl. Phys.* **116**, 113506 (2014).
140. M. Feneberg, M.Winkler, J. Klamser et al., Anisotropic optical properties of semipolar AlGaN layers grown on m-plane sapphire, *Appl. Phys. Lett.* **106**, 182102 (2015).
141. R. Pässler, Dispersion-related description of temperature dependencies of band gaps in semiconductors, *Phys. Rev. B* **66**, 085201 (2002).
142. T. Onuma, S. F. Chichibu, A. Uedono et al., Radiative and nonradiative processes in strain-free AlxGa1-xN films studied by time-resolved photoluminescence and positron annihilation techniques, *J. Appl. Phys.* **95**, 2495 (2004).
143. N. Nepal, J. Li, M. L. Nakami et al., Exciton localization in AlGaN alloys, *Appl. Phys. Lett.* **88**, 062103 (2006).
144. M. Feneberg, S. Osterburg, M. F. Romero et al., Optical properties of magnesium doped $Al_xGa_{1-x}N$ (0.61 ≤x ≤ 0.73), *J. Appl. Phys.* **116**, 143103 (2014).
145. S. Kurai, H. Miyake, K. Hiramatsu et al., Microscopic potential fluctuations in Si-doped AlGaN epitaxial layers with various AlN molar fractions and Si concentrations, *J. Appl. Phys.* **119**, 025707 (2016).
146. A. Sedhain, L. Du, J. H. Edgar et al., The origin of 2.78 eV emission and yellow coloration in bulk AlN substrates, *Appl. Phys. Lett.* **95**, 262104 (2009).
147. M. Bickermann, B. M. Epelbaum, O. Filip et al., Faceting in AlN bulk crystal growth and its impact on optical properties of the crystals, *Phys. Status Solidi (c)* **9**, 449 (2012).
148. T. Koyama, M. Sugawara, T. Hoshi et al., Relation between Al vacancies and deep emission bands in AlN epitaxial films grown by NH3-source molecular beam epitaxy, *Appl. Phys. Lett.* **90**, 241914 (2007).
149. J. A. Freitas Jr., J. C. Culbertson, M. A. Mastro et al., Structural and optical properties of thick freestanding AlN films prepared by hydride vapor phase epitaxy, *J. Crystal Growth* **350**, 33 (2012).
150. M. Strassburg, J. Senawiratne, N. Dietz et al., The growth and optical properties of large, high-quality AlN single crystals, *J. Appl. Phys.* **96**, 5870 (2004).
151. M. Bickermann, B. M. Epelbaum, O. Filip et al., Point defect content and optical transitions in bulk aluminum nitride crystals, *Phys. Status Solidi (b)* **246**, 1181 (2009).
152. T. Schulz, M. Albrecht, K. Irmscher et al., Ultraviolet luminescence in AlN, *Phys. Status Solidi (b)* **248**, 1513 (2011).
153. G. A. Slack, L. J. Schowalter, D. Morelli et al., Some effects of oxygen impurities on AlN and GaN, *J. Crystal Growth* **246**, 287 (2002).
154. J. M. Mäki, I. Makkonen, F. Tuomisto et al., Identification of the V_{Al}-O_N defect complex in AlN single crystals, *Phys. Rev. B* **84**, 081204(R) (2011).
155. R. Collazo, J. Xie, B. E. Gaddy et al., On the origin of the 265nm absorption band in AlN bulk crystals, *Appl. Phys. Lett.* **100**, 191914 (2012).
156. C. Stampfl and C. G. Van de Walle, Theoretical investigation of native defects, impurities, and complexes in aluminum nitride, *Phys. Rev. B* **65**, 155212 (2002).

157. K. Laaksonen, M. G. Ganchenkova, and R. M. Nieminen, Vacancies in wurtzite GaN and AlN, *J. Phys.: Condens. Matter* **21**, 015803 (2009).

158. Q. Yan, A. Janotti, M. Scheffler et al., Origins of optical absorption and emission lines in AlN, *Appl. Phys. Lett.* **105**, 111104 (2014).

159. N. T. Son, A. Gali, A. Szabo et al., Defects at nitrogen site in electron-irradiated AlN, *Appl. Phys. Lett.* **98**, 242116 (2011).

160. T. Nagashima, Y. Kubota, T. Kinoshita et al., Structural and optical properties of carbon-doped AlN substrates grown by hydride vapor phase epitaxy using AlN substrates prepared by physical vapor transport, *Appl. Phys. Express* **5**, 125501 (2012).

161. B. E. Gaddy, Z. Bryan, I. Bryan et al., Vacancy compensation and related donor-acceptor pair recombination in bulk AlN, *Appl. Phys. Lett.* **103**, 161901 (2013).

162. B. E. Gaddy, Z. Bryan, I. Bryan et al., The role of the carbon-silicon complex in eliminating deep ultraviolet absorption in AlN, *Appl. Phys. Lett.* **104**, 202106 (2014).

163. K. B. Nam, M. L. Nakarmi, J. Y. Lin et al., Deep impurity transitions involving cation vacancies and complexes in AlGaN alloys, *Appl. Phys. Lett.* **86**, 222108 (2005).

164. M. L. Nakarmi, N. Nepal, J. Y. Lin et al., Photoluminescence studies of impurity transitions in Mg-doped AlGaN alloys, *Appl. Phys. Lett.* **94**, 091903 (2009).

165. L. F. J. Piper, T. D. Veal, C. F. McConville et al., Origin of the *n*-type conductivity of InN: The role of positively charged dislocations, *Appl. Phys. Lett.* **88**, 252109 (2006).

166. R. E. Jones, K. M. Yu, S. X. Li et al., Evidence for *p*-type doping of InN, *Phys. Rev. Lett.* **96**, 125505 (2006).

167. T. L. Tansley and C. P. Foley, Infrared absorption in indium nitride, *J. Appl. Phys.* **60**, 2092 (1986).

168. V. Yu. Davydov, A. A. Klochikhin, R. P. Seisyan et al., Absorption and emission of hexagonal InN. Evidence of narrow fundamental band gap, *Phys. Status Solidi (b)* **229**, R1 (2002).

169. S. H. Wei, X. Nie, I. G. Batyrev et al., Breakdown of the band-gap-common-cation rule: The origin of the small band gap of InN, *Phys. Rev. B* **67**, 165209 (2003).

170. P. Rinke, M. Scheffler, A. Qteish et al., Band gap and band parameters of InN and GaN from quasiparticle energy calculations based on exact-exchange density-functional theory, *Appl. Phys. Lett.* **89**, 161919 (2006).

171. T. Inushima, M. Higashiwaki, T. Matsui et al., Electron density dependence of the electronic structure of InN epitaxial layers grown on sapphire (0001), *Phys. Rev. B* **72**, 0852010 (2005).

172. B. Arnaudov, T. Paskova, P. P. Paskov et al., Energy position of near-band0edge emission spectra of InN epitaxial layers with different doping levels, *Phys. Rev. B* **69**, 115216 (2004).

173. M. Feneberg, J. Däubler, K. Thonke et al., Mahan excitons in degenerate wurtzite InN: Photoluminescence spectroscopy and reflectivity, *Phys. Rev. B* **77**, 245207 (2008).

174. S. Zhao, Z. Mi, M. G. Kibria et al., Understanding the role of Si doping on surface charge and optical properties: Photoluminescence study of intrinsic and Si-doped InN nanowires, *Phys. Rev. B* **85**, 245313 (2012).

175. S. Zhao, S. Fathololoumi, K. H. Bevan et al., Tuning the surface charge properties of epitaxial InN nanowires, *Nano Lett.* **12**, 2877 (2012).

176. Z. Mi and S. Zhao, Extending group-III nitrides to the infrared: Recent advances in InN, *Phys. Status Solidi (b)* **252**, 1050 (2015).

177. L. H. Dmowski, M. Baj, X. Q. Wang et al., Advantage of In- over N-polarity for disclosure of *p*-type conduction in InN:Mg, *J. Appl. Phys.* **115**, 173704 (2014).

178. S. Khromov, P. O. Å. Persson, X. Wang et al., Correlation between switching to *n*-type conductivity and structural defects in highly Mg-doped InN, *Appl. Phys. Lett.* **106**, 232102 (2015).

179. M. Y. Xie, N. Ben Sedrine, S. Schöche et al., Effect of Mg doping on the structural and free-charge carrier properties of InN films, *J. Appl. Phys.* **115**, 163504 (2014).

180. S. Zhao, X. Liu and Z. Mi, Photoluminescence properties of Mg-doped InN nanowires, *Appl. Phys. Lett.* **103**, 203113 (2013).

181. X. Wang, S. B. Che, Y. Ishitani et al., Growth and properties of Mg-doped In-polar InN films, *Appl. Phys. Lett.* **90**, 201913 (2007).

182. T. Hirasaki, K. Asano, M. Banno et al., Growth of thick InGaN layers by tri-halide vapor phase epitaxy, *Jpn. J. Appl. Phys.* **53**, 05FL02 (2014).

183. P. Schley, R. Goldhahn, A. T. Winzer et al., Dielectric function and Van Hove singularities for In-rich InxGa1-xN alloys: Comparison of N- and metal-face materials, *Phys. Rev. B* **75**, 205204 (2007).

184. E. Sakalaus, Ö. Tuna, A. Kraus et al., Dielectric function and bowing parameters of InGaN alloys, *Phys. Status Solidi (b)* **249**, 485 (2012).

185. P. G. Moses and C. G. Van de Walle, Band bowing and band alignment in InGaN alloys, *Appl. Phys. Lett.* **96**, 021908 (2010).

186. S. F. Chichibu, A. Uedono, T. Onuma et al., Origin of defect-insensitive emission probability in In-containing (Al,In,Ga)N alloy semiconductors, *Nature Mat.* **5**, 810 (2006).

4

Electronic and Transport Properties of III-Nitride Semiconductors

Yuh-Renn Wu

4.1 Introduction

The III-nitride semiconductors have become one of the most important semiconductors in the world due to the rapid growth of solid state lighting, laser, and power electronic applications [1–14]. The III-nitrides including alloys of AlN, InN, and GaN are direct bandgap materials, which span the entire visible emission spectrum including Deep UV to infrared wavelength. Besides the excellent electronic properties, the GaN and AlN have a relatively wider bandgap and very good thermal properties and a reasonable electron effective mass, which make them very suitable for power electronic applications [15–25]. In this chapter, we will introduce the basic electronic properties of nitride-based materials.

4.2 Band Structure of Wurtzite III-N Semiconductors

The III-nitride semiconductor materials are basically wurtzite structures, with hexagonal symmetrical properties. The band structures are basically constructed by sp3 orbitals. Since they are all direct bandgap materials, the conduction band is mainly constructed by s orbital, while the valance bands are mainly constructed by $p_x, p_y,$ and p_z orbitals ($|X\rangle, |Y\rangle$, and $|Z\rangle$ states). Figure 4.1 shows that the complete band structures of AlN, GaN, and InN are calculated by the First Principle theory. The valence band of GaN is mainly

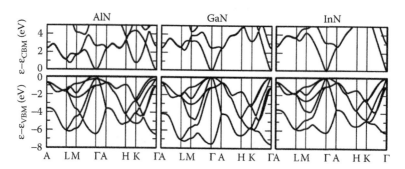

FIGURE 4.1 The calculated band structure of nitride materials with OPEx. For ease of comparison the valence bands have been aligned at the valence-band maximum and the conduction bands at the minimum. (From Delaney, K.T. et al., *Appl. Phys. Lett.*, 94, 191109, 2009. With permission.)

constructed by $|X \pm iY\rangle$ and $|Z\rangle$, which are mainly called the CH1, CH2, and CH3 bands. The CH1 and CH2 bands are equal to heavy hole band and light hole band but are mainly constructed by $|X\rangle$ and $|Y\rangle$ orbitals. Since they are mainly constructed by these two states, the effective masses are quite large for the $|X\rangle$ state along k_y and k_z direction and small for the k_x direction. The same situation is also found for $|Y\rangle$ state where the effective masses along k_x and k_z direction are quite large and small for k_y direction. Therefore, for the *E-k* dispersion relation away from Γ center, we will note that the HH band has very large effective mass along all directions and LH has large effective mass along k_z direction and small effective mass along k_x and k_y directions contributed by $|X\rangle$ and $|Y\rangle$ state, respectively. These are due to the crystal symmetric. Unfortunately, the effective mass of HH band is quite large, which is close to 1.1–1.8 m_0. This leads to a very low mobility for the hole transport. The light hole band has a much smaller effective mass (~0.20 m_0) in the k_x and k_y directions. If we want the light hole band to be the top valence band, strain engineering is needed for lifting the light hole band. No matter what strain is applied on GaN, the effective masses along the k_z direction for both heavy and light hole bands (CH1 and CH2) are very large. Therefore, the mobility of GaN/InN-based alloy is much smaller, which is not good especially for light-emitting diodes (LEDs) cases. Only the third CH3 band which is composed of $|Z\rangle$ states has a much smaller effective mass along the k_z direction. However, only the AlN and AlGaN alloys with high aluminum compositions have $|Z\rangle$ state on top, which might give a higher hole mobility. However, for the light-emitting applications, the $|Z\rangle$-liked state will emit TM polarized light, which is bad for surface-emitting devices.

On the other hand, the effective electron masses of GaN, AlN, and InN are much consistent in each direction since they are direct bandgap materials composed by $|S\rangle$-orbital. Due to the hcp crystal symmetric, the effective masses of GaN along kz and kx-ky directions are slightly different. The $m_\perp^*$ is around 0.21 m_0 and $m_\parallel^*$ is around 0.20 m_0. This is simply within the error bar ranges. This small effective mass compared with Si (1.12 eV) with much larger bandgap (3.4 eV) makes GaN to be an excellent candidate for high-power and high-frequency devices. The second valley of GaN device is the L-M valley. The valley height is around 1.1 to 2 eV according to different calculations and simulation result. The height of L-M valley would be very critical for high field transport. As we know, at the high electric field condition, the electron will gain its energy. As the electron is accelerated by the electric field, it also suffers the scattering loss, especially from the polar optical phonon scattering. However, if the field is too high where the electron's energy is close to the second L-M valley, the nonequivalent inter-valley scattering will be turned on so that the electron will be scattered into the second valley. The much larger effective mass of the second valley will slow down the average electron speed and cause the negative differential mobility at the high field region. Therefore, the height of valley separation will decide when the electron speed will start to be saturated. Although the position of the second valley limits the saturation velocity, it also prevents the electron's energy over the bandgap to induce the impact ionization scattering for device breakdown. Therefore, the breakdown field is much larger than expectation at this bandgap.

Except the bandgap issues, the wurtzite nitride-based material is also the polar material. The GaN, AlN, and InN have a spontaneous polarization. If they are under strain, the piezoelectric polarization will also be induced in these materials. These will lead to many interesting device properties, which cause many different issues for different applications. In the following subsections, we will discuss how the strain and crystal orientation affect the band structure, electronic, and polarization properties.

4.3 The Strain and Piezoelectric Polarization in Nitride-Based Materials

4.3.1 Strain Induced for the *c*-Plane Structure

As we know, nitride-based materials are wurtzite structures. The lattice sizes of GaN, AlN, and InN are 3.11 Å, 3.189 Å, and 3.53 Å, respectively. The lattice mismatch is quite large, especially between InN and GaN. Therefore, it could cause strain when different alloys are pseudomorphically grown. When the alloy $In_xGa_{1-x}N$ or $Al_xGa_{1-x}N$ is made, the lattice constant can be obtained by linear interpolation, where

$$a_{InGaN} = xa_{InN} + (1-x)a_{GaN} \tag{4.1}$$

For the *c*-plane case, ideally, when the layer material is grown on top of the buffer material, it will suffer the lateral strain due to the lattice mismatch. The lateral strain induced in the layer can be calculated by

$$\epsilon_{\parallel} = \frac{a_s - a_L}{a_L} = \epsilon_{xx} = \epsilon_{yy} \tag{4.2}$$

To obtain the strain induced in the growth direction, we need to understand the strain and stress relation first. For the 6×6 strain stress tensor of wurtzite structure is shown below

$$\begin{pmatrix} Xx \\ Yy \\ Zz \\ Yz \\ Zx \\ Xy \end{pmatrix} = \begin{pmatrix} c_{11} & c_{12} & c_{13} & 0 & 0 & 0 \\ c_{12} & c_{11} & c_{13} & 0 & 0 & 0 \\ c_{13} & c_{13} & c_{33} & 0 & 0 & 0 \\ 0 & 0 & 0 & c_{44} & 0 & 0 \\ 0 & 0 & 0 & 0 & c_{44} & 0 \\ 0 & 0 & 0 & 0 & 0 & c_{66} \end{pmatrix} \begin{pmatrix} e_{xx} \\ e_{yy} \\ e_{zz} \\ e_{yz} \\ e_{zx} \\ e_{xy} \end{pmatrix} \tag{4.3}$$

where:

Xx, Yy, and Zz are the normal stress in each direction and Xy, Yz, Zx the shear stress

c_{ij} is the elastic coefficient

e_{ij} is the strain element

1, 2, 3, 4, 5, and 6 are representing xx, yy, zz, yz, zx, and xy, respectively

Note that $e_{xx} = \epsilon_{xx}$ but $e_{xy} = 2\epsilon_{xy}$. When the material is grown along *c*-axis, the stress along the growth direction is 0. Therefore, we can obtain

$$Zz = 0 = c_{13}e_{xx} + c_{13}e_{yy} + c_{33}e_{zz} = 0 \tag{4.4}$$

The strain along the *z*-direction is

$$e_{zz} = \epsilon_{zz} = -\frac{c_{13}}{c_{33}}(e_{xx} + e_{yy}) = -\frac{2c_{13}}{c_{33}}\epsilon_{\parallel} \tag{4.5}$$

After obtaining the strain, the piezoelectric induced polarization charge p^{pz} can be calculated by

$$p^{pz} = \begin{pmatrix} p_x^{pz} \\ p_y^{pz} \\ p_z^{pz} \end{pmatrix} = \begin{pmatrix} 0 & 0 & 0 & 0 & e_{15} & 0 \\ 0 & 0 & 0 & e_{15} & 0 & 0 \\ e_{31} & e_{31} & e_{33} & 0 & 0 & 0 \end{pmatrix} \begin{pmatrix} \epsilon_{xx} \\ \epsilon_{yy} \\ \epsilon_{zz} \\ \epsilon_{yz} \\ \epsilon_{zx} \\ \epsilon_{xy} \end{pmatrix} \tag{4.6}$$

Except the piezoelectric polarization charges induced at the strained layer, the nitride-based material also has spontaneous polarization. Therefore, at the interface, it will induce the polarization charge

$$\Delta P_z = P_L^{pz} + P_L^{sp} - P_S^{sp} \tag{4.7}$$

where P_S^{sp} is the spontaneous polarization of reference layer and P_L^{pz} and P_L^{sp} are the piezoelectric and spontaneous polarization charges, respectively. The polarization difference induced at the interface is quite large. The typical value is around $1 \times 10^{13} \, cm^{-2}$ to $3 \times 10^{13} \, cm^{-2}$. The polarization induced polar changes will induce a strong electric field along the c-axis and cause a serious band bending. For the light-emitting device application, this strong electric field usually leads to the quantum-confined Stark effect (QCSE) and reduces the radiative recombination efficiency. However, for the power electronic device, the polarization field will induce the 2DEG in the channel and reduce the access resistance, which are very promising properties for power electronic devices.

4.3.2 Strain Induced on the Semipolar Plane and Nonpolar Plane Structures

When the growth plane is inclined from the c-plane by an angle θ, the total polarization discontinuity for a coherent layer of $In_xGa_{1-x}N$ on GaN induced by the anisotropic strain can be calculated based on the anisotropic linear elasticity. To deal with this problem, if we define ϕ as the angle between the θ rotating axis (x'-axis) and x-axis as shown in Figure 4.2, we rotate the coordinates by θ with the rotating axis as defined by Figure 4.2 we can derive the rotation matrix U as:

$$U = \begin{pmatrix} \cos\phi & -\sin\phi & 0 \\ \cos\theta\sin\phi & \cos\theta\cos\phi & -\sin\theta \\ \sin\theta\sin\phi & \sin\theta\cos\phi & \cos\theta \end{pmatrix} \tag{4.8}$$

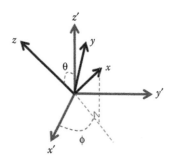

FIGURE 4.2 Coordinate systems used in calculations of strain induced polarization: *xyz*—natural coordinate system associated with *c* axis; *x'y'z'*—coordinate system related to the layer surface normal. (From Yuji, Z. et al., *Jpn. J. Appl. Phys*, 53, 100206, 2014. Quantum confined stark.)

Due to the crystal symmetric, we can neglect the effect of ϕ and assume it is zero. Then the rotation matrix becomes

$$U = \begin{pmatrix} 1 & 0 & 0 \\ 0 & \cos\theta & -\sin\theta \\ 0 & \sin\theta & \cos\theta \end{pmatrix} \tag{4.9}$$

The relations between the original coordinate system and the rotated coordinate system for the wave vectors, strain tensors, and elastic stiffness constants are expressed as

$$k'_i = \sum_\alpha U_{i\alpha} k_\alpha \tag{4.10}$$

$$\epsilon'_{ij} = \sum_{\alpha\beta} U_{i\alpha} U_{j\beta} \epsilon_{\alpha\beta} \tag{4.11}$$

$$C'_{ijkl} = \sum_{\alpha\beta\gamma\delta} U_{i\alpha} U_{j\beta} U_{k\gamma} U_{l\delta} C_{\alpha\beta\gamma\delta} \tag{4.12}$$

Here C_{ijkl} is the 9×9 tensor instead of the 6×6 elastic constant tensor mentioned earlier i, j, k, and l are x, y, and z, respectively. The relationship between 9×9 tensor to 6×6 tensor is that $C_{xxxx} = C_{11}$, $C_{xxyy} = C_{12}$, ..., and so on. Therefore, in the inclined coordinate, we can get the strain induced by the semipolar lateral direction as

$$\epsilon'_\parallel = \frac{a'_s - a'_L}{a'_L} = \epsilon_{x'x'} = \epsilon_{y'y'} \tag{4.13}$$

Then we can get

$$\epsilon'_\perp = \epsilon'_{zz} = -\frac{2C'_{13}}{C'_{33}} \epsilon'_\parallel = -\frac{2C'_{13}}{C'_{33}} \epsilon'_{xx} \tag{4.14}$$

and C'_{13} and C'_{33} can be obtained from Equation 4.12. After obtaining ϵ'_{xx}, ϵ'_{yy}, and ϵ'_{zz}, we can obtain the 6 (or 9) strain components (ϵ_{xx}, ϵ_{yy}, ϵ_{zz}, ϵ_{yz}, ϵ_{zx}, and ϵ_{xy}) from Equation 4.11. Note that since it is an inclined substrate, the shear strain component will be induced, and will be zero. By Equation 4.6, we can then calculate the piezoelectric polarization in the inclined plane. After we obtain the piezoelectric polarization value at the original coordinate, P^{pz}, we need to multiply the $\cos\theta$ to 3 components (p_x^{pz}, p_y^{pz}, p_z^{pz}) and sum them. Romanov et al. [32] had derived the final simple equation to obtain the piezoelectric polarization.

$$\begin{aligned} P_{L_{z'}}^{pz} = e_{31}\cos\theta\epsilon_{x'x'} &+ \left(e_{31}\cos^3\theta + \frac{e_{33} - e_{15}}{2}\sin\theta\sin2\theta \right)\epsilon_{y'y'} \\ &+ \left(\frac{(e_{31} + e_{15})}{2}\sin\theta\sin2\theta + e_{33}\cos^3\theta \right)\epsilon_{z'z'} \\ &+ \left[(e_{31} - e_{33})\cos\theta\sin2\theta + e_{15}\sin\theta\cos2\theta \right]\epsilon_{y'z'} \end{aligned} \tag{4.15}$$

The total polarization difference at the strained layer and the substrate layer (InGaN/GaN or AlGaN/GaN) can be calculated by

$$\Delta P'_z = P_{L_{z'}}^{pz} + \left(P_L^{sp} - P_S^{sp} \right)\cos\theta \tag{4.16}$$

where P_L^{sp} and P_S^{sp} are the spontaneous polarization of the InGaN layer and the substrate layer, respectively. From these equations, we can know that the spontaneous polarization will simply be cosθ in respect to the inclined angle θ. However, the piezoelectric polarization is not so simple. Due to the crystal symmetric and induced shear strain the polarization value is not changed linearly with cosθ but is complicated as shown in Equation 4.15. In Romanov's paper, it shows that the $\Delta P_z'$ will change the sign near 45° due to the induced shear strain to change the piezoelectric polarization value. In addition to the change in the total polarization discontinuity, $\Delta P_z'$, the band structure and light emission polarization properties will also be modified due to the anisotropic strain in coherently stressed nonpolar and semipolar heterostructures. In the next section, we will discuss the influence of strain to bandstructure modification.

4.3.3 The Band Structure Modification due to the Strain

As we have discussed, the characterization of AlN, GaN, and InN bulk material has been calculated and understood well. However, when these materials are grown under the strain condition, the band structure will be modified due to the strain. In the earlier section, we have discussed that the strain will modify the polarization value. Beside this, the strain will also modify the bandstructure especially in the valence band, where P_x, P_y, and P_y orbital are very sensitive to the direction of strain. Therefore, the band structure, as well as the light emission polarization properties will be affected due to the strain, especially in the semipolar plane where the anisotropic strain is in coherently stressed nonpolar and semipolar heterostructures.

To understand the influence of strain on the bandstructure modification in the *c*-plane, nonpolar plane, and semipolar plane, we need to consider the following issues: (1) the changes of the VB states mixing due to the induced strain; (2) the energy separation of the first VB (CH1) and the second VB (CH2) (3) If the QW structure is formed, the effective mass differences along the vertical confined direction of CH1 to CH2 bands will also need to be studied, since they will strongly affect the quantum confinement. For example, if the effective mass of CH1 band along the *z'* direction ($m_{z'}^*$) is smaller than the CH2 band, the energy separation of CH1 and CH2 bands will become smaller because the CH1 band will have a larger confinement energy. For some conditions, the CH2 can be higher than CH1 band to be the dominated band.

As we mentioned, the VB structure of wurtzite GaN is developed from the on the $|X\rangle$, $|Y\rangle$, $|Z\rangle$ states, which correspond to the nitrogen p_x, p_y and p_z orbitals, respectively. The six band structure bases of the wurtzite structure including spins can be expressed as $|u_1\rangle$ to $|u_6\rangle$ and are $\left(1/\sqrt{2}\right)|X+iY,\uparrow\rangle$, $\left(1/\sqrt{2}\right)|X+iY,\downarrow\rangle$, $|Z,\uparrow\rangle$, $|Z,\downarrow\rangle$, $\left(1/\sqrt{2}\right)|X-iY,\uparrow\rangle$, $\left(1/\sqrt{2}\right)|X-iY,\downarrow\rangle$, where the polarization of the emission light is strongly affected by these bases. For the *c*-plane structure in GaN and InN, the upper two bands are basically $|u_1\rangle$ and $|u_6\rangle$ ($|X \pm iY\rangle$ states), where the $|X\rangle$ and $|Y\rangle$ are equally mixed. However, for the AlN, the upper band becomes $|Z\rangle$ states. When the InGaN-based alloy grown on the *c*-plane GaN substrate or buffer layer, the equi-biaxial compressive stress for an InGaN layer does not change the separation of the top two VBs $|u_{1,6}\rangle$ and $|u_{2,5}\rangle$ ($|X\rangle$ and $|Y\rangle$ states are moving up equally and the $|Z\rangle$ state is moving down. Therefore, the dominated band structure would still be $|X \pm iY\rangle$ states. For the light-emitting device, the emission strength of the polarization direction $\hat{e}$ would be the inner product of $\langle S|\hat{e}\cdot\nabla|X \pm iY\rangle$. Therefore, for the dominate $|X \pm iY\rangle$ states, the light polarization will be mainly parallel to X and Y direction, which is the TE polarized light. This would be good for surface-emitting devices. For the AlN, the top band is $|Z\rangle$ states. For the AlGaN alloy mixed with AlN and GaN, the $|Z\rangle$ states will be the top valence band when Al composition is larger than 20%. However, when the AlGaN material is under strain, the condition will change. Since the lateral compress strain will move the $|Z\rangle$ state down the tensile strain will lift up the $|Z\rangle$ states. Therefore, it is important to know how strain will affect the shift of valence band states, especially how it will affect the TM and TE polarized light emission strength. In the semipolar plane and nonpolar plane, due to the wurtzite structure, the anisotropic strain will have an anisotropic effect to the Z and X-Y states, which has a more complicated behavior to band structures.

To understand the detailed influence of the strain, the $6 \times 6\,k \cdot p$ method is usually used to analyze these problems.[41] The 6×6 Hamiltonian can be expressed as

$$H^v u = \begin{pmatrix} F & 0 & -H^* & 0 & K^* & 0 \\ 0 & G & \Delta & -H^* & 0 & K^* \\ -H & \Delta & \lambda & 0 & I^* & 0 \\ 0 & -H & 0 & \lambda & \Delta & I^* \\ K & 0 & I & \Delta & G & 0 \\ 0 & K & 0 & I & 0 & F \end{pmatrix} \begin{pmatrix} |u_1\rangle \\ |u_2\rangle \\ |u_3\rangle \\ |u_4\rangle \\ |u_5\rangle \\ |u_6\rangle \end{pmatrix} \tag{4.17}$$

where

$$F = \Delta_1 + \Delta_2 + \lambda + \theta$$

$$G = \Delta_1 - \Delta_2 + \lambda + \theta$$

$$\lambda = \left(\frac{\hbar^2}{2m_0} \right) \left[A_1 k_z^2 + A_2 \left(k_x^2 + k_y^2 \right) \right] + D_1 \epsilon_{zz} + D_2 \left(\epsilon_{xx} + \epsilon_{yy} \right)$$

$$\theta = \frac{\hbar^2}{2m_0} \left[A_3 k_z^2 + A_4 \left(k_x^2 + k_y^2 \right) \right] + D_3 \epsilon_{zz} + D_4 \left(\epsilon_{xx} + \epsilon_{yy} \right) \tag{4.18}$$

$$K = \frac{\hbar^2}{2m_0} A_5 \left(k_x + ik_y \right)^2 + D_5 \left(\epsilon_{xx} - \epsilon_{yy} + 2i\epsilon_{xy} \right)$$

$$H = \frac{\hbar^2}{2m_0} i \left[A_6 k_z \left(k_x + ik_y \right) + A_7 \left(k_x + ik_y \right) \right] + i D_6 \left(\epsilon_{xz} + i\epsilon_{yz} \right)$$

$$I = \frac{\hbar^2}{2m_0} i \left[A_6 k_z \left(k_x + ik_y \right) - A_7 \left(k_x + ik_y \right) \right] + i D_6 \left(\epsilon_{xz} + i\epsilon_{yz} \right)$$

$$\Delta = \sqrt{2} \Delta_3,$$

D_1–D_6 are the deformation potentials and A_1–A_7 are the fitting parameters to the VB structure. k_i and ϵ_{ij} ($i, j = x, y, z$) are the wave vector and the strain tensor. Δ_1 is the crystal-field energy and Δ_2 and Δ_3 are the spin-orbit energy parameters. These parameters can be found in References [27–30] or Table 1.2 in Chapter 1. Note that there are many theoretical calculations which provide different values for A_1 to A_7. Particularly, the deformation potentials D_1 to D_6 are very different from these studies. However, the trend of applied structure to the band structures shift are the same. If we looked at the 6×6 matrix for the Γ center $k_x = k_y = k_z = 0$ for the c-plane layer and the shear strain component $\epsilon_{xy} = \epsilon_{yz} = \epsilon_{zx} = 0$. In addition, $\epsilon_{xx} = \epsilon_{yy} = \epsilon_{\parallel}$ and $\epsilon_{zz} = \epsilon_{\perp}$, we can obtain the matrix to be

$$H^v u = \begin{pmatrix} F & 0 & 0 & 0 & 0 & 0 \\ 0 & G & \Delta & 0 & 0 & 0 \\ 0 & \Delta & \lambda & 0 & 0 & 0 \\ 0 & 0 & 0 & \lambda & \Delta & 0 \\ 0 & 0 & 0 & \Delta & G & 0 \\ 0 & 0 & 0 & 0 & 0 & F \end{pmatrix} \begin{pmatrix} |u_1\rangle \\ |u_2\rangle \\ |u_3\rangle \\ |u_4\rangle \\ |u_5\rangle \\ |u_6\rangle \end{pmatrix} \tag{4.19}$$

and

$$F = \Delta_1 + \Delta_2 + (D_1 + D_3)\epsilon_{zz} + (D_2 + D_4)(\epsilon_{xx} + \epsilon_{yy})$$

$$G = \Delta_1 - \Delta_2 + (D_1 + D_3)\epsilon_{zz} + (D_2 + D_4)(\epsilon_{xx} + \epsilon_{yy})$$

$$\lambda = D_1\epsilon_{zz} + D_2(\epsilon_{xx} + \epsilon_{yy})$$

$$\theta = D_3\epsilon_{zz} + D_4(\epsilon_{xx} + \epsilon_{yy}) \qquad (4.20)$$

$$K = H = I = 0$$

$$\Delta = \sqrt{2}\Delta_3$$

We can find that the $|u_1\rangle$ and $|u_2\rangle$ state ($|X \pm iY\rangle$) will basically move in the same direction when the strain is applied and $|u_3\rangle$ state ($|Z\rangle$) will move in a different direction or magnitude. If we neglect the coupling term of Δ, the relative shift of $|u_1\rangle$ and $|u_2\rangle$ state to $|u_3\rangle$ due to the strain would be the term $\theta = D_3\epsilon_{zz} + D_4(\epsilon_{xx} + \epsilon_{yy})$. The D_3 values are usually positive and D_4 values are usually negative even from different sources [33]. Therefore, for the compress lateral strain where ϵ_{xx} and ϵ_{yy} are smaller than 0 and ϵ_{zz} is larger than 0. We can find that $\theta > 0$ and $|u_1\rangle$ and $|u_2\rangle$ state move up more than $|u_3\rangle$ state. that is the $|X\rangle$ and $|Y\rangle$ will move up and $|Z\rangle$ state will move down respectively. However, if the layer is under tensile strain, the relative movement will switch where the $|Z\rangle$ will move up and $|X\rangle$ and $|Y\rangle$ will move down. Therefore, for the AlGaN grown on GaN buffer layer, where the AlGaN layer is suffering the tensile strain, the $|Z\rangle$ state will move up. At the low Al composition case, the tensile strain will move up the $|Z\rangle$ state to be the top band. At high Al composition, the tensile strain will move up $|Z\rangle$ state even more. On the other hand, if the AlGaN layer is grown on AlN buffer layer or grown on high Al composition layer; where the compressive strain has occurred, the $|Z\rangle$ state will move down and $|X\rangle$ and $|Y\rangle$ will move up. When the AlGaN layer is grown on AlN buffer layer under the compressive strain $|X\rangle$ and $|Y\rangle$ will remain to be the top valence band when the Al composition is below 60%–70%. Of course, when the quantum confined effect is considered in QW, the movement is even more complicated. Using the finite difference method to solve the band structure in the QW case is most common method to analyze the strain in quantum confined systems.

For the semipolar plane case, when the crystal orientation rotates to the primed system, k_x, k_y and k_z need to be rotated as well in the Hamiltonian. The relation of the new coordinates and the old coordinates can be rotated by the rotation matrix U_{ij} as mentioned in Equation 4.8. The relation of k to k' can be expressed as

$$k_i' = \sum_\alpha U_{i\alpha} k_\alpha \qquad (4.21)$$

If x-axis is the rotation axis, at inclined angle θ

$$k_x' = k_x$$

$$k_y' = \cos\theta k_y - \sin\theta k_z \qquad (4.22)$$

$$k_z' = \sin\theta k_y + \cos\theta k_z$$

after replacing the k_x', k_y' and k_z' into the 6×6 Hamiltonian and strain term as derived in Equation 4.14. We can solve the Hamiltonian by an eigenvalue solver. For simpler nonpolar cases, the growth direction becomes the original y direction and the *c*-plane and *a*-plane become the lateral directions. At the

Γ center, Hamiltonion in the Equation 4.19 will be similar. However, here the ϵ_{xx} and ϵ_{zz} will be equal to $\epsilon_{\parallel}$ and ϵ_{yy} and will become $\epsilon_{\perp}$. If we check the Equations 4.19 and 4.20, it will become

$$H^v u = \begin{pmatrix} F & 0 & 0 & 0 & K^* & 0 \\ 0 & G & \Delta & 0 & 0 & K^* \\ 0 & \Delta & \lambda & 0 & 0 & 0 \\ 0 & 0 & 0 & \lambda & \Delta & 0 \\ K & 0 & 0 & \Delta & G & 0 \\ 0 & K & 0 & 0 & 0 & F \end{pmatrix} \begin{pmatrix} |u_1\rangle \\ |u_2\rangle \\ |u_3\rangle \\ |u_4\rangle \\ |u_5\rangle \\ |u_6\rangle \end{pmatrix} \tag{4.23}$$

$$F = \Delta_1 + \Delta_2 + (D_1 + D_2 + D_3 + D_4)\epsilon_{\parallel} + (D_2 + D_4)(\epsilon_{\perp})$$

$$G = \Delta_1 - \Delta_2 + (D_1 + D_2 + D_3 + D_4)\epsilon_{\parallel} + (D_2 + D_4)(\epsilon_{\perp})$$

$$K = D_5(\epsilon_{xx} - \epsilon_{yy}) = D_5(\epsilon_{\parallel} - \epsilon_{\perp})$$

$$\lambda = (D_1 + D_2)\epsilon_{\parallel} + D_2(\epsilon_{\perp})$$

$$\theta = D_3\epsilon_{zz} + D_4(\epsilon_{xx} + \epsilon_{yy}) \tag{4.24}$$

$$K = H = I = 0$$

$$\Delta = \sqrt{2}\Delta_3$$

As you can see in Equation 4.24, at Γ center, the nonzero term K appears at the nonpolar case. The K term is the coupling between the $|u_1(|X + iY, \uparrow\rangle)$ and $|u_5(|X - iY, \uparrow\rangle)$ or the coupling between the $|u_2(|X + iY, \downarrow\rangle)$ and $|u_6(|X - iY, \downarrow\rangle)$. Basically, this term will break the symmetric and decouple $|X + iY\rangle$ into pure $|X\rangle$ and $|Y\rangle$ state. For the c-plane case, the $|Z\rangle$ state suffers the tensile strain and it will move down. Therefore, in this case, the $|Y\rangle$ state ($|Z'\rangle$ states) is along the growth and suffers the tensile strain and $|X\rangle$ state has the compressive strain as the c-plane case. The $|Z\rangle$ state (now the Y' state) also suffers compressive strain. Therefore, $|X\rangle$ (X') state will move up as the first top band. The $|Z\rangle$ state (Y' state) also moves up to be the second subband and the orginial $|Y\rangle$ state (Z') move down to be the third band as shown in Figure 4.3c. For the semipolar plane case, it is more complicated. Therefore, we can solve these problems by solving $k.p$ Hamiltonian numerically.

Figure 4.3 shows the calculated E-k relation of the bulk $In_{0.2}Ga_{0.8}N$ material under the same lateral compressive strain in different growth orientations from our group published in Reference [31]. Note that this result does not reflect the influence of the quantum confined effect but only considers the influence of strain deformation potential. The red, green, and blue (RGB) colors mean the strength of $|X'$-, $|Y'$-, and $|Z'\rangle$-like states in the primed system, respectively. If the states are mixed, the color will change according to the RGB strength.

Figure 4.3a shows the E-k relation of the bulk c-plane InGaN material without strain. As we can see at the center of the Γ valley, the top two bands are mixed with $|X\rangle$ and $|Y\rangle$ states, which are usually denoted as $|X \pm iY\rangle$ states, while the $|Z\rangle$ state is the third VB. As we have demonstrated in Equations 4.4 and 4.5, when we apply the biaxial compressive strain in the lateral direction as shown in Figure 4.3b, the $|X \pm iY\rangle$ states remain as the top two bands and the $|Z\rangle$ state, however, moves to even lower position. For the m-plane case as shown in Figure 4.3c, as we have mentioned in Equation 4.14. The top band becomes $|X'\rangle$ state. For the semipolar plane case, such as $(11\bar{2}2)$ as shown in Figure 4.3d at $\theta = 58°$ the top VB will be still mainly to be (X') state. The second and third states are basically mixed with $|X$ and $|Z\rangle$-like states (Y' and Z') depending on the rotation angle. Hence, to realize an accurate estimate of the VB energies for different template orientation angles, especially for QWs, we need to solve the $k \cdot p$ equations for each case and include quantum confinement. More detail can be found in Reference [31,33].

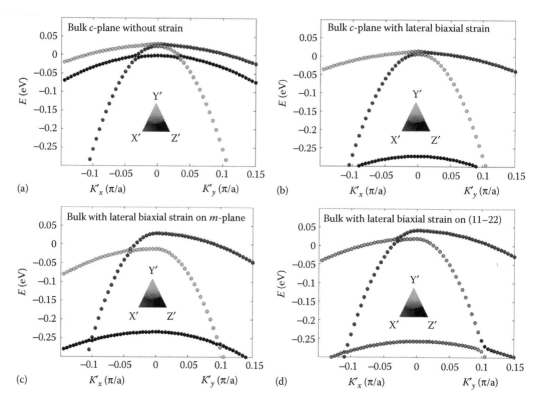

FIGURE 4.3 (a) The calculated valence band *E-k* relationship for bulk *c*-plane In$_{0.2}$Ga$_{0.8}$N without strain. (b–d) are the calculated *E-k* relationship for bulk In$_{0.2}$Ga$_{0.8}$N under biaxial strain for the *c*-plane, *m*-plane, and (1122) plane, respectively. The RGB color represents the strength of X′, Y′, and Z′ states, respectively. Note that for *c*-plane, X′ = X, Y′ = Y, and Z′ = Z. (From Yuji, Z. et al., *Jpn. J. Appl. Phys*, 53, 100206, 2014. With permission.)

In addition, the effective mass in the different direction will also change with the strain applied. In the quantum well condition, the freedom of carrier is limited to the lateral direction and 2D density of state should be considered. Therefore, the m$^*_{DOS,2D}$ for a 2D system should be

$$m_{DOS,2D} = \sqrt{m^*_{x'} m^*_{y'}} \tag{4.25}$$

For the transport problem, the effective mass for the mobility calculation should be relied on the effective mass of the transport direction (m$^*_T$), and the mobility would be decided by

$$\mu = \frac{e\tau}{m^*_T} \tag{4.26}$$

However, the mobility will also rely on the relaxation time $\langle \tau \rangle$ and in the next section, we will discuss how the τ is calculated in the nitride system in the next section.

4.4 Transport Properties of III-N Semiconductor

As we have mentioned, the wideband material such as GaN and AlN are very potential candidates for high power and high-frequency transistor. Due to the wide bandgap, the breakdown electric field of GaN could be at least higher than 3 MV/cm. And the breakdown electric field of AlN could be over 10 MV/cm. Since GaN is a direct bandgap material, the electron effective mass is around 0.2 m$_0$, which

TABLE 4.1 The Basic Parameters of Nitride Materials

	GaN	AlN	InN
Bandgap (eV)	3.4	6.2	0.7
$m^*_{e\perp}$	0.21	0.32	0.07
$m^*_{e,\parallel}$	0.20	0.30	0.07
$m^*_{hh,\parallel}$	1.87	2.68	1.61
$m^*_{hh,\perp}$	1.87	2.68	1.61
$m^*_{lh,\parallel}$	0.14	0.26	0.11
$m^*_{lh,\perp}$	0.14	0.26	0.11
P_{sp}	−0.034	−0.090	−0.042
P_{sp} bowing term (alloy to GaN)		0.021	0.037
e_{13}	0.73	1.5606	0.73
e_{33}	−0.49	−0.5361	−0.49
e_{15}/e_{14}	−0.40	0.4176	−0.40

Source: Ambacher, O. et al., *J. Phys. Condens. Matter*, 14, 3399–3434, PII S0953-8984(02)29173-0, 2002; Ambacher, O. et al., *J. Appl. Phys.*, 87, 334–344, 2000; Vurgaftman, I. and Meyer, J.R., *J. Appl. Phys.*, 94, 3675–3696, 2003; Vurgaftman, I., Meyer, J.R., and Ram-Mohan, L.R., *J. Appl. Phys.*, 89, 5815–5875, 2001.

is very competitive to Si-based technology. The calculation and experimentshow that GaN's mobility can be over 1500 cm²/Vs under the 2DEG form at the AlN/GaN interface. Currently, the nitride-based LEDs have been widely used in solid state lighting. Therefore, the technology for growing GaN-based epistar has become more mature than any other semiconductor materials, except the Si industry. Therefore, this makes nitride-based material to be the most potential semiconductor material to be applied in the field where Si cannot be used, especially for high-power applications. Although GaN has a relatively small electron effective mass at this wide bandgap, the hole-effective mass is relatively high. As shown in Table 4.1, the CH1 band (HH)'s effective mass is close to 1.81 m_0. Although in different crystal orientation, the effective mass will vary from 1.1 to 1.8 m_0. All these ranges are still quite high for the hole transport. Therefore, even in the low doping condition, the hole mobility is estimated to be around 100 cm²/Vs. The typical mobility of heavy p-type-doped GaN material is only around 2–10 cm²/Vs. Currently, GaN material is mainly used for HFET transistor. The major applications are high-power and high-frequency devices due to their wide bandgap. Although some efforts have been put on HBT devices, the high activation energy of Mg and low hole mobility limit the device performance.

4.4.1 AlGaN/GaN Heterojunction Field Effect Transistor

AlGaN/GaN heterojunction field effect transistor (some also called high electron mobility transistor, HEMT) has attracted significant attention due to their potential in high-power applications. As mentioned earlier, due to the large bandgap, GaN-based HFET has a very large breakdown electric field (>3 MV/cm). For a proper design, the operation voltage could be very high (>1000 V). In addition, GaN material can be operated in very high temperatures due to its wide bandgap, crystal stability, and strength. Besides these material advantages, the most attractive feature is the formation of positive polar charges at AlGaN/GaN interface, which attracts a huge amount of 2DEG to GaN channel at the AlGaN/GaN.

As shown in Figure 4.4, the AlGaN/GaN interface is formed and the surface is in contact with metal. The Schottky contact will be formed at the metal-semiconductor interface. According to the Equation 4.7, for the AlGaN/GaN interface grown in the *c*-axis direction, the polarization charge at the interface would be

$$\Delta P_z = P_S^{sp} - P_L^{pz} - P_L^{sp} \tag{4.27}$$

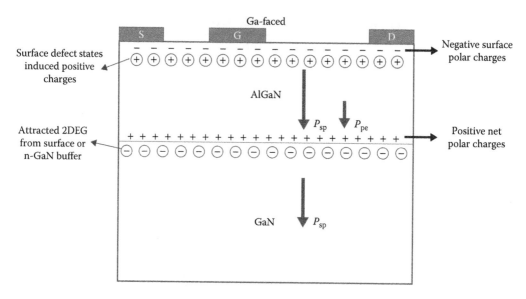

FIGURE 4.4 The formation of 2DEG at AlGaN/GaN interface.

For the AlGaN/GaN interface, according to Reference [27] and Table 4.1, the derive ΔP_z will be

$$P_L^{pz}\left(\frac{Al_xGa_{1-x}N}{GaN}\right)=\left[-0.0525x+0.0282x(1-x)\right]\left(C\cdot m^{-2}\right)$$

$$P_L^{sp}\left(Al_xGa_{1-x}N\right)=-0.090x-0.034(1-x)+0.021x(1-x)$$

$$\Delta P_z(x)=+0.090x-0.021x(1-x)+0.0525x-0.0282x(1-x)$$

$$=0.1425x-0.0492x(1-x)=0.0933x+0.0492x^2$$

For example, for a 30% Al composition AlGaN grown on the GaN, the spontaneous and piezoelectric polarization field would be 0.0562 C/m². The electric field would be 6.2 MV/cm, which is already larger than the breakdown electric field. At this large polarization field as shown in Figure 4.5a without carrier screening, the potential difference at 20 nm AlGaN layer could be over 10 eV, which is very large. In addition, at the AlGaN and GaN interface, the calculated $\Delta P_z(0.3)$ would be close to $2.0 \times 10^{13}\,cm^{-2}$. This is a very large sheet charge density at AlGaN/GaN interface. Some studies show that the measured polarization charge might be smaller due to the nonlinear effect or random alloy fluctuation [4,34]. However, the sheet charge density over $10^{13}\,cm^{-2}$ has been confirmed. At this large polarization field, the system will try to minimize their energy. As shown in Figure 4.5b, At the AlGaN/GaN interface, the formed positive polarized sheet charge will attract almost same amount 2DEG either from bulk-ntype GaN buffer layer or from top AlGaN/air surface to screen this large electric field induced band bending. For the AlGaN/air or AlGaN/metal surface, the physical behavior would be more complicated. For the AlGaN layer, including the spontaneous polarization and piezoelectric polarization, the surface polarization charge at air/AlGaN interface is negative charges. These charges will be screened by oppositely charged surface defects [35–37] or the absorbed charges from the environment such as ionized gas, and so on as shown in Figure 4.5b. For the metal/AlGaN interface, due to the high surface state density, the Fermi-level pinning is very clear. For the GaN, the Fermi level is usually pinned at the 1.1 eV below the conduction band and for AlGaN, the typical pinning position is 1.3 to 1.5 eV. Therefore, the influence of metal work function is weak since the defects, polarization charge, and metal electron gas will find the most balanced position.

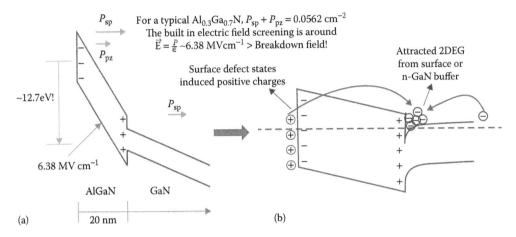

For a typical $Al_{0.3}Ga_{0.7}N$, $P_{sp} + P_{pz} = 0.0562$ cm^{-2}
The built in electric field screening is around
$\vec{E} = \frac{P}{\epsilon} \sim 6.38$ MVcm^{-1} > Breakdown field!

FIGURE 4.5 (a) The potential band bending caused by the polarization field without any screening. For the $Al_{0.3}Ga_{0.7}N$ layer, the built-in electric field could be 6.2 MV/cm, which is larger than the breakdown electric field. If the AlGaN layer is 20 nm, the potential difference at two sites could be 12.7 eV. (b) The system cannot tolerate this large electric field. The surface defect state will induce positive defect charge to compensate this large electric field. The disappeared electron charges might go to AlGaN/GaN interface to form the 2DEG.

When the surface Fermi level is pinning at defect center, the positive polarization charge at AlGaN/GaN interface will start to put down the potential. When the AlGaN layer is thick enough, the conduction band at GaN is lower than the Fermi level, the 2DEG will start to be attracted to the interface. Figure 4.6 shows the AlGaN thickness versus 2DEG density. When the AlGAN is more than 3 nm, the 2DEG will start to accumulate. When the layer is thicker and thicker, the increase rate will start to saturate since the attracted 2DEG will screen all the polar charges. When the 2DEG is formed at the AlGaN/GaN interface due to the attraction of positive fixed polar charge, it will significantly reduce the access resistance in the channel. As shown in Figure 4.4, the typical AlGaN/GaN HFET constructed with a drain, source, and gate region. Since the 2DEG is induced by polarization charges instead of doping, there is no ionized impurity scattering in the channel. In addition, due to the high sheet charge density (>10^{13} cm^{-2}) as shown in Figure 4.6, the carrier will screen the potential fluctuation caused by dislocation and defects. Therefore, the mobility of 2DEG in the GaN channel could be over 1,500 cm^2/Vs at this high-carrier density condition. This would be excellent for high-power and high-speed applications due

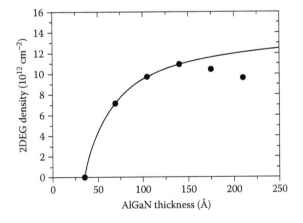

FIGURE 4.6 Room temperature 2DEG density measured as a function of $Al_{0.34}Ga_{0.64}N$ barrier thickness. (From Ibbetson, J.P. et al., *Appl. Phys. Lett.*, 77, 250–252, 2000. With permission.)

to the low scattering rate with high speed and huge carrier density for high-current density. Therefore, many works have been put to push GaN HEMT device in RF applications [16,19,21,39–42]. However, due to the scaling ability in Si continues to improve and the 10 nm gate length CMOS technology has started to ship, many RF technology has been replaced by Si CMOS. The possible applications of nitride in RF applications are mainly limited to military or RF base station. On the other hand, the high-power applications are the major area for potential applications. Due to the large sheet carrier density and high breakdown field, the GaN HFET has been a very strong candidate for these applications. Due to the increasing demands on power convertors such as electric cars, solar power plant, smart grid, and so on, we will need a high-current density power convertor with very high PAE, where GaN has demonstrated their ability to meet these requirements.

4.4.2 Current Collapse and Surface Passivation

As we have mentioned, the surface state at the air/AlGaN surface will not only neutralize the negative polarization charge, it will also act as a defect trapping center. The electrons, to neutralize the negative polar charge, will leave from the surface to either AlGaN/GaN interface to be the source of 2DEG or other places. Since these surface states are empty, they are easy to trap electrons again. Therefore, when the device is under operation especially at high voltage, the electron leaking from gate contact will start to leak through the surface and will be trapped by these empty defects. Since these kinds of hopping behavior through defect state is very low, it is hard to be reversed. As a result, the trapped electron at surfaces will expel the 2DEG in the channel and increase the channel resistance as shown in Figure 4.7. Therefore, the surface treatment has been a critical issue for AlGAN/GaN HFET applications. Researchers found that using the SiN$_x$ to passivate the AlGaN surface has the better performance compared to other passivation materials [42–49]. However, the trapping of the electron on the surface state can only be minimized. It cannot be totally removed. In addition, although SiN$_x$ is a good material for passivation, it is not good to act as an insulator. The typical electron affinity of SiN is around 2.9 to 3.0 eV that cannot form a large Schottky barrier, to stop the leakage of current. However, other large band-gap materials such as SiO$_2$ are Al$_2$O$_3$ are not good passivation materials [42–49]. Therefore, multilayer

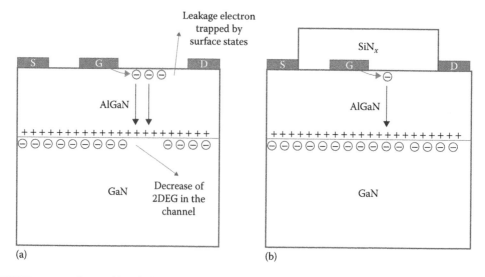

(a)   (b)

FIGURE 4.7 A Schema of how leakage trapping at the surface influences the 2DEG in the channel and leads to the current collapse. In (a), there is no SiNx passivated at AlGaN/air interface. The surface state density is much higher. Carriers are easier to be trapped at the surface, pulled up the surface potential and decrease the 2DEG in the Channel. In (b), surface states decrease so that the leakage current and the decrease of 2DEG is less.

structure for MOISFET configuration is used to reduce the gate leakage. For example, metal/Al$_2$O$_3$/SiNx/AlGaN/GaN MOISFET has been used to improve the leakage and reduce the surface trapping at the same time. Besides using the oxide as the gate insulator, the p-GaN layer below the gate metal is also proposed to be a method to improve the leakage issues. Researchers found that putting a p-GaN layer on top of AlGaN layer can improve the leakage behavior. It suggests that the Mg acceptor will fill those trap centers first and them form the hole carriers. The p-GaN layer can also prevent the electron leakage since it increases the barrier for the electron to be injected into the Channel.

4.4.3 Passive Mode and Enhance Mode

Due to the existence of 2DEG, the channel in the AlGaN/GaN HFET is already turned on at the 0 V gate voltage. Therefore, to turn off the GaN HFET, the negative bias is needed. For the typical AlGaN/GaN HFET, the typical turn-off voltage is −3 to −5 V. If the oxide insulator is used, due to the increased thickness and reduced capacitance, the applied negative bias could be even higher. The 2DEG in the channel provides low access resistance to the channel. However, the device needs to be operated in the passive mode. However, there is a critical demand for enhanced mode HFET for safety issues. The device needs to be normally off when the gate is unbiased. To achieve this, the 2DEG in the channel needs to be depleted at 0 V. This contradicts the advantage of 2DEG. If we deplete the 2DEG in the whole channel, the access resistance will increase significantly so that the one advantage of AlGaN/GaN HFET will be lost. Therefore, the best condition is that we only deplete the 2DEG under the metal gate.

As shown in Figure 4.6, the AlGaN layer thickness is below 4 nm, the 2DEG in the channel will be significantly depleted. Therefore, the simple idea is to recess the gate as shown in Figure 4.8a [50–53]. When the AlGaN layer under the metal gate region is recessed, the 2DEG below the recess region will be depleted. Therefore, the Vth will move toward the positive voltage. However, using recess technology will reduce the AlGaN barrier thickness and increase the possibility of tunneling leakage. The recess gate structure is often used in high-speed RF applications since it will increase the gate control ability, and reduce the short channel effects. However, with the high-power applications where the large drain bias is applied, the leakage issue will be a serious problem. Some papers even recess the gate to the GaN layer and use the oxide as the gate insulator to improve the leakage issues [54,55].

The other way to achieve the enhanced mode is using the p-GaN layer as shown in Figure 4.8b [44,50,56,57]. As mentioned, the activated p-GaN can passivate the surface. It will also ping the Fermi level at the valence band edge. Therefore, this lifts up the conduction band potential and induces the depletion field. The p-GaN layer will also reduce the 2DEG in the whole channel. Therefore, etching the

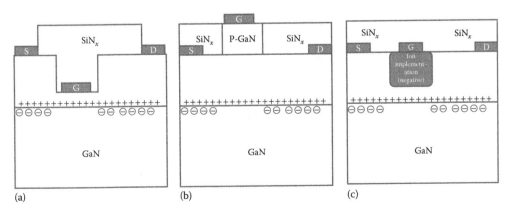

FIGURE 4.8 (a) The recess gate structure. (b) p-GaN layer HFET. (c) Using ion implementation technology to implement negative ionized charges.

p-GaN layer in the ungated region would be a proper choice to keep the high 2DEG and high-current density. The third method is to ion implement under the metal gate region. Materials such as CF4 [58] treatment is widely used to provide the negative ion. However, how to avoid the deterioration of crystal quality of AlGaN layer would be critical.

4.4.4 Mobility and Velocity-Field Curve

The main application for AlGaN HFET is focusing on the high-power and high-speed devices. As we can notice, the typical operation voltage at high-power electronics is over 100 V or even 1,000 V. It can be expected that the electric field in the AlGaN HFET would be very high. For example, the average electric field for 1 μm gate length at $V_D = 100$ V would be 10^6 V/cm, which is near the breakdown field. Actually, when the device is operated at saturation region, most electric fields will be fallen at the drain side near the gate edge. Therefore, the local electric field could be even higher. Figure 4.9 shows the 2D plot potential of 5 μm gate length AlGaN/HFET and electric field (E_x) along the channel at $V_D = 20$ V and $V_G = -3$ V. Although the average electric field at this 5 μm gate length case is 4×10^4 V/cm. The electric field peak in the channel could be as high as 6×10^5 V/cm. Therefore, the breakdown voltage in the device will actually happen earlier. If we assume the breakdown field to happen at electric field equal to 3×10^6 V/cm, we will predict that breakdown voltage is over 1,000 V. However, the breakdown will appear near 100 V in this case. Therefore, the actual electric in the channel is much higher than the average electric field.

At this high electric field, the mobility in the channel will no longer be constant. For example, if the mobility is always 1,500 cm²/Vs at a high electric field and the electron velocity is equal to $v = \mu \bar{E}$. The electron speed would be near 1.5×10^9 cm/s, which is impossible since this is almost 1/20 of the speed of light. Actually, under this high field condition, the scattering rate in the device will increase. The equivalent effective mass at high energy region will also become much larger to slow down the electron speed. Therefore, the carrier velocity at the high field region is no longer a linear effect to the electric field.

To study this problem, Monte Carlo simulation [59–64] has been the major tool in modeling these transport issues. In the Monte Carlo simulation, the carrier is considered as a point particle whose scattering rates are given by the Fermi's golden rule expressions. Carriers transport in Monte Carlo techniques is

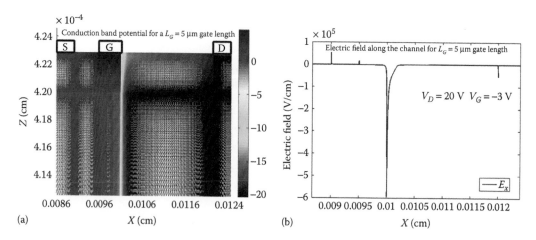

FIGURE 4.9 (a) The 2D conduction band potential distribution for a AlGaN/GaN HFET with $L_G = 5$ μm at $V_G = -3$ V, $V_D = 20$ V. (b) The electric field (E_x) along the AlGaN/GaN interface for the same case.

viewed as a series of free flight and scattering events. During each free flight time, the scattering rates will be calculated, and the scattering events are treated as instantaneous and decided by the probability with random number. The simulation involves the following stages:

1. Calculate the different scattering rates for intra and inter subband scattering. If there is intervalley scattering such as Γ to L-M valley. This is also considered for the nonequivalent intervalley scattering.
2. Particle injection into the channel: The particles are injected with a pre-chosen distribution of carrier momenta or decided by thermal distribution.
3. Free flight of carriers: The scattering event is considered to be instantaneous. Between scattering processes, the electron simply acclerates in the electric field with the change of k vector by $F = \hbar dk/dt$.
4. Scattering Events: A specific prescription is used in Monte Carlo method to determine the time between scattering events. Scattering rates of different scattering mechanism are calculated by Fermi's golden rule. The free flight time is decided by the total scattering rate. At the end of a free flight, scattering occurs, which alters the flight pattern of the electron.
5. Selection of scattering events: Random choice of scattering events according to the scattering rates calculated in step 4. This choice is again based on the Monte Carlo method.
6. State of electron after scattering: Use Monte Carlo to determine the momentum of electron immediately after the collision has occurred. The duration is assumed to be zero. To determine the final state of electron, we needed detailed information on the scattering process.

Steps 3–6 are repeated until the end of the simulation. The Monte Carlo simulation takes the following scattering mechanisms into account:

1. Polar optical phonon absorption and emission;
2. Acoustic phonon scattering;
3. Interface roughness scattering;
4. Equivalent and nonequivalent inter-valley scattering;
5. Alloy scattering, and;
6. Charged dislocation scattering.
7. Ionized impurity scattering if the doping exists.

The scattering rate can be calculated by Fermi's golden rule, which is given by

$$M_{k,k'} = k \left| H_{\text{pert}} \right| k' \tag{4.28}$$

$$W_i(k,k') = \frac{2\pi}{\hbar} \sum_{k'} \left| M_{k,k'} \right|^2 \delta(E_f - E_i) \tag{4.29}$$

where:
H_{pert} is the perturbation Hamiltonian for the interaction
k' is the index over all final states

The resultant wave functions from solution of Schrodinger equations are then used to calculate overlap integrals, $I_{k,k'}$, which are used in the calculation of the intersubband matrix elements.

$$I_{k,k'} = \int_{\text{Entire device}} \phi_k^*(z) \phi_{k'}(z) dz \tag{4.30}$$

134 *Handbook of GaN Semiconductor Materials and Devices*

In the confined three-dimensional or quasi two-dimensional structures, the matrix elements used in Equation 4.30 would be slightly different. The matrix element is calculated by integrating an overlap integral term with the 3D matrix element. For example, for phonon scattering, we have

$$I_{k,k'}(q) = \int\limits_{\text{Entire device}} \phi_k^*(z)\phi_{k'}(z)\,dz \tag{4.31}$$

$$|M_{k,k'}|^2 = |M(Q,q)|^2 |I_{kk'}|^2\,dq$$

where Q and q are components of *phonon* wave vectors parallel and perpendicular to the heterointerface, respectively.

In the quasi-2D Monte Carlo approach, the electron is treated as a particle in the parallel direction. In the vertical direction, the electron is treated as an envelope wave function. For the heterostructure like AlGaN/GaN or InN/GaN, the wave function is localized at the heterojunction interface. For higher subband, the envelope will spread into GaN or AlGaN region. To calculate the scattering rate for a state that overlaps several different regions, we need to use a reasonable averaging procedure.

For different purposes, the Monte Carlo Simulation is separated into steady state Monte Carlo simulation and ensemble Monte Carlo simulation. For the steady state simulation, one electron is injected into the device for a very long period of simulation for certain electric field until it reaches the steady state velocity. The resultant E-v curve is used to deduce the low-field drift mobility μ_l and be used to solve the drift-diffusion equation. Figure 4.10 shows the calculated v-E field curve calculated by the steady state Monte Carlo simulation results [65,66] under different doping density.

As shown in Figure 4.10, at low-field region, the velocity field curve is linear. When the electric field is near 3×10^4 V/cm, there is a kink point. Since the polar optical phonon's energy is near 90 meV, the polar

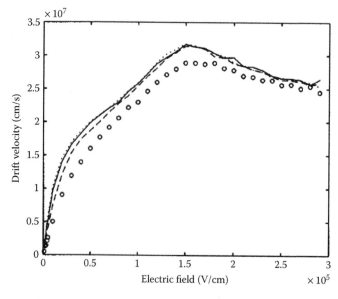

FIGURE 4.10 Doping dependence of velocity-field characteristics of GaN, $T = 300$ K. Dotted line, $N_d = n = 10^{16}$ cm^{-3}, without ionized impurity scattering; line, $N_d = n = 10^{16}$ cm^{-3}; dashed line, $N_d = n = 10^{17}$ cm^{-3}; circles, $N_d = n = 10^{18}$ cm^{-3}. (From Bhapkar, U.V. and Shur, M.S., *J. Appl. Phys.*, 82, 1649–1655, 1997; Zhang, Y. and Singh, J., *J. Appl. Phys.*, 89, 386–389, 2001. With permission.)

optical phonon emission scattering just turns on at this electric condition. This increases the scattering rate significantly to cause this nonlinear *v-E* curve [67–70].

For ensemble Monte Carlo simulation, a large number of electrons (typically ~200 k) are injected into real device potential profiles to calculate the average transient time of the electrons through the potential profile. The distribution of carrier density and carrier velocity information can then be a feedback to the Poisson solver to recalculate the potential. These two solvers can couple together and self-consistently solve the problem. However, the convergence issue is another factor to be considered.

4.4.5 Field Plate for High Power Applications

As mentioned in earlier section, the electric field in the nitride HFET is not uniformly distribution distributed in the channel. The electric field is usually from near the drain side with a limited region. This depletion region area side would be the key factor to limit the breakdown field. The way to increase the depletion area size would be to simply reduce the 2DEG density and increase the sheet resistance in the channel so that more potential will drop in all areas to make the electric field distributed at wide region. However, the whole channel resistance is increased and the current will drop significantly. Therefore, the required high-current density at high voltage cannot be achieved. The alternated way to improve this problem is to add an extra gate as the field plate. The field plate has a thickness insulator layer thickness as shown in Figure 4.11. Therefore, it pinches off the channel at a much larger negative bias. At the small drain bias, the channel is pinched off by the original gate. For example, if the *Vth* of the original gate is −5 V and $V_G = 0$ V, when V_D increases to 5 V, and the V_{GD} is close to −5 V, the 2DEG in the channel under the gate edge near the drain side is depleted. Therefore, the current starts to saturate and the saturation behavior can be observed. For the field plate, whether its bias is controlled by the source or gate voltage, it has a much thicker layer thickness. The typical thickness could be 60–100 nm SiN or Al_2O_3 material. Therefore, the pinched off bias could be <−30 to −50 V or even smaller bias. When the drain bias goes to 50 V, V_{GD} or V_{SD} would be <−50 V. Therefore, the 2DEG in the channel under the field plate region will start to become depleted. Then the electric field will be fallen in the wider region and make the breakdown voltage larger. If we want to apply an even large drain bias, multiple field plate could be used with different layer thickness. However, in the high-frequency applications, the equivalent gate length would be larger so that delay of response would be expected.

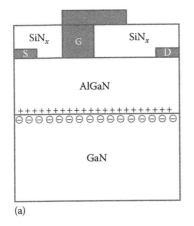

(a)

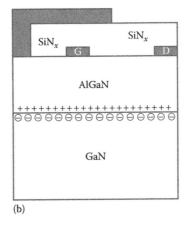

(b)

FIGURE 4.11 The design concept of field plate. (a) The field plate that connects to the gate. (b) The field plate that connects to the source.

4.5 Summary

In this chapter, we basically discuss the material properties of nitrides including the band structure modification through strain, alloy, and quantum system. The transport issues of GaN have also been discussed. In this chapter, the design issue for AlGaN/GaN HFET has been discussed in detail. The influences of surface trap to the leakage, the polarization effects, surface passivation, and velocity saturation issues are the key factors to be overcome.

References

1. Le, L. et al., Utilization of polarization-inverted AlInGaN or relatively thinner AlGaN electron blocking layer in InGaN-based blue–violet laser diodes. *J. Vac. Sci. Technol. B*, **33**(1):011209, 2015.
2. Matin, K., Y. Zheng, and A. Bar-Cohen, Numerical modeling and simulation of laser diode diamond microcoolers. In *2014 IEEE Intersociety Conference on Thermal and Thermomechanical Phenomena in Electronic Systems (ITherm)*, Orlando, FL, May 2014.
3. Lermer, T. et al., Gain of blue and cyan InGaN laser diodes. *Appl. Phys. Lett.*, **98**(2):021115, 2011.
4. Wu, C.-K., C.-K. Li, and Y.-R. Wu, Percolation transport study in nitride based LED by considering the random alloy fluctuation. *J. Comput. Electron.*, **14**(2):416–424, 2015.
5. Yang, T.-J. et al., The influence of random indium alloy fluctuations in indium gallium nitride quantum wells on the device behavior. *J. Appl. Phys.*, **116**(11):113104, 2014.
6. Chen, C.-Y. and Y.-R. Wu, Studying the short channel effect in the scaling of the AlGaN/GaN nanowire transistors. *J. Appl. Phys.*, **113**(21):214501, 2013.
7. Li, C.-K. et al., Three dimensional numerical study on the efficiency of a core-shell InGaN/GaN multiple quantum well nanowire light-emitting diodes. *J. Appl. Phys.*, **113**(18):183104, 2013.
8. Wu, Y.-R. et al., Analyzing the physical properties of InGaN multiple quantum well light emitting diodes from nano scale structure. *Appl. Phys. Lett.*, **101**(8):083505, 2012.
9. Li, C.-K. and Y.-R. Wu, Current spreading effect in vertical GaN/InGaN LEDs. *Proceedings of the SPIE 7939, Gallium Nitride Materials and Devices VI, p. 79392K* (March 03, 2011); doi:10.1117/12.877663; http://dx.doi.org/10.1117/12.877663.
10. Zhao, Y. et al., Atomic-scale nanofacet structure in semipolar (20 bar 2} bar 1}) and (20 bar 2} 1) InGaN single quantum wells. *Appl. Phys. Express*, **7**(2):025503, 2014.
11. Masui, H. et al., Light-polarization characteristics of electroluminescence from InGaN/GaN light-emitting diodes prepared on (1122)-plane GaN. *J. Appl. Phys.*, **100**(11):113109, 2006.
12. Masui, H. et al., Polarized light emission from nonpolar InGaN light-emitting diodes grown on a bulk m-plane GaN Substrate. *Jpn. J. Appl. Phys.*, **44**(43):L1329–L1332, 2005.
13. Mukai, T., K. Takekawa, and S. Nakamura, InGaN-based blue light-emitting diodes grown on epitaxially laterally overgrown GaN substrates. *Jpn. J. Appl. Phys.*, **37 Part 2**(7B), 1998.
14. Nakamura, S. et al., Superbright green InGaN single-quantum-well-structure light-emitting diodes. *Jpn. J. Appl. Phys.*, **34 Part 2**(10B), 1995.
15. Chakraborty, A. et al., Interdigitated multipixel arrays for the fabrication of high-power light-emitting diodes with very low series resistances. *Appl. Phys. Lett.*, **88**(18):181120, 2006.
16. Palacios, T. et al., AlGaN/GaN high electron mobility transistors with InGaN back-barriers. *IEEE Electron Dev. Lett.*, **27**(1):13–15, 2006.
17. Rajan, S. et al., Electron mobility in graded AlGaN alloys. *Appl. Phys. Lett.*, **88**(4): p. 042103, 2006 doi: http://dx.doi.org/10.1063/1.2165190.
18. Chakraborty, A. et al., Defect reduction in nonpolar a-plane GaN films using in situ SiN_x nanomask. *Appl. Phys. Lett.*, **89**(4):041903, 2006.
19. Palacios, T. and U.K. Mishra, Improved technology for high frequency AlGaN/GaN HEMTs. *ONR CANE/MURI Rev.*, 2005.

20. Shen, L. et al., Unpassivated GaN/AlGaN/GaN power high electron mobility transistors with dispersion controlled by epitaxial layer design. *J. Electron. Mater.*, **33**(5):422–425, 2004.

21. Singh, M., et al. Monte Carlo study of noise scaling in AlGaN/GaN HFETs. In *27th International Conference on the Physics of Semiconductors*, Flagstaff, AZ, 2004.

22. Chini, A. et al., 2.1 A/mm current density AlGaN/GaN HEMT. *Electro. Lett.*, **39**(7):625–626, 2003.

23. Mishra, U.K., P. Parikh, and Y.F. Wu, AlGaN/GaN HEMTs—An overview of device operation and applications. *Proc. IEEE*, **90**:1022–1031, 2002.

24. Wu, Y.F. et al., Very-high power density AlGaN/GaN HEMTs. *IEEE Trans. Electron Dev.*, **48**:586–590, 2001.

25. Mishra, U.K. et al., GaN microwave electronics. *IEEE Trans. Microwave Theory Tech.*, **46**(6):756–761, 1998.

26. Delaney, K.T., P. Rinke, and C.G. Van de Walle, Auger recombination rates in nitrides from first principles. *Appl. Phys. Lett.*, **94**(19):191109, 2009.

27. Ambacher, O. et al., Pyroelectric properties of Al(In)GaN/GaN hetero- and quantum well structures. *J. Phys. Condens. Matter*, 14:3399–3434, PII S0953-8984(02)29173-0, 2002.

28. Ambacher, O. et al., Two dimensional electron gases induced by spontaneous and piezoelectric polarization in undoped and doped AlGaN/GaN heterostructures. *J. Appl. Phys.*, **87**(1):334–344, 2000.

29. Vurgaftman, I. and J.R. Meyer, Band parameters for nitrogen-containing semiconductors. *J. Appl. Phys.*, **94**(6):3675–3696, 2003.

30. Vurgaftman, I., J.R. Meyer, and L.R. Ram-Mohan, Band parameters for III–V compound semiconductors and their alloys. *J. Appl. Phys.*, **89**(11):5815–5875, 2001.

31. Yuji, Z. et al., Valence band states and polarized optical emission from nonpolar and semipolar III–nitride quantum well optoelectronic devices. *Jpn. J. Appl. Phys*, **53**(10):100206, 2014.

32. Romanov, A.E. et al., Strain-induced polarization in wurtzite III-nitride semipolar layers. *J. Appl. Phys.*, **100**(2):023522, 2006.

33. Huang, H.-H. and Y.-R. Wu, Light emission polarization properties of semipolar InGaN/GaN quantum well. *J. Appl. Phys.*, **107**(5):053112, 2010.

34. Yang, T.-J., J.S. Speck, and Y.-R. Wu, Influence of nanoscale indium fluctuation in the InGaN quantum-well LED to the efficiency droop with a fully 3D simulation model. In *SPIE OPTO*, San Francisco, CA, February 2014.

35. Eller, B.S., J. Yang, and R.J. Nemanich, Polarization effects of GaN and AlGaN: Polarization bound charge, band bending, and electronic surface states. *J. Electron. Mater.*, **43**(12):4560–4568, 2014.

36. Segev, D. and C.G.V.d. Walle, Origins of fermi-level pinning on GaN and InN polar and nonpolar surfaces. *EPL (Europhysics Letters)*, **76**(2):305, 2006.

37. Van de Walle, C.G. and D. Segev, Microscopic origins of surface states on nitride surfacesa. *J. Appl. Phys.*, **101**(8):081704, 2007.

38. Ibbetson, J.P. et al., Polarization effects, surface states, and the source of electrons in AlGaN/GaN heterostructure field effect transistors. *Appl. Phys. Lett.*, **77**(2):250–252, 2000.

39. Palacios, T., et al. Use of multichannel heterostrucures to improve the access resistance and f_T linearilty in GaN-based HEMTs. In *62nd Device Research Conference*, Notre Dame, IN, June 2004.

40. Palacios, T. et al., AlGaN/GaN HEMTs with an InGaN-based back-barrier. In *63rd Device Research Conference*, 181–182, San Diego, CA, June 2005.

41. Higashiwaki, M., T. Matsui, and T. Mimura, 30-nm-gate AlGaN/GaN MIS-HFETs with 180 GHz f_T. In *Device Research Conference*, State College, PA, June 2006.

42. Shen, L. et al., Unpassivated high power deeply recessed GaNHEMTs with fluorine-plasma surface treatment. *IEEE Electron Dev. Lett.*, **27**(4):214–216, 2006.

43. Arulkumaran, S. et al., Surface passivation effects on AlGaN/GaN high-electron-mobility transistors with SiO2,Si3N4, and silicon oxynitride. *Appl. Phys. Lett.*, **84**(4):613–615, 2004.

44. Coffie, R. et al., P-capped GaN-AlGaN-GaN high-electron mobility transistors (HEMTs). *IEEE Electron Dev. Lett.*, **23**(10):588–590, 2002.

45. Hashizume, T., et al. Surface passivation of GaN and GaN/AlGaN heterostructures by dielectric films and its application to insulated-gate heterostructure transistors. *J. Vac. Sci. Technol. B: Microelectron Nanometer. Struct.*, **21**(4):1828–1838, 2003.

46. Shen, L. et al., High-power polarization-engineered GaN/AlGaN/GaN HEMTs without surface passivation. *IEEE Electron Dev. Lett.*, **25**(1):7–9, 2004.

47. Koley, G. and M.G. Spencer, On the origin of the two-dimensional electron gas at the AlGaN/GaN heterostructure interface. *Appl. Phys. Lett.*, **86**(4), 2005.

48. Higashiwaki, M., T. Matsui, and T. Mimura, AlGaN/GaN MIS-HFETs with f_T of 163 GHz using cat-CVD SiN gate-insulating and passivation layers. *IEEE Electron Dev. Lett.*, **27**(1):16–18, 2006.

49. Tan, W.S. et al., Surface leakage currents in SiN_xPassivated AlGaN/GaN HFETs. *IEEE Electron Dev. Lett.*, **27**(1):1–3, 2006.

50. Lin, Y.J. and Y.L. Chu, Effects of the thickness of capping layers on electrical properties of Ni ohmic contacts on p-AlGaN and p-GaN using an ohmic recessed technique. *Semicond. Sci. Technol.*, **21**(8):1172–1175, 2006.

51. Saito, W. et al., Recessed-gate structure approach toward normally off high-voltage AlGaN/GaN hemt for power electronics applications. *IEEE Trans. Electron Dev.*, **53**(2):356–362, 2006.

52. Shen, L. et al., High performance deeply-recessed GaN power HEMTs without surface passivation. *Electron. Lett.*, **42**(9):555–556, 2006.

53. Wang, W.K. et al., Performance enhancement by using the n^+-GaN cap layer and gate recess technology on the AlGaN–GaN HEMT fabrication. *IEEE Electron Dev. Lett.*, **26**(1):5–7, 2005.

54. Kambayashi, H. et al., Over 100 A operation normally-off AlGaN/GaN hybrid MOS-HFET on Si substrate with high-breakdown voltage. *Solid-State Electron.*, **54**(6):660–664, 2010.

55. Matocha, K. and R.J. Gutmann, High-voltage normally off GaN MOSFETs on sapphire substrates. *IEEE Electron Dev. Lett.*, **52**(1):6–10, 2005.

56. Hilt, O., et al. Normally-off AlGaN/GaN HFET with p-type Ga Gate and AlGaN buffer. In *2010 22nd International Symposium on Power Semiconductor Devices & IC's (ISPSD)*, Hiroshima, Japan, June 2010.

57. Coffie, R. et al., Unpassivated p-GaN/AlGaN/GaN HEMTs with 7.1 W/mm at 10 GHz. *Electron. Lett.*, **39**(19):1419–1420, 2003.

58. Cai, Y. et al., Control of threshold voltage of AlGaN/GaN HEMTs by fluoride-based plasma treatment: From depletion mode to enhancement mode. *IEEE Trans. Electron Dev.*, **53**(9):2207–2215, 2006.

59. Jacobini, C. and L. Reggiani, The Monte Carlo method for the solution of charge transport in semiconductors with applications to covalent materials. *Rev. Mod. Phys.*, **55**(3):645–705, 1983.

60. Mosko, M. and A. MoskoVA, Ensemble Monte-Carlo simulation of electron-electron scattering–improvements of conventional methods. *Phys. Rev. B*, **44**(19):10794–10803, 1991.

61. Kosina, H. and S. Selberherr, A hybrid device simulator that combines Monte Carlo and drift-diffusion analysis. *IEEE Trans. Computer-Aided Design*, **13**(2):201–210, 1994.

62. Li, T., R.P. Joshi, and C. Fazi, Monte Carlo evaluations of degeneracy and interface roughness effects on electron transport in AlGaN–GaN heterostructures. *J. Appl. Phys.*, **88**(2):829–837, 2000.

63. Singh, M. and J. Singh, Design of high electron mobility devices with composite nitride channels. *J. Appl. Phys.*, **94**(4):2498–2506, 2003.

64. Singh, J., Velocity overshoot effects and transit times in III–V nitride HFETs: A Monte Carlo study. In *62nd Device Research Conference*, Notre Dame, IN, June 2004.

65. Bhapkar, U.V. and M.S. Shur, Monte Carlo calculation of velocity-field characteristics of wurtzite GaN. *J. Appl. Phys.*, **82**(4):1649–1655, 1997.

66. Zhang, Y. and J. Singh, Monte Carlo studies of two dimensional transport in GaN/AlGaN transistors: Comparison with transport in AlGaAs/GaAs channels. *J. Appl. Phys.*, **89**(1):386–389, 2001.

67. Wu, Y.-R. and J. Singh, Transient study of self-heating effects in AlGaN/GaN HFETs: Consequence of carrier velocities, temperature, and device performance. *J. Appl. Phys.*, **101**:113712, 2007.

68. Singh, M., Y.-R. Wu, and J. Singh, Velocity overshoot effects and scaling issues in III–V nitrides. *IEEE Trans. Electron Devices*, **52**(3):311–316, 2005.

69. Yuh-Renn, W., M. Singh, and J. Singh, Device scaling physics and channel velocities in AlGaN/GaN HFETs: Velocities and effective gate length. *IEEE Trans. Electron Dev.*, **53**(4):588–593, 2006.

70. Yuh-Renn, W., M. Singh, and J. Singh, Sources of transconductance collapse in III–V nitrides - consequences of velocity-field relations and source/gate design. *IEEE Trans. Electron Dev.*, **52**(6):1048–1054, 2005.

II

Growth and Processing

5

Growth Technology for GaN and AlN Bulk Substrates and Templates

Michael Slomski,
Lianghong Liu,
John F. Muth,
and Tania Paskova

5.1 Nitride Applications and Templates versus Bulk Substrates

5.1.1 Nitride Applications

Gallium nitride (GaN) and its close chemical counterparts-aluminum nitride (AlN), indium nitride (InN), and their ternary alloys Al(Ga, In)N-form the III-V nitride-based semiconductor family (often referred to as nitrides). The most established applications of III-nitride semiconductors are optoelectronics and electronics, but applications such as spintronics, photovoltaics, and sensors are recently under extensive research as well. Detailed summaries of their applications and market analysis can be found in various publications.[1–3]

From the application point of view, optoelectronics has the largest market due to the fact that devices emitting from ultraviolet to red can be fabricated by using the nitride semiconductors. They have a direct band-gap, and the band-gap energy of the Al(Ga, In)N varies in a wide range between 6.2 and 0.7 eV depending on their alloy composition.[4] The interest in nitrides for electronic application is also very strong, motivated by their unique properties. In particular, GaN is one of the few materials with both a high breakdown voltage and a high saturation velocity.[5] The nitrides are well suited for high-temperature applications because of their wide bandgaps and low intrinsic carrier concentrations. The electron mobility in GaN is quite high considering the magnitude of the bandgap and is even higher in selectively doped AlGaN/GaN heterostructures where two-dimensional electron gases (2DEG) may form, producing high sheet carrier densities. The strong piezoelectric and polarization effects presented in the AlGaN/GaN structures enhance these densities. These properties make the nitrides the best semiconductor system for high-frequency electronic applications. Finally, GaN possesses a very high breakdown field, allowing the devices to support large voltages for high-power operation.[6]

5.1.2 Nitride Bulk Substrates

Despite this incredible potential of the nitride materials, device performance and the fundamental understanding of the properties are still limited by the material quality.[7] Unfortunately, due to high melting temperatures and very high dissociation pressure, bulk nitrides cannot be grown by using conventional methods used for GaAs or Si, such as the Czochraslski or Bridgman techniques. Bulk GaN and AlN crystals have only recently become commercially available, and mostly with limited size. Even so, the price of these materials is still beyond the cost-sensitive customers.[8,9]

Most applications of nitrides have relied on heteroepitaxy. A wide variety of materials have been studied for substrates for nitride epitaxy including insulating metal oxides, metals, metal nitrides, and other semiconductors. However, despite the significant success in GaN heteroepitaxy on many different substrates, the mismatch in thermal expansion coefficients and lattice parameters leads to some degradation of the film quality. A thorough discussion of substrates for GaN heteroepitaxy is given in an earlier review by Liu and Edgar.[7] The growth process and device fabrication technology have been described in many reviews such as those by Ambacher,[10] Gibart et al.,[11] Sheu and Chi,[12] Pearton et al.,[13,14] Morkoç et al.,[15,16] Monemar,[17] J. W. Orton and Foxon,[18] Munoz et al.,[19] Steigerwald et al.,[20] Bickermann and Paskova.[21]

The first GaN-based light-emitting diodes (LEDs) were fabricated on sapphire with typical dislocation density higher than 10^9 cm^{-2}. Before that, good luminescence efficiency, regardless of the semiconductor, was thought to require very low dislocation densities, less than 10^6 (cm^{-2}).[22] Researchers were surprised to measure excellent luminescence efficiency from GaN-based LEDs despite dislocation densities four orders of magnitude higher. As devices other than LEDs have been developed using GaN, it has become apparent that the conventional wisdom on the requirements for device fabrication and epitaxial growth still has merit. It became clear, in agreement with the prior knowledge, that high dislocation densities are detrimental to the performance of devices with more sophisticated structures such as laser diodes (LDs), or for devices requiring large areas or operating at greater power densities.[23]

In the case of LDs, it is very well known that the lifetime is largely increased when LDs are fabricated on the lower dislocation region realized by epitaxial lateral overgrowth (ELO) technique, which can produce GaN with dislocation density down to mid-10^6 cm^{-2}. Dislocations can cause deterioration in the operation of quantum-well-based LDs mainly by three mechanisms[24]: (i) serving as nonradiative recombination centers for electrons and holes leading to heat generation instead of optical emission; (ii) introducing fast diffusion along the dislocation lines, smearing out quantum wells and *p-n* junctions; and (iii) disturbing the epitaxial growth, so that atomically flat structures cannot be obtained. The power, lifetime, and reliability of LDs fabricated on bulk GaN with even lower dislocation were improved greatly. Thus, bulk GaN substrates with dislocation density less than 10^5 cm^{-2}, preferably less than 10^3 cm^{-2} are highly desirable for reliable high power LDs.

For short wavelength violet LEDs, dislocations can also noticeably affect performance. This influence may result from longer diffusion length for low In-composition of InGaN alloy. It was predicted that when the diffusion length is longer than 500 nm, the value similar to those for other conventional LED materials, dislocation densities must be lower than 10^6 cm^{-2}. In contrast, the diffusion length is as low as 50 nm in blue InGaN/GaN LEDs, which can allow the dislocation densities up to 10^9 cm^{-2} without acting as recombination center.[25] On the other hand, white LEDs based on UV LEDs plus RGB phosphors will be competing against blue LEDs with yellow phosphors in a variety of applications with its advantages. In this case, bulk nitride substrates with low dislocation densities may help to realize high-efficiency white LEDs. It was confirmed that threading dislocations could cause leakage current[26] and aging of high power LEDs.[27] As Tsao et al.[28] pointed out, the small-area chip (0.25 mm^2) scenario for solid-state lighting commercialization may require 10^6 cm^{-2} or less dislocation density.

Hence, development of alternative substrate technology could potentially open a huge market for bulk nitrides.

In the case of electronic devices, dislocations also have a deterioration effect on the device performance. It was revealed that screw dislocation lines could act as a channel for the reverse-bias leakage current in Schottky diodes[29,30] by promoting the metal diffusion along the lines.[31] Yamada et al.[32] found that the reverse bias leakage current of *p-n* junction devices was reduced by more than two orders of magnitude by decreasing the threading dislocation density down to 10^6 cm^{-2} when employing ELO technique in comparison with growing directly on a sapphire substrate. McCarthy et al.[33] compared the performance of GaN heterojunction bipolar transistors (HBTs) fabricated on MOCVD GaN templates and ELO materials. It was identified that the dislocations were the primary source of collector-emitter leakage and it was reduced by four orders of magnitude for devices on ELO materials. It is well known that the key for GaN HEMT with high performance is low leakage in the buffer layer, which demands highly resistive epilayers with low defect density. Moreover, the traps associated with dislocations in the buffer layer also cause a current collapse in AlGaN/GaN FETs.[16] Additionally, studies on device reliability over their lifetime indicate that the dislocations also result in aging degradation of the electronic devices.[34] In summary, the effect on both device performance, lifetime, and reliability is significant, which could explain the demand for high-quality, low-defect-density GaN substrates.

Furthermore, it is important to underline that bulk nitride substrates offer a few more advantages in addition to lower dislocation density: (i) conductive GaN substrates allow vertical device geometries, therefore, higher current flow and higher yield for device fabrication; (ii) the same cleaving property between substrate and epilayers makes it easy for fabricating laser mirrors for resonance oscillation; (iii) higher thermal conductivity improves heat dissipation for high power devices; (iv) various crystallographic orientations, which allows device alignment with different polarities; and last (v) with respect to the device growth process, bulk nitride substrates eliminate the steps of low-temperature nucleation layer growth, temperature ramping, and several micron buffer layer growth, therefore, reduce material usage, increase throughput, and reduce cost.

5.2 Nitride Growth Methods

This section will focus on the various growth methods that have been intensively studied in the last two decades, including hydride vapor phase epitaxy (HVPE), sublimation, ammonothermal growth, high-pressure solution growth (HPSG), and the combination of growth methods. Various techniques were used for bulk crystal growth, template growth, single substrate growth, boule growth, and seeded regrowth.

5.2.1 Sublimation

Sublimation is almost exclusively used for growth of AlN. This method is usually carried out in a furnace with a temperature gradient, which provides the driving force for the growth. There are two major types of furnaces: resistively heated and RF heated furnaces. In order to achieve considerable growth rates, the sublimation growth of AlN requires high temperature (>1,800°C, and typically about 2,200°C). Such a high temperature along with the high reactivity of Al vapor could create severe problems for the stability of the furnace fixture, which could be degraded quickly or even melt. In order to reduce the temperature, various approaches were proposed by the research groups working on this topic, including different reactor configurations with various heating systems (a few examples shown in Figure 5.1a–d).

AlN sublimation, a kind of physical vapor transport growth, is a relatively simple process. Initially, kinetic studies, performed by Liu[35] and Noveski[36] demonstrated that the growth rate was limited by the mass transport in the gas phase at a pressure >200 torr and temperature range of 1,700°C ~ 2,400°C. Reducing pressure was effective in increasing the growth rate since it increased not only the partial

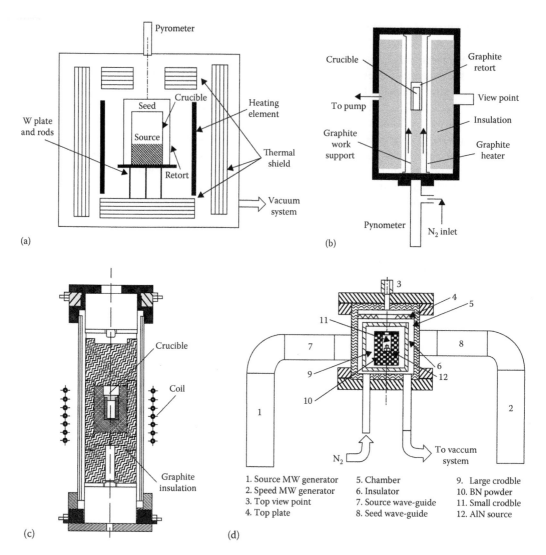

FIGURE 5.1 AlN sublimation reactors with various heating systems: (a) resistively heating system using tungsten (Reprinted from *J. Cryst. Growth*, 220, Liu, L. and Edgar, J. H., Transport efforts in the sublimation growth of aluminum nitride, 243–253, Copyright 2000, with permission from Elsevier.); (b) resistively heating system using graphite (Reprinted from *J. Mater. Sci. Eng. B*, 117, Liu, B. et al., Free nucleation of aluminum nitride single crystals in HPBN crucible by sublimation, 99–105, Copyright 2005, with permission from Elsevier.); (c) induction heating system (Reprinted from *J. Cryst. Growth*, 264, Noveski, V. et al., Mass transfer in AlN crystal growth at high temperature, 369–378, Copyright 2004, with permission from Elsevier.); (d) microwave heating system (Reprinted from *J. Cryst. Growth*, 262, Zhuang, D. et al., Bulk AlN crystal growth by direct heating of the source using microwaves, 168–174, Copyright 2004, with permission from Elsevier.)

pressure of aluminum but also the diffusivity. Yet, the growth was limited by the stability of crucible under high aluminum partial pressure. Nevertheless, later studies revealed that the surface kinetics would take effect at low pressure.[37] At that time, the rate controlling step of growth was the availability of active nitrogen adatoms rather than that of aluminum at crystal surface due to the low sticking coefficient of N_2. Currently, most of the reported growth rates are still below 1 mm/hr, which is limited by the stability of crucible at high temperature and low N_2 pressure.

AlN growth was usually conducted in a crucible inside a furnace. The selection of suitable crucible materials has been a major challenge at such high temperature. The properties and stabilities of different materials for crucibles or coating such as W, graphite, BN, SiC, TaC, TaN, NbC, Ta, Ta2C, Re, ZrO2, ZrC, ZrN, HfN, HfC and so on, under various sublimation conditions have been reviewed by Zhuang et al.,[38] Dalmau et al.,[39] Slack et al.,[40] and Epelbaum et al.[41] The following major conclusions were drawn: (1) The crystal color, growth habit, and morphology were affected by the choice of crucible materials. The density of extended defects and their distribution in the crystals remained unaffected; (2) Pure W is quite a durable crucible material for AlN sublimation growth at medium temperatures. It also ensures high purity crystal growth. At temperatures above 2,300°C, the W crucibles suffer from attack by aluminum vapor that eventually leads to failure. Nevertheless, the free nucleation in W crucible has very high density and is difficult to control; (3) TaN, TaC, and NbC may be the best choices considering the crucible lifetime, material compatibility, and crystal quality; however, the cost is high. In addition, the coated graphite components are not suitable as the coatings tend to spall off; (4) BN caused highly anisotropic growth rates along different crystallographic directions. The crystals produced are highly striated; (5) Other materials may offer inferior performance due to the reaction with aluminum or sublimation at the typical crystal growth temperatures; (6) The thermal expansion also needs to be considered for the case of crystal growth contacting the crucible walls. Materials with similar thermal expansion coefficient to that of AlN are preferred in order to reduce the stress during cooling down.

Other furnace fixtures along with the crucible and source themselves can cause profound contamination in AlN during growth. The unintentional impurity incorporation in bulk AlN crystals is an important issue, as impurities not only can change physical, optical (i.e., UV transparency), and electrical properties, but also can change the growth mechanism and initiate crystal defects. For example, Brenner et al.[42] revealed that boron substitutional impurities can enhance binding and incorporation of growth species onto steps. In addition to forming crystal defects, impurities can also change the crystal morphology as a consequence of surfactant effects and changes in the Al-N thermodynamics.[43] Oxygen in AlN crystals is difficult to avoid since the commercial AlN sources can contain as high as 1.0 wt%. This oxygen contamination from the source material was found to be the most challenging problem because of oxygen-assisted transport of aluminum at relatively low temperature of 1,700°C–1,800°C. The oxygen impurity can occur as a substitution impurity on nitrogen sites at concentrations up to 1×10^{21} cm^{-3}.[44] At even higher concentration, it eventually produces Al_2O_3 inclusions. The oxygen in the source is transported presumably as AlO and Al_2O with AlON during heating-up stage.[45] A subsequent growth of dense AlN on such powder layer generated at the heating-up stage was possible, but resulted in a very small grain size of 0.1–0.3 mm.[46] For improving the nucleation control, a temperature gradient reversal method was used, that is, in the beginning of the process the source was kept at a lower temperature, and the oxygen concentration can be reduced to 10^{19} cm^{-3}.[46,47] Intensive studies of AlN properties revealed that the oxygen impurity has a significant effect on various material properties, such as thermal conductivity, luminescence, optical absorption, UV transparency, and so on.[48-52]

The free-nucleation growth of AlN on graphite walls was reported by Tanaka et al.[53] to result in three morphologies at different temperatures: flat plates at 2,100°C; needles with hexagonal cross-section at 2,000°C; and needles with rectangular cross-sections at 1,900°C. The yellow color of crystals was attributed to blue light adsorption by the oxygen impurity. This growth habit of crystal shape dependence on temperature was also observed by Epelbaum et al.[54] and Singh et al.[55] Edgar et al.[56-58] studied the self-nucleation on both W and BN crucible in a resistively heated furnace with W or graphite heating elements. The largest growth rate of 0.4 mm/hr was found in BN crucible, though it resulted in thin platelets. The self-seeding produces crystals of highest perfection, lowest stress, and lowest Si and C impurity content compared to seeded growth on SiC. The crystal quality, (as revealed by Raman scattering spectroscopy—full width at half maximum (FWHM) of the high-frequency peak (E1(LO)+A1(LO))—and X-ray diffraction—FWHM of the 0002 reflection) was found higher in the freely nucleated crystals than that in material grown by seeded growth.[59] The crystals grown by Raghothamachar et al.[60] also displayed high quality, with a dislocation density of about 10^3 cm^{-2}. It is important to note that usually AlN single crystals grow in random crystallographic orientations and the crystallite size is on the order of mm. Different shapes of AlN single crystals, such as boules, platelets, prisms, needles, and whiskers, have been reported by different researchers. The large diameter (cm size) crystals by self-nucleation were, unfortunately, polycrystalline, though the growth rate was as high as 1 mm/hr.

In order to increase the crystal size and control the shape of free nucleation, the Crystal IS research team, which continues the pioneering work of Slack et al.,[61] utilized a cone-shaped crucible. Ideally, a single nucleus forms at the tip end of the crucible at a high temperature such as about 2,250°C and evolves into a large single crystal grain. The single grains become larger and begin to dominate as the boule grows, while the surface area occupied by the polycrystalline region decreases. Typically, however, several nuclei form and compete with each other as the growth proceeds. The growth rate was reported as high as to 0.9 mm/hr,[62] with the largest size recently claimed to be two inches. A corresponding step flow growth mechanism, which originated from the screw dislocation, was also proposed by the same research group.[63,64] Screw dislocations dominated and no edge dislocations were observed in these large single crystal grains, though the screw dislocation densities were below 5×10^4 cm^{-2}. The characterization results indicated that the dislocation density was 800–1,000 cm^{-2} and FWHM of the 0002 rocking curve was less than 10 arcsec.[65]

Using a substrate as a seed crystal is an attractive alternative since the crystal orientation, crystal polarity, and the initial nucleation can be better controlled. Moreover, the large diameter seed crystals up to the available size of SiC may greatly facilitate scaling to produce large crystals. Single crystalline 6H-SiC (0001) wafers were used due to the relatively small lattice mismatch to AlN (0.9%) and their high-temperature stability. The growth of AlN on SiC was first reported by Balkas et al.[66] The results from TEM analyses revealed the absence of line and planar defects; Raman and X-ray diffraction studies also indicated an excellent crystalline quality of the grown AlN crystal. However, the crystals contained small (~2 × 2 mm^2), individual hexagonal sub-grains and cracks upon cooling. Edgar et al.[58,67-71] conducted extensive research work on AlN seeded growth on SiC. Typically, the source temperature was above 1,800°C, the source-substrate temperature gradient was less than 5°C/mm, and the ambient pressure varied in the range of 100–800 torr. Significant difficulties were encountered when 6H-SiC substrates were used as seed crystals for bulk AlN growth, such as: decomposition of 6H-SiC at high temperatures, isolated random nucleation of AlN and subsequent individual sub-grain growth, cracks in the AlN caused by stress from the mismatch of thermal expansion between AlN and SiC. Direct growth on as-received Si-face SiC resulted in formation of discontinuous hexagonal subgrains of 1-mm size. The growth of AlN on carbon-polarity substrates was almost completely absent.[58] Such incomplete coalescence was caused by decomposition of the 6H-SiC in the presence of AlN, forming hillocks on the surface. Subsequently, the AlN tends to nucleate on these SiC hillocks. In some areas, sub-grain boundaries were formed as the AlN islands grew and merged together. In other regions, the AlN islands stopped growing laterally, and simply started growing vertically (along the [0001] direction) leaving deep, irregular crevasses between the AlN islands.[67] Using high-temperature MOCVD, grown AlN effectively improved the wetting of SiC and resulted in 2D growth mode during sublimation.[68,69] The AlN epilayer promoted layer-by-layer growth forming a continuous single grain film, however, the crystal always suffered from cooling-down cracking. For thick AlN films, the AlN was under biaxial compressive stress on the surface but tensile stress at the AlN/SiC interface. Large tensile stresses at the interface of the AlN/SiC heterostructure, up to 1 GPa, were detected by the frequency shift of the Raman E1(TO) mode. The frequency shift along with stress had an approximate linear distribution along the depth of sample. An AlN crystal thickness greater than 2 mm was estimated as necessary to completely eliminate cracking. In order to solve the cracking problem, a deposition of pure AlN was proposed on a thick layer (0.1–0.5 mm) of (AlN)x(SiC)1-x alloy film on SiC by sublimation from a mixture of AlN-SiC powders at 1,800°C.[70] The intermediate properties of an (AlN-SiC) alloy layer greatly reduced cracking in the final AlN crystal. In addition, the presence of silicon and carbon in the source material reduced or eliminated the decomposition of the SiC substrate during sublimation crystal growth, enabling longer duration crystal growth. By employing this technique, high-quality colorless AlN single crystals up to 1 cm in diameter were produced. Analysis by X-ray diffraction indicated that the AlN single crystal was of comparable quality to the original 6H-SiC (0001) substrate.[71] For AlN single crystals seeded on 6H-SiC substrates, the FWHM of X-ray rocking curve for asymmetric peaks had a larger value than that of the symmetric peaks (i.e., the (0002) peak). Such disparity between the FWHM of the symmetric and asymmetric peaks is commonly observed for the group III nitrides and has been attributed to the presence of significantly more dislocations of edge type rather than screw type.[72,73]

The growth of AlN on AlN seed grown by self-nucleation was also attempted. Schlesser et al.[74] grew cm-size single crystals at 2,100°C–2,300°C on AlN seeds grown by vaporization of Al in a nitrogen atmosphere. Seeded growth occurred preferentially in the *c*-direction. The FWHM of 0002 rocking curve was only 25 arcsec, indicating high quality of crystals grown by seeded homogrowth. Noveski et al.[47,75] developed a process for continuous growth on previously deposited materials. By using the inverted temperature gradient method, part of the grown materials were removed, and the renucleation of AlN was suppressed on the previously grown materials, which were exposed to air during AlN source replenishment in order to maintain constant source-seed distance. Through a multiple regrowth run, cm size single crystal grains were demonstrated. The same research group continued extensive work using AlN for seeded growth in both c- and m-orientation;[76,77] however, the seed expansion to 2 inch is extremely challenging due to the process repeatability issue.

5.2.2 Hydride Phase Vapor Epitaxy

Unlike sublimation, HVPE growth has been used for the growth of all three nitride compounds: AlN, GaN, and InN, as well as for their ternary alloys.

5.2.2.1 HVPE Growth of GaN

HVPE is an attractive technique for bulk GaN growth due to its ability to produce thick layers of high quality with high growth rates and relatively low cost. The thick layers produced by HVPE can be used as quasi-bulk substrates after delamination, or to obtain GaN boules which can be sliced into GaN substrates. Nearly all of the GaN substrates in today's market are fabricated by the HVPE technique, owing to its favorable processing conditions. Atmospheric pressure and relatively low processing temperature as compared to other techniques create a cost-effective growth method when paired with its high growth rate above 100 µm/h.[78] The HVPE technique is able to obtain these high growth rates due to its high surface migration rates for the halide species.

The HVPE technique has proven that it is capable of producing *n*-type, *p*-type, and semi-insulating material. It is not possible to achieve completely undoped material, as HVPE grown material always exhibits unintentional *n*-type doping due to background doping by silicon and oxygen introduced form the quartz within the reactor or process gases. Intentional *n*-type doping is usually achieved with silicon and oxygen, although doping with Ge and Sn has been accomplished. Si doping is usually performed by introducing a flow gas of Silane (SiH_4), disilane (Si_2H_6), or dichlorosilane (SiH_2Cl_2).[79]

HVPE growth optimizations have been reported in numerous publications, including several review papers.[78–81] For successful HVPE GaN growth, the following major conclusions could be summarized: (1) An inert ambient atmosphere is needed and is more effective for GaN growth than hydrogen atmosphere, the latter being widely used for other III-V systems as well. (2) In HVPE-GaN growth, NH_3 should be used as the source of the group V element rather than a nitrogen halide (NCl_3), which is highly explosive. Moreover, the thermal dissociation of NH_3 results in formation of N_2 molecules which are extremely stable and nonreactive at the temperatures used, while the thermal dissociation of $AsCl_3$, in GaAs growth, results in the formation of As_2 and As_4 molecules, which remain typically volatile and chemically reactive in the next step of the growth reaction. (3) The development of the last-running process for very thick samples is difficult to perform, due to massive condensation of reaction species on the reactor walls. The GaN reaction tends to create huge amounts of by-products such as NH_3Cl, $GaCl_3$, and $GaCl_3 \cdot NH_3$.

The kinetics of the process and growth mechanisms occurring at the solid/vapor interface during HVPE growth of GaN were simulated by several models,[82–84] assuming a surface process involving the following steps: (a) adsorption of NH_3 molecules, (b) adsorption of N atoms forming NGaCl, and (4) decomposition of the NGaCl via different desorption mechanisms. Two of them were suggested by analogy with the GaAs model: desorption forming HCl and desorption forming $GaCl_3$. One more $GaCl_2$ desorption mechanism was suggested for the GaN growth based on the experimental results.[84] Statistical treatment of the dynamic equilibrium between the adsorbed and gas-phase species allowed explicit expressions of the growth rates via the different pathways. There are several detailed kinetics considerations published in the literature.[81] The overall dependence of growth rate upon temperature is generally agreed upon among different researchers, although details are partly in contradiction with each other. Growth rate dependence on the temperature shows two regions, a low- and high-temperature region with different trends.[80] The low-temperature regime shows an increasing growth rate with temperature, while at high temperatures a maximum is reached at around 950°C–1,080°C and growth rate decreases with further temperature increase.

Growth in the high-temperature region has seen the most experimental studies due to larger practical interest stemming from the faster growth rates and better material quality obtained. Variations in the temperature value that achieve the fastest growth rate have been attributed to differences in reactor design and growth conditions. At optimum temperature, the growth rate has been shown to be affected by HCl flow rate and the distance of the substrate from the gas mixing point. The growth rate is shown to increase linearly with increasing HCl flow rate and decrease exponentially with substrate distance from mixing point. High growth rate can, however, interfere with the crystalline quality, as increasing the mixing space can improve the uniformity of the growth rate of the substrate and be helpful in avoiding polycrystalline deposition.[85–88] Crystalline quality is also dependent on factors, which do not affect the flow rate, including NH_3 flow rate and substrate orientation.[80]

Reactors used for the HVPE growth of GaN are based on the same initial concept and are broken down into vertical and horizontal geometries as shown in Figure 5.2a–d.[80,89,90] Vertical and horizontal reactors of similar designs are used for both GaN and AlN growth. The principle of both types of reactors are similar: reactant gases are introduced at optimum distances to ensure mixing and optimal gas distribution, temperature is maintained in each zone by the use of resistive heating, and the sample holder is oriented in a certain position or is rotated to obtain uniform growth on the films.

In the horizontal reactor, HCl gas is introduced upstream to react with a liquid Ga pool in order to form GaCl and H_2 at about 850°C. Further downstream, a second zone can be used for other metallic sources like In or Al, or other dopants typically maintained at a different temperature. Past this point, NH_3 is introduced into a zone in the temperature range of 1,000°C–1,060°C, where it mixes with GaCl. The GaCl and NH_3 mixture reaches the substrate surface holder to form GaN, at temperatures ranging from 950°C–1,150°C. The sample holder can be oriented parallel, inclined, or perpendicular to the gas flow direction depending on which gives the most uniform deposition for a given reactor design. Most commonly, gas inlet tubes deliver reactant gases to the mixing point using separated quartz liners, although a coaxial arrangement has been shown to improve the uniformity of the layers by achieving better mixing of the reactant gases.[87,91] For growth of compound nitrides, a reactor with two temperature zones—one for the metal container ensuring the reaction of HCl with the metal, often called source zone and one zone where the growth is to take place, called the growth zone—are sufficient (Figure 5.2a). For growth of ternary alloys or for doping, one more temperature zone is needed for the second metal container. Some advanced reactors contain up to five temperature zones (Figure 5.2b) for ensuring higher flexibility for growing various nitrides and/or stabilizing the temperature profile at the edges. Typically, each temperature zone is controlled independently.

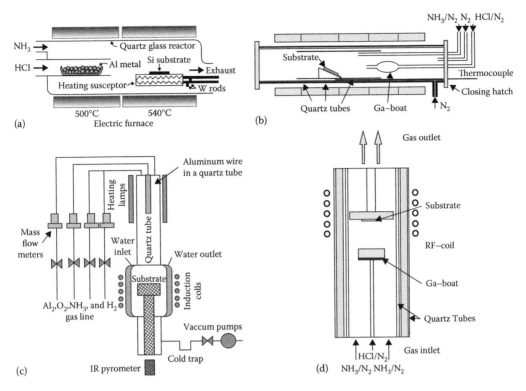

FIGURE 5.2 HVPE reactors with different design: (a) Horizontal reactor with two resistive heating temperature zones used for AlN growth at temperatures up to 1,410°C (From Kumagai, Y. et al.: Characterization of a freestanding AlN substrate prepared by HVPE, *Phys. Stat. Sol. C.* 2008. 1512. Copyright Wiley-VCH Verlag GmbH & Co. KGaA. Reproduced with permission.); (b) horizontal reactor with 5 resistive heating temperature zones, allowing doping and/or ternary nitride growth; [21] (c) vertical reactor with top-down gas source low with two temperature zones with both resistive and induction heating (From Coudurier, N. et al.: Effects of the V/III ratio on the quality of AlN grown on (0001) sapphire by HTHVPE. *Phys. Status solid C.* 2013. 10. 362. Copyright Wiley-VCH Verlag GmbH & Co. KGaA. Reproduced with permission.); (d) vertical reactor with RF heating.

The vertical reactor design offers benefits of improvements in film uniformity by facilitating easier substrate rotation. The substrate is held flat on a susceptor that is perpendicular to the gas flow direction (Figure 5.2c). One important modification is the implementation of a technique of lowering the substrate holder isothermally into a dump tube, in which there is a counterflow of NH$_3$ in a N$_2$ carrier gas.[80] This modification allows for an abrupt interruption of the growth, which enables the change of the gas flows in the main tube or to slowly cool down the sample by further lowering the holder. The vertical reactor design can be further modified by supplying the process gases through the bottom inlet flange, while the top flange can be lifted for loading and unloading of substrates (Figure 5.2d). Reasons for the use of the inverted design include retaining the benefits of the vertical design while providing the possibility of raising the substrate holder and also minimization of solid particle contamination on the growing surface.

5.2.2.2 HVPE Growth of AlN

The HVPE growth of AlN is similar to that of GaN—the gas mixture produced from HCl passing over the Al metal in the source zone is introduced to growth zone along with NH$_3$. AlCl$_3$, which is compatible with quartz is dominantly generated over AlCl, could be utilized to grow crystalline AlN at relatively low source temperatures such as 500°C. The comprehensive thermodynamic analysis of HVPE growth of AlN was reported by Kumagai et al.[92,93]

The initial efforts on AlN HVPE growth resulted in thin epilayers, whose quality was not comparable to that by other methods such as MOCVD.[94,95] Researchers from TDI have investigated AlN HVPE growth for both thin templates[96–98] and bulk crystals.[99,100] Both SiC and Si have been used as substrates with no buffer layer. The deposition was conducted at temperatures in the range of 1,000°C–1,100°C in a horizontal atmospheric hot-wall reactor, with a growth rate up to 60 μm/hr. The thick films separated from substrates have been used for seeded growth to further improve crystal quality. The typical diameter of the grown AlN boules was about 1 inch with largest one up to 1.75 inches, with resistivity exceeding 10^8 Ωcm. However, the FWHM of (0002) ω-scan was ~700 arcsec, proving the challenge to obtain high-quality crystals. AlN dislocation density reduction with increasing thickness was much less noteworthy than that of GaN, that is, the dislocation density of ~10^8 cm^{-2} still preserved at the surface for 200-μm-thick film.[101] This indicates the weak interaction (coalescence) of AlN grains at low-temperature HVPE as expected.

In order to further improve the quality, HVPE at a high temperature above 1,200°C has been intensively studied recently. Different reactor configurations for high-temperature HVPE have been adopted, as shown in Figure 5.2. With MOCVD grown AlN buffer layer on sapphire, Gong et al.[102] grew 12.4 μm AlN with 102 FWHM of 600 arcsec at 1,450°C. Kumagai's group has achieved 85 μm/hr growth rate at growth temperature as high as 1,400°C, though the FWHM of the (0002) rocking curve was in the order of 100 arcmin.[103] The result also confirmed that the HVPE AlN was under AlCl$_3$ mass transportation limited process even at high temperature. Later, the group has achieved FWHM of 002 rocking curve of 295 arcsec for 83-μm-thick layer grown at 1,450°C under a high growth rate of 57 μm/hr.[104] The results clearly demonstrated that higher growth rates resulted in smaller FWHM of the rocking curves due to enhanced coalescence. However, the cracking is still inevitable for thick AlN layer. Pons's group found that the insertion of a low-temperature nucleation layer from 650°C to 850°C can improve the crystallinity of the HT AlN layer deposited at 1,400°C, as well as its morphology.[105] With further optimization, the group achieved 350 arcsec for FWHM of the 002 rocking curve.[106] The optimal V/III ratios were also reported under specific growth conditions,[90,107] although the group did not observe strong growth rate dependency on V/III or NH$_3$, in contrast to what Yamane et al. observed.[93]

Recently, Nitride Solutions Inc. has made significant progress on AlN template growth. The company is producing 1 ~ 4 μm high-quality AlN templates on sapphire. The FWHM of 002 and 102 rocking curves are below 250 and 450 arcsec, respectively. The UV transparency of 2 μm films is greater than 80% with a low level of C and O concentration. In order to further improve the film thickness and quality without cracking, patterned substrates have been used. By adopting grooved patterns on SiC, the critical thickness of AlN was increased significantly, and FWHM of 100 rocking curves reached 833 arcsec for 22-μm-thick film[108] and 525 arcsec for 35–μm-thick film.[109] The more than two-order dislocation density reduction by utilizing LEO with proper patterns is significant in AlN HVPE growth.[110] However, the issues encountered in GaN HVPE such as cracking during growth, cool-down cracking, and surface morphology degrading are still expected in HVPE AlN for thick free-standing bulk growth. The so-called "weak-interface" method, widely adopted in GaN HVPE, was employed by Kumagai et al.[111,112] to achieve thick AlN, followed by self-separation during cool-down. A 85 μm thick free-standing AlN was separated from sapphire with interfacial voids. The same group also used Si as a substrate and obtained 112 μm free standing AlN by etching away the silicon substrate. However, the Si impurity level in AlN was as high as 10^{19} due to contamination from the substrate used.[91]

5.2.2.3 HVPE Growth of AlGaN Templates

The growth of low-defect AlGaN used as template (or bulk) substrates is highly desirable for fabrication of advanced AlGaN-based power devices, high-frequency transistors, and UV LEDs. While, other growth methods have been tried such as solution growth technique[113], the well-established HVPE technology developed for both AlN and GaN is currently the most successful in growing AlGaN templates up to 2-mm-thick bulk substrates, as reported in patent application.[114]

The thermodynamic analysis of AlGaN HVPE has been reported by Koukitu et al.,[115] and the effect of H_2 was studied in detail later.[116] It was found that the AlGaN growth was in thermo-equilibrium and the dependency of AlGaN composition was calculated on the growth conditions, which agreed with the experimental results.[117]

In HVPE growth of AlGaN, *in situ* formation of $AlCl_3$ and $GaCl$ by passing HCl through metal aluminum and gallium pools, are utilized to provide the III-species. A wide range of $Al_xGa_{1-x}N$ with composition x varying from 0 to 1 has been successfully grown by HVPE. One of the big challenges for AlGaN HVPE is the composition uniformity across the template radius and thickness, as well as the cracking. The uniformity complications can result from (i) temperature nonuniformity across the wafer due to different driving force for AlN and GaN deposition, since the response of the Gibbs energy on temperature is totally in reverse for AlN and GaN formation; (ii) different diffusivities of $AlCl_3$ and $GaCl$ in the gas phase, leading to different composition across the wafer even in reactor equipped with a premixed $AlCl_3$ and $GaCl$ source; (iii) preferred incorporation of gallium over aluminum on non-c-plane facets during coalescence across the growth thickness.

Different substrates have been used for AlGaN template growth by HVPE. Sapphire is the most commonly used substrate,[116–120] while SiC[121] has also been investigated. In order to reduce cracking, increase growth thickness, and achieve lower defect density, trench and honeycomb patterned sapphire both with AlN buffer layers have also been used, and 40 μm $Al_{0.45}Ga_{0.55}N$[122] and 20 μm thick $Al_{0.3}Ga_{0.7}N$ has been demonstrated.[123]

5.2.3 Ammonothermal Growth

The ammonothermal growth process is conducted by transporting the mass from high- to low- temperature zones and has been reported to possess several advantages,[80] including: (i) lower dissolution density due to solubilizing in the higher temperature zone, (ii) lower probability of cubic formation, (iii) absence of disastrous effects of the basic solution on the autoclave materials, which results in safety improvements and cost effectiveness. Compared to other techniques, ammonothermal growth is characterized by relatively low growth rate capacity; however, it requires relatively low temperature and reasonable pressure that allows controllable recrystallization enabling high-quality material.

The ammonothermal technique employs a solvothermal process, which allows solubilization of polycrystalline III-nitride nutrient or feedstock in supercritical ammonia under high pressure by utilizing a solubilizing agent or mineralizer.[78] The type of mineralizer determines the main type of the ammonothermal approach, which in turn determines their advantages and disadvantages. Mineralizers are divided into three groups: (i) basic mineralizers, which introduce NH_2^- ions to the solution, (ii) acidic mineralizers, which introduce NH_4^+ ions, and (iii) neutral mineralizers which introduce neither NH_2^- ions nor NH_4^+ ions. Dissolution of III-nitrides using potassium azide (KN_3) or potassium amide (KNH_2) ammonothermal-basic solutions have been suggested by several groups[124–127] to occur through the formation of soluble intermediate compounds, which subsequently decompose with temperature. Other crystalline mechanisms, which involve forward solubility of nitride species, have been reported[124,128,129] for ammonothermal-acidic solutions, which utilize ammonium halide mineralizers.

The quality of ammonothermally grown GaN material still requires detailed evaluation, as optimization of the growth conditions and producing a systematic series for thorough analysis will take some time. The best results in terms of crystallinity were reported for self-prepared seed and multiple regrowth processes.[130] The first experimental results indicate that the properties of the ammonothermal GaN material are strongly influenced by the quality of the seeds used. With further research, ammonothermal growth is very promising for its ability to be scalable by supporting the growth of large crystals and multiple simultaneous seeded growths.

5.2.4 Solution Growth at High Pressure and High Temperature

Solution growth at high pressure and high temperature (HPSG) has been proven as a viable technique for many semiconductors. Despite over two decades of effort with GaN semiconductors, equilibrium solution growth proved to be challenging due to its high melting temperature and high associated equilibrium N_2 pressure (> 45 kbar).[78] Due to the very low solubility of nitrogen in Ga of only ~1%, the growth rate is very low and crystal sizes have been limited to several millimeter long prisms or 15 mm × 20 mm lateral-size platelets for about 200 h growth time.[131]

HPSG has produced the highest structural quality GaN material with dislocation density lower than 100 cm^{-2}; however, the dominating lateral growth mode observed under these conditions favors higher levels of background impurity incorporation. While the dominating impurity, oxygen (~10^{19} cm^{-3}), leads to a high n-type carrier density that is beneficial for many devices, the ability to control the carrier density is important for most applications. HPSG has also been shown capable of producing AlGaN bulk crystals of high crystalline quality by providing both Ga and Al sources and carefully controlling the nitrogen solubility and temperature.

5.2.5 Mixed Growth Methods

Another approach that recently attracted significant attention is the so-called "seeded growth" method. The approach aims at combining the high growth rate and higher crystalline quality achievable by the homoepitaxial regrowth employing GaN seeds instead of using sapphire to avoid the initial highly defective interface region. Several alternatives of seeded growth have been elaborated recently by researches from Unipress. Namely, seeds produced either by HPSG, HVPE, or ammonothermal growth methods have been used for regrowth by HVPE or vice versa.

The first type of seeded growth listed above was reported first due to the early availability of high-quality HPSG crystals, although of small form factor. The main challenge of this seeded growth was the expansion of the seed crystal.[132,133] In HVPE growth of GaN, crystal expansion can be controlled through physical-chemical gradients and concentrations, which result in stabilization of crystal planes under the appropriate growth conditions. For *in*-plane expansion, size expansion will depend on the anisotropy of the growth rates for different crystallographic planes. Growth parameters, such as temperature, pressure, growth rate (source species flux) or the ratio between the nitrogen and the gallium species, can influence the stabilization of one crystal plane over another. Additionally, the geometric configuration of the seed in the system may contribute to the flux of the growth species from additional sources, such as surface migration. As shown in the ELOG growth of GaN,[134,135] inclined crystalline facets can be formed under certain growth conditions that give undesirable crystalline morphologies. An important part of the growth process development is to identify growth parameters that give a planar (as opposed to prismatic) crystal, and then find the way to extend the growth in directions normal to those edge planes. One issue that has been observed with seed expansion is the variation in impurity content and carrier concentration. The physical and electrical properties of GaN grown by HVPE are very sensitive to the orientation of the crystallization front. This could present a challenge for bulk growth, particularly for undoped and semi-insulating material, where a low background concentration of shallow impurities, such as oxygen, is desired.

With the availability of larger wafers, produced by HVPE, the regrowth by HVPE was optimized and proven successful in maintaining the seed wafers size during the HVPE seeded growth.[136] Even the very early HVPE regrowth on HVPE free-standing crystals showed that the material produced during regrowth possesses significantly improved structural characteristics. In particular, the dislocation density was found to be noticeably reduced. The analysis of the defect density in the regrown HVPE-GaN with different thicknesses has shown that the trend of decreasing the defect density with increasing thickness remains.[137] The high purity of the crystals and the high optical quality have been proven reproducible. In particular, the low-temperature photoluminescence (PL) spectra of the regrown GaN films showed comparable narrow (~2 meV) exciton peaks as in the PL spectrum of the seed. However, based on different studies performed by variety of techniques, such as X-ray diffraction, PL, and Raman scattering spectroscopies,[136,137] it is clear that both the residual strain and curvature (radius of about 10 m in the best case) of the seed was reproduced, which remains the main undesirable characteristics of these materials.

The most recent availability of low-dislocation-density wafers grown by ammonothermal method open a new possibility for homoepitaxial seeded growth. The HVPE seeded growth on such seeds has also been proven successful in reproducing the seed characteristics in regard to very low curvature (radius close to 100 m) and low dislocation density (in the angle of 10^4 cm^{-2}).[138] Moreover, in contrast to the ammonothermal seed characteristics, the HVPE-GaN wafer grown on ammonothermal seed possessed high purity and transparency, which are typical characteristics of the HVPE growth. Thus, the combination of the two methods seems to enable the advantages of both in the regrown material.

5.3 Properties of Nitride Substrates

The three main characteristics that are particularly important for substrate applications and determine the demand for homoepitaxial growth on bulk nitride substrate over the widely established heteroepitaxial growth on sapphire are the following: low dislocation density, controllable electrical conductivity in a wide range, and high thermal conductivity. All characteristics are strongly dependent on the growth methods and process strategies.

5.3.1 Defect Density

The dislocation density and distribution represent especially important characteristics. Currently, the lowest reported values for AlN and GaN substrates are in the range of 10^2–10^4 cm^{-2} for AlN grown by sublimation[139] and for GaN grown by ammonothermal method.[140] Unfortunately, the low dislocation density for both materials are accompanied by relatively high point defect densities.

There are two primary approaches toward decreasing the dislocation density in HVPE-GaN: (i) ELOG-type approach, employing stripe or star patterning and (ii) increasing GaN thickness with the ultimate goal to grow boules that allow wafer slicing. The ELOG approaches can provide areas of very low defect density, where the lateral growth mode dominates. However, the approach results in nonuniform lateral distribution of dislocation across the wafer surface which requires a very precise alignment of the subsequent technology steps for device structure growth.

The second approach of increasing the layer thickness allows for annihilation of mixed dislocations with opposite Burger's vectors[141] and/or by point defect-assisted dislocation climb leads to a substantial decrease of dislocation density along the growth direction. This method has the advantage of uniform lateral dislocation density as shown in the cathodoluminescence(CL) panchromatic images for GaN layers with 30 and 600 mm thickness, shown in the inset of Figure 5.3. Figure 5.3 shows the trend of dislocation density versus GaN layers thickness, summarizing results obtained by different characterization techniques,[79,141-143] namely by transmission electron microscopy (TEM), CL topography,[142,143] atomic force microscopy and different etching procedures.[144,145] The experimental data shows a monotonic decrease of dislocation density with GaN layer thickness over several orders of magnitude, consistent with the theoretical predictions.

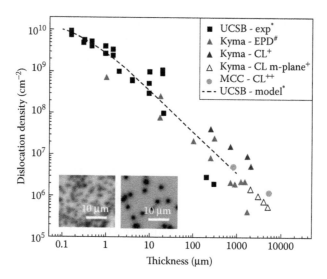

FIGURE 5.3 Combined graph of experimental data and theoretical predictions for dislocation density as a function of the GaN layer thickness. *Mathis et al. [141]; #Hanser et al. [142]; +Paskova et al. [79]; ++Fujimori et al. [143]. Inset: Representative panchromatic CL images visualizing in-plane dislocation distribution in GaN layer grown by HVPE with thickness of about 150 μm (left) and 700 μm (right).

5.3.2 Doping and Electrical Conductivity

All bulk AlN is known to have a substantial amount of residual impurities and point defects, which are responsible for the coloring of the bulk crystals and for reducing its transparency. Various technological strategies have been explored with the ultimate goal to reduce the background impurity concentration with modest success. Efforts for intentional doping of bulk AlN crystals to systematically alter the electrical properties of the material are rare although doping by Si has been reported.[146,147]

GaN substrates have been demonstrated with *n*-type and *p*-type conductivity, as well as semi-insulating. All undoped GaN still retains *n*-type conductivity due to unavoidable oxygen and silicon residual impurities. The lowest silicon and oxygen concentrations vary for the different growth techniques, growth reactor ambient, sources and growth procedures. The lowest residual impurity levels have been reported in the range of 10^{15} cm^{-3} for HVPE growth.[21]

Intentional *n*-type doping has been reported by various impurities, attempting controllable doping in a wide range. Currently, the most developed approach for *n*-type doping was reported for O and Si impurities introduced by gaseous or solid sources. Typical gaseous Si dopant precursors for HVPE growth include silane, disilane, and dichlorosilane. Alternatively, Si doping can be achieved in HVPE by exposing Si to the HCl gas to form $SiCl_x$ which then transports to and reacts with the growing film. HVPE grown bulk GaN doped with Si by using SiH_4 was reported to result in material with carrier concentration in the range of 1.3×10^{16}–8.3×10^{18} cm^{-3} while maintaining high crystalline and optical quality.[148] Alternative doping with oxygen, using gaseous source of high quality was also reported to achieve carrier concentration up to 8.5×10^{18} cm^{-3}. Temperature dependence of Hall carrier concentration and mobility of two such representative samples doped with Si (n = 7.3×10^{16} cm^{-3}) and O (n = 2×10^{17} cm^{-3}) are shown in Figure 5.4a and b, respectively, showing similar trends as the ones for undoped material with the offset related to the higher free carrier concentrations.

P-type doping of GaN bulk material was reported by both HVPE and ammonothermal techniques. Today Mg is the element proven to be the most efficient and controllable in achieving *p*-type conductivity. Despite that, high hole concentrations are challenging due to the high thermal activation energy of Mg, 150–200 meV,[149] which results in only a few percent of the Mg acceptor ionization at room temperature.

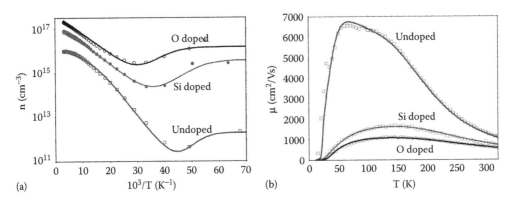

FIGURE 5.4 (a) Temperature dependence of (a) free electron concentration and (b) mobility as determined by Hall effect measurements for three representative samples, undoped, Si doped, and O doped. (Courtesy of David Look.)

Additionally, post-growth annealing is required for removing the hydrogen passivation of the Mg dopant. Other elements such as Zn and Cd have also been used but with even less activation efficiency due to their larger thermal activation energy and much less activation achievable at room temperature. The doping with Mg in HVPE was done by either using a separate temperature zone and source bath or by mixing the Mg in the Ga source. The HCl gas, reacting with Mg can for MgCl which is transported to the growth zone.[150] The role of H carrier gas was reported in a controversial way. Although one would expect that it is preferable to use inert carrier gas to avoid Mg-H passivation, the experimental results have shown the preferable behavior of the passivation during growth to prevent the incorporation of residual donor impurities.[151] Thus, despite the technological breakthroughs in growing *p*-type GaN by HVPE, the reports are very limited, and the exact governing mechanisms remain unclear.

Semi-insulating GaN bulk substrates are currently only possible by doping with compensating elements and were reported grown by HVPE and ammonothermal growth methods. The ammonothermal approach has used various compensating elements such as Fe, Mg, Zn, Cd[130]. The most promising approach suggested so far in HVPE growth is doping with Fe using a metalorganic source. The material was reported with high resistivity (larger than 10^6 Ωcm). Reports on this topic are relatively limited, and the compensation mechanisms are not fully understood, but the growth of semi-insulating GaN with high quality was proven successful.

5.3.3 Thermal Conductivity

One of the largest drawbacks to using heteroepitaxially grown GaN for power device applications is the low thermal conductivity of most of the typically used foreign substrates. High power and current in devices generate a substantial amount of Joule heating that must be removed in order to prevent failures associated with operation outside of the devices temperature range. In order to remove the heat generated during operation, the substrate must have thermal conductivity equal to or greater than the material of the device in order to effectively transfer heat to a heat sink or to be radiated to air. Typical thermal conductivities of foreign substrates used for epitaxial growth such as sapphire (~23 W/mK) and silicon (~110 W/mK) are lower than that of undoped or lightly doped GaN and act as a thermal barrier for heat flow out of the device. The most appropriate substrate for GaN would be an undoped native GaN substrate, which would provide high thermal conductivity (>200 W/mK). Foreign substrates such as SiC would provide high thermal conductivity needed for thermal management; however, the need for a buffer, the misfit dislocation generation, and the thermal barrier resistance may negate their benefits.

A thermal conductivity of ~150 Wm^{-1}K^{-1} at room temperature was typically reported for thin GaN epilayers with dislocation densities typically in the range of 10^8–10^9 cm^{-2}, which is relatively lower compared to that of SiC.[48,152-154] For bulk GaN, recent major advances in GaN crystal growth have resulted in significantly improved material properties of GaN substrates.[44,78,155-157] Slack et al. reported 227 Wm^{-1}K^{-1} for an undoped GaN plate grown by hydride vapor phase epitaxy (HVPE),[44] estimated using a steady-state heat flow method. For low threading dislocation density bulk GaN of 5×10^5 cm^{-2}, Mion et al. demonstrated an improvement in the thermal conductivity of HVPE bulk GaN, reaching 230 Wm^{-1}K^{-1} for Fe-doped GaN, determined through the 3ω method.[155] For unintentionally doped GaN, Richter et al. achieved a thermal conductivity of ~290 Wm^{-1}K^{-1} [156], while Shibata et al. reported a GaN thermal conductivity as high as 380 Wm^{-1}K^{-1} [157], both determined using the laser flash method. Both studies are consistent with the theory from Witek, suggesting a maximum GaN thermal conductivity of 410 Wm^{-1}K^{-1} [158]. Despite the large variation in the recently reported numbers, most likely related to different ensemble of residual and native point defects as a result of specific backgrounds and sources used in the different growth techniques, the numbers are consistently reported larger than thought for a long time. Therefore, bulk GaN substrates could offer an interesting alternative to SiC from the perspective of thermal management.

Temperature dependence of thermal conductivity of doped GaN was reported in several publications;[152-154, 159-161] however, the material studied was either of low crystalline quality (possessing high dislocation density [152-154]), or of only single representative sample doped with either Mg, or O, or the Fe was compared with undoped material.[159-161] We have recently studied thermal conductivity of bulk GaN at elevated temperatures, being most relevant for high power device at operation conditions. Figure 5.5a shows the temperature dependence of thermal conductivity, measured by 3ω methods in the temperature range of 300–460 K for two bulk GaN samples grown by HVPE on sapphire (provided by Kyma Technology) and on ammonothermal bulk GaN substrates (provided by Unipress) after removing the substrates for both. The two samples have dislocation density in the rage of 10^6 cm^{-2} and 10^4 cm^{-2}, and slightly different thermal conductivity of 265 and 279 W/m.K at RT, respectively. The latter is a bit surprising, considering that most theoretical models predict no significant effect of the dislocation density below 10^8cm^{-2} on the thermal conductivity, which calls for a more precise analysis of additional factors that may play a role.[153,155] The dependence of temperature is proportional to $T^{-\varepsilon}$ with $\varepsilon = 1.43$ and 1.38, respectively, for the representative samples. This slope is close to previously estimated slope of −1.43 and is characteristic of pure adamantine crystals in the temperature range below the Debye temperature.[162]

The thermal conductivity of bulk AlN reported so far varies greatly, which was attributed to the difference in the purity and microstructure of AlN.[163,164] The largest reported thermal conductivity value to date is 340 W/m.K at room temperature by Rojo et al.[49] One of the problems caused by extensive unavoidable oxygen impurity was a significant reduction of intrinsic thermal conductivity. The thermal

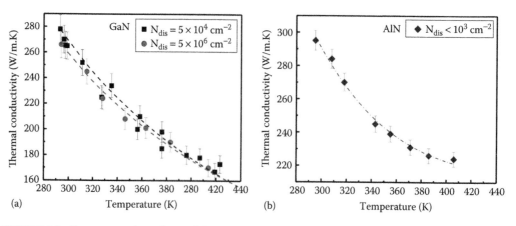

FIGURE 5.5 Temperature dependence of thermal conductivity of: (a) free standing GaN grown by HVPE on sapphire with dislocation density of about 10^6 cm^{-2} and on ammonothermal bulk GaN with dislocation density of about 10^4 cm^{-2}; (b) bulk AlN grown by modified HVPE method with dislocation density $<10^3$ cm^{-2}.

conductivity at room temperature of several samples was analyzed as a function of oxygen concentration, resulting in a trend that extrapolated to 320 W/m.K for pure AlN. A temperature dependence of thermal conductivity of AlN bulk samples grown by modified HVPE (provided by Nitride Solutions) in the temperature region of 300–460 K is shown in Figure 5.5b. The thermal conductivity at RT was determined 295 W/m.K among the best reported so far. The temperature dependence slope was determined to be –1.11, which is lower than the values of –1.21 estimated by Slack et al.[44] explained by acoustic phonon transport, and is indicative of more complex scattering mechanism in play.

The existing data on thermal conductivity of binary compound nitrides reveal the complexity of this property and its complex interplay with the different defects present in the materials grown by different techniques, on different substrates, and with different growth optimizations. The thermal transport in the ternary alloys such as AlGaN and AlInN are even more complicated due to the additional effect of alloy scattering, which is plausibly expected to dominate. Nevertheless, the defect contributions are likely to be significant in a similar way as proven to in the binary nitrides. However, despite the significant practical importance of the AlGaN alloy system its thermal properties remain the least explored and clarified.

Figure 5.6 visualizes the theoretical predictions based on the virtual-crystal model.[165] The experimental data published by two other groups are shown in Figure 5.6 (black symbols), as well as the experimental data collected on AlGaN layers produced by Kyma Technology (red symbols). One can see that while all the experimental values for the alloys are generally below the values for the AlN and GaN, there is quite a large scattering and deviation from the theoretically predicted numbers. Also, it is visible that the values for Kyma material are a bit higher than the predicted ones. A potential explanation could be related to the dislocation density. These initial results were collected for layers with thickness of about 1.5 μm, 10 μm and 25 μm while the other experimental values are for thinner layers with the thickness of about 0.5 μm. The results from TEM study show a noticeable reduction of dislocation density of about one order of magnitude with increasing thickness from 1 to 25 μm, which could explain the higher thermal conductivity, but more statistics is needed for samples with systematically varied parameters in order to extract reliable trends and to develop a correct alloy model.

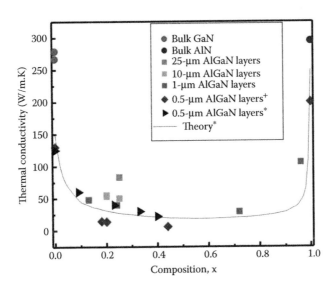

FIGURE 5.6 Thermal conductivity as a function of Al content, x, in $Al_xGa_{1-x}N$. For comparison purposes experimental data is shown for bulk GaN (dot symbols), AlN (dot symbols) and AlGaN (square symbols) grown by HVPE, thin AlGaN layers on sapphire from Reference 166 (rhombic symbols) and from Reference 165 (triangle symbols). The theoretical curve was calculated from the virtual-crystal model at RT. (From Daly, B. C. et al., *J. Appl. Phys.*, 92, 3820, 2002. With permission.)

5.4 Conclusions

The nitride substrate technologies have made a remarkable progress in the last two decades, though their development still lags behind the nitride thin-film device technology. Recent advances in nitride bulk crystal growth led to a significant improvement of material quality, which allowed better understanding of their properties and more correct estimation of their fundamental properties. Yet, growth related defects, doping, and thermal conductivity remain not fully elucidated, and more research is needed in order to reveal their interplay and to be fully controlled.

Acknowledgments

Support by National Science Foundation under grant DMR-1207075 and CBET-1336464 is acknowledged.

References

1. J. Y. Duboz, GaN as seen by the industry, *Phys. Stat. Sol. (a)*, 176, 5 (1999).
2. K. Rammohan et al., Progress of GaN Transistors, *Power Electronics Europe* 2015, 3, http://www.power-mag.com/pdf/issuearchive/77.pdf.
3. Strategies Unlimited, Gallium Nitride – 2003, Technology Status, Applications, and Market Forecasts, June 2003.
4. V. Y. Davydov et al., Band gap of hexagonal InN and InGaN alloys, *Phys. Stat. Sol. (b)*, 240, 787–795 (2002).
5. S. J. Pearton, C. R. Abernathy, M. E. Overberg, G. T. Thaler, A. H. Onstine, B. P. Gila, F. Ren, B. Lou, and J. Kim, New applications for gallium nitride, *Mater. Today*, 24–31 (2002).
6. M. Kuzuhara and H. Tokuda, Low-loss and high voltage III-nitride transistors for power switching applications, *IEEE Trans. Electron. Dev.*, 62, 405 (2015).
7. L. Liu and J. H. Edgar, Substrates for gallium nitride epitaxy, *Mater. Sci. Eng. R*, 37, 61–127 (2002).
8. R. Stevenson, GaN substrates offer high performance at a price, *Compound Semiconductor Magazine* 2004, 10, http://compoundsemiconductor.net/articles/ magazine/10/7/1.
9. K. Evans, Native substrates spar with the established technology, *Compound Semiconductor Magazine* 2004, 10, http://compoundsemiconductor.net/articles/magazine/ 10/10/4/1.
10. O. Ambacher, Growth and applications of group III-nitrides, *J. Phys. D: Appl. Phys.*, 31, 2653–2710 (1998).
11. B. Beaumont, P. Vennegues, and P. Gibart, Epitaxial lateral overgrowth of GaN, *Phys. Stat. Sol. (b)*, 227, 1–43 (2001); P. Gibart, Metal organic vapour phase epitaxy of GaN and lateral overgrowth, *Rep. Prog. Phys.*, 67, 667–715 (2004).
12. J. K. Sheu and G. C. Chi, The doping process and dopant characteristics of GaN, *J. Phys: Condens. Matter*, 14, R657–R702 (2002).
13. S. J. Pearton, J. C. Zolper, R. J. Shul, and F. Ren, GaN: Processing, defects and devices, *J. Appl. Phys.*, 86, 1–70 (1999).
14. S. J. Pearton, F. Ren, A. P. Zhang, and K. O. Lee, Fabrication and performance of GaN electronic devices, *Mat. Sci. Eng. R*, 30, 55–212 (2000).
15. H. Morkoc, R. Cingolani, and B. Gil, Polarization effects in nitride semiconductor device structures and performane of MODFETs, *Solid-State Electron.*, 43, 1753 (1999).
16. H. Morkoc and L. Liu, GaN-based MODFET and HBT, In *Nitride Semiconductors Handbook on Materials and Devices*, Ruterana, P., Albrecht, M., and Neugebauer, J., (eds.). Wiley-VCH: Weinheim, Germany, 547–626 (2004).
17. B. Monemar and G. Pozina, Group III-nitride based hetero and quantum structures, *Prog. Quant. Electron.*, 24, 239–290 (2000).

18. J. W. Orton and C. T. Foxon, Group III nitride semiconductor for short wavelength light-emitting devices, *Rep. Prog. Phys.*, 61, 1–75 (1998).

19. E. Munoz, E. Monroy, J. L. Pau, F. Calle, F. Omnes, and P. Gibart, III-nitrides and UV detection, *J. Phys: Condens. Matter*, 13, 7115–7137 (2001).

20. D. A. Steigerwald, J. C. Bhat, D. Collins, R. M. Fletcher, M. O. Holcomb, M. J. Ludowise, P. S. Martin, and S. L. Rudaz, Illumination with solid state lighting technology, *IEEE IEEE J. Sel. Top. Quant.*, 8, 310–320 (2002).

21. M. Bickermann and T. Paskova, Vapor transport growth of wide bandgap materials. In *Handbook of Crystal Growth*, Nishinaga, T. and Rudolph, P., (eds.). Elsevier: Amsterdam, the Netherlands, Vol. 2, pp. 621–669, (2015).

22. S. D. Lester, F. A. Ponce, M. G. Craford, and D. A. Steigerwald, High dislocation densities in high efficiency GaN-based light-emitting diodes, *Appl. Phys. Lett.* 66, 1249–1251 (1995).

23. S. Nakamura, InGaN-based violet LDs, *Semicond. Sci. Technol.*, 14, R27 1249–1251 (1999).

24. S. Poroski et al., Blue lasers on high pressure grown GaN single crystal substrates, *Europhys. News*, 35 (3), (2004).

25. T. Sugahara, H. Sato, M. Hao, Y. Naoi, S. Kurai, S. Tottori, K. Yamashita, K. Nishio, L. T. Romano, and S. Sakai, Direct evidence that dislocations are non-radiative recombination centers in GaN, *Jpn. J. Appl. Phys.* 37, 398 (1999).

26. X. A. Cao, J. A. Teetsov, F. Shahedipour-Sandvik, and S. D. Arthur, Microstructural origin of leakage current in GaN/InGaN ligh-emitting diodes, *J. Cryst. Growth*, 264, 172–177 (2004).

27. Z. Z. Chen, J. Zhao, Z. X. Qin, X. D. Hu, T. J. Yu, Y. Z. Tong, Z. J. Yang, X. Y. Zhou, G. Q. Yao, B. Zhang, and G. Y. Zhang, Study on the stability of the high-brightness with LED, *Phys. Stat. Sol. (b)*, 241, 2664–2667 (2004).

28. J. Y. Tsao, Solid-state lighting: Lamps, chips and materials for tomorrow, *IEEE Cir. Dev.*, 20, 28–37 (2004).

29. J. W. P. Hsu and M. J. Manfra, Direct imaging of reverse-bias leakage through pure screw dislocations in GaN films grown by MBE on GaN templates, *Appl. Phys. Lett.*, 81, 79–81 (2002).

30. J. W. P. Hsu, M. J. Manfra, D. V. Lang, S. Richter, S. N. G. Chu, A. M. Sergent, R. N. Kleiman, L. N. Pfeiffer, and R. J. Molnar, Inhomogeneous spatial distribution of reverse bias leakage in GaN Schottky diodes, *Appl. Phys. Lett.*, 78, 1685 (2001).

31. C. Y. Hsu, W. H. Lan, and Y. C. Wu, Effect of thermal annealing of Ni/Au ohmic contact on the leakage current of GaN based LEDs, *Appl. Phys. Lett.*, 83, 2447–2449 (2003).

32. A. Yamada, Y. Kawaguchi, and T. Yokogawa, Reduction of leakage current of p-n junction by using air-bridged lateral epitaxial growth technique, *Phys. Stat. Sold. (c)*, 7, 2494–2497 (2003).

33. L. S. McCarthy, I. P. Smorchkova, H. Xing, P. Kozodoy, P. Fini, J. Limb, D. L. Pulfrey, J. S. Speck, M. J. W. Rodwell, S. P. DenBaars, and U. K. Mishra, GaN GHBT: Toward an RF device, *IEEE Trans. Electron Dev.*, 48, 543–551 (2001).

34. F. Rampazzo, R. Pierobon, D. Pacetta, C. Gaquiere, D. Theron, B. Boudart, G. Meneghesso, and E. Zanoni, Hot carrier aging degradation phenomena in GaN based MESFETs, *Microelectron. Reliab.*, 44, 1375–1380 (2004).

35. L. Liu and J. H. Edgar, Transport efforts in the sublimation growth of aluminum nitride, *J. Cryst. Growth*, 220, 243–253 (2000).

36. V. Noveski, R. Schlesser, S. Mahajan, S. Beaudoin, and Z. Sitar, Mass transfer in AlN crystal growth at high temperature, *J. Cryst. Growth*, 264, 369–378 (2004).

37. B. Liu, J. H. Edgar, B. Raghothamachar, M. Dudley, J. Y. Lin, H. X. Jiang, A. Sarua, and M. Kuball, Free nucleation of aluminum nitride single crystals in HPBN crucible by sublimation. *J. Mater. Sci. Eng. B*, 117, 99–105 (2005).

38. D. Zhuang, J. H. Edgar, B. Liu, H. E. Huey, H. X. Jiang, J. Y. Lin, M. Kuball, F. Mogal, J. Chaudhuri, and Z. Rek, Bulk AlN crystal growth by direct heating of the source using microwaves, *J. Cryst. Growth*, 262, 168–174 (2004).

39. R. Dalmau, B. Raghothamachar, M. Dudley, R. Schlesser, and Z. Sitar, Crucible selection in AlN bulk crystal growth, *Mat. Res. Soc. Symp. Proc.*, 798, Y2.9.1 (2004).

40. G. A. Slack, J. Whitlock, K. Morgan, and L. J. Schowalter, Properties of crucible materials for bulk growth of AlN, *Mat. Res. Soc. Symp. Proc.*, 798, Y10.74.1 (2004).

41. B. M. Epelbaum, D. Hofmann, M. Bickermann, and A. Winnacker, Sublimation growth of bulk AlN crystals: Materials compatibility and crystal quality, *Mater. Sci. Forum*, 389–393, 1445–1448 (2002).

42. D. W. Brenner, R. Schlesser, Z. Sitar, R. Dalmau, R. Collazo, and Y. Li, Model for the influence of boron impurities on the morphology of AlN grown by PVT, *Surf. Sci.*, 560, L202–L206 (2004).

43. B. Liu, D. Zhuang, and J. H. Edgar, AlN Grown by Sublimation, In *Vacuum Science and Technology: Nitrides as Seen by the Technology*, Tanya, P. and Monemar, B., (eds.). Research Signpost: Trivandrum, India 59–78 (2002).

44. G. A. Slack, L. J. Schowalter, D. Morelli, and J. A. Freitas Jr., Some effects of oxygen impurities on AlN and GaN, *J. Cryst. Growth*, 246, 287–298 (2002).

45. S. Y. Karpov, A. V. Kulik, I. N. Przhevalskii, M. S. Ramm, and Y. N. Makarov, Role of oxygen in AlN sublimation growth, *Phys. Stat. Sol. (c)*, 1989–1992 (2003).

46. M. Bickermann, B. M. Epelbaum, and A. Winnacker, PVT growth of bulk AlN crystals with low oxygen contamination, *Phys. Stat. Sol. (c)*, 0, 1993–1996 (2003).

47. V. Noveski, R. Schlesser, B. Raghothamachar, M. Dudley, S. Mahajan, S. Beaudoin, and Z. Sitar, Seeded growth of bulk AlN crystals and grain evolution in polycrystalline AlN boules, *J. Cryst. Growth*, 279, 13–19 (2005).

48. G. A. Slack, Nonmetallic crystals with high thermal conductivity, *J. Phys. Chem. Solids*, 34, 321 (1973).

49. J. C. Rojo, L. J. Schowalter, K. Morgan, D. I. Florescu, F. H. Pollak, B. Raghothamachar, and M. Dudley, Single crystal AlN substrate preparation from bulk crystals, *Mat. Res. Symp. Proc.*, 680, E2.1.1. (2001).

50. P. Lu, R. Collazo, E. F. Dalmau, G. Durkaya, N. Dietz, and Z. Sitar, Different optical absorption edges in AlN bulk crystals grown in m- and c-orientation, *APL*, 93, 131922 (2008).

51. Q. Yan, A. Janotti, M. Scheffler, and C. Van de Walle, Origins of optical absorption and emission lines in AlN, *APL*, 105, 111104 (2014).

52. M. Bickermann, B. Epelbaum, O. Filip, P. Heimann, S. Nagata, and A. Winnacker, UV transparent single crystalline bulk AlN substrates, *Phys. Status Solidi.*, C7, 21 (2010).

53. M. Tanaka, S. Nakahata, K. Sogabe, H. Nakata, and M. Tobioka, *Jpn. J. Appl. Phys.*, 36, L1062 (1998).

54. B. M. Epelbaum, C. Seitz, A. Magerl, M. Bickermann, and A. Winnacker, Natural growth habit of bulk AlN crystals, *J. Cryst. Growth*, 265, 577–581 (2004).

55. N. B. Singh, A. Berghmans, H. Zhang, T. Wait, R. C. Clarke, J. Zingaro, and J. C. Golombeck, PVT growth of large AlN crystals, *J. Cryst. Growth*, 250, 107–112 (2003).

56. J. H. Edgar, L. H. Robins, S. E. Coatney, L. Liu, K. Ignatiev, and Z. Rek, A comparison of aluminum nitride freely nucleated and seeded on 6H-SiC, *Mater. Sci. Forum*, 338–342, 1599–1602 (2000).

57. J. H. Edgar, L. Liu, B. Liu, D. Zhuang, J. Chaudhuri, M. Kuball, S. Rajasingam, Bulk AlN crystal growth: Self-seeding and seeding on 6H-SiC substrates, *J. Cryst. Growth*, 246, 187–193 (2002).

58. Y. Shi, Z. Y. Xie, L. Liu, B. Liu, J. H. Edgar, and M. Kuball, Influence of buffer layer and 6H-SiC substrate polarity on the nucleation of AlN grown by the sublimation sandwich technique, *J. Cryst. Growth*, 233, 177–186 (2001).

59. M. Kuball, J. M. Hayes, Y. Shi, and J. H. Edgar, Phonon lifetimes in bulk AlN and their temperature dependence, *Appl. Phys. Lett.*, 77, 1958–1960 (2000).

60. B. Raghothamachar, W. M. Vetter, M. Dudley, R. Dalmau, R. Schlesser, Z. Sitar, E. Michael, and J. W. Kolis, Synchrotron white beam topography characterization of PVT grown AlN and ammonothermal GaN, *J. Cryst. Growth*, 246, 271–280 (2002).

61. G. A. Slack and T. F. McNelly, Ain single crystals, *J. Cryst. Growth*, 42, 560–563 (1977).

62. J. C. Rojo, G. A. Slack, K. Morgan, L. J. Schowalter, and M. Dudley, Growth of self-seeded AlN by sublimation-recondensation and substrate preparation, *Mat. Res. Soc. Symp.*, 639, G1.10.2 (2001).

63. L. J. Schowalter, J. C. Rojo, N. Yakovlev, Y. Shusterman, K. Dovidenko, R. Wang, I. Bhat, and G. A. Slack, Preparation and characterization of single crystal AlN substrates, *MRS Internet J. Nitride Semicond. Res.*, 5S1, W6.7 (2000).

64. N. L. Yakovlev, J. C. Rojo, and L. J. Schowalter, Morphology of facets on vapor-grown AlN crystals, *Surf. Sci.*, 493, 519–525 (2001).

65. B. Raghothamachar, M. Dudley, J. C. Rojo, K. Morgan, and L. J. Schowalter, x-ray characterization of bulk AlN single crystals grown by the sublimation technique, *J. Cryst. Growth*, 250, 244–250 (2003).

66. C. M. Balkas, Z. Sitar, T. Zheleva, L. Bergman, R. Nemanich, and R. F. Davis, Sublimation growth and characterization of bulk AlN single crystals, *J. Cryst. Growth*, 179, 363–370 (1997).

67. B. Liu, Y. Shi, L. Liu, J. H. Edgar, and D. N. Braski, Surface morphology and composition characterization at the initial stages of AlN crystal growth, *Mat. Res. Soc. Symp. Proc.*, 639, G3.13.1 (2001).

68. L. Liu, D. Zhuang, B. Liu, Y. Shi, J. H. Edgar, S. Rajasingam, and M. Kuball, Characterization of AlN crystals grown by sublimation, *Phys. Stat. Sol. (a)*, 188, 769–773 (2001).

69. L. Liu, B. Liu, Y. Shi, and J. H. Edgar, Growth mode and defects in AlN sublimed on (0001) 6H-SiC substrates, *MRS Internet J. Nitride Semicond. Res.*, 6, 7 (2001).

70. Y. Shi, B. Liu, L. Liu, J. H. Edgar, H. M. Meyer III, E. A. Payzant, L. R. Walker, N. D. Evans, J. G. Swadener, and J. Chaudhuri, Initial nucleation study and new technique for sublimation growth of AlN on SiC substrate, *Phys. Stat. Sol. (a)*, 188, 757–762 (2001).

71. Y. Shi, B. Liu, L. Liu, J. H. Edgar, E. A. Payzant, J. M. Hayes, and M. Kuball, New technique for sublimation growth of AlN single crystals, *MRS Nitride Semicond. Res.*, 6, 5 (2001).

72. B. Heying, X. H. Xu, S. Keller, Y. Li, D. Kapolnek, B. P. Keller, S. P. DenBaars, and J. S. Speck, Role of threading dislocation structure on the x-ray diffraction peak widths in epitaxial GaN films, *Appl. Phys. Lett.*, 68, 463 (1996).

73. T. Metzger, R. Höpler, E. Born, S. Christiansen, M. Albrecht, H. P. Strunk, O. Ambacher, M. Stutzmann, R. Stömmer, and H. Göbel, Coherent X-Ray Scattering Phenomenon in Highly Disordered Epitaxial AlN Films, *Phys. Status Solidi A*, 162, 529–535 (1997).

74. R. Schlesser, R. Dalmau, and Z. Sitar, Seeded growth of AlN bulk single crystals by sublimation, *J. Cryst. Growth*, 241, 416–420 (2002).

75. V. Noveski, R. Schlesser, J. Freitas, Jr. S. Mahajan, S. Beaudoin, and Z. Sitar, Vapor phase transport of AlN in an RF heated reactor: Low and high temperature studies, *Mat. Res. Soc. Symp. Proc.*, 798, Y2.8.1 (2004).

76. D. Zhuang, Z. G. Herro, R. Schlesser, and Z. Sitar, Seeded growth of AlN single crystals by physical vapor transport, *J. Cryst. Growth*, 287, 372 (2006).

77. P. Lu, R. Collazo, R. F. Dalmau, G. Durkaya, N. Dietz, B. Raghothamachar, M. Dudley, and Z. Sitar, Seeded growth of AlN bulk crystals in m- and c-orientation, *J. Cryst. Growth*, 312, 58 (2010).

78. T. Paskova and K. R. Evans, GaN substrates – Progress, status, and prospects, *IEEE J. Sel. Quant. Electron.*, 15, 4 (2009).

79. T. Paskova, D. A. Hanser, and K. R. Evans, GaN substrates for III-nitride devices, *Proc. IEEE*, 98, 7 (2010).

80. B. Feigelson and T. Paskova, Growth of bulk GaN crystals, In Comprehensive Semiconductor Science and Technology, Bhattacharya, P., Fornari, R., and Kamimura, H., (eds.). Elsevier: Amsterdam, the Netherlands, Vol. 3, pp. 232–281 (2011).

81. T. Paskova and B. Monemar, HVPE-GaN quasi-substrates for nitride device structures, In Low Dimensional Nitride Semiconductors, Gil, B., (ed.). Oxford University Press: Oxford, UK, pp. 79–104, (2002).

82. E. Aujol, J. Napierala, A. Trassoudaine, E. Gil-Lafon, and R. Cadoret, Hydrogen and nitrogen ambient effects on epitaxial growth of GaN by hydride vapor epitaxy, *J. Cryst. Growth*, 222, 538–548 (2001).

83. R. Cadoret, Growth Mechanisms of (00.1) GaN Substrates in the hyrdride vapor-phase method: Surface diffusion, spiral growth, H_2 and $GaCl_3$ mechanisms. *J. Cryst. Growth*, 205, 123–135 (1999).

84. A. Trassaudaine, E. Aujol, R. Cadoret, T. Paskova, and B. Monemar, A new mechanism in the growth process of GaN by HVPE. *Materials Research Society Symposium Proceedings*, G.3.2, Boston, MA, (2000).

85. T. Paskova, E. Svedberg, A. Henry, I. G. Ivanov, R. Yakimova, and B. Monemar, Thick GaN layers grown on a-plane sapphire substrates by hydride vapor epitaxy. *Physica Scripta*, T79, 67–70 (1999a).

86. A. Shintani and S. Minagawa, Kinetics of the epitaxial growth of GaN using Ga, HCl and NH_3, *J. Cryst. Growth*, 22, 1 (1974).

87. W. Seifert, R. Franzheld, E. Butter, H. Sobotta, and V. Reide, On the origin of free carriers in high-conducting n-GaN. *Cryst. Res. Technol.*, 18, 383 (1983).

88. S. S. Liu and D. A. Stevenson, Growth kinetics and catalytic effects in the vapor phase epitaxy of gallium nitride. *J. Electrochem. Soc.*, 125, 1161 (1978).

89. N. Coudurier, R. Boichot, V. Fellmann, A. Claudel, E. Blanquet, A. Crisci, S. Coindeau, D. Pique, and M. Pons, Effects of the V/III ratio on the quality of AlN grown on (0001) sapphire by HTHVPE, *Phys. Status Solid C*, 10, 362 (2013).

90. Y. Kumagai, T. Nagashima, H. Murakami, K. Takada, and A. Koukitu, Characterization of a free-standing AlN substrate prepared by HVPE, *Phys. Stat. Sol. C*, 5, 1512 (2008).

91. S. A. Safvi, N. R. Perkins, M. N. Horton, R. Matyi, and T. F. Kuech, Effect of reactor geometry and growth parameters on the uniformity and material properties of GaN/sapphire grown by hydride vapor-phase epitaxy, *J. Cryst. Growth*, 182, 233–240 (1997)

92. Y. Kumagai, T. Yamane, T. Miyajia, H, Murakami, Y. Kangawa, and A. Koukitu, Hydride vapor phase epitaxy of AlN: Thermodynamic analysis of aluminum source and its application to growth, *Phys. Stat. Sol. (c)*, 0, 2498–2501 (2003).

93. T. Yamane, H. Murakami, Y. Kangawa, Y. Kumagai, and A. Koukitu, Growth of thick AlN layer on sapphire (001) substrate using HVPE, *Phys. Stat. Sol. C*, 2, 2062 (2005).

94. W. M. Yim, E. J. Stofko, P. J. Zanzucchi, J. I. Pankove, M. Ettenberg, and S. L. Gilbert, Epitaxially grown AlN and its optical bandgap, *J. Appl. Phys.*, 44, 292–295 (1973).

95. A. V. Dobrynin, G. A. Naida, and V. A. Novoselov, *Phys. Stat. Sol. (a)*, 104, K47 (1987).

96. Y. Melnik, D. Tsvetkov, A. Pechnikov, I. Nikitina, N. Kuznetsov, and V. Dmitriev, Characterization of AlN/SiC epitaxial wafers fabricated by HVPE, *Phys. Stat. Sol. (a)*, 188, 463–466 (2001).

97. O. Y. Ledyaev, A. E. Cherenkov, A. E. Nikolaev, I. P. Nikitina, N. I. Kuznetsov, and M. S. Dunaevski, Properties of AlN layers grown on SiC substrates in wide temperature range by HVPE, *Phys. Stat. Sol. (c)*, 0, 474–478 (2002).

98. A. S. Usikov et al., Material quality improvement for homoepitaxial GaN and AlN layers grown on sapphire-based templates, *Phys. Stat. Sol. (c)*, 2580–2584 (2004).

99. A. Nikolaev, I. Nikitina, A. Zubrilov, M. Mynbaeva, Y. Melnik, and V. Dmitriev, AlN wafers fabricated by HVPE, *MRS Internet J. Nitride Res.* W6.5 (1999).

100. Y. Melnik et al., AlN substrates: Fabrication via vapor phase growth and characterization, *Phys. Stat. Sol. (a)*, 200, 22–25 (2003).

101. M. Albrecht, I. P. Nikitina, A. E. Nikolaev, Yu. V. Melnik, V. A. Dmitriev, and H. P. Strunk, Dislocation reduction in AlN and GaN bulk crystals grown by HVPE, *Phys. Stat. Sol. (a)*, 176, 453–458 (1999).

102. X. Gong, K. Xu, J. Huang, T. Liu, G. Ren, J. Wang, and J. Zhang, Evolution of the surface morphology of AlN epitaxial film by HVPE, *J. Cryst. Growth*, 409, 100 (2015).

103. T. Nagashima, M. Harada, H. Yanagi, Y. Kumagai, A. Koukitu, and K. Takada, High speed epitaxial growth of AlN above 1200°C by HVPE, *J. Cryst. Growth*, 300, 42 (2007).

104. T. Nagashima, M. Harada, H. Yanagi, H. Fukuyama, Y. Kumagai, A. Koukitu, and K. Takada, Improvement of AlN crystalline quality with high epitaxial growth rates by HVPE, *J. Cryst. Growth*, 305, 355 (2007).

105. M. Balaji et al., Effects of AlN nucleation layers on the growth of AlN films using high temperature HVPE, *J. Alloy. Compd.*, 526, 103 (2012).

106. M. Pons, R. Boichot, N. Coudurier, A. Claudel, E. Blanquet, S. Lay, F. Mercier, and D. Pique, HT CVD of AlN, growth and evaluation, *Surf. Coat. Technol.*, 230, 111 (2013).

107. A. Claudel et al., Influence of the V/III ratio in the gas phase on thin epitaxial AlN layers grown on (0001) sapphire by high temperature HVPE, *Thin Solid Films*, 573, 140 (2014).

108. K. Okumura, T. Nomura, H. Miyake, K. Hiramatsu, and O. Eryuu, HVPE growth of AlN on trench-patterned 6H-SiC substrates, *Phys. Status Solidi.C*, 8, 467 (2011).

109. K. Fujita, K. Okuura, H. Miyake, K. Hiramatsu, and H. Hirayama, HVPE growth of thick AlN on trench-patterned substrate, *Phys. Status Solidi.C*, 8, 1483 (2011).

110. D. Kamber, Y. Wu, E. Letts, S. DenBaars, J. Speck, and S. Nakamura, LEO of AlN on patterned SiC substrates by HVPE, *APL*, 90, 122116 (2007).

111. Y. Kumagai, J. Tajima, M. Ishizuki, T. Nagashima, H. Murakami, K. Takada, and A. Koukitu, Self separation of a thick AlN layer from a sapphire substrate via interfacial voids formed by the decomposition of sapphire, *Appl. Physics Exp.*, 1, 045003 (2008).

112. Y. Kumagai, Y. Enatsu, M. Ishizuki, Y. Kubota, J. Tajima, T. Nagashima, H. Murakami, K. Takada, and A. Koukitu, Investigation of void formation beneath thin AlN layers by decomposition of sapphire substrates for self-separation of thick AlN layers grown by HVPE, *J. Cryst. Growth*, 312, 2530 (2010).

113. P. Geiser, J. Jun, S. M. Kazakov, P. Wagli, J. Parpinski, B. Batlogg, and L. Klemm, AlxGa1-xN bulk single crystals, *App. Phys. Lett.*, 86, 081908 (2005).

114. M. Miyanaga, N. Mizuhara, S. Tanizaki, I. Satoh, and H. Nakahata, AlGaN bulk crystal manufacturing method and AlGaN substrate manufacturing method, US Patent 2011/0114016A1.

115. A. Koukitu, J. Kikuchi, Y. Kangawa, and Y. Kumagai, Thermodynamic analysis of AlGaN HVPE growth, *J.C.G.* 281, 47 (2005).

116. H. Murakami, J. Kikuchi, Y. Kumagai, and A. Koukitu, Thermodynamic study on the role of hyroden during hydride vapor phase epitaxy of AlGaN, *Phys. Stat. Sol. (c)*, 3, 1457 (2006).

117. A. Koukitum, F. Satoh, T. Yamane, H. Murakami, and Y. Kumagai, HVPE growth of AlGaN ternary alloy using AlCl3 and GaCl, *J. Cryst. Growth*, 305, 335 (2007).

118. H. Ahn et al., Growth of thick AlGaN by mixed-source HVPE, *Appl. Surf. Sci.*, 243, 178 (2005).

119. K. S. Jang, K. H. Kim, S. L. Hwang, H. S. Jeon, H. S. Ahn, M. Yang, W. J. Choi, S. W. Kim, Y. Honda, M. Yamaguchi, N. Sawaki, J. Yoo, S. M. Lee, M. Koike, Characterization of AlGaN, Te-doped GaN and Mg-doped GaN grown by hydride vapor phase epitaxy, *Phys. Stat. Sol.(C)*, 4, 133–136 (2007).

120. V. Soukhoveev, O. Kovalenkov, L. Shapovalova, V. Uvantsov, A. Usikov, V. Dmitriev, V. Davydov, and A. Smirnov, AlGaN epitaxial layers grown by HVPE on sapphire substrates, *Phys. Stat. Sol. (c)*, 3, 1483 (2006).

121. A. Usikov, V. Soukhoveev, L. Shapovalova, A. Syrkin, O. Kovalenkov, A. Volkova, V. Sizov, V. Ivantsov, and V. Dmitriev, New results on HVPE growth of AlN, GaN, InN and their alloys, *Phys. Stat. Sol. (c)*, 5, 1825 (2008).

122. S. Hagedorn, E. Richter, U. Zeimer, and M. Weyers, HVPE growth of thick Al0.45Ga0.55N layers on trench patterned sapphire substrates, *Phys. Status Solidi. C*, 10, 355 (2013).

123. E. Richter, S. Fleischmann, D. Goran, S. Hagedorn, W. John, A. Mogilatenko, D. Prasai, U. Zeimer, M. Weyers, and G. Trankle, HVPE of c-plane AlGaN Layers on patterned sapphire substrates, *J. Elec. Mater.*, 43, 814 (2014).

124. D. R. Ketchum and J. W. Kolis, Crystal growth of gallium nitride in supercritical ammonia, *J. Cryst. Growth*, 222, 431–434, (2001).

125. R. Dwilinski, R. Doradzinski, J. Garczynski, L. Sierzputowski, J. M. Baranowski, and M. Kaminska, AMMONO method for GaN and AlN production, *Diamond Rel. Mater.*, 7, 1348–1350, (1998).

126. B. Wang and M. Callahan, Transport growth of GaN crystals by the ammonothermal technique using various nutrients, *J. Cryst. Growth*, 291, 455–460, (2006).

127. T. Hashimoto, K. Fujito, F. Wu, B. Haskell, P. T. Fini, J. S. Speck, and S. Nakamura, Structural characterization of thick GaN films grown on free-standing GaN seeds by ammonothermal method using basic ammonia, *Jpn. J. Appl. Phys.*, 44 (25), L797–L799, (2005).

128. T. Fukuda and D. Ehrentraut, Prospects for ammonothermal growth of large GaN crystals, *J. Cryst. Growth*, 305, 304–310, (2007).

129. M. P. D'Evelyn, H. C. Hong, D.-S. Park, H. Lu, E. Kaminsky, R. R. Melkote, P. Perlin, M. Lesczynski, S. Porowski, and R. J. Molnar, Bulk GaN crystal growth by the high-pressure ammonothermal method, *J. Cryst. Growth*, 300, 11–16 (2007).

130. R. Dwilinski, R. Doradzinski, J. Garczynski, L. P. Sierzputowski, A. Puchalski, Y. Kanbara, K. Yagi, H. Minakuchi, and H. Hayashi, Excellent crystallinity of truly bulk ammonothermal GaN, *J. Cryst. Growth*, 310, 3911–3916 (2008).

131. M. Bockowski, Review: Bulk growth of gallium nitride: Challenges and difficulties, *Cryst. Res. Technol.*, 42 (12), 1162–1175 (2007).

132. B. Łucznik, B. Pastuszka, I. Grzegory, M. Boćkowski, G. Kamler, E. Litwin-Staszewska, S. Porowski, Deposition of thick GaN layers by HVPE on the pressure grown GaN substrates, *J. Cryst. Growth*, 281, 38–46 (2005).

133. I. Grzegory, B. Łucznik, M. Boćkowski, B. Pastuszka, M. Kryśko, G. Kamler, G. Nowak, S. Porowski, Growth of bulk GaN by HVPE on pressure grown seeds, *Proc SPIE*, 6121, 612107 (2006).

134. R. F. Davis, A. M. Roskowski, E. A. Preble, J. S. Speck, B. Heying, J. A. Freitas, E. R. Glaser, W. E. Carlos, Gallium nitride materials - progress, status, and potential roadblocks, *Proc IEEE*, 90, 993–1005 (2002).

135. K. Motokia, T. Okahisaa, S. Nakahataa, N. Matsumotoa, H. Kimurab, H. Kasaib, K. Takemotob, K. Uematsua, M. Uenoa, Y. Kumagaic, A. Koukituc, H. Sekic, Growth and characterization of freestanding GaN substrates, *J. Cryst. Growth*, 237, 912–921 (2002).

136. B. Lucznik, B. Pasturszka, G. Kamler, I. Grzegory, and S. Porowski, In Technology of *Gallium Nitride Crystal Growth*, Ehrentraut, D., Meissner, E., Bockowski, M., (eds.). Growth of Bulk GaN Crystals by HVPE on Single Crystalline GaN seeds, Springer series in materials science, Springer: Berlin, Germany, Vol. 133, 61–78 (2009).

137. T. Paskova, P. P. Paskov, J. Birch, E. Valcheva, M. Abrashev, S. Tungasmita, B. Monemar, HVPE regrowth on free-standing GaN quasi-substrate, *IPAP Conf. Ser.*, C1, 19–22 (2000).

138. T. Sochacki, Z. Bryan, M. Amilusik, R. Collazo, B. Lucznik, J. L. Weyher, G. Nowak, B. Sadovyi, G. Kamler, R. Kucharski, Preparation of Free-Standing GaN Substrates from Thick GaN Layers Crystallized by Hydride Vapor Phase Epitaxy on Ammonothermally Grown GaN Seeds, *Appl. Phys. Express*, 6, 075504 (2013).

139. Hexatech Inc. AlN Technology Overview. 2011. http://www.hexatechinc.com/aln-technology-overview.html.

140. T. Sochacki, M. Amilusik, M. Fijalkowski, M. Iwinska, B. Lucznik, J. L. Weyher, G. Kamler, R. Kucharski, I. Grzegory, and M. Bockowski, Examination of defects and the seed's critical thickness in HVPE-GaN growth on ammonothermal GaN seed, *Phys. Status Solidi. B*, 252, 1172–1179 (2015). doi:10.1002/pssb.201451604.

141. S. K. Mathis, A. E. Romanov, L. F. Chen, G. E. Beltz, W. Pompe, and J. S. Speck, Modeling of Threading Dislocation Reduction in Growing GaN Layers, *Phys. Status Solidi. A*, 179, 125–145 (2000).

142. D. Hanser, L. Liu, E. A. Preble, D. Thomas, and M. Williams, Growth and fabrication of 2 inch free-standing GaN substrates via the boule growth method, *Mater. Res. Soc. Symp. Proc.*, 798, Y2.1.1 (2004).

143. T. Fujimori, HVPE and ammonothermal GaN substrates for high performance devices, *Plenary Talk at Int Workshop on Nitride Semiconductors*, October 6–10, Montreux, Switzerland, (2008).

144. T. Hino, S. Tomiya, T. Miyajima, K. Yanashima, S. Hashimoto, and M. Ikeda, Characterization of threading dislocations in GaN epitaxial layers, *Appl. Phys. Lett.*, 76, 3421–3423 (2000).

145. J. L. Weyher, L. Macht, G. Kamber, J. Borysiuk, and I. Grzegory, Characterization of GaN single crystals by defect-selective etching, *Phys. Status Solidi. C*, 3, 821–826 (2003).

146. Y. Taniyasu, M. Kasu, and T. Makimoto, Electrical conduction properties of n-type Si-doped AlN with high electron mobility (>100cm^{2}V^{-1}s^{-1}), *Appl. Phys. Lett.*, 85, 4672–4674 (2004).

147. T. Schulz, K. Irmscher, M. Albrecht, C. Hartmann, J. Wollweber, and R. Fornari, n-type conductivity in sublimation-grown AlN bulk crystals, *Phys. Status Solidi. RRL*, 1, 147–149 (2007).

148. M. Slomski, P. P. Paskov, J. H. Leach, J. F. Muth, and T. Paskova, Thermal conductivity of bulk GaN grown by HVPE: Effect of Si doping, *Phys. Stat. Solidi B*, 10.1002/pssb.201600713 (2017)

149. W. Götz, R.S Kern, C.H Chen, H. Liu, D.A Steigerwald, R.M Fletcher, Hall-effect characterization of III–V nitride semiconductors for high efficiency light emitting diodes, *Mater. Sci. Eng. B*, 59, 211–217 (1999)

150. A. Usikov, O. Kovalenkov, V. Ivantsov, V. Sukhoveev, V. Dmitriev, N. Schmidt, D. Poloskin, V. Petrov, V. Ratnikov, p-Type GaN epitaxial layers and AlGaN/GaN heterostructures with high hole concentration and mobility grown by HVPE, *Mater. Res. Soc. Symp. Proc.*, 831, 453–457 (2005).

151. J. Neugebauer and C. G. Van der Walle, Hydrogen in GaN: Novel Aspects of a Common Impurity, *Phys. Rev. Lett.*, 75, 4452–4455 (1995).

152. D. I. Florescu, V. M. Asnin, F. H. Pollak, R. J. Molnar, and C. E. C. Wood, High spatial resolution thermal conductivity and Raman spectroscopy investigation of hydride vapor phase epitaxy grown n-GaN/sapphire (0001): Doping dependence, *J. Appl. Phys.*, 88, 3295 (2000).

153. J. Zou, D. Kotchetkov, A. A. Balandin, D. I. Florescu and F. H. Pollak, Thermal conductivity of GaN films: Effect of impurities and dislocations, *J. Appl. Phys.*, 92, 2534 (2002).

154. D. Kotchetkov, J. Zou, A. A. Balandin, D. I. Florescu, and F. H. Pollak, Effect of dislocations on thermal conductivity of GaN layers, *Appl. Phys. Lett.*, 79, 4316–4318 (2001).

155. C. Mion, J. F. Muth, E. A. Preble, and D. Hanser, Accurate dependence of GaN thermal conductivity on dislocation density, *Appl. Phys. Lett.*, 89, 092123 (2006).

156. E. Richter, M. Gründer, B. Schineller, F. Brunner, U. Zeimer, C. Netzel, M. Weyers, and G. Tränkle, GaN boules grown by high rate HVPE, *Phys. Stat. Sol. (c)*, 8, 5 (2011).

157. H. Shibata, Y. Waseda, K. Shimoyama, K. Kiyomi, and H. Nagaoka, Gallium nitride based material and method of manufacturing the same, US 2009/0081110A1 (2009).

158. A. Witek, Some aspects of thermal conductivity of isotopically pure diamond—A comparison with nitrides, *Diam. Relat. Mater.*, 7, 962 (1998).

159. A. Jezowski, B. A. Danilchenko, M. Bockowski, I. Grzegory, S. Krukowski, T. Suski, and T. Paszkiewicz, Thermal conductivity of GaN crystals in 4.2 -300 K range, *Sol. State Commun.*, 128, 69–72 (2003).

160. K. Jagannadham, E. A. Berkman, and N. Elmasry, Thermal conductivity of semi-insulating, p-type, and n-type GaN films on sapphire, *J. Vac. Sci. Technol. A*, 26, 375–379 (2008).

161. R. B. Simon, J. Anaya, and M. Kuball, Thermal conductivity of bulk GaN --Effects of oxygen, magnesium doping, and strain field compensation, *Appl. Phys. Lett.*, 105, 202105 (2014).

162. C. Mion, J. F. Muth, E. A. Preble, and D. Hanser, Accurate dependence of gallium nitride thermal conductivity on dislocation density, *Appl. Phys. Lett.*, 89, 092123 (2006).

163. W. Li and N. Mingo, Thermal conductivity of bulk and nanowire InAs, AlN, and BeO polymorphs from first principles, *J. Appl. Phys.*, 114, 183505 (2013).

164. G. A. Slack, Nonmetallic crystals with high thermal conductivity, *J. Phys. Chem. Sol.*, 34 (2), 321–335 (1973).

165. W. Liu and A. A. Balandin, Thermal conduction in Al$_x$Ga$_{1-x}$N alloys and thin films, *JAP*, 97, 073710 (2005).

166. B. C. Daly, H. J. Maris, A. V. Nurmikko, M. Kuball, and J. Han. Optical pump-and-probe measurement of the thermal conductivity of nitride thin films, *J. Appl. Phys.*, 92, 3820 (2002).

6

III-Nitride Metalorganic Vapor-Phase Epitaxy

Daniel D. Koleske

6.1 Introduction to the MOVPE Growth Technique

The metalorganic vapor-phase epitaxy (MOVPE) growth technique is well suited for the group III nitrides for reasons that will become obvious in this review. It is the growth technique of choice for visible LED manufacturing since both the GaN film of sufficient thickness can be formed on sapphire, SiC or Si, along with the active light-emitting region. Numerous other optical and electronic devices have also been grown using MOVPE but our main focus for this review will be LED growth as this was the primary driver of the majority of research into group III-nitride growth.

Similar to the MOVPE growth of the other III-V semiconductors, III-N semiconductors use trimethyl or triethyl sources of gallium, indium, and aluminum. The unique aspect of III-N growth is the use of ammonia NH_3, as the hydride precursor. During III-N growth, NH_3 is used in great excess (large V/III ratios) to achieve the best films. Typically, for GaN, V/III ratios are ~1,000–2,000, while for InGaN, V/III ratios are ~10,000 and for AlN, V/III ratios are ~1–100.

Based on the equilibrium constant for NH_3 dissociation, N_2 and H_2 should be favored over NH_3 above 200°C, implying that the III-Ns could be grown at temperatures as low as 200°C. Yet, to produce surface N of sufficient quantities, the surface must be heated to above 500°C–1,000°C for GaN. This is because NH_3 must be catalytically dissociated at the surface at high temperatures to break the three ~4 eV N-H bonds (Koleske et al., 1998a). Curiously, the bond energies of N≡N and H–H are significantly and slightly larger, respectively, than the N–H bonds, suggesting that once N_2 is formed, it no longer participates in the growth process while H_2 plays a unique role reacting with adsorbed metals and scavenging adsorbed impurities.

The high growth temperature required for the growth of GaN provides an additional benefit. This benefit is a gas-phase diffusion mechanism for the Ga atoms which becomes active at temperatures >900°C (Koleske et al., 2014a). This gas-phase transport mechanism allows for the growth of large grained GaN with low enough dislocation densities to be useful for LED and HEMT device fabrication provided the correct nucleation layer (NL) is first established. Without this gas-phase transport mechanism, it is unlikely that many of the breakthroughs discussed in the upcoming sections would have occurred as rapidly as they did.

6.2 A Brief GaN History

GaN was first synthesized at the University of Chicago in 1932 (Johnson et al., 1932). Interestingly, after describing the synthesis, the authors described that GaN sublimes above 800°C in H_2 and they also are the first to state GaN's well-known chemical inertness and thermal stability (Johnson et al., 1932). These hallmarks of GaN, its chemical inertness, and thermal stability result from the strong covalent and partially ionic bonding present in the material (Koleske et al., 1998a). The strong bonding produces the large bandgap of 3.39 eV, leading to its use for both visible-wavelength optical and high-power electronic devices.

Synthesis of GaN in the 1960s and 1970s was conducted with hydride vapor phase epitaxy (HVPE) (Maruska and Tietjen, 1969; Ejder, 1974; Madar et al., 1977), using NH_3 with either metallic gallium (Ejder, 1974) or reactor generated gallium chlorides (Elwell and Elwell, 1988). This early HVPE work on GaN established crystallographic orientations of GaN on sapphire (Madar et al., 1977) and other surfaces (Liu and Edgar, 2002), while today it is considered to be more of a bulk growth technique. Using HVPE, Maruska and Tietjen were able to make colorless GaN and verify GaN's direct band-gap energy of 3.39 eV (Maruska and Tietjen, 1969). Later p-type doping work by Maruska led to the first working GaN-diode (Maruska et al., 1973; Maruska and Rhines, 2015). At this time, the material quality was severely limited, primarily because of large densities of defects, such as dislocations from the heteroepitaxy and the low purity of the chemicals used for growth, especially NH_3 with high levels of H_2O, which lead to high n-type carrier concentrations (Maruska and Tietjen, 1969; Elwell and Elwell, 1988). Over time, the chemical purity improved and bulk GaN substrates, although extremely difficult to grow (Karpiński et al., 1984), are currently more available (Dwiliński et al., 2010).

6.2.1 First III-N MOVPE Work

The first reported growth of III-N materials using MOVPE was by Hal Manasevit and coworkers in the late 1960s (Manasevit, 1972, 1981). In his early work, Manasevit used TMAl, NH_3, and a high flow rate of H_2 carrier gas to grow AlN at ~1,200°C on Si, SiC, sapphire, and $MgAl_2O_4$ surfaces (Manasevit, 1972). For GaN growth, TMGa was used at 925°C–975°C on c-plane sapphire, resulting in c-plane GaN rotated by 30 degrees with respect to the in-plane sapphire lattice (Manasevit, 1972). The 6 μm thick GaN films had background carrier concentrations above 10^{19} cm^{-3} and a mobility of 60 cm^2/Vs and were found to be resistant to acids and bases, but could be destroyed by heating H_2 to 1,050°C (Manasevit, 1972), confirming the earlier observations of Johnson et al. (1932). While MOVPE proved useful for III-N materials growth, issues with high defect density, reagent impurities, and rough film growth, continued to limit the usefulness of this material system.

The pivotal breakthrough for all III-N research occurred when then graduate student Hiroshi Amano under the direction of Prof. Isamu Akasaki at Nagoya University invented nucleation layers (NLs) for GaN on sapphire growth. Prior to this breakthrough, the GaN films grown on sapphire were rough looking like *frosted glass* (Amano et al., 1986). However, by first growing AlN NLs at a lower growth temperature, the GaN films layers appeared smooth and transparent (Amano et al., 1986). These low-temperature AlN layers resulted in narrower XRD linewidths for the GaN compared to films without the AlN layers (Amano et al., 1986). If AlN temperature was too high, cracks were observed along the <11–20> direction; if the AlN temperature was too low, pits formed in the GaN, suggesting that optimization of the AlN buffer was required (Amano et al., 1986).

Further breakthroughs were made by Shuji Nakamura at Nichia Chemical Corporation starting ~1989, which included development of low-temperature GaN buffer layers (Nakamura, 1991c), design of a novel two flow MOVPE system to reduce parasitic gas-phase chemistry and improve growth rate (Nakamura et al., 1991a), and *in situ* observations of growth (Nakamura, 1991a, 1991b), all of which were reported in 1991. The key to these breakthroughs was experimenting with the design of the reactors (Amano et al., 1986; Nakamura et al., 1991a), especially on how to suppress the gas-phase prereaction in the boundary layer as described in Section 6.2.2.

Other early groups working on the III-N materials include Matsuoka et al. (1990, 1992), Yoshimoto et al. (1991), and Takao et al. (1989), all working on establishing InGaN growth conditions using lower growth temperatures (500°C–800°C) in the absence of H_2 at high V/III ratios. Also of note is the work of Asif Khan and coworkers at Honeywell to develop AlGaN materials (Khan et al., 1983) ultimately leading to the first AlGaN/GaN heterojunctions with a mobility of 620 cm^2/Vs at room temperature (Khan et al., 1991).

6.2.2 Reactor Design

After increasing the maximum operating temperature, reactors used for III-As and III-P growth were suitable for the initial exploration of III-N growth. Growth efficiency in these systems was poor due to gas-phase parasitic chemistry which decreased the quantity of reactants reaching the surface. Recognizing this fact, Amano et al. developed an atmospheric pressure MOVPE reactor with increased flow to improve growth efficiency (Amano et al., 1986). Later, Nakamura developed a revolutionary design where the MOs and NH_3 precursors were delivered from a side quartz tube and pushed against the surface vertically using high flow rates of N_2 and H_2 carrier gas. This atmospheric pressure reactor produced GaN films on sapphire of unprecedented quality as witnessed by the high electron mobility of 900 cm^2/Vs at room temperature and 3,000 cm^2/Vs at 77 K (Nakamura et al., 1992b). Based on the chemical flow rates used and growth rates for GaN of 4.5 μm/hr, it can be speculated that Nakamura's reactor design was very efficient. The high vertical flow rate also referred to as the *pressing flow* likely suppressed the gas-phase parasitic reactions and thinned the boundary layer. Whatever the exact advantages of this reactor design, it allowed Nakamura to rapidly explore III-N growth, ultimately leading to the first candela-class, nitride-based blue LEDs (Nakamura et al., 1994).

Over the years, MOVPE reactor designs for the III-N's have evolved from single 1–2″ wafer machines to production platforms able to grow on numerous 2″, 4″, and 6″ wafers at once. For these designs, the MOs and NH$_3$ are kept separate to avoid the gas-phase parasitic reactions until they reach the heated surface (Kadinski et al., 2004). Other means to thin the boundary layer include rotating the wafer carrier at high-speeds to produce a viscous drag on the vertical gas flow, resulting in pulling it horizontally across the wafer surface (Coltrin et al., 1989). In horizontal reactors, another option to avoid gas-phase parasitic reactions is to reduce the reagent residence time by increasing the flow rates through the reactor (Matsumoto et al., 2008). These high flow rates decrease reagent residence time in the reactor resulting in high growth rates and reduced particulate formation (Matsumoto et al., 2008). For shorter distances between the gas inlet and growth zone, Dauelsberg and coworkers found that GaN growth efficiency in a close coupled showerhead and horizontal reactors was heavily dependent on gas-phase particulate formation with the particles being pushed to forming soot on the reactor top wall due to thermo-phoretic forces (Dauelsberg et al., 2007, 2011). The degree of gas-phase particulate formation could be reduced by heating the top reactor wall (Dauelsberg et al., 2007, 2011). All of these design changes essentially reduce gas-phase particulate formation, improving overall reagent usage, and decreased cost of ownership.

Finally, for group III-N growth higher pressure should stabilize and suppress thermal decomposition (Ambacher et al., 1996). Initial GaN growth by Amano et al. and Nakamura was at atmospheric pressure (Amano et al., 1986; Nakamura et al., 1991a). Higher growth pressure is desired to overcome the high equilibrium-vapor-pressure of nitrogen over the solid phase, possibly leading to higher growth temperatures for the solid and better crystallinity (Matsuoka, 2011). Recently, several groups have developed above atmospheric pressure growth with the goal of further material improvement (Atsushi et al., 1999; Woods and Dietz, 2006; Matsuoka et al., 2011). Above atmospheric pressure, growth has been shown to produce GaN with lower dislocation density compared to atmospheric (Atsushi et al., 1999) by promoting larger lateral growth rates through the enhanced gas-phase transport of Ga atoms (Mitchell et al., 2001). Obviously, the possible benefits of above atmospheric pressure MOVPE growth of III-N materials need further exploration.

6.2.3 Quest for the Blue LED Kicked Off the III-N Revolution

The development of a blue wavelength LEDs has been the primary driver of the III-N revolution, ultimately garnering the 2014 Nobel Prize in Physics for Akasaki, Amano, and Nakamura (Akasaki, 2015; Amano, 2015; Nakamura, 2015). However, before igniting this revolution, many breakthroughs were required in the materials quality, materials reproducibility, p-type doping and activation, active QW materials growth, and electrical confinement of carriers. Each of these breakthroughs and the role played by MOVPE growth are described in Sections 6.3 through 6.7.

Initial material improvement is described in Section 6.3 which includes the development and use of low-temperature AlN and GaN NLs. Simultaneous to the improved materials was the development of *in situ* growth monitoring described in Section 6.4. The pinnacle achievement for *in situ* monitoring was the rapid development of LEDs on Si substrates as described in Section 6.7.2. For diodes, both n- and p-type doping are required as described in Section 6.5. Initially, p-type doping of GaN resulted in resistive material until activation schemes were developed. This was followed by InGaN material development as described in Section 6.6 to red shift the light emission from the bandgap of GaN (364 nm) into the blue (450–470 nm) and green (505–540 nm) wavelength regimes. Confining carriers in the InGaN layers by capping with an AlGaN electron blocking layer (EBL) is described in Section 6.7. Once the GaN, InGaN, and AlGaN materials evolved to exceed Candela-class brightness (Nakamura et al., 1994), LEDs became attractive for general-purpose lighting. However, using LEDs to replace the incumbent incandescent and fluorescent lighting would require substantial development (Haitz and Tsao, 2011). Progress on increased luminous efficacy soon followed (Krames et al., 2007), leading to the current availability of LED lights.

6.3 GaN Heteroepitaxy and the Use of Nucleation Layers

The essential breakthrough in group III-N materials occurred from the intentional use of lower temperature AlN and GaN buffer or NLs (Amano et al., 1986; Nakamura, 1991c). Before NLs, GaN growth on sapphire or other substrates resulted in the morphology shown in Figure 6.1a. These GaN films grown directly on sapphire were generally rough with hexagonal morphology. These films were likely N-polar since the sapphire at high temperature was first exposed to NH_3 before the TMGa. The use of AlN NLs as demonstrated by Amano et al. (1986) produced smooth, mirror-like GaN similar to the morphology shown in Figure 6.1b.

Typically, in other material systems, *buffer layers* were used to bridge small lattice constant differences or to bury impurities at the starting surface. However, a large in-plane lattice difference of ~14% exists between GaN and sapphire, even with 30-degree rotation lattices taken into account. This large difference in lattice constants would suggest that a dislocation needs to form every seventh atomic spacing, which at a first glance should negate sapphire as a logical substrate for GaN growth. Yet, as pointed out by Green et al., how well materials wet and grown is perhaps more important than the usual lattice matching criterion applied to semiconductor growth, especially for the growth of materials with radically different chemical composites, such as nitrides on oxides (Green et al., 1970). This difference in the GaN epitaxial growth mode is exemplified in Figure 6.2a and b which show 15 minutes of GaN growth on sapphire. For this growth mode, the GaN is aligned to the underlying sapphire but is composed of individual grains, which once coalesced reduces the GaN dislocation density as discussed in Section 6.3.2.

The question often arises as to why sapphire is used for GaN growth at all. One answer to this question is that sapphire, somehow, works. Part of that answer lies in the experience gained by researchers in growing GaN on sapphire from early on (Maruska and Tietjen, 1969; Manasevit, 1972). Another answer is that sapphire, unlike many other substrates, is chemical stability at the high temperatures used for GaN growth, is not overly corroded in the H_2 or NH_3 environments, is reasonably chemically pure, is polished to a smooth finish, and is obtainable at a reasonable cost. Preparation of the sapphire surface now requires no *ex situ* chemical cleaning before loading into the MOVPE system.

After loading into the growth chamber, the sapphire is simply heated in H_2 to 1,000°C–1,100°C to remove any residual atmospheric and hydrocarbon impurities, and possibly improve step edge definition (Ribič and Bratina, 2007). However, excessive annealing in H_2 at temperatures >1,200°C can improve surface crystallography but results in undesirable step bunching (Ribič and Bratina, 2007), and surface pitting. Some degree of sapphire annealing has been shown to improve the GaN grain alignment (Wickenden et al., 1999), likely by aligning GaN grains to the sapphire step edges. After annealing the sapphire, it is exposed to NH_3 at a lower temperature (500°C–950°C) to nitride the surface. Depending on the extent of this

(a)   (b)

FIGURE 6.1 Nomarski microscope images of GaN on sapphire grown (a) without and (b) with a low-temperature deposited AlN nucleation layer prior to the GaN growth.

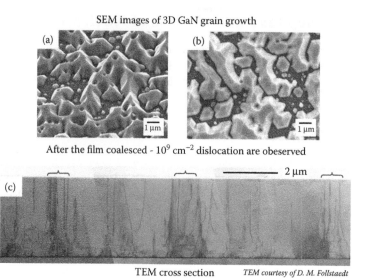

FIGURE 6.2 (a) Tilted and (b) planar SEM images of 15 minutes of high-temperature GaN growth on a GaN nucle-ation layer on sapphire. For these two images the growth was intentionally forced to grow with a 3D morphology by reducing the growth temperature and NH$_3$ flow. In (c) is shown a TEM cross-section of the GaN film after full coales-cence. The dislocations are visible as lines in the image and there are regions close to the GaN sapphire interface where dislocations are bent parallel to the growth direction. These dislocations are bent as a result of the 3D morphology.

nitridation, the nitride layer has been described as a thin oxynitride layer or one to several monolayers of AlN, with the main effect being modification to the surface energy (Uchida et al., 1996; Hashimoto et al., 1999a). How exactly these annealing and nitridation steps restructure the sapphire surface prior to the NL growth is uncertain; however, the sequential exposure to H$_2$ and NH$_3$ gases at high temperature followed by cooling for the NL growth likely leaves the surface partly or mostly hydrogen terminated.

Sections 6.3.2 through 6.3.4 discuss the three most used NL schemes for high-temperature GaN growth on sapphire and SiC surfaces, but first, several unique properties of high-temperature GaN MOVPE growth are discussed as they ultimately influence how the high-temperature GaN layers grow on the NLs.

6.3.1 High-Temperature GaN Growth

There are several unique features of high-temperature MOVPE growth of GaN. These include a Ga gas-phase transport diffusion mechanism, thermal decomposition, the catalytic nature of ammonia disso-ciation, the presence of gas-phase parasitic chemistry, and the GaN films usefulness despite high defect and dislocation density. These unique features dictate not only GaN MOVPE growth but also influence the growth of the low composition alloys with indium and aluminum.

6.3.1.1 Gas-Phase Diffusive Transport of Ga Atoms

The most important difference between GaN MOVPE growth and AlN and InN MOVPE growth is the ability for Ga to diffuse through the gas phase at high temperature. This growth mechanism was first uncovered by Mitchell, Coltrin, and Han, in lateral transport studies of epitaxial lateral overgrowth (ELO) (Mitchell et al., 2001). Before this work the exact nature of the growth enhancement was unclear. It was speculated that the growth enhancement during the lateral overgrowth was the result of Ga atoms landing on the oxide or nitride masked regions and diffusing across the surface to incorporate prefer-entially at the growth edge. To discern whether surface diffusion or gas-phase diffusion leads to the growth enhancement, this group cleverly etched deep trenches into the masked regions to interrupt the surface diffusion pathway. After growth, the resulting profiles were virtually identical independent of the trench features, indicating that a gas-phase lateral transport mechanism was responsible for the lateral overgrowth (Mitchell et al., 2001).

Both the gas-phase and surface diffusion mechanisms along with examples are shown in Figure 6.3. These two mechanisms influence the growth morphology and can be discerned from one another fitting the power spectral density (PDS) or $g(q)$ measured from AFM images (Koleske et al., 2014a). Figure 6.3a shows the gas-phase diffusion mechanism which involves the evaporation of Ga atoms from the surface and their recondensation elsewhere on the surface. This smoothing mechanism from the AFM analysis has also been referred to as the evaporation/recondensation mechanism. This transport mechanism dominates at temperatures >900°C, has a power law dependence of 2 in the PSD, and allows Ga gas-phase diffusion across the surface laterally up to 1–10's of microns (Koleske et al., 2014a). The other smoothing mechanism depicted in Figure 6.3b is the more common surface diffusion mechanism. The surface diffusion mechanism dominates below 800°C, has a power law dependence of 4 in the PSD, and for GaN involves diffusion lengths of 10–100's of nanometers (Koleske et al., 2014a). Between 800°C and 900°C, both mechanisms operate and are also influenced to a lesser degree by the carrier gas (N_2 or H_2), growth rates, and other factors.

Strong evidence for the evaporation/recondensation mechanism is observed in AFM images of high-temperature MOVPE growth of GaN on sapphire. Three different-sized AFM scans of GaN films grown on sapphire are shown in Figure 6.3c–e. Note that these AFM images have progressively larger RMS roughness as the scan size increases. The reason for the increasing roughness is because the RMS roughness is calculated from the square root of the sum of the PSD measured at each reciprocal length scale, q. The PSDs, for each image shown in Figures 6.3c–e, are shown in Figure 6.3f. Each PSD as shown in Figure 6.3f has a power law dependency close to $n = 2$, indicating that the high-temperature GaN films are smoothed by the evaporation/recondensation mechanism (Koleske et al., 2014a). Hydrogen plays a special role in this mechanism since it has been shown to increase the rate at which Ga atoms desorb from the surface; likely as intact Ga-H species (Koleske et al., 2001).

In addition to smoothing high-temperature GaN layers, the evaporation/recondensation mechanism is also responsible for improving GaN material during lateral overgrowth of GaN on masked regions (Pierre, 2004). In 1997, Nam et al. reported dislocation reduction by several orders of magnitude when GaN was overgrown on a mask material (Nam et al., 1997), with the implication that this material would be useful for high-power electronic devices, such as lasers. Further exploration of mask geometry led to a better understanding of the GaN shape evolution (Marchand et al., 1998; Kazumasa, 2001), and especially how it depended on the MOVPE growth conditions and overgrowth geometry (Kazumasa, 2001; Coltrin and Mitchell, 2003; Du et al., 2005; Sun et al., 2008). Specifically, growth factors such as the temperature and pressure generated either lateral overgrowth, triangular features, or hexagonal pyramids depending on the mask geometry (Kazumasa, 2001). Later, generalized theorems were developed to describe facets evolution specifically how certain facets grow to extinction leaving the slower growing facets to dominate the GaN geometry (Coltrin and Mitchell, 2003; Du et al., 2005). Because the growth is dominated by surface kinetics, Sun et al. were able to develop kinetic-Wulff plots to correlate growth dynamics to growth features, which included anisotropic nucleation islands, highly striated surfaces, and pentagonal or triangular pits (Sun et al., 2008).

While not discussed here, the surface diffusion mechanism (shown in Figure 6.2b) is essential to explaining lower temperature growth, including GaN NL growth, high-temperature AlN growth, and mounding behavior observed in InGaN underlayer growth as discussed in Section 6.6.3.

6.3.1.2 Thermal Decomposition

As first mentioned by Johnson et al. (1932), GaN decomposes at temperatures well below its melting temperature (Koleske et al., 2001). For this strongly bonded semiconductor, the presence of a thermal decomposition mechanism might aid in bringing the growth closer to equilibrium, particularly at the surface. Many studies on GaN thermal decomposition have been reported, and these studies are summarized in Ref. (Koleske et al., 2001). Higher annealing pressure was shown to increase the rate of GaN decomposition

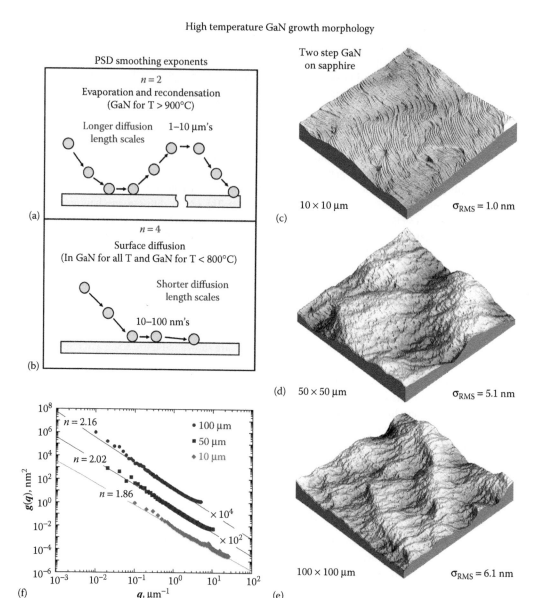

FIGURE 6.3 Growth of morphology of the high-temperature MOVPE GaN films. In (a) is shown the evaporation and recondensation smoothing mechanism, which allows large gas-phase diffusion lengths for the gallium atoms and mainly dominates at temperatures above 900°C. In (b) is shown the surface diffusion smoothing mechanism, which is shorter range and mainly operates at temperatures below 800°C. In (c–e) are AFM images of MOVPE grown GaN over different length scales. In (f) is the power spectral density, $g(q)$ plotted as a function of the recipro-cal length, q. The power law fit through the data in (f) has an $n \sim 2$ indicating that the high-temperature smoothing of the GaN film is controlled by the evaporation and recondensation smoothing mechanism. (From Koleske, D.D. et al., *J. Cryst. Growth*, 391, 85–96, 2014b. With permission.)

in H_2 when studied in a quartz tube MOVPE system (Koleske et al., 1998b). One caveat to this study is that the GaN surface temperature could be higher at higher pressure when measured by pyrometry, since as the reactor pressure increases the surface temperature also increases (Creighton et al., 2006).

From the previous decomposition studies summarized in Koleske et al. (2001), the measured activation energies for GaN ranged from 0.34 to 3.62 eV and fell into four distinctive groups, indicating that up to four different mechanisms may limit the GaN decomposition rate. Of note are two proposed decomposition mechanisms where hydrogen is added to N atoms at lower temperatures or higher annealing pressures (Koleske et al., 2001). Metallic gallium desorption in pure H_2 also had a 10x higher pre-exponential factor compared to desorption in N_2 with a similar desorption activation energy, suggesting the possible formation and desorption of a Ga-H bonded species (Koleske et al., 2001). GaN decomposition studies with small amounts of added NH_3 dramatically decreased the GaN decomposition rate by 100x compared to just flowing N_2; however, as the NH_3 concentration over the surface increased the GaN decomposition rate increased slightly (Koleske et al., 1999b). Similar reductions in the decomposition rate in NH_3 compared to vacuum have also been reported by Karpov et al. (2000). Using ammonia flows similar to those used for growth, the GaN decomposition rate was shown to be up to ¼ of the growth rate when TMGa was added back to the reactor (Koleske et al., 1999b). The addition of NH_3 certainly stabilizes the growth surface and certainly incorporates N into the growing lattice. However, the exact nature of NH_3 catalytic dissociation remains unclear.

6.3.1.3 Catalytic Ammonia Dissociation

Following purely thermodynamic assessment of NH_3 dissociation, the products N_2 and H_2 should be favored above 200°C based on the equilibrium constant (Haynes and Lide, 2011). However, because the N–H bond energies are ~4 eV and are comparable with the Ga–N bond strength, high temperatures are required to catalytically dissociate NH_3 to produce active N species on the growth surface (Koleske et al., 1998a). After high-temperature (2,000–3,200 K) gas-phase NH_3 pyrolysis experiments, Davidson et al. proposed a complicated reaction mechanism for NH_3 decomposition comprising 9 intermediate or final species that participated in 21 different chemical reactions (Davidson et al., 1990). These studies provided rate coefficients for the reactive intermediates required for III-N growth such as N, NH, and NH_2 (Davidson et al., 1990). Experiments in quartz tubes by Ban and later by Liu and Stevenson indicated that NH_3 pyrolysis was minimal under conventional MOVPE growth conditions, unless a surface or catalyst was placed in the tube (Ban, 1972; Liu and Stevenson, 1978). Other experiments at GaN growth temperatures and atmospheric pressures indicated that the conversion of NH_3 was less than 25% and the extent of pyrolysis was reduced as the pressure was reduced (Monnery et al., 2001). Despite the apparent lack of NH_3 chemical reactivity, GaN NLs are grown at a temperature as low as 500°C (Wickenden et al., 1994); made possible from the increased NH_3 or NH_x fragment residence time at the surface at these lower temperatures.

Based on these studies, it would appear that large amounts of NH_3 are required to generate sufficient surface reactive nitrogen species (N, NH, NH_2) (Parikh and Adomaitis, 2006) for growth. Because such an excess of NH_3 is used, many researchers have assumed that growth occurs under N-rich conditions. However, because of the catalytic NH_3 dissociation and unknown N surface coverage, the growth may occur under metal-rich conditions, similar to MBE (Tarsa et al., 1997). Occasionally, metallic droplets are observed after growth, suggesting a slightly metal-rich surface. A likely scenario is that the N is incorporated into the lattice to pin the metal atoms into the growing lattice. This might make the rates of breaking the last N–H bond and N_2 reformation and desorption crucial to the growth process, making the ill-understood chemical processes that occur in the selvedge region substantially more interesting and dynamic than previously thought.

Some evidence to support N–H bonds being intact late in the N incorporation is that hydrogen desorption from a Ga rich surface is typically complete above ~500°C (Chiang et al., 1995; Shekhar and Jensen, 1997), suggesting that surface H generated from NH_3 dissociation is easily desorbed and that there is a high degree of surface NH_3 dissociation. Predosing a GaN surface with D followed by NH_3 exposure results in desorption of deuterated ammonia species (Chiang et al., 1995) suggesting reversible NH_3 surface dissociation and desorption. Likewise, Bartram and Creighton found both reversible and irreversible NH_3 decomposition by dosing with GaN with $N^{15}H_3$ and subsequent TPD experiments (Bartram and Creighton, 1998). In addition, they observed $N^{15}N^{14}$ desorption well below the onset of GaN sublimation with the N^{14} atom coming from the subsurface, implying lattice-bonded nitrogen can be removed when the surface is nitrogen rich (Bartram and Creighton, 1998). So with significant NH_3 and N_2 reformation reactions occurring during growth, to what extent is the surface saturated with active N or NH_x fragments? King et al. have suggested that at higher temperatures the hydrogen desorption and ammonia desorption/decomposition are fast enough that only partial surface coverage can be maintained (King et al., 1999). If the turnover rate of NH_3 and desorption rates of H_2 and N_2 are faster than the incorporation of N into the lattice, then surface diffusion (and the gas-phase transport for Ga) is solely limited to the metal atoms (Koleske et al., 1998a) with N atoms locking the metal atoms into the lattice. This rapid exchange of active N species and Ga diffusion would then explain the apparent step flow growth observed at high temperatures (Ramana Murty et al., 2000) since Ga atoms would not incorporate until they reach the step edge.

Finally, an irony of the catalytic dissociation of NH_3 is that while the gas-phase V/III ratio is often reported for MOVPE growth, the actual surface V/III ratio might be significantly lower. For InGaN growth, the common assumption is that the high V/III ratios are required since NH_3 dissociation becomes extremely sluggish at a lower temperature. Evidence to suggest this sluggish chemistry is that the indium incorporation can be dramatically reduced by adding even small amounts of H_2 to the growth (Piner et al., 1997; Koleske et al., 2014b). Interestingly, the dissociation of NH_3, which produces hydrogen as a byproduct, does not apparently change the indium incorporation into the InGaN film (Piner et al., 1997). So H_2 with a bond strength of 4.5 eV is preferentially dissociated compared to three N–H bonds with bond strengths of ~4.0 eV. This implies that NH_3 dissociation does not supply significant surface H during InGaN growth otherwise indium might be completely removed from the surface. In addition, higher NH_3 flows incorporate more indium into the InGaN compared to lower flows (Keller et al., 1997b), again implying no significant impact of H generated during NH_3 dissociation on surface indium removal. This suggests that at low temperatures significant numbers of N–H bonds remain intact until N–Ga or N–In bonds are formed. Further complicating any clear picture of the catalytic dissociation of NH_3 and how N is incorporated into the lattice is the role of the gas-phase chemistry and adduct formation in or near the boundary layer.

6.3.1.4 Gas-Phase Parasitic Chemistry

Compared to NH_3 dissociation, the effects of gas-phase parasitic chemistry and its limits on III-N growth are better understood. While 100% conversion of reactants to products for growth would be the ideal situation, flowing reactive molecules over a heated surface creates the potential for chemistry to occur in the gas-phase. Besides reducing the reagent flow reaching the surface, the particles generated in the gas-phase can land on the surface creating surface defects or change alloy concentrations away from their gas-phase compositions (Han et al., 1998).

Creighton et al. observed a thin layer of particles generated in the boundary layer using *in situ* laser light scattering in an inverted stagnation flow reactor (Creighton et al., 2002). The particle layer was located 6 mm from the surface; a location well predicted as a balance between the thermophoretic and convective forces. The generation of the gas-phase particles was estimated to be 20%–80% of the input TMAl, suggesting the potential for a highly inefficient growth process (Creighton et al., 2002). Later work also showed particles generated when using TMGa and TMIn, with the rate of particle generation decreasing as the growth temperature decreases (Creighton et al., 2008a). When using TMIn for InGaN growth, some of the generated gas-phase particles might form after the indium has interacted and desorbed from the surface (Koleske et al., 2014b), which may explain the more metallic nature of the captured InN gas-phase particles (Creighton et al., 2008a).

The particle formation is suspected to initiate from adducts formed between the MOs and NH_3 (Thon and Kuech, 1996). Early on the adduct was suspected to be a $[(CH_3)_2Ga:NH_2]_3$ complex formed after methane elimination (Thon and Kuech, 1996). Later work using infrared spectroscopy measured the equilibrium constant for the formation and reversible dissociation of $(CH_3)_3Ga:NH_3$ and $(CH_3)_3In:NH_3$ adducts (Creighton and Wang, 2005). As these simple adducts are heated, the elimination of methane leads to stable dimers and trimers (Mihopoulos et al., 1998), which then nucleate particle growth. Since nonideality in growth rates and alloy composition are clear indications of parasitic chemical reactions, gas-phase parasitic sink terms must be included to accurately predict the experimental growth rates (Mihopoulos et al., 1998; Coltrin et al., 2006). To account for the particle generation, Coltrin et al. developed a nine-step reaction mechanism describing the Ga-precursor decomposition, Al-adduct formation and methane elimination, particle nucleation, and particle growth to achieve good agreement with AlGaN growth rate data in a rotating disk reactor (Coltrin et al., 2006). Understanding and modeling the parasitic chemistry can potentially help with developing growth conditions, and reactor designs, or MO injection schemes that reduce the parasitic loss of MO.

For preventing extensive parasitic chemistry, most commercial reactors keep the MO's separate from the NH_3 (to avoid adduct formation) until they are introduced close to the growth surface. Another method is to heat the reactor walls to reverse the dimer or trimer adduct formation (Creighton and Wang, 2005). For AlGaN, the lower growth pressure is typically used to reduce parasitic chemistry and improve material quality (Allerman et al., 2004). Chen et al. showed using a horizontal reactor that GaN growth efficiency at atmospheric pressure was comparable to that of GaAs; however, growth pressures of 250 torr were required to achieve reasonable AlGaN growth rates (Chen et al., 1996). Nakamura et al. reported that lowering the TMAl partial pressure was effective in preventing the gas-phase parasitic reactions (Nakamura et al., 1998). Finally, instead of keeping the MO's and NH_3 spatially separated, several groups have separated them temporarily using a pulsed approach leading to reduced parasitic chemistry and enhanced surface diffusion of the metal atoms (Fareed et al., 2003; Zhang et al., 2003).

6.3.1.5 Material Defects

The largest metric of the material quality is the overall material defectivity, especially for semiconductors. The most obvious defects in GaN heteroepitaxial films are the dislocations that initiate at the sapphire/GaN interface. Above this interface, the dislocations annihilate to some degree, but the rest remains and threads vertically up through the films (Hiramatsu et al., 1991). Each of these regions is shown in the cross-section TEM in Figure 6.2c, where a high density of defects/dislocations are observed near the GaN/sapphire interface. Above that, interface dislocations undergo annihilation through the bending of the dislocations laterally on the triangular facets, increasing the possibility for interaction with a dislocation with opposite Burger's vectors (Follstaedt et al., 2002), and the extent to which this occurs can be purposefully increased by lowering the NH_3 flow rate and temperature (Koleske et al., 2002a) as shown in Figure 6.2a and b. Also evident in Figure 6.2c are that the threading dislocations (TDs) form bundles denoted by the parenthesis marks which are believed to originate during grain coalescence (Fini et al., 1998).

Once dislocations were identified as a major contributor to material quality, accurate means were sought to quantify them and determine their influence on material properties (Weimann et al., 1998; Atsushi et al., 1999; Koleske et al., 2002a; Dai et al., 2009). The high dislocation density in GaN was initially thought to be due to the large lattice mismatch between the substrate and the growing GaN film; however, later work suggested that the highly defective NLs and unique grain growth mechanism (Akasaki et al., 1989) were the cause for the resulting dislocation density (Fini et al., 1998; Moram et al., 2010).

The most visually satisfying way to image dislocations was using transmission electron microscopy (TEM) (Wu et al., 1996). Plan-view TEM images provide a more accurate measure for vertically TDs since they typically appear as a line segment in the image, while cross-section TEM views present difficulties during quantification due to uncertainties in the foil thickness. Imaging of all three dislocation types (pure edge, pure screw, and mixed) in plan-view foils can require changes away from the strict g·b imaging criteria and can be accomplished by tipping the sample in the TEM (Follstaedt et al., 2003). The most obvious change in dislocation densities is observed in cross-section TEM images of laterally overgrown GaN either on masked GaN (Zheleva et al., 1997) or from sapphire posts (Follstaedt et al., 2002). Dislocation density can also be estimated from XRD which is faster and nondestructive; however, simplifying assumptions have to be made in order to extract dislocation density estimations from X-ray diffraction (XRD) linewidths (Lee et al., 2005; Moram and Vickers, 2009). As first discussed by Heying et al., several reflections (minimum of two) are required to estimate the screw and edge component dislocation densities (Heying et al., 1996). Dislocation cores have also been imaged using atomic force microscopy (AFM) although clear identification of the edge component dislocations can be difficult. Several etching (Stocker et al., 1998; Hino et al., 2000) or dislocation decoration (Oliver et al., 2006) techniques have been developed to accentuate imaging and better resolve the dislocation types. Because edge dislocations act as nonradiative recombination centers (NRCs), Cathodoluminescence (CL) has also been used to estimate dislocation density by counting the number of dark spots in images (Tomoya et al., 1998; Cherns et al., 2001). Recently, nondestructive electron beam channeling contrast imaging has been developed to measure dislocation densities using a conventional backscatter detector in a commercial scanning electron microscope (Kamaladasa et al., 2011).

While all the above-mentioned methods can be used to measure or estimate material dislocation density, they do not stand on an equal quantitative footing. TEM is clearly the most accurate and can image all three types of dislocations (Follstaedt et al., 2003); however, it can only measure selected, small regions. AFM can underestimate dislocation density if imaging conditions are not correctly established due to tip size or imaging artifacts. XRD measures the entire GaN film volume, and therefore, can overestimate the dislocation density at the top of the film. XRD also requires assumptions of how the measured reflections are correlated to the dislocation density (Lee et al., 2005) and based on these assumptions the screw and edge component dislocations are estimated. Finally, CL only images dislocations or defects that contain a nonradiative component such as threading edge dislocations (Tomoya et al., 1998; Cherns et al., 2001). Also during InGaN or AlGaN growth on GaN, dislocations can be generated once the critical thickness is exceeded (Lee et al., 2004b; Holec et al., 2008). For example, misfit dislocations were generated between the GaN barriers and InGaN QWs when the indium content was 22% (Holec et al., 2008).

6.3.2 Low-Temperature AlN NLs on Sapphire

Amano et al. produced smooth GaN films using MOVPE by first depositing a low-temperature AlN NL (Amano et al., 1986). This work followed Yoshida et al. who previously used AlN NLs to improve GaN films on sapphire using molecular beam epitaxy (Yoshida et al., 1983). Besides producing smoother GaN films, the AlN NL improved other GaN properties such as, improved carrier mobility (Amano et al., 1988), improved PL intensity (Amano et al., 1988), and ultimately lower dislocation density (Akasaki et al., 1989). Dissection of the growth process revealed that the optimal AlN NLs provided enough GaN nuclei to grow both vertically and laterally across the surface, resulting in a smooth film (Akasaki et al., 1989; Hiramatsu et al., 1991). TEM and SEM studies of the GaN growth on the AlN NL showed an initially high density of GaN nuclei, followed by a geometric down selection of GaN nuclei that were crystallographic aligned to the c-axis (Hiramatsu et al., 1991). As growth progressed, the selected GaN nuclei had a high lateral growth velocity resulting in trapezoid islands on the surface (Hiramatsu et al., 1991). Ultimately, these trapezoidal islands coalesce at the same vertical height resulting in the observed smooth morphology (Hiramatsu et al., 1991). As discussed in the previous section, TDs are annihilated near the AlN/GaN interface, resulting in GaN films with 10x fewer dislocations compared to GaN films without the AlN NL. As many have observed, the rate of film coalescence can be controlled by many factors including, growth temperature (Figge et al., 2000), NH_3 flow rate (Figge et al., 2000; Koleske et al., 2002a), and the use of N_2 or H_2 during growth (Cho et al., 2007).

When optimal AlN NLs are deposited on the sapphire, the GaN has a Ga-polar orientation as evident by smooth films (as shown in Figure 6.1b) and inertness to etching in hot KOH (Collazo et al., 2006). The AlN and GaN polarity appears to be influenced by the degree of nitridation and AlN NL growth conditions {Collazo, 2006 #869}. Collazo et al. have shown that the polarity of the AlN NLs (either N-polar or Al-polar) directly carries over to the GaN film orientation (Collazo et al., 2006). The surface is excessively nitrided compared to the Al-polar films to achieve N-polar films (Collazo et al., 2006).

One major issue with the use of AlN NLs for GaN growth are the potential issues with reproducibility especially when insulating GaN films are desired for HEMT structures (Twigg et al., 1999, 2001). These growth issues are likely related to prior growth runs in the same reactor and subtle changes to the initial nitridation or AlN film growth have been shown to change the AlN NL structure (Twigg et al., 2001). Specifically, high-resistivity GaN films grew on large-grained AlN NLs with a smooth AlN/sapphire interface (Twigg et al., 2001). Repeating the same recipe in the reactor over time resulted in low-resistivity GaN films. Examination of the AlN NLs using TEM showed the AlN to have a much finer grain structure and a rougher AlN/sapphire interface (Twigg et al., 2001). While these lower resistivity films had a lower overall dislocation density, the small-grained AlN allowed oxygen from the sapphire to autodope the GaN near the GaN/AlN interface (Twigg et al., 2001). This buried conductive layer can be easily detected using capacitance–voltage measurements (Twigg et al., 1999; Cheong et al., 2000; Wickenden et al., 2000) and the oxygen at the interface can be measured using SIMS (Xu et al., 2000).

How the AlN NL nucleates the GaN nuclei at high temperature or how a change in the AlN NL grain structure would influence the number density of GaN nuclei formatted is unclear; however, annealed AlN NLs have been shown to roughen and develop surface undulations likely to relieve layer stress (Lorenz et al., 2000). Another possible nucleation site might be the protruding AlN grains observed by Twigg et al. (2001). These surface undulations or protrusions could provide the nucleation sites for GaN growth with grain selection and density determined within the first seconds of high-temperature growth. Once established, the Ga species preferentially incorporate on the GaN grains with few additional grains formed as suggested in Figure 6.2a and b. This growth process would eventually lead to a planar film once the film has coalesced since the nascent grains all grow at the same lateral and vertical rates.

A final twist to low-temperature deposited AlN NLs is the more recent use of magnetron sputtering to deposit the AlN prior to the growth of GaN (Paskova et al., 2012). This process might be especially useful for growth on substrates other than sapphire such as Si (111) (Takaya et al., 2013). Using this process Lee et al. have reported a TEM measured dislocation density reduction from 6.8×10^7 cm^{-2} to 2.6×10^7 cm^{-2} using reactive plasma deposited AlN on a patterned sapphire substrate (Lee et al., 2014). Using a sputter-deposited approach for AlN on sapphire (or other substrates) might help alleviate some of the above-mentioned reproducibility issues for MOVPE AlN NL growth in conventional reactors.

6.3.3 Low-Temperature GaN NLs on Sapphire

Starting in 1990, Nakamura worked on GaN NLs to generate a different way to produce high-quality GaN films on sapphire (Johnstone, 2007). At about same time, Dennis Wickenden at the Applied Physics Laboratory was also using lower temperature GaN buffer layers to improve the quality of GaN films grown using MOVPE (Wickenden et al., 1991). Interestingly, the patent for highly resistive GaN NLs was awarded to Ted Moustakas at Boston University who had used MBE to deposit low-temperature GaN layers (100°C–400°C) before the high-temperature growth at 600°C–900°C (Moustakas, 1997). Initial reports by Nakamura suggested GaN material grown on the GaN NLs (Nakamura, 1991c; Nakamura et al., 1992b) were as good if not slightly better than AlN NLs (Amano et al., 1986), however Nakamura's novel reactor design might have also played a role in the GaN material improvement.

The low-temperature GaN NL has been described as containing small grains (Wickenden et al., 1994), amorphous regions (Sugiura et al., 1997), and stacking faults or zinc blende material (Lorenz et al., 2000). It also contains high concentrations of H and C from the incomplete dissociation of the Ga precursor (Ambacher et al., 1996). The GaN NL is relatively smooth after deposition as observed in TEM images (Lorenz et al., 2000), AFM images (Koleske et al., 2004), and from optical reflectance measurements (Koleske et al., 2005). However, upon annealing to high temperature, the GaN NL roughens (Lorenz et al., 2000; Lada et al., 2003; Koleske et al., 2004) and the optical reflectance signal decreases due to a loss of material and roughening (Koleske et al., 2004, 2005). Initially, it was suspected that the GaN NL underwent a solid-state sintering since there were indications from absorption measurements of an increase in the grain size (Wickenden et al., 1994). Instead of solid-phase recrystallization, wurtzite GaN grains were observed in TEM images to form on top of the original GaN NL (Lorenz et al., 2000; Lada et al., 2003). The formation of the wurtzite GaN grains was suspected to be driven through Ga atom desorption and re-adsorption on the NL surface (Lorenz et al., 2000). Conversely, Sugiura et al. proposed that the wurtzite GaN grains developed at the sapphire/GaN interface and grow in size, as GaN material is lost to the gas phase (Sugiura et al., 1997).

To prove that the GaN nuclei formed as a result of a gas-phase driven re-crystallization, Koleske et al. grew a series of GaN NLs all to the same thickness and then annealed these NLs to high temperature, stopping at different points along the annealing schedule to freeze in the evolved structure (Koleske et al., 2004). AFM images of the annealed GaN NLs showed different degrees of high-temperature GaN nuclei formation and when annealed long enough, a total disappearance of the GaN nuclei (Koleske et al., 2004). One AFM image measured after annealing the NL to 1,000°C is shown in Figure 6.4a

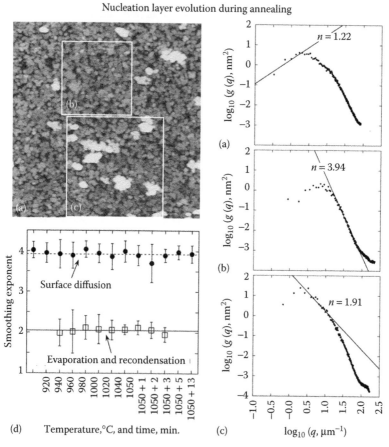

FIGURE 6.4 AFM and power spectral density analysis applied to GaN nucleation layer evolution. The $3 \times 3\,\mu m$ AFM image shows the nascent high-temperature GaN grains (whiter regions) forming on top of the lower temperature deposited GaN (darker grey). Power spectral density, $g(q)$, analysis is shown in plots (a–c) for the different regions in the AFM image. In the image (b), the power law fit is $n \sim 4$, indicating that the low-temperature GaN NL is smoothed by the surface diffusion mechanism. In the image (c), which has several high-temperature GaN grains, the power law fit is $n \sim 2$, indicating that these high-temperature GaN grains are smoothed by the evaporation/recondensation mechanism. In (d) the power law fits are summarized for each step along the GaN NL annealing schedule. This work illustrates the importance of the gas-phase transport mechanism for reforming the wurtzite GaN grains on the NL. (From Koleske, D.D. et al., *J. Cryst. Growth*, 273, 86–99, 2004. With permission.)

(Koleske et al., 2004). In this AFM image, the nascent GaN nuclei can be seen as the lighter regions while the surrounding GaN NL is darker. Following the PSD analysis described in Section 6.3.1 and shown in Figure 6.2, the surface diffusion and evaporation/recondensation smoothing mechanisms were selectively measured. The PSD for each of the boxed regions is shown in Figure 6.4a–c, where the PSD is calculated for (a) the whole AFM image, (b) a sub region without any GaN nuclei, and (c) a subregion with several high-temperature GaN nuclei. Applying the power law fits to the (b) subregion image shows that the original GaN NL is smoothed by surface diffusion mechanism ($n = 4$), while the (c) subregion image shows that the GaN nuclei are smoothed by the evaporation/recondensation mechanism ($n = 2$). In Figure 6.4d, the power law fits are summarized for the entire NL annealing schedule. The data in Figure 6.4 show that upon heating, the GaN NL decomposes [130] and the desorbed Ga atoms are transported through the gas-phase to eventually incorporate into the growing wurtzite GaN nuclei. From the AFM measurements, only a third of the initial NL volume ends up in the GaN nuclei, suggesting a net loss of material during the annealing step (Koleske et al., 2004).

Like the other growth parameters, the GaN NL growth conditions need to be optimized in each reactor. By varying the NL thickness, Nakamura settled on an optimal layer thickness of 20 nm (Nakamura, 1991c) and controlling the NL thickness to less than 0.1 nm is possible using *in situ* reflectometry (Kobayashi et al., 1998; Koleske et al., 2004). The exact NL thickness and transformation to GaN nuclei depend on the optimal number density of GaN nuclei that form and the complementary lateral growth rate required for film coalescence. It has previously been seen that lower GaN nucleation density correlates with reduced dislocation density in GaN films (Hashimoto et al., 1999b; Moran et al., 2004). Because not all of the NL volume is transformed into GaN nuclei, the nuclei density appears to scale with the square of the initial NL thickness (Koleske et al., 2007). Using multiple NL growth and annealing cycles, Lang et al. have achieved extremely sparse nuclei densities that resulted in GaN films with dislocation densities of 1×10^8 cm^{-2} (Lang et al., 2005, 2006). By adding H$_2$ during the NL annealing process, Han and coworkers showed an increased nuclei spatial orientation that initially leads to rougher films with better quality when fully coalesced (Han et al., 1997).

6.3.4 AlN NLs on SiC

The advantage of growing AlN (0001) on 6H-SiC (0001) is that the in-plane lattice mismatch is ~1% (Weeks et al., 1995). Using bulk SiC substrates obtained from Cree Research, Weeks and coworkers at NCSU deposited high-temperature AlN on the SiC to act as a NL for subsequent GaN growth (Weeks et al., 1995). The resulting GaN film was smooth, showed strong near band edge emission in PL, and had a dislocation density of ~1×10^9 cm^{-2} (Weeks et al., 1995). Conductive AlGaN NLs were also developed on SiC so that the n-type SiC could act as the bottom contact layer for blue LEDs (Doverspike et al., 1997). Later AFM work by Toshio et al. of GaN grown on these high-temperature AlN NLs showed a twilled step structure with no evidence of spiral growth originating from screw dislocations (Toshio et al., 1998). TEM cross-section measurements for GaN grown on AlN on SiC revealed the absence of screw dislocations in the AlN and a low screw dislocation density near the AlN/GaN interface (Koleske et al., 2002b) consistent with the observations of Toshio et al. (1998).

Unlike sapphire, nitridation of SiC prior to the NL growth generates an amorphous SiN layer, which is detrimental to further epitaxy. To prevent SiN layer formation, the TMAl can be first introduced into the reactor shortly before the NH$_3$ is introduced. Another consideration in using SiC as a substrate for GaN growth was the quality of the surface polish and subsequent damage incurred to the top layer. Different AlN NL growth temperatures were required for optimal GaN growth on 6H–SiC (on axis and 3.5-degree off-axis) with a rougher polish compared to 4H–SiC with a smoother polish (Koleske et al., 2002b). For achieving optimal electrical properties, an AlN NL temperature of 1,080°C was used for the planar and 3.5-degree off-axis 6H–SiC, while an AlN NL temperature of 980°C was used for 4H–SiC (Koleske et al., 2002b). AFM images suggest that the AlN morphology influences the GaN film structure and subsequent electron mobility (Koleske et al., 2002b).

6.4 *In Situ* Monitoring

Nakamura initially described the benefit of using *in situ* diagnostics by monitoring changes in the IR radiation transmission intensity (Nakamura, 1991a, 1991b). Nakamura found that decreases in the IR intensity corresponded to increased film roughness while increases in the IR intensity corresponded to increased film smoothness. Through the use of GaN NLs (Nakamura, 1991c), Nakamura had previously identified that smoother GaN films correlated with better materials properties (Nakamura, 1991a, 1991b), leading to the use of *in situ* IR transmission intensity as a means to prescreen GaN material for further analysis.

Over time, the instrumentation of these reflectance-based techniques become more sophisticated (Breiland and Killeen, 1994, 1995; Killeen and Breiland, 1994; Breiland et al., 1999), allowing monitoring of growth rate, alloy concentrations, and film uniformity. In addition to the optical reflectance techniques, a second essential *in situ* technique used directed laser beams to monitor film strain. This latter technique allowed the rapid development of GaN growth on Si (111) surface as reported by Dadgar, Krost and coworkers (Dadgar et al., 2000, 2006; Krost et al., 2005b). Details of additional *in situ* growth monitoring techniques are described elsewhere (Kuech, 2014). Of recent interest for LED fabrication is an *in situ* PL measurement that measures InGaN MQW emission wavelengths, ultimately allowing for adjustments of the growth conditions to correct the emission wavelength (Prall et al., 2014).

Unlike MBE growth, which can use both photon and electron-based monitoring techniques, only photons can be used during MOVPE growth because of the higher pressure. Typically, visible wavelength photons are used, although X-ray sources could be used but are too low in intensity to convey real-time information except in special circumstances (Yamamoto et al., 2014). This leaves IR to UV wavelengths for the III-N *in situ* monitoring of pyrometry, reflectance, and strain measurements described in Sections 6.4.1 through 6.4.3.

6.4.1 Pyrometry

The MOVPE growth temperature is routinely controlled using proximity thermocouples, optical pyrometry, or applied heater power. More often a combination of these techniques is used. While pyrometry might not be thought of as an *in situ* monitoring technique, knowing the temperature during InGaN QW growth is essential to achieve the correct LED wavelength, since a 1°C temperature difference can shift the emission wavelength by 1–2 nm (Creighton et al., 2006, 2008b). Complicating this seemingly simple measurement for III-N growth is the transparent nature of both the film and the substrate. This means that instead of measuring the surface temperature, the pyrometer is measuring the susceptor or carrier temperature which depending on growth pressure can differ by ~70°C (Creighton et al., 2006). Emissivity changes to the susceptor from deposits can also influence the measured growth temperature (Breiland, 2003). Finally, the pyrometer should be calibrated against a black body source or some other self-consistent means to account for geometric sighting issues and radiation loss factors that occur during the pyrometry measurement (Breiland, 2003).

Emissivity-correcting pyrometers (ECPs) were developed (Breiland, 2003) to address some of these issues. For this technique, the emissivity is measured using the reflectance from a separate light source reflecting from the surface (Breiland, 2003). The ECP technique was shown to provide more accurate temperature measurements of III-V material when the substrates were opaque. For III-N materials two wavelength ranges can be used to measure the surface temperature, one <400 nm where the GaN film is opaque at ~1,000°C (Creighton et al., 2006) and the other in the mid-IR where the sapphire is opaque (Creighton et al., 2008b). Both of these wavelength ranges present challenges for routine operation. For the 400 nm ECP, the light generated at InGaN growth temperatures is low, requiring special counting techniques and photomultiplier tubes (Creighton et al., 2006). For the 400 nm ECP the noise is less than 0.1°C above 1,000°C, but as the temperature decreases to 700°C–800°C, the temperature noise is ~1°C (Creighton et al., 2006). On the other hand, the signal is plentiful in the mid-IR and is transparent to the rotation–vibration absorption bands of NH_3 near ~7–8 μm. The thermal emission reflected back onto the surface can also be used to measure the emissivity, however slight changes to the ambient temperature and accurate reflectivity measurement make this technique somewhat difficult for routine implementation (Creighton et al., 2008b). A prototype mid-IR pyrometer was developed and tested resulting in improved control of the PL wavelength for InGaN MQWs (Creighton et al., 2008b).

Besides pyrometry, direct measurements of the optical absorption edge can provide a measure of the temperature provided the dependence of the absorption edge on temperature is known. This technique has demonstrated better than 1°C sensitivity but can be sensitive to thin-film interference, which causes the apparent temperature to oscillate (Johnson and Tiedje, 1997). This oscillation can be removed from the temperature measurement by using the width of the adsorption knee to correct temperature errors (Johnson and Tiedje, 1997). Commercial instruments based on this technique are available from k-Space Associates (Harris et al., 2007).

6.4.2 Optical Reflectance

Next to pyrometry, optical reflectance is widely used as *in situ* diagnostic, because the equipment is relatively easy to implement and the results can be modeled or fit to the reflectance waveform (Breiland and Killeen, 1995). Initially, single wavelength reflectance waveforms were monitored resulting in growth rate and optical constant information for each deposited layer (Breiland and Killeen, 1994). Implementation of this technique has been described by Breiland et al. and consists of measuring chopped light from a light source reflected from the surface in a near normal incidence geometry (Breiland and Killeen, 1994). Such an approach can usually be combined with temperature measurement to produce an emissivity-corrected pyrometer (Breiland and Killeen, 1994). Breiland and Killeen developed a virtual interface model allowing extraction of information on the growth that requires no prior knowledge other than the starting reflectance of the substrate (Breiland and Killeen, 1995). Because of the richness of the measurement, in many cases, just being able to monitor the reflectance waveform in a *finger print* mode gives the MOVPE operator confidence that the desired material is being grown.

An example optical reflectance waveform measured for the two-step GaN growth process on sapphire (Nakamura, 1991c) is shown in Figure 6.5. The waveform was measured at 550 nm and is shown in solid line, while the growth temperature is shown in dashed line. The best understood growth signatures from this waveform such as the NL thickness (Koleske et al., 2004), the 3D to 2D recovery time (Figge et al., 2000; Balmer et al., 2002; Koleske et al., 2002a), the bulk GaN growth rate (Breiland and Killeen, 1995), and the bulk layer roughness (Balmer et al., 2002; Chaowang and Ian, 2007), and uniformity regions of the waveform are denoted in black. Secondary features of the waveform are labeled blue and include temperature-dependent optical constants for sapphire, GaN NL, and bulk GaN, and the extent of the NL decomposition (Koleske et al., 2005). Finally, features that physically influence the waveform are labeled in red and include the NL roughness and the 3D islanding. Methods to extract this roughening information from the waveform have not yet been developed, though determining the GaN grain size might be possible using multiwavelength reflectivity measurements. Since the reflectance signal decreases to near zero during the 3D-roughening phase, measurements of the diffuse reflected light intensity might also provide information during this growth step.

As previously mentioned, Nakamura suggested that his choice of growth conditions was guided by IR transmission measurements when trying to maximize the smoothness of the GaN films (Nakamura, 1991a). Later Balmer et al. used an effective medium approach to model the surface roughness to explain the measured decrease in the reflectance waveform during the 3D–2D growth transition (Balmer et al., 2002). Optical reflectance has also been used to measure GaN decomposition resulting in a measured activation energy of 2.7 eV for both NLs and bulk films (Koleske et al., 2003). During these decomposition studies, the pre-exponential for NL decomposition was measured to be 4–9x larger than for bulk GaN, which might be expected from the more defective NLs (Koleske et al., 2003). An analytical model was developed to fit the reflectance waveforms during NL annealing which included the temperature

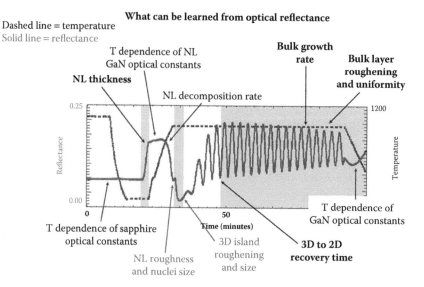

FIGURE 6.5 *In situ* monitoring of a typical GaN on sapphire growth run. The growth temperature changes are shown by the dashed line and the *in situ* optical reflectance at 550 nm is shown as the solid line. As the growth progresses, lots of information are conveyed about the growth rates, morphology, film uniformity, and optical constants.

dependence of refraction indices and the decomposition rates measured under varying or constant temperature (Koleske et al., 2005). Using optical reflectance, Fitouri et al. quantified the SiN treatment process and found that an effective medium approximation could be used to describe the growth rate and refractive indices for GaN growth after SiN treatment (Fitouri et al., 2005).

6.4.3 Stress Monitoring (GaN on Si)

With a potentially large difference in the coefficients of thermal expansion (COE) between III-N materials and many substrates, it has become increasingly important to manage film strain during growth. If not managed properly wafer bow can lead to cracks, delamination, or wafers so warped they cannot be lithographically patterned or processed. Several *in situ* techniques have been developed to monitor the wafer curvature using single (Belousov et al., 2004) or multiple (Gillard and Nix, 1993; Floro et al., 1996) laser beams. For the single laser beam technique, a position sensitive detector is used to measure the laser reflection during the rotation (Belousov et al., 2004). For the multiple laser beam technique, (Floro et al., 1996; Chason et al., 1999) the reflected laser beam positions are tracked by a CCD camera and the wafer curvature can be calculated from the beam separation (Floro et al., 1996). The measured laser deflection along with an application of the Stoney equation (Stoney, 1909) can be used to determine the wafer curvature. If the III-N alloy concentration and thickness are known, the strain state of the film can be calculated.

This technique was used to measure the stress evolution for GaN films growth on sapphire (Amano et al., 1998; Hearne et al., 1999). Hearne et al. found that even though GaN has a 16% compressive lattice mismatch to sapphire, at high-temperature GaN grows in tension and only after cooling to room temperature is the GaN film put into compression (Hearne et al., 1999). In other experiments, the GaN strain state could be controlled by inserting AlN interlayers between the GaN layers to reduce this tensile stress and put the film back into compression (Amano et al., 1998). Later, the critical layer thicknesses for

GaN strain relaxation on AlGaN layers were measured using the *in situ* strain measurement by gradually increasing the thicknesses at a fixed aluminum concentration and capping each AlGaN layer with a standard GaN layer (Lee et al., 2004b). One anomalous feature of the multiple beam laser technique on transparent films and substrates is that the reflection is susceptible to thickness gradients that can steer the laser beam away from the ideal specular condition leading to thickness induced oscillations in the measured strain curve (Breiland et al., 2004). Also, for films with nonuniform thickness, the Stoney equation cannot be used quantitatively.

The growth of InGaN-based LEDs on Si (111) is the most significant application of *in situ* strain measurements as spearheaded by Dadgar and Krost (Dadgar et al., 2003, 2006; Krost et al., 2005a, 2005b). Using the *in situ* curvature measurement technique, the impact of 3D-island coalescence, layer thermal mismatch, and doping on the collective film/substrate strain state could be monitored and controlled through the insertion of AlN and AlGaN strain relieving layers (Dadgar et al., 2003; Krost et al., 2005a), ultimately resulting in III-N films on Si(111) with near zero wafer curvature. Recently, with the aid of *in situ* strain monitoring, crack-free, uniform InGaN/GaN LEDs with a strain-engineered buffer layer were grown on an 8-inch diameter Si (111) substrate (Lee et al., 2015). Also with knowledge of the lattice-mismatch, the Stoney equation, and the strain state of GaN on sapphire, the AlGaN and InGaN alloy concentration can be determined during growth (Brunner et al., 2008a).

6.5 Doping the III-N's

In any semiconductor system, the better the control over the n-type and p-type doping the greater the portfolio of device structures that can be made. As a first step to controlled doping, the background impurities and electrically contributing point defects must be at low concentrations or controlled to some degree. Early GaN films typically had n-type background carrier concentration presumably due to oxygen impurities and intrinsic defects (Ambacher, 1998). However, as chemical purity improved and the use of NLs reduced dislocation density, controlled n- and p-type doping in the III-N system soon followed.

6.5.1 n-Type Doping (Si, Ge)

In most III-V semiconductors, silicon incorporates onto the Group III site making it a donor except at higher doping levels where carrier compensation occurs. Silicon presents very little memory effect due to its low vapor pressure. Silicon is typically delivered using silane (SiH_4) or disilane (Si_2H_6), and both dopant chemicals have shown good incorporation over wide temperature ranges for GaN growth (Wickenden et al., 1995b; Eiting et al., 1998). Silicon incorporation depends linearly on the flow from the gas-phase source (Koide et al., 1991; Nakamura et al., 1992a; Wickenden et al., 1995b) and using Si_2H_6 Wickenden et al. achieved doping levels from 1×10^{17} cm^{-3} to 4×10^{19} cm^{-3} with doping efficiency slightly greater than 0.5 (Wickenden et al., 1995b). This doping efficiency for Si_2H_6 might be expected due to the stepwise decomposition of Si_2H_6 to adsorbed SiH_x species with remaining $SiH_{x \geq 3}$ species desorbing into the gas phase (Jasinski and Gates, 1991). By comparing SIMS measurements to carrier concentration measured with variable temperature Hall Effect measurements the single donor/single acceptor compensation was <0.3 and the carrier activation energy was 34 meV at 6×10^{16} cm^{-3} (Wickenden et al., 1995a). As the Si concentration increased, the carrier activation energy decreased, indicating a transition from hopping to band-to-band conduction (Wickenden et al., 1995a). Gemanium has also been successfully used as an n-type dopant in GaN (Nakamura et al., 1992a).

For high Si-doping levels in GaN, an extra tensile strain is added, which can lead to cracking. This tensile strain is likely due to the smaller ionic radii of Si compared to Ga (Hageman et al., 1999). However, first-principles calculations show that only negligible changes occur as Si is substituted for Ga in the lattice, suggesting that cracks do not form during growth but during cooling to room temperature (Romano et al., 2000). The threshold for crack formation was found to occur for a 2.0 μm thick film with a Si concentration of 2×10^{19} cm^{-3} (Romano et al., 2000). To reduce this tensile film strain, Fritze et al. have reported on the use of germane and isobutylgermane to achieve Ge-doping levels up to 2.9×10^{20} cm^{-3} and crack-free GaN films (Fritze et al., 2012).

For AlGaN and AlN, the Si carrier activation energy increases as the Al concentration increases as expected from the increase in bandgap (Taniyasu et al., 2002; Thapa et al., 2008). Eiting et al. measured room temperature electron concentrations as high as 8.8×10^{18} cm^{-3} and a mobility of 110 cm^2/Vs in 0.1 μm thick Al$_{0.10}$Ga$_{0.90}$N (Eiting et al., 1998). In 70% aluminum AlGaN, Zhu et al. achieved Si-doping levels 6.0×10^{19} cm^{-3}, resulting in an electron concentration of 3.3×10^{19} cm^{-3} and mobility of 25 cm^2/Vs at room temperature (Zhu et al., 2004). Lastly, n-type AlN has been achieved with electron concentrations of 1×10^{17} cm^{-3} at Si-doping levels of 3×10^{19} cm^{-3}, indicating a carrier activation energy of ~100 eV (Taniyasu et al., 2002). Bulk AlN substrates with dislocation densities of ~10^6 cm^{-2} doped with Si-doping concentrations of 3×10^{17} cm^{-3} resulted in an AlN film with a mobility of 426 cm^2/Vs (Taniyasu et al., 2006a).

Optically, silicon doping of GaN and InGaN films also increases band-edge (BE) emission intensity and decreases the yellow band emission (Nakamura et al., 1993a). For example, Nakamura mentions that the BE emissions of Si-doped InGaN films were about 36 times stronger than that of undoped InGaN films and 20 times stronger than that of the blue emission (at 450 nm) of Mg-doped p-type GaN films (Nakamura et al., 1993a). Silicon doping of the GaN barriers in InGaN/GaN single and multiquantum wells also influences the growth morphology and PL emission intensity (Keller et al., 1998a), providing another possible way to tune the optical quality of InGaN quantum wells (QWs).

6.5.2 p-Type Doping with Mg and Mg Activation

Initially, Mg-doped GaN films were highly resistive despite doping into the material with little n-type background carrier concentrations (Akasaki et al., 1989; Nakamura, 1991c). Somehow the Mg doping was making the GaN go from being weakly n-type to highly resistive, suggesting that the Mg was not acting as a p-type dopant, not sitting on the right lattice site, had too high of an activation energy, or possibly some other mechanism was preventing dopant activation. Similar highly resistive Mg-doped GaN films had been observed during HVPE growth (Maruska et al., 1973; Maruska and Rhines, 2015), and only weak electroluminescence was apparent after applying 150 V across point contacts (Maruska and Rhines, 2015). Subjecting Mg:GaN to a low energy electron beam irradiation produced modest p-type conductivity (Amano et al., 1989) and this procedure was later replicated and extended to achieve record high acceptor densities of 3×10^{18} cm^{-3} in GaN (Nakamura et al., 1991b). Thermal annealing of Mg-doped GaN in flowing N$_2$ by Nakamura et al. resulted in p-type GaN, while annealing in H$_2$ or NH$_3$ returned or kept the GaN resistive, suggesting that hydrogen was a possible compensator to the desired p-type activity (Nakamura et al., 1992c).

When GaN is doped with Mg, H creates a complex with Mg, rendering the Mg doping inactive. To identify the H lattice location Wampler and coworkers used a combination of ion channeling and density functional theory to calculate total energies for various configurations and charge states of hydrogen (Wampler et al., 2001). Using H$_2$ gas charging to passivate the Mg, H was found to reside at the transverse antibonding site to Mg with a binding energy of ~0.7 eV (Wampler et al., 2001). Hydrogen was found to readily diffuse in GaN p/n junctions at temperatures ranging from 200°C to 310°C (Seager et al., 2002), suggesting the high annealing temperatures >600°C (Nakamura et al., 1992c) required for Mg activation was not due to dissociation of the H-Mg complex. Instead, Myers et al. found the high temperature was required to overcome the surface activation barrier for hydrogen release (Myers et al., 2001). By annealing in D$_2$, the hydrogen release from the surface was accelerated (Wampler and Myers, 2003) and thermal annealing in vacuum, N$_2$, and O$_2$ showed that hydrogen release is rate limited by surface desorption (Myers et al., 2004). Depending on the annealing condition, H_2 was found to form and desorb with rates proportional to the square of H surface coverage, or if N–H and O–H species were involved, linear with the H coverage (Myers et al., 2004). Further experiments by Lu et al. have confirmed that annealing Mg-doped GaN in mixed N_2 and O_2 provided significant improvements in Mg activation, ultimately producing hole concentrations of 9×10^{17} cm^{-3} with resistivities of 0.62 Ω-cm (Lu et al., 2013). Mg-doped GaN films can be activated in the MOVPE system during the cool down by annealing for several minutes at ~800°C in N_2 (Bour et al., 1999) or through post-growth anneals. No matter how it is implemented, the Mg activation step is critical in achieving high hole concentrations required for working diodes (Doo-Hyeb et al., 1999; Harima et al., 1999; Svensk et al., 2007).

Compared to other III-V semiconductors, the acceptor density is limited to ~1×10^{18} cm^{-3} despite doping levels approaching 1×10^{20} cm^{-3}. Temperature-dependent Hall Effect measurements and defect spectroscopy identified energy levels in Mg-doped GaN that were 170 meV (Gotz et al., 1996), 135–155 meV (Kim et al., 2000), and 112–190 eV (Kozodoy et al., 2000) above the valance band depending on doping level. These experimental values for the Mg acceptor level are consistent with the wide bandgap nature of GaN as derived from theory (Van de Walle et al., 1998). The Mg acceptor level also scales with bandgap for the III-N alloys where the Mg level gets deeper as aluminum is added and shallower as indium is added (Lee et al., 2004a). Ideal Mg concentrations in GaN are around 3×10^{19} cm^{-3} and above this doping level the Mg auto-compensates (Kaufmann et al., 2000; Kozodoy et al., 2000) leading a reduction in the hole concentration (Obloh et al., 1998). These larger Mg concentrations in GaN have been suspected of producing Mg-rich planar defects (Liliental-Weber et al., 1999) and inversion domains (Romano et al., 1996, 2001; Romano and Myers, 1997; Vennegues et al., 2000).

For MOVPE growth, the Mg dopant bis-(cyclopentadienyl)-magnesium, Cp_2Mg, is used although the methyl substituted, $(MeCp)_2Mg$, can also be used with the advantage that it has a higher vapor pressure (Kozodoy et al., 1998). Since the Mg precursor is solid when used for MOVPE growth, it has been mixed with various low-vapor pressure solutions to keep the precursor from sintering and allowing for more uniform pickup and delivery out of the bubbler (Qi et al., 2001). One potential issue with these low-vapor pressure solutions is the potential for them to be pushed out of the bubbler at high bubbler flow rates which might cause clogging of the downstream bubbler lines and mass flow controllers.

To achieve reasonable Mg-doping levels, Kozodoy et al. used near atmospheric-pressure growth to achieve [Mg] of 1×10^{19} cm^{-3}, which produced acceptor concentrations of 8×10^{17} cm^{-3} (Kozodoy et al., 1998). The amount of Mg incorporated was independent of the growth rate and V/III ratio but as the growth pressure was reduced a higher V/III ratio was required to maintain low resistivity p-type films (Kozodoy et al., 1998), suggesting a lower chemical reactivity of NH_3 at this lower pressure. At 76 Torr, x10 more magnesium was incorporated into the film (Kozodoy et al., 1998), with this increase being attributed to reduced parasitic adduct formation at low pressures (Wang and Creighton, 2004). Although the acceptor concentration increased at low growth rates and higher V/III ratios at low pressure, the maximum concentration obtained was only one-quarter of the concentration obtained at atmospheric pressure (Kozodoy et al., 1998). The increase in acceptor concentration was assigned to reduced compensation of the GaN film by residual donor defects, suspected to be nitrogen vacancies (Kozodoy et al., 1998).

In Figure 6.6 illustrates typical PL and Hall-effect data for Mg-doped GaN films. For this work $(Mecp)_2Mg$ was used as the Mg precursor, with the p-type GaN layer grown at 970°C and 200 torr in a Veeco D125 short jar MOVPE system. The GaN growth rate was ~0.6 μm/hr using a $(Mecp)_2Mg$ flow rate of ~30 sccm (corresponds to 0.16 μmoles/min). Using these growth conditions ~3×10^{19} cm^{-3} Mg atoms are incorporated into the lattice resulting in routine hole concentrations of $3 - 5 \times 10^{17}$ cm^{-3}. The growth temperature of 970°C is chosen since it allows enough thermal budget for blue LED growth by maintaining the QW emission intensity. Lower p-side growth temperatures are required for longer wavelength LEDs however lower hole concentrations have been observed at these lower growth temperatures compared to those used for blue wavelength LEDs (Lee et al., 2007).

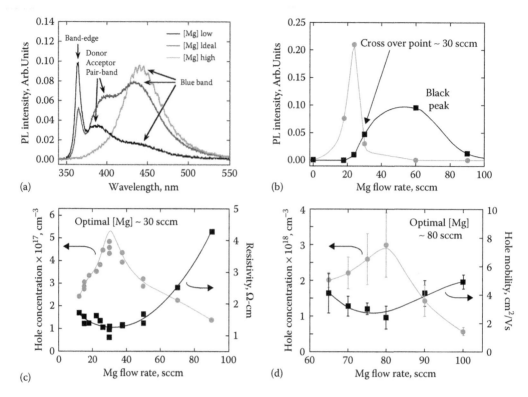

FIGURE 6.6 (a) Plots of PL intensity for Mg concentrations in GaN that are too low (strong band-edge emission and weak DAP), ideal (higher DAP), and too high (dominated by blue band). The PL peaks correspond to the band-edge intensity near 364 nm, the donor-acceptor pair-band (DAP) near 380 nm, and the blue band near 440 nm. (b) PL intensity at different cp_2Mg-doping flow rates for the DAP peak and black peak. A change over in which peak (DAP or black) has maximum intensity occurs near a Mg flow rate of 30 sccm. (c) The hole concentration (left axis) and layer resistivity (right axis) as a function of Mg flow rate. The maximum hole concentration and minimum resistivity occur at a Mg flow rate of 30 sccm. (d) The hole concentration (left axis) and hole mobility (right axis) as a function of Mg flow rate for 7%–8% indium concentration InGaN. The maximum hole concentration occurs at a Mg flow rate of 80 sccm.

Representative room-temperature PL scans of the Mg-doped GaN are shown in Figure 6.6a. These films were activated in the MOVPE system by annealing for five minutes at 820°C in N_2 after growth. The three common PL features observed when GaN is doped with Mg are the band edge emission peak at 364 nm, a broader donor-acceptor pair (DAP) band peak around 380–390 nm, and an even broader blue band emission peak around 440 nm. The origin of the blue band emission at 440 nm is not entirely clear; however, it has been attributed to possible Mg defect levels or Mg clustering that results when too much Mg is incorporated into the GaN film (Kaufmann et al., 2000; Kozodoy et al., 2000; Qu et al., 2003; Iida et al., 2010). The relative intensity of these three common PL features depends on the Mg-doping level. At low Mg levels, the PL luminescence is dominated by band-edge emission at 364 nm as shown in the blue scan in Figure 6.6a. As the Mg flow increases the DAP band increases in intensity and the band-edge intensity decreases as shown by comparing the too low and ideal scans. Finally, as the Mg-doping level increases further, the blue band near 440 nm increases in intensity along with a corresponding decrease in the DAP band. The measured PL intensities are plotted in Figure 6.6b for the DAP and black peaks. At lower Mg flow rates the DAP peaks at about 12 sccm and decreases above that value, while the black peak has a maximum near 40–60 sccm. Of interest is the point where the DAP and black peaks cross over in intensity which is near 30 sccm as shown in Figure 6.6b.

The hole concentration (filled circles) is plotted in Figure 6.6c for Mg-doped GaN after activation in the MOVPE reactor. The peak hole concentrations occur at a Mg flow rate ~30 sccm which is the same position as the DAP and blue peak cross over point shown in Figure 6.6b. Note that the peak in the hole concentration results in a resistivity of ~1 Ω-cm (filled squares). The peak in the hole concentration occurs near a doping concentration of ~3 × 10^{19} cm^{-3} with increased or decreased [Mg] flows resulting in lower hole concentrations (Obloh et al., 1998). Similar strong blue luminescence has also been reported for heavily Mg-doped GaN (Kaufmann et al., 2000; Kozodoy et al., 2000; Qu et al., 2003; Iida et al., 2010).

As the indium concentration is increased in InGaN, the Mg induced hole activation energy lowers compared to that of GaN (Kumakura et al., 2000a; Pantha et al., 2009). By lowing the activation energy, higher hole concentrations are observed as shown in Figure 6.6d. For this plot, the hole concentration from Hall effect is plotted versus the Mg flow rate for In$_{0.08}$GaN$_{0.92}$N grown at a temperature of 830°C. Near a Mg flow rate of 80 sccm, the average hole concentration of ~3 × 10^{18} cm^{-3} is achieved. The measured mobility is also plotted and decreases from ~10 cm^2/Vs typically observed for GaN to around ~3 cm^2/Vs for the In$_{0.08}$GaN$_{0.92}$N. The higher Mg flow rate might be required to overcome indium surface site blocking; however, at this growth temperature the adsorption and desorption of indium from the surface are quite high so the steady-state surface coverage is uncertain. Typically, because the InGaN films are grown in the absence of H$_2$, the p-type InGaN is activated as grown and does not require further annealing in N$_2$.

Other groups have also achieved higher hole concentrations in p-type InGaN. For example, Pantha et al. studied Mg doping of In$_{0.35}$Ga$_{0.65}$N films and found a lowering of the activation energy from ~170 meV measured for GaN to ~43 meV (Pantha et al., 2009). For Mg-doped In$_{0.22}$Ga$_{0.78}$N a room temperature resistivity of 0.4 Ω-cm and a hole concentration ~5 × 10^{18} cm^{-3} was achieved (Pantha et al., 2009). Higher hole concentrations of 6.7 × 10^{18} cm^{-3} were achieved in In$_{0.14}$Ga$_{0.86}$N after optimizing the Mg-doping concentration, growth rate, and temperature (Kumakura et al., 2000a, 2000b). Using In$_{0.22}$Ga$_{0.75}$N/GaN superlattices, Kumakura et al. achieved hole concentrations of 2.8 × 10^{19} cm^{-3}, finding that the hole concentrations for the superlattices were larger than InGaN bulk layers with the same average indium mole fraction (Kumakura et al., 2000c). The origin of holes in Mg-doped In$_x$Ga$_{1-x}$N films is somewhat mysterious. Unlike in Mg-doped GaN, Zvanut et al. using EPR were not able to correlate the number of Mg-related acceptor states in p-type InGaN to the hole density, and instead measured a decrease in the Mg acceptor density as the indium concentration increased (Zvanut et al., 2012), leaving the origin of the increased hole concentrations in p-type InGaN somewhat of a mystery.

Mg doping of AlGaN is also important since it is used as an EBL in LEDs. As expected the acceptor activation energy increases when AlGaN is doped with Mg. The activation energy for Mg-doped Al$_{0.08}$Ga$_{0.92}$N was estimated to be about 35 meV deeper than that in GaN (Tanaka et al., 1994). Using atmospheric pressure MOVPE, the net acceptor concentration increased in Al$_{0.15}$Ga$_{0.85}$N with increasing [Cp$_2$Mg]/[III] and saturated at about 7 × 10^{18} cm^{-3} while the Mg concentration did not saturate until 5 × 10^{19} cm^{-3} (Suzuki et al., 1998). For Al$_{0.15}$Ga$_{0.85}$N the acceptor activation energy was estimated to be 250 meV at room temperature (Suzuki et al., 1998). PL measurements by Kinoshita et al. suggested that Mg doping at high V/III ratio can effectively suppress self-compensation by the formation of nitrogen vacancy complexes (Kinoshita et al., 2013). Using this approach, they were able to measure 1.3 × 10^{17} cm^{-3} holes in Al$_{0.7}$Ga$_{0.3}$N (Kinoshita et al., 2013). Kawanishi et al. have reported finding holes in AlGaN doped with carbon (Kawanishi and Tomizawa, 2012).

One way to overcome the large activation energy of acceptors in III-N materials is through the use of binary and ternary superlattices (SLs) which has been shown to enhance the acceptor activation by more than one order of magnitude (Schubert et al., 1996). Kozodoy and coworkers grew Mg-doped AlGaN/GaN resulting in measured hole concentration at room temperature over 2.5 × 10^{18} cm^{-3} more than 10 times that obtained in bulk AlGaN layers (Kozodoy et al., 1999). The temperature dependence of the resistivity provided evidence of the formation of a confined hole gas whose origin was attributed to valence band bending from the piezoelectric and spontaneous polarization fields (Kozodoy et al., 1999). Even higher hole concentrations were achieved using a $Al_{0.63}Ga_{0.37}N/Al_{0.51}Ga_{0.49}N$ superlattice where Zheng et al. were able to reach hole concentrations of 3.5×10^{18} cm^{-3} at room temperature; a 10x improvement in conductivity compared with that of conventional SLs (Zheng et al., 2016).

Codoping schemes have also been proposed as another way to potentially lower the hole activation energy and increase hole concentrations. This method relies on the simultaneous placement of the counter dopant atoms in close proximity to each other to create a substitutional complex that in theory lowers the activation energy. By codoping GaN with Mg with oxygen the hole concentrations were increased to 2×10^{18} cm^{-3} and the resistivity of codoped layers decreased from 8 to 0.2 Ω-cm (Korotkov et al., 2001, 2002). Variable temperature Hall effect measurements showed that the acceptor activation energy decreases from 170 to 135 meV upon oxygen codoping (Korotkov et al., 2001, 2002). In another study GaN was codoped with Mg and Si, and the blue luminescence intensity was shown to decrease as the Si doping increased (Han et al., 2003), suggesting a reduction in the native defect formation and its influence on Fermi energy on defect stability in GaN (Han et al., 2003).

To study how the carrier concentration, carrier type, and compensation change, as GaN is codoped, we separately doped and codoped GaN films grown on sapphire using cp$_2$Mg and SiH$_4$. For the SiH$_4$-doped GaN a [Si] ~2×10^{19} cm^{-3} was used resulting in room temperature n-type carrier concentration of 1.7×10^{19} cm^{-3}, a resistivity ρ, of 0.0033 Ω-cm, and a mobility μ, of 115 cm^2/Vs. In the same material doped with a [Mg] ~3×10^{19} cm^{-3}, the GaN was p-type with a carrier concentration of 3.1×10^{17} cm^{-3}, a $\rho = 2.5$ Ω-cm, and a $\mu = 8.0$ cm^2/Vs. Interestingly, when the GaN was codoped with both dopants (i.e., [Si] ~2×10^{19} cm^{-3} and [Mg] ~3×10^{19} cm^{-3}) the film measured p-type with a carrier concentration of 2.8×10^{17} cm^{-3}, a $\rho = 3.4$ Ω-cm, and a $\mu = 6.5$ cm^2/Vs. Note that the codoped p-type sample is slightly more compensated (lower free carriers, higher resistivity, and lower mobility) than the p-type film with just Mg doping. One expectation from this codoping experiment might be that since Si produces more free carriers than Mg, the film should end up n-type instead of the measured p-type. However, this is not the case, since the [Mg] is larger than the [Si] resulting in the Fermi level being pinned closer to the valence band than the conduction band. Only when, [Si] > [Mg], would this material be expected to turn n-type. On the other hand, if [Si] $\approx$ [Mg], the material should be closer to insulating with the Fermi level pinned somewhere between the valence and conduction bands. The results discussed here are similar to the work of Mita et al. who reported that intentionally Si-doped GaN became semi-insulating when the carbon concentration exceeded that of silicon (Mita et al., 2008) since carbon has been shown to produce a deep acceptor. Still, instead of producing insulating material by trying to balance the n-type and p-type behavior with dopants, it is better to have low intrinsic carrier concentrations and pin the Fermi level near mid-bandgap using carbon and iron doping as discussed in Section 6.5.3.

Finally, magnesium doping of III-N material can result in a memory effect where Mg is still incorporated into the growing film even after the MO source is shut off. The memory effect was speculated to originate from the Mg precursor adsorption onto the walls or stainless tubing in the reactor. Kuech et al. tested this assumption by introducing baffles to vary the surface area of the growth chamber and were able to show that quartz was a source of adsorption sites, not the stainless steel of the gas injection manifold (Kuech et al., 1988). This memory effect can be reduced somewhat if reactor purging is performed (Rask et al., 1988). Contributing to the Mg memory effect is the Lewis acid–base complexes that form between NH$_3$ and cp$_2$Mg which are reversible when heated (Wang and Creighton, 2004) but may condense out on cooler parts of the MOVPE reactor. For n-type material grown on Mg-doped GaN, a long tail of Mg has been shown to profile into the n-type material, thereby preventing a sharp interface p/n junction (Huili et al., 2003). Huili et al. showed that this Mg tailing into Si-doped GaN was shown to be as large as ~115 nm/decade after shutoff to reach 1×10^{19} cm^{-3} and 750 nm/decade to reach 1×10^{17} cm^{-3} region (Huili et al., 2003). By experimenting with surface cleans and growing Si-doped films in a *Mg-free* reactor, they were able to show that the majority of the memory effect Mg-doping profile was due to Mg adsorption in the reactor, however a slight effect of residual Mg on the surface was also observed (Huili et al., 2003). Obviously, the Mg memory effectively limits the type of p- on n-type junctions that can be grown using MOVPE.

6.5.3 Unintentional Doping (C, O, Si, H) and Insulating Material (Fe Doping)

Since the metalorganics are transported to the growth surface with some degree of their hydrocarbon ligands still attached (Parikh and Adomaitis, 2006; Danielsson et al., 2016), the incorporation of carbon and hydrogen into the III-N films has always been an important consideration. In addition, the growth chamber constructed of quartz or stainless steel, with stainless steel gas manifolds, and high-temperature heater parts might also contribute to Si, O, and other metals being unintentionally incorporated during growth. In this section, several of the major contaminants are discussed along with their influence on the film properties.

Carbon is the most likely and most important unintentionally incorporated impurity during MOVPE growth. In GaN, carbon acts as a deep acceptor to compensate residual n-type background doping however it also produces trapping type defects responsible for current collapse in HEMTs (Klein et al., 2001). The carbon concentration in GaN films can be controlled through the choice of growth conditions (Parish et al., 2000; Koleske et al., 2002c; Wickenden et al., 2004) as detailed in the next several paragraphs, although where carbon sits in the lattice dictates its influence on the electrical properties rather than its overall concentration.

Studies of carbon incorporation as a function MOVPE growth conditions were first reported by Parish et al. (2000) and later by Koleske et al. (2002c). Both of these studies and later ones find that carbon incorporation is enhanced as the growth temperature (Parish et al., 2000; Koleske et al., 2002c), growth pressure (Koleske et al., 2002c; Wickenden et al., 2004; Kato et al., 2007; Luong et al., 2015), or V/III ratio (Parish et al., 2000; Koleske et al., 2002c; Kato et al., 2007) are reduced. Although carbon incorporation is from MO sources, surprisingly, there is only a weak dependence on TMGa flow rate (Parish et al., 2000; Koleske et al., 2002c), where a fourfold increase in TMGa flow results in only a 30% increase in the carbon concentration (Koleske et al., 2002c). The carbon concentration also increased by a factor of 3.5 when the H_2 flow rate was increased from 1 to 6.4 SLM; however, this change in the carbon concentration might be more related to a reduction in GaN growth temperature from the higher heat capacity of H_2 compared to N_2 (Koleske et al., 2002c). Parish et al. also reported that carbon incorporation also increases as the Al concentration is increased in AlGaN films (Parish et al., 2000).

All of these dependencies of carbon incorporation might be expected from the TMGa dissociation on the surface; however, the exact chemical form of the carbon doping fragment remains uncertain. Naively, one might expect the last Ga-CH_3 bond might remain intact as it adsorbs on the surface. With the $-CH_3$ fragment on the surface, an addition of a surface H could lead to methane production and desorption, while further dissociation of the $-CH_3$ fragment might lead to carbon incorporation. The hydrogenation of adsorbed methyl fragments might follow the same re-hydrogenation conditions that remove surface N atoms and favor NH_3 reformation and GaN decomposition (Koleske et al., 2001). Two recent studies might contradict this simple picture of carbon doping of GaN. The first is a theoretical study where Danielsson et al. were able to calculate that during GaN growth the main gallium precursor shifts from Ga-CH_3 at low temperatures to atomic Ga at higher temperatures and the carbon doping shifts from $-CH_3$ groups at low temperatures to C_2H_x at higher temperatures (Danielsson et al., 2016). The second is a comprehensive study by Li et al. using various alkanes (CH_4, C_3H_8, and i-C_4H_{10}), an alkene (C_2H_4), an alkyne (C_2H_2), and trimethylamine [$N(CH_3)_3$] to dope GaN (Li et al., 2015). The unsaturated hydrocarbons C_2H_4 and C_2H_2 were found to produce high carbon-doping concentrations (10^{18}–10^{19} cm^{-3}) with less of an effect on the GaN crystal quality than the saturated ones (Li et al., 2015). Recent work by Lundin et al. confirmed propane (C_3H_8) is an effective carbon dopant over a wide range of MOVPE growth conditions for GaN (Lundin et al., 2016). For carbon doping of AlGaN, Kawanishi et al. showed p-type doping (p ~ 3.2 × 10^{18} cm^{-3}) of $Al_{0.10}Ga_{0.90}N$ using CBr_4 (Kawanishi and Tomizawa, 2012).

In their study of carbon incorporation to produce insulating films, Wickenden et al. suggested that carbon may incorporate differently along dislocation cores compared to bulk incorporation (Wickenden et al., 2004). This suggestion was also proposed by Klein et al. in current collapse studies of AlGaN/GaN HEMTs, where the deepest trap was correlated to a carbon-related defect, while the mid-gap trap was

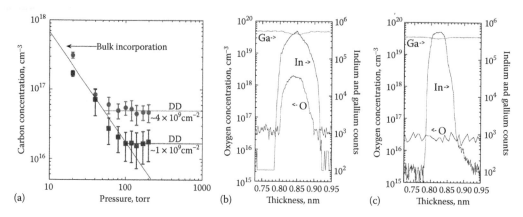

FIGURE 6.7 (a) SIMS measured carbon incorporation into GaN as a function of pressure grown on GaN templates with two different dislocation densities as measured using XRD. (From Lee, S.R. et al., *Appl. Phys. Lett.*, 86, 241904, 2005. With permission.) (b) SIMs measured oxygen concentration in a 3% indium InGaN thick layer before changing out the NH_3 purifier. The indium and gallium counts are plotted on the right axis to denote the location of the InGaN film. (c) is the same as (b) after changing the NH_3 purifier and repeating the same growth run.

thought to be associated with grain boundaries or dislocations (Klein et al., 2001). Additional evidence for differences in carbon incorporation along dislocations compared to bulk is shown in Figure 6.7a. For this work, two GaN templates were grown with 4×10^9 cm^{-2} and 1×10^9 cm^{-2} dislocation densities (Lee et al., 2005) following the procedures described in Koleske et al. (2002a). The growth pressure was varied on each dislocation density template, and the carbon concentration measured using SIMS. At 20 torr the carbon concentration is $>10^{17}$ cm^{-3}; however, it drops rapidly as the growth pressure is increased as shown by the solid black line in Figure 6.7a. For the GaN template with the 4×10^9 cm^{-2} dislocation density, the carbon concentration drops to 5×10^{16} cm^{-3} for pressures >50 torr and remains constant. For the lower dislocation density template, the carbon concentration drops to 1.6×10^{16} cm^{-3} and remains constant for pressures >100 torr. These results suggest that at low pressure the carbon concentration is dominated by bulk incorporation into the GaN, while as the pressure is increased a constant amount of carbon is incorporated into the GaN which might depend more on the GaN dislocation density. This may be one of the reasons it is easier to get highly resistive GaN films at higher dislocation densities compared to lower dislocation densities (Wickenden et al., 2004).

In addition to different incorporation rates in the bulk or at dislocation cores, carbon and other impurities incorporate differentially on different surface facets. SIMS analysis revealed that C, O, and Al incorporate in GaN more readily on the N-polar surface compared to the Ga-polar surface (Sumiya et al., 2000; Lin et al., 2013) and similar results have been observed for HVPE growth on N-polar bulk GaN crystals (Tuomisto et al., 2005). Xu et al. found that carbon incorporation was largest on nonpolar GaN (11-20), followed by the (0001) and semipolar (11-22) facets (Xu et al., 2010).

From a theoretical point of view, the carbon incorporated at a nitrogen site has been calculated to act as a deep acceptor in InN, GaN, and AlN (Duan and Stampfl, 2009; Lyons et al., 2014), while the carbon on the cation site acts like a shallow donor in InN and GaN, but behaves like a DX center in AlN (Lyons et al., 2014). Similar results were found for carbon sitting on the N site for InGaN monolayer QWs, where it was calculated to act as an acceptor over a wide range of Fermi Energy values (Alfieri et al., 2015). Therefore, carbon could compensate n-type carriers residing in the InGaN QWs or act as a nonradiative recombination center, reducing the efficiency of the radiative recombination.

Silicon is the n-type dopant used to generate donors in III-N materials and usually supplied by SiH_4 or Si_2H_6. Silicon impurity incorporation has been shown to depend inversely on the growth rate, especially if the Si impurity is coming from the growth chamber (Koleske et al., 2002c). In addition, unintentional

Si doping can come from the MO sources, SiC-coated graphite susceptors, substrates (Si and SiC), wall coating previously doped with Si, leaks in the Si switching valve stems, or pieces inside the MOVPE system. From the work of Koleske et al., it was shown that the precursors (TMGa and NH_3), the quartz liner, and potential leaky valve stems in the Si-doping line were the likely culprits for the background Si-doping levels (Koleske et al., 2002c). The SiC-coated graphite-coated susceptor also has potential to be corroded in the high-temperature NH_3/H_2 mixture and therefore cannot be ruled out as a possible source of Si background doping (Koleske et al., 2002c).

Like Silicon, *Oxygen* is also an n-type impurity that might have been responsible for the large n-type carrier background in the early days of GaN growth (Maruska and Rhines, 2015). Oxygen concentrations weakly follow the same trends as carbon (Parish et al., 2000; Wickenden et al., 2004), suggesting similar chemistry is responsible for both incorporation and removal of oxygen from the GaN surface. In studies of the carrier gas, oxygen incorporation was found to be lower when using hydrogen (Parish et al., 2000), suggesting hydrogen dissociation aids in removing oxygen-containing species from the surface similar to hydrocarbon fragments (Koleske et al., 2002c). Lower growth temperature has also been found to increase background oxygen incorporation (Lin et al., 2013). Intentional oxygen doping of AlN NLs has been conducted by Kuhn and Scholz using 1% oxygen in nitrogen to improve GaN film properties by providing a partial reduction in the available GaN nucleation sites on the AlN NL (Kuhn and Scholz, 2001).

Oxygen impurities can come from several sources, including precursors, carrier gases, chamber walls, residual H_2O or O_2 during loading, and the substrate materials (Al_2O_3 sapphire). Oxygen is known to incorporate into metalorganics by inserting between the metal atom and the methyl or ethyl group, that is, $(CH_3)_2-M-OOCH_3$ (Alexandrov et al., 1987) to produce a peroxide or monoxide species. Oxygen is also present in NH_3 in the form of CO, CO_2, and H_2O, which is removed using a point-of-use purifier before flowing into the growth chamber. Leaks in the manifold and reactor can potentially introduce ambient gases (O_2 and H_2O), although the manifold gas line is usually maintained above the atmospheric pressure, which can reduce this effect. Oxygen leached out of quartz through the interaction of liquid Ga on the SiO_2 could also contaminate films; however, no noticeable change was observed in [O] concentration (i.e., $>2.0\_10^{17}$ cm^{-3}) for a stainless steel reaction chamber with and without a quartz liner (Koleske et al., 2002c).

An example of oxygen impurity incorporation in InGaN films grown using a spent and new NH_3 purifier is shown in Figure 6.7b and c, respectively. In Figure 6.7b and c, SIMS profiles are shown for a 120 nm thick 3% indium InGaN film sandwiched between high-temperature GaN layers. Note that in Figure 6.7b the oxygen (left axis) background is $\sim 3 \times 10^{16}$ cm^{-3} in the GaN film but increases to $\sim 2 \times 10^{18}$ cm^{-3} when the growth temperature is lowered to 850°C for the growth of the 3% indium concentration InGaN layer. While the oxygen peak coincides with the increase in the indium ion counts (plotted along the right axis) samples grown at 850°C without indium also showed a similar increase in the oxygen signal, suggesting oxygen-containing species were not being removed by the NH_3 purifier. Immediately after replacing the NH_3 purifier, the same growth run was repeated, and the SIMS scan for this run is shown in Figure 6.7c. In this scan, the oxygen level is constant at $\sim 2.5 \times 10^{16}$ cm^{-3} for both the InGaN growth and high-temperature GaN growth. These two growths used the same MOs, N_2, and H_2 purifiers, and growth chamber conditions, suggesting that having a working NH_3 purifier is important for reducing oxygen in lower temperature III-N films. This work also suggests that the H_2 etching mechanism responsible for removing hydrocarbon fragments on the GaN surface might be quite efficient at removing residual oxygen-containing species during high-temperature growth (Koleske et al., 2001, 2002c).

As GaN grows and coalesces on sapphire, oxygen from the sapphire can dope the GaN near the GaN/sapphire interface (Xu et al., 2000). This oxygen auto-doping is more of an issue when GaN NLs are used compared to AlN NL since the GaN NL decomposes forming discrete grains which can uncover large regions of the sapphire surface (Koleske et al., 2002a) as shown in Figure 6.2. For AlN NLs, the extent of oxygen diffusion has been shown to depend on the porosity of the AlN NL and is prevented to some degree if the AlN NL is composed of larger grains (Twigg et al., 2001). Oxygen diffusion through GaN films grown on $LiAlO_2$ has also been observed in SIMS measurements suggesting continuous oxygen diffusion from the substrate through the growing GaN (Mauder et al., 2011). The oxygen auto-doped region near the GaN/sapphire interface has been shown to produce a buried conductive layer which can lead to a parallel conduction pathway in either lightly doped GaN (Look et al., 1997; Hsu et al., 2000; Xu et al., 2000) or in AlGaN/GaN 2DEG layers (Twigg et al., 2001; Wickenden et al., 2004).

Hydrogen is plentiful in III-N materials, especially in Mg-doped materials. For most common impurities in the III-N materials, hydrogens influence on the materials properties is least known and (except for Mg-doped GaN) it appears to be rather innocuous in terms of its effect on electrical and structural properties. Hydrogen has been reported to be a donor impurity in both GaN (Wright et al., 2003) and InN (Janotti and Van de Walle, 2008). As discussed in Section 6.5.2, hydrogen is known to effectively passivate Mg dopants (Van de Walle and Neugebauer, 2004), which then require annealing to activate the Mg-H complex (Nakamura et al., 1992c; Myers et al., 2001). Because GaN and AlN NLs are grown at low temperature they likely contain large quantities of hydrogen (Ambacher et al., 1996), which leads to a more rapid thermal decomposition when heated compared to the high-temperature material to form the wurtzite GaN nuclei (Koleske et al., 2003) or relax as in AlN NLs (Lorenz et al., 2000). Annealing out the hydrogen in these NLs plays a key role in their restructuring and material quality prior to the high-temperature GaN growth.

Iron is used as a dopant to achieve highly resistive GaN layers and to overcome the residual n-type background commonly caused from Si or O impurities (Heikman et al., 2002; Polyakov et al., 2004). Heikman et al. used ferrocene (cp_2Fe) to dope GaN up to concentrations of 1.7×10^{19} cm^{-3} (Heikman et al., 2002). The Fe source showed a linear dependence on the precursor partial pressure and was insensitive to growth temperature, pressure, and NH_3 partial pressure. Similar to Mg doping, memory effects were also observed using cp_2Fe which can be partly alleviated through annealing of the chamber and changing out the growth liner (Heikman et al., 2002; Rudziński et al., 2006). When doped into GaN, the Fe creates a deep acceptor level creating a highly resistive film (i.e., 7×10^9 Ω/square) at an Fe doping a level of 1.3×10^{19} cm^{-3} (Heikman et al., 2002). Confirming the deep acceptor nature of Fe in GaN, Polyakov et al. showed that the Fermi level was pinned at 0.5 eV below the conduction band for Fe-doped GaN films (Polyakov et al., 2004). Using defect selective etching, Rudziński et al. also noticed that Fe doping increases the TD density that might also contribute to the increased resistivity.

Other metals can also be incorporated from the surrounding chamber or heater materials. This includes component materials in stainless steel which has a balance of Fe and may contain the elements Cr, Ni, Mo, Mg, Si, N, C, P, and S in decreasing order of concentration. In addition, refractory metals (i.e., Mo, TZM, W, and Re) and high-temperature ceramics (alumina, BN, SiC, and graphite) are used as heating elements, electrical insulators, heat shields, spindles, and other internal reactor parts. Also, the vacuum chamber pieces need to be sealed using traditional ultra-high vacuum techniques, which includes Ag-coated Cu gaskets sealed on knife edge seals. As a point of reference, we have recently conducted SIMS studies to look for potential contamination from these metals and found no evidence for increased Co, Ti, Fe, Ni, Cu, or Mo in GaN or InGaN films above the SIMS background detection limits.

6.5.4 Point Defect Incorporation

It is well known that chemical impurities are incorporated during MOVPE growth, however harder to detect point defects, such as vacancies, interstitials, antisite defects also occur during the growth of III-N films. Less work has been done in this area because specialized spectroscopies, such as positron annihilation, deep-level transient spectroscopy (DLTS), or deep-level optical spectroscopy (DLOS), are required in conjunction with a complete knowledge of the impurity concentrations, and optical and structural properties before point defect concentrations can be used as an explanation for any data trend. A summary of defects related properties is described by Reshchikov and Morkoç (2005).

Initial studies by Look et al. used high-energy (0.7–1 MeV) electron irradiation of GaN grown on sapphire to produce shallow donors attributed to N-vacancies and acceptors attributed to N interstitials (Look et al., 1997). Later using photoionization spectroscopy, Klein et al. uncovered two deep trap states responsible for current collapse in AlGaN/GaN high electron mobility transistors (Klein et al., 2001). In this work, the deeper trap was associated with a carbon-related defect, and the mid-gap trap was associated with grain boundaries or dislocations in the GaN material (Klein et al., 2001).

Defects that act as NRCs in InGaN MQW and LED materials have also been identified using DLOS in conjunction with other techniques (Armstrong et al., 2012a, 2012b, 2014, 2015). In MQWs, higher defect concentrations were found in the first QW, and as the number of QWs decreased, the DLOS-measured defect concentration decreased (Armstrong et al., 2012a). Using DLOS and lighted capacitance–voltage measurements, Armstrong et al. determined that TDs strongly influence point defect incorporation in InGaN/GaN LEDs with both defects being detrimental to LED efficiency (Armstrong et al., 2012b). Green wavelength MQWs have been shown to have increased nonradiative defect concentrations compared to blue wavelength MQWs (Armstrong et al., 2014). The increased defect concentrations in the green MQWs also reduced the peak IQE compared to blue MQWs and shifted the peak IQE to higher laser pump power (Armstrong et al., 2014). Lastly, using DLOS and differential carrier lifetime measurements, the defect concentrations were measured in a single blue QW LED with and without a dilute $In_xGa_{1-x}N$ (x ~ 0.03) underlayer (UL) (Armstrong et al., 2015). It was found that the inclusion of the UL below the InGaN QW significantly improved LED radiative efficiency by 3.9 times compared to the LED without an UL, despite both LEDs having the same radiative recombination rates (Armstrong et al., 2015). The decrease in radiative efficiency was attributed to the concentration difference of a near-mid-gap defect state, which was measured to be 3.4 times lower for the LED on the InGaN UL (Armstrong et al., 2015).

In addition to DLOS measurements of defects, positron annihilation spectroscopy has been used to measure negatively charged defects in GaN and InGaN materials (Saarinen et al., 1997; Chichibu et al., 2006). Using this technique, Saarinen et al. identified Ga vacancies at concentrations 10^{17}–10^{18}cm^{-3} in both GaN bulk crystals and epitaxial layers (Saarinen et al., 1997). These Ga vacancies were also shown to increase from 10^{16} to 10^{19} cm^{-3} when the V/III molar ratio increases from 1,000 to 10,000 (Saarinen et al., 1998), suggesting that growth conditions directly influence vacancy formation rates. This same study revealed that the creation of Ga vacancies is accompanied by the decrease in the free electron concentration from 10^{20} to 10^{16} cm^{-3}, indicating that Ga vacancies may act as compensating centers (Saarinen et al., 1998). In studies with GaN films with grain sizes varying from 0.2 to 2–5 µm, Oila et al. found the Ga vacancy concentration to be independent of the grain size, suggesting that Ga vacancies exist in the grain interior (Oila et al., 2003). In addition, shallow positron traps were attributed to negatively charged edge-type dislocations, which define the grain boundaries (Oila et al., 2003). Chichibu et al. suggested that these same Ga vacancies were related to NRCs in GaN (Chichibu et al., 2006) and similar cation vacancies have been identified in InGaN films along with cation and nitrogen vacancy clusters (Uedono et al., 2013, 2014). Clearly, these nonradiative point defects decrease device efficiency, and a thorough understanding of how they are generated during MOVPE growth is still lacking.

6.6 InGaN

Ideally, if indium could be added to GaN in a continuous manner without creating too many defects, optical-emitting devices could be constructed covering all of the visible wavelengths (400–650 nm) and reaching into the near-IR region (~650–1400 nm). Yet achieving these longer wavelengths with the good optical quality material is difficult for reasons that will become evident in this section.

Generally, the MOVPE growth of InGaN is conducted at 600°C–900°C, which is a substantially lower growth temperature than GaN. This lower growth temperature is required to incorporate the indium into the growing GaN before it desorbs from the surface. The desorption activation energy for indium from the surface is ~2.0–2.5 eV (Rongen et al., 1995; Koblmüller et al., 2007; Choi et al., 2008; King et al., 2008) although lower values of 0.8 eV have been measured (Schenk et al., 1999). Certainly, the indium desorption activation energy is lower than that for Ga desorption (2.7 eV) (Koleske et al., 2001). At the lower growth temperature used for InGaN, the catalytic dissociation of the NH_3 is reduced; partly due to the reduced temperature and partly due to the lower catalytic dissociation rates of indium compared to gallium. Therefore, more NH_3 is needed for the growth of InN or InGaN compared to GaN, resulting in a V/III ratio often in the range 10,000–100,000 for InGaN growth. If the MO fluxes are too high (Markus et al., 2013), the NH_3 activation is too low (Kim et al., 2003), or the growth temperature is too low, grayish InGaN films result. The grayish appearance is due to metallic clustering which if due to indium metal shows up in XRD as a peak near $2\theta = 33.0°$ (Thaler et al., 2010). InGaN is also grown in the absence of hydrogen carrier gas (Piner et al., 1997) since even small amounts of H_2 during growth decreases the indium incorporation (Piner et al., 1997; Koleske et al., 2014b).

6.6.1 InGaN MOVPE Growth

The initial work by Matsuoka and coworkers established some necessary growth parameters for MOVPE growth of InGaN (Matsuoka et al., 1990, 1992; Yoshimoto et al., 1991). In comparison to GaN growth, these researchers found lower growth temperatures and high V/III ratios were required for the best quality InGaN. In addition, a linear relationship between the solid-phase indium incorporation to the gas-phase indium to gallium ratio was found at 500°C; however, as the InGaN growth temperature was increased to 700°C and 800°C, this relationship becomes highly sublinear with only a small fraction of the supplied indium incorporated into the growing solid (Matsuoka et al., 1992). At these higher temperatures, the main reason for the decreased indium incorporation is the higher vapor pressure of indium compared to gallium (Matsuoka et al., 1992), which results in indium desorption instead of incorporation. Despite the difficulty incorporating indium at a higher temperature, this growth condition produced InGaN material with the best PL intensity (Yoshimoto et al., 1991).

Following this work, Nakamura grew InGaN material at temperatures between 780°C and 830°C using high indium source flow rates and high growth rates (Nakamura and Mukai, 1992). With these growth conditions, sharp band edge emission was observed between 400 and 445 nm with very little deep level emission (Nakamura and Mukai, 1992). Si doping of the InGaN was later shown to further increase the relative PL intensity (Nakamura et al., 1993a). Initial, LED structures by Nakamura et al. used relatively thick double heterostructures with 20 nm thick InGaN regions (Nakamura et al., 1993c). No mention was made of employing a GaN capping layer before increasing the growth temperature to complete the p-GaN growth at 1,020°C, although it is highly likely that the InGaN material was capped with at least a thin GaN to preserve the InGaN layer (Johnstone, 2007). GaN capping InGaN QWs prior to heating was later mentioned by several groups who also reported that not capping the InGaN QW results in reduced indium in the QW (Tzu-Chi and Wei, 2001; Olaizola et al., 2002).

Improvements in blue LED intensity by Nakamura and coworkers at Nichia would continue for the next several years where output powers climbed from 125 μW (Nakamura et al., 1993c) to 1.5 mW within a years' time (Nakamura et al., 1994). For green wavelengths the use of 3 nm thick, single QWs are mentioned for the first time resulting in 3 mW output at 520 nm (Nakamura et al., 1995). Later laser diode structures were developed with more complicated structures using 26 periods, 2.5 nm InGaN QWs, and 5.0 nm thick InGaN barriers (Nakamura et al., 1996). Further work showed that the highest PL emission intensity and narrowest linewidths were obtained by using thin 2.0–3.0 nm QWs with GaN barriers of ~5.0 nm (Kozodoy et al., 1997). These designs produced 445 nm LEDs with output power of 2.2 mW at 20 mA (Kozodoy et al., 1997).

To determine the indium concentration in thin films or QWs two primary techniques are used, XRD and photoluminescence (PL). An example of an XRD radial diffraction scan (ω/2θ) for a 5 period InGaN/GaN MQW structure is shown in Figure 6.8a. The data for this scan is shown in red, and the dynamic diffraction simulation is shown in blue. The MQW parameters derived from the fit are shown

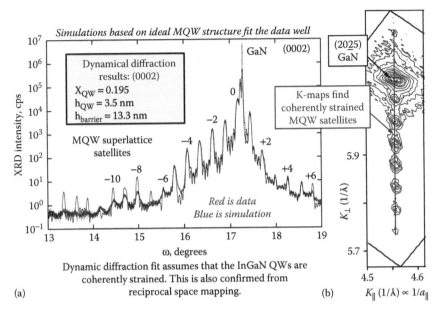

FIGURE 6.8 (a) X-ray diffraction analysis of green wavelength five period InGaN/GaN MQW structure. The red data are the measured scan and the blue solid line is the dynamic diffraction simulation (fit), which assumes that the InGaN QWs are coherently strained to the underlying GaN template. The fit parameters are listed in the yellow box. The superlattice peaks are labeled (0, ±2, ±4, etc...). (b) is a reciprocal space map around the (20–25) reflection, which also indicates that the InGaN MQWs are coherently strained to the underlying GaN template since all of the diffraction peaks have nearly the same parallel in-plane lattice constant.

in the yellow box with x_{QW} being the indium fraction in the QW, h_{QW} being the height (thickness) of the QW, and $h_{barrier}$ being the height of the GaN barriers. The superlattice peaks, from the periodic placement of the QW+barrier, are labeled (0, ±2, ±4, etc.). Near the main GaN diffraction peak, the fringing between superlattice peaks provides information on the entire MQW thickness. Note that the dynamic diffraction analysis overestimates the peak intensity for superlattice peaks greater than ±6. The primary reasons for this intensity overestimation are differences in the indium grading on the top and bottom InGaN/GaN interfaces and lateral film-thickness variations (Lee et al., 2012). Usually, square indium concentration profiles are used to model the indium concentration changes; however, similar fits can also be obtained if trapezoidal profiles are used instead. Since both square and trapezoidal profiles result in good fits, they both provide measures of the average indium concentration, although the square profile might slightly underestimate the peak indium concentration modeled by the trapezoidal profile.

One main assumption of the dynamic diffraction analysis shown in Figure 6.8a is that the InGaN QW is coherently strained to the underlying GaN lattice. (The software does allow for the full or partial strain relaxation of the InGaN QW if required.) Confirmation that the QWs are coherently strained is shown in the reciprocal space map measured around the (20–25) reflection is shown in Figure 6.8b. The solid red line goes through the main GaN and superlattice peaks indicating that layers all have the same in-plane lattice constant, proving that the MQW are coherently strained to the underlying GaN.

Figure 6.9 shows several experiments to study indium incorporation as a function of growth conditions. For these studies, the combined MO flow rates ranged from ~30 to 200 μmoles/min and used 15 SLM NH$_3$ and 10 SLM N$_2$ at 300 torr in a D125 Veeco short-jar MOVPE system. After the InGaN QW growth, the QW is capped with a ~1.0 nm thick GaN cap layer at the QW temperature followed by ramping to 850°C and cooling back to the QW growth temperature while continuously growing the GaN barrier layer. Five-period InGaN/GaN QW-barrier structure was grown, and the indium concentration

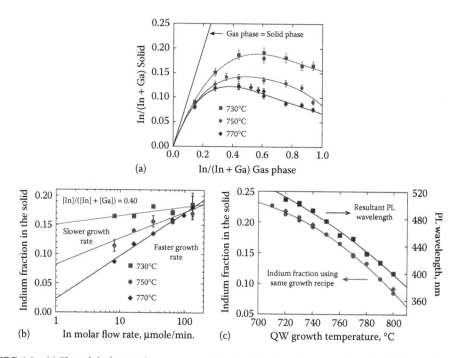

(a)

(b)

(c)

FIGURE 6.9 (a) The solid-phase indium incorporation is plotted as a function of the [In]/([In]+[Ga]) gas-phase ratio for three different QW growth temperatures. For the higher indium fractions the TMGa flow was reduced and the QW growth time increased to achieve the same QW thickness. (b) At fixed [In]/([In]+[Ga]) = 0.40 the MO flow rates were varied at three different temperatures and the QW growth time adjusted to produce QWs ~ 3.0 nm thick. (c) The indium fraction in the solid and the resultant PL wavelength for MQWs grown with the same recipe except for the QW growth temperature.

in the solid was determined from the dynamic diffraction analysis of the $\omega/2\theta$ XRD scan. For these samples, the InGaN QW thickness is ~3.0 nm and the GaN barriers' thickness is ~10 nm.

Similar to the results of Matsuoka, Yoshimoto, and coworkers (Matsuoka et al., 1990, 1992; Yoshimoto et al., 1991), the indium fraction incorporated into the solid in Figure 6.9a is significantly less than the indium fraction in the gas phase. The solid black line denotes the expected condition where the solid-phase [In]/([In]+[Ga]) is equal to the gas-phase [In]/([In]+[Ga]). Three different growth temperatures were used to produce MQWs which emitted in the deep blue (~440 nm) at 770°C, cyan (~480–490 nm) at 750°C, and green (520–530 nm) at 730°C. The three curves show that even when the same TMIn and TMGa flows are used, the amount of indium that gets incorporated into the solid decreases as the QW temperature increases. The reason for this temperature dependence on the indium incorporation is because the steady-state indium coverage, θ_{In}, is dose and desorption rate limited. At 770°C, a peak solid-phase indium incorporation of 0.12 was obtained at a gas-phase fraction of 0.30–0.42, while at lower QW growth temperatures, the peak indium incorporation occurs at higher gas-phase indium fractions (i.e., ~0.45 at 750°C and 0.55 at 730°C). It was deduced from Figure 6.9a that the best [In]/([In]+[Ga]) gas-phase ratio was between 0.4 and 0.6 for the growth of blue to green wavelength QWs.

The effect of increasing the QW growth rate is shown in Figure 6.9b. For these data, the [In]/([In]+[Ga]) gas-phase ratio was kept constant at 0.40. To keep the InGaN MQW structures the same the QW growth time was adjusted to achieve the 3 nm thick InGaN QWs. Note in Figure 6.9b that as the indium flow and the gallium flow increase, the QW growth rate increases and the solid-phase indium concentration increases. Similar results were observed by Keller et al. where the indium incorporation increased as the QW growth rate was increased by increasing the TMGa flow rate (Keller et al., 1997a). These data suggest the indium incorporation is mainly driven by the TMGa flow rate with the indium atoms essentially getting buried in the lattice before they can desorb. Another observation from the data is that at the fastest QW growth rates (highest indium flow rates) the solid-phase indium incorporation reaches a constant value approaching ~0.20. This limit on indium incorporation is due to the coherency strain limit (Masaya et al., 1997; Liliental-Weber et al., 2001; Pereira et al., 2002), which is discussed in more detail in Section 6.6.4.

As expected, if the MQW growth conditions are kept the same, and only the QW growth temperature is changed, the indium incorporation increases as the growth temperature decreases as shown in Figure 6.9c. The resulting PL wavelength plotted by the solid blue squares follows the band-gap bowing parameters established by Wu et al. (2002) if the redshift due to the polarization fields and slight increase in the QW thickness as the growth temperature is decreased are taken into account (Monemar and Pozina, 2000).

From Figure 6.9a–c it is evident that the indium incorporation can be controlled through the indium flow rate and the QW growth temperature. This sets up an interesting question on how to best control the indium, either indium flow rate or temperature and which method produces the brightest MQWs. To settle this question, the same InGaN/GaN MQW structures were grown at QW temperatures ranging from 680°C to 780°C with the indium flow rate adjusted to achieve ~0.15 indium fractions in the QW. For this study, the necessary TMIn flow rates were varied from 2.5 µmoles/min at 680°C to 160 µmoles/min at 780°C while the TMGa was kept constant at 18 µmoles/min. The TMIn flow rates used are plotted in Figure 6.10a as the squares along the left axis. The resultant indium fraction in the QW is plotted as circles along the right axis. Note as the QW growth temperature is increased, an exponential increase in the TMIn flow is required to maintain the 0.15 indium fraction in the QW. Achieving the 0.15 indium fraction at 780°C was difficult, and only indium fractions of 0.12–0.14 were achieved.

The PL intensity of the 450 nm emitting InGaN QWs was measured using a cw 325 nm HeCd laser at a low excitation power of <5 mW. The PL intensity is shown in Figure 6.10b for the selected samples from Figure 6.10a with the peak PL intensity occurring at 760°C. At lower temperatures, the QW emission intensity is reduced suggesting a possible increase in NRCs. At 780°C the PL intensity is also reduced likely because of reduced carrier confinement especially if indium diffuses into the GaN barrier (Koleske et al., 2015). The obvious conclusion from this study is that InGaN films should be grown at higher temperatures, with high TMIn flow rates to achieve the best optical quality. Similar results have been published by Hammersley et al. where higher QW growth temperatures resulted in higher

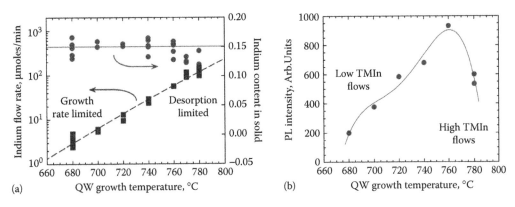

FIGURE 6.10 (a) The indium flow rate required to produce InGaN QWs at 450 nm is plotted versus the QW growth temperature. The TMGa flow rate was kept constant at 18 µmoles/min and the resulting indium incorporated into the solid is plotted along the right axis. (b) The corresponding 450 nm PL intensity from select growth runs shown in (a) is plotted.

IQE and reduced nonradiative point defects (Hammersley et al., 2016). In addition to higher growth temperatures, higher growth pressures have been shown to increase indium incorporation resulting in better optical quality (Daisuke et al., 2010). Similar results have been observed for high-pressure growth of InN (Dietz et al., 2008; Durkaya et al., 2009).

Finally, InGaN can be thermally sensitive after growth especially if higher temperature layers are required to complete the device structure, such as the p-side for a LED. Previous studies of InGaN thermal stability have been examined in MQW structures (McCluskey et al., 1998; Chuo et al., 2001; Jacobs et al., 2003) and single InGaN single layers (Chalker et al., 2000; Sakuta et al., 2005; Thaler et al., 2010). These studies find indium and gallium interdiffusion after annealing (McCluskey et al., 1998; Chuo et al., 2001; Jacobs et al., 2003), voids and metallic indium inclusions (Jacobs et al., 2003), and loss of indium (Chalker et al., 2000) through diffusion (Sakuta et al., 2005). In a study by Thaler et al., thick InGaN layers were sandwiched between GaN layers and annealed resulting in decreases in the PL intensity at much lower temperatures compared to the onset of changes in the XRD intensity (Thaler et al., 2010). This work suggests that point defects that quench luminescence are generated well before the onset of structural decomposition. For an InGaN film with 18% indium, the temperature-dependent InGaN decomposition yielded an activation energy of 0.87 ± 0.07 eV, which is similar to the activation energy for bulk InN decomposition (Thaler et al., 2010).

6.6.2 InGaN/GaN Interfaces

MOVPE of indium incorporation is also influenced by the temporal gas switching. This includes, the time for the indium to build up on the surface, the maximum coverage, θ, that builds up on the surface, and depleting the indium reservoir after turning off the indium source. This gas switching effect mainly impacts shorter-timed growths such as InGaN QW and to a lesser extent bulk-like films, although it can influence the sharpness of the interface. Since better PL intensity is obtained for higher temperature InGaN growth as shown in Figure 6.10, in the situations that follow, high indium flow rates are used with the indium desorption controlling the indium fraction in the solid.

For QW growth on c-plane GaN, a gas switching schematic is shown in Figure 6.11a along with two proposed situations of indium surface coverage, θ_{In}. As shown on the top of Figure 6.11a, the TMGa and TMIn are both switched on at the same time for the growth of the InGaN QW. After the QW growth is finished, the TMIn is turned off, and the QW is capped with a thin GaN layer before temperature ramps or other growth steps are conducted. For the first situation, labeled (1) the indium surface coverage, θ_{In}, reaches ~0.15 of a full monolayer, which leads to a 0.15 indium fraction in the InGaN QW. For the second situation, labeled (2) the θ_{In} is ~0.25, which should lead to a 0.20 indium fraction in the InGaN QW. The reduced indium fraction for (2) is due to the coherency strain limit (Masaya et al., 1997; Liliental-Weber et al., 2001; Pereira et al., 2002). Jiang et al. have measured indium surface coverages up to a maximum of θ_{In} ~0.26 by *in situ* XRD (Jiang et al., 2006).

When both the TMGa and TMIn are switched on, it takes time for the metal atoms to cover the surface. Ignoring loss of MO from gas-phase prereaction, it is assumed that once the MO fragment is through the boundary layer, its sticking coefficient on the surface is unity. This means that the surface coverage of indium, θ_{In}, or gallium, θ_{Ga}, will depend on the number of available surface sites and the relative rate of desorption from the surface. Since the indium desorption rate is orders of magnitude higher than the gallium desorption rate, it will take longer to reach a steady state θ_{In} than a steady state θ_{Ga}. Since the desorption rate of Ga from the surface is negligible, the QW growth rate depends mainly on the TMGa flow rate. The slower buildup of indium at the surface means that θ_{In} follows the well-known Langmuir adsorption isotherm (Langmuir, 1918), which is depicted by the solid line in Figure 6.11a. The flowing NH_3 then locks the surface gallium and indium into the growing InGaN lattice. Because of the finite time required to buildup the steady state θ_{In}, there will be an increasing indium concentration grade through the GaN/InGaN interface. The width of this concentration grade can be several monolayers (Lee et al., 2012).

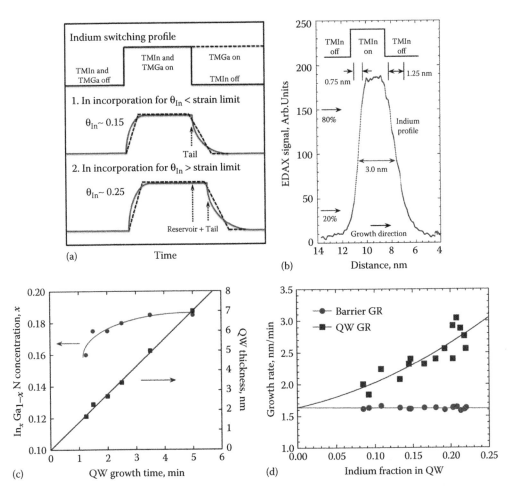

FIGURE 6.11 (a) Schematic of the TMIn and TMGa switching profile and the indium surface coverage for situation (1) where the maximum indium surface coverage, θ_{In}, is 0.15 of a monolayer and situation (2) where the maximum $\theta_{In} \sim 0.25$ of a monolayer. For the InGaN QW growth both the TMIn and TMGa are turned on, whereas for the GaN-capping layer only the TMGa is on. After the TMIn is switched off, there is an indium tail into the GaN barrier for both situations (1) and (2). In addition, there is an additional surface indium reservoir that must be depleted in situation (2). (b) EDX measured indium signal for a green wavelength QW along with the TMIn switching profile. The bottom of the QW (near 11 nm) has an interface width of 0.75 nm as measured from the increase in the EDX signal from 20% to 80% of the full intensity. Measured in a similar way, the top of the QW (near 7 nm) has an interface width of 1.25 nm. (c) XRD measured indium concentration (circles) and QW thickness (squares) for MQWs where the QW growth time was varied. (d) Calculated InGaN QW and GaN barrier growth rates based on recipe switching times. This plot is from the same data set shown in Figure 6.9c for InGaN/GaN MQWs where only the QW growth temperature was varied.

After the QW growth is complete, the TMIn is shut off, and there is a reservoir of indium on the surface corresponding to either situation (1) $\theta_{In} \sim 0.15$ or situation (2) $\theta_{In} \sim 0.25$. With the TMGa still flowing, the indium reservoir both desorbs and is incorporated into the top regions of the InGaN QW. For situation (1) the residual indium tails into the growing GaN barrier, resulting in a graded top InGaN/GaN interface, similar to the lower GaN/InGaN interface (Lee et al., 2012). For situation (2) the indium continues to incorporate into the QW for a period time followed by the same grading of indium into the InGaN/GaN top interface. Because of the increased θ_{In} in situation (2) and the lower growth temperature required to achieve these coverages, it will take a longer time to incorporate the remaining θ_{In}. This should result in a larger grading of the top InGaN/GaN interface compared to the bottom GaN/InGaN interface (Lee et al., 2012). The black-dashed lines on top of the indium profiles in Figure 6.11a denote the expected indium concentration as a function of growth time.

Evidence for leading and trailing indium grades in InGaN MQWs has been observed in atom beam tomography (ABT) profiles (Ren et al., 2015), and TEM studies (Massabuau et al., 2015), and XRD diffraction scan modeling (Lee et al., 2012). Further evidence for the nonabruptness of the MQW interfaces is shown in Figure 6.11b where the indium profile for a MQW structure was measured using energy dispersive X-ray (EDX) spectroscopy in a STEM. The MQW consisted of five InGaN QWs with a 0.20 indium fraction surrounded with GaN barriers (Koleske et al., 2015) and the measured EDX signal was averaged over all five QWs resulting in the profile shown in Figure 6.11b. From this indium profile, the bottom GaN/InGaN interface width is ~0.75 nm while the top InGaN/GaN interface width is ~1.25 nm. The larger width of the top QW interface is expected to be based on situation (2) described in Figure 6.11a and observed in the TEM images shown in Koleske et al. (2015). One slight complication to this simple explanation for the indium profiles is the increased multiple step bunching observed in 0.20 indium fraction QWs (Koleske et al., 2010; Lee et al., 2012), which could contribute to the graded profile observed at the InGaN/GaN interfaces.

More evidence of a finite time to achieve θ_{In} was reported by Hoffmann et al. who changed the effective thickness of the QW from 1.1 to 3.8 nm by varying the QW growth time from 30 to 100 s (Hoffmann et al., 2010). By observing less indium in the QW for shorter growth times, they concluded that there was an effective turn on time for indium incorporation into the QW (Hoffmann et al., 2010). A similar observation is shown in Figure 6.11c where the indium solid fraction is plotted along the left axis versus the QW growth time. Since the TMGa flow rate dictates the QW growth rate, the QW thickness increases linearly with time as shown by the solid blue squares in Figure 6.11c. For the 1-minute QW growth time, the indium concentration is ~16% which increases to ~18% as the QW growth time increases, which indicates the time frame needed for θ_{In} to reach steady-state coverage.

One additional example of the indium grading into the GaN barrier is shown in Figure 6.11d. The MQW data for Figure 6.11a is the same MQWs previously shown in Figure 6.9c, except the InGaN QW and GaN barrier growth rates are plotted as a function of the indium fraction in QW. For this series of MQWs, the QW was grown for 1.25 minutes, the GaN cap for 0.5 minutes, and the temperature ramp for completion of the GaN was an additional 6.0 minutes. Assuming the QW occurred only over the 1.25 minutes and the GaN barrier over the 6.5 minutes, the QW and barrier growth rates were calculated and plotted in Figure 6.11d. Note that the barrier growth rate (circles) is very constant while the InGaN QW growth rate (blue squares) gradually increases as the indium concentration in the QW increases. The reason for this apparent increase in the InGaN QW growth rate is partly due to the assumption that the QW is only grown over the programmed 1.25 minutes. Instead, the QW continues to grow until the indium reservoir is consumed. In other words, the actual QW growth time exceeds the allotted 1.25 minutes, especially as the QW growth temperature decreases.

Several studies have explored ways to thwart the expected Langmuir adsorption of indium to produce a much squarer indium profile. Indium preflows were used by Muthusamy-Senthil and coworkers to get θ_{In} to a higher steady-state value before QW growth (Muthusamy Senthil et al., 2008). While this predosing approach did not affect the interface abruptness as intended, it influenced the initial growth surface resulting in a 30% enhancement in the IQE for deep green (>530 nm) MQWs (Muthusamy Senthil et al., 2008). In another study of Deng et al. also, using an indium predose before QW growth resulted in 0.16 indium concentration at the lower QW interface, which increased to 0.19 at the top interface (Deng et al., 2014).

Instead of tailoring the InGaN QW distribution to make it more square, other groups have contemplated restructuring the indium distribution to improve the electron-hole wavefunction overlap and increase the spontaneous emission efficiency (Choi et al., 2003; Takeyoshi et al., 2003; Zhao et al., 2011). Takeyoshi et al. reported changing the indium composition in the InGaN QW from square to trapezoidal resulting in a 40% higher emission intensity (Takeyoshi et al., 2003). Following similar thinking, Choi et al. reported similar results with triangular shaped MQWs grown by grading indium composition resulting in LEDs with narrower linewidths, lower operation voltages, and a 2x output power increase (Choi et al., 2003). Further evidence of the improved wave function overlap was observed in the EL peak energy being nearly independent of the injection current and temperature (Choi et al., 2003). Unfortunately, the main difficulty in explicitly tailoring the indium distribution is the time required to reach steady state θ_{In} and proving that the desired indium distribution has been achieved.

Finally, the maximum θ_{In} that can be achieved for reasonable MOVPE growth conditions has been measured by Jiang et al. (2006). The buildup of indium on the surface was shown to follow the Langmuir isotherm with θ_{In} dependent on the surface temperature (Jiang et al., 2006). This group found that the maximum indium coverage on GaN (0001) surface is about ~0.25 ML with an estimated error of 20% (Jiang et al., 2006). Northrup has suggested that higher indium coverages can be obtained on semipolar surfaces, due to the reduced repulsive interaction between indium atoms on the (11–22) surface compared to the (10–10) surface (Northrup, 2009). This increased indium incorporation has led to greater indium incorporation on some semipolar planes (Tim et al., 2012; Zhao et al., 2012). Increased indium incorporation on N-polar GaN has also been shown at higher growth temperatures than for Ga-polar GaN surface, allowing for the growth of green to red (633 nm) LEDs (Kanako et al., 2015). Despite the increase in the indium incorporation, the bulk strain relaxation criterion still holds, and if the bulk strain is exceeded, defects will be formed to relieve the strain energy (Feezell et al., 2013; Wu et al., 2014) as discussed in Section 6.6.4.

6.6.3 InGaN Morphology

During InGaN growth on GaN, rich and complex surface morphologies evolve that may actually improve or enhance the QW light emission properties. These morphologies might be coupled with carrier localization phenomena, which have been postulated to reduce carrier recombination at dislocations. These carrier localizations have been proposed to occur within quantum dots (Narukawa et al., 1997), composition fluctuations (Gerthsen et al., 2000), well-width fluctuations coupled to the strong polarization fields (Graham et al., 2005), and on localized In-N-In chains or clusters (Chichibu et al., 2006). An anti-localization theory also suggests that there is an energetic screening of carriers away from dislocations via the creation of a higher energy landscape around the dislocation cores (Hangleiter et al., 2005). One final suggestion is that localization is not required since the minority carrier diffusion length is so small in the III-nitrides (Speck and Rosner, 1999). While these different perspectives for III-N blue LED efficiencies have been proposed, there is yet no definitive proof to rule in favor of one mechanism over the others.

The most obvious InGaN morphology feature is the V-type defects, which are typically co-located with TDs (Wu et al., 1998). As the growth temperature is lowered, the c-plane growth rate becomes larger than the lateral growth rate (Mitchell et al., 2001; Coltrin and Mitchell, 2003), resulting in inverted pyramidal features composed of six (10–11) planes, which when imaged in TEM cross-section appear as a V-shape (Wu et al., 1998). While V-defects are typically found around dislocation cores, their presence has also been detected in solar cell MQW structures grown on low dislocation density substrates (Young et al., 2013), suggesting that other types of defects may also initiate their presence. When higher temperature p-type GaN is grown on a MQW with V-defects, the V-defects fill in resulting in a smooth surface (Sharma et al., 2000).

In addition to V-defects, InGaN grown on GaN develops a hillock/valley or mounding morphology. Initial observations of this morphology by Keller et al. showed that nanoscale islands formed around dislocations with a screw component and predosing the GaN surface with disilane enhanced this morphology (Keller et al., 1998b). This hillock morphology was attributed to a transition in the thermodynamic mode of film growth, as 3D islands nucleated on the cores of screw-type dislocations

(Miraglia et al., 2003). Oliver and coworkers showed that this transition from 2D to 3D growth increased as the NH_3 flow increased and tentatively attributed this change to a decrease in the barrier to adatom incorporation at step edges (Oliver et al., 2004). Increasing the NH_3 partial pressure by increasing the reactor pressure also had the same effect on the 2D to 3D growth transition (Oliver et al., 2004). Further studies revealed that this transition could also be achieved by increasing the TMIn flux and increasing the growth rate (Oliver et al., 2005). By increasing the temperature after the InGaN QW growth, van der Laak et al. observed gross well-width fluctuations linked to an overall 3D morphology consisting of a network of interlinking InGaN strips (van der Laak et al., 2007). Interestingly, if these well-width fluctuations are capped with GaN, the surface morphology became smooth with no indication of the underlying morphology (van der Laak et al., 2007). The network of interlinking InGaN strips was shown to correlate well with the alignment of GaN step edges, and they suggested that the InGaN network forms via the decomposition of In-rich regions near step edges (van der Laak et al., 2007) and these networks can be partly filled by further annealing (Oliver et al., 2009). Similar networks have also been found in blue and UV-emitting QWs (Oliver et al., 2008). It has been suggested that these structures help with carrier localization, since thinner QW regions have shorter wavelength emission, while thicker QW regions have longer wavelength emission (Sonderegger et al., 2006).

The InGaN hillock/valley or mounding coincides with the reduction in growth temperature to <900°C. At this temperature in the N_2 growth ambient, the evaporation/recondensation smoothing mechanism shuts down, and only the surface diffusion smoothing mechanism will be active (Koleske et al., 2014a). As mentioned in Section 6.3.1 these two mechanisms operate over different length scales with the surface diffusion leading to characteristic mounding at length scales of tens to hundreds of nanometers (Koleske et al., 2014a), which are observed in many AFM images of InGaN growth on GaN. In addition, roughening of the InGaN surface might offer slight reductions in the overall strain state of the InGaN (Koleske et al., 2010). Changes in the surface step structure from single-layer step heights to multiple-layer step heights have been observed for single-QW and MQW growth (Koleske et al., 2010). This change in step heights increases as the growth temperature is decreased, as the indium is increased, or as the number of QWs increases (Koleske et al., 2010) and possibly indicate the start of the 2D to 3D morphology changes observed by Oliver et al. (2004, 2005). The increased frequency of multistep heights is a way to partially accommodate the in-plane film strain, but this restructuring is insufficient to trigger bulk strain relaxation (Koleske et al., 2010). This intrinsic multiple-layer step restructuring, when coupled with the strong piezoelectric fields present in the wurtzite group III-nitrides (Graham et al., 2005), could explain the enhanced carrier localization in InGaN QWs (Koleske et al., 2010). Galtrey et al. have observed evidence in atom beam tomography (ABT) measurements that the top InGaN QW interface is rougher than the bottom interface (Galtrey et al., 2008a), consistent with increased presence of multiheight steps after even single QWs (Koleske et al., 2010).

Finally, larger macroscopic defects also develop during InGaN growth including inclusions embedded within V-defects (Florescu et al., 2003; Ting et al., 2003; Kumar et al., 2007), metallic clusters, and ring-like defects (Massabuau et al., 2012). As more QW periods were added, Ting et al. found that embedded inclusions resulted in progressive deterioration of the surface morphology especially at higher indium concentrations (Ting et al., 2003). The generation of these clusters within the V-defects was believed to be driven by localized strain-energy variations at the apex of V-defects (Florescu et al., 2003). Fortunately, fixing this surface morphology was accomplished by raising the GaN barrier growth temperature to 900°C or adding H_2 to the GaN barrier growth (Ting et al., 2003). More recently Massabuau et al. have studied the origin of trench defects which appear as ring-like defects on thick InGaN films or InGaN/GaN MQWs with higher indium concentrations (Massabuau et al., 2012). Massabuau et al. proposed that the trench defects consist of stacking faults lying in the basal plane terminated by a vertical stacking mismatch boundary, which resulted in a partially relaxed InGaN film within the enclosed region (Massabuau et al., 2012). These defects are also observed on MQWs that have high quantum efficiencies, so any deleterious nature of these defects is not understood (Massabuau et al., 2012), but ultimately decreases the total emitting area. Additional work on how these defects are generated and propagate will be required, especially how the increased strain in the InGaN films generates these defects.

6.6.4 Implications of InGaN Strain Relaxation

Unlike the other III-V semiconductors the small size of the N atom results in increased strain on the lattice, which has a larger effect for InGaN films on GaN compared to AlGaN films on GaN. In fact, InGaN films with only 0.20 indium fraction have a ~2% larger in-plane lattice constant compared to GaN. This results in a critical thickness for $In_{0.20}Ga_{0.80}N$ of only 6–7 nm (Holec et al., 2008). This low critical thickness rules out any useful device structure requiring >0.20 indium fractions even on low defect density GaN films.

In the early studies of InGaN growth on GaN, many researchers reported the observation of two XRD diffraction or PL peaks. Initially, this was believed to be due to a solid-phase miscibility gap (Ho and Stringfellow, 1996), generated because of the large difference in the interatomic spacing between GaN and InN. This phenomenon has also referred to as *composition pulling* and *phase separation* with the InGaN film devolving into two InGaN films; one with a higher and one with a lower indium concentration. Using a regular solution model Matsuoka determined that InGaN was always predicted to have an unstable mixing region at temperatures lower than 3,000°C (Matsuoka, 1997) and the results of Ho and Stringfellow (1996) essentially support this viewpoint.

Kawaguchi et al. report this *composition pulling effect* was observed for InGaN films grown on GaN and AlGaN to be a stronger effect with increasing lattice mismatch, with the crystalline quality being good near the InGaN/GaN interface and decreasing in quality as the film thickness and/or the indium mole fraction increased (Kawaguchi et al., 1996). This *composition pulling effect* was believed to be due to the exclusion of indium atoms from the InGaN lattice during the early growth stages, which in turn, reduced the deformation energy from the large lattice mismatch (Hiramatsu et al., 1997). TEM images revealed that the InGaN crystal quality degraded as edge dislocations propagating through the InGaN ultimately resulting in a pitted surface (Hiramatsu et al., 1997). Single-phase InGaN films were reported by El-Masry et al. up to an indium fraction of 0.28, but for higher indium fractions, the films spinodally decomposed to multiphase InGaN (El-Masry et al., 1998). Many other instances of InGaN films with indium-rich regions were reported after annealing (McCluskey et al., 1998; Romano et al., 1998) and using low-pressure growth (Tran et al., 1997; Chen et al., 2004).

Later, the increased InGaN film strain, compared to the GaN lattice was considered as an alternative explanation to the miscibility gap (Scholz et al., 1997). From XRD reciprocal space maps and TEM cross-sections, it was clear that InGaN formed a coherently strained layer close to the InGaN/GaN interface, which relaxed after reaching the critical thickness (Masaya et al., 1997; Liliental-Weber et al., 2001; Pereira et al., 2002). An illustration of a ~100 nm thick InGaN film is shown in Figure 6.12a, which shows a lower coherently strained InGaN layer, and a top strain relaxed InGaN layer with a rougher surface. Experimentally, InGaN was grown on GaN at 760°C with increasing TMIn flow rates at a fixed TMGa flow. Using XRD analysis, the indium fractions and strain states could be measured for both the coherently strained and strain-relaxed layers depicted in Figure 6.12a. As the [In]/([In]+[Ga]) gas-phase ratio increases, the coherently strained InGaN (blue squares) reaches a maximum indium fraction of ~0.20. For the top, strain-relaxed, InGaN layer, the indium fraction in the solid increases reaching a solid-phase fraction of ~0.40 for a gas-phase ratio of 0.57. So even at high gas-phase indium fractions (here 0.57), indium only incorporated up to ~0.20 until the InGaN film strain relaxed.

An elegant *in situ* observation of InGaN strain relaxation during growth was reported by Richard et al. using real-time XRD reciprocal space mapping (Richard et al., 2010). They show that after the first 3 nm of growth the indium concentration reaches an indium fraction of $x = 0.26$ and remains constant (Richard et al., 2010). After reaching a thickness of 7 nm a second peak emerges with a strain relaxed indium fraction of $x = 0.76$; a concentration three times the coherently strained InGaN (Richard et al., 2010). This higher indium composition film initiates when the smaller area relaxed islands absorb residual indium rejected from the larger surface area of the coherent film (Richard et al., 2010).

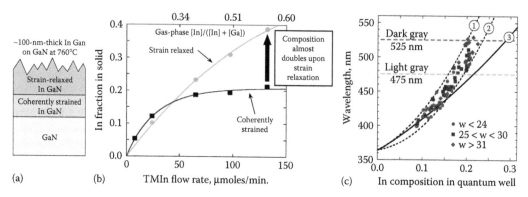

FIGURE 6.12 (a) Depiction of the InGaN growth on GaN showing a coherently strained region and the strain-relaxed layer (filled circles). (b) X-ray measurement of the solid-phase indium versus the TMIn flow rate for both the coherently strained InGaN (filled squares and curve) and the strain relaxed InGaN (filled circles and curve). The TMGa flow was fixed at 98 μmoles/min and the gas-phase composition is marked on the top of the figure. (c) Measure of the resulting photoluminescence peak wavelength as a function of the indium composition in the quantum well (QW) measured by X-ray diffraction fit using dynamic diffraction analysis software. The solid circles denote QWs widths less than 24 Å, solid squares denote QW widths between 25 and 30 Å, and the solid diamonds denote QW widths greater than 31 Å. The black solid line denotes the bandgap bowing from (Wu et al., 2002), while the black-dashed lines fit the lower and upper bounds of wavelengths that result from each measured indium composition. The light gray- and dark gray-dashed lines denote the 475 and 525 nm wavelengths, respectively, and suggest by their intersection with the black-dashed lines the possible indium concentrations required to reach these wavelengths.

As the InGaN film grows thicker the indium concentration in the strain relaxed region decreases closer to the gas-phase concentration near 0.57 (Richard et al., 2010). This kinetic mechanism of forming the strain relaxed inhomogeneity was shown to be strongly coupled to surface morphology, and should be cumulative in multilayer structures such as MQWs (Richard et al., 2010).

Despite the coherency strain limit of no more than ~0.20 for InGaN growth on c-plane GaN, green InGaN MQWs can be grown as shown in Figure 6.12c. Plotted in Figure 6.12c is PL wavelength measured using a cw 325 HeCd laser plotted as a function of the indium concentration in the InGaN QW as measured using XRD. The dashed lines labeled 1 and 2 are the maximum and minimum wavelength achieved for each indium concentration. The solid line labeled 3 is the bulk PL emission wavelength based on the band bowing formula of Wu et al. (2002). Note that compared to the bulk InGaN curve 3, there is a substantial redshift in the emission wavelength for the InGaN QWs due to the strong polarization fields that bend the band structure leading to the longer wavelength emission (Monemar and Pozina, 2000). So, for the c-plane, the strong polarization fields definitely help in achieving longer wavelength emission, however as the current density increases the emission intensity blue shifts due to band filling. Even for green wavelength, the MQWs TEM studies have identified large well-width fluctuations and misfit dislocations generated in the QWs that thread to the sample surface (Costa et al., 2006). On some semi-polar GaN planes, higher indium incorporation has been observed, resulting in longer wavelength LEDs (Natalie et al., 2008; Sato et al., 2008; Masui et al., 2010); however, misfit dislocations formed by basal or nonbasal plane slip systems can also occur on these nonpolar or semi-polar structures resulting in reduced emission efficiency (Shan Hsu et al., 2012). Developing a strain-relaxed, low-defect density InGaN template (Miura et al., 2008) would facilitate indium fractions greater than 0.20, possibly leading to yellow and red wavelength LEDs in the nitrides.

6.6.5 Methods for Manipulating InGaN QW Brightness

For improving the brightness of InGaN QW emission, it should be expected that the InGaN QW growth conditions are most critical. This was previously discussed in Section 6.6.1, showing that the brightest

QW were grown using high temperatures and high-indium flow rates. Of course, more indium can be incorporated into the QW by lowering the growth temperature at the risk of reducing emission intensity.

While the QW growth is important, results by many groups also suggest that the growth of the materials surrounding the InGaN QWs is just as important as the QW growth itself. These growth parameters include the growth of GaN or InGaN barrier layers, the use of alloy or superlattice layers beneath the QWs, capping the QWs with GaN or more recently AlGaN layers prior to heating, and finally artificially structuring the QW using temperature annealing or hydrogen dosing schemes. These optimization schemes and their influence on the QW emission intensity are described in the following paragraphs.

Wider bandgap *barrier layers* separate the InGaN QWs and provide the needed confinement of the electrons and holes. While single QW or double heterostructure III-N LED devices were made first by Nakamura and coworkers (Nakamura et al., 1993b), these days most LED structures consist of MQW structures with the number of QWs ranging from a few to many. There are several possible advantages for using MQWs instead of SQWs for LEDs. First, for c-plane GaN LEDs, the electron and hole wavefunction overlap is maximized for InGaN QW thicknesses of ~2–3 nm which leaves less volume for recombination while MQWs would increase this emission volume. Second, as discussed in Section 6.6.1, MQWs can be easily characterized using XRD to determine the indium composition and QW and barrier thicknesses. Third, the MQW emission intensity increases rather dramatically as the number of QWs increases. This increase in the QW emission intensity is shown by the open circles in Figure 6.13a for increasing numbers of QWs emitting at 440 nm. For this measurement, a low power cw HeCd laser was used, and so the reason for the increase in the PL intensity is due to the reduction in NRCs as the number of QWs increases (Armstrong et al., 2012a). For this plot, the absorption of the laser as a function of thickness, was taken into account to normalize the PL intensity. Note that the PL intensity increases by over three orders of magnitude in going from a single QW to 10 MQWs. Also, note that the PL intensity scales as the number of QWs to the 4.2 power which indicates that the QWs do not contribute equally to the overall emission intensity and that NRCs are gradually removed as the number of QWs increases. Similar decreases in the number of nonradiative defects were also measured using DLOS studies on MQW LEDs (Armstrong et al., 2012a). Finally, David et al. have shown that no matter how many QWs are grown, only the QWs nearest the p-layer emit significant light under electrical pumping (David et al., 2008). This observation suggests that many of the underlying QW structures are inherently unnecessary but somehow improve the emission intensity of the topmost QWs closest to the p/n junction.

In addition to the number of InGaN/GaN MQWs used, GaN barriers' growth conditions are also important. These include the thickness of the GaN cap layer, the GaN barrier growth temperature and

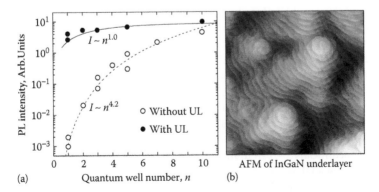

FIGURE 6.13 (a) PL intensity versus the quantum well number, *n*, with (filled circles) and without (open circles) the InGaN UL. (b) 3 × 3 µm AFM image of a 170 nm thick InGaN underlayer (UL) with ~3% indium concentration.

ramp rates, GaN barrier thickness, and the Si-doping level. GaN capping of the InGaN QW prior to any temperature changes will lead to better retention of the indium fraction in the QW. The growth of a more defective or too thin of a GaN cap layer leads to indium out diffusion or desorption (Jian-Shihn et al., 1997) and a reduction in the QW thickness (Tzu-Chi and Wei, 2001). Increased retention of indium was confirmed by Pendlebury et al. after capping the InGaN QWs with a 3.0 nm thick GaN layer before increasing the temperature to complete the GaN barrier (Pendlebury et al., 2007). Higher temperature GaN barriers have been shown to improve PL or LED emission intensity (Olaizola et al., 2002; Wu et al., 2003) and improve the surface morphology through the reduction in number of V-type defects (Florescu et al., 2003; Ting et al., 2003; Kumar et al., 2007). The amount of indium lost from the QW based on higher GaN barrier growth temperature was shown to be related to the temperature ramp rate, with faster ramp rates leading to increased indium loss (Wang et al., 2007). Also, the GaN barrier thickness has been shown to influence the IQE of indium-graded InGaN/GaN QWs with thinner barriers leading to improved IQE compared to thicker barriers (Wei et al., 2013). With increased thickness of the GaN barrier, the abruptness of the interface between InGaN and GaN layers deteriorated resulting in reduced PL intensity and broader emission (Dong-Joon et al., 2001). Massabuau et al. warned that for optimal QW efficiency, it is important to grow the active region using a growth method, which leads to QW interfaces that are as abrupt as possible (Massabuau et al., 2015).

Silicon doping of the GaN barriers has also been shown to influence the QW emission intensity, purportedly by screening the polarization fields (Fiorentini et al., 1999) although the doping levels required to achieve this effect need to be ~10^{20} cm^{-3} (Young et al., 2016). Si doping of GaN barriers by Keller et al. showed higher PL intensity in the SQWs placed on pretreated Si dosed GaN surfaces compared to no pretreatment (Keller et al., 1998a). Other more recent studies have shown improved luminescence properties of MQW with Si delta doping of the GaN barriers (Kwon et al., 2005), improved light output power from LEDs with increased Si doping in the GaN barrier layers (Xia et al., 2012), and a 17%, increase in LED light output with light Si doping (Zhiting et al., 2015). Zhiting et al. cautioned that too high of Si doping can interfere with interface quality and results in current leakage and a decline in luminous efficiency (Zhiting et al., 2015).

Since residual indium can grade into the GaN barrier as discussed in Section 6.6.2, several groups have used InGaN barriers with lower indium concentrations to separate the InGaN QWs. Chichibu et al. mention the use of Si-doped and undoped InGaN barriers for MQW laser diode structures, resulting in a smaller Stokes-like shift observed in the Si-doped barriers and a reduction in threshold power density required for stimulated emission (Chichibu et al., 1998). For green/yellow wavelength InGaN/GaN MQWs, Hwang et al. used InGaN layers embedded in the barriers as a strain-controlling interlayer to obtain a 552 nm emitting LED with an external quantum efficiency at 23.5% at 20 mA (Hwang et al., 2013) and Zhiting et al. have also observed improved InGaN QW emission intensity (Zhiting et al., 2016).

Underlayers (ULs) are often used under MQW or LED structures to improve overall emission efficiency. Many different types of ULs have been developed including, low-temperature GaN layers, thick low indium concentration InGaN layers, or InGaN/GaN or AlGaN/GaN short-period superlattices. The exact mechanism for their effectiveness is not entirely known, but a reduction in nonradiative recombination defects has been observed for SQW LEDs grown with and without InGaN ULs (Armstrong et al., 2015). The UL may act as a sequestration area for point defects (Akasaka et al., 2004, 2006), a region to mitigate the possible impurity incorporation as a result of growth interruption (Akasaka et al., 2004, 2006), or as a strain relief layer to decrease cation vacancies. Although not as widely discussed as some of the other growth optimization procedures, the use of these layers to improve InGaN QW emission intensity is widespread.

A properly designed InGaN UL can dramatically improve the QW emission intensity. Evidence for this improvement using a UL is shown in Figure 6.13a, where the PL intensity is plotted as a function of increasing numbers as filled circles. Similar to the open circle MQWs previously discussed, the same MQW structures were grown on a 170 nm thick ~3% indium concentration InGaN UL. For this plot, the same 325 cw HeCd laser was used, and the laser power attenuation was taken into account to normalize the PL intensities. Note that the PL intensity is ~1,000 times larger for the SQW grown on the InGaN UL compared to no UL. As the number of QWs increases, the difference in PL intensity between the MQWs with and with the UL decreases rapidly. Note too that for MQWs on the InGaN ULs, the PL intensity, I, scales directly with the number of QWs, n, (i.e., $I \sim n^{1.0}$), signifying that each QW adds linearly to the PL intensity. This differs from the MQWs grown without the UL, where $I \sim n^{4.2}$, suggesting that lower lying QWs barely contribute to the overall PL intensity. Eventually, after the growth of 10 MQWs with and without ULs approach each other in PL intensity. One way to think about this observation is that for the MQWs without the UL, the lower QWs essentially act as ULs to gradually improve the PL emission intensity, while for MQWs with the UL this gradual improvement is not required. This observation may provide an explanation as to why early LEDs were typically composed of many MQWs even though only the top most QWs participated in the LED light emission (David et al., 2008). These days the number of QWs needed for good optical emission has been significantly reduced.

The morphology of the InGaN UL is shown in Figure 6.13b by a 3×3 µm AFM image. The InGaN has a mounded structure (Koleske et al., 2014a) with the peaks corresponding to classic growth around screw dislocations (Burton et al., 1949). This InGaN layer was grown at 870°C using high indium flow rate and an InGaN growth rate of 350 nm/hr. Based on the peak in the power spectral density near 5–10 µm^{-1}, the roughening of the InGaN UL occurs over lengths of ~100–200 nm (Koleske et al., 2014a). There is also an increase in the number of double step heights compared to the starting GaN template (Koleske et al., 2010). The predominantly outward step edges formed with this morphology would provide some outward strain relaxation for InGaN QWs that are grown on top. Any slight relaxation during the InGaN QW growth might either incorporate more indium or have a decreased likelihood of forming point or vacancy related defects. The slight strain relaxation that these ULs provide has been measured with XRD (Lee et al., 2012) and many groups refer to the InGaN UL as a strain relief layer (Nanhui et al., 2006, 2007; Ali et al., 2013; Li et al., 2013).

The first mention of InGaN or superlattice ULs is from patents issued to LumiLeds (Goetz et al., 2002, 2003) where they are referred to as smoothing layers. Soon reports from Akasaka et al. showed that when 50 nm thick InGaN ULs were used the radiative efficiency of InGaN MQWs could be improved to 71% (Akasaka et al., 2004), by reducing NRCs (Akasaka et al., 2005). Using GaN and InGaN ULs for violet wavelength LEDs, the EL intensity was shown to be five times larger when using InGaN ULs compared to GaN ULs (Akasaka et al., 2006). Leem et al. used 30 nm thick $In_{0.03}Ga_{0.97}N$ UL to reduce peak wavelength shift as the current was increased (Leem et al., 2008) and Törmä et al. used dilute indium concentration InGaN ULs and reported an almost 50% increase in the electroluminescence EL (Törmä et al., 2008). The insertion of an $In_{0.03}Ga_{0.97}N$ UL resulted in MQWs with reduced recombination lifetime, increased spatial homogeneity in the QW luminescence, and partial strain relaxation as evidenced by reciprocal space mapping (Li et al., 2013). IQE improvements have been reported using ULs ranging in indium concentration from 1% to 5% (Nanhui et al., 2006, 2007; Son et al., 2006; Leem et al., 2008; Wang et al., 2010; Christian et al., 2015; Davies et al., 2016). Even the use of a single violet-emitting QW at the bottom of QWs was suspected of improving light emission for orange-emitting LEDs (Huang et al., 2006).

In addition to bulk InGaN layers, InGaN/GaN superlattice ULs have also been reported to improve EL and reduce wavelength shift as the current was increased (Nanhui et al., 2006, 2007; Leem et al., 2008; Okada et al., 2015). Using atom probe tomography, Larson et al. reported finding a 21 period InGaN/GaN superlattice in a commercial OSRAM LED structure (Larson et al., 2013). Some groups have also used AlN/GaN multilayers (Ali et al., 2013) and n-$Al_{0.06}Ga_{0.94}N$/n-GaN strained-layer superlattices (SLSs) as ULs (Cheng-Liang et al., 2006; Ali et al., 2013).

As mentioned in Section 6.6.2, capping the InGaN QWs first with GaN prior to ramping the growth temperature for the remainder of the GaN barrier allows the QW to retain much of the deposited indium. More recently the Toshiba group has reported using *AlGaN interlayers* (ILs) to cap InGaN QWs prior to heating (Hashimoto et al., 2013, 2014; Shinji et al., 2013; Jong-Il et al., 2014). AlGaN capping and annealing the InGaN QWs allowed the Toshiba group to achieve EQEs of ~20% for yellow (560 nm)

LEDs (Hashimoto et al., 2014) and ~3% for red (629 nm) LEDs (Jong-Il et al., 2014). The Toshiba group proposed possible reasons for the improved efficiency with AlGaN ILs, including improved hetero-interfaces, higher InGaN crystalline quality with fewer nonradiative defects, and a polarization field shift of electron and hole wavefunctions all of which have been confirmed (Koleske et al., 2015). Two other important factors for achieving these longer wavelengths were uncovered by Koleske et al. (2015). First, using the AlGaN IL allows for the growth of the InGaN QW well below typical green wavelength growth temperatures to achieve indium concentrations up to ~0.25 (Koleske et al., 2015). Second, annealing the IL-capped QW prior to the GaN barrier growth improves the AlGaN IL smoothness as determined by AFM, improves the InGaN/AlGaN/GaN interface quality as determined from STEM images and XRD, and increases the radiative efficiency by reducing nonradiative defects as determined by time-resolved PL measurements (Koleske et al., 2015). Finally, the AlGaN IL increases the spontaneous and piezoelectric polarization-induced electric fields acting on the InGaN QW, providing an additional red-shift to the emission wavelength as determined from modeling of the experimental data (Koleske et al., 2015).

Hydrogen dosing of InGaN QWs would seem to be counterproductive since hydrogen is well known to decrease indium incorporation in InGaN films (Piner et al., 1997). If hydrogen is added to the gas flow before completely capping the InGaN QWs, gaps are produced in the InGaN QWs as measured using atom probe tomography (Hu et al., 2012). These same *gappy* QWs can also be produced using a temperature bounce procedure described by Oliver et al. (2013). Using atom probe tomography, Ren et al. observed these discontinuous InGaN QWs made by dosing the InGaN QW with H_2 as the GaN cap layer is grown (Ren et al., 2015). Relative to continuous InGaN QWs grown using no H_2 during the GaN capping layer, the hydrogen dosed GaN cap layers had a 50% increase in PL emission intensity (Ren et al., 2015). Similar to the discontinuous, temperature-bounced QWs of Oliver et al. (2013), adding H_2 during the QW capping possibly removes more defective or indium-rich regions leading to a reduction in the nonradiative defects and an increase in the radiative recombination (Ren et al., 2015).

Attempts to use H_2 to produce lower indium concentration InGaN barrier layers while maintaining the same TMGa and TMIn flow rates resulted in MQWs with decreased PL intensity (Koleske et al., 2014b). Using XRD and dynamic diffraction analysis, the indium concentration in both the InGaN QW and InGaN barrier could be measured as a function of the H_2 flow rate. Kinetic analysis of the steady state θ_{In} suggested that hydrogen enhances indium surface desorption through the formation of more volatile indium-hydride species (Koleske et al., 2014b). Similar results were found by Czernecki et al. who found that H_2 dosing reduced the QW thickness, decreased the QW indium concentration, and decreased the barrier growth rate (Czernecki et al., 2015).

Another way to produce discontinuous InGaN MQWs is to *temperature ramp* the QW prior to the growth of the GaN barrier. Referred to as the *temperature-bounce* method, it has been extensively studied by the Cambridge group (Oliver et al., 2013). This work stemmed from early work where gross well-width fluctuations were observed in uncapped InGaN QWs, resulting in an overall 3D morphology consisting of a network of interlinking InGaN strips (van der Laak et al., 2007). The InGaN network appears to correlate to GaN template step edges, suggesting that the network structure forms via the decomposition of indium-rich regions arising at step edges (van der Laak et al., 2007). GaN-capped QWs grown with this method had QWs with gross fluctuations in width when viewed in TEM cross-sections (van der Laak et al., 2007).

Olivier et al. have shown that LEDs with these discontinuous QWs have higher quantum efficiencies than continuous QWs (Oliver et al., 2013). Two different *temperature-bounce* methods were shown to provide better IQEs. The first is a two temperature growth, where after the QW growth the temperature is ramped and the GaN barrier is grown at a higher temperature. The second is single temperature growth, where after the QW growth the temperature is *bounced* and then returned to the QW temperature for the growth of the barrier layer (Oliver et al., 2013). Both methods produced similar QW width fluctuations (Oliver et al., 2013). Similarly, structured QWs were previously reported but without detail on the exact growth procedure (Zhu et al., 2006; Carvalho et al., 2015). Previous TEM results by

Gerthsen et al. have also observed inhomogeneous indium distribution with nanometer-sized, indium-rich agglomerates, and 10 nm composition fluctuations; however, no information was provided on the exact growth conditions (Gerthsen et al., 2001). Galtrey et al. also report seeing similar structures in commercially available UV LEDs with an efficiency of 67% and it was suggested that these discontinuous InGaN QWs played a large role in the overall efficiency of these LEDs (Galtrey et al., 2008b).

Each of these growth techniques can be used to ostensibly improve the overall QW emission efficiency, likely through an increase in the degree of carrier localization and a decrease in NRCs. What does not appear to matter is how the increased localization or decrease in NRCs is achieved since a variety of growth methods and MQW structures can be used to realize efficient light emission. While these approaches work to improve IQE and EQE at lower carrier densities, the overall efficiency decreases at high carrier densities where other carrier loss mechanisms dominate, such as Auger recombination, and the overall emission efficiency decreases (Shen et al., 2007).

6.7 Aluminum Containing Nitrides

Unlike GaN and InGaN growth, AlN and AlGaN are typically grown at lower pressure to reduce the gas-phase parasitic chemistry discussed in Section 6.3.1. The gas-phase parasitic reactions remove aluminum from the gas phase resulting in AlGaN alloys with reduced aluminum fraction (Allerman et al., 2004) with the prospect of these particles landing on the surface to create large scale growth defects. Chen et al. report growing AlN at 250 torr in a horizontal reactor but as the pressure was increased the growth rate did decrease due to gas-phase parasitic chemistry (Chen et al., 1996). In addition, AlN and AlGaN do not show evidence for gas-phase etching and decomposition reactions normally observed for GaN growth at high temperature. This means that lower NH_3 flows can be used (Allerman et al., 2004) to enhance adatom surface diffusion (Fareed et al., 2005; Choi et al., 2009). Such little NH_3 is required for growth that alternating pulses of NH_3 and MOs can be used to enhance Al adatom migration, resulting in improved AlGaN quality and smoother surfaces (Fareed et al., 2005; Choi et al., 2009). Also due to the materials stability, AlN has been grown at temperatures up to 1,600°C (Imura et al., 2006b) resulting in smooth surface with improved material quality suggesting that enhanced Al mobility on the surface is essential for improved material (Fujimoto et al., 2006; Imura et al., 2006a; Amano, 2011). This progress has resulted in UV LEDs (Amano et al., 2008; Khan et al., 2008) with wavelengths down to 210 nm (Taniyasu et al., 2006b).

In this section, the growth of AlN will be covered briefly followed by a discussion on using thin AlGaN and AlN layers for strain relief and dislocation filtering in GaN materials. Through the use of these layers, GaN can be grown directly on Si substrates without severe cracking. Also, AlGaN growth is discussed as it is routinely used as an electron blocking layer (EBL) in InGaN-based LEDs to prevent electron overflow into the p-type region. Other structures including superlattices, alloy graded layers, and InAlN are also discussed in relationship to how they are used as an EBLs in LEDs.

6.7.1 AlN Growth

Since AlN is more thermally stable than the other III-N compounds, it can be grown at substantially higher temperatures, provided the MOVPE system is designed with this in mind. Yasuo et al. reported AlN growth on sapphire at growth temperatures as high as 1,300°C at low V/III ratio resulting in smooth films (Yasuo and Ako, 1996). Later, this same group reported using a thin first AlN layer grown at 1,250°C followed by a second layer grown at 1,350°C (Yasuo et al., 1997). Smooth AlN layers were grown on sapphire at temperatures between 1,300°C and 1,600°C resulting in a dislocation density of less than 2×10^9 cm^{-2} (Imura et al., 2006b). Other groups have also reported the growth of AlN on sapphire at high temperature (Fujimoto et al., 2006; Brunner et al., 2008b; Kröncke et al., 2013), resulting in films with smoother surfaces, reduced etch pit density and narrower XRD linewidths. Sakai et al. have reported that high-quality GaN can be grown on high-temperature AlN films on SiC, resulting in GaN films with a dislocation density of 5×10^7 cm^{-2} (Sakai et al., 2002).

Proof that AlN is more thermally stable has been investigated by Yoshinao et al. by heating AlN in H_2, He, and NH_3 flows (Yoshinao et al., 2008). In flowing H_2, slight thermal decomposition occurred above 1,200°C, which only become significant until 1,400°C. Thermal decomposition in He or $H_2 + NH_3$ up to 1,400°C did not produce noticeable AlN decomposition (Yoshinao et al., 2008). Also, the Al-polarity AlN layer decomposition rate was less than the N-polarity AlN layer at each temperature (Yoshinao et al., 2008). Following similar decomposition trends as GaN (Koleske et al., 2001), AlN decomposes more readily in H_2 at lower temperatures than decomposition in N_2, suggesting a H_2 etching mechanism akin to GaN and possibly a gas-phase transport of Al atoms at high temperature.

6.7.2 AlN and AlGaN Interlayers for Strain and Dislocation Reduction

Many groups have reported using thin AlN, AlGaN, or superlattice interlayers (ILs) to reduce dislocation density and strain in GaN and AlGaN layers (Amano et al., 1998, 1999; Motoaki et al., 1998; Koleske et al., 1999a; Waldrip et al., 2001; Wang et al., 2002). Initially, for GaN, the growth temperature was lowered to grow another AlN nucleation layer followed by heating and restarting of the GaN growth (Amano et al., 1998). With the advent of *in situ* stress measurements, the insertion of the low-temperature AlN ILs was shown to be effective in reducing the GaN film tensile stress and cracking (Amano et al., 1998). Low-temperature GaN ILs also reduced the density of TDs and decrease the tensile stress that normally occurs during high-temperature GaN growth (Amano et al., 1998). Both the low-temperature AlN and GaN ILs also reduced the etch pit density of the top GaN layer (Motoaki et al., 1998). Using this approach, Amano et al. report crack-free, low dislocation density AlGaN on sapphire over the entire compositional range (Amano et al., 1999).

The exact nature of the dislocation density reduction using AlN ILs was later investigated using Hall effect, XRD, and TEM cross-section measurements (Koleske et al., 1999a). From this study, it was shown that as the number of AlN IL/HT GaN layers increases, the electron mobility increased from 440 to 725 cm^2/Vs (Koleske et al., 1999a). TEM measurements revealed a significant reduction in the screw dislocation density for GaN films grown on the AlN IL, while the symmetric and off-axis X-ray linewidths increased as the number of AlN IL/HT GaN layers increase, indicating that overall alignment of the larger GaN grains was increasing (Koleske et al., 1999a). The increase in mobility as the number of AlN ILs increase was attributed to a decreased acceptor concentration possibly associated with the reduction in screw dislocations (Koleske et al., 1999a).

By providing a means for controlling GaN and AlGaN film strain, Waldrip et al. demonstrated 60 pairs, crack-free, AlGaN/GaN distributed Bragg reflectors (DBRs) that had a 99% reflectivity at a peak wavelength of 380 nm (Waldrip et al., 2001). Later Wang et al. used AlN/AlGaN superlattices (SLs) to reduce TD density to grow thick AlGaN on sapphire (Wang et al., 2002). A similar dislocation reduction was observed for low-temperature AlGaN ILs for GaN films where the AlGaN IL layer was shown to block TD from coming into the underlying GaN film (Tsai and Gong, 2004). Instead of reducing the GaN dislocation density, the ultimate utility of AlN and AlGaN ILs was applied to achieve crack-free GaN on Si.

6.7.3 Cracking and Stress Management—GaN on Si (111)

The major advantage of growing compound semiconductors on Si (111) or Si (100) substrates is the availability of large area wafers and the potential for matching CMOS-compatible circuitry with the functionality that the III-V device provides. As mentioned in Section 6.4.3, *in situ* monitoring of strain evolution was critical for the rapid development of GaN growth on Si (111). This is due to the large lattice mismatch between the GaN and Si of 17% and coefficient of thermal expansion (CTE) mismatch of 56% which severely limits GaN thickness unless AlN or AlGaN strain relieving layers are judiciously placed in the structure. Even after the cracking is prevented, the high dislocation density hampered the development of useful devices on Si (111) until silicon nitride auto-masking techniques were applied to reduce the dislocation density (Dadgar et al., 2003). Several groups have pursued GaN-based LEDs and power electronic devices on Si substrates due to their compatibility with existing process lines used in the electronics industry (Zhu et al., 2013).

Initially, GaN growth on Si (111) used low-temperature GaN NLs using the standard two-step growth technique (Ishikawa et al., 1998). During the heating, the GaN NL on Si liberated Ga atoms from the NL decomposition interacted with the Si surface resulting in a wavy surface appearance (Ishikawa et al., 1998). Even an intermediate temperature GaN growth did not improve the observed surface swelling and EDX analysis showed large amounts of Ga located in regions where the film had swelled (Ishikawa et al., 1998). Kobayashi et al. tried an aluminum oxide compound layer as a linking layer between the GaN and the Si (111) substrate, resulting in films with a rough morphology mainly because the aluminum oxide was composed of different grain heights (Kobayashi, et al. 1998). Also, epitaxial cubic SiC buffer layers were grown on either carbonized or uncarbonized Si (111) substrates at low temperature for the growth of hexagonal GaN (Boo et al., 1998). Later Wei et al. deposited cubic and hexagonal GaN on 3C-SiC on Si (100) resulting in a 91% cubic component GaN under the optimized growth conditions (Wei et al., 2000).

To reduce the melt-back etching of Si by Ga metal at high temperature, Hiroyasu et al. instead used AlN as the linking layer between the Si substrate and the GaN (Hiroyasu et al., 1999). Similar to the high-temperature growth of AlN on SiC mentioned in Section 6.3.4 the use of high-temperature AlN and AlGaN were used to seal the Si surface and prevent the Ga metal etchback of the Si substrate. The only caveat to this approach is to introduce the TMAl precursor first before the NH_3 (Hiroyasu et al., 1999), to avoid excessive nitridation of the Si surface. The AlN layer also provides a compressive layer on which to place the GaN which grows under tension (Hearne et al., 1999). With this approach, Hiroyasu et al. reported a final GaN surface without cracks or pits when they used an 80 nm thick AlN at 1,100°C followed by an AlGaN layer with 27% aluminum concentration (Hiroyasu et al., 1999). Other groups followed this initial success with Lahrèche et al. achieving dislocations in the low 10^{10} cm^{-2} range (Lahrèche et al., 2000), using AlN ILs (Bläsing et al., 2002), or AlGaN ILs (Han et al., 2001), or AlN/AlGaN superlattices (Chen et al., 2002) to manage the film strain as monitored using an *in situ* optical stress sensor (Krost et al., 2005b).

After managing the film strain on Si, the GaN dislocation densities were typically $>10^{10}$ cm^{-2} making these films marginally useful. To decrease the dislocation density Dadgar et al. used *in situ* deposited SiN auto-masking layers to restart the GaN growth and filter the dislocation density (Dadgar et al., 2003). By optimizing the SiN growth time, the dislocation induced etch pit density was decreased to 1.3×10^9 cm^{-2} (Dadgar et al., 2003), making these GaN films useful for LEDs and HEMT structures. Other methods to reduce dislocation densities in the GaN films on Si (111) included low-temperature AlN IL (Zhang et al., 2004) and multiple AlN ILs between the GaN layers (Cong et al., 2005).

The growth of GaN on the more technologically important Si (100) surface was reported by Dadgar et al. by breaking the normal fourfold symmetry of the (100) surface using a four degrees off-oriented Si (100) substrate (Dadgar et al., 2007). After the growth of 100 nm of GaN on this miscut substrates, a rather smooth surface was formed with only a few misoriented crystallites (Dadgar et al., 2007). The growth of GaN on Si (110) showed a nearly threefold reduction in the full width at half maximum (FWHM) of the (1–100) omega-scan, suggesting even better GaN grain alignment compared to Si (111) (Dadgar et al., 2007).

Possibly a better approach to GaN growth on Si is to use sputter deposited AlN NLs following the discussion in Section 6.3.2. Recently, Takaya et al. have used reactive-sputter-deposited AlN intermediate layer (sp-AlN) at 350°C to successfully grow GaN on Si (111) resulting in films with dislocations of about 10^9 cm^{-2} (Takaya et al., 2013). Reactive magnetron sputtering of AlN on Si (100) resulted in c-axis oriented (0001) possibly making this a suitable approach for GaN films (Zhang et al., 2005). Utilizing high-resolution XRD, Lee et al. were able to show that the GaN crystal quality was improved on the sputter deposited AlN NL compared to the conventional low-temperature GaN NLs (Lee et al., 2014), making this approach attractive for GaN growth on all Si surfaces.

6.7.4 Al-based Electron Block Layers

To improve the efficiency of MQW LEDs, the EBL has played an important role in confining electrons to the MQW region and preventing them from leaking into the p-side of the device (Schubert, 2006). Typically, thin AlGaN EBL layers are used with Al fractions of ≤0.20. To best confine the electrons on the n-type side of the junction, the AlGaN is p-type doped with Mg, which presents several growth issues with this layer. First, as mentioned in Section 6.5.2, Mg is already a deep acceptor in GaN which becomes deeper in the valence band as aluminum is added. As a result, higher Mg-doping levels are required with consequential lower crystal quality in this layer. Second, the growth conditions required for optimal AlGaN growth, such as higher temperature in H_2, are quite different from the InGaN MQW growth conditions at a lower temperature without H_2. So a compromise is required for optimal MQWs and AlGaN EBL growth requiring a transition in growth conditions between the two layers. Sometimes this compromise is an undoped, last GaN barrier placed between the last QW and the EBL. Other material compositions have been used as EBLs including lattice-matched InAlN, quaternary AlInGaN, superlattice structures, and graded or tapered structures, which are highlighted in the following paragraphs.

In most LED structures, *AlGaN* EBLs are used to prevent electron overshoot into the p-side. However, as pointed out by Liu et al. the effectiveness of conventional EBLs at blocking electrons is still questionable due to the serious negative band-bending between the last GaN barrier and the AlGaN EBL (Liu et al., 2012). Xia et al. pointed out that in some cases LEDs without the AlGaN EBL perform better than structures with the EBL and suggested that some care must be taken in both its overall growth and LED design considerations (Xia et al., 2013). The EBL can also influence EQE with higher EQEs found at lower current when AlGaN EBLs are used while at higher current higher EQE was observed without EBLs (Han et al., 2009).

Nevertheless, AlGaN EBL is found in many LED structures with the most important growth parameters being the aluminum concentration, AlGaN thickness, AlGaN growth temperature and Mg-doping level. For EBLs, aluminum concentrations are ≤0.20 and should depend on the indium concentration in the InGaN QWs. Using modeling, Li et al. found that moderate Al fractions were most effective for low indium fraction InGaN QWs, while higher Al fractions were found to be harmful for high indium fraction QWs (Li et al., 2016). Kim et al. found the highest EL intensity for AlGaN EBLs containing 0.12 aluminum while higher aluminum percentages resulted in issues with carrier transport (Kim et al., 2012).

For blue wavelength LEDs, the luminescence efficiency increased as the Mg concentration in the AlGaN EBL was increased with the increased Mg concentration improving hole injection and carrier concentration (Wu and Chen, 2008). Also, the Mg diffusion into the InGaN QW growth region was suspected by Stephan et al. who suggested that the Mg diffusion can be reduced by growing a 30 nm thick Mg-doped AlGaN EBL rather than a thinner EBL (Stephan et al., 2003). Another method to suppress Mg diffusion into the QWs is to increase the UID GaN region or to grow the Mg-doped AlGaN layer at lower temperatures (Stephan et al., 2003). The use of lower EBL growth temperatures of 850°C in N_2 was found by Ru-Chin et al. to result in three times greater EL compared to AlGaN EBLs grown at 1,050°C in N_2 or H_2 (Ru-Chin et al., 2003). Higher Mg concentrations closest to the active region were also shown to improve hole injection resulting in a higher EL efficiency (Stephan et al., 2004; Kohler et al., 2005).

Taking advantage of the strong polarization gradients in the III-N materials can lead to bulk electron or hole distributions without the use of impurity doping (Rajan et al., 2006). This led to the consideration of using a *graded AlGaN* EBL first modeled by Simon et al. (2010) and then demonstrated by Zhang et al. for Mg-doped AlGaN graded from 30% to 0% on Al-polar AlN templates on sapphire (Zhang et al., 2010). Later LED results using a graded AlGaN EBL purportedly had a better blocking capability and a greatly improved hole injection efficiency (Zhang and Yao, 2011). However later modeling by Piprek suggested that the graded AlGaN EBL provided similar hole injection rates as a single aluminum concentration AlGaN EBL (Piprek, 2012). Piprek also suggested that polarization doping is less advantageous for Ga-polar LEDs compared to N-polar LEDs since for Ga-polar LEDs this approach generates a valence band barrier at the MQW/AlGaN interface which effectively blocks holes from entering the MQW (Piprek, 2012). Later work using AlGaN/GaN superlattice structures with either modulation doping of Mg (Liu et al., 2016) or grading of the aluminum concentration (Park et al., 2013) also claimed higher hole concentrations and better hole injection. Undoubtedly, holes generated through the use of polarization could become important for higher aluminum concentration LEDs, but much work still needs to be done to explore this possibility.

Certain physical and growth related properties of *InAlN* potentially make it a more ideal EBL layer than AlGaN. First, InAlN has a wider energy bandgap than GaN, a large conduction-band offset, and small valence-band offset, which would allow better hole injection and more effectively block electrons from entering the p-type side (Zhang et al., 2012). Second, InAlN is grown at lower temperatures than AlGaN making it ideal for the thermal preservation of the InGaN QWs. Third, InAlN with an ~0.18 indium fraction should lattice match the underlying GaN, reducing applied strain to the LED. Using an $In_{0.18}Al_{0.82}N$ EBL in blue LEDs, Choi et al. increased the EL compared to similar LEDs grown without an EBL or with an $Al_{0.2}Ga_{0.8}N$ EBL, suggesting improved electron confinement by the InAlN EBL (Choi et al., 2010). LEDs with InAlN EBLs grown at 780°C produced higher emission efficiency compared to InAlN EBLs grown at 840°C, suggesting better preservation of the InGaN QWs (Alec et al., 2010). Kim et al. also showed improved IQE for green wavelength LEDs using lower temperature InAlN which caused less thermal degradation of the green active region (Kim et al., 2010).

One issue in the growth of InAlN is the observation of unintentional Ga incorporation into the InAlN even though the Ga source is not on. Zhu et al. found as much as 45% gallium incorporation into AlInN films under a very indium-rich growth environment with no gallium precursor flowing (Zhu et al., 2012). The source of the unintentional gallium incorporation was suspected to be from the GaN template, which was enhanced by adding H_2, suggesting decomposing GaN as the gallium source (Zhu et al., 2012). Later work showed possible sources of the unintentional gallium to be the growth chamber walls, deposits on the wafer carrier, and cooler parts of the MOVPE system (Choi et al., 2014; Kim et al., 2014, 2015). The transport of gallium to the growth surface was speculated to result in the formation of an In–Ga eutectic from the residual chamber Ga atoms and the decomposed TMIn precursor (Choi et al., 2014). Experiments where the growth chamber was intentionally coated with TMGa and TMIn prior to InAlN growth suggested some sort of interaction between liberated gas-phase gallium atoms and metallic indium generated gas-phase gallium which was eventually observed in the InAlN (Choi et al., 2014; Kim et al., 2014, 2015). Clearly, this work calls into question the compositional assignment of InAlN films in much previous work as well as suggesting some care must be taken when growing InAlN films.

6.8 Summary and Remaining Questions

The MOVPE growth considerations for group III nitrides were reviewed, herein, with emphasis on the chemical and physical aspects of growth required for visible LEDs. The groundwork laid by Amano and Akasaki (Amano et al., 1986) and the revolutionary progress of Nakamura as well as many other researchers not only generated early favorable results but also stimulated further advances from many enthusiastic researchers. Astonishing progress in device advances and performance has often outpaced general understanding of the scientific underpinnings. In summary, remaining questions are posed relating to the growth mechanisms, improving the material quality, and origin and correlation of point defects to growth conditions.

Ultimately, GaN growth was not about finding a lattice-matched substrate, but about inventing AlN (Amano et al., 1986) and GaN (Nakamura, 1991c) NLs. This invention revolutionized the nitrides and may apply to future materials that do not have a lattice matched substrate. Sapphire works for GaN growth because it is thermally stable, can be slightly nitrided, and produces films under compressive stress after growth. In addition, the high-temperature MOVPE growth of GaN provides a gas-phase transport mechanism (discussed in Section 6.3.1) that allows the development of GaN with large grains that laterally coalesce resulting in a reduced dislocation density. The growth to the extinction of the c-plane under these high-temperature conditions (Du et al., 2005) leads to the smooth c-plane surface. Once the GaN grains are established the gas-phase Ga transport is similarly independent of substrate, meaning that growth on SiC, sapphire, or Si(111) substrates results in similar dislocation densities in the mid to high 10^8 cm^{-2}. This gas-phase transport mechanism is a fortuitous advantage when GaN is grown using MOVPE.

One other fortuitous advantage of GaN is the relative lack of sensitivity to dislocation density (for LED and HEMT devices) despite the high dislocation density. Semiconductors in the traditional III-V materials with dislocation densities of 10^8–10^9 cm^{-2} would be rendered useless and optically dead. While a variety of localization (Narukawa et al., 1997; Gerthsen et al., 2000; Graham et al., 2005; Chichibu et al., 2006), antilocalization (Hangleiter et al., 2005), or short carrier lifetimes (Speck and Rosner, 1999) are used to explain away carrier recombination at the dislocation cores, it is still not clear how GaN is immune from such a large dislocation density compared to other III-V semiconductors. Furthermore, any brightness achieved in blue wavelength LEDs is purposefully built into the QWs, aided by the careful construction of the material surrounding the QWs as discussed in Section 6.6.5. Bright QWs were developed through Edisonian optimization, which is good for LED manufacturing, and suggesting that many different growth approaches exist to accomplish bright LEDs. Details behind the driving physics are still lacking since no unifying theory yet explains all active regions.

Lastly, understanding point defects and their influence on device properties is extremely difficult in this materials system. This is of course due to the numerous materials properties, including grain structure and boundaries, dislocation density and type, impurities, and morphological properties that first need to be fully deconvolved before assigning any materials properties as solely due to point defects. With the advent of larger, single crystal GaN substrates, several of the materials properties such as grain structure, grain boundaries, and dislocation density can be removed as contributing factors. However, growth morphology and device properties on these bulk single crystals can be influenced by the manufacturer and method of manufacturing (Kizilyalli et al., 2015), suggesting more substrate and growth development will be required for the further advancement of devices using GaN homoepitaxial growth.

Acknowledgment

A great number of people need to be thanked for their direct or indirect input into this chapter. Although initially trained as a surface scientist, I learned the art of MOVPE crystal growth from Dr. Alma Wickenden at the Naval Research Laboratory (NRL) starting in 1995. Others at NRL including Kurt Gaskill, Richard Henry, Mark Twigg, Jaime Freitas, Mohammad Fatemi, Steve Binari, and Vic Bermudez taught me valuable characterization and interpretation skills. After moving to Sandia National Laboratories in 2001, I benefitted greatly from conversations and collaborations with Andy Allerman, Christine Mitchell, Jeff Figiel, Mike Coltrin, Randy Creighton, Steve Lee, Bob Biefeld, Art Fischer, Mary Crawford, Steve Kurtz, Jon Wierer, David Follstaedt, Andy Armstrong, Jeff Kempisty, Ganesh Subramanian, George Wang, Jeff Tsao, Jerry Simmons, Sam Myers, Alan Wright, Carl Seager, Greg Peake, and Jeff Cederberg. I also thank Bob Biefeld and Brendan Gunning for thoroughly reading and commenting on this manuscript.

This work was funded by the Laboratory Directed Research and Development program at Sandia National Laboratories. Sandia National Laboratories is a multiprogram laboratory managed and operated by Sandia Corporation, a wholly owned subsidiary of Lockheed Martin Corporation, for the U.S. Department of Energy's National Nuclear Security Administration under contract DE-AC04-94AL85000.

References

Akasaka T, Gotoh H, Kobayashi Y, Nakano H, Makimoto T (2006) InGaN quantum wells with small potential fluctuation grown on InGaN underlying layers. *Applied Physics Letters* 89:101110.
Akasaka T, Gotoh H, Nakano H, Makimoto T (2005) Blue-purplish InGaN quantum wells with shallow depth of exciton localization. *Applied Physics Letters* 86:191902.

Akasaka T, Gotoh H, Saito T, Makimoto T (2004) High luminescent efficiency of InGaN multiple quantum wells grown on InGaN underlying layers. *Applied Physics Letters* 85:3089–3091.

Akasaki I (2015) Nobel Lecture: Fascinated journeys into blue light. *Reviews of Modern Physics* 87:1119–1131.

Akasaki I, Amano H, Koide Y, Hiramatsu K, Sawaki N (1989) Effects of AlN buffer layer on crystallographic structure and on electrical and optical properties of GaN and $Ga_{1-x}Al_xN$ $(0 < x \leq 0.4)$ films grown on sapphire substrate by MOVPE. *Journal of Crystal Growth* 98:209–219.

Alec MF, Kewei WS, Reid J, Fernando AP, Jae-Hyun R, Hee Jin K, Suk C, Seong-Soo K, Russell DD (2010) Effect of growth temperature on the electron-blocking performance of InAlN layers in green emitting diodes. *Applied Physics Express* 3:031003.

Alexandrov YA, Vyshinskii NN, Kokorev VN, Alferov VA, Chikinova NV, Makin GI (1987) On the possibility of complex formation of Group IIIB organic compounds with molecular oxygen. *Journal of Organometallic Chemistry* 332:259–269.

Alfieri G, Tsutsumi T, Micheletto R (2015) Electronic properties of substitutional impurities in InGaN monolayer quantum wells. *Applied Physics Letters* 106:192102.

Ali AH, Abu Bakar AS, Egawa T, Hassan Z (2013) InGaN-based multi-quantum well light-emitting diode structure with the insertion of superlattices under-layer. *Superlattices and Microstructures* 60:201–207.

Allerman AA, Crawford MH, Fischer AJ, Bogart KHA, Lee SR, Follstaedt DM, Provencio PP, Koleske DD (2004) Growth and design of deep-UV (240-290 nm) light emitting diodes using AlGaN alloys. *Journal of Crystal Growth* 272:227–241.

Amano H (2011) Impact of high-temperature metalorganic vapor phase epitaxial growth of AlGaN-based UV-A, UV-B and UV-C quantum wells on the improvement of their internal quantum efficiency. In: *17th International Conference on Microscopy of Semiconducting Materials 2011* (Walther T, Midgley PA, eds). Cambridge, UK: IPO Publishing.

Amano H (2015) Nobel Lecture: Growth of GaN on sapphire via low-temperature deposited buffer layer and realization of p-type GaN by Mg doping followed by low-energy electron beam irradiation. *Reviews of Modern Physics* 87:1133–1138.

Amano H, Akasaki I, Hiramatsu K, Koide N, Sawaki N (1988) Eeffects of the buffer layer in metalorganic vapor-phase epitaxy of GaN on sapphire substrate. *Thin Solid Films* 163:415–420.

Amano H, Imura M, Iwaya M, Kamiyama S, Akasaki I (2008) AlN and AlGaN by MOVPE for UV Light Emitting Devices. *Materials Science Forum* 590:175–210.

Amano H, Iwaya M, Hayashi N, Kashima T, Nitta S, Wetzel C, Akasaki I (1999) Control of dislocations and stress in AlGaN on sapphire using a low temperature interlayer. *Physica Status Solidi (B)* 216:683–689.

Amano H, Iwaya M, Kashima T, Katsuragawa M, Akasaki I, Han J, Hearne S, Floro JA, Chason E, Figiel J (1998) Stress and defect control in GaN using low temperature interlayers. *Japanese Journal of Applied Physics* 37:L1540–L1542.

Amano H, Kito M, Hiramatsu K, Akasaki I (1989) P-type conduction in Mg-doped GaN treated with low-energy electron beam irradiation (LEEBI). *Japanese Journal of Applied Physics* 28:L2112.

Amano H, Sawaki N, Akasaki I, Toyoda Y (1986) Metalorganic vapor-phase epitaxial-growth of a high-quality GaN film using an AlN buffer layer. *Applied Physics Letters* 48:353–355.

Ambacher O (1998) Growth and applications of Group III-nitrides. *Journal of Physics D: Applied Physics* 31:2653.

Ambacher O, Brandt MS, Dimitrov R, Metzger T, Stutzmann M, Fischer RA, Miehr A, Bergmaier A, Dollinger G (1996) Thermal stability and desorption of Group III nitrides prepared by metal organic chemical vapor deposition. *Journal of Vacuum Science & Technology B: Microelectronics and Nanometer Structures Processing, Measurement, and Phenomena* 14:3532–3542.

Armstrong A, Henry TA, Koleske DD, Crawford MH, Lee SR (2012a) Quantitative and depth-resolved deep level defect distributions in InGaN/GaN light emitting diodes. *Optics Express* 20:A812–A821.

Armstrong A, Henry TA, Koleske DD, Crawford MH, Westlake KR, Lee SR (2012b) Dependence of radiative efficiency and deep level defect incorporation on threading dislocation density for InGaN/GaN light emitting diodes. *Applied Physics Letters* 101:162102.

Armstrong AM, Bryant BN, Crawford MH, Koleske DD, Lee SR, Wierer Jr JJ (2015) Defect-reduction mechanism for improving radiative efficiency in InGaN/GaN light-emitting diodes using InGaN underlayers. *Journal of Applied Physics* 117:134501.

Armstrong AM, Crawford MH, Koleske DD (2014) Contribution of deep-level defects to decreasing radiative efficiency of InGaN/GaN quantum wells with increasing emission wavelength. *Applied Physics Express* 7:032101.

Atsushi W, Hirokazu T, Toshiyuki T, Hiroyuki O, Kiyofumi C, Hiroshi A, Takayuki K, Ryo N, Isamu A (1999) Correlation between dislocation density and the macroscopic properties of GaN grown by metalorganic vapor phase epitaxy. *Japanese Journal of Applied Physics* 38:L1159.

Balmer RS, Pickering C, Pidduck AJ, Martin T (2002) Modelling of high temperature optical constants and surface roughness evolution during MOVPE growth of GaN using in-situ spectral reflectometry. *Journal of Crystal Growth* 245:198–206.

Ban VS (1972) Mass spectrometric studies of vapor-phase crystal growth II. *Journal of the Electrochemical Society* 119:761–765.

Bartram E, Creighton JR (1998) GaN CVD reactions: Hydrogen and ammonia decomposition and the desorption of Gallium Michael. *MRS Online Proceedings Library Archive* 537:G3.68.

Belousov M, Volf B, Ramer JC, Armour EA, Gurary A (2004) In situ metrology advances in MOCVD growth of GaN-based materials. *Journal of Crystal Growth* 272:94–99.

Bläsing J, Reiher A, Dadgar A, Diez A, Krost A (2002) The origin of stress reduction by low-temperature AlN interlayers. *Applied Physics Letters* 81:2722–2724.

Boo J-H, Ustin SA, Ho W (1998) Growth of hexagonal GaN thin films on Si(1 1 1) with cubic SiC buffer layers. *Journal of Crystal Growth* 189–190:183–188.

Bour DP, Connell GAN, Scifres DR (1999) In-situ acceptor activation in group III-V nitride compound semiconductors. US Patent 5,926,250, filed September 12, 1997, and issued July 20, 1999.

Breiland WG (2003) *Reflectance-Correcting Pyrometry in Thin Film Deposition Applications.* Albuquerque, NM: Sandia National Laboratories, p. 85, SAND2003-1868.

Breiland WG, Coltrin ME, Creighton JR, Hou HQ, Moffat HK, Tsao JY (1999) Organometallic vapor phase epitaxy (OMVPE). *Materials Science and Engineering: R-Reports* 24:241–274.

Breiland WG, Killeen KP (1994) Spectral reflectance as an in-situ monitor for MOCVD. In: *Diagnostic Techniques for Semiconductor Materials Processing* (Glembocki OJ, Pang SW, Pollak FH, Crean GM, Larrabee G, eds), pp. 99–104. Boston, MA: Material Research society.

Breiland WG, Killeen KP (1995) A virtual interface method for extracting growth-rates and high-temperature optical-constants from thin semiconductor-films using in-situ normal incidence reflectance. *Journal of Applied Physics* 78:6726–6736.

Breiland WG, Lee SR, Koleske DD (2004) Effect of diffraction and film-thickness gradients on wafer-curvature measurements of thin-film stress. *Journal of Applied Physics* 95:3453–3465.

Brunner F, Knauer A, Schenk T, Weyers M, Zettler JT (2008a) Quantitative analysis of in situ wafer bowing measurements for III-nitride growth on sapphire. *Journal of Crystal Growth* 310:2432–2438.

Brunner F, Protzmann H, Heuken M, Knauer A, Weyers M, Kneissl M (2008b) High-temperature growth of AlN in a production scale 11 × 2′ MOVPE reactor. *Physica Status Solidi (c)* 5:1799–1801.

Burton WK, Cabrera N, Frank FC (1949) Role of dislocations in crystal growth. *Nature* 163:398–399.

Carvalho D, Morales FM, Ben T, García R, Redondo-Cubero A, Alves E, Lorenz K, Edwards PR, O'Donnell KP, Wetzel C (2015) Quantitative chemical mapping of InGaN quantum wells from calibrated high-angle annular dark field micrographs. *Microscopy and Microanalysis* 21:994–1005.

Chalker PR, Morrice D, Joyce TB, Noakes TCQ, Bailey P, Considine L (2000) Indium segregation in MOCVD InGaN layers studied by medium energy ion scattering. *Diamond and Related Materials* 9:520–523.

Chaowang L, Ian MW (2007) Quantitative simulation of in situ reflectance data from metal organic vapour phase epitaxy of GaN on sapphire. *Semiconductor Science and Technology* 22:629.

Chason EH, Floro JA, Seager CH, Sinclair MB (1999) Measurement of the curvature of a surface using parallel light beams. US Patent 5,912,738, filed November 25, 1996, and issued June 15, 1999.

Chen CH, Liu H, Steigerwald D, Imler W, Kuo CP, Craford MG, Ludowise M, Lester S, Amano J (1996) A study of parasitic reactions between NH_3 and TMGa or TMAl. *Journal of Electronic Materials* 25:1004–1008.

Chen CQ, Zhang JP, Gaevski ME, Wang HM, Sun WH, Fareed RSQ, Yang JW, Khan MA (2002) AlGaN layers grown on GaN using strain-relief interlayers. *Applied Physics Letters* 81:4961–4963.

Chen ZZ, Qin ZX, Hu XD, Yu TJ, Yang ZJ, Tong YZ, Ding XM, Zhang GY (2004) Study of photoluminescence and absorption in phase-separation InGaN films. *Physica B: Condensed Matter* 344:292–296.

Cheng-Liang W, Jyh-Rong G, Ming-Fa Y, Bor-Jen W, Wei-Tsai L, Tai-Yuan L, Chung-Kwei L (2006) Improvement in the characteristics of GaN-based light-emitting diodes by inserting AlGaN-GaN short-period superlattices in GaN underlayers. *IEEE Photonics Technology Letters* 18:1497–1499.

Cheong MG, Kim KS, Oh CS, Namgung NW, Yang GM, Hong C-H, Lim KY et al. (2000) Conductive layer near the GaN/sapphire interface and its effect on electron transport in unintentionally doped n-type GaN epilayers. *Applied Physics Letters* 77:2557–2559.

Cherns D, Henley SJ, Ponce FA (2001) Edge and screw dislocations as nonradiative centers in InGaN/GaN quantum well luminescence. *Applied Physics Letters* 78:2691–2693.

Chiang CM, Gates SM, Bensaoula A, Schultz JA (1995) Hydrogen desorption and ammonia adsorption on polycrystalline GaN surfaces. *Chemical Physics Letters* 246:275–278.

Chichibu S, Cohen DA, Mack MP, Abare AC, Kozodoy P, Minsky M, Fleischer S et al. (1998) Effects of Si-doping in the barriers of InGaN multiquantum well purplish-blue laser diodes. *Applied Physics Letters* 73:496–498.

Chichibu SF, Uedono A, Onuma T, Haskell BA, Chakraborty A, Koyama T, Fini PT et al. (2006) Origin of defect-insensitive emission probability in In-containing (Al, In, Ga) N alloy semiconductors. *Nature Materials* 5:810–816.

Cho YS, Hardtdegen H, Kaluza N, Steins R, Heidelberger G, Lüth H (2007) The growth mechanism of GaN with different H_2/N_2 carrier gas ratios. *Journal of Crystal Growth* 307:6–13.

Choi RJ, Hahn YB, Shim HW, Han MS, Suh EK, Lee HJ (2003) Efficient blue light-emitting diodes with InGaN/GaN triangular shaped multiple quantum wells. *Applied Physics Letters* 82:2764–2766.

Choi S, Kim HJ, Ryou J-H, Dupuis RD (2009) Digitally alloyed modulated precursor flow epitaxial growth of $Al_xGa_{1-x}N$ layers with AlN and $Al_yGa_{1-y}N$ monolayers. *Journal of Crystal Growth* 311:3252–3256.

Choi S, Kim HJ, Kim SS, Liu J, Kim J, Ryou JH, Dupuis RD, Fischer AM, Ponce FA (2010) Improvement of peak quantum efficiency and efficiency droop in III-nitride visible light-emitting diodes with an InAlN electron-blocking layer. *Applied Physics Letters* 96:221105.

Choi S, Kim HJ, Lochner Z, Kim J, Dupuis RD, Fischer AM, Juday R et al. (2014) Origins of unintentional incorporation of gallium in AlInN layers during epitaxial growth, part I: Growth of AlInN on AlN and effects of prior coating. *Journal of Crystal Growth* 388:137–142.

Choi S, Kim T-H, Wolter S, Brown A, Everitt HO, Losurdo M, Bruno G (2008) Indium adlayer kinetics on the gallium nitride (0001) surface: Monitoring indium segregation and precursor-mediated adsorption. *Physical Review B* 77:115435.

Christian GM, Hammersley S, Davies MJ, Dawson P, Kappers MJ, Massabuau FCP, Oliver RA, Humphreys CJ (2015) Room temperature PL efficiency of InGaN/GaN quantum well structures with prelayers as a function of number of quantum wells. *Physica Status Solidi* (c) 13:248–251.

Chuo C-C, Lee C-M, Chyi J-I (2001) Interdiffusion of In and Ga in InGaN/GaN multiple quantum wells. *Applied Physics Letters* 78:314–316.

Collazo R, Mita S, Aleksov A, Schlesser R, Sitar Z (2006) Growth of Ga- and N- polar gallium nitride layers by metalorganic vapor phase epitaxy on sapphire wafers. *Journal of Crystal Growth* 287:586–590.

Coltrin ME, Kee RJ, Evans GH (1989) A mathematical-model of the fluid-mechanics and gas-phase chemistry in a rotating-disk chemical vapor-deposition reactor. *Journal of the Electrochemical Society* 136:819–829.

Coltrin ME, Mitchell CC (2003) Mass transport and kinetic limitations in MOCVD selective-area growth. *Journal of Crystal Growth* 254:35–45.

Coltrin ME, Randall Creighton J, Mitchell CC (2006) Modeling the parasitic chemical reactions of AlGaN organometallic vapor-phase epitaxy. *Journal of Crystal Growth* 287:566–571.

Cong G, Lu Y, Peng W, Liu X, Wang X, Wang Z (2005) Design of the low-temperature AlN interlayer for GaN grown on Si (1 1 1) substrate. *Journal of Crystal Growth* 276:381–388.

Costa PMFJ, Datta R, Kappers MJ, Vickers ME, Humphreys CJ, Graham DM, Dawson P, Godfrey MJ, Thrush EJ, Mullins JT (2006) Misfit dislocations in In-rich InGaN/GaN quantum well structures. *Physica Status Solidi (A)* 203:1729–1732.

Creighton JR, Breiland WG, Coltrin ME, Pawlowski RP (2002) Gas-phase nanoparticle formation during AlGaN metalorganic vapor phase epitaxy. *Applied Physics Letters* 81:2626–2628.

Creighton JR, Breiland WG, Koleske DD, Thaler G, Crawford MH (2008b) Emissivity-correcting mid-infrared pyrometry for group-III nitride MOCVD temperature measurement and control. *Journal of Crystal Growth* 310:1062–1068.

Creighton JR, Coltrin ME, Figiel JJ (2008a) Observations of gas-phase nanoparticles during InGaN metal-organic chemical vapor deposition. *Applied Physics Letters* 93:171906.

Creighton JR, Koleske DD, Mitchell CC (2006) Emissivity-correcting near-UV pyrometry for group-III nitride OMVPE. *Journal of Crystal Growth* 287:572–576.

Creighton JR, Wang GT (2005) Reversible Adduct Formation of Trimethylgallium and Trimethylindium with Ammonia. *The Journal of Physical Chemistry A* 109:133–137.

Czernecki R, Grzanka E, Smalc-Koziorowska J, Grzanka S, Schiavon D, Targowski G, Plesiewicz J et al. (2015) Effect of hydrogen during growth of quantum barriers on the properties of InGaN quantum wells. *Journal of Crystal Growth* 414:38–41.

Dadgar A, Blasing J, Diez A, Alam A, Heuken M, Krost A (2000) Metalorganic chemical vapor phase epitaxy of crack-free GaN on Si (111) exceeding 1 mu m in thickness. *Japanese Journal of Applied Physics Part 2 - Letters* 39:L1183–L1185.

Dadgar A, Hums C, Diez A, Blaesing J, Krost A (2006) Growth of blue GaN LED structures on 150-mm Si(111). *Journal of Crystal Growth* 297:279–282.

Dadgar A, Poschenrieder M, Reiher A, Blasing J, Christen J, Krtschil A, Finger T, Hempel T, Diez A, Krost A (2003) Reduction of stress at the initial stages of GaN growth on Si(111). *Applied Physics Letters* 82:28–30.

Dadgar A, Schulze F, Wienecke M, Gadanecz A, Bläsing J, Veit P, Hempel T, Diez A, Christen J, Krost A (2007) Epitaxy of GaN on silicon impact of symmetry and surface reconstruction. *New Journal of Physics* 9:389.

Dai Q, Schubert MF, Kim MH, Kim JK, Schubert EF, Koleske DD, Crawford MH et al. (2009) Internal quantum efficiency and nonradiative recombination coefficient of GaInN/GaN multiple quantum wells with different dislocation densities. *Applied Physics Letters* 94:111109.

Daisuke I, Kensuke N, Takafumi M, Motoaki I, Satoshi K, Hiroshi A, Isamu A, Akira B, Takashi U (2010) Growth of GaInN by raised-pressure metalorganic vapor phase epitaxy. *Applied Physics Express* 3:075601.

Danielsson O, Li X, Ojamae L, Janzen E, Pedersen H, Forsberg U (2016) A model for carbon incorporation from trimethyl gallium in chemical vapor deposition of gallium nitride. *Journal of Materials Chemistry C* 4:863–871.

Dauelsberg M, Brien D, Püsche R, Schön O, Yakovlev EV, Segal AS, Talalaev RA (2011) Investigation of nitride MOVPE at high pressure and high growth rates in large production reactors by a combined modelling and experimental approach. *Journal of Crystal Growth* 315:224–228.

Dauelsberg M, Martin C, Protzmann H, Boyd AR, Thrush EJ, Käppeler J, Heuken M, Talalaev RA, Yakovlev EV, Kondratyev AV (2007) Modeling and process design of III-nitride MOVPE at near-atmospheric pressure in close coupled showerhead and planetary reactors. *Journal of Crystal Growth* 298:418–424.

David A, Grundmann MJ, Kaeding JF, Gardner NF, Mihopoulos TG, Krames MR (2008) Carrier distribution in (0001)InGaN/GaN multiple quantum well light-emitting diodes. *Applied Physics Letters* 92:053502.

Davidson DF, Kohse-Höinghaus K, Chang AY, Hanson RK (1990) A pyrolysis mechanism for ammonia. *International Journal of Chemical Kinetics* 22:513–535.

Davies MJ, Hammersley S, Massabuau FC-P, Dawson P, Oliver RA, Kappers MJ, Humphreys CJ (2016) A comparison of the optical properties of InGaN/GaN multiple quantum well structures grown with and without Si-doped InGaN prelayers. *Journal of Applied Physics* 119:055708.

Deng Z, Jiang Y, Wang W, Cheng L, Li W, Lu W, Jia H, Liu W, Zhou J, Chen H (2014) Indium segregation measured in InGaN quantum well layer. *Scientific Reports* 4:6734.

Dietz N, Alevli M, Atalay R, Durkaya G, Collazo R, Tweedie J, Mita S, Sitar Z (2008) The influence of substrate polarity on the structural quality of InN layers grown by high-pressure chemical vapor deposition. *Applied Physics Letters* 92:41911.

Dong-Joon K, Yong-Tae M, Keun-Man S, Seong-Ju P (2001) Effect of barrier thickness on the interface and optical properties of InGaN/GaN multiple quantum wells. *Japanese Journal of Applied Physics* 40:3085.

Doo-Hyeb Y, Lachab M, Hao M, Sugahara T, Takenaka H, Naoi Y, Sakai S (1999) Investigation on the p-type activation mechanism in Mg-doped GaN films grown by metalorganic chemical vapor deposition. *Japanese Journal of Applied Physics* 38:631–634.

Doverspike K, Bulman GE, Sheppard ST, Kong HS, Leonard M, Dieringer H, Weeks TW et al. (1997) Status of nitride based light emitting and laser diodes on SiC. *MRS Proceedings* 482:1169–1178.

Du D, Srolovitz DJ, Coltrin ME, Mitchell CC (2005) Systematic prediction of kinetically limited crystal growth morphologies. *Physical Review Letters* 95:155503.

Duan XM, Stampfl C (2009) Defect complexes and cluster doping of InN: First-principles investigations. *Physical Review B* 79:035207.

Durkaya G, Alevli M, Buegler M, Atalay R, Gamage S, Kaiser M, Kirste R, Hoffmann A, Jamil M, Ferguson I (2009) Growth temperature-phase stability relation in In$_{1-x}$Ga$_x$N epilayers grown by high-pressure CVD. *MRS Proceedings* 1202:1202–I1205–1221.

Dwiliński R, Doradziński R, Garczyński J, Sierzputowski L, Kucharski R, Zając M, Rudziński M, Kudrawiec R, Serafińczuk J, Strupiński W (2010) Recent achievements in AMMONO-bulk method. *Journal of Crystal Growth* 312:2499–2502.

Eiting CJ, Grudowski PA, Dupuis RD (1998) P- and N-type doping of GaN and AlGaN epitaxial layers grown by metalorganic chemical vapor deposition. *Journal of Electronic Materials* 27:206–209.

Ejder E (1974) Growth and morphology of GaN. *Journal of Crystal Growth* 22:44–46.

El-Masry NA, Piner EL, Liu SX, Bedair SM (1998) Phase separation in InGaN grown by metalorganic chemical vapor deposition. *Applied Physics Letters* 72:40–42.

Elwell D, Elwell MM (1988) Crystal growth of gallium nitride. *Progress in Crystal Growth and Characterization* 17:53–78.

Fareed Q, Gaska R, Shur MS (2003) Migration enhanced metal organic chemical vapor deposition of AlN/GaN/InN-based heterostructures. In: *Semiconductor Device Research Symposium, 2003 International*, pp. 402–403. Piscataway, NJ: IEEE.

Fareed RSQ, Zhang JP, Gaska R, Tamulaitis G, Mickevicius J, Aleksiejunas R, Shur MS, Khan MA (2005) Migration enhanced MOCVD (MEMOCVD (TM)) buffers for increased carrier lifetime in GaN and AlGaN epilayers on sapphire and SiC substrate. *Physica Status Solidi C* 2:2095–2098.

Feezell DF, Speck JS, DenBaars SP, Nakamura S (2013) Semipolar (20-2-1) InGaN/GaN light-emitting diodes for high-efficiency solid-state lighting. *Journal of Display Technology* 9:190–198.

Figge S, Böttcher T, Einfeldt S, Hommel D (2000) In situ and ex situ evaluation of the film coalescence for GaN growth on GaN nucleation layers. *Journal of Crystal Growth* 221:262–266.

Fini P, Wu X, Tarsa EJ, Golan Y, Srikant V, Keller S, Denbaars SP, Speck JS (1998) The effect of growth environment on the morphological and extended defect evolution in GaN grown by metalorganic chemical vapor deposition. *Japanese Journal of Applied Physics* 37:4460.

Fiorentini V, Bernardini F, Della Sala F, Di Carlo A, Lugli P (1999) Effects of macroscopic polarization in III-V nitride multiple quantum wells. *Physical Review B* 60:8849–8858.

Fitouri H, Benzarti Z, Halidou I, Boufaden T, Jani BE (2005) Laser-reflectometry monitoring of the GaN growth by MOVPE using SiN treatment: study and simulation. *Physica Status Solidi (A)* 202:2467–2473.

Florescu DI, Ting SM, Ramer JC, Lee DS, Merai VN, Parkeh A, Lu D, Armour EA, Chernyak L (2003) Investigation of V-defects and embedded inclusions in InGaN/GaN multiple quantum wells grown by metalorganic chemical vapor deposition on (0001) sapphire. *Applied Physics Letters* 83:33–35.

Floro JA, Chason E, Lee SR (1996) Real time measurement of epilayer strain using a simplified wafer curvature technique. In: *Diagnostic Techniques for Semiconductor Materials Processing Ii* (Pang SW, Glembocki OJ, Pollak FH, Celii FG, SotomayorTorres CM, eds), pp. 491–496. Boston, MA: Material Research society.

Follstaedt DM, Missert NA, Koleske DD, Mitchell CC, Cross KC (2003) Plan-view image contrast of dislocations in GaN. *Applied Physics Letters* 83:4797–4799.

Follstaedt DM, Provencio PP, Missert NA, Mitchell CC, Koleske DD, Allerman AA, Ashby CIH (2002) Minimizing threading dislocations by redirection during cantilever epitaxial growth of GaN. *Applied Physics Letters* 81:2758–2760.

Fritze S, Dadgar A, Witte H, Bügler M, Rohrbeck A, Bläsing J, Hoffmann A, Krost A (2012) High Si and Ge n-type doping of GaN doping - Limits and impact on stress. *Applied Physics Letters* 100:122104.

Fujimoto N, Kitano T, Narita G, Okada N, Balakrishnan K, Iwaya M, Kamiyama S et al. (2006) Growth of high-quality AlN at high growth rate by high-temperature MOVPE. *Physica Status Solidi C* 3:1617–1619.

Galtrey MJ, Oliver RA, Kappers MJ, Humphreys CJ, Clifton PH, Larson D, Saxey DW, Cerezo A (2008a) Three-dimensional atom probe analysis of green- and blue-emitting $In_xGa_{1-x}N$/GaN multiple quantum well structures. *Journal of Applied Physics* 104:013524.

Galtrey MJ, Oliver RA, Kappers MJ, McAleese C, Zhu D, Humphreys CJ, Clifton PH, Larson D, Cerezo A (2008b) Compositional inhomogeneity of a high-efficiency $In_xGa_{1-x}N$ based multiple quantum well ultraviolet emitter studied by three dimensional atom probe. *Applied Physics Letters* 92:041904.

Gerthsen D, Hahn E, Neubauer B, Rosenauer A, Schon O, Heuken M, Rizzi A (2000) Composition fluctuations in InGaN analyzed by transmission electron microscopy. *Physica Status Solidi A* 177:145–155.

Gerthsen D, Neubauer B, Rosenauer A, Stephan T, Kalt H, Schon O, Heuken M (2001) InGaN composition and growth rate during the early stages of metalorganic chemical vapor deposition. *Applied Physics Letters* 79:2552–2554.

Gillard VT, Nix WD (1993) Study of the dislocation processes associated with strain relaxation in Si-Ge Heteroepitaxial films. *Zeitschrift Fur Metallkunde* 84:874–880.

Goetz W, Camras M, Gardner N, Kern R, Kim A, Stockman S (2002) Indium gallium nitride smoothing structures for III-nitride devices. U.S. Patent 6,489,696, filed March 29, 2001, and issued November 21, 2002, p. 15.

Goetz WK, Camras MD, Gardner NF, Kern RS, Kim AY, Stockman SA (2003) Indium gallium nitride smoothing structures for III-nitride devices. U.S. Patent 6,635,904, filed March 29, 2001, and issued October 21, 2003, p. 15.

Gotz W, Johnson NM, Walker J, Bour DP, Street RA (1996) Activation of acceptors in Mg-doped GaN grown by metalorganic chemical vapor deposition. *Applied Physics Letters* 68:667–669.

Graham DM, Soltani-Vala A, Dawson P, Godfrey MJ, Smeeton TM, Barnard JS, Kappers MJ, Humphreys CJ, Thrush EJ (2005) Optical and microstructural studies of InGaN/GaN single-quantum-well structures. *Journal of Applied Physics* 97:103508.

Green AK, Dancy J, Bauer E (1970) Insignificance of Lattice Misfit for Epitaxy. *Journal of Vacuum Science & Technology* 7:159–163.

Hageman PR, Devillers MAC, Zauner ARA, Kirilyuk V, Bouwens WS, Crane RCM, Larsen PK (1999) A study on the silane doping of hetero-epitaxial MOCVD grown GaN. *Physica Status Solidi (B)* 216:609–613.

Haitz R, Tsao JY (2011) Solid-state lighting: 'The case' 10 years after and future prospects. *Physica Status Solidi (A)* 208:17–29.

Hammersley S, Kappers MJ, Massabuau FCP, Sahonta S-L, Dawson P, Oliver RA, Humphreys CJ (2016) Effect of QW growth temperature on the optical properties of blue and green InGaN/GaN QW structures. *Physica Status Solidi (C)* 13:209–213.

Han B, Gregie JM, Wessels BW (2003) Blue emission band in compensated GaN:Mg codoped with Si. *Physical Review B* 68:045205.

Han J, Figiel JJ, Crawford MH, Banas MA, Bartram ME, Biefeld RM, Song YK, Nurmikko AV (1998) OMVPE growth and gas-phase reactions of AlGaN for UV emitters. *Journal of Crystal Growth* 195:291–296.

Han J, Ng T-B, Biefeld RM, Crawford MH, Follstaedt DM (1997) The effect of H_2 on morphology evolution during GaN metalorganic chemical vapor deposition. *Applied Physics Letters* 71:3114–3116.

Han J, Waldrip KE, Lee SR, Figiel JJ, Hearne SJ, Petersen GA, Myers SM (2001) Control and elimination of cracking of AlGaN using low-temperature AlGaN interlayers. *Applied Physics Letters* 78:67–69.

Han SH, Lee DY, Lee SJ, Cho CY, Kwon MK, Lee SP, Noh DY, Kim DJ, Kim YC, Park SJ (2009) Effect of electron blocking layer on efficiency droop in InGaN/GaN multiple quantum well light-emitting diodes. *Applied Physics Letters* 94:231123.

Hangleiter A, Hitzel F, Netzel C, Fuhrmann D, Rossow U, Ade G, Hinze P (2005) Suppression of nonradiative recombination by V-shaped pits in GaInN/GaN quantum wells produces a large increase in the light emission efficiency. *Physical Review Letters* 95:127402.

Harima H, Inoue T, Nakashima S, Ishida M, Taneya M (1999) Local vibrational modes as a probe of activation process in p-type GaN. *Applied Physics Letters* 75:1383–1385.

Harris JJ, Thomson R, Taylor C, Barlett D, Campion RP, Grant VA, Foxon CT, Kappers MJ (2007) Evaluation of sapphire substrate heating behaviour using GaN band-gap thermometry. *Journal of Crystal Growth* 300:194–198.

Hashimoto R, Hwang J, Saito S, Nunoue S (2013) High-efficiency green-yellow light-emitting diodes grown on sapphire (0001) substrates. *Physica Status Solidi (C)* 10:1529–1532.

Hashimoto R, Hwang J, Saito S, Nunoue S (2014) High-efficiency yellow light-emitting diodes grown on sapphire (0001) substrates. *Physica Status Solidi (C)* 11:628–631.

Hashimoto T, Terakoshi Y, Yuri M, Ishida M, Imafuji O, Sugino T, Itoh K (1999a) Quantitative study of nitridated sapphire surfaces by x-ray photoelectron spectroscopy. *Journal of Applied Physics* 86:3670–3675.

Hashimoto T, Yuri M, Ishida M, Terakoshi Y, Imafuji O, Sugino T, Itoh K (1999b) Reduction of threading dislocations in GaN on sapphire by buffer layer annealing in low-pressure metalorganic chemical vapor deposition. *Japanese Journal of Applied Physics* 38:6605.

Haynes WM, Lide DR (2011) *CRC Handbook of Chemistry and Physics: A Ready-Reference Book of Chemical and Physical Data*. Boca Raton, FL: CRC Press.

Hearne S, Chason E, Han J, Floro JA, Figiel J, Hunter J, Amano H, Tsong IST (1999) Stress evolution during metalorganic chemical vapor deposition of GaN. *Applied Physics Letters* 74:356–358.

Heikman S, Keller S, DenBaars SP, Mishra UK (2002) Growth of Fe doped semi-insulating GaN by metalorganic chemical vapor deposition. *Applied Physics Letters* 81:439–441.

Heying B, Wu XH, Keller S, Li Y, Kapolnek D, Keller BP, DenBaars SP, Speck JS (1996) Role of threading dislocation structure on the x-ray diffraction peak widths in epitaxial GaN films. *Applied Physics Letters* 68:643–645.

Hino T, Tomiya S, Miyajima T, Yanashima K, Hashimoto S, Ikeda M (2000) Characterization of threading dislocations in GaN epitaxial layers. *Applied Physics Letters* 76:3421–3423.

Hiramatsu K, Itoh S, Amano H, Akasaki I, Kuwano N, Shiraishi T, Oki K (1991) Growth-mechanism of GaN grown on sapphire with AlN buffer layer by MOVPE. *Journal of Crystal Growth* 115:628–633.

Hiramatsu K, Kawaguchi Y, Shimizu M, Sawaki N, Zheleva T, Davis RF, Tsuda H, Taki W, Kuwano N, Oki K (1997) The composition pulling effect in MOVPE grown InGaN on GaN and AlGaN and its TEM characterization. *MRS Internet Journal of Nitride Semiconductor Research* 2:e6.

Hiroyasu I, Guang-Yuan Z, Naoyuki N, Takashi E, Takashi J, Masayoshi U (1999) GaN on Si Substrate with AlGaN/AlN Intermediate Layer. *Japanese Journal of Applied Physics* 38:L492.

Ho I-H, Stringfellow GB (1996) Solid phase immiscibility in GaInN. *Applied Physics Letters* 69:2701–2703.

Hoffmann V, Netzel C, Zeimer U, Knauer A, Einfeldt S, Bertram F, Christen J, Weyers M, Tränkle G, Kneissl M (2010) Well width study of InGaN multiple quantum wells for blue–green emitter. *Journal of Crystal Growth* 312:3428–3433.

Holec D, Zhang Y, Rao DVS, Kappers MJ, McAleese C, Humphreys CJ (2008) Equilibrium critical thickness for misfit dislocations in III-nitrides. *Journal of Applied Physics* 104:123514.

Hsu JWP, Lang DV, Richter S, Kleiman RN, Sergent AM, Molnar RJ (2000) Nature of the highly conducting interfacial layer in GaN films. *Applied Physics Letters* 77:2873–2875.

Hu Y-L, Farrell RM, Neufeld CJ, Iza M, Cruz SC, Pfaff N, Simeonov D et al. (2012) Effect of quantum well cap layer thickness on the microstructure and performance of InGaN/GaN solar cells. *Applied Physics Letters* 100:161101.

Huang C-F, Tang T-Y, Huang J-J, Shiao W-Y, Yang CC, Hsu C-W, Chen LC (2006) Prestrained effect on the emission properties of InGaN/GaN quantum-well structures. *Applied Physics Letters* 89:051913.

Huili X, Green DS, Haijiang Y, Mates T, Kozodoy P, Keller S, Denbaars SP, Mishra UK (2003) Memory effect and redistribution of Mg into sequentially regrown GaN layer by metalorganic chemical vapor deposition. *Japanese Journal of Applied Physics, Part 1* 42:50–53.

Hwang J, Hashimoto R, Saito S, Nunoue S (2013) Effects of local structure on optical properties in green-yellow InGaN/GaN quantum wells. In: *Gallium Nitride Materials and Devices VIII*, pp. G-1–6. San Francisco, CA: SPIE.

Iida D, Tamura K, Iwaya M, Kamiyama S, Amano H, Akasaki I (2010) Compensation effect of Mg-doped a- and c-plane GaN films grown by metalorganic vapor phase epitaxy. *Journal of Crystal Growth* 312:3131–3135.

Imura M, Nakano K, Fujimoto N, Okada N, Balakrishnan K, Iwaya M, Kamiyama S et al. (2006a) High-temperature metal-organic vapor phase epitaxial growth of AlN on sapphire by multi transition growth mode method varying V/III ratio. *Japanese Journal of Applied Physics Part 1* 45:8639–8643.

Imura M, Nakano K, Kitano T, Fujimoto N, Okada N, Balakrishnan K, Iwaya M et al. (2006b) Micro-structure of thick AlN grown on sapphire by high-temperature MOVPE. *Physica Status Solidi (A)* 203:1626–1631.

Ishikawa H, Yamamoto K, Egawa T, Soga T, Jimbo T, Umeno M (1998) Thermal stability of GaN on (1 1 1) Si substrate. *Journal of Crystal Growth* 189–190:178–182.

Jacobs K, Daele BV, Leys MR, Moerman I, Tendeloo GV (2003) Effect of growth interrupt and growth rate on MOVPE-grown InGaN/GaN MQW structures. *Journal of Crystal Growth* 248:498–502.

Janotti A, Van de Walle CG (2008) Sources of unintentional conductivity in InN. *Applied Physics Letters* 92:032104.

Jasinski JM, Gates SM (1991) Silicon chemical vapor deposition one step at a time: fundamental studies of silicon hydride chemistry. *Accounts of Chemical Research* 24:9–15.

Jian-Shihn T, Jan-Dar G, Shih-Hsiung C, Ming-Shiann F, Chun-Yen C (1997) Investigation of the indium atom interdiffusion on the growth of GaN/InGaN heterostructures. *Japanese Journal of Applied Physics* 36:1728.

Jiang F, Wang R-V, Munkholm A, Streiffer SK, Stephenson GB, Fuoss PH, Latifi K, Thompson C (2006) Indium adsorption on GaN under metal-organic chemical vapor deposition conditions. *Applied Physics Letters* 89:161915.

Johnson SR, Tiedje T (1997) Effect of substrate thickness, back surface texture, reflectivity, and thin film interference on optical band-gap thermometry. *Journal of Crystal Growth* 175:273–280.

Johnson WC, Parsons JB, Crew MC (1932) Nitrogen compounds of gallium III Gallic nitride. *The Journal of Physical Chemistry* 36:2651–2654.

Johnstone B (2007) *Brilliant!: Shuji Nakamura and the Revolution in Lighting Technology.* Amherst, NY: Prometheus Books, Publishers.

Jong-Il H, Rei H, Shinji S, Shinya N (2014) Development of InGaN-based red LED grown on (0001) polar surface. *Applied Physics Express* 7:071003.

Kadinski L, Merai V, Parekh A, Ramer J, Armour EA, Stall R, Gurary A, Galyukov A, Makarov Y (2004) Computational analysis of GaN/InGaN deposition in MOCVD vertical rotating disk reactors. *Journal of Crystal Growth* 261:175–181.

Kamaladasa RJ, Liu F, Porter LM, Davis RF, Koleske DD, Mulholland G, Jones KA, Picard YN (2011) Identifying threading dislocations in GaN films and substrates by electron channelling. *Journal of Microscopy* 244:311–319.

Kanako S, Tomoyuki T, Jung-Hun C, Shigeyuki K, Takashi H, Ryuji K, Takashi M (2015) Red to blue wavelength emission of N-polar (000-1) InGaN light-emitting diodes grown by metalorganic vapor phase epitaxy. *Applied Physics Express* 8:061005.

Karpiński J, Jun J, Porowski S (1984) Equilibrium pressure of N_2 over GaN and high pressure solution growth of GaN. *Journal of Crystal Growth* 66:1–10.

Karpov SY, Talalaev RA, Makarov YN, Grandjean N, Massies J, Damilano B (2000) Surface kinetics of GaN evaporation and growth by molecular-beam epitaxy. *Surface Science* 450:191–203.

Kato S, Satoh Y, Sasaki H, Masayuki I, Yoshida S (2007) C-doped GaN buffer layers with high breakdown voltages for high-power operation AlGaN/GaN HFETs on 4-in Si substrates by MOVPE. *Journal of Crystal Growth* 298:831–834.

Kaufmann U, Schlotter P, Obloh H, Kohler K, Maier M (2000) Hole conductivity and compensation in epitaxial GaN: Mg layers. *Physical Review B* 62:10867–10872.

Kawaguchi Y, Shimizu M, Hiramatsu K, Sawaki N (1996) The composition pulling effect in InGaN growth on the GaN and AlGaN epitaxial layers grown by MOVPE. *MRS Online Proceedings Library Archive* 449:89.

Kawanishi H, Tomizawa T (2012) Carbon-doped p-type (0001) plane AlGaN (Al?=?6-55%) with high hole density. *Physica Status Solidi B-Basic Solid State Physics* 249:459–463.

Kazumasa H (2001) Epitaxial lateral overgrowth techniques used in group III nitride epitaxy. *Journal of Physics: Condensed Matter* 13:6961.

Keller S, Chichibu SF, Minsky MS, Hu E, Mishra UK, DenBaars SP (1998a) Effect of the growth rate and the barrier doping on the morphology and the properties of InGaN/GaN quantum wells. *Journal of Crystal Growth* 195:258–264.

Keller S, Keller BP, Kapolnek D, Mishra UK, DenBaars SP, Shmagin IK, Kolbas RM, Krishnankutty S (1997a) Growth of bulk InGaN films and quantum wells by atmospheric pressure metalorganic chemical vapour deposition. *Journal of Crystal Growth* 170:349–352.

Keller S, Mishra UK, Denbaars SP (1997b) Flow modulation epitaxy of indium gallium nitride. *Journal of Electronic Materials* 26:1118–1122.

Keller S, Mishra UK, Denbaars SP, Seifert W (1998b) Spiral growth of InGaN nanoscale islands on GaN. *Japanese Journal of Applied Physics* 37:L431.

Khan A, Balakrishnan K, Katona T (2008) Ultraviolet light-emitting diodes based on group three nitrides. *Nature Photonics* 2:77–84.

Khan MA, Skogman RA, Schulze RG, Gershenzon M (1983) Properties and ion implantation of $Al_xGa_{1-x}N$ epitaxial single crystal films prepared by low pressure metalorganic chemical vapor deposition. *Applied Physics Letters* 43:492–494.

Khan MA, Van Hove JM, Kuznia JN, Olson DT (1991) High electron mobility $GaN/Al_xGa_{1-x}N$ hetero-structures grown by low-pressure metalorganic chemical vapor deposition. *Applied Physics Letters* 58:2408–2410.

Killeen KP, Breiland WG (1994) In-situ spectral reflectance monitoring of III-V epitaxy. *Journal of Electronic Materials* 23:179–183.

Kim DJ, Ryu DY, Bojarczuk NA, Karasinski J, Guha S, Lee SH, Lee JH (2000) Thermal activation energies of Mg in GaN: Mg measured by the Hall effect and admittance spectroscopy. *Journal of Applied Physics* 88:2564–2569.

Kim HJ, Choi S, Kim S-S, Ryou J-H, Yoder PD, Dupuis RD, Fischer AM, Sun K, Ponce FA (2010) Improvement of quantum efficiency by employing active-layer-friendly lattice-matched InAlN electron blocking layer in green light-emitting diodes. *Applied Physics Letters* 96:101102.

Kim J, Ji M-H, Detchprohm T, Dupuis RD, Fischer AM, Ponce FA, Ryou J-H (2015) Effect of Group-III precursors on unintentional gallium incorporation during epitaxial growth of InAlN layers by metalorganic chemical vapor deposition. *Journal of Applied Physics* 118:125303.

Kim J, Lochner Z, Ji M-H, Choi S, Kim HJ, Kim JS, Dupuis RD et al. (2014) Origins of unintentional incorporation of gallium in InAlN layers during epitaxial growth, part II: Effects of underlying layers and growth chamber conditions. *Journal of Crystal Growth* 388:143–149.

Kim K-H, Lee S-W, Lee S-N, Kim J (2012) Effect of p-$Al_xGa_{1-x}N$ electron blocking layer on optical and electrical properties in GaN-based light emitting diodes. *Journal of Vacuum Science & Technology B* 30:061204.

Kim S, Lee K, Lee H, Park K, Kim C-S, Son S-J, Yi K-W (2003) The influence of ammonia pre-heating to InGaN films grown by TPIS-MOCVD. *Journal of Crystal Growth* 247:55–61.

King SW, Carlson EP, Therrien RJ, Christman JA, Nemanich RJ, Davis RF (1999) X-ray photoelectron spectroscopy analysis of GaN/(0001)AlN and AlN/(0001)GaN growth mechanisms. *Journal of Applied Physics* 86:5584–5593.

King SW, Davis RF, Nemanich RJ (2008) Kinetics of Ga and In desorption from (7×7) Si(1 1 1) and (3×3) 6H-SiC(0 0 0 1) surfaces. *Surface Science* 602:405–415.

Kinoshita T, Obata T, Yanagi H, Inoue S-I (2013) High p-type conduction in high-Al content Mg-doped AlGaN. *Applied Physics Letters* 102:012105.

Kizilyalli IC, Edwards AP, Aktas O, Prunty T, Bour D (2015) Vertical power p-n diodes based on bulk GaN. *IEEE Transactions on Electron Devices* 62:414–422.

Klein PB, Binari SC, Ikossi K, Wickenden AE, Koleske DD, Henry RL (2001) Current collapse and the role of carbon in AlGaN/GaN high electron mobility transistors grown by metalorganic vapor-phase epitaxy. *Applied Physics Letters* 79:3527–3529.

Kobayashi Y, Akasaka T, Kobayashi N (1998) In-situ monitoring of GaN MOVPE by shallow-angle reflectance using ultraviolet light. *Journal of Crystal Growth* 195:187–191.

Koblmüller G, Gallinat CS, Speck JS (2007) Surface kinetics and thermal instability of N-face InN grown by plasma-assisted molecular beam epitaxy. *Journal of Applied Physics* 101:083516.

Kohler K, Stephan T, Perona A, Wiegert J, Maier M, Kunzer M, Wagner J (2005) Control of the Mg doping profile in III-N light-emitting diodes and its effect on the electroluminescence efficiency. *Journal of Applied Physics* 97:104914.

Koide N, Kato H, Sassa M, Yamasaki S, Manabe K, Hashimoto M, Amano H, Hiramatsu K, Akasaki I (1991) Doping of GaN with Si and properties of blue M/I/N/N_ GaN LED with Si-doped N+-layer by MOVPE. *Journal of Crystal Growth* 115:639–642.

Koleske DD, Coltrin ME, Allerman AA, Cross KC, Mitchell CC, Figiel JJ (2003) In situ measurements of GaN nucleation layer decompostion. *Applied Physics Letters* 82:1170–1172.

Koleske DD, Coltrin ME, Cross KC, Mitchell CC, Allerman AA (2004) Understanding GaN nucleation layer evolution on sapphire. *Journal of Crystal Growth* 273:86–99.

Koleske DD, Coltrin ME, Lee SR, Thaler G, Cross KC, Russell MJ (2007) Understanding GaN nucleation layer evolution on sapphire and its impact on GaN dislocation density. In: *Proceedings of the SPIE*, pp. 68410H–68412. Piscataway, NJ: IEEE.

Koleske DD, Coltrin ME, Russell MJ (2005) Using optical reflectance to measure GaN nucleation layer decomposition kinetics. *Journal of Crystal Growth* 279:37–54.

Koleske DD, Fischer AJ, Allerman AA, Mitchell CC, Cross KC, Kurtz SR, Figiel JJ, Fullmer KW, Breiland WG (2002a) Improved brightness of 380 nm GaN light emitting diodes through intentional delay of the nucleation island coalescence. *Applied Physics Letters* 81:1940–1942.

Koleske DD, Fischer AJ, Bryant BN, Kotula PG, Wierer JJ (2015) On the increased efficiency in InGaN-based multiple quantum wells emitting at 530–590 nm with AlGaN interlayers. *Journal of Crystal Growth* 415:57–64.

Koleske DD, Henry RL, Twigg ME, Culbertson JC, Binari SC, Wickenden AE, Fatemi M (2002b) Influence of AlN nucleation layer temperature on GaN electronic properties grown on SiC. *Applied Physics Letters* 80:4372–4374.

Koleske DD, Lee SR, Crawford MH, Cross KC, Coltrin ME, Kempisty JM (2014a) Connection between GaN and InGaN growth mechanisms and surface morphology. *Journal of Crystal Growth* 391:85–96.

Koleske DD, Lee SR, Thaler G, Crawford MH, Coltrin ME, Cross KC (2010) Indium induced step transformation during InGaN growth on GaN. *Applied Physics Letters* 97:071901.

Koleske DD, Twigg ME, Wickenden AE, Henry RL, Gorman RJ, Freitas JA, Fatemi M (1999a) Properties of Si-doped GaN films grown using multiple AlN interlayers. *Applied Physics Letters* 75:3141–3143.

Koleske DD, Wickenden AE, Henry RL (1999b) GaN Decomposition in Ammonia. *MRS Online Proceedings Library Archive* 595:F99W93.64.

Koleske DD, Wickenden AE, Henry RL, Culbertson JC, Twigg ME (2001) GaN decomposition in H_2 and N_2 at MOVPE temperatures and pressures. *Journal of Crystal Growth* 223:466.

Koleske DD, Wickenden AE, Henry RL, DeSisto WJ, Gorman RJ (1998a) Growth model for GaN with comparison to structural, optical, and electrical properties. *Journal of Applied Physics* 84:1998.

Koleske DD, Wickenden AE, Henry RL, Twigg ME (2002c) Influence of MOVPE growth conditions on carbon and silicon concentrations in GaN. *Journal of Crystal Growth* 242:55–69.

Koleske DD, Wickenden AE, Henry RL, Twigg ME, Culbertson JC, Gorman RJ (1998b) Enhanced GaN decomposition in H_2 near atmospheric pressures. *Applied Physics Letters* 73:2018–2020.

Koleske DD, Wierer Jr JJ, Fischer AJ, Lee SR (2014b) Controlling indium incorporation in InGaN barriers with dilute hydrogen flows. *Journal of Crystal Growth* 390:38–45.

Korotkov RY, Gregie JM, Wessels BW (2001) Electrical properties of p-type GaN:Mg codoped with oxygen. *Applied Physics Letters* 78:222–224.

Korotkov RY, Gregie JM, Wessels BW (2002) Codoping of wide gap epitaxial III-Nitride semiconductors. *Opto-Electronics Review* 10:243–249.

Kozodoy P, Abare A, Sink RK, Mack M, Keller S, DenBaars SP, Mishra UK, Steigerwald D (1997) MOCVD growth of high output power ingan multiple quantum well light emitting diode. *MRS Proceedings* 468:481.

Kozodoy P, Hansen M, DenBaars SP, Mishra UK (1999) Enhanced Mg doping efficiency in Al0.2Ga0.8N/GaN superlattices. *Applied Physics Letters* 74:3681–3683.

Kozodoy P, Keller S, DenBaars S, Mishra UK (1998) MOVPE growth and characterization of Mg-doped GaN. *Journal of Crystal Growth* 195:265–269.

Kozodoy P, Xing HL, DenBaars SP, Mishra UK, Saxler A, Perrin R, Elhamri S, Mitchel WC (2000) Heavy doping effects in Mg-doped GaN. *Journal of Applied Physics* 87:1832–1835.

Krames MR, Shchekin OB, Mueller-Mach R, Mueller GO, Zhou L, Harbers G, Craford MG (2007) Status and future of high-power light-emitting diodes for solid-state lighting. *Journal of Display Technology* 3:160–175.

Kröncke H, Figge S, Aschenbrenner T, Hommel D (2013) Growth of AlN by pulsed and conventional MOVPE. *Journal of Crystal Growth* 381:100–106.

Krost A, Dadgar A, Schulze F, Blasing J, Strassburger G, Clos R, Diez A, Veit P, Hempel T, Christen J (2005b) In situ monitoring of the stress evolution in growing group-III-nitride layers. *Journal of Crystal Growth* 275:209–216.

Krost A, Schulze F, Dadgar A, Strassburger G, Haberland K, Zettler T (2005a) Simultaneous measurement of wafer curvature and true temperature during metalorganic growth of group-III nitrides on silicon and sapphire. *Physica Status Solidi B-Basic Solid State Physics* 242:2570–2574.

Kuech T (2014) *Handbook of Crystal Growth: Thin Films and Epitaxy*. Amsterdam, The Netherlands: Elsevier Science.

Kuech TF, Wang PJ, Tischler MA, Potemski R, Scilla GJ, Cardone F (1988) The control and modeling of doping profiles and transients in MOVPE growth. *Journal of Crystal Growth* 93:624–630.

Kuhn B, Scholz F (2001) An oxygen doped nucleation layer for the growth of high optical quality GaN on sapphire. *Physica Status Solidi (A)* 188:629–633.

Kumakura K, Makimoto T, Kobayashi N (2000a) Activation energy and electrical activity of Mg in Mg-doped $In_xGa_{1-x}N$ (x < 0.2). *Japanese Journal of Applied Physics Part 2 - Letters* 39:L337–L339.

Kumakura K, Makimoto T, Kobayashi N (2000b) High hole concentrations in Mg-doped InGaN grown by MOVPE. *Journal of Crystal Growth* 221:267–270.

Kumakura K, Makimoto T, Kobayashi N (2000c) Efficient hole generation above 10^{19} cm^{-3} in Mg-doped InGaN/GaN superlattices at room temperature. *Japanese Journal of Applied Physics Part 2 - Letters* 39:L195–L196.

Kumar MS, Park JY, Lee YS, Chung SJ, Hong CH, Suh EK (2007) Effect of barrier growth temperature on morphological evolution of green InGaN/GaN multi-quantum well heterostructures. *Journal of Physics D: Applied Physics* 40:5050.

Kwon M-K, Park I-K, Beak S-H, Kim J-Y, Park S-J (2005) Improvement of photoluminescence by Si delta-doping in GaN barrier layer of GaN/$In_xGa_{1-x}N$ multi-quantum wells. *Physica Status Solidi (A)* 202:859–862.

Lada M, Cullis AG, Parbrook PJ (2003) Effect of anneal temperature on GaN nucleation layer transformation. *Journal of Crystal Growth* 258:89–99.

Lahrèche H, Vennéguès P, Tottereau O, Laügt M, Lorenzini P, Leroux M, Beaumont B, Gibart P (2000) Optimisation of AlN and GaN growth by metalorganic vapour-phase epitaxy (MOVPE) on Si (1 1 1). *Journal of Crystal Growth* 217:13–25.

Lang T, Odnoblyudov M, Bougrov V, Sopanen M (2005) MOCVD growth of GaN islands by multistep nucleation layer technique. *Journal of Crystal Growth* 277:64–71.

Lang T, Odnoblyudov M, Bougrov V, Suihkonen S, Sopanen M, Lipsanen H (2006) Morphology optimization of MOCVD-grown GaN nucleation layers by the multistep technique. *Journal of Crystal Growth* 292:26–32.

Langmuir I (1918) The adsorption of gases on plane surfaces of glass, mica, and platinum. *Journal of the American Chemical Society* 40:1361–1403.

Larson DJ, Prosa TJ, Olson D, Lefebvre W, Lawrence D, Clifton PH, Kelly TF (2013) Atom probe tomography of a commercial light emitting diode. *Journal of Physics: Conference Series* 471:012030.

Lee C-Y, Tzou A-J, Lin B-C, Lan Y-P, Chiu C-H, Chi G-C, Chen C-H, Kuo H-C, Lin R-M, Chang C-Y (2014) Efficiency improvement of GaN-based ultraviolet light-emitting diodes with reactive plasma deposited AlN nucleation layer on patterned sapphire substrate. *Nanoscale Research Letters* 9:1–6.

Lee S-J, Song J-C, Park H-J, Park J-B, Jeon S-R, Lee C-R, Jeon D-W, Baek JH (2015) High Brightness, Large Scale GaN Based Light-Emitting Diode Grown on 8-Inch Si Substrate. *ECS Journal of Solid State Science and Technology* 4:Q92–Q95.

Lee SN, Son J, Sakong T, Lee W, Paek H, Yoon E, Kim J, Cho YH, Nam O, Park Y (2004a) Investigation of optical and electrical properties of Mg-doped p-In$_x$Ga$_{1-x}$N, p-GaN and p-Al$_y$Ga$_{1-y}$N grown by MOCVD. *Journal of Crystal Growth* 272:455–459.

Lee SR, Koleske DD, Crawford MH, Wierer Jr JJ (2012) Effect of interface grading and lateral thickness variation on x-ray diffraction by InGaN/GaN multiple quantum wells. *Journal of Crystal Growth* 355:63–72.

Lee SR, Koleske DD, Cross KC, Floro JA, Waldrip KE, Wise AT, Mahajan S (2004b) In situ measurements of the critical thickness for strain relaxation in AlGaN/GaN heterostructures. *Applied Physics Letters* 85:6164–6166.

Lee SR, West AM, Allerman AA, Waldrip KE, Follstaedt DM, Provencio PP, Koleske DD, Abernathy CR (2005) Effect of threading dislocations on the Bragg peakwidths of GaN, AlGaN, and AlN heterolayers. *Applied Physics Letters* 86:241904.

Lee W, Limb J, Ryou JH, Yoo D, Ewing MA, Korenblit Y, Dupuis RD (2007) Nitride-based green light-emitting diodes with various p-type layers. *Journal of Display Technology* 3:126–132.

Leem SJ, Shin YC, Kim KC, Kim EH, Sung YM, Moon Y, Hwang SM, Kim TG (2008) The effect of the low-mole InGaN structure and InGaN/GaN strained layer superlattices on optical performance of multiple quantum well active layers. *Journal of Crystal Growth* 311:103–106.

Li T, Wei QY, Fischer AM, Huang JY, Huang YU, Ponce FA, Liu JP, Lochner Z, Ryou J-H, Dupuis RD (2013) The effect of InGaN underlayers on the electronic and optical properties of InGaN/GaN quantum wells. *Applied Physics Letters* 102:041115.

Li X, Danielsson Ö, Pedersen H, Janzén E, Forsberg U (2015) Precursors for carbon doping of GaN in chemical vapor deposition. *Journal of Vacuum Science & Technology B* 33:021208.

Li X, Zhao DG, Jiang DS, Chen P, Liu ZS, Zhu JJ, Yang J et al. (2016) The effectiveness of electron blocking layer in InGaN-based laser diodes with different indium content. *Physica Status Solidi (A)* 213:2223–2228.

Liliental-Weber Z, Benamara M, Swider W, Washburn J, Grzegory I, Porowski S, Dupuis RD, Eiting CJ (1999) Ordering in bulk GaN: Mg samples: Defects caused by Mg doping. *Physica B-Condensed Matter* 273:124–129.

Liliental-Weber Z, Benamara M, Washburn J, Domagala JZ, Bak-Misiuk J, Piner EL, Roberts JC, Bedair SM (2001) Relaxation of InGaN thin layers observed by x-ray and transmission electron microscopy studies. *Journal of Electronic Materials* 30:439–444.

Lin Z, Zhang J, Cao R, Ha W, Zhang S, Chen X, Yan J et al. (2013) Effect of growth temperature on the impurity incorporation and material properties of N-polar GaN films grown by metal-organic chemical vapor deposition. *Journal of Crystal Growth* 384:96–99.

Liu L, Edgar JH (2002) Substrates for gallium nitride epitaxy. *Materials Science and Engineering: R-Reports* 37:61–127.

Liu S, Stevenson D (1978) Growth kinetics and catalytic effects in the vapor phase epitaxy of gallium nitride. *Journal of Electrochemical Society* 125:1161–1169.

Liu Z, Ma J, Yi X, Guo E, Wang L, Wang J, Lu N, Li J, Ferguson I, Melton A (2012) p-InGaN/AlGaN electron blocking layer for InGaN/GaN blue light-emitting diodes. *Applied Physics Letters* 101:261106.

Liu Z, Yi X, Yu Z, Yuan G, Liu Y, Wang J, Li J, Lu N, Ferguson I, Zhang Y (2016) Impurity resonant states p-type doping in wide-band-gap nitrides. *Scientific Reports* 6:19537.

Look DC, Reynolds DC, Hemsky JW, Sizelove JR, Jones RL, Molnar RJ (1997) Defect donor and acceptor in GaN. *Physical Review Letters* 79:2273–2276.

Lorenz K, Gonsalves M, Kim W, Narayanan V, Mahajan S (2000) Comparative study of GaN and AlN nucleation layers and their role in growth of GaN on sapphire by metalorganic chemical vapor deposition. *Applied Physics Letters* 77:3391–3393.

Lu W, Aplin D, Clawson AR, Yu PKL (2013) Effects of the gas ambient in thermal activation of Mg-doped p-GaN on Hall effect and photoluminescence. *Journal of Vacuum Science & Technology A* 31:011502.

Lundin WV, Sakharov AV, Zavarin EE, Kazantsev DY, Ber BY, Yagovkina MA, Brunkov PN, Tsatsulnikov AF (2016) Study of GaN doping with carbon from propane in a wide range of MOVPE conditions. *Journal of Crystal Growth* 449:108–113.

Luong TT, Ho YT, Tran BT, Woong YY, Chang EY (2015) Barrier strain and carbon incorporation-engineered performance improvements for AlGaN/GaN high electron mobility transistors. *Chemical Vapor Deposition* 21:33–40.

Lyons JL, Janotti A, Van de Walle CG (2014) Effects of carbon on the electrical and optical properties of InN, GaN, and AlN. *Physical Review B* 89:035204.

Madar R, Michel D, Jacob G, Boulou M (1977) Growth anisotropy in GaN-Al$_2$O$_3$ system. *Journal of Crystal Growth* 40:239–252.

Manasevit HM (1972) The use of metalorganics in the preparation of semiconductor materials: Growth on insulating substrates. *Journal of Crystal Growth* 13–14:306–314.

Manasevit HM (1981) Recollections and reflections of MOCVD. *Journal of Crystal Growth* 55:1–9.

Marchand H, Ibbetson JP, Fini PT, Keller S, DenBaars SP, Speck JS, Mishra UK (1998) Mechanisms of lateral epitaxial overgrowth of gallium nitride by metalorganic chemical vapor deposition. *Journal of Crystal Growth* 195:328–332.

Markus P, Abdul K, Michael K (2013) Surface transitions during InGaN growth on GaN(0001) in metal-organic vapor phase epitaxy. *Japanese Journal of Applied Physics* 52:08JB23.

Maruska HP, Rhines WC (2015) A modern perspective on the history of semiconductor nitride blue light sources. *Solid-State Electronics* 111:32–41.

Maruska HP, Stevenson DA, Pankove JI (1973) Violet luminescence of Mg-doped GaN. *Applied Physics Letters* 22:303–305.

Maruska HP, Tietjen JJ (1969) The preparation and properties of vapor-deposited single-crystalline GaN. *Applied Physics Letters* 15:327–329.

Masaya S, Yasutoshi K, Kazumasa H, Nobuhiko S (1997) Metalorganic vapor phase epitaxy of thick InGaN on sapphire substrate. *Japanese Journal of Applied Physics* 36:3381.

Massabuau FCP, Davies MJ, Blenkhorn WE, Hammersley S, Kappers MJ, Humphreys CJ, Dawson P, Oliver RA (2015) Investigation of unintentional indium incorporation into GaN barriers of InGaN/GaN quantum well structures. *Physica Status Solidi (B)* 252:928–935.

Massabuau FC-P, Sahonta S-L, Trinh-Xuan L, Rhode S, Puchtler TJ, Kappers MJ, Humphreys CJ, Oliver RA (2012) Morphological, structural, and emission characterization of trench defects in InGaN/GaN quantum well structures. *Applied Physics Letters* 101:212107.

Masui H, Nakamura S, DenBaars SP, Mishra UK (2010) Nonpolar and semipolar III-nitride light-emitting diodes: Achievements and challenges. *IEEE Transactions on Electron Devices* 57:88–100.

Matsumoto K, Tokunaga H, Ubukata A, Ikenaga K, Fukuda Y, Tabuchi T, Kitamura Y, Koseki S, Yamaguchi A, Uematsu K (2008) High growth rate metal organic vapor phase epitaxy GaN. *Journal of Crystal Growth* 310:3950–3952.

Matsuoka T (1997) Calculation of unstable mixing region in wurtzite In$_{1-x-y}$Ga$_x$Al$_y$N. *Applied Physics Letters* 71:105–106.

Matsuoka T, Liu Y, Kimura T, Zhang Y, Prasertsuk K, Katayama R (2011) Paving the way to high-quality indium nitride: the effects of pressurized reactor. In: *SPIE OPTO*, p. 19. San Francisco, CA: SPIE.

Matsuoka T, Tanaka H, Sasaki T, Katsui K (1990) Wide-gap semiconductor (In, Ga)N. In: *Proceedings of the 16th International Symposium on GaAs and Related Compounds*, Karuizawa, Japan, pp. 141–146. Bristol: Institute of Physics.

Matsuoka T, Yoshimoto N, Sasaki T, Katsui A (1992) Wide-gap semiconductor InGaN and InGaAlN grown by MOVPE. *Journal of Electronic Materials* 21:157–163.

Mauder C, Reuters B, Behmenburg H, De Souza RA, Woitok JF, Chou MMC, Heuken M, Kalisch H, Jansen RH (2011) Mechanisms of impurity incorporation during MOVPE growth of m-plane GaN layers on LiAlO2. *Physica Status Solidi (C)* 8:2050–2052.

McCluskey M, Romano L, Krusor B, Bour D, Johnson N, Brennan S (1998) Phase separation in InGaN/ GaN multiple quantum wells. *Applied Physics Letters* 72:1730–1732.

Mihopoulos TG, Gupta V, Jensen KF (1998) A reaction-transport model for AlGaN MOVPE growth. *Journal of Crystal Growth* 195:733–739.

Miraglia PQ, Preble EA, Roskowski AM, Einfeldt S, Lim SH, Liliental-Weber Z, Davis RF (2003) Helical-type surface defects in InGaN thin films epitaxially grown on GaN templates at reduced temperatures. *Thin Solid Films* 437:140–149.

Mita S, Collazo R, Rice A, Dalmau RF, Sitar Z (2008) Influence of gallium supersaturation on the properties of GaN grown by metalorganic chemical vapor deposition. *Journal of Applied Physics* 104:013521.

Mitchell CC, Coltrin ME, Han J (2001) Mass transport in the epitaxial lateral overgrowth of gallium nitride. *Journal of Crystal Growth* 222:144–153.

Miura A, Nagai T, Senda R, Kawashima T, Iwaya M, Kamiyama S, Amano H, Akasaki I (2008) Realization of low-dislocation-density, smooth surface, and thick GaInN films on m-plane GaN templates. *Journal of Crystal Growth* 310:3308–3312.

Monemar B, Pozina G (2000) Group III-nitride based hetero and quantum structures. *Progress in Quantum Electronics* 24:239–290.

Monnery WD, Hawboldt KA, Pollock AE, Svrcek WY (2001) Ammonia pyrolysis and oxidation in the claus furnace. *Industrial & Engineering Chemistry Research* 40:144–151.

Moram MA, Sadler TC, Häberlen M, Kappers MJ, Humphreys CJ (2010) Dislocation movement in GaN films. *Applied Physics Letters* 97:261907.

Moram MA, Vickers ME (2009) X-ray diffraction of III-nitrides. *Reports on Progress in Physics* 72:036502.

Moran B, Wu F, Romanov AE, Mishra UK, Denbaars SP, Speck JS (2004) Structural and morphological evolution of GaN grown by metalorganic chemical vapor deposition on SiC substrates using an AlN initial layer. *Journal of Crystal Growth* 273:38–47.

Motoaki I, Tetsuya T, Shigeo Y, Christian W, Hiroshi A, Isamu A (1998) Reduction of etch pit density in organometallic vapor phase epitaxy-grown GaN on sapphire by insertion of a low-temperature-deposited buffer layer between high-temperature-grown GaN. *Japanese Journal of Applied Physics* 37:L316.

Moustakas TD (1997) Highly insulating monocrystalline gallium nitride thin films. U.S. Patent 5,686,738, filed January 13, 1995, and issued November 11, 1997.

Muthusamy Senthil K, Jae Young P, Yong Seok L, Sang Jo C, Chang-Hee H, Eun-Kyung S (2008) Improved internal quantum efficiency of green emitting InGaN/GaN multiple quantum wells by in preflow for InGaN well growth. *Japanese Journal of Applied Physics* 47:839.

Myers SM, Vaandrager BL, Wampler WR, Seager CH (2004) Influence of ambient on hydrogen release from p-type gallium nitride. *Journal of Applied Physics* 95:76–83.

Myers SM, Wright AF, Petersen GA, Wampler WR, Seager CH, Crawford MH, Han J (2001) Diffusion, release, and uptake of hydrogen in magnesium-doped gallium nitride: Theory and experiment. *Journal of Applied Physics* 89:3195–3202.

Nakamura F, Hashimoto S, Hara M, Imanaga S, Ikeda M, Kawai H (1998) AlN and AlGaN growth using low-pressure metalorganic chemical vapor deposition. *Journal of Crystal Growth* 195:280–285.

Nakamura S (1991a) Analysis of real-time monitoring using interference effects. *Japanese Journal of Applied Physics Part 1* 30:1348–1353.

Nakamura S (1991b) Insitu monitoring of GaN growth using interference effects. *Japanese Journal of Applied Physics Part 1* 30:1620–1627.

Nakamura S (1991c) GaN growth using GaN buffer layer. *Japanese Journal of Applied Physics Part 2* 30:L1705–L1707.

Nakamura S (2015) Nobel lecture: Background story of the invention of efficient blue InGaN light emitting diodes. *Reviews of Modern Physics* 87:1139–1151.

Nakamura S, Harada Y, Seno M (1991a) Novel metalorganic chemical vapor-deposition system for GaN growth. *Applied Physics Letters* 58:2021–2023.

Nakamura S, Mukai T (1992) High-quality InGaN films grown on GaN films. *Japanese Journal of Applied Physics* 31:L1457.

Nakamura S, Mukai T, Senoh M (1992a) Si-doped and Ge-doped GaN films grown with GaN buffer layers. *Japanese Journal of Applied Physics Part 1* 31:2883–2888.

Nakamura S, Mukai T, Senoh M (1992b) Insitu monitoring and hall measurements of GaN grown with GaN buffer layers. *Journal of Applied Physics* 71:5543–5549.

Nakamura S, Mukai T, Senoh M (1993a) Si-doped InGaN films grown on GaN films. *Japanese Journal of Applied Physics* 32:L16.

Nakamura S, Mukai T, Senoh M (1994) Candela-class high-brightness InGaN/AlGaN double-heterostructure blue-light-emitting diodes. *Applied Physics Letters* 64:1687–1689.

Nakamura S, Mukai T, Senoh M, Iwasa N (1992c) Thermal annealing effects on p-type Mg-doped GaN films. *Japanese Journal of Applied Physics Part 2* 31:L139–L142.

Nakamura S, Senoh M, Iwasa N, Nagahama S-I, Yamada T, Mukai T (1995) Superbright green InGaN single-quantum-well-structure light-emitting diodes. *Japanese Journal of Applied Physics* 34:L1332.

Nakamura S, Senoh M, Mukai T (1991b) Highly P-typed Mg-doped GaN films grown with GaN buffer layers. *Japanese Journal of Applied Physics, Part 2* 30:L1708–1711.

Nakamura S, Senoh M, Mukai T (1993b) High-power InGaN/GaN double-heterostructure violet light emitting diodes. *Applied Physics Letters* 62:2390–2392.

Nakamura S, Senoh M, Mukai T (1993c) P-GaN/N-InGaN/N-GaN double-heterostructure blue-light-emitting diodes. *Japanese Journal of Applied Physics* 32:L8.

Nakamura S, Senoh M, Nagahama S-I, Iwasa N, Yamada T, Matsushita T, Kiyoku H, Sugimoto Y (1996) InGaN-based multi-quantum-well-structure laser diodes. *Japanese Journal of Applied Physics* 35:L74.

Nam O-H, Bremser MD, Zheleva TS, Davis RF (1997) Lateral epitaxy of low defect density GaN layers via organometallic vapor phase epitaxy. *Applied Physics Letters* 71:2638–2640.

Nanhui N, Huaibing W, Jianping L, Naixin L, Yanhui X, Jun H, Jun D, Guangdi S (2006) Improved quality of InGaN/GaN multiple quantum wells by a strain relief layer. *Journal of Crystal Growth* 286:209–212.

Nanhui N, Huaibing W, Jianping L, Naixin L, Yanhui X, Jun H, Jun D, Guangdi S (2007) Enhanced luminescence of InGaN/GaN multiple quantum wells by strain reduction. *Solid-State Electronics* 51:860–864.

Narukawa Y, Kawakami Y, Funato M, Fujita S, Fujita S, Nakamura S (1997) Role of self-formed InGaN quantum dots for exciton localization in the purple laser diode emitting at 420 nm. *Applied Physics Letters* 70:981–983.

Natalie F, Hitoshi S, Hisashi M, Steven PD, Shuji N (2008) Increased polarization ratio on semipolar (1122) InGaN/GaN light-emitting diodes with increasing indium composition. *Japanese Journal of Applied Physics* 47:7854.

Northrup JE (2009) GaN and InGaN(1122) surfaces: Group-III adlayers and indium incorporation. *Applied Physics Letters* 95:133107.

Obloh H, Bachem KH, Kaufmann U, Kunzer M, Maier M, Ramakrishnan A, Schlotter P (1998) Self-compensation in Mg doped p-type GaN grown by MOCVD. *Journal of Crystal Growth* 195:270–273.

Oila J, Saarinen K, Wickenden AE, Koleske DD, Henry RL, Twigg ME (2003) Ga vacancies and grain boundaries in GaN. *Applied Physics Letters* 82:1021–1023.

Okada N, Kashihara H, Sugimoto K, Yamada Y, Tadatomo K (2015) Controlling potential barrier height by changing V-shaped pit size and the effect on optical and electrical properties for InGaN/GaN based light-emitting diodes. *Journal of Applied Physics* 117:025708.

Olaizola SM, Pendlebury ST, Neill JPO, Mowbray DJ, Cullis AG, Skolnick MS, Parbrook PJ, Fox AM (2002) Influence of GaN barrier growth temperature on the photoluminescence of InGaN/GaN heterostructures. *Journal of Physics D: Applied Physics* 35:599.

Oliver RA, Kappers MJ, Humphreys CJ (2008) Gross well-width fluctuations in InGaN quantum wells. *Physica Status Solidi (C)* 5:1475–1481.

Oliver RA, Kappers MJ, Humphreys CJ, Briggs GAD (2004) The influence of ammonia on the growth mode in InGaN/GaN heteroepitaxy. *Journal of Crystal Growth* 272:393–399.

Oliver RA, Kappers MJ, Humphreys CJ, Briggs GAD (2005) Growth modes in heteroepitaxy of InGaN on GaN. *Journal of Applied Physics* 97:013707.

Oliver RA, Kappers MJ, Sumner J, Datta R, Humphreys CJ (2006) Highlighting threading dislocations in MOVPE-grown GaN using an in situ treatment with SiH_4 and NH_3. *Journal of Crystal Growth* 289:506–514.

Oliver RA, Massabuau FC-P, Kappers MJ, Phillips WA, Thrush EJ, Tartan CC, Blenkhorn WE et al. (2013) The impact of gross well width fluctuations on the efficiency of GaN-based light emitting diodes. *Applied Physics Letters* 103:141114.

Oliver RA, Sumner J, Kappers MJ, Humphreys CJ (2009) Morphological changes of InGaN epilayers during annealing assessed by spectral analysis of atomic force microscopy images. *Journal of Applied Physics* 106:054319.

Pantha BN, Sedhain A, Li J, Lin JY, Jiang HX (2009) Electrical and optical properties of p-type InGaN. *Applied Physics Letters* 95:261904.

Parikh RP, Adomaitis RA (2006) An overview of gallium nitride growth chemistry and its effect on reactor design: Application to a planetary radial-flow CVD system. *Journal of Crystal Growth* 286:259–278.

Parish G, Keller S, Denbaars SP, Mishra UK (2000) SIMS investigations into the effect of growth conditions on residual impurity and silicon incorporation in GaN and $Al_xGa_{1-x}N$. *Journal of Electronic Materials* 29:15–20.

Park JH, Yeong Kim D, Hwang S, Meyaard D, Fred Schubert E, Dae Han Y, Won Choi J, Cho J, Kyu Kim J (2013) Enhanced overall efficiency of GaInN-based light-emitting diodes with reduced efficiency droop by Al-composition-graded AlGaN/GaN superlattice electron blocking layer. *Applied Physics Letters* 103:061104.

Paskova T, Preble EA, Clites TI, Hanser AD, Evans KR (2012) Group III nitride templates and related heterostructures, devices, and methods for making them. U.S. Patent 20120235161, filed May 6, 2008, and issued September 20, 2012.

Pendlebury ST, Parbrook PJ, Mowbray DJ, Wood DA, Lee KB (2007) InGaN/GaN quantum wells with low growth temperature GaN cap layers. *Journal of Crystal Growth* 307:363–366.

Pereira S, Correia MR, Pereira E, O'Donnell KP, Alves E, Sequeira AD, Franco N, Watson IM, Deatcher CJ (2002) Strain and composition distributions in wurtzite InGaN/GaN layers extracted from x-ray reciprocal space mapping. *Applied Physics Letters* 80:3913–3915.

Pierre G (2004) Metal organic vapour phase epitaxy of GaN and lateral overgrowth. *Reports on Progress in Physics* 67:667.

Piner EL, Behbehani MK, ElMasry NA, McIntosh FG, Roberts JC, Boutros KS, Bedair SM (1997) Effect of hydrogen on the indium incorporation in InGaN epitaxial films. *Applied Physics Letters* 70:461.

Piprek J (2012) AlGaN polarization doping effects on the efficiency of blue LEDs. In: *SPIE OPTO*, pp. 82620E–82611. International Society for Optics and Photonics. San Francisco, CA: SPIE.

Polyakov AY, Smirnov NB, Govorkov AV, Pearton SJ (2004) Properties of Fe-doped semi-insulating GaN structures. *Journal of Vacuum Science and Technology B* 22:120–125.

Prall C, Ruebesam M, Weber C, Reufer M, Rueter D (2014) Photoluminescence from GaN layers at high temperatures as a candidate for in situ monitoring in MOVPE. *Journal of Crystal Growth* 397:24–28.

Qi YD, Musante C, Lau KM, Smith L, Odedra R, Kanjolia R (2001) OMVPE growth of P-type GaN using solution Cp_2Mg. *Journal of Electronic Materials* 30:1382–1386.

Qu BZ, Zhu QS, Sun XH, Wan SK, Wang ZG, Nagai H, Kawaguchi Y, Hiramatsu K, Sawaki N (2003) Photoluminescence of Mg-doped GaN grown by metalorganic chemical vapor deposition. *Journal of Vacuum Science & Technology A* 21:838–841.

Rajan S, DenBaars SP, Mishra UK, Xing H, Jena D (2006) Electron mobility in graded AlGaN alloys. *Applied Physics Letters* 88:042103.

Ramana Murty MV, Fini P, Stephenson GB, Thompson C, Eastman JA, Munkholm A, Auciello O, Jothilingam R, DenBaars SP, Speck JS (2000) Step bunching on the vicinal GaN(0001) surface. *Physical Review B* 62:R10661–R10664.

Rask M, Landgren G, Andersson SG, Lundberg A (1988) Abrupt p-type doping transition using bis-(cyclopentadienyl)-magnesium in metal-organic vapor-phase epitaxy of GaAs. *Journal of Electronic Materials* 17:311–314.

Ren X, Riley JR, Koleske DD, Lauhon LJ (2015) Correlated high-resolution x-ray diffraction, photoluminescence, and atom probe tomography analysis of continuous and discontinuous InxGa1–xN quantum wells. *Applied Physics Letters* 107:022107.

Reshchikov MA, Morkoç H (2005) Luminescence properties of defects in GaN. *Journal of Applied Physics* 97:061301.

Ribič PR, Bratina G (2007) Behavior of the (0 0 0 1) surface of sapphire upon high-temperature annealing. *Surface Science* 601:44–49.

Richard M-I, Highland MJ, Fister TT, Munkholm A, Mei J, Streiffer SK, Thompson C, Fuoss PH, Stephenson GB (2010) In situ synchrotron x-ray studies of strain and composition evolution during metal-organic chemical vapor deposition of InGaN. *Applied Physics Letters* 96:051911.

Romano LT, Kneissl M, Northrup JE, Van de Walle CG, Treat DW (2001) Influence of microstructure on the carrier concentration of Mg-doped GaN films. *Applied Physics Letters* 79:2734–2736.

Romano LT, McCluskey MD, Krusor BS, Bour DP, Chua C, Brennan S, Yu KM (1998) Phase separation in annealed InGaN/GaN multiple quantum wells. *Journal of Crystal Growth* 189–190:33–36.

Romano LT, Myers TH (1997) The influence of inversion domains on surface morphology in GaN grown by molecular beam epitaxy. *Applied Physics Letters* 71:3486–3488.

Romano LT, Northrup JE, Okeefe MA (1996) Inversion domains in GaN grown on sapphire. *Applied Physics Letters* 69:2394–2396.

Romano LT, Van de Walle CG, Ager JW, Götz W, Kern RS (2000) Effect of Si doping on strain, cracking, and microstructure in GaN thin films grown by metalorganic chemical vapor deposition. *Journal of Applied Physics* 87:7745–7752.

Rongen R, Leys MR, Hall PJ, Es CM, Vonk H, Wolter JH (1995) Investigations on indium phosphide grown by chemical beam epitaxy. *Journal of Electronic Materials* 24:1391–1398.

Ru-Chin T, Chun-Ju T, Shyi-Ming P, Chang-Cheng C, Sheu JK, Ching-En T, Te-Chung W, Gou-Chung C (2003) Improvement of near-ultraviolet InGaN-GaN light-emitting diodes with an AlGaN electron-blocking layer grown at low temperature. *IEEE Photonics Technology Letters* 15:1342–1344.

Rudziński M, Desmaris V, van Hal PA, Weyher JL, Hageman PR, Dynefors K, Rödle TC, Jos HFF, Zirath H, Larsen PK (2006) Growth of Fe doped semi-insulating GaN on sapphire and 4H-SiC by MOCVD. *Physica Status Solidi (C)* 3:2231–2236.

Saarinen K, Laine T, Kuisma S, Nissilä J, Hautojärvi P, Dobrzynski L, Baranowski JM et al. (1997) Observation of native Ga vacancies in GaN by positron annihilation. *Physical Review Letters* 79:3030–3033.

Saarinen K, Seppälä P, Oila J, Hautojärvi P, Corbel C, Briot O, Aulombard RL (1998) Gallium vacancies and the growth stoichiometry of GaN studied by positron annihilation spectroscopy. *Applied Physics Letters* 73:3253–3255.

Sakai M, Ishikawa H, Egawa T, Jimbo T, Umeno M, Shibata T, Asai K et al. (2002) Growth of high-quality GaN films on epitaxial AlN/sapphire templates by MOVPE. *Journal of Crystal Growth* 244:6–11.

Sakuta H, Kawano Y, Yamanaka Y, Kurai S, Taguchi T (2005) Diffusion of In atoms in InGaN ultra-thin films during post-growth thermal annealing by high-resolution Rutherford backscattering spectrometry. *Physica Status Solidi (C)* 2:2407–2410.

Sato H, Chung RB, Hirasawa H, Fellows N, Masui H, Wu F, Saito M et al. (2008) Optical properties of yellow light-emitting diodes grown on semipolar (11-22) bulk GaN substrates. *Applied Physics Letters* 92:221110.

Schenk HPD, de Mierry P, Laügt M, Omnès F, Leroux M, Beaumont B, Gibart P (1999) Indium incorporation above 800°C during metalorganic vapor phase epitaxy of InGaN. *Applied Physics Letters* 75:2587–2589.

Scholz F, Sohmer A, Off J, Syganow V, Dörnen A, Im JS, Hangleiter A, Lakner H (1997) In incorporation efficiency and composition fluctuations in MOVPE grown GaInN/GaN hetero structures and quantum wells. *Materials Science and Engineering: B* 50:238–244.

Schubert EF (2006) *Light-Emitting Diodes.* Cambridge: Cambridge University Press.

Schubert EF, Grieshaber W, Goepfert ID (1996) Enhancement of deep acceptor activation in semiconductors by superlattice doping. *Applied Physics Letters* 69:3737–3739.

Seager CH, Myers SM, Wright AF, Koleske DD, Allerman AA (2002) Drift, diffusion, and trapping of hydrogen in p-type GaN. *Journal of Applied Physics* 92:7246–7252.

Shan Hsu P, Hardy MT, Young EC, Romanov AE, DenBaars SP, Nakamura S, Speck JS (2012) Stress relaxation and critical thickness for misfit dislocation formation in (10-10) and (30-31) InGaN/GaN heteroepitaxy. *Applied Physics Letters* 100:171917.

Sharma N, Thomas P, Tricker D, Humphreys C (2000) Chemical mapping and formation of V-defects in InGaN multiple quantum wells. *Applied Physics Letters* 77:1274–1276.

Shekhar R, Jensen KF (1997) Temperature programmed desorption investigations of hydrogen and ammonia reactions on GaN. *Surface Science* 381:L581–L588.

Shen YC, Mueller GO, Watanabe S, Gardner NF, Munkholm A, Krames MR (2007) Auger recombination in InGaN measured by photoluminescence. *Applied Physics Letters* 91:141101.

Shinji S, Rei H, Jongil H, Shinya N (2013) InGaN light-emitting diodes on c-face sapphire substrates in green gap spectral range. *Applied Physics Express* 6:111004.

Simon J, Protasenko V, Lian C, Xing H, Jena D (2010) Polarization-induced hole doping in wide–band-gap uniaxial semiconductor heterostructures. *Science* 327:60–64.

Son JK, Lee SN, Sakong T, Paek HS, Nam O, Park Y, Hwang JS, Kim JY, Cho YH (2006) Enhanced optical properties of InGaN MQWs with InGaN underlying layers. *Journal of Crystal Growth* 287:558–561.

Sonderegger S, Feltin E, Merano M, Crottini A, Carlin JF, Sachot R, Deveaud B, Grandjean N, Ganière JD (2006) High spatial resolution picosecond cathodoluminescence of InGaN quantum wells. *Applied Physics Letters* 89:232109.

Speck JS, Rosner SJ (1999) The role of threading dislocations in the physical properties of GaN and its alloys. *Physica B: Condensed Matter* 273–274:24–32.

Stephan T, Köhler K, Kunzer M, Schlotter P, Wagner J (2003) Electroluminescence efficiency of InGaN light emitting diodes: Dependence on AlGaN:Mg electron blocking layer width and Mg doping profile. *Physica Status Solidi (C)* 7:2198–2201.

Stephan T, Koehler K, Maier M, Kunzer M, Schlotter P, Wagner J (2004) Influence of Mg doping profile on the electroluminescence properties of GaInN multiple-quantum-well light-emitting diodes. In: *Light-Emitting Diodes: Research, Manufacturing, and Applications VIII*, pp. 118–126. San Francisco, CA: SPIE.

Stocker DA, Schubert EF, Redwing JM (1998) Crystallographic wet chemical etching of GaN. *Applied Physics Letters* 73:2654–2656.

Stoney GG (1909) The tension of metallic films deposited by electrolysis. *Proceedings of the Royal Society of London Series A* 82:172–175.

Sugiura L, Itaya K, Nishio J, Fujimoto H, Kokubun Y (1997) Effects of thermal treatment of low-temperature GaN buffer layers on the quality of subsequent GaN layers. *Journal of Applied Physics* 82:4877–4882.

Sumiya M, Yoshimura K, Ohtsuka K, Fuke S (2000) Dependence of impurity incorporation on the polar direction of GaN film growth. *Applied Physics Letters* 76:2098–2100.

Sun Q, Yerino CD, Ko TS, Cho YS, Lee I-H, Han J, Coltrin ME (2008) Understanding nonpolar GaN growth through kinetic Wulff plots. *Journal of Applied Physics* 104:093523.

Suzuki M, Nishio J, Onomura M, Hongo C (1998) Doping characteristics and electrical properties of Mg-doped AlGaN grown by atmospheric-pressure MOCVD. *Journal of Crystal Growth* 189–190:511–515.

Svensk O, Suihkonen S, Lang T, Lipsanen H, Sopanen M, Odnoblyudov MA, Bougrov VE (2007) Effect of growth conditions on electrical properties of Mg-doped p-GaN. *Journal of Crystal Growth* 298:811–814.

Takao N, Takeshi K, Hiroyuki M, Osamu O (1989) Properties of Ga$_{1-x}$In$_x$N films prepared by MOVPE. *Japanese Journal of Applied Physics* 28:L1334.

Takaya Y, Tomoyuki T, Yoshio H, Masahito Y, Hiroshi A (2013) Growth of GaN on Si(111) substrates via a reactive-sputter-deposited AlN intermediate layer. *Japanese Journal of Applied Physics* 52:08JB16.

Takeyoshi O, Yoshimasa U, Eun-Kyung S, Hyung-Jae L, Takayuki S, Shigefusa FC (2003) Improved emission efficiency in InGaN/GaN quantum wells with compositionally-graded barriers studied by time-resolved photoluminescence spectroscopy. *Japanese Journal of Applied Physics* 42:L1369.

Tanaka T, Watanabe A, Amano H, Kobayashi Y, Akasaki I, Yamazaki S, Koike M (1994) p-type conduction in Mg-doped GaN and Al0.08Ga0.92N grown by metalorganic vapor phase epitaxy. *Applied Physics Letters* 65:593–594.

Taniyasu Y, Kasu M, Kobayashi N (2002) Intentional control of n-type conduction for Si-doped AlN and Al$_x$Ga$_{1-x}$N (0.42 <= x < 1). *Applied Physics Letters* 81:1255–1257.

Taniyasu Y, Kasu M, Makimoto T (2006a) Increased electron mobility in n-type Si-doped AlN by reducing dislocation density. *Applied Physics Letters* 89:182112.

Taniyasu Y, Kasu M, Makimoto T (2006b) An aluminium nitride light-emitting diode with a wavelength of 210 nanometres. *Nature* 441:325–328.

Tarsa EJ, Heying B, Wu XH, Fini P, DenBaars SP, Speck JS (1997) Homoepitaxial growth of GaN under Ga-stable and N-stable conditions by plasma-assisted molecular beam epitaxy. *Journal of Applied Physics* 82:5472–5479.

Thaler GT, Koleske DD, Lee SR, Bogart KHA, Crawford MH (2010) Thermal stability of thin InGaN films on GaN. *Journal of Crystal Growth* 312:1817–1822.

Thapa SB, Hertkorn J, Scholz F, Prinz GM, Feneberg M, Schirra M, Thonke K, Sauer R, Biskupek J, Kaiser U (2008) MOVPE growth of high quality AlN layers and effects of Si doping. *Physica Status Solidi (C)* 5:1774–1776.

Thon A, Kuech TF (1996) High temperature adduct formation of trimethylgallium and ammonia. *Applied Physics Letters* 69:55–57.

Tim W, Lukas S, Carsten N, Jens R, Veit H, Simon P, Arne K, Markus W, Ulrich S, Michael K (2012) Indium incorporation and emission wavelength of polar, nonpolar and semipolar InGaN quantum wells. *Semiconductor Science and Technology* 27:024014.

Ting SM, Ramer JC, Florescu DI, Merai VN, Albert BE, Parekh A, Lee DS et al. (2003) Morphological evolution of InGaN/GaN quantum-well heterostructures grown by metalorganic chemical vapor deposition. *Journal of Applied Physics* 94:1461–1467.

Tomoya S, Hisao S, Maosheng H, Yoshiki N, Satoshi K, Satoru T, Kenji Y, Katsushi N, Linda TR, Shiro S (1998) Direct evidence that dislocations are non-radiative recombination centers in GaN. *Japanese Journal of Applied Physics* 37:L398.

Törmä PT, Svensk O, Ali M, Suihkonen S, Sopanen M, Odnoblyudov MA, Bougrov VE (2008) Effect of InGaN underneath layer on MOVPE-grown InGaN/GaN blue LEDs. *Journal of Crystal Growth* 310:5162–5165.

Toshio N, Tetsuya A, Naoki K (1998) Step-flow metalorganic vapor phase epitaxy of GaN on SiC substrates. *Japanese Journal of Applied Physics* 37:L459.

Tran CA, Karlicek RF, Schurman M, Salagaj T, Thompson A, Stall R (1997) Phase separation in bulk InGaN and quantum wells grown by low pressure MOCVD. In: *Vertical-Cavity Lasers, Technologies for a Global Information Infrastructure, WDM Components Technology, Advanced Semiconductor Lasers and Applications, Gallium Nitride Materials, Processing, and Devi*, pp. 29–30. Piscataway, NJ: IEEE.

Tsai YL, Gong JR (2004) Influence of low-temperature AlGaN intermediate multilayer structures on the growth mode and properties of GaN. *Optical Materials* 27:425–428.

Tuomisto F, Saarinen K, Lucznik B, Grzegory I, Teisseyre H, Suski T, Porowski S, Hageman PR, Likonen J (2005) Effect of growth polarity on vacancy defect and impurity incorporation in dislocation-free GaN. *Applied Physics Letters* 86:031915.

Twigg ME, Henry RL, Wickenden AE, Koleske DD, Culbertson JC (1999) Nucleation layer microstructure, grain size, and electrical properties in GaN grown on a-plane sapphire. *Applied Physics Letters* 75:686–688.

Twigg ME, Koleske DD, Wickenden AE, Henry RL, Binari SC (2001) Correlation between nucleation layer structure, dislocation density, and electrical resistivity for GaN films grown on a-plane sapphire by metalorganic vapor phase epitaxy. *Applied Physics Letters* 79:4322–4324.

Tzu-Chi W, Wei IL (2001) Influence of barrier growth temperature on the properties of InGaN/GaN quantum well. *Japanese Journal of Applied Physics* 40:5302.

Uchida K, Watanabe A, Yano F, Kouguchi M, Tanaka T, Minagawa S (1996) Nitridation process of sapphire substrate surface and its effect on the growth of GaN. *Journal of Applied Physics* 79:3487–3491.

Uedono A, Ishibashi S, Oshima N, Suzuki R, Sumiya M (2014) (Invited) Point defect characterization of group-III nitrides by using monoenergetic positron beams. *ECS Transactions* 61:19–30.

Uedono A, Tsutsui T, Watanabe T, Kimura S, Zhang Y, Lozac, Apos, HM, Sang LW, Ishibashi S, Sumiya M (2013) Point defects introduced by InN alloying into $In_xGa_{1-x}N$ probed using a monoenergetic positron beam. *Journal of Applied Physics* 113:123502.

Van de Walle CG, Neugebauer J (2004) First-principles calculations for defects and impurities: Applications to III-nitrides. *Journal of Applied Physics* 95:3851–3879.

Van de Walle CG, Stampfl C, Neugebauer J (1998) Theory of doping and defects in III-V nitrides. *Journal of Crystal Growth* 189:505–510.

van der Laak NK, Oliver RA, Kappers MJ, Humphreys CJ (2007) Characterization of InGaN quantum wells with gross fluctuations in width. *Journal of Applied Physics* 102:013513.

Vennegues P, Benaissa M, Beaumont B, Feltin E, De Mierry P, Dalmasso S, Leroux M, Gibart P (2000) Pyramidal defects in metalorganic vapor phase epitaxial Mg doped GaN. *Applied Physics Letters* 77:880–882.

Waldrip KE, Han J, Figiel JJ, Zhou H, Makarona E, Nurmikko AV (2001) Stress engineering during metalorganic chemical vapor deposition of AlGaN/GaN distributed Bragg reflectors. *Applied Physics Letters* 78:3205–3207.

Wampler WR, Myers SM (2003) Hydrogen release from magnesium-doped GaN with clean ordered surfaces. *Journal of Applied Physics* 94:5682–5687.

Wampler WR, Myers SM, Wright AF, Barbour JC, Seager CH, Han J (2001) Lattice location of hydrogen in Mg doped GaN. *Journal of Applied Physics* 90:108–117.

Wang GT, Creighton JR (2004) Complex formation between magnesocene ($MgCp_2$) and NH_3: Implications for p-type doping of group III nitrides and the Mg memory effect. *Journal of Physical Chemistry A* 108:4873–4877.

Wang H-M, Zhang J-P, Chen C-Q, Fareed Q, Yang J-W, Khan MA (2002) AlN/AlGaN superlattices as dislocation filter for low-threading-dislocation thick AlGaN layers on sapphire. *Applied Physics Letters* 81:604–606.

Wang J, Wang L, Zhao W, Zou X, Luo Y (2010) Study on internal quantum efficiency of blue InGaN multiple-quantum-well with an InGaN underneath layer. *Science China Technological Sciences* 53:306–308.

Wang Y, Pei XJ, Xing ZG, Guo LW, Jia HQ, Chen H, Zhou JM (2007) Effects of barrier growth temperature ramp-up time on the photoluminescence of InGaN/GaN quantum wells. *Journal of Applied Physics* 101:033509.

Weeks TW, Bremser MD, Ailey KS, Carlson E, Perry WG, Davis RF (1995) GaN thin-films deposited via organometallic vapor-phase epitaxy on alpha(6H)-SiC(0001) using high-temperature monocrystalline AlN buffer layers. *Applied Physics Letters* 67:401–403.

Wei CH, Xie ZY, Li LY, Yu QM, Edgar JH (2000) MOCVD growth of cubic GaN on 3C-SiC deposited on Si (100) substrates. *Journal of Electronic Materials* 29:317–321.

Wei L, Li-Hong Z, Fan-Ming Z, Ling Z, Wei-Cui L, Xiao-Ying L, Bao-Lin L, Zhe-Chuan F (2013) Influence of GaN barrier thickness on optical properties of in-graded InGaN/GaN multiple quantum wells. *Applied Physics Express* 6:081001.

Weimann NG, Eastman LF, Doppalapudi D, Ng HM, Moustakas TD (1998) Scattering of electrons at threading dislocations in GaN. *Journal of Applied Physics* 83:3656–3659.

Wickenden AE, Gaskill DK, Koleske DD, Doverspike K, Simons DS, Chi PH (1995a) On compensation and impurities in state-of-the-art GaN epilayers grown on sapphire. *MRS Proceedings* 395:679.

Wickenden DK, Kistenmacher TJ, Bryden WA, Morgan JS, Estes Wickenden A (1991) The effect of self nucleation layers on the Mocvd growth of gallium nitride on sapphire. *MRS Proceedings* 221:167.

Wickenden AE, Koleske DD, Henry RL, Gorman RJ, Culbertson JC, Twigg ME (1999) The impact of nitridation and nucleation layer process conditions on morphology and electron transport in GaN epitaxial films. *Journal of Electronic Materials* 28:301–307.

Wickenden AE, Koleske DD, Henry RL, Gorman RJ, Twigg ME, Fatemi M, Freitas JA, Moore WJ (2000) The influence of OMVPE growth pressure on the morphology, compensation, and doping of GaN and related alloys. *Journal of Electronic Materials* 29:21–26.

Wickenden AE, Koleske DD, Henry RL, Twigg ME, Fatemi M (2004) Resistivity control in unintentionally doped GaN films grown by MOCVD. *Journal of Crystal Growth* 260:54–62.

Wickenden AE, Rowland LB, Doverspike K, Gaskill DK, Freitas JA, Simons DS, Chi PH (1995b) Doping of gallium nitride using disilane. *Journal of Electronic Materials* 24:1547–1550.

Wickenden AE, Wickenden DK, Kistenmacher TJ (1994) The effect of thermal annealing on GaN nucleation layers deposited on (0001) sapphire by metalorganic chemical vapor deposition. *Journal of Applied Physics* 75:5367–5371.

Woods V, Dietz N (2006) InN growth by high-pressures chemical vapor deposition: Real-time optical growth characterization. *Material Science and Engineering B-Solid State Materials for Advanced Technology* 127:239–250.

Wright AF, Seager CH, Myers SM, Koleske DD, Allerman AA (2003) Hydrogen configurations, formation energies, and migration barriers in GaN. *Journal of Applied Physics* 94:2311–2318.

Wu F, Zhao Y, Romanov A, DenBaars SP, Nakamura S, Speck JS (2014) Stacking faults and interface roughening in semipolar (20-2-1) single InGaN quantum wells for long wavelength emission. *Applied Physics Letters* 104:151901.

Wu GA, Chen SH (2008) Blue InGaN/GaN light-emitting diodes using Mg-doped AlGaN electron-blocking barriers. *Journal of the Korean Physical Society* 52:1570–1574.

Wu J, Walukiewicz W, Yu KM, Ager JW, Haller EE, Lu H, Schaff WJ (2002) Small band gap bowing in In1−xGaxN alloys. *Applied Physics Letters* 80:4741–4743.

Wu LW, Chang SJ, Su YK, Chuang RW, Wen TC, Kuo CH, Lai WC, Chang CS, Tsai JM, Sheu JK (2003) Nitride-based green light-emitting diodes with high temperature GaN barrier layers. *IEEE Transactions on Electron Devices* 50:1766–1770.

Wu XH, Brown LM, Kapolnek D, Keller S, Keller B, DenBaars SP, Speck JS (1996) Defect structure of metal-organic chemical vapor deposition-grown epitaxial (0001) GaN/Al$_2$O$_3$. *Journal of Applied Physics* 80:3228–3237.

Wu XH, Elsass CR, Abare A, Mack M, Keller S, Petroff PM, DenBaars SP, Speck JS, Rosner SJ (1998) Structural origin of V-defects and correlation with localized excitonic centers in InGaN/GaN multiple quantum wells. *Applied Physics Letters* 72:692–694.

Xia CS, Li ZS, Lu W, Zhang ZH, Sheng Y, Da Hu W, Cheng LW (2012) Efficiency enhancement of blue InGaN/GaN light-emitting diodes with an AlGaN-GaN-AlGaN electron blocking layer. *Journal of Applied Physics* 111:094503.

Xia CS, Simon Li ZM, Sheng Y (2013) On the importance of AlGaN electron blocking layer design for GaN-based light-emitting diodes. *Applied Physics Letters* 103:233505.

Xu SR, Hao Y, Zhang JC, Cao YR, Zhou XW, Yang LA, Ou XX, Chen K, Mao W (2010) Polar dependence of impurity incorporation and yellow luminescence in GaN films grown by metal-organic chemical vapor deposition. *Journal of Crystal Growth* 312:3521–3524.

Xu XL, Beling CD, Fung S, Zhao YW, Sun NF, Sun TN, Zhang QL et al. (2000) Formation mechanism of a degenerate thin layer at the interface of a GaN/sapphire system. *Applied Physics Letters* 76:152–154.

Yamamoto T, Iida D, Kondo Y, Sowa M, Umeda S, Iwaya M, Takeuchi T, Kamiyama S, Akasaki I (2014) In situ X-ray diffraction monitoring of GaInN/GaN superlattice during organometalic vapor phase epitaxy growth. *Journal of Crystal Growth* 393:108–113.

Yasuo O, Ako H (1996) Growth of high-quality AlN and AlN/GaN/AlN heterostructure on sapphire substrate. *Japanese Journal of Applied Physics* 35:L1013.

Yasuo O, Hiroaki Y, Rie S (1997) Growth of high-quality AlN, GaN and AlGaN with atomically smooth surfaces on sapphire substrates. *Japanese Journal of Applied Physics* 36:L1565.

Yoshida S, Misawa S, Gonda S (1983) Improvements on the electrical and luminescent properties of reactive molecular beam epitaxially grown GaN films by using AlN-coated sapphire substrates. *Applied Physics Letters* 42:427–429.

Yoshimoto N, Matsuoka T, Sasaki T, Katsui A (1991) Photoluminescence of InGaN films grown at high temperature by metalorganic vapor phase epitaxy. *Applied Physics Letters* 59:2251–2253.

Yoshinao K, Jumpei T, Masanari I, Toru N, Hisashi M, Kazuya T, Akinori K (2008) Self-separation of a thick AlN layer from a sapphire substrate via interfacial voids formed by the decomposition of sapphire. *Applied Physics Express* 1:045003.

Young NG, Farrell RM, Hu YL, Terao Y, Iza M, Keller S, DenBaars SP, Nakamura S, Speck JS (2013) High performance thin quantum barrier InGaN/GaN solar cells on sapphire and bulk (0001) GaN substrates. *Applied Physics Letters* 103:173903.

Young NG, Farrell RM, Oh S, Cantore M, Wu F, Nakamura S, DenBaars SP, Weisbuch C, Speck JS (2016) Polarization field screening in thick (0001) InGaN/GaN single quantum well light-emitting diodes. *Applied Physics Letters* 108:061105.

Zhang BS, Wu M, Liu JP, Chen J, Zhu JJ, Shen XM, Feng G et al. (2004) Reduction of tensile stress in GaN grown on Si(1 1 1) by inserting a low-temperature AlN interlayer. *Journal of Crystal Growth* 270:316–321.

Zhang JP, Wang HM, Sun WH, Adivarahan V, Wu S, Chitnis A, Chen CQ et al. (2003) High-quality AlGaN layers over pulsed atomic-layer epitaxially grown AlN templates for deep ultraviolet light-emitting diodes. *Journal of Electronic Materials* 32:364–370.

Zhang JX, Cheng H, Chen YZ, Uddin A, Yuan S, Geng SJ, Zhang S (2005) Growth of AlN films on Si (100) and Si (111) substrates by reactive magnetron sputtering. *Surface and Coatings Technology* 198:68–73.

Zhang L, Ding K, Yan JC, Wang JX, Zeng YP, Wei TB, Li YY, Sun BJ, Duan RF, Li JM (2010) Three-dimensional hole gas induced by polarization in (0001)-oriented metal-face III-nitride structure. *Applied Physics Letters* 97:062103.

Zhang Y-Y, Yao G-R (2011) Performance enhancement of blue light-emitting diodes with AlGaN barriers and a special designed electron-blocking layer. *Journal of Applied Physics* 110:093104.

Zhang YY, Zhu XL, Yin YA, Ma J (2012) Performance enhancement of near-UV light-emitting diodes with an InAlN/GaN superlattice electron-blocking layer. *IEEE Electron Device Letters* 33:994–996.

Zhao H, Liu G, Zhang J, Poplawsky JD, Dierolf V, Tansu N (2011) Approaches for high internal quantum efficiency green InGaN light-emitting diodes with large overlap quantum wells. *Optics Express* 19:A991–A1007.

Zhao Y, Yan Q, Huang C-Y, Huang S-C, Shan Hsu P, Tanaka S, Pan C-C et al. (2012) Indium incorporation and emission properties of nonpolar and semipolar InGaN quantum wells. *Applied Physics Letters* 100:201108.

Zheleva TS, Nam O-H, Bremser MD, Davis RF (1997) Dislocation density reduction via lateral epitaxy in selectively grown GaN structures. *Applied Physics Letters* 71:2472–2474.

Zheng TC, Lin W, Liu R, Cai DJ, Li JC, Li SP, Kang JY (2016) Improved p-type conductivity in Al-rich AlGaN using multidimensional Mg-doped superlattices. *Scientific Reports* 6:21897.

Zhiting L, Haiyan W, Yunhao L, Meijuan Y, Wenliang W, Guoqiang L (2016) Influence of In content in InGaN barriers on crystalline quality and carrier transport of GaN-based light-emitting diodes. *Journal of Physics D: Applied Physics* 49:115112.

Zhiting L, Rui H, Guoqiang L, Shuguang Z (2015) Effect of Si doping in barriers of InGaN/GaN multiple quantum wells on the performance of green light-emitting diodes. *Japanese Journal of Applied Physics* 54:022102.

Zhu D, Kappers MJ, Costa PMFJ, McAleese C, Rayment FDG, Chabrol GR, Graham DM et al. (2006) A comparative study of near-UV emitting InGaN quantum wells with AlGaN and AlInGaN barriers. *Physica Status Solidi (A)* 203:1819–1823.

Zhu D, Wallis DJ, Humphreys CJ (2013) Prospects of III-nitride optoelectronics grown on Si. *Reports on Progress in Physics* 76:106501.

Zhu JJ, Fan YM, Zhang H, Lu GJ, Wang H, Zhao DG, Jiang DS et al. (2012) Contribution of GaN template to the unexpected Ga atoms incorporated into AlInN epilayers grown under an indium-very-rich condition by metalorganic chemical vapor deposition (MOCVD). *Journal of Crystal Growth* 348:25–30.

Zhu K, Nakarmi ML, Kim KH, Lin JY, Jiang HX (2004) Silicon doping dependence of highly conductive n-type Al0.7Ga0.3N. *Applied Physics Letters* 85:4669–4671.

Zvanut ME, Willoughby WR, Koleske DD (2012) The source of holes in p-type $In_xGa_{1-x}N$ films. *Journal of Applied Physics* 112:086102.

7

Molecular Beam Epitaxial Growth of III-Nitride Nanowire Heterostructures and Emerging Device Applications

Shizhao Fan,
Songrui Zhao,
Faqrul A. Chowdhury,
Renjie Wang,
and Zetian Mi

7.1 Introduction

In the past decade, tremendous progress has been made in the growth/synthesis and characterization of III-nitride nanowires, including InN, GaN, AlN, and their alloys (Bertness et al. 2010; Lin et al. 2010; Hestroffer et al. 2011; Jenichen et al. 2011; Waag et al. 2011; González-Posada et al. 2012; Albert et al. 2013; Jahangir et al. 2013; Katsumi et al. 2013; Nguyen et al. 2013; Sarwar et al. 2015b; Zhao et al. 2016). Compared to their conventional planar counterparts, nearly dislocation and strain-free III-nitride nanowires can be epitaxially grown on Si, sapphire, SiO_x and other foreign substrates, due to the efficient strain relaxation associated with the large surface-to-volume ratio (Zhao et al. 2013a). Moreover, recent studies, both theoretically and experimentally, have shown that the group III substitutional Mg-doping has significantly lower surface formation energy compared to that in the bulk region (Zhao et al. 2013b; Zhao et al. 2015a), thereby leading to enhanced dopant incorporation and efficient p-type conduction that is often difficult to achieve in III-nitride planar structures. It has been shown that the room-temperature emission wavelengths of III-nitride nanowires can be tuned from the deep ultraviolet (UV) (~207 nm for AlN), through the deep visible, to the near-infrared spectral range (~1.9 µm for InN) by varying the alloy compositions (Wu 2009; César et al. 2011). Recently, it has also been discovered that III-nitrides are the only semiconductors whose energy band edges can straddle water redox potentials under deep visible and even possibly near-infrared light irradiation, which is essential for the efficient generation of solar fuels through water splitting and CO_2 reduction (Moses et al. 2010; Kibria et al. 2016b).

In this chapter, we describe some of the major advances made by the molecular beam epitaxy (MBE) growth of III-nitride nanowire heterostructures, including InN, In-rich InGaN, AlN, and Al-rich AlGaN nanowires. Their emerging device applications, such as color-tunable light-emitting diode (LED) arrays, deep UV LEDs and lasers, and solar fuel devices are also presented. The growth mechanism of III-nitride nanowires by MBE and the realization of high-quality In-rich InGaN and Al-rich AlGaN nanowire heterostructures are presented in Section 7.2. Some of the recent demonstrations of full-color single nanowire LEDs on a single chip and high-efficiency AlN nanowire LEDs and AlGaN nanowire lasers operating in the UV-C band are also discussed. The use of InGaN nanowires for solar-to-hydrogen conversion, including both photocatalytic and photoelectrochemical (PEC) water splitting is presented in Section 7.3. Moreover, their integration with Si solar cells to achieve high-efficiency double-band photoelectrodes is also discussed. Finally, conclusions are made in Section 7.4.

7.2 MBE Growth and Characterization of III-Nitride Nanowires

In this section, we describe some of the recent advances made by the MBE growth of III-nitride nanowire arrays, including the catalyst-free self-organized growth and selective area epitaxy of InGaN and AlGaN nanowires on Si and sapphire substrates. With the use of nanowire structures, efficient p-type conduction has been demonstrated for InN and AlN that was not previously possible for planar structures. Moreover, the formation of nearly dislocation-free III-nitride film structures through controlled nanowire coalescence is also presented. Their emerging applications in full-color display and lighting and deep UV optoelectronics are also discussed.

7.2.1 Growth Mechanism

7.2.1.1 Spontaneous Formation of III-Nitride Nanowires

To date, a large number of III-nitride nanowire structures by MBE are spontaneously formed through a self-organized process under nitrogen-rich conditions on Si substrate, driven by the anisotropy of various surface properties such as chemical potential and sticking coefficient. Different from the growth of III-nitride epilayers, wherein the growth process is largely determined by the incorporation of atoms directly impinged from effusion cells, both impinged atoms and migrated (diffused) atoms along the substrate and nanowire lateral surface play important roles in the nucleation and formation of III-nitride nanowires. Consequently, the spontaneous formation of MBE-grown III-nitride nanowires is also referred to as a diffusion-driven process (Debnath et al. 2007).

Additionally, detailed studies have further indicated the existence of a thin amorphous interface, that is, SiN_x layer between the nanowire and Si substrate (Stoica et al. 2008; Chang et al. 2009; Landré et al. 2009; Knelangen et al. 2010), which could suggests the absence of epitaxial relationship with the underlying substrate. This has been further confirmed by the direct growth of GaN, InGaN/GaN dot-in-a-wire, and InN nanowire structures on the SiO_x template (Zhao et al. 2013a). Importantly, it is noted that the nanowire structures exhibit comparable and/or better structural and optical properties compared to the nanowire structures grown directly on Si (Zhao et al. 2013a). These studies essentially lead to more choices of substrates for MBE-grown III-nitride nanowire structures, and thus, render MBE-grown III-nitride nanowires promising candidates for a wide range of functional electronic and photonic devices. Recently, III-nitride nanowire LEDs grown directly on SiO_x template has been transferred onto Cu and other flexible substrates to achieve enhanced performance (Nguyen et al. 2014). Moreover, laterally arranged multicolor InGaN/GaN dot-in-nanowire LEDs have been demonstrated on large area Si substrate by using a three-step growth process on patterned SiO_x/Si substrate (Wang et al. 2014b).

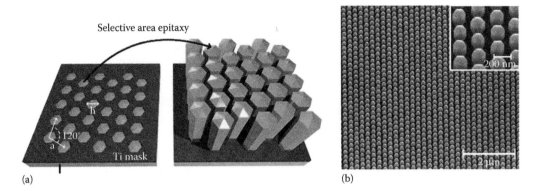

FIGURE 7.1 (a) Left: Arrays of nanoscale opening apertures defined by e-beam lithography process on a Ti mask on planar GaN template on *c*-plane sapphire substrate. Right: Schematic of the selective area epitaxy of GaN nanowire arrays on the nano-patterned substrate. (b) 45°-tilted view SEM image showing GaN nanowire arrays selectively grown in the opening apertures. Inset: High-magnification SEM image of GaN nanowire arrays. (From Le, B. et al., *Advanced Materials*, 28, 8446–8454, 2016. With permission.)

7.2.1.2 Selective Area Epitaxy

The MBE growth of III-nitride nanowires has also been performed on patterned substrates, including Si, GaN, and sapphire (Kishino et al. 2008; Bertness et al. 2010; Ishizawa et al. 2011; Le et al. 2016; Ra et al. 2016). In this growth process, a mask layer such as SiO_x, SiN_x, or Ti, is typically first deposited on the substrate, followed by e-beam lithography techniques to open nanoholes, illustrated in Figure 7.1a (Le et al. 2016). Due to the different kinetics of atoms on the mask layer surface and the substrate surface, nucleation process takes place preferentially in the nanoholes (Kishino et al. 2008; Bertness et al. 2010; Bengoechea-Encabo et al. 2011; Schumann et al. 2011; Kamimura et al. 2012; Gačević et al. 2015; Yamano et al. 2015). Therefore, in this growth process, the nanowire diameter can be precisely determined by the nanohole size. Shown in Figure 7.1b is the scanning electron microscope (SEM) image of GaN nanowire arrays grown on patterned GaN template on the sapphire substrate (Le et al. 2016), which are vertically aligned along the *c*-axis and show a high degree of size uniformity. Efforts have also been devoted to understanding the growth mechanism. Through detailed studies of the growth of GaN nanowires on GaN template with TiN_x mask, Gačević et al. have suggested a two-stage mechanism including the selective area kinetics-driven stage and the free surface energy minimization stage (Gačević et al. 2015). Yamano et al. have further demonstrated the controlled growth on the top-down-etched GaN nanopillars on GaN template (Yamano et al. 2015). In this work, the nanowires were formed only on GaN nanopillars due to the shadow effect, and the overgrown nanowire size was determined by the GaN nanopillar size. This demonstration of excellent control of the formation site, size, and spacing of III-nitride nanowire structures may render III-nitride nanostructures promising candidates for many device applications. For example, by precisely controlling the nanowire formation site, size, and spacing, high-quality single photon sources and photonic crystal lasers have been demonstrated (Ishizawa et al. 2011; Holmes et al. 2014).

7.2.2 InN and In-Rich InGaN Nanowires

7.2.2.1 InN Nanowires

To date, the MBE growth of InN nanowire structures has been performed on Si (Johnson et al. 2004; Shen et al. 2006; Stoica et al. 2006; Calleja et al. 2007a; Harui et al. 2008; Chang et al. 2009; Richter et al. 2009; Segura-Ruiz et al. 2009; Werner et al. 2009; Cuscó et al. 2010; Calarco et al. 2012; Zhao et al. 2012a; Sarwar et al. 2015a), GaN template (Satoshi et al. 2008; Wang et al. 2015a), and sapphire

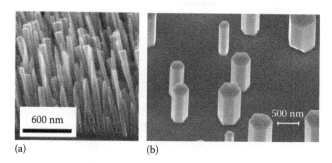

FIGURE 7.2   (a) SEM image of nominally non-doped tapered InN nanowires on Si (From Stoica, T. et al., *Nano Letters*, 6, 1541–1547, 2006. With permission.) (b) SEM image of non-tapered hexagonal InN nanowires on Si. (From Zhao, S. et al., *Nano Letters*, 12, 2877–2882, 2012a. With permission.)

(Nishikawa et al. 2007). During the early studies, the MBE-grown InN nanowires have tapered morphology, with the nanowire top considerably larger than the nanowire root, as illustrated in Figure 7.2a. Detailed characterizations of their optical and electrical properties have further indicated that such tapered (both nominally nondoped and Mg-doped) InN nanowires possess n-type degenerate characteristics. Characterizations on their electrical transport properties with single nanowire back gate field-effect transistor devices suggest average electron concentrations on the order of 10^{18} cm^{-3}, or higher, and mobility values in the range of 76 to 760 cm^2/Vs, drastically smaller than theoretical predictions and nondoped InN epilayers (Chang et al. 2005; Calleja et al. 2007b; Richter et al. 2009; Werner et al. 2009; Koley et al. 2011). Moreover, high surface electron density, on the order of 10^{11-14} cm^{-3}, has been derived. This high surface electron density consequently pins the Fermi level deep above the conduction band at the surface. The high background electron concentration and Fermi-level pinning at the surface, largely determine their optical properties. For example, regarding to the photoluminescence characteristics, the photoluminescence peak energies are generally larger than the bandgap energy of InN, with nearly no dependence on temperature and excitation power, due to a Mahan exciton emission mechanism (Johnson et al. 2004; Shen et al. 2006; Stoica et al. 2006; Feneberg et al. 2008; Segura-Ruiz et al. 2009; Zhao et al. 2012b). Moreover, coupled optical phonon modes have also been observed in Raman spectroscopy due to the presence of surface electron accumulation layer (Lazić et al. 2008; Cuscó et al. 2009; Cuscó et al. 2010; Jeganathan et al. 2010; Schäfer-Nolte et al. 2010; Domènech-Amador et al. 2012). Additionally, a fast initial decay of carrier lifetime, on the order of a picosecond, has been measured for small-diameter (~30 nm) nanowires, manifesting the influence of surface on the carrier relaxation dynamics (Chang and Gwo 2009; Ahn et al. 2012).

Recently, with the improved MBE growth process, nontapered InN nanowires with hexagonal shape have been demonstrated on Si, illustrated in Figure 7.2b. The valence band spectrum from angle-resolved X-ray photoelectron spectroscopy (XPS) experiment is shown in Figure 7.3a. It is seen that the near-surface Fermi level is located about 0.5 eV above valence band maximum (VBM), for a bandgap energy of 0.65 eV, which indicates that the near-surface Fermi level is not pinned in the conduction band and there is no surface electron accumulation. As a comparison, the InN nanowires with tapered morphology have also been investigated. The SEM of such nanowires is shown in the inset of Figure 7.3b. As illustrated in Figure 7.3b, the near-surface Fermi level is pinned about 0.35 eV in the conduction band. Compared to nontapered hexagonal InN nanowires, the tapered InN nanowires are generally grown under nonoptimized conditions, and n-type defects such as nitrogen vacancies have been found to exist (Stampfl et al. 2000; Zhao et al. 2012a). Detailed first principle calculations have further indicated a much lower formation energy of defect donors at the surface compared to that of bulk, rendering the preferential incorporation of defect donors at the surface (Zhao et al. 2012a). This, as a consequence, leads to the surface electron accumulation and Fermi-level pinning at InN surface. It is, thus, strongly

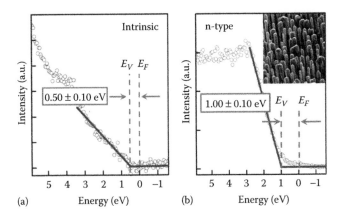

FIGURE 7.3 Valence band spectra of (a) intrinsic InN nanowires and (b) n-type InN nanowires. The inset of (b) shows the corresponding SEM image. (From Mi, Z. and Zhao, S. et al., *Basic Solid State Physics* (*B*), 252, 1050–1062, 2015. With permission.)

suggested that the surface electron accumulation and Fermi-level pinning at InN surface are largely due to the presence of defects and impurities, and *not* intrinsic properties of InN. This conclusion is further supported by the experiments performed on cleaved InN epilayers, wherein the absence of surface electron accumulation and Fermi-level pinning in the conduction band have been suggested (Wu et al. 2008; Ebert et al. 2011).

Electrical transport properties of such hexagonal InN nanowires have also been investigated by nano-probing technique (Zhao et al. 2013c). Illustrated in Figure 7.4a, a very low electron concentration on the order of 10^{13} cm^{-3} is achieved for large-radius nanowires. Moreover, large electron mobility ~12,000 cm^2/Vs has been derived for low electron concentrations, shown in Figure 7.4b. It is observed that the low electron concentration is associated with a large nanowire radius, which suggests that surface scattering plays an important role in the average electron mobility. With such a low level of background electron concentration, the optical properties have been found to be drastically different from n-type degenerate InN nanowires. For example, free-excitation emission can be measured at low temperatures (Zhao et al. 2012a,b). Moreover, phonon replica has also been measured (Zhao et al. 2012c).

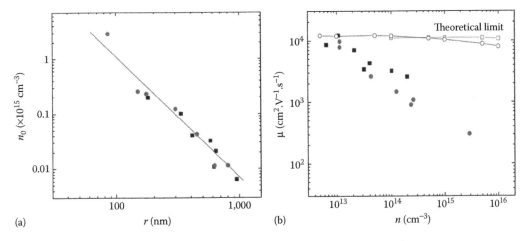

FIGURE 7.4 Electrical properties of intrinsic InN nanowires with (a) the derived electron concentration versus nanowire radius and (b) the derived electron mobility versus electron concentration. The line in (a) is a linear fit. The open symbols in (b) are theoretical predictions. (From Mi, Z. and Zhao, S., *Basic Solid State Physics* (*B*), 252, 1050–1062, 2015. With permission.)

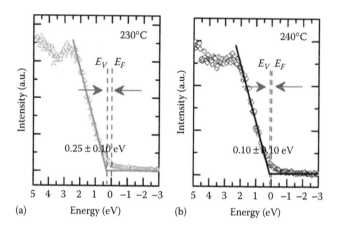

FIGURE 7.5 p-Type surface measured from Mg-doped InN nanowires with (a) Mg cell temperature of 230°C and (b) of 240°C (The higher the Mg cell temperature, the higher the Mg doping concentration). (From Zhao, S. et al., *Nano Letters*, 13, 5509–5513, 2013b. With permission.)

In addition, phonon-plasma coupling mode is suggested to be absent in micro Raman spectroscopy studies (Zhao et al. 2012c).

With the minimization of background electron doping concentration and elimination of surface electron accumulation, p-type InN nanowires have been demonstrated through direct Mg doping (Zhao et al. 2013b; Le et al. 2014; Le et al. 2015). As illustrated in Figure 7.5, instead of the presence of surface electron accumulation, p-type surface is measured from XPS, which is in direct contrast to the previously reported Mg-doped InN epilayers (Yoshikawa et al. 2010). The realization of p-type surface in Mg-doped InN nanowires is attributed to the enhanced surface doping due to the lower In-substitutional Mg-doping formation energy at surface, which is similar to the previously discussed lower formation energy of defect donors (Zhao et al. 2012a; Zhao et al. 2013b). It is thus suggested that such an effect of enhanced surface doping in nanostructures could be a very promising approach to address the p-type doping challenge of III-nitrides. Another example is the achievement of efficient p-type conduction in Mg-doped Al(Ga)N nanowires, which will be discussed in Section 7.2.3.

The device schematic of a p-type InN single nanowire transistor is shown in Figure 7.6a. The source-drain current versus source-drain voltage under varying back-gate voltages is shown in Figure 7.6b. It is seen that the channel conduction increases considerably with more negative back-gate voltages, providing an unambiguous evidence for the achievement of p-type conduction. Field-effect mobility around 100 cm²/Vs and free hole concentration on the order of 10^{16} cm^{-3} have been derived. Detailed studies have further indicated an ambipolar transport behavior from such p-type InN nanowire transistors, consistent with the absence of Fermi-level pinning (Le et al. 2015).

Another evidence for the achievement of p-type InN nanowires is the demonstration of light-emitting diodes (LEDs; Le et al. 2014). The device structure is illustrated in Figure 7.7a, consisting of p-, i-, and n-InN segments embedded in a single nanowire. The electrodes are fabricated by standard e-beam lithography processing. The Si and Mg doping concentrations are ~5 × 10^{17} cm^{-3} and ~4 × 10^{19} cm^{-3}, respectively. The *I-V* characteristics measured at 77 K is shown in Figure 7.7b, with the electroluminescence spectrum shown in the inset of Figure 7.7b. It is seen that an electroluminescence peak around 0.7 eV, corresponding to ~1.8 μm in wavelength, was measured, providing clear evidence for the achievement of p-type conduction.

7.2.2.2 In-Rich InGaN Nanowires

High-quality In-rich InGaN materials are essentially required for application in high-efficiency deep visible LEDs and lasers, multijunction photovoltaic devices, and solar fuel production. However, the

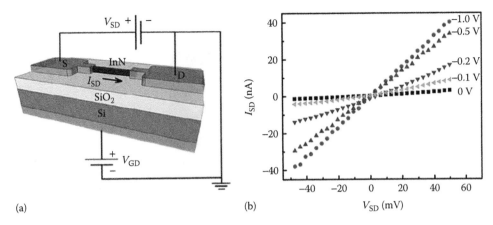

(a) (b)

FIGURE 7.6 (a) Schematic of p-type InN nanowire transistor. (b) The source-drain current (I_{SD}) versus source-drain voltage (V_{SD}) under different back-gate voltages. (From Zhao, S. et al., *Nano Letters*, 13, 5509–5513, 2013b. With permission.)

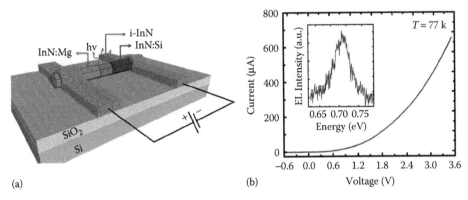

(a) (b)

FIGURE 7.7 (a) Schematic of single InN nanowire LED. (b) The *I-V* characteristics and electroluminescence spectrum (inset) measured at 77 K. The electroluminescence spectrum was at an injection of 500 μA under a continuous-wave biasing. (From Le, B. H. et al., *Applied Physics Letters*, 105, 231124, 2014. With permission.)

epitaxy of In-rich InGaN alloys has remained challenging, due to the large lattice mismatch between InN and GaN (~11%) and the lack of suitable substrates. Moreover, the presence of In phase separation limits the wavelength tunability of InGaN alloys. The high growth temperature of the conventional metalorganic chemical vapor phase epitaxial deposition (MOCVD) growth process also leads to enhanced In desorption at the growth front (Doppalapudi et al. 1998). Compared to MOCVD, MBE offers several advantages for the growth of In-rich InGaN, including a relatively low growth temperature and the use of plasma nitrogen sources to generate atomic nitrogen species independent of the growth temperature. In this regard, significant progress has been made in the MBE growth of InGaN nanowire heterostructures. In what follows, the recent advances of InGaN/GaN well/disk/dot-in-nanowires, core-shell nanowire heterostructures, and nearly homogeneous $In_{0.5}Ga_{0.5}N$ nanowire crystals are described.

The InGaN/GaN well/disk/dot-in-nanowire heterostructures have been developed to improve carrier confinement and achieve high-efficiency light emission. Kikuchi et al. have shown that InGaN/GaN LEDs can operate in the deep visible wavelength range of (530–645 nm) (Kikuchi et al. 2004). Phosphor-free white LEDs with the incorporation of multiple InGaN disks in GaN nanowires have also been realized by Guo et al (Guo et al. 2011). The incorporation of multiple, vertically aligned self-organized

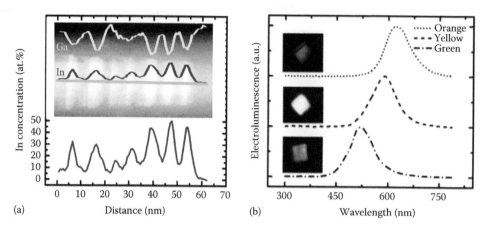

FIGURE 7.8 (a) EDXS profiles showing the variation of In and Ga compositions along the growth direction. (b) Room temperature electroluminescence spectra of LEDs with different In compositions in the quantum dot active regions. The corresponding optical images of the red, orange, yellow, and green LEDs are shown in the insets. (From Nguyen, H. P. T. et al., *Nano Letters*, 11, 1919–1924, 2011b. With permission.)

InGaN quantum dots in nearly defect-free GaN nanowires can significantly enhance the radiative electron-hole recombination efficiency (Nguyen et al. 2011a, b; Nguyen et al. 2012). Illustrated in Figure 7.8a are the energy-dispersive X-ray spectroscopy (EDXS) profiles for In and Ga compositions in InGaN/GaN dot-in-nanowire heterostructures. By varying the In composition and/or quantum dot size, emission wavelengths across nearly the entire visible spectral range can be achieved. Shown in Figure 7.8b are the electroluminescence emission spectra of multicolor InGaN/GaN dot-in-nanowire LEDs grown on Si substrate (Chang et al. 2010; Nguyen et al. 2011a, b; Nguyen et al. 2012). Blue color single photon source has also been demonstrated from single InGaN/GaN dot-in-nanowire structures under electrical injection (Deshpande et al. 2013).

The presence of surface states and defects in In-rich InGaN nanowires and the resulting nonradiative surface recombination generally lead to extremely short carrier lifetime in nanowire structures, severely limiting the performance of nanowire LEDs (Parkinson et al. 2009; Chang et al. 2012; Joyce et al. 2013). Zhang et al. have calculated the carrier injection efficiency of typical InGaN nanowire LEDs, which is well below 10%, due to the presence of surface recombination (Zhang et al. 2014). The extremely low carrier injection efficiency, together with the associated junction heating effect, explains the commonly reported very low output power of nanowire LEDs. In this regard, various surface passivation techniques have been developed to break the carrier injection bottleneck of nanowire LEDs (Chen et al. 1996; Martinez et al. 2000; Chevtchenko et al. 2007; Tajik et al. 2012; Jahangir et al. 2013; Nguyen et al. 2013; Nguyen et al. 2015). The incorporation of an *in situ* grown large bandgap AlGaN shell has shown to be one of the most promising and effective approaches to suppress nonradiative surface recombination and to enhance carrier injection efficiency (Nguyen et al. 2013; Nguyen et al. 2015; Wang et al. 2015b). The core-shell heterostructures can be spontaneously formed by growing InGaN/AlGaN quantum dot heterostructures (Nguyen et al. 2015), due to the reduced Al atom migration and the preferential incorporation of Al in the near-surface region of nanowires compared to In and Ga atoms. Shown in Figure 7.9 is the elemental mapping of the group III elements (Ga, In, and Al) of the InGaN/AlGaN dot-in-nanowire core-shell heterostructures. Vertically aligned InGaN quantum dots are formed at the center of the nanowire and are surrounded by Al-rich AlGaN shell structures. With the incorporation of such core-shell heterostructures, the carrier lifetime is enhanced from 0.3 to 4.5 ns (Nguyen et al. 2015). Detailed studies have further shown that the output power of InGaN/GaN nanowire LEDs can be enhanced by two to three orders of magnitude by incorporating a large bandgap AlGaN shell structure. Shown in

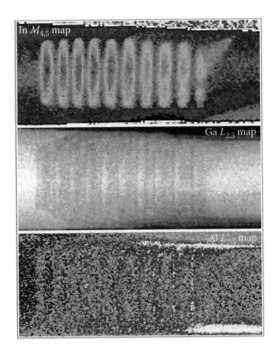

FIGURE 7.9 Elemental mapping result showing distributions of In, Ga, and Al elements within the nanowire LED structure. (From Nguyen, H. P. T. et al., *Scientific Reports*, 5, 7744, 2015. With permission.)

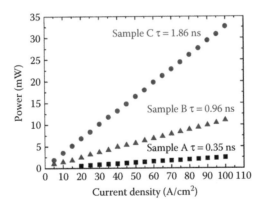

FIGURE 7.10 Light output power of AlInGaN LEDs versus injection current. The device area is 1 mm × 1 mm. The measurements were performed under pulsed biasing conditions (1% duty cycle). The measured carrier lifetimes of three InGaN/GaN nanowire LED samples with various large-bandgap AlInGaN shell structures are also shown. (From Wang, R. et al., *Applied Physics Letters*, 106, 261104, 2015b. With permission.)

Figure 7.10 is the output power versus injection current for a 1 × 1 mm² nanowire LED device without any packaging or other approaches of light extraction enhancement (Wang et al. 2015). An output power over 30 mW has been achieved. The device can operate efficiently in the entire visible spectral range by varying the alloy compositions of the dots.

For solar fuel device applications, the presence of a large bandgap shell structure is often detrimental for efficient charge carrier extraction from the nanowire lateral surfaces, which is required for efficient water splitting and CO_2 reduction reactions. Recently, In-rich InGaN nanowire arrays with In compositions over 50% have been synthesized by MBE, exhibiting nearly homogeneous distribution of indium across the InGaN nanocrystal (Fan et al. 2016). The suppression of severe phase separation is realized by using a closely packed GaN nanowire template on Si as nucleation sites for InGaN, adopting an InGaN/GaN multiple cycle growth technique to counteract the long-distance diffusion of indium in nanowires, and optimizing the combined effect of growth temperature and nitrogen flow rate to finely control the growth kinetics. Based on the analysis by Tourbot et al., the plastic strain relaxation could lead to the formation of a relatively homogeneous In-rich InGaN alloy on GaN (Tourbot et al. 2012). Fan et al. have further investigated the MBE growth of In-rich InGaN nanowires at reduced temperature (~520°C–550°C) using closely packed GaN nanowire template as the nucleation layer. Nucleation on the coarse GaN nanowire template may induce plastic relaxation at the initial stage of InGaN growth. The beam flux equivalent pressures for In and Ga are 8×10^{-8} Torr and 2×10^{-8} Torr, respectively. The nitrogen flow rate is in the range of 0.4–0.5 standard cubic centimeter per minute (sccm), which allows a sufficient diffusion length of indium adatoms on the nanowire surface for uniform incorporation at such low growth temperatures. As the growth proceeds, the constant annealing of the underlying grown InGaN crystal may induce spinodal decomposition of In-N and long-distance diffusion of indium toward the surface of nanowires, considering the relatively low nitrogen vapor pressure in the vicinity of the growth front. For counteracting such a thermodynamic impact, an InGaN/GaN multiple cycle growth technique is used, wherein a thin GaN capping layer of 10–20 nm is deposited at the same growth temperature after each InGaN segment is grown.

As shown in the SEM image of Figure 7.11, the InGaN nanowires exhibit heights of ~1 μm and diameters of ~300 nm (Fan et al. 2017). In the bottom right corner of the displayed region, the closely packed GaN nanowire template of ~100 nm in height is exposed due to the removal of InGaN nanowires by cleavage. Detailed studies on single nanowires using scanning transmission electron spectroscopy (STEM) and electron energy loss spectroscopy (EELS) are shown in Figure 7.12 (Fan et al. 2016). The relatively uniform intensity in the STEM high-angle annular dark-field (HAADF) image of atomic number contrast confirms the absence of a pure GaN nanowire shell. EELS analysis reveals that the indium molar fraction can reach 50% in the core of the nanowire, while slightly lower indium composition of ~30% is obtained at the surface of the nanowire. The high-resolution STEM high-angle annular dark-field imaging (HAADF) image in Figure 7.12c further indicates that the InGaN nanocrystal is defect-free with a notable atomic-ordering feature. In contrast to the InGaN/GaN core-shell nanowires which exhibit growth direction in the *c*-direction and top facet of *c*-planes, the fast Fourier transform of the In-rich InGaN image in Figure 7.12c suggests that the growth direction of this specific InGaN nanowire

FIGURE 7.11   A 45° tilted SEM image of $In_{0.5}Ga_{0.5}N$ nanowires grown on closely packed GaN nanowire template. (From Fan, S. et al., *Advanced Energy Materials*, 1600952, 2017. With permission.)

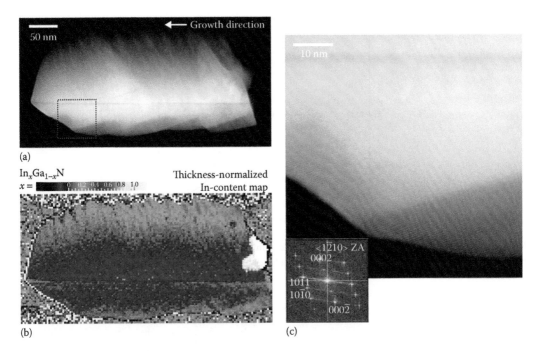

(a)

(b)

(c)

FIGURE 7.12 (a) The STEM–HAADF image of an InGaN nanowire viewed along (1-210) zone-axis, showing evidence of contrast at sidewalls due to the strong surface texturing. (b) The thickness-projected indium-content map derived using STEM–EELS spectrum. Relatively uniform distribution of indium was observed in the nanowire (~50% in the bulk, ~30% at the near-surface region). (c) High-magnification image of the region in the red dashed box in (a) highlighting the presence of atomic ordering along *c*-axis direction. Additional inset shows the image's FFT confirming the growth direction of <10$\bar{1}$1>. (From Fan, S. et al., *APL Materials*, 4, 076106, 2016a. With permission.)

is <10$\bar{1}$1>. A majority of other nanowires examined in STEM also exhibit a non-*c*-plane growth axis (such as <10$\bar{1}$2> or <10$\bar{1}$0>), further confirming the plastic relaxation of such nanowires. The corresponding photoluminescence emission spectrum is shown in Figure 7.13 (Fan et al. 2016). The In-rich InGaN nanowire arrays with ~50% indium by MBE (Sample C) show significantly enhanced emission intensity compared with InGaN nanowires of lower indium compositions (Samples A and B). The spectral linewidth (full-width-at-half-maximum–FWHM) is ~70 nm, which is among the narrowest values measured from InGaN materials in this wavelength range.

7.2.2.3 Single InGaN Nanowires

Recent selective area epitaxy studies have shown that the alloy compositions of InGaN nanowires depend not only on the growth parameters, but also on the nanowire size and spacing. Sekiguchi et al. have demonstrated that multicolor light emission can be achieved from ensemble InGaN nanowires grown on the same substrate by varying their lateral sizes (Sekiguchi et al. 2010), which takes advantage of the shadowing effect of neighboring nanowires to alter the alloy compositions. More recently, it has been shown that the alloy compositions of single InGaN nanowires can also be controllably varied by simply changing their diameters (Ra et al. 2016). Illustrated in Figure 7.14a, single InGaN/GaN dot-in-nanowire structures are grown on GaN template on the sapphire substrate using the technique of selective area epitaxy by MBE. Opening sizes in the range of 80 nm to nearly 2 μm are created on the Ti mask on GaN template. SEM images of single nanowire structures with different diameters are shown in Figure 7.14b. The nanowires have hexagonal morphology and Ga-polarity on their top surface.

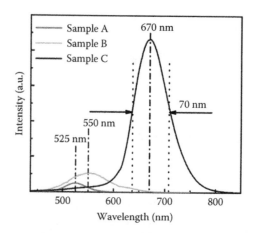

FIGURE 7.13 Room-temperature photoluminescence emission of In$_x$Ga$_{1-x}$N nanowires with three different indium molar fractions. (From Fan, S. et al., *APL Materials*, 4, 076106, 2016a. With permission.)

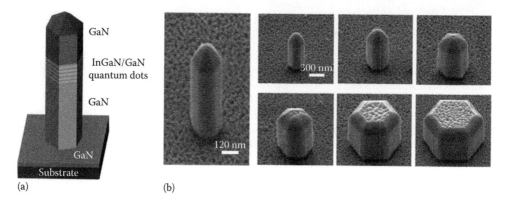

FIGURE 7.14 (a) Schematic of a single InGaN/GaN dot-in-nanowire structure grown on GaN template on sapphire substrate. (b) 30° tilted SEM image of single InGaN/GaN nanowires with various diameters. (From Ra, Y.-H. et al., *Nano Letters*, 16, 4608–4615, 2016. With permission.)

Room-temperature photoluminescence emission spectra of single InGaN/GaN dot-in-nanowire structures have also been measured, as shown in Figure 7.15. The nanowires have different diameters and are grown on the same substrate in a single epitaxial step. It is seen that the photoluminescence emission shows a consistent blueshift with increasing nanowire diameter. The emission wavelength can be continuously tuned from 640 nm to 465 nm by changing the nanowire lateral sizes from 150 nm to ~2 μm. The underlying mechanism has been studied. During single nanowire epitaxy, the atom incorporation at the growth front consists of atoms migrated from the lateral surfaces as well as directly impinged atoms. Due to the use of relatively high growth temperature, the diffusion length of Ga atoms (~1 μm) is significantly larger than that of In atoms (~100 nm) (Sekiguchi et al. 2010). The small In atom diffusion length is due to thermal desorption. Since the In atom diffusion length is smaller than the nanowire diameters, their incorporation at the nanowire growth front is expected to be significantly reduced for nanowires with larger diameters. On the other hand, the incorporation of Ga atoms is expected to show a small, or negligible dependence on nanowire sizes in the range of interest in this study. Therefore, the overall In composition in the nanowire structures is expected to decrease significantly with increasing nanowire diameter, thereby leading to a consistent blueshift with increasing size. These results are

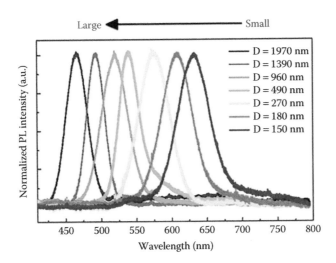

FIGURE 7.15 Normalized PL spectra of InGaN/GaN single nanowires with different diameters (D) measured at room-temperature, showing a blueshift in the emission peak with increasing nanowire diameter. (From Ra, Y.-H. et al., *Nano Letters*, 16, 4608–4615, 2016. With permission.)

distinctly different from the epitaxy of planar heterostructures, wherein alloy compositions are only determined by the global growth parameters and expected to be uniform across the wafer. It is also worth mentioning that in ensemble nanowire arrays, a redshift in the emission wavelength has often been measured with increasing nanowire diameter, which is directly related to the shadowing effect of neighboring nanowires (Kishino et al. 2013; Kishino et al. 2015; Sekiguchi et al. 2010). The beam shadowing effect, however, is not present in single nanowire epitaxy.

Detailed structural characterization has further revealed the correlation between the composition of InGaN quantum dots and the lateral size of InGaN nanowires during single nanowire epitaxy. Illustrated in Figure 7.16a are the STEM–HAADF atomic number contrast images of different diameter nanowires grown on the same substrate. The high-resolution EELS mapping of indium of the quantum dot active regions in each nanowire are shown in Figure 7.16b. The regions with dark intensity are In-free or In-deficient, whereas the bright regions are rich in indium. For nanowires with small diameters (active regions A and B), InGaN quantum dots are positioned at the center of nanowires and are vertically aligned along the *c*-axis. However, with increasing nanowire diameter (active regions C and D), InGaN quantum dots are formed on the semipolar planes of nanowires. The strong dependence of indium incorporation on nanowire diameter provides clear evidence that, during single nanowire epitaxy, the atom incorporation depends strongly on the nanowire size. Moreover, the formation of InGaN quantum dots on the semipolar planes toward the outer diameters provides unambiguous evidence that indium incorporation is limited by indium atom supplied from lateral diffusion. Further shown in Figure 7.16c, the indium concentration of InGaN quantum dots decreases drastically with increasing nanowire diameter, which is consistent with the blue-shift of the photoluminescence emission peak measured in Figure 7.15.

Full-color single InGaN/GaN dot-in-nanowire LEDs have also been demonstrated on a single chip by simply varying nanowire diameters in a single epitaxy step, illustrated in Figure 7.17a. The nanowire diameters are varied from 200 to 600 nm. Each nanowire consists of ~0.44 μm n-GaN, six InGaN/GaN quantum dots, and 0.15 μm p-GaN. Shown in Figure 7.17b is the SEM image of InGaN/GaN nanowire arrays with four different diameters (~220, 320, 420, and 630 nm). Subsequently, single InGaN/GaN nanowire LEDs are fabricated, and the performance of each single nanowire device is characterized. The devices exhibit excellent current-voltage and light-current characteristics.

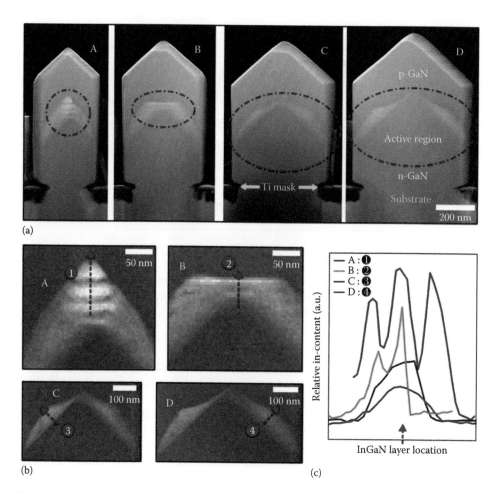

FIGURE 7.16 (a) An STEM–HAADF image for InGaN/GaN dot-in-nanowire structures with different diameters grown on GaN template on sapphire substrate along <11$\bar{2}$0> direction. The active region of nanowires with diameters of ~320 nm, ~420 nm, ~500 nm, and ~595 nm are labelled as A, B, C, and D, respectively. (b) High-resolution STEM–EELS maps of the In-distribution of active regions A, B, C, and D normalized to the sample thickness. Line profiles were integrated along areas as marked by the dashed red line in each active region. (c) Elemental profiles of relative In-content derived from EELS analysis along line 1 in active region A, line 2 in active region B, line 3 in active region C, and line 4 in active region D, showing higher In-content in smaller diameter wires. (From Ra, Y.-H. et al., *Nano Letters*, 16, 4608–4615, 2016. With permission.)

Illustrated in Figure 7.17c, single nanowire LEDs with diameters of 220, 320, 420, and 630 nm exhibit peak emission wavelengths of 659, 625, 526, and 461 nm, respectively (Ra et al. 2016). The selective area epitaxy of single nanowire heterostructures offers a unique approach to realize multiwavelength nanoscale optoelectronic devices with potential application for imaging, projection display, sensing, communication, and medical diagnostics on a single chip.

7.2.3 AlN and Al-Rich AlGaN Nanowires

In this section, we discuss the recent advances in the MBE growth and characterization of Al(Ga)N nanowires and their device applications including high-efficiency deep UV LEDs and electrically injected ultralow threshold deep UV lasers (Carnevale et al. 2012; Wang et al. 2013; Kent et al. 2014; Li et al. 2015; Zhao et al. 2015a, c, d).

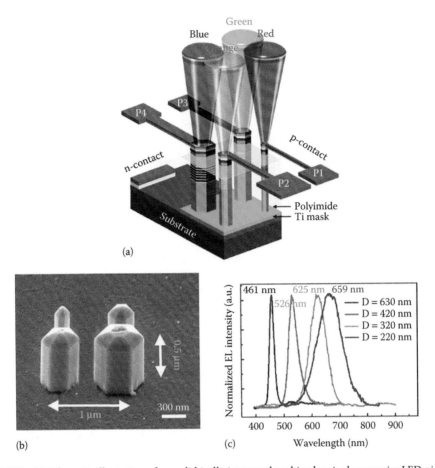

FIGURE 7.17 (a) Schematic illustration of monolithically integrated multi-color single nanowire LED pixels on a single chip. (b) SEM image of single InGaN/GaN dot-in-nanowires with different diameters grown by selective area epitaxy. (c) Electroluminescence spectra of single nanowire LEDs with different diameters. (From Ra, Y.-H. et al., *Nano Letters*, 16, 4608–4615, 2016. With permission.)

7.2.3.1 AlN Nanowires

AlN holds the key to achieving high-efficiency deep UV optoelectronics. In the past several years, significant progress has been made on the growth and characterization of AlN nanowires (Liu et al. 2001; Tang et al. 2001; Zhang et al. 2001; Balasubramanian et al. 2004; Zhao et al. 2005; Landré et al. 2010a; Kenry et al. 2012; Zhao et al. 2015a). Various techniques have been developed including the template-confined method (Liu et al. 2001; Zhang et al. 2001), *dc* arc plasma method (Balasubramanian et al. 2004), and CVD techniques (Tang et al. 2001; Zhao et al. 2005). Limited by the uncontrolled impurity incorporation, these AlN nanowires only exhibit defect-related emission. Recently, with the use of MBE, AlN nanowires with significantly improved quality have been demonstrated (Landré et al. 2010a, b; Wang et al. 2014; Connie et al. 2015; Zhao et al. 2015a, b; Mi et al. 2016). In what follows, we discuss the growth and characterization of AlN nanowires grown by MBE.

The growth of AlN nanowires by MBE has been achieved through a diffusion-driven process under nitrogen-rich conditions without any foreign catalyst. In addition, to promote the formation of AlN nanowires, GaN nanowire template is used (Wang et al. 2014; Connie et al. 2015; Zhao et al. 2015a, b; Mi et al. 2016). The use of GaN nanowire template can provide an additional degree of freedom in controlling the AlN nanowire size, density, and morphology. It has been found that by controlling the growth

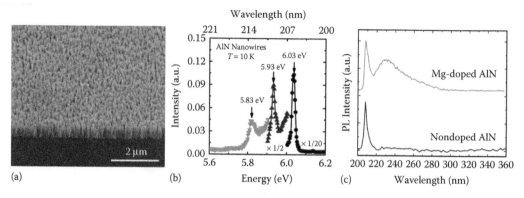

FIGURE 7.18 (a) SEM image of AlN nanowire arrays grown on Si substrate. (From Zhao, S. et al., *Scientific Reports*, 5, 8332, 2015a. With permission.) (b) Photoluminescence spectra of AlN nanowires measured at 10 K, showing the free-exciton emission peak around 6.03 eV and phonon replicas with an energy separation of around 100 meV. (From Connie, A. T. et al., *Applied Physics Letters*, 106, 213105, 2015. With permission.) (c) Photoluminescence spectra of Mg-doped AlN nanowires measured at room temperature. The Photoluminescence spectrum of nondoped AlN nanowires is also shown for a comparison. (From Connie, A. T. et al., *Applied Physics Letters*, 106, 213105, 2015. With permission.)

temperature of GaN nanowire template, the density of AlN nanowires can be well controlled (Mi et al. 2016). Shown in Figure 7.18a is an SEM image of AlN nanowires grown under optimized conditions. It is seen that large-scale highly dense and uniform AlN nanowires are formed on Si substrate, which is essential for the fabrication of large area devices.

Optical properties of such AlN nanowires have also been investigated by photoluminescence experiments. The spectrum measured at 10 K is shown in Figure 7.18b. A strong photoluminescence peak at 6.03 eV (205.6 nm) is observed, which is attributed to free-exciton emission (Wang et al. 2014). Moreover, phonon replicas associated with the free-exciton emission peak have also been measured, suggesting superior material quality (Wang et al. 2014). Detailed studies have further indicated that such AlN nanowires possess an internal quantum efficiency of around 80% (Wang et al. 2014), comparable to the values measured from state-of-the-art InGaN blue LEDs.

The growth and characterization of Mg-doped AlN nanowires have also been investigated (Connie et al. 2015; Zhao et al. 2015a). In order to enhance Mg incorporation, a lower substrate temperature is used. With optimized growth parameters, Mg-doped AlN nanowires can exhibit similar morphology as that seen from nondoped AlN nanowires (Figure 7.18a). The room-temperature photoluminescence spectrum of Mg-doped AlN nanowires is shown in Figure 7.18c. It is seen that besides the band edge emission at 207 nm (~6 eV), another low energy peak at 230 nm (~5.4 eV) is clearly measured. This peak energy separation (~0.6 eV) is consistent with the Mg activation energy in AlN (Taniyasu et al. 2006; Nakarmi et al. 2009). In addition, it is seen that the Mg acceptor-related emission peak is highly asymmetric, which partly overlaps with the band edge emission peak.

Recently, the first AlN nanowire LEDs have been demonstrated (Zhao et al. 2015a, b). A schematic of the device structure is shown in Figure 7.19a, which consists of n-GaN/p-AlGaN contact layers, and AlN p-i-n junction. Figure 7.19b shows the *I-V* characteristics of AlN nanowire LEDs measured under different temperatures. At room temperature the turn-on voltage is only ~5.5 V, much lower compared to the previously measured 20–30 V for *c*-plane AlN LEDs (Taniyasu et al. 2006). Furthermore, for a forward current of 20 mA (device size: 300 μm × 300 μm), the forward voltage is only ~7 V, which corresponds to an electrical efficiency of ~85% for an emission wavelength at ~207 nm. Moreover, only a very small reverse bias current is measured at room temperature, with a rectification ratio of ~10^6 at ±8 V (inset of Figure 7.19b). Excellent *I-V* characteristics have also been measured at low temperatures, which is

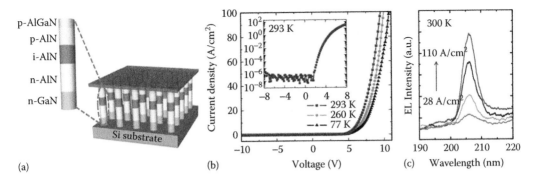

FIGURE 7.19 (a) Schematic of AlN nanowire LEDs on Si substrate. (From Zhao, S. et al., *Scientific Reports*, 5, 8332, 2015a; Zhao, S. et al, *Nano Letters*, 15, 7006, 2015b. With permission.) (b) *I-V* characteristics of AlN nanowire LEDs measured at different temperatures. Inset: room-temperature *I-V* curve in a semi-logarithmic scale. (c) Electroluminescence spectrum of AlN nanowire LEDs measured at room temperature.

attributed to the much improved p-type conduction through Mg impurity band (Connie et al. 2015). The room-temperature electroluminescence spectra are shown in Figure 7.19c, and strong emission around 207 nm is observed.

Compared to AlN LEDs with planar structures, AlN nanowire LEDs have shown much improved electrical performance. The Mg-doping mechanism in AlN nanowires has been investigated through detailed ab-initio calculations. It is found that, compared to AlN planar structures the Al-substitutional Mg dopant formation energy in AlN nanowires is significantly reduced, due to much more efficient strain relaxation (Zhao et al. 2015a). This can lead to a dramatic enhancement of Mg incorporation into AlN nanowires and the formation of Mg impurity band. Detailed electrical transport characterizations have further indicated that the conduction of p-type AlN nanowires is dominant through an impurity band conduction mechanism, with an activation energy of only 20–30 meV, which is significantly smaller compared to the ionization energy of Mg acceptors (Connie et al. 2015; Zhao et al. 2015a). Such a surface-enhanced doping mechanism has also been previously observed in Si and Ge nanowires (Xie et al. 2009), Si- and Mg-doped InN nanowires (Zhao et al. 2012a; Zhao et al. 2013b), and Si-doped GaN nanowires (Fang et al. 2015).

7.2.3.2 AlGaN Nanowires

With the use of CVD technique, ternary AlGaN nanowires on a variety of substrates, including Si, SiO_x, and sapphire, have been demonstrated (He et al. 2011; Chen et al. 2014a, b; Chen et al. 2015; Sivadasan et al. 2015; Ye et al. 2015). However, the emission peaks have been limited in the UV-A to visible spectral range, due to the presence of extensive defects. Improved material quality has been achieved with the use of MBE (Ristic et al. 2002; Bertness et al. 2006; Park et al. 2006; Carnevale et al. 2012; Carnevale et al. 2013; Pierret et al. 2013a; Kent et al. 2014; Zhao et al. 2015a, d). Under nitrogen-rich conditions and with the use of GaN nanowire template, spontaneously formed nearly defect-free ternary AlGaN nanowires with tunable emission wavelength covering UV-A, B, and C bands have been demonstrated (Carnevale et al. 2012; Carnevale et al. 2013; Pierret et al. 2013a, b; Kent et al. 2014; Zhao et al. 2015a, d). Moreover, Al-rich AlGaN shell and AlGaN nanoclusters have also been found to be present in such MBE-grown AlGaN nanowires (Wang et al. 2012; Pierret et al. 2013b, c; Wang et al. 2013; Zhao et al. 2015c), which can provide superior charge carrier confinement and enable high-efficiency deep UV optoelectronic devices.

Al-rich AlGaN shell has been commonly observed from ternary AlGaN nanowires, due to the smaller diffusion length of Al atoms, compared to Ga atoms (Iliopoulos and Moustakas 2002). Illustrated in Figure 7.20a is the EELS spectrum imaging of Ga and Al signals, Al accumulates on the nanowire

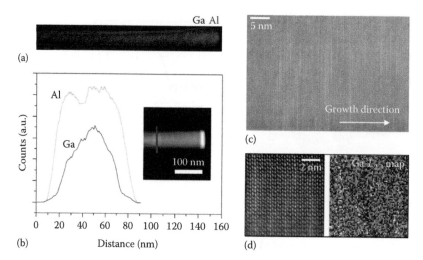

FIGURE 7.20 (a) Color-coded EELS spectrum imaging map showing Ga (blue) and Al (red) distribution in a single AlGaN nanowire. (From Li, K. H. et al., *Nature Nanotechnology*, 10, 140, 2015. With permission.) (b) EDXS line scans of Al and Ga across the nanowire radial direction, with the inset showing the scan direction by the red line. (From Li, K. H. et al., *Nature Nanotechnology*, 10, 140, 2015. With permission.) (c) A higher magnification STEM–HAADF image. The bright striations correspond to the Ga-rich AlGaN quantum dots/dashes. (From Zhao, S. et al., *Nano Letters*, 15, 7801–7807, 2015d. With permission.) (d) Ga-elemental mapping and the concurrently acquired ADF signal from EELS-SI at atomic resolution. The corresponding STEM image is also shown. (From Zhao, S. et al., *Nano Letters*, 15, 7801–7807, 2015d. With permission.)

sidewall, formatting Al-rich AlGaN shell. Figure 7.20b shows the EDXS line scans of Ga and Al across the nanowire radial direction, and the aggregation of Al on the nanowire sidewall is also clearly seen. The presence of a large bandgap AlGaN shell can suppress nonradiative surface recombination and lead to high-efficiency nanowire optoelectronic devices.

Furthermore, compositional modulations in AlGaN nanowires have been observed (Pierret et al. 2013a, b, c; Zhao et al. 2015c, d). Shown in Figure 7.20c is a high-magnification high-angle dark-field image. It is seen that extensive atomic-scale Ga-/Al-rich modulations, appearing as brighter and darker regions along the growth direction are clearly measured. More importantly, the Ga-rich AlGaN (brighter region) is not continuous along the lateral direction (perpendicular to the growth direction). Such Ga-rich AlGaN regions have sizes varying from single atomic layer from 0.25 nm to 2 nm along the growth direction, and lateral sizes varying from 2 to 5 nm. A Ga-signal profile for such modulation is shown in Figure 7.20d, with the corresponding high-magnification high-angle annular dark-field image also shown. The localized variations in the Ga-concentration are estimated to be 5–10 at .%. Such AlGaN quantum-dot-like nanostructures can lead to strong quantum confinement of charge carriers (Zhao et al. 2015d).

With the use of reduced nitrogen flow rate, AlGaN nanowires with tunable emission at room temperature have been demonstrated (Zhao et al. 2016b). The photoluminescence spectra of AlGaN nanowires with different Al contents are shown in Figure 7.21a. It is seen that by changing alloy composition, the emission wavelength can be widely tuned in the deep UV band. Due to the presence of Al-rich AlGaN shell and the spontaneously formed AlGaN quantum-dot-like nanostructures, such AlGaN nanowires can also possess a high internal quantum efficiency. As illustrated in Figure 7.21b, an internal quantum efficiency of nearly 80% has been measured for a sample emitting around 265 nm (Zhao et al. 2015d). The room temperature and low-temperature photoluminescence spectra are also shown in the inset of Figure 7.21b.

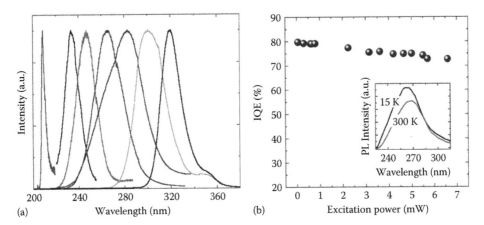

(a) Wavelength (nm)   (b) Excitation power (mW)

FIGURE 7.21 (a) Emission spectra of AlGaN nanowires with different Al contents measured at room temperature. (b) Internal quantum efficiency versus excitation power for a sample emitting around 265 nm at room temperature. (From Zhao, S. et al., *Nano Letters*, 15, 7801–7807, 2015d. With permission.)

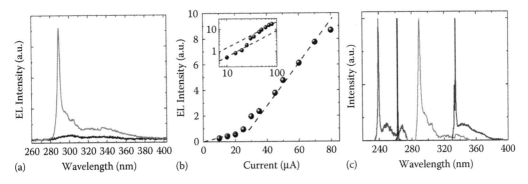

(a) Wavelength (nm)   (b) Current (µA)   (c) Wavelength (nm)

FIGURE 7.22 (a) Lasing spectrum at 80 µA for a 289 nm AlGaN nanowire laser measured at room temperature. The background emission at 10 µA is also shown (black curve). (From Zhao, S. et al., *Applied Physics Letters*, 107, 043101, 2015c. With permission.) (b) The *L-I* curve of the 289 nm laser. The inset shows the *L-I* curve in a logarithmic scale. (From Zhao, S. et al., *Nano Letters*, 15, 7801–7807, 2015d. With permission.) (c) AlGaN nanowire lasers at different lasing wavelengths. The lasing peaks at 262 nm and 334 nm are measured at cryogenic temperatures. (From Le, B. H. et al., *Applied Physics Express*, 8, 061001, 2015; Zhao, S. et al., *Applied Physics Letters*, 107, 043101, 2015c; Zhao, S. et al., *Nano Letters*, 15, 7801–7807, 2015d. With permission.)

Electrically injected lasers have also been realized in the deep UV band with AlGaN nanowire structures, for the first time (Zhao et al. 2015c, d). The laser cavity is formed due to the light Anderson localization effect, and the optical loss through the Si substrate is minimized due to the tapered nanowire morphology. The electroluminescence spectra for a 289 nm AlGaN nanowire laser device are shown in Figure 7.22a. It is seen that at a low injection current of 10 µA (below threshold), only a broad peak is measured. With the increase of injection current (80 µA, above threshold), a sharp lasing peak appears. The inset of Figure 7.22b shows the light output curve in a logarithmic scale. A clear s-shape can be seen, which corresponds to spontaneous emission, super-linear amplified spontaneous emission, and lasing, providing unambiguous evidence for the achievement of lasing. Figure 7.22b shows the light output versus current, and a clear threshold around 30 µA is measured. This threshold current is orders of magnitude lower compared to the previously reported planar AlGaN quantum well lasers operating in the UV-A band at room temperature (Yoshida et al. 2008a, b). Such a low lasing threshold is attributed to the nearly defect-free AlGaN nanowires, the presence of quantum-dot-like nanostructures, and the high Q optical cavity (Zhao et al. 2015c). Moreover, with tuning the Al content, electrically injected AlGaN nanowire lasers at different wavelengths can also be realized (Li et al. 2015; Zhao et al. 2015c, d), illustrated in Figure 7.22c.

7.2.4 Controlled Coalescence of III-Nitride Nanowire Arrays

One grand challenge for III-nitride planar devices is the presence of large dislocation and defect densities, due to the large lattice and thermal mismatches between the grown epilayer and commonly available substrates. In this regard, various epitaxial lateral overgrowth (ELO) techniques, which generally involve the use of a patterned dielectric layer on the substrate, have been developed (Nishinaga et al. 1988; Akira et al. 1997; Nam et al. 1997). In this process, the initial growth only occurs selectively in the openings, followed by the lateral growth over the masked region until coalescence. For such epilayers, the portion grown above the openings has remained defective, while the portion laterally grown above the masked region can have a very low defect density. The average dislocation densities are generally on the order of 10^6–10^8 cm^{-2} (Imer et al. 2006; Ni et al. 2006; Kawashima et al. 2007). As an alternative solution, coalescence from spontaneously formed GaN nanowires has also been previously studied (Bertness et al. 2008; Consonni et al. 2009; Kato et al. 2009; Li et al. 2009; Dogan et al. 2011; Jenichen et al. 2011; Grossklaus et al. 2013; Fan et al. 2014). Due to the lack of control of the nanowire size, height, morphology, and twist/tilt in crystal orientation, the coalescence of such randomly misoriented nanowires generally leads to the presence of strain, inversion domains, low-angle grain boundaries and networks of structural defects including dislocations at the coalescence boundary. As a consequence, the realization of any practical devices has not been possible. Compared to these approaches, coalescence from nanowires on patterned substrate through a selective area epitaxy process could provide a viable path to realize large-scale defect free AlGaN thin films. In what follows, we discuss the recent advances on the coalesced III-nitride thin films from nanowires grown by selective area growth (SAG).

In general, when two misoriented nanowires coalesce at a grain boundary with a tilting angle θ, such low-angle grain boundaries will result in a periodic network of edge dislocations, preventing the realization of large-scale defect-free thin films from spontaneously formed nanowires. However, as illustrated in Figure 7.23a, if the nanowires are well aligned, that is, they have nearly identical orientation relationship with respect to the substrate, the formation of dislocations during the coalescence process is not energetically favorable, such that any small degree of distortion can be accommodated, rendering defect free film structures. In this regard, coalescence of SAG AlGaN nanowires have been investigated (Le et al. 2016). Shown in Figure 7.23b is a schematic of the formation of AlGaN film structures through coalesced AlGaN nanowires. The AlGaN nanowires are grown on Ti-patterned GaN template on sapphire substrate. It is seen that as the growth proceeds, due to the lateral growth of the nanowires, coalescence occurs. Figure 7.23c shows the coalesced AlGaN nanowire structure viewed from the top surface, highlighting the well-ordered semipolar planes.

Detailed studies have further indicated that such coalesced AlGaN film structures can exhibit a room-temperature IQE in the range of 60%–80% for emission wavelengths in the UV-A band (Le et al. 2016). This high IQE is consistent with the absence of dislocations in the coalesced film structures. The electrical properties of the coalesced semipolar p-Al$_{0.35}$Ga$_{0.65}$N layers, with a Mg doping concentration in the range of 10^{20} cm^{-3}, have also been investigated (Le et al. 2016). At room temperature, Hall mobility and hole concentration are measured to be 8.85 cm^2/V·s and 7.4×10^{18} cm^{-3}, respectively.

AlGaN UV LEDs based on the coalesced semipolar AlGaN film structures have also been demonstrated (Le et al. 2016). The device structure consists of p-/n-GaN contact layers and AlGaN double heterojunctions. Detailed studies on the structural properties have been performed. Figure 7.24a shows the cross-section low-magnification high-angle annular dark-field image. Figure 7.24b shows the

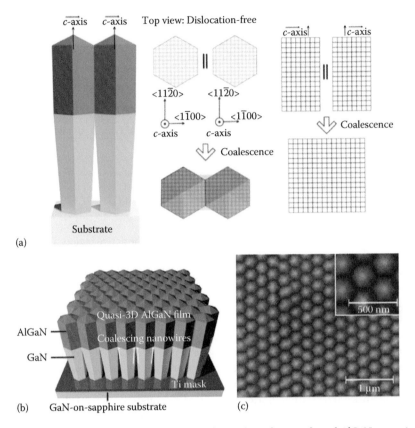

(a)

(b)

(c)

FIGURE 7.23 (a) Schematic of the formation of defect-free epilayer from coalesced AlGaN nanowires by selective area epitaxy. (b) Schematic of the coalesced AlGaN film on top of GaN nanowire template. (c) Top view SEM image of coalesced AlGaN film structure. (From Le, B. et al., *Advanced Materials*, 28, 8446–8454, 2016. With permission.)

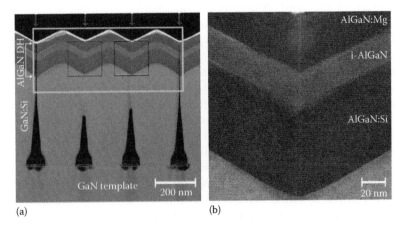

(a)

(b)

FIGURE 7.24 (a) A low-magnification HAADF–STEM image of GaN/AlGaN nanowire LED structure. (b) High-resolution HAADF–STEM image taken from the active region, highlighting the abrupt interface without any dislocation/stacking faults. (From Le, B. et al., *Advanced Materials*, 28, 8446–8454, 2016. With permission.)

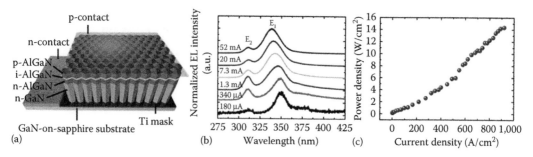

FIGURE 7.25 (a) Schematic of AlGaN UV LEDs fabricated from coalesced AlGaN film structures. (b) The EL spectra measured under different injection currents. (c) The light output power versus injection current. (From Le, B. et al., *Advanced Materials*, 28, 8446–8454, 2016. With permission.)

high-resolution high-angle annular dark-field image taken from the active region. Detailed studies have further confirmed that the coalesced AlGaN structures are free of dislocations (Le et al. 2016).

The schematic of the fabricated device is shown in Figure 7.25a. The device has a turn-on voltage ~3.3 V. The operation voltage is ~4.4 V for a current density of 100 A/cm² (CW), which is comparable to, or better than AlGaN quantum well LEDs operating at similar wavelengths (Adivarahan et al. 2001; Jeon et al. 2004; Peng et al. 2004). Illustrated in Figure 7.25b, strong emission around 340 nm is measured. The very weak emission peak at 310 nm is attributed to electron overflow and emission from the p-AlGaN layer (Le et al. 2016). The measured output power versus current is shown in Figure 7.25c. The output power increases near-linearly with the increase of injection current, and an output power density of ~15 W/cm² is measured at room temperature at an injection current density of 900 A/cm². Evidently, the controlled coalescence of nanowire arrays through selective area epitaxy offers a powerful platform to realize high-efficiency planar UV photonic devices, including LEDs, lasers, and photodetectors.

7.3 Solar Water Splitting on III-Nitride Nanowire Arrays

The chemical transformation of sunlight, water, and carbon dioxide into energy-rich fuels, commonly known as artificial photosynthesis, is perceived to be one of the key sustainable energy technologies in the future energy arena (Tachibana et al. 2012). One of the key steps in artificial photosynthesis is proton reduction (hydrogen generation) through solar water splitting, which can be realized at the solid-liquid interface of nanocrystals immersed in aqueous solutions via two strategies, including the PEC approach and the photochemical (or photocatalytic) approach (Walter et al. 2010). In PEC water splitting either one or both of the electrodes can be semiconductor photocatalyst in order to capture solar energy, and a metallic conductor is commonly used for transporting the photo-generated carriers between the electrodes. Effective carrier separation and conduction in this system demands the application of external bias and highly conductive electrolyte and substrate. Photochemical dissociation of water, on the other hand, is a spontaneous and wireless approach of harnessing solar energy in which the counter electrode is mounted on the photocatalyst surface in the form of micro/nano-electrode, commonly defined as co-catalysts (Bolton and Hall, 1979; Chen et al. 2010; Walter et al. 2010).

In the past four decades, intensive research effort has been devoted to metal oxides for solar water splitting, including TiO_2 (Chen and Mao 2007), WO_3 (Su et al. 2010), $BiVO_4$ (Kim and Choi 2014; Shi et al. 2014), and α-Fe_2O_3 (Kim et al. 2013; Liu et al. 2016). However, the large bandgaps of metal oxides (>2.1 eV) severely limit light absorption. Moreover, the electronic band structure of metal oxides does not straddle the water redox potential, shown in Figure 7.26 (Chowdhury et al. 2015), and consequently, an external bias is required to drive water splitting, thereby leading to reduced efficiency. On the other hand, semiconductors including Si and conventional III–V alloys such as GaAs, InP, and $GaInP_2$, have valence band edges that are more negative than the potential for oxygen evolution reaction (E_0(OER)).

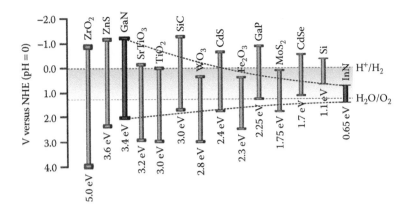

FIGURE 7.26 Energetic position of band edges (versus NHE at pH~0) and associated bandgaps (eV) of commonly used photocatalysts. The oxidation and reduction potentials of water are shown in horizontal dotted line. The band edge positions of In$_x$Ga$_{1-x}$N as a function of indium incorporation (0-1 from left to right) varies according to curved dotted lines. (From Chowdhury, F. A. et al., *APL Materials*, 3, 104408, 2015. With permission.)

It, thus, requires an external bias to drive water oxidation, the first and foremost step of solar water splitting. These materials also suffer from poor stability in aqueous solution. For these reasons, the performance of solar water splitting devices is still much inferior to that of solar cells (Lee et al. 2012; Hu et al. 2013a; Krawicz et al. 2013; Gao et al. 2014; Hu et al. 2014; Hettick et al. 2015; MacLeod et al. 2015; Standing et al. 2015; Sun et al. 2015; Gao et al. 2016). In this context, III-nitride semiconductors offer several fundamental advantages for solar water splitting. The energy bandgap of InGaN can be tuned from 0.65 eV (InN) to 3.4 eV (GaN) by simply varying the alloy compositions, which nearly covers the entire solar spectrum. III-nitrides also have high light absorption coefficients and excellent carrier transport properties (Bandić et al. 2000; Makimoto et al. 2001; Hahn et al. 2011), promising multijunction solar cells with an efficiency up to 62% (De Vos 1992). More importantly, the CBM and VBM of InGaN are suitable for water splitting with an indium composition up to 50%, which corresponds to an energy bandgap of ~1.75 eV (Wu 2009). III-nitride nanowires also offer efficient charge carrier separation and a large number of reaction sites on their lateral surfaces. Recent studies have further shown that, when grown under nitrogen-rich conditions, the nonpolar lateral surfaces of III-nitride nanowires are terminated by nitrogen that passivates the nanowire surfaces against attack by air/aqueous electrolytes (Kibria et al. 2016a).

7.3.1 Photochemical Water Splitting

Some of the initial studies of photocatalytic water splitting on III-nitrides were performed on GaN powders (Kida et al. 2006; Maeda et al. 2007; Maeda 2011), which, however, exhibit very low quantum efficiency. The photocatalytic water splitting efficiency depends critically on the efficiency of charge carrier spatial separation and extraction, which requires the dimension of the photocatalytic materials to be comparable to, or smaller than the carrier diffusion distance of photogenerated carriers. In addition, the presence of any nonradiative recombination centers should be minimized. Therefore, the development of single crystalline nanostructured photocatalysts, through precise engineering of the morphology and bandgap, is of great interest. In this context, GaN nanowires grown by MBE on Si substrate have been explored to demonstrate photochemical neutral pH water splitting (Wang et al. 2011). In a nanowire photocatalyst, wherein the carrier transport is no longer diffusion limited, the photogenerated holes and electrons can migrate toward the nanowire/electrolyte interface and take part in water splitting reactions. Moreover, Rh/Cr$_2$O$_3$ core-shell hydrogen evolution co-catalyst nanoparticles are coupled to the nitride surface for efficient separation of charge carriers (Maeda and Domen 2010; Wang et al. 2011).

By incorporating defect-free InGaN quantum dots and quantum wells in GaN nanowires grown by MBE, neutral pH water splitting under visible light irradiation has also been achieved using multiband InGaN/GaN heterostructures under blue and green light irradiation (up to 560 nm), the longest wavelength ever reported for spontaneous overall water splitting on any photocatalyst materials (Kibria et al. 2013).

Recently, it has been recognized that Fermi-level pinning and the resulting surface band bending creates an additional energy barrier for charge carrier transport to the photocatalyst surface (Kibria et al. 2014). For example, the presence of an upward band bending in an n-type semiconductor inhibits the extraction of electrons and therefore suppresses proton reduction reaction. On a p-type semiconductor photocatalyst surface, the presence of a downward band bending prevents the efficient extraction of holes and therefore limits the H_2O oxidation reaction. Unique to the nonpolar GaN surfaces is that the occupied surface states are bunched near the band edge or positioned outside of the bandgap. Recent studies have further shown that the near-surface band bending can be precisely tuned through controlled Mg-dopant incorporation in GaN and InGaN nanowire arrays (Kibria et al. 2014; Kibria et al. 2015). The downward surface band bending for a p-type GaN (or InGaN) nanowire can be minimized by optimizing the Mg doping distribution in nanowire structures. Illustrated in Figure 7.27a is the H_2 evolution rates for different GaN: Mg nanowire samples. The H_2 generation rate shows a significant increase with Mg cell temperature (Mg concentration). The low photocatalytic activity under low Mg doping concentration is due to the presence of a downward band bending that suppresses H_2O oxidation reaction. With increased Mg dopant incorporation, the downward surface band bending is reduced, leading to efficient extraction of charge carriers and accelerated water oxidation and proton reduction. The absorbed photon conversion efficiency can reach ~50% for GaN nanowires with optimum Mg-doping levels. Shown in Figure 7.27b is the evolution of H_2 and O_2 as a function of irradiation time for the optimized Mg-doped GaN sample in pure water. The H_2/O_2 ratio is nearly 2:1, suggesting a balanced redox reaction. The turnover number, which is defined as the ratio of the amount of gas generated to the amount of nanowire photocatalyst, exceeds 10,000 during 22 hr reaction, confirming the excellent stability of Mg-doped GaN nanowire photocatalyst. More recently, the atomic origin of the long-term stability and high efficiency of such InGaN nanowire arrays for photocatalytic overall water splitting has been revealed (Kibria et al. 2016b). It has been observed that the nonpolar lateral surfaces of InGaN nanowires grown by MBE are terminated by nitrogen, which can passivate the Ga(In)N nanowires against attack by aqueous electrolyte (oxidation and photocorrosion).

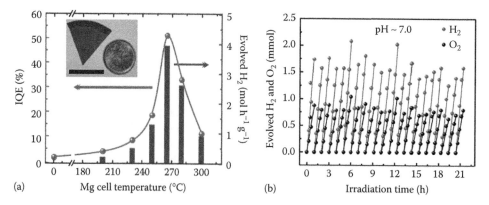

FIGURE 7.27   (a) H_2 evolution rate and internal quantum efficiency for different GaN:Mg samples in overall pure water splitting. Reactions were performed using ~0.387 mg GaN nanowire catalyst in pure water under illumination of a Xenon lamp (300 W). The sample size is shown in the inset. Scale bar 2 cm. (b) Repeated cycles of overall pure water splitting for GaN:Mg: 265°C, showing the stability of the nanowires. (From Kibria, M. G. et al., *Nature Communications*, 5, 3825, 2014. With permission.)

By optimizing the surface electronic properties of InGaN nanowire arrays through controlled Mg-dopant incorporation, an absorbed photon conversion efficiency ~69% has been achieved under visible light irradiation (Chowdhury et al. 2015; Kibria et al. 2015). The energy conversion efficiency for GaN/InGaN nanowire double band photocatalysts reaches ~7.5% for light illumination in the wavelength range of 200–475 nm. The solar-to-hydrogen efficiency, however, is still relatively low (~1.8%), which is limited by the absorption of photons in the deep visible and near-infrared spectral range. Further enhancement in the solar-to-hydrogen efficiency is under investigation by using In-rich InGaN nanowires to extend the absorption edge toward longer wavelengths.

7.3.2 PEC Water Splitting

The first demonstration of PEC water splitting on GaN was reported by Kocha, S. et al. in 1995 (Kocha et al. 1995). In their experiment, the flat-band potential of n-doped GaN thin film shifted from −1.5 V versus saturated calomel electrode (SCE) to −0.85 V versus SCE as the pH of electrolyte was varied from 14 to 2. The onset potential of photocurrent, however, was observed to be 0.5 V more positive than the corresponding flat-band potential. After loading RuO_2 co-catalyst on the surface of GaN, zero-bias water splitting has been demonstrated. Etching of GaN in PEC water splitting has also been reported (Huygens et al. 2000), which is attributed to the poor material quality as well as the Ga-polarity of the samples (Weyher et al. 1997; Yoshida 1997). It has been found that stable PEC operation on GaN can be realized in hydrogen halide acids. Later, Fujii and Ohkawa have confirmed that the oxidation reaction on n-type GaN surface is due to simultaneous oxygen generation and GaN photocorrosion (Fujii and Ohkawa, 2005). Research on the improvement of PEC performance by GaN and InGaN nanostructures in hydrogen halides has witnessed steady progress, but the photocurrent density in the potential window of 0–1.2 V versus normal hydrogen electrode (NHE) has hitherto remained less than 2 mA/cm^2 under AM1.5G one sun illumination (Ebaid et al. 2015; Tao et al. 2016). Recently, it has been found that NiO_x, as a co-catalyst, can stabilize GaN for PEC operation in strong basic solutions and stable photocurrent of over 500 hr has been reported (Ohkawa et al. 2013).

Based on the state-of-the-art oxygen evolution reaction cocatalysts and the electrocatalytic performance of Pt for hydrogen evolution, the minimum photovoltage required to drive an electrode current density of 20 mA/cm^2 is in the range of 1.7–1.9 V (Walter et al. 2010; Hu et al. 2013b). Therefore, to effectively manage photons in the solar spectrum, dual light absorbers with an energy bandgap configuration of 1.75 eV for the top light absorber and 1.12 eV for the bottom light absorber are required to sustain a matched photocurrent density of ~22 mA/cm^2 for unassisted water splitting under AM1.5G illumination of 100 mW/cm^2, corresponding to a solar-to-hydrogen (STH) efficiency of 27%. The dual light absorber can be realized by pairing a large bandgap photoanode with a Si-based photocathode. Another approach is to use a semitransparent single junction solar cell to assist a Si photoelectrode. The simplest and most scalable design is based on the monolithic integration of a top light absorber with Si substrates as the bottom light absorber. The MBE growth of InGaN nanowires on Si substrates provides a nearly ideal monolithically integrated dual light absorber, that is, InGaN nanowires with 52% of indium (1.75 eV) and Si (1.12 eV). In what follows, PEC properties of (In)GaN nanowires grown by MBE and their integration with Si solar cell wafer for solar hydrogen generation are presented.

7.3.2.1 InGaN Nanowire Photocathode

The application of III-nitrides as photocathodes for water splitting has been hindered by the difficulty in realizing p-doped InGaN. Although Aryal et al. have demonstrated that Mg-doped InGaN can function as a photocathode for hydrogen generation in HBr solution, the photocurrent density can only reach 1 mA/cm^2 at an electric bias of 1.2 V versus the Pt counter electrode under AM1.5G illumination of 132 mW/cm^2 (Aryal et al. 2010). Nakamura et al. have further shown that downward band bending in an undoped GaN epilayer can be enabled by using an undoped GaN/AlN (~5 nm)/n-doped GaN tunnel junction and that, under the illumination of 110 mW/cm^2 from a Xenon lamp, a saturated photocurrent density of ~0.5 mA/cm^2 has been observed at 0 V versus NHE in 0.5 M H_2SO_4 (Nakamura et al. 2014). In this context, we have realized a p-type InGaN nanowire photocathode on n-type Si substrate. In this work, to enable the tunneling transfer of photogenerated holes from p-type InGaN nanowires toward the Si substrate, n[+]-GaN nanowires were grown on the n[+]-Si substrate first, followed by the growth of n[++]-GaN (20 nm), InGaN (4 nm), and p[++]-GaN (20 nm) in sequence to form a polarization-enhanced tunnel junction. Pt nanoparticle co-catalyst was formed on the lateral surface of InGaN nanowires by a photodeposition process (Wang et al. 2010). The linear scanning voltammetric (LSV) curves of the InGaN nanowire photocathode is shown in Figure 7.28 (Fan et al. 2015). The photocurrent density reaches ~1.8 mA/cm^2 at 0 V versus NHE under AM1.5G illumination of 130 mW/cm^2. The PEC performance is stable in 1.0M HBr as well as in 1.0M HNO_3 solution.

Fan et al. have further demonstrated an adaptive double-junction photocathode by integrating such nanowire arrays on a planar Si solar cell wafer (Fan et al. 2015). Schematically shown in Figure 7.29a, the InGaN/Si photocathode consists of a planar n[+]-p Si solar cell wafer and tunnel junction nanowire arrays with an n[+]-GaN bottom segment and p-InGaN top segment along the axial direction of each nanowire. The InGaN nanowire segment on top has an absorption edge of ~510 nm, while the underlying n[+]-p Si junction absorbs deep visible light and infrared light photons up to 1100 nm. Herein, the n[++]-GaN/ InGaN/p[++]-GaN polarization-enhanced tunnel junction is used to connect the top InGaN and the bottom Si light absorbers, which can facilitate the transfer of photogenerated holes in InGaN toward the n[+]-GaN bottom segment to recombine with photogenerated electrons from the n[+]-p Si junction. As shown by the STEM images of an InGaN tunnel junction nanowire after photodeposition in Figure 7.29b–d, Pt nanoparticles are uniformly distributed on the lateral surfaces of the top InGaN segments and the bottom n[+]-GaN segment, which indicates that both the top InGaN and the bottom GaN/Si light absorbers can drive proton reduction. Under illumination, the photogenerated electrons in the p-InGaN nanowire top light absorber migrate toward the nanowire/electrolyte interface due to the downward band bending, while the photogenerated electrons in the n[+]-p Si bottom light absorber inject into the n[+]-Si bottom nanowire segment by taking advantage of the negligible conduction band offset between the n[+]-Si emitter and the n[+]-GaN nanowire base (Guha and Bojarczuk 1998). The majority of the injected electrons can drive proton reduction on the lateral surfaces of the n[+]-GaN nanowire base with a small portion of

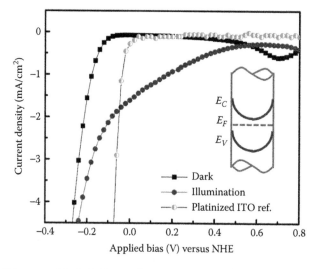

FIGURE 7.28 The LSV curves of the InGaN tunnel junction nanowire photocathode under dark (dark squares) and 1.3 sun of AM1.5G illumination (dark circles). Indium tin oxide (ITO) substrate (12 Ω cm) with 5 μg/cm^2 Pt as a reference, represented by the curve with light half-filled circles. The inset shows the lateral band diagram of the n[+]-GaN bottom nanowire segment. (From Fan, S. et al., *Nano Letters*, 15, 2721–2726, 2015. With permission.)

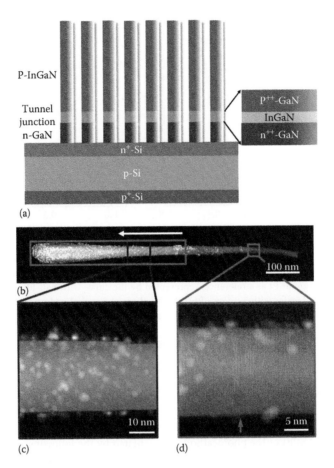

FIGURE 7.29   (a) The schematic of the photocathode formed by InGaN tunnel junction nanowires on n⁺-p Si solar cell substrate. (b), (c) STEM images of an InGaN tunnel junction nanowire with Pt nanoparticles. The white arrow indicates the growth direction. Two additional STEM images show that Pt nanoparticles are uniformly distributed (b) on the p-InGaN top nanowire segment and (c) on the GaN bottom nanowire segment. The gray arrow indicates the tunnel junction. (From Fan, S. et al., *Nano Letters*, 15, 2721–2726, 2015. With permission.)

the electrons recombining with holes from the p-InGaN in the adaptive tunnel junction. Such a unique design can surpass the restriction of current matching in conventional dual absorber devices and at the same time provide energetic photogenerated electrons from each light absorber to the HER catalysts, which is made possible by the unique lateral carrier extraction of the 1D nanowire structure.

The accumulation of photogenerated holes in p-InGaN and their transport across the tunnel junction can promote the flux of photogenerated electrons in the n⁺-p Si junction toward the n⁺-GaN bottom nanowire segment. Therefore, interfacial resistance originated from the n⁺-Si/n⁺-GaN heterointerface and the tunnel junction is expected to be lower in the presence of the p-InGaN top light absorber. Moreover, the InGaN tunnel junction nanowire arrays play a critical role in enhancing the performance of the underlying Si solar cell due to the enhanced light absorption by nanowire light trapping effect, efficient carrier extraction from the large lateral surface of nanowires, and reduced surface recombination by extracting the photogenerated electrons from the planar Si into n⁺-GaN nanowires.

After photodeposition of Pt nanoparticles, the PEC performance of the InGaN/Si double-band adaptive junction photocathode was measured, shown in Figure 7.30a (Fan et al. 2015). It is seen that the electrical bias (versus NHE) required for a photocurrent density of –2 mA/cm² shifts negatively from

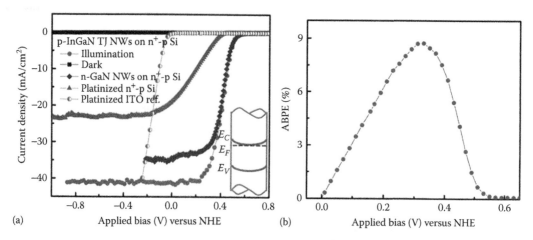

FIGURE 7.30 (a) The LSV curves of InGaN tunnel junction nanowires grown on n$^+$-p Si substrate under dark (fully filled black squares) and under AM1.5G illumination of 130 mW/cm^2 (fully filled gray circles). ITO substrate (12 Ω cm) with 5 µg/cm^2 Pt as a reference, represented by the curve with half-filled gray circles. The LSV curves of n$^+$-p Si solar cell wafer with 5 µg/cm^2 Pt (fully filled gray triangles) and n$^+$-GaN nanowires grown on n$^+$-p Si substrate (fully filled dark gray diamonds) are also shown for comparison. The inset shows the lateral band diagram of the n$^+$-GaN bottom nanowire segment. (b) The ABPE of the photocathode formed by InGaN tunnel junction nanowires on n$^+$-p Si substrate. (From Fan, S. et al., *Nano Letters*, 15, 2721–2726, 2015. With permission.)

−0.08 V on the InGaN tunnel junction nanowire photocathode to 0.5 V on the InGaN/Si adaptive tunnel junction photocathode, and more importantly, a saturated photocurrent density of −40.5 mA/cm^2 at 0.23 V versus NHE is observed under AM1.5G illumination of 130 mW/cm^2. The applied bias photon-to-current efficiency (ABPE) of the photocathode is calculated by the following equation assuming an ideal counter electrode.

$$\mathrm{ABPE} = \frac{\left| J\,[\mathrm{mA/cm^2}] \right| \times \left(1.23\mathrm{V} - |V_b| \right)}{130\,[\mathrm{mW/cm^2}]} \times 100\% \tag{7.1}$$

where V_b is the applied bias versus NHE. The maximal ABPE reaches 8.7% at 0.33 V versus NHE, shown in Figure 7.30b.

Another two types of photocathodes, n$^+$-p Si solar cell substrates and n$^+$-GaN nanowires grown on the n$^+$-p Si solar cell substrate were studied, both of which were deposited with Pt cocatalyst prior to PEC experiment. The n$^+$-GaN nanowires on n$^+$-p Si solar cell exhibited a similar onset potential and reduced saturated photocurrent density of −35 mA/cm^2. The compromised PEC performance in the absence of p-InGaN top nanowire segments and the tunnel junction indicates the increase in interfacial resistance and the reduction of electron migration toward the lateral surface of nanowires. In addition, the photocathode of the n$^+$-p Si solar cell exhibited a much positive onset potential (−0.36 V versus NHE for −2 mA/cm^2) and a low saturated photocurrent density of −23 mA/cm^2. The drastic improvement on PEC performance of n$^+$-p Si solar cell photocathode by the MBE growth of n$^+$-GaN nanowires is due to the increased light absorption in Si due to the anti-reflection effect of vertically aligned nanowire arrays, the effective carrier separation realized by the injection of photogenerated electrons from Si to GaN nanowire, and the reduced hydrogen evolution overpotential at the (In)GaN/Pt/electrolyte interface compared to that at the Si/Pt/electrolyte interface. In comparison with the state-of-art Si-based photocathodes (Boettcher et al. 2011; Esposito et al. 2013; Seger et al. 2013; Ji et al. 2014), the onset potential of 0.55 V versus NHE (for −0.5 mA/cm^2) is among the most positive values, and the saturated photocurrent density is among the highest values after normalized to one sun.

7.3.2.2 InGaN Nanowire Photoanodes

An InGaN/Si tandem photoanode can be realized by MBE growth of In-rich InGaN nanowires on a p⁺-n Si substrate with a Si tunnel junction. Schematic of the proposed InGaN nanowire/Si double-band photoanode is illustrated in Figure 7.31a. The Si tunnel junction can be formed by fabricating a thin layer of n⁺-Si above the p⁺-Si emitter in the n-type Si solar cell. An n⁺-GaN bottom nanowire segment is grown first as a hole blocking layer to separate the photogenerated holes in InGaN from the Si tunnel junction. The migration of photogenerated holes in n-type InGaN nanowires toward the nanowire/electrolyte interface is driven by the upward band bending at the lateral nanowire surface. The photogenerated holes from Si and photogenerated electrons from InGaN recombine at the Si tunnel junction, which satisfies the current matching condition under standard AM1.5G illumination and can add up the photovoltage from both light absorbers. Besides, the nanowire light trapping effect can enhance light absorption in the InGaN nanowires as well as in the underlying p⁺-n Si junction.

The room-temperature photoluminescence spectrum of InGaN nanowires is shown in Figure 7.31b. The emission peak is at 725 nm, corresponding to a bandgap of about 1.7 eV. Two types of photoanodes, including InGaN nanowires grown on n⁺-Si substrates and InGaN nanowires grown on n⁺-p Si solar cell substrates with the presence of the Si tunnel junction, were tested in 1 M HBr solution. Shown in Figures 7.32a, b, the InGaN nanowire photoanode exhibited an onset potential of 0.55 V and reached a photocurrent density of 2 mA/cm² at 0.76 V, with an applied bias photon-to-current (ABPE) efficiency of 0.7%, while the InGaN/Si double-band photoanode exhibited an onset potential of 0.2 V and the saturated photocurrent density reached 16.3 mA/cm² with an ABPE of 8.3%. Under AM1.5G illumination of 100 mW/cm², such a photocurrent density is significantly higher than other InGaN-based photoelectrodes or photovoltaic devices. By comparing the two photoanodes, efficient tunneling recombination of photogenerated electrons in InGaN with the photo-generated holes in the buried p⁺-n Si junction was validated by the significant cathodic shift of the onset potential and the dramatically improved fill factor of the polarization curve. However, an electric bias of 0.2–0.3 V was required to drive PEC hydrogen generation. One major cause of photovoltage loss could be the inevitable formation of a resistive SiN$_x$ layer (~2 nm) at the GaN/Si heterointerface during the self-assembled nucleation process of InGaN nanowires in MBE.

The onset potential needs to be further improved to enable zero bias water splitting by the double-band photoanode. The photovoltage from the Si bottom cell is about 0.5 V. With conductive passivation, the photovoltage of Si is expected to exceed 0.7 V. The deposition of a conductive layer such as titanium nitride to protect Si from the formation of a SiN$_x$ layer at the nanowire/Si interface could also improve

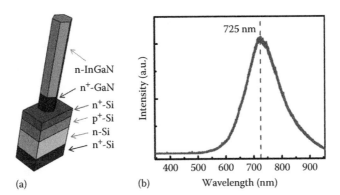

(a)   (b)

FIGURE 7.31 (a) Schematic of InGaN/Si nanowire double-band photoanode. The Si solar cell consists of an n⁺-Si back field layer, a n-Si thick base layer, a p⁺-Si emitter layer, and an n⁺-Si ultrathin layer to form Si tunnel junction with the p⁺-Si emitter. Each nanowire includes an n⁺-GaN bottom segment to block photogenerated holes in InGaN, and an n-InGaN top segment as the top light absorber. (b) Photoluminescence spectrum of InGaN nanowires measured at room temperature. (From Fan, S. et al., *Advanced Energy Materials*, 1600952, 2017. With permission.)

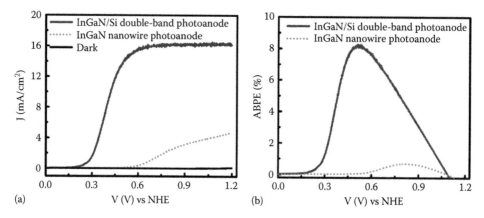

FIGURE 7.32 (a) LSV scans of the InGaN/Si double-band photoanode (gray solid curve) and the InGaN nanowire photoanode (light gray dotted curve) under illumination. The black curve represents dark current density from the InGaN nanowire photoanode. The scan rate is 20 mV/sec. (b) The ABPE of the InGaN nanowire photoanode (gray solid curve) and the InGaN/Si double-band photoanode (light gray dotted curve). (From Fan, S. et al., *Advanced Energy Materials*, 1600952, 2017. With permission.)

the PEC performance. The photovoltage of InGaN nanowires can also be improved by forming a p⁺-n InGaN lateral junction. Moreover, the charge separation efficiency at the InGaN/electrolyte interface also needs to be improved by passivating the surface of InGaN nanowires.

7.4 Summary

In summary, we have provided an overview of the recent advances made on III-nitride nanowires, with a special focus on the MBE growth of InN, AlN, In-rich InGaN, and Al-rich AlGaN nanowire structures. By optimizing the epitaxial growth conditions, efficient p-type conduction has been demonstrated for InN and AlN nanowires. The incorporation of quantum-confined nanostructures in InGaN and AlGaN nanowire arrays has led to high-efficiency full-color LEDs on a single chip as well as electrically pumped semiconductor lasers operating in the UV-B and UV-C bands. III-nitride nanowires have also emerged as a new generation of photocatalysts for high-efficiency artificial photosynthesis, due to their tunable energy bandgap and distinct photocatalytic properties. Moreover, the direct integration of III-nitride nanowire photocatalysts with Si solar cell wafers offers a new avenue for realizing high efficiency, low cost, scalable, and stable production of solar fuels.

References

Adivarahan, V., A. Chitnis, J. P. Zhang, M. Shatalov, J. W. Yang, G. Simin, M. A. Khan, R. Gaska, and M. S. Shur. Ultraviolet light-emitting diodes at 340 nm using quaternary AlInGaN multiple quantum wells. *Applied Physics Letters* 79(25): 4240–4242 (2001).

Ahn, H., C.-C. Yu, P. Yu, J. Tang, Y.-L. Hong, and S. Gwo. Carrier dynamics in InN nanorod arrays. *Optics Express* 20(2): 769–775 (2012).

Akira, U., S. Haruo, S. Akira, and A. A. Yamaguchi. Thick GaN epitaxial growth with low dislocation density by hydride vapor phase epitaxy. *Japanese Journal of Applied Physics* 36(7B): L899 (1997).

Albert, S., A. Bengoechea-Encabo, M. A. Sánchez-García, X. Kong, A. Trampert, and E. Calleja. Selective area growth of InGaN/GaN nanocolumns by molecular beam epitaxy on gan-buffered Si (111): From ultraviolet to infrared emission. *Nanotechnology* 24(17): 175303 (2013).

Aryal, K., B. N. Pantha, J. Li, J. Y. Lin, and H. X. Jiang. Hydrogen generation by solar water splitting using P-InGaN photoelectrochemical cells. *Applied Physics Letters* 96: 052110 (2010).

Balasubramanian, C., V. P. Godbole, V. K. Rohatgi, A. K. Das, and S. V. Bhoraskar. Synthesis of nanowires and nanoparticles of cubic aluminium nitride. *Nanotechnology* 15(3): 370–373 (2004).

Bandić, Z. Z., P. M. Bridger, E. C. Piquette, and T. C. McGill. The values of minority carrier diffusion lengths and lifetimes in GaN and their implications for bipolar devices. *Solid-State Electronics* 44(2): 221–228 (2000).

Bengoechea-Encabo, A., F. Barbagini, S. Fernandez-Garrido, J. Grandal, J. Ristic, M. A. Sanchez-Garcia, E. Calleja. Understanding the selective area growth of GaN nanocolumns by MBE using Ti Nanomasks. *Journal of Crystal Growth* 325(1): 89–92 (2011).

Bertness, K. A., A. Roshko, L. M. Mansfield, T. E. Harvey, and N. A. Sanford. Mechanism for spontaneous growth of GaN nanowires with molecular beam epitaxy. *Journal of Crystal Growth* 310(13): 3154–3158 (2008).

Bertness, K. A., A. Roshko, N. A. Sanford, J. M. Barker, and A. V. Davydov. Spontaneously grown GaN and AlGaN nanowires. *Journal of Crystal Growth* 287(2): 522–527 (2006).

Bertness, K. A., A. W. Sanders, D. M. Rourke, T. E. Harvey, A. Roshko, J. B. Schlager, and N. A. Sanford. Controlled nucleation of GaN nanowires grown with molecular beam epitaxy. *Advanced Functional Materials* 20(17): 2911–2915 (2010).

Boettcher, S. W., E. L. Warren, M. C. Putnam, E. A. Santori, D. Turner-Evans, M. D Kelzenberg, and M. G. Walter. Photoelectrochemical hydrogen evolution using Si microwire arrays. *Journal of the American Chemical Society* 133(5): 1216–1219 (2011).

Bolton, J. R. and D. O. Hall. Photochemical conversion and storage of solar energy. *Annual Review of Energy* 4(1): 353–401 (1979).

Calarco, R. InN nanowires: Growth and optoelectronic properties. *Materials* 5(11): 2137–2150 (2012).

Calleja, E., J. Grandal, M. A. Sánchez-García, M. Niebelschütz, V. Cimalla, and O. Ambacher. Evidence of electron accumulation at nonpolar surfaces of InN nanocolumns. *Applied Physics Letters* 90(26): 262110 (2007b).

Calleja, E., J. Ristić, S. Fernández-Garrido, L. Cerutti, M. A. Sánchez-García, J. Grandal, A. Trampert et al. Growth, morphology, and structural properties of group-III-nitride nanocolumns and nanodisks. *Physica Status Solidi* (b) 244(8): 2816–2837 (2007b).

Carnevale, S. D., T. F. Kent, P. J. Phillips, M. J. Mills, S. Rajan, and R. C. Myers. Polarization-induced Pn diodes in wide-band-gap nanowires with ultraviolet electroluminescence. *Nano Letters* 12(2): 915–920 (2012).

Carnevale, S. D., T. F. Kent, P. J. Phillips, A. T. Sarwar, C. Selcu, R. F. Klie, and R. C. Myers. Mixed polarity in polarization-induced p-n junction nanowire light emitting diodes. *Nano Letters* 13(7): 3029–3035 (2013).

César, M., Y. Ke, W. Ji, H. Guo, and Z. Mi. Band gap of $In_xGa_{1-x}N$: A first principles analysis. *Applied Physics Letters* 98(20): 202107 (2011).

Chang, C. C., C. Y. Chi, M. Yao, N. Huang, C. C. Chen, J. Theiss, A. W. Bushmaker et al. Electrical and optical characterization of surface passivation in gaas nanowires. *Nano Letters* 12(9): 4484–4489 (2012).

Chang, C.-Y., G.-C. Chi, W.-M. Wang, L.-C. Chen, K.-H. Chen, F. Ren, and S. J. Pearton. Transport properties of InN nanowires. *Applied Physics Letters* 87(9): 093112 (2005).

Chang, Y.-L., F. Li, A. Fatehi, and Z. Mi. Molecular beam epitaxial growth and characterization of non-tapered InN nanowires on Si (111). *Nanotechnology* 20(34): 345203 (2009).

Chang, Y.-L., J. L. Wang, F. Li, and Z. Mi. High efficiency green, yellow, and amber emission from InGaN/GaN dot-in-a-wire heterostructures on Si (111). *Applied Physics Letters* 96(1): 013106 (2010).

Chang, Y.-M. and S. Gwo. Carrier and phonon dynamics of wurtzite InN nanorods. *Applied Physics Letters* 94(7): 071911 (2009).

Chen, F., X. Ji, Z. Lu, Y. Shen, and Q. Zhang. Shape-controlled synthesis of one-dimensional $Al_xGa_{1-x}N$ nanotower arrays: Structural characteristics and growth mechanism. *Journal of Physics D: Applied Physics* 47(6): 065311 (2014a).

Chen, F., X. Ji, Z. Lu, Y. Shen, and Q. Zhang. Structural and raman properties of compositionally tunable $Al_xGa_{1-x}N$ (0.66≤X≤1) nanowires. *Materials Science and Engineering: B* 183: 24–28 (2014b).

Chen, F., X. Ji, and Q. Zhang. Morphology-controlled synthesis and structural characterization of ternary $Al_xGa_{1-x}N$ nanostructures by chemical vapor deposition. *CrystEngComm* 17(6): 1249–1257 (2015).

Chen, W. D., X. Q. Li, L. H. Duan, X. L. Xie, and Y. D. Cui. Photoluminescence enhancement of $(NH_4)_2S_x$ passivated Inp surface by rapid thermal annealing. *Applied Surface Science* 100–101: 592–595 (1996).

Chen, X. and S. S. Mao. Titanium dioxide nanomaterials: Synthesis, properties, modifications, and applications. *Chemical Reviews* 107(7): 2891–2959 (2007).

Chen, X., S. Shen, L. Guo, and S. S. Mao. Semiconductor-based photocatalytic hydrogen generation. *Chemical Reviews* 110(11): 6503–6570 (2010).

Chevtchenko, S. A., M. A. Reshchikov, Q. Fan, X. Ni, Y. T. Moon, A. A. Baski, and H. Morkoç. Study of SiN_x and SiO_2 passivation of GaN surfaces. *Journal of Applied Physics* 101(11): 113709 (2007).

Chowdhury, F. A., Z. Mi, M. G. Kibria, and M. L. Trudeau. Group III-nitride nanowire structures for photocatalytic hydrogen evolution under visible light irradiation. *APL Materials* 3(10): 104408 (2015).

Connie, A. T., S. Zhao, S. M. Sadaf, I. Shih, Z. Mi, X. Du, J. Lin, and H. Jiang. Optical and electrical properties of Mg-doped AlN nanowires grown by molecular beam epitaxy. *Applied Physics Letters* 106(21): 213105 (2015).

Consonni, V., M. Knelangen, U. Jahn, A. Trampert, L. Geelhaar, and H. Riechert. Effects of nanowire coalescence on their structural and optical properties on a local scale. *Applied Physics Letters* 95(24): 3 (2009).

Cuscó, R., E. Alarcón-Lladó, J. Ibáñez, T. Yamaguchi, Y. Nanishi, and L. Artús. Raman scattering study of background electron density in Inn: A hydrodynamical approach to the LO-phonon-plasmon coupled modes. *Journal of Physics: Condensed Matter* 21(41): 415801 (2009).

Cuscó, R., N. Domenech-Amador, L. Artu´s, T. Gotschke, K. Jeganathan, T. Stoica, and R. Calarco. Probing the electron density in undoped, Si-doped, and Mg-doped Inn nanowires by means of Raman scattering. *Applied Physics Letters* 97(22): 221906 (2010).

Debnath, R. K., R. Meijers, T. Richter, T. Stoica, R. Calarco, and H. Lüth. Mechanism of molecular beam epitaxy growth of GaN nanowires on Si (111). *Applied Physics Letters* 90(12): 123117 (2007).

Deshpande, S., J. Heo, A. Das, and P. Bhattacharya. Electrically driven polarized single-photon emission from an InGaN quantum dot in a GaN nanowire. *Nature Communications* 4: 1675 (2013).

De Vos, A. *Endoreversible Thermodynamics of Solar Energy Conversion.* John Wiley & Sons: Weinheim, Germany, 1992.

Dogan, P., O. Brandt, C. Pfuller, J. Lahnemann, U. Jahn, C. Roder, A. Trampert, L. Geelhaar, and H. Riechert. Formation of high-quality GaN microcrystals by pendeoepitaxial overgrowth of GaN nanowires on Si (111) by molecular beam epitaxy. *Crystal Growth and Design* 11(10): 4257–4260 (2011).

Domènech-Amador, N., R. Cuscó, R. Calarco, T. Yamaguchi, Y. Nanishi, and L. Artús. Double resonance raman effects in Inn nanowires. *Physica Status Solidi (RRL) – rapid research letters* 6(4): 160–162 (2012).

Doppalapudi, D., S. Basu, K. Ludwig, and T. Moustakas. Phase separation and ordering in InGaN alloys grown by molecular beam epitaxy. *Journal of Applied Physics* 84(3): 1389–1395 (1998).

Ebaid, M., J.-H. Kang, S.-H. Lim, J.-S. Ha, J. K. Lee, Y.-H. Cho, and S.-W. Ryu. Enhanced solar hydrogen generation of high density, high aspect ratio, coaxial InGaN/GaN multi-quantum well nanowires. *Nano Energy* 12: 215–223 (2015).

Ebert, P., S. Schaafhausen, A. Lenz, A. Sabitova, L. Ivanova, M. Dahne, Y. L. Hong, S. Gwo, and H. Eisele. Direct measurement of the band gap and fermi level position at InN (1120). *Applied Physics Letters* 98(6): 062103 (2011).

Esposito, D. V, I. Levin, T. P. Moffat, and A. A. Talin. H_2 evolution at Si-based metal–insulator–semiconductor photoelectrodes enhanced by inversion channel charge collection and H spillover. *Nature Materials* 12(6): 562–568 (2013).

Fan, S., B. AlOtaibi, S. Y Woo, Y. Wang, G. A. Botton, and Z. Mi. High efficiency solar-to-hydrogen conversion on a monolithically integrated InGaN/GaN/Si adaptive tunnel junction photocathode. *Nano Letters* 15(4): 2721–2726 (2015).

Fan, S., I. Shih, and Z. Mi. A monolithically integrated InGaN nanowire/Si tandem photoanode approaching the ideal bandgap configuration of 1.75/1.13 eV. *Advanced Energy Materials*: 1600952 (2017). doi: 10.1002/aenm.201600952.

Fan, S., S. Y. Woo, S. Vanka, G. A. Botton, and Z. Mi. An $In_{0.5}Ga_{0.5}N$ nanowire photoanode for harvesting deep visible light photons. *APL Materials* 4: 076106 (2016).

Fan, S., S. Zhao, X. Liu, and Z. Mi. Study on the coalescence of dislocation-free GaN nanowires on Si and SiO_x. *Journal of Vacuum Science and Technology B* 32(2): 5 (2014).

Fang, Z., E. Robin, E. Rozas-Jimenez, A. Cros, F. Donatini, N. Mollard, J. Pernot, and B. Daudin. Si donor incorporation in GaN nanowires. *Nano Letters* 15(10): 6794–6801 (2015).

Feneberg, M., J. Däubler, K. Thonke, R. Sauer, P. Schley, and R. Goldhahn. Mahan excitons in degenerate wurtzite Inn: Photoluminescence spectroscopy and reflectivity measurements. *Physical Review B* 77(24): 245207 (2008).

Fujii, K. and K. Ohkawa. Photoelectrochemical properties of P-Type GaN in comparison with N-Type GaN. *Japanese Journal of Applied Physics* 44(7L): L909 (2005).

Gačević, Ž., D. G. Sanchez, and E. Calleja. Formation mechanisms of GaN nanowires grown by selective area growth homoepitaxy. *Nano Letters* 15(2): 1117–1121 (2015).

Gao, L., Y. Cui, R. H. J. Vervuurt, D. van Dam, R. P. J. van Veldhoven, J. P. Hofmann, and A. A. Bol. High-efficiency InP-based photocathode for hydrogen production by interface energetics design and photon management. *Advanced Functional Materials* 26(5): 679–686 (2016).

Gao, L., Y. Cui, J. Wang, A. Cavalli, A. Standing, T. T. T. Vu, and M. A. Verheijen. Photoelectrochemical hydrogen production on In P nanowire arrays with molybdenum sulfide electrocatalysts. *Nano Letters* 14(7): 3715–3719 (2014).

González-Posada, F., R. Songmuang, M. D. Hertog, and E. Monroy. Room-temperature photodetection dynamics of single GaN nanowires. *Nano Letters* 12(1): 172–176 (2012).

Grossklaus, K. A., A. Banerjee, S. Jahangir, P. Bhattacharya, and J. M. Millunchick. Misorientation defects in coalesced self-catalyzed GaN nanowires. *Journal of Crystal Growth* 371: 142–147 (2013).

Guha, S. and N. A. Bojarczuk. Ultraviolet and violet GaN light emitting diodes on silicon. *Applied Physics Letters* 72(4): 415–417 (1998).

Guo, W., A. Banerjee, P. Bhattacharya, and B. S. Ooi. In GaN/GaN disk-in-nanowire white light emitting diodes on (001) silicon. *Applied Physics Letters* 98(19): 193102 (2011).

Hahn, C., Z. Zhang, A. Fu, C. H. Wu, Y. J. Hwang, D. J. Gargas, and P. Yang. Epitaxial growth of InGaN nanowire arrays for light emitting diodes. *ACS Nano* 5(5): 3970–3976 (2011).

Hannah, J. J., J. D. Callum, G. Qiang, H. Hoe Tan, J. Chennupati, L.-H. James, M. Herz Laura, and B. J. Michael. Electronic properties of GaAs, InAs and InP nanowires studied by terahertz spectroscopy. *Nanotechnology* 24(21): 214006 (2013).

He, C., Q. Wu, X. Wang, Y. Zhang, L. Yang, N. Liu, Y. Zhao, Y. Lu, and Z. Hu. Growth and characterization of ternary AlGaN alloy nanocones across the entire composition range. *ACS Nano* 5: 1291 (2011).

Hestroffer, K., C. Leclere, C. Bougerol, H. Renevier, and B. Daudin. Polarity of GaN nanowires grown by plasma-assisted molecular beam epitaxy on Si (111). *Physical Review B* 84(24): 245302 (2011).

Hettick, M., M. Zheng, Y. Lin, C. M. Sutter-Fella, J. W. Ager, and A. Javey. Nonepitaxial thin-film In p for scalable and efficient photocathodes. *The Journal of Physical Chemistry Letters* 6(12): 2177–2182 (2015).

Nguyen, H. P. T., K. Cui, S. Zhang, S. Fathololoumi, and Z. Mi. Full-color InGaN/GaN dot-in-a-wire light emitting diodes on silicon. *Nanotechnology* 22(44): 445202 (2011a).

Nguyen, H. P. T., M. Djavid, S. Y. Woo, X. Liu, A. T. Connie, S. Sadaf, Q. Wang et al. Engineering the carrier dynamics of InGaN nanowire white light-emitting diodes by distributed p-AlGaN electron blocking layers. *Scientific Reports* 5: 7744 (2015).

Nguyen, H. P. T., Q. Wang, and Z. Mi. Phosphor-free InGaN/GaN dot-in-a-wire white light-emitting diodes on copper substrates. *Journal of Electronic Materials* 43(4): 868–872 (2014).

Nguyen, H. P. T., S. Zhang, K. Cui, X. Han, S. Fathololoumi, M. Couillard, G. A. Botton, and Z. Mi. P-type modulation doped InGaN/GaN dot-in-a-wire white-light-emitting diodes monolithically grown on Si (111). *Nano Letters* 11(5): 1919–1924 (2011b).

Nguyen, H. P. T., S. Zhang, K. Cui, A. Korinek, G. A. Botton, and Z. Mi. High-efficiency InGaN/GaN dot-in-a-wire red light-emitting diodes. *IEEE Photonics Technology Letters* 24(4): 321–323 (2012).

Holmes, M. J., K. Choi, S. Kako, M. Arita, and Y. Arakawa. Room-temperature triggered single photon emission from a III-Nitride site-controlled nanowire quantum dot. *Nano Letters* 14(2): 982–986 (2014).

Hu, S., C.-Y. Chi, K. T. Fountaine, M. Yao, H. A. Atwater, P. D. Dapkus, N. S. Lewis, and C. Zhou. Optical, electrical, and solar energy-conversion properties of gallium arsenide nanowire-array photoanodes. *Energy and Environmental Science* 6(6): 1879–1890 (2013a).

Hu, S., M. R. Shaner, J. A. Beardslee, M. Lichterman, B. S. Brunschwig, and N. S. Lewis. Amorphous Tio2 coatings stabilize Si, Gaas, and gap photoanodes for efficient water oxidation. *Science* 344(6187): 1005–1009 (2014).

Hu, S., C. Xiang, S. Haussener, A. D. Berger, and N. S. Lewis. An analysis of the optimal band gaps of light absorbers in integrated tandem photoelectrochemical water-splitting systems. *Energy and Environmental Science* 6(10): 2984–2993 (2013b).

Huygens, I., K. Strubbe, and W. P. Gomes. Electrochemistry and photoetching of N-GaN. *Journal of the Electrochemical Society* 147(5): 1797–1802 (2000).

Iliopoulos, E. and T. D. Moustakas. Growth kinetics of AlGaN films by plasma-assisted molecular-beam epitaxy. *Applied Physics Letters* 81(2): 295 (2002).

Imer, B. M., F. Wu, S. P. DenBaars, and J. S. Speck. Improved quality (1120) a-plane GaN with sidewall lateral epitaxial overgrowth. *Applied Physics Letters* 88(6): 061908 (2006).

Ishizawa, S., K. Kishino, R. Araki, A. Kikuchi, and S. Sugimoto. Optically pumped green (530–560 Nm) stimulated emissions from InGaN/GaN multiple-quantum-well triangular-lattice nanocolumn arrays. *Applied Physics Express* 4(5): 055001 (2011).

Jahangir, S., M. Mandl, M. Strassburg, and P. Bhattacharya. Molecular beam epitaxial growth and optical properties of red-emitting (λ = 650 nm) InGaN/GaN disks-in-nanowires on silicon. *Applied Physics Letters* 102(7): 071101 (2013).

Jeganathan, K., V. Purushothaman, R. K. Debnath, R. Calarco, and H. Luth. Raman scattering on intrinsic surface electron accumulation of InN nanowires. *Applied physics letters* 97(9): 093104 (2010).

Jenichen, B., O. Brandt, C. Pfuller, P. Dogan, M. Knelangen, and A. Trampert. Macro- and micro-strain in GaN nanowires on Si (111). *Nanotechnology* 22(29): 5 (2011).

Ji, L., M. D. McDaniel, S. Wang, A. B. Posadas, X. Li, H. Huang, J. C. Lee. A silicon-based photocathode for water reduction with an epitaxial srtio3 protection layer and a nanostructured catalyst. *Nature Nanotechnology* 10(1): 84–90 (2014).

Johnson, M. C., C. J. Lee, E. D. Bourret-Courchesne, S. L. Konsek, S. Aloni, W. Q. Han, and A. Zettl. Growth and morphology of 0.80 eV photoemitting indium nitride nanowires. *Applied Physics Letters* 85(23): 5670–5672 (2004).

Joyce, H. J., C. J. Docherty, Q. Gao, H. H. Tan, C. Jagadish, J. Lloyd-Hughes, L. M. Herz and M. B. Johnston. Electronic properties of GaAs, InAs and InP nanowires studied by terahertz spectroscopy. *Nanotechnology* 24(21): 214006 (2013).

Kamimura, J., K. Kishino, and A. Kikuchi. Photoluminescence properties of selectively grown Inn microcrystals. *Physica Status Solidi (RRL)-Rapid Research Letters* 6(4): 157–159 (2012).

Kato, K., K. Kishino, H. Sekiguchi, and A. Kikuchi. Overgrowth of GaN on be-doped coalesced GaN nanocolumn layer by rf-plasma-assisted molecular-beam epitaxy-formation of high-quality GaN microcolumns. *Journal of Crystal Growth* 311(10): 2956–29561 (2009).

Kawashima, T., T. Nagai, D. Iida, A. Miura, Y. Okadome, Y. Tsuchiya, M. Iwaya et al. Reduction in defect density over whole area of (100)m -plane GaN using one-sidewall seeded epitaxial lateral overgrowth. *Physica Status Solidi (b)* 244(6): 1848–1852 (2007).

Kishino, K., T. Hoshino, S. Ishizawa, and A. Kikuchi. Selective-area growth of GaN nanocolumns on titanium-mask-patterned silicon (111) substrates by RF-plasma-assisted molecular-beam epitaxy. *Electronics Letters* 44(13): 819–821 (2008).

Kishino, K., K. Nagashima, and K. Yamano. Monolithic integration of InGaN-based nanocolumn light-emitting diodes with different emission colors. *Applied Physics Express* 6(1): 012101 (2013).

Kishino, K., A. Yanagihara, K. Ikeda, and K. Yamano. Monolithic integration of four-colour InGaN-based nanocolumn LEDs. *Electronics Letters* 51(11): 852–854 (2015).

Kenry, K.-T. Y. and S. F. Yu. AlN nanowires: Synthesis, physical properties, and nanoelectronics applications. *Journal of Materials Science* 47(14): 5341–5360 (2012).

Kent, T. F., S. D. Carnevale, A. T. Sarwar, P. J. Phillips, R. F. Klie, and R. C. Myers. Deep ultraviolet emitting polarization induced nanowire light emitting diodes with $Al_xGa_{1-x}N$ active regions. *Nanotechnology* 25(45): 455201 (2014).

Kibria, M., F. A. Chowdhury, S. Zhao, B. AlOtaibi, M. L. Trudeau, H. Guo, and Z. Mi. Visible light-driven efficient overall water splitting using p-type metal-nitride nanowire arrays. *Nature Communications* 6: 6797 (2015).

Kibria, M. G, H. P. T. Nguyen, K. Cui, S. Zhao, D. Liu, H. Guo, and M. L. Trudeau. One-step overall water splitting under visible light using multiband InGaN/GaN nanowire heterostructures. *ACS Nano* 7(9): 7886–7893 (2013).

Kibria, M. G. and Z. Mi. Artificial photosynthesis using metal/nonmetal-nitride semiconductors: Current status, prospects, and challenges. *Journal of Materials Chemistry A* 4: 2801–2820 (2016b).

Kibria, M. G., R. Qiao, W. Yang, I. Boukahil, X. Kong, F. A. Chowdhury, M. L. Trudeau et al. Mi atomic-scale origin of long-term stability and high performance of P-GaN nanowire arrays for photocatalytic overall pure water splitting. *Advanced Materials* 28: 8388–8397 (2016a).

Kibria, M. G., S. Zhao, F. A. Chowdhury, Q. Wang, H. P. T. Nguyen, M. L. Trudeau, H. Guo, and Z. Mi. Tuning the surface fermi level on p-type gallium nitride nanowires for efficient overall water splitting. *Nature Communications* 5: 3825 (2014).

Kida, T., Y. Minami, G. Guan, M. Nagano, M. Akiyama, and A. Yoshida. Photocatalytic activity of gallium nitride for producing hydrogen from water under light irradiation. *Journal of Materials Science* 41(11): 3527–3534 (2006).

Kikuchi, A., M. Kawai, M. Tada, and K. Kishino. InGaN/GaN multiple quantum disk nanocolumn light-emitting diodes grown on (111) Si substrate. *Japanese Journal of Applied Physics, Part 2: Letters* 43(12A): L1524–L1526 (2004).

Kim, J. Y., G. Magesh, D. H. Youn, J.-W. Jang, J. Kubota, K. Domen, and J. S. Lee. Single-crystalline, wormlike hematite photoanodes for efficient solar water splitting. *Scientific Reports* 3: 2681 (2013).

Kim, T. W. and K.-S. Choi. Nanoporous bivo4 photoanodes with dual-layer oxygen evolution catalysts for solar water splitting. *Science* 343(6174): 990–994 (2014).

Knelangen, M., V. Consonni, A. Trampert, and H. Riechert. In situ analysis of strain relaxation during catalyst-free nucleation and growth of GaN nanowires. *Nanotechnology* 21(24): 245705 (2010).

Kocha, S. S., M. W. Peterson, D. J. Arent, J. M. Redwing, M. A. Tischler, and J. A. Turner. Electrochemical investigation of the gallium nitride-aqueous electrolyte interface. *Journal of the Electrochemical Society* 142(12): L238–L240 (1995).

Koley, G., Z. Cai, E. B. Quddus, J. Liu, M. Qazi, and R. A. Webb. Growth direction modulation and diameter-dependent mobility in InN nanowires. *Nanotechnology* 22(29): 295701 (2011).

Krawicz, A., J. Yang, E. Anzenberg, J. Yano, I. D. Sharp, and G. F. Moore. Photofunctional construct that interfaces molecular cobalt-based catalysts for H2 production to a visible-light-absorbing semiconductor. *Journal of the American Chemical Society* 135(32): 11861–11868 (2013).

Landré, O., C. Bougerol, H. Renevier, and B. Daudin. Nucleation mechanism of GaN nanowires grown on (111) Si by molecular beam epitaxy. *Nanotechnology* 20(41): 415602 (2009).

Landré, O., V. Fellmann, P. Jaffrennou, C. Bougerol, H. Renevier, A. Cros, and B. Daudin. Molecular beam epitaxy growth and optical properties of AlN nanowires. *Applied Physics Letters* 96(6): 061912 (2010).

Landré, O., V. Fellmann, P. Jaffrennou, C. Bougerol, H. Renevier, and B. Daudin. Growth mechanism of catalyst-free [0001] GaN and AlN nanowires on Si by molecular beam epitaxy. *Physica Status Solidi (c)* 7(7–8): 2246–2248 (2010).

Lazić, S., E. Gallardo, J. M. Calleja, F. Agulló-Rueda, J. Grandal, M. A. Sánchez-Garcia, and E. Calleja. Raman scattering by longitudinal optical phonons in InN nanocolumns grown on Si (111) and Si (001) substrates. *Physica E: Low-dimensional Systems and Nanostructures* 40(6): 2087–2090 (2008).

Le, B. H., S. Zhao, X. Liu, S. Y. Woo, G. A. Botton, and Z. Mi. Controlled coalescence of AlGaN nanowire arrays: An architecture for nearly dislocation-free planar ultraviolet photonic device applications. *Advanced Materials* 28: 8446–8454 (2016).

Le, B. H., S. Zhao, N. H. Tran, and Z. Mi. Electrically injected near-infrared light emission from single InN nanowire PIN diode. *Applied Physics Letters* 105(23): 231124 (2014).

Le, B. H., S. Zhao, N. H. Tran, T. Szkopek, and Z. Mi. On the fermi-level pinning of InN grown surfaces. *Applied Physics Express* 8(6): 061001 (2015).

Lee, M. H., K. Takei, J. Zhang, R. Kapadia, M. Zheng, Y.-Z. Chen, J. Nah. p-Type InP nanopillar photocathodes for efficient solar-driven hydrogen production. *Angewandte Chemie* 124(43): 10918–10922 (2012).

Li, K. H., X. Liu, Q. Wang, S. Zhao, and Z. Mi. Ultralow-threshold electrically injected AlGaN nanowire ultraviolet lasers on Si operating at low temperature. *Nature Nanotechnology* 10: 140 (2015).

Li, Q., Y. Lin, J. R. Creighton, J. J. Figiel, and G. T. Wang. Nanowire-templated lateral epitaxial growth of low-dislocation density nonpolar a-plane GaN on r-plane sapphire. *Advanced Materials* 21(23): 2416–2420 (2009).

Lin, H.-W., Y.-J. Lu, H.-Y. Chen, H.-M. Lee, and S. Gwo. InGaN/GaN nanorod array white light-emitting diode. *Applied Physics Letters* 97(7): 073101 (2010).

Liu, G., S. Ye, P. Yan, F.-Q. Xiong, P. Fu, Z. Wang, Z. Chen, J. Shi, and C. Li. Enabling an integrated tantalum nitride photoanode to approach the theoretical photocurrent limit for solar water splitting. *Energy and Environmental Science* 9(4): 1327–1334 (2016).

Liu, J., X. Zhang, Y. Zhang, R. He, and J. Zhu. Novel synthesis of AlN nanowires with controlled diameters. *Journal of Materials Research* 16: 3133 (2001).

MacLeod, B. A., K. X. Steirer, J. L. Young, U. Koldemir, A. Sellinger, J. A Turner, T. G. Deutsch, and D. C. Olson. Phosphonic acid modification of gainp2 photocathodes toward unbiased photoelectrochemical water splitting. *ACS applied materials and interfaces* 7(21): 11346–11350 (2015).

Maeda, K. Photocatalytic water splitting using semiconductor particles: History and recent developments. *Journal of Photochemistry and Photobiology C: Photochemistry Reviews* 12(4): 237–268 (2011).

Maeda, K. and K. Domen. Photocatalytic water splitting: Recent progress and future challenges. *The Journal of Physical Chemistry Letters* 1(18): 2655–2661 (2010).

Maeda, K., K. Teramura, N. Saito, Y. Inoue, and K. Domen. Photocatalytic overall water splitting on gallium nitride powder. *Bulletin of the Chemical Society of Japan* 80(5): 1004–1010 (2007).

Makimoto, T., K. Kumakura, and N. Kobayashi. High current gains obtained by InGaN/GaN double heterojunction bipolar transistors with P-InGaN base. *Applied Physics Letters* 79(3): 380–381 (2001).

Martinez, G. L., M. R. Curiel, B. J. Skromme, and R. J. Molnar. Surface recombination and sulfide passivation of GaN. *Journal of Electronic Materials* 29(3): 325–331 (2000).

Mi, Z. and S. Zhao. Extending group-III nitrides to the infrared: Recent advances in InN. *Basic Solid State Physics (B)* 252(5): 1050–1062 (2015).

Mi, Z., S. Zhao, S. Y. Woo, M. Bugnet, M. Djavid, X. Liu, J. Kang et al. Molecular beam epitaxial growth and characterization of Al(Ga)N nanowire deep ultraviolet light emitting diodes and lasers. *Journal of Physics D-Applied Physics* 49(36): 364006 (2016).

Moses, P. G. and C. G. Van de Walle. Band bowing and band alignment in InGaN alloys. *Applied Physics Letters* 96(2): 021908 (2010).

Nakamura, A., K. Fujii, M. Sugiyama, and Y. Nakano. A nitride based polarization-engineered photocathode for water splitting without a p-type semiconductor. *Physical Chemistry Chemical Physics* 16(29): 15326–15330 (2014).

Nakarmi, M. L., N. Nepal, J. Y. Lin, and H. X. Jiang. Photoluminescence studies of impurity transitions in Mg-doped AlGaN alloys. *Applied Physics Letters* 94(9): 091903 (2009).

Nam, O.-H., M. D. Bremser, T. S. Zheleva, and R. F. Davis. Lateral epitaxy of low defect density GaN layers via organometallic vapor phase epitaxy. *Applied Physics Letters* 71(18): 2638–2640 (1997).

Nguyen, H. P. T., S. Zhang, A. T. Connie, M. G. Kibria, Q. Wang, I. Shih, and Z. Mi. Breaking the carrier injection bottleneck of phosphor-free nanowire. Nano Letters 13(11): 5437–5442 (2013).

Ni, X., Ü. Özgür, Y. Fu, N. B., J. Xie, A. A. Baski, H. Morkoç, and Z. Liliental-Weber. Defect reduction in (1120) a-Plane GaN by two-stage epitaxial lateral overgrowth. *Applied Physics Letters* 89(26): 262105 (2006).

Nishikawa, S., Y. Nakao, H. Naoi, T. Araki, H. Na, and Y. Nanishi. Growth of InN nanocolumns by Rf-Mbe. *Journal of Crystal Growth* 301–302(4): 490–495 (2007).

Nishinaga, T., Nakano, T., Zhang, S. Epitaxial lateral overgrowth of GaAs by LPE. *Japanese Journal of Applied Physics* 27(6A), L964 (1988).

Ohkawa, K., W. Ohara, D. Uchida, and M. Deura. Highly stable GaN photocatalyst for producing H2 gas from water. *Japanese Journal of Applied Physics* 52(8S): 08JH04 (2013).

Park, Y. S., B. R. Hwang, J. C. Lee, H. Im, H. Y. Cho, T. W. Kang, J. H. Na, and C. M. Park. Self-assembled $Al_xGa_{1-x}N$ nanorods grown on Si (001) substrates by using plasma-assisted molecular beam epitaxy. *Nanotechnology* 17(18): 4640–4643 (2006).

Parkinson, P., H. J. Joyce, Q. Gao, H. H. Tan, X. Zhang, J. Zou, C. Jagadish, L. M. Herz and M. B. Johnston. Carrier lifetime and mobility enhancement in nearly defect-free core–shell nanowires measured using time-resolved terahertz spectroscopy. *Nano Letters* 9(9): 3349–3353 (2009).

Peng, H., E. Makarona, Y. He, Y.-K. Song, A. V. Nurmikko, J. Su, Z. Ren et al. Ultraviolet light-emitting diodes operating in the 340 nm wavelength range and application to time-resolved fluorescence spectroscopy. *Applied Physics Letters* 85(8): 1436–1438 (2004).

Pierret, A., C. Bougerol, M. den Hertog, B. Gayral, M. Kociak, H. Renevier, and B. Daudin. Structural and optical properties of $Al_xGa_{1-x}N$ nanowires. *Physica Status Solidi (RRL) - Rapid Research Letters* 7(10): 868–873 (2013b).

Pierret, A., C. Bougerol, B. Gayral, M. Kociak, and B. Daudin. Probing alloy composition gradient and nanometer-scale carrier localization in single AlGaN nanowires by nanocathodoluminescence. *Nanotechnology* 24(30): 305703 (2013a).

Pierret, A., C. Bougerol, S. Murcia-Mascaros, A. Cros, H. Renevier, B. Gayral, and B. Daudin. Growth, structural and optical properties of AlGaN nanowires in the whole composition range. *Nanotechnology* 24(11): 115704 (2013c).

Ra, Y.-H., R. Wang, S. Y. Woo, M. Djavid, S. M. Sadaf, J. Lee, G. A. Botton, and Z. Mi. Full-color single nanowire pixels for projection displays. *Nano Letters* 16(7): 4608–4615 (2016).

Richter, T., H. Lüth, T. Schäpers, R. Meijers, K. Jeganathan, S. E. Hernandez, R. Calarco, and M. Marso. Electrical transport properties of single undoped and n-type doped InN nanowires. *Nanotechnology* 20(40): 405206 (2009).

Ristic, J., M. A. Sanchez-Garcia, E. Calleja, J. Sanchez-Paramo, J. M. Calleja, U. Jahn, and K. H. Ploog. AlGaN nanocolumns grown by molecular beam epitaxy: Optical and structural characterization. *Physica Status Solidi (a)* 192: 60 (2002).

Salwar, A. T. M. G, S. D. Carnevale, T. F. Kent, M. R. Laskar, B. J. May, and R. C. Myers. Molecular beam epitaxy of Inn nanowires on Si. *Journal of Crystal Growth* 428: 59–70 (2015a).

Salwar, A. T. M. G., S. D. Carnevale, F. Yang, T. F. Kent, J. J. Jamison, D. W. McComb, and R. C. Myers. Semiconductor nanowire light-emitting diodes grown on metal: A direction toward large-scale fabrication of nanowire devices. *Small* 11(40): 5402–5408 (2015b).

Satoshi, H., T. Hidetoshi, A. Takanobu, M. Hideto, H. Kazumasa, A. Tsutomu, and N. Yasushi. Transmission electron microscopy characterization of position-controlled InN nanocolumns. *Japanese Journal of Applied Physics* 47(7R): 5330 (2008).

Schäfer-Nolte, E., T. Stoica, T. Gotschke, F. Limbach, E. Sutter, P. Sutter, D. Grützmacher, and R. Calarco. Enhanced light scattering of the forbidden longitudinal optical phonon mode studied by micro-raman spectroscopy on single InN nanowires. *Nanotechnology* 21(31): 315702 (2010).

Schumann, T., T. Gotschke, F. Limbach, T. Stoica, and R. Calarco. Selective-area catalyst-free Mbe growth of GaN nanowires using a patterned oxide layer. *Nanotechnology* 22(9): 095603 (2011).

Seger, B., T. Pedersen, A. B Laursen, P. C. Vesborg, O. Hansen, and I. Chorkendorff. Using TiO_2 as a conductive protective layer for photocathodic H_2 evolution. *Journal of the American Chemical Society* 135(3): 1057–1064 (2013).

Segura-Ruiz, J., N. Garro, A. Cantarero, C. Denker, J. Malindretos, and A. Rizzi. Optical studies of Mbe-grown InN nanocolumns: Evidence of surface electron accumulation. *Physical Review B* 79(11): 115305 (2009).

Sekiguchi, H., K. Kishino, and A. Kikuchi. Emission color control from blue to red with nanocolumn diameter of InGaN/GaN nanocolumn arrays grown on same substrate. *Applied Physics Letters* 96(23): 231104 (2010).

Seong-Ran, J., M. Gherasimova, Z. Ren, J. Su, G. Cui, J. Han, H. Peng. High performance algainn ultra-violet light-emitting diode at the 340 nm wavelength. *Japanese Journal of Applied Physics* 43(11A): L1409 (2004).

Shen, C.-H., H.-Y. Chen, H.-W. Lin, S. Gwo, A. A. Klochikhin, and V. Yu. Davydov. Near-infrared photoluminescence from vertical inn nanorod arrays grown on silicon: Effects of surface electron accumulation layer. *Applied Physics Letters* 88(25): 253104 (2006).

Shi, X., I. Y. Choi, K. Zhang, J. Kwon, D. Y. Kim, J. K. Lee, S. H. Oh, J. K. Kim, and J. H. Park. Efficient photoelectrochemical hydrogen production from bismuth vanadate-decorated tungsten trioxide helix nanostructures. *Nature Communications* 5: (2014).

Sivadasan, A. K., A. Patsha, S. Polaki, S. Amirthapandian, S. Dhara, A. Bhattacharya, B. K. Panigrahi, and A. K. Tyagi. Optical properties of monodispersed AlGaN nanowires in the single-prong growth mechanism. *Crystal Growth and Design* 15(3): 1311–1318 (2015).

Stampfl, C., C. Van de Walle, D. Vogel, P. Krüger, and J. Pollmann. Native defects and impurities in InN: First-principles studies using the local-density approximation and self-interaction and relaxation-corrected pseudopotentials. *Physical Review B* 61(12): R7846 (2000).

Standing, A., S. Assali, L. Gao, M. A Verheijen, D. van Dam, Y. Cui, P. H. Notten, J. E. Haverkort, and E. P. Bakkers. Efficient water reduction with gallium phosphide nanowires. *Nature Communications* 6: 7284 (2015).

Stoica, T., R. J. Meijers, R. Calarco, T. Richter, E. Sutter, and H. Lüth. Photoluminescence and intrinsic properties of MBE-grown InN nanowires. *Nano Letters* 6(7): 1541–1547 (2006).

Stoica, T., E. Sutter, R. J. Meijers, R. K. Debnath, R. Calarco, H. Lüth, and D. Grützmacher. Interface and wetting layer effect on the catalyst-free nucleation and growth of GaN nanowires. *Small* 4(6): 751–754 (2008).

Su, J., X. Feng, J. D. Sloppy, L. Guo, and C. A. Grimes. Vertically aligned Wo$_3$ nanowire arrays grown directly on transparent conducting oxide coated glass: Synthesis and photoelectrochemical properties. *Nano Letters* 11(1): 203–208 (2010).

Sun, K., Y. Kuang, E. Verlage, B. S. Brunschwig, C. W. Tu, and N. S. Lewis. Sputtered niox films for stabilization of P+ N-InP photoanodes for solar-driven water oxidation. *Advanced Energy Materials* 5(11): 1402276 (2015).

Tachibana, Y., L. Vayssieres, and J. R. Durrant. Artificial photosynthesis for solar water-splitting. *Nature Photonics* 6(8): 511–518 (2012).

Tajik, N., C. M. Haapamaki, and R. R. LaPierre. Photoluminescence model of sulfur passivated P-Inp nanowires. *Nanotechnology* 23(31): 315703 (2012).

Tang, C. C., S. S. Fan, M. L. De La Chapelle, and P. Li. Silica-assisted catalytic growth of oxide and nitride nanowires. *Chemical Physics Letters* 333: 12 (2001).

Taniyasu, Y., M. Kasu, and T. Makimoto. An aluminium nitride light-emitting diode with a wavelength of 210 nanometres. *Nature* 441(7091): 325–328 (2006).

Tao, T., T. Zhi, B. Liu, M. Li, Z. Zhuang, J. Dai, and Y. Li. Significant improvements in InGaN/GaN nano-photoelectrodes for hydrogen generation by structure and polarization optimization. *Scientific Reports* 6: 20218 (2016).

Tourbot, G., C. Bougerol, F. Glas, L. F. Zagonel, Z. Mahfoud, S. Meuret, P. Gilet et al. Growth mechanism and properties of InGaN insertions in GaN nanowires. *Nanotechnology* 23(13): 135703 (2012).

Waag, A., X. Wang, S. Fündling, J. Ledig, M. Erenburg, R. Neumann, and M. A. Suleiman. The nanorod approach: GaN nanoleds for solid state lighting. *Physica Status Solidi (c)* 8(7–8): 2296–2301 (2011).

Walter, M. G., E. L. Warren, J. R. McKone, S. W. Boettcher, Q. Mi, E. A. Santori, and N. S. Lewis. Solar water splitting cells. *Chemical Reviews* 110(11): 6446–6473 (2010).

Wang, D., A. Pierre, M. G. Kibria, K. Cui, X. Han, K. H. Bevan, and H. Guo. Wafer-level photocatalytic water splitting on GaN nanowire arrays grown by molecular beam epitaxy. *Nano letters* 11(6): 2353–2357 (2011).

Wang, K., T. Araki, T. Yamaguchi, Y. T. Chen, E. Yoon, and Y. Nanishi. InN nanocolumns grown by molecular beam epitaxy and their luminescence properties. *Journal of Crystal Growth* 430: 93–97 (2015a).

Wang, Q., A. T. Connie, H. P. Nguyen, M. G. Kibria, S. Zhao, S. Sharif, I. Shih, and Z. Mi. Highly efficient, spectrally pure 340 nm ultraviolet emission from Alxga(1)-Xn nanowire based light emitting diodes. *Nanotechnology* 24(34): 345201 (2013).

Wang, Q., H. P. T. Nguyen, K. Cui, and Z. Mi. High efficiency ultraviolet emission from Al$_x$Ga$_{1-x}$N core-shell nanowire heterostructures grown on Si (111) by molecular beam epitaxy. *Applied Physics Letters* 101(4): 043115 (2012).

Wang, Q., S. Zhao, A. T. Connie, I. Shih, Z. Mi, T. Gonzalez, M. P. Andrews et al. Optical properties of strain-free AlN nanowires grown by molecular beam epitaxy on Si substrates. *Applied Physics Letters* 104(22): 223107 (2014a).

Wang, R., X. Liu, I. Shih, and Z. Mi. High efficiency, full-color AlInGaN quaternary nanowire light emitting diodes with spontaneous core-shell structures on Si. *Applied Physics Letters* 106(26): 261104 (2015b).

Wang, R., H. P. T. Nguyen, A. T. Connie, J. Lee, I. Shih, and Z. Mi. Color-tunable, phosphor-free InGaN nanowire light-emitting diode arrays monolithically integrated on silicon. *Optics Express* 22(S7): A1768–A1775 (2014b).

Werner, F., F. Limbach, M. Carsten, C. Denker, J. Malindretos, and A. Rizzi. Electrical conductivity of InN nanowires and the influence of the native indium oxide formed at their surface. *Nano Letters* 9(4): 1567–1571 (2009).

Weyher, J. L., S. Müller, I. Grzegory, and S. Porowski. Chemical polishing of bulk and epitaxial GaN. *Journal of Crystal Growth* 182(1): 17–22 (1997).

Wu, C.-L., H.-M. Lee, C.-T. Kuo, C.-H. Chen, and S. Gwo. Absence of fermi-level pinning at cleaved nonpolar InN surfaces. *Physical review letters* 101(10): 106803 (2008).

Wu, J. When group-III nitrides go infrared: New properties and perspectives. *Journal of Applied Physics* 106(1): 011101 (2009).

Xie, P., Y. Hu, Y. Fang, J. Huang, and C. M. Lieber. Diameter-dependent dopant location in silicon and germanium nanowires. *Proceedings of the National Academy of Sciences of the United States of America* 106(36): 15254–15258 (2009).

Yamano, K., K. Kishino, H. Sekiguchi, T. Oto, A. Wakahara, and Y. Kawakami. Novel selective area growth (SAG) method for regularly arranged AlGaN nanocolumns using nanotemplates. *Journal of Crystal Growth* 425: 316–321 (2015).

Ye, F., X.-M. Cai, X. Zhong, H. Wang, X.-Q. Tian, D.-P. Zhang, P. Fan. The role of Ga in the chemical vapor deposition growth of ternary $Al_xGa_{1-x}N$ nanowires. *Journal of Alloys and Compounds* 620: 87–90 (2015).

Yoshida, H., Y. Yamashita, M. Kuwabara, and H. Kan. A 342-nm ultraviolet AlGaN multiple-quantum-well laser diode. *Nature Photonics* 2(9): 551–554 (2008a).

Yoshida, H., Y. Yamashita, M. Kuwabara, and H. Kan. Demonstration of an ultraviolet 336 nm AlGaN multiple-quantum-well laser diode. *Applied Physics Letters* 93(24): 241106 (2008b).

Yoshida, S. Electrochemical etching of a conductive GaN crystal for patterning. *Journal of Crystal Growth* 181(3): 293–296 (1997).

Yoshikawa, A., X. Wang, Y. Ishitani, and A. Uedono. Recent advances and challenges for successful p-type control of InN films with Mg acceptor doping by molecular beam epitaxy. *Physica Status Solidi (a)* 207(5): 1011–1023 (2010).

Zhang, S., A. T. Connie, D. A. Laleyan, H. P. T. Nguyen, Q. Wang, J. Song, I. Shih, and Z. Mi. On the carrier injection efficiency and thermal property of InGaN/GaN axial nanowire light emitting diodes. *IEEE Journal of Quantum Electronics* 50(6): 483–490 (2014).

Zhang, Y., J. Liu, R. He, Q. Zhang, X. Zhang, and J. Zhu. Synthesis of aluminum nitride nanowires from carbon nanotubes. *Chemistry of Materials* 13: 3899 (2001).

Zhao, C., T. K. Ng, R. T. ElAfandy, A. Prabaswara, G. B. Consiglio, I. A. Ajia, and I. S. Roqan. Droop-free, reliable, and high-power InGaN/GaN nanowire light-emitting diodes for monolithic metal-optoelectronics. *Nano Letters* 16(7): 4616–4623 (2016a).

Zhao, Q., H. Zhang, X. Xu, Z. Wang, J. Xu, D. Yu, G. Li, and F. Su. Optical properties of highly ordered AlN nanowire arrays grown on sapphire substrate. *Applied Physics Letters* 86(19): 193101 (2005).

Zhao, S., A. T. Connie, M. H. Dastjerdi, X. H. Kong, Q. Wang, M. Djavid, S. Sadaf et al. Aluminum nitride nanowire light emitting diodes: Breaking the fundamental bottleneck of deep ultraviolet light sources. *Scientific Reports* 5: 8332 (2015a).

Zhao, S., M. Djavid, and Z. Mi. A surface emitting, high efficiency near-vacuum ultraviolet light source with aluminum nitride nanowires monolithically grown on silicon. *Nano Letters* 15: 7006 (2015b).

Zhao, S., S. Fathololoumi, K. Bevan, D. Liu, M. G. Kibria, Q. Li, G. T. Wang, H. Guo, and Z. Mi. Tuning the surface charge properties of epitaxial InN nanowires. *Nano Letters* 12(6): 2877–2882 (2012a).

Zhao, S., X. Liu, S. Y. Woo, J. Kang, G. A. Botton, and Z. Mi. An electrically injected AlGaN nanowire laser operating in the ultraviolet-c band. *Applied Physics Letters* 107(4): 043101 (2015c).

Zhao, S., B. H. Le, D. P. Liu, X. D. Liu, M. G. Kibria, T. Szkopek, H. Guo, and Z. Mi. P-type InN nanowires. *Nano Letters* 13(11): 5509–5513 (2013b).

Zhao, S., M. G. Kibria, Q. Wang, H. P. T. Nguyen, and Z. Mi. Growth of large-scale vertically aligned GaN nanowires and their heterostructures with high uniformity on Sio X by catalyst-free molecular beam epitaxy. *Nanoscale* 5(12): 5283–5287 (2013a).

Zhao, S., Z. Mi, M. G. Kibria, Q. Li, and G. T. Wang. Understanding the role of Si doping on surface charge and optical properties: Photoluminescence study of intrinsic and Si-doped InN nanowires. *Physical Review B* 85(24): 245313 (2012b).

Zhao, S., O. Salehzadeh, S. Alagha, K. L. Kavanagh, S. P. Watkins, and Z. Mi. Probing the electrical transport properties of intrinsic InN nanowires. *Applied Physics Letters* 102(7): 073102 (2013c).

Zhao, S., Q. Wang, Z. Mi, S. Fathololoumi, T. Gonzalez, and M. P. Andrews. Observation of phonon sideband emission in intrinsic InN nanowires: A photoluminescence and micro-raman scattering study. *Nanotechnology* 23(41): 415706 (2012c).

Zhao, S., S. Y. Woo, M. Bugnet, X. Liu, J. Kang, G. A. Botton, and Z. Mi. Three-dimensional quantum confinement of charge carriers in self-organized AlGaN nanowires: A viable route to electrically injected deep ultraviolet lasers. *Nano Letters* 15(12): 7801–7807 (2015d).

Zhao, S., S. Y. Woo, S. M. Sadaf, Y. Wu, A. Pofelski, D. A. Laleyan, R. Rashid et al. Molecular beam epitaxy growth of Al-rich AlGaN nanowires for deep ultraviolet optoelectronics. *APL Materials* 4: 086115 (2016b).

8

Advanced Optoelectronic
Device Processing

Fengyi Jiang

Over the last 20 years, great advancements have been made in nitride-based devices. The achievement can be credited to the development of material growth, as well as the innovation of device fabrication technologies. Among those nitride devices, GaN-based LED is the most mature product, which has been widely used in human life. This chapter will focus on the device fabrication process of GaN-based LEDs in the aspect of LED structure and related key technologies.

Various types of GaN LED devices have been commercialized for different applications. But the majority of them have the device structure similar to one of the three types: conventional lateral structure,[1] thin film vertical structure[2] and flip-chip structure.[3]

As illustrated in Figure 8.1a, the lateral structure is mostly based on sapphire substrate, both n-electrode and p-electrode are on the same side, the current flows downwards across the quantum wells (QWs) and then laterally through n-GaN. With reflector at the backside of sapphire substrate, the light emitted from the QWs can be extracted from the top face and the four sidewalls. For the vertical structure with n-electrode and p-electrode on two opposite sides, current flows vertically through the QWs. As the reflector is at the bottom surface of p-GaN and the epilayer is very thin (~3 μm), the sidewall emission can be neglected, and thus, the light can be extracted only from the top face. Highly directional light can be generated from the vertical LEDs. And for the flip-chip structure, which is almost an inverted structure compared to lateral structure with similar current flowing path but different light extraction faces, reflector is on the top surface of p-GaN, and light is extracted through the surface of the sapphire substrate and the four side walls. The fabrication processes and related techniques of the three type structures are introduced in detail, respectively.

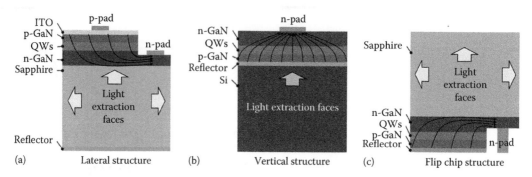

FIGURE 8.1 Schematic illustration of GaN-based LED with (a) lateral structure, (b) vertical structure, and (c) flip-chip structure.

8.1 Lateral LEDs Structure and Processing

Sapphire is the most widely used substrate for GaN-based LEDs with excellent mechanical property and chemical stability. Consequently, the original substrate for GaN is hard to be removed during device fabrication process. The conventional GaN LED structure, namely the lateral structure is commonly used in LED chip fabrication process, with relatively low difficulty.

Typically, the epitaxial structure of a GaN LED consists of an n-GaN layer grown on sapphire substrate, a p-GaN layer on the top surface and InGaN QWs lay between n-GaN and p-GaN, as illustrated in Figure 8.2.

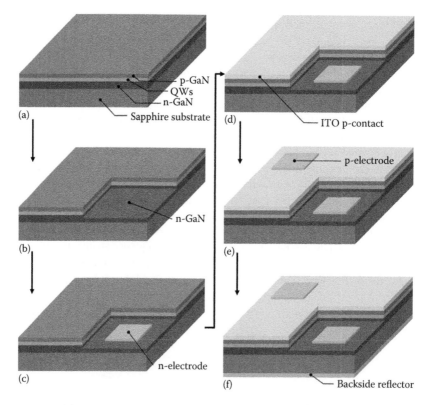

FIGURE 8.2 Device fabrication processes of lateral LEDs grown on sapphire substrate (a) epi-structure, (b) etching off p-GaN for n-contact, (c) deposition of n-pad, (d) deposition of transparent p-contact, (e) deposition of p-pad, and (f) deposition of reflector.

The chip fabrication process is made up by following steps: (1) etching of p-GaN to expose n-GaN for n-contact deposition, (2) deposition of n-contact pad for wire bonding, (3) covering the surface of p-GaN with transparent conductive p-contact for better current spreading, (4) deposition of p-pad for wire bonding, and (5) deposition of reflector on the backside of sapphire substrate for a higher light extraction efficiency.

8.1.1 Selective Etching

With an insulated sapphire substrate and p-GaN on the top surface, a selective etching conducted on p-GaN is essential to expose a region for n-contact deposition.

Compared with other conventional III-V semiconductors, III-nitrides have much higher bond energies, resulting in excellent chemical stability and good mechanical properties, which is advantageous for application of GaN-based devices in harsh environments and under extreme operating conditions. However, it also causes great impediments to device processing.

Under room temperature, wet etching of high-quality GaN is very difficult, and very slow etching rate can be observed in some alkalis. Table 8.1 lists the etch rates and etching planes of GaN in various chemicals.[4] Elevating the temperature in wet etching process can enhance the etching rates, but the no effective photolithography mask can sustain high-temperature alkalis. Therefore, the selective etching process for GaN devices cannot be conducted by wet etching.

Based on the wet etching process, photoelectrochemical (PEC) etching was developed to enhance etching rate at room temperature under the assistance of photons. Figure 8.3 illustrates the PEC process of n-GaN: photons with energy greater than the bandgap energy of GaN is illuminating on the GaN. The following oxidation reactions are supposed to be responsible for the decomposition of GaN[5]:

TABLE 8.1 Etch Rates and Etching Planes of GaN in Various Chemicals

Chemical	Temperature (°C)	Etch Rate (μm/min)	Etching Planes
CH_3COOH	30	<0.001	None
HCl	30	<0.001	None
HNO_3	81	<0.001	None
H_3PO_4	108–195	0.013–0.32	$\{10\bar{1}2\},\{10\bar{1}3\}$
H_2SO_4	93	<0.001	None
KOH molten	150–247	0.003–2.3	$\{10\bar{1}0\},\{10\bar{1}1\}$
50% KOH in H_2O	83	<0.001	None
10%–50% KOH in $C_2H_4(OH)_2$	90–182	0.0015–1.3	$\{10\bar{1}0\}$
50% NaOH in H_2O	100	<0.001	None
20% NaOH in $C_2H_4(OH)_2$	178	0.67–1.0	None

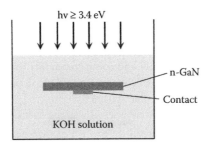

FIGURE 8.3 Schematic illustration of PEC of GaN.

$$2GaN + 6h^+ \rightarrow 2Ga^{3+} + N_2 \uparrow$$

$$2Ga^{3+} + 6OH^- \rightarrow Ga_2O_3 + 3H_2O$$

The illuminating photons provide energy to generate holes and electrons in GaN, then the holes act as oxidizing agent and dissociate GaN into its component elements Ga^{3+}, and finally, Ga^{3+} reacts with the OH^- in the solution and forms soluble Ga_2O_3.

At the interface between n-type GaN and KOH solution, the energy band bends upward, which drives hole to the surface and electrons into the bulk.[6] With the continuing consumption of holes, electrons will accumulate in bulk and recombine with photogenerated holes resulting in hole-lacking and etching rate slowing down. A contact can be deposited on the surface of GaN, which can conduct electrons into the solution and enhance etching rate. Contrarily, if GaN is p-type, the band will bend downward and drive holes into the bulk and prevent etching; thus p-type GaN is unable to be etched using PEC process.

In conclusion, the wet etching processes (include PEC), are not suitable for selective etching for p-GaN. Although GaN perfect lattice is hard to be etched by wet etching, dislocations are able to be etched in the etching solution. Therefore, wet etching can be used to reveal defects in GaN.[7-9]

Since wet etching of GaN is difficult, the dry etching methods which use plasma to etch GaN are developed.[10] The plasma consisting of atoms, radicals, and ions is generated by applying RF field on the reactant gas. The active species in the plasma are accelerated by induced field to the GaN surface and react with the regions not protected by masks, reaction byproducts desorb from the surface and being pumped away continuously. The most common dry plasma etching techniques for GaN are reactive ion etching (RIE) and inductively coupled plasma etching (ICP). For RIE, the reactant gas is normally chlorine-based, such as CCl_4, $SiCl_4$, and Cl_2. The etching rate can be increased by increasing the magnitude of RF power which may also damage the surface of GaN.[11,12] ICP is a promising choice for GaN etching with high etching rat, low damage, and high anisotropic.[13] Hence, the selective etching of p-GaN is normally carried by ICP.

8.1.2 Ohmic Contact to GaN

After finishing the selective etching of p-GaN, n-electrode is deposited on the exposed n-GaN region via electron beam evaporation. The electrode and GaN form a metal-semiconductor (MS) junction. The MS junction can be either Shockley type or ohmic type, which depends on the work function of contact metal and semiconductor.[14] Figure 8.4 illustrates four cases of band structure when a metal is contacted with GaN. (a) For n-GaN with Femi level near conduction band, if its work function is smaller than that of the metal, its conduction band will bend upward and form a barrier for electrons transport from metal to n-GaN. (b) If n-GaN has a work function greater than metal, the band will bend downward and electrons can flow fluently from metal to n-GaN. (c) For p-GaN with Femi level near valence band, if its work function is lower than that of the metal, its conduction band will bend downward and form a barrier for holes transport from metal to p-GaN. And (d) if p-GaN has a work function greater than metal, the band will bend upward and form no barrier for holes transportation. In the case of making n-electrode or p-electrode for LED devices with low working voltage, the contacts should be ohmic contact. Without considering surface states, metals with low work function are preferred for n type contact, and high work functions are preferred for p type contact.

Table 8.2 lists the work functions of GaN and some commonly used contact materials. The n-type GaN have a relatively low work function which is around 4.1 eV, doping level may slightly change the value but not too much. Among the listed materials, Al, Ti, Ag, and Cr can be selected to be n-contact material as their work functions are relatively low. Al was the first choice of n-contact metal. Ohmic contact can be easily formed between Al and n-GaN, but the contact may degrade as Al can be easily oxidized. Ti and Ag are even more unstable than Al. In this context, multiple layers of contact metals are developed instead of a single layer metal. Many concerns are considered when designing the multiple

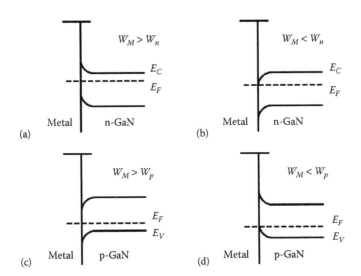

FIGURE 8.4 Schematic band structure of MS junction for the cases of (a) metal's work function higher than n-GaN, (b) metal's work function lower than n-GaN, (c) metal's work function higher than p-GaN, and (d) metal's work function lower than p-GaN.

TABLE 8.2   Work Functions of Related Materials for GaN LED Device

Materials	n-GaN	Al	Ti	Ag	Cr
Work function (eV)	~4.1	4.3	4.3	4.3	4.5
Materials	p-GaN	Pt	Ni	Au	ITO
Work function (eV)	~7.5	5.7	5.2	5.1	4.8

layers metal contact: low work function metals should be selected as the contacting layer, and noble metals can be used to protect the contact metals and improve the stability. A simple way to make a stable n-contact is using Ti/Al bilayer as n-contact pad, whereas Ti acts as contact layer and Al acts as capper layer. Ti and Al may form Al_3Ti alloy which can prevent the contact layer from oxidizing.[15] Another important issue that should be considered is the stickiness of the n-contact pad. With Al as the capper layer, wires are hard to be bonded to n-contact pad as the surface is easily oxidized. Therefore, an Au layer acts as the bonding layer is introduced as the capper, and a blocking layer is introduced between Au and Al to prevent Au diffuse to Al/Ti. The commonly used structures of n-contact metal multiple layers are Ti/Al/Ti/Au, Al/Ti/Au, Ti/Al/Ni/Au and Ti/Al/Pt/Au.[16–18]

The p-type ohmic contact has always been a problem for GaN LED device fabrication.[19] There are two main reasons: the first one is that the work function of p-GaN is as high as 7.5 eV, which is much higher than that of most metals, resulting in a high barrier for hole transportation; the second reason is the hole concentration of p-GaN is relatively low. Among the materials listed in Table 8.2, Pt seems to be the most appropriate one with high work function and noble chemical stability. Using Pt as the p-GaN contact metals can obtain a low specific contact resistance and high stability ohmic contact.[20] However, because the p-GaN layer is very thin and its electrical conductivity is low, a small size of p-pad is not good for current spreading. But large size of p-pad may decrease the light extraction efficiency. Therefore, a transparent conductive p-contact covered the whole p-GaN surface is adopted as current spreading layer in the chip fabrication process. The most common transparent conductive material is indium tin oxides (ITO).[21,22] ITO has a high electrical conductivity and high light transmission factor which is promising for current spreading and light extraction. However, with a low work function of 4.8 eV, ohmic contact is difficult to be formed. The solution is introducing a thin Ni layer of few angstroms thick between the

GaN and ITO. The Ni layer can be oxidized into transparent NiO layer, and easily form ohmic contact with p-GaN. Both high electrical conductivity and high light extraction efficiency can be achieved.[23-26]

A highly Mg-doped layer on the top surface can help to reduce contact resistance. Prior to the Ni contacting layer deposition, the GaN wafer should be annealed to remove H atom for p-GaN and active the Mg doping. After completing the transparent conductive ITO contact layer covering the whole p-GaN surface, a small size of p-pad for wire bonding, normally Au, is deposited on ITO, finishing the metal contact process.

8.1.3 Reflector

The sapphire substrate of a conventional lateral LED functions as the carrier of the chip, where the backside is mounted on the heat sink. Therefore, the backside of sapphire should be covered with reflective mirror to enhance light extraction efficiency. Both Ag and Al have very high reflectance near the visible spectrum range; thus, they are selected as the reflective metals.[23] As shown in Figure 8.5, the reflectance of Ag is higher than that of Al for visible light, but it drops sharply in UV light range. Therefore, Al is often used as a reflector in UV LEDs and Ag is often used as the reflector in blue and green LEDs. Besides the mirror layer, an Au cap layer is often deposited above to protect the reflective mirror from oxidizing.

Instead of using metals as a reflector, distributed Bragg reflector (DBR) is also widely used in lateral LEDs. The DBR structure is generally formed from a repeated periodical stack of alternating high and low index layers, the thickness of each layer is a one-quarter wavelength of the light. As the refraction indexes of the two materials are different, Fresnel reflection will occur at the interfaces. Better reflection effect can be obtained if the difference of refractive index is larger. Usually, the difference of the refraction index between the two materials is not large so that the reflection effect is not good at each interface. Increasing the periods of DBRs can effectively increase the total reflectivity to close to 100%. The two materials of DBR for nitride LEDs are normally SiO_2 and TiO_2.[27,28] With proper design, the reflectivity of the DBR can be higher than that of the metallic reflectors. The DBR can also be integrated in the epi-structure during epitaxial growth, in the form of $Al_{0.2}Ga_{0.8}N/GaN$ repeating layers.[29-31]

The advantage of lateral LED structure is that the fabrication process is relatively simple, and the manufacturing cost is low. However, with both n- and p-contact on one side of the LED structure, the current is flowing laterally through n-GaN, resulting in current crowding, which is an important issue for lateral LEDs. Inhomogeneous current distribution largely affects the efficiency of lateral LEDs. The current crowding may also cause regional temperature raise up and limit the driven current

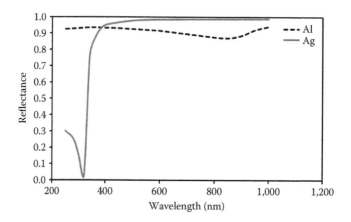

FIGURE 8.5   Dependence of reflectance on wavelength.

density; thus, lateral LED is not suitable for high power applications. And moreover, the heat generated by the device is mainly conducted away through the sapphire substrate, which has a poor thermal conductivity. Thus, the operation condition of lateral LEDs is restricted at relatively low temperature with low input power.

8.2 Vertical LEDs Structure and Processing

It is known that GaN LEDs can be grown on sapphire, SiC, and Si substrate. Sapphire and SiC are hard to be removed due to their excellent chemical stability and mechanical properties; thus, they are often processed to lateral LEDs in the early days. Unlike them, the Si substrate LEDs can be processed to vertical thin film structure easily as Si can be easily removed by wet etching. The mature Si-based IC technology brings convenient transfer technology for GaN film grown on Si substrate.

The process of fabricating vertical LEDs grown on Si substrate is schematically illustrated in Figure 8.6. Starting from (a) a LED structure grown on Si substrate, (b) a reflective p-contact is deposited on the p-GaN surface, (c) then it bonded to a new Si carrier or metal carrier via metal bonding technique, (d) the original Si substrate is removed by wet etching, leaving n-GaN expose to the air, (e) n-GaN surface is roughed by wet etching, and (f) n-electrode deposited on n-GaN.[32]

Some of the techniques used in lateral LEDs can be shared in thevertical LEDs fabrication process, which we will not discuss here anymore. The following new techniques used in vertical LEDs fabrication process will be focused.

8.2.1 Complementary Contact

The epilayer thickness of a vertical thin-film LED is normally 2 ~ 3 μm, while the length and width are in the magnitude of hundreds of micrometers. With reflective mirror on the p-GaN surface, majority of the light is emitted from the n-GaN surface. The sidewall emission can be neglected as the epilayer is very thin compared with the size of the LED chip. Two problems will be raised by the n-contact deposited on the top surface: current crowding and light blocking. As illustrated in Figure 8.7a, in a conventional vertical structure LED, current prefers to flow through the paths with low serial resistance, resulting in high current density under n-contact region. Nonuniform current distribution may largely affect device performance of vertical LEDs, and the situation will become even worse when injection

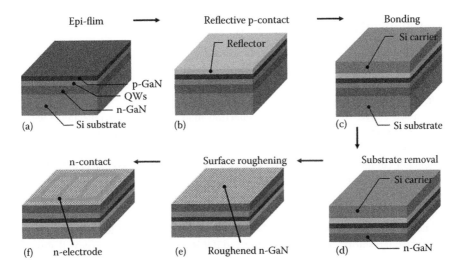

FIGURE 8.6 Device fabrication processes of vertical LEDs grown on Si substrate (a)–(f) described in text.

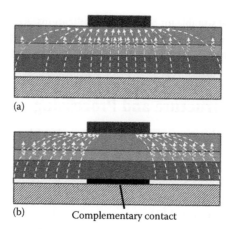

(a)

(b)   Complementary contact

FIGURE 8.7   Light extraction and current spreading in vertical LED (a) without complementary contact and (b) with complementary contact.

level is high. Moreover, the emitted light under n-contact region is hindered by n-contact, which cannot be extracted from the LED device, leading to low extraction efficiency.

A complementary contact is designed to solve the problems of current crowding and light blocking.[33] The idea is to form a high resistive contact on the p-GaN surface with the same geometry of n-contact, which is so-called "complementary contact." The high resistive contact can be achieved by deposition of dielectric materials like SiO_2 or SiN_x, forming a high resistive media between p-GaN and p-contact. Complementary contact can also be formed by deposing metal with unmatched work function as the media instead of dielectric materials. As illustrated in Figure 8.7b, the current is forced to flow away from the region under n-contact and distribute more uniformly across the junction, relieving current crowding. And because of no current flowing through the region under n-contact, no light will be emitted from the junction area under n-contact, and thus no light will be blocked by the n-contact.

Further investigation of device performance reveals that complementary contact is an effective technique for improving light extraction efficiency. With complementary contact, light extraction efficiency can be enhanced by 10%–40%; the actual value is depending on the coverage ratio of n-contact.

8.2.2 High Reflectivity Reflector with Low Resistivity

Unlike lateral LEDs with reflective mirrors on the backside of sapphire, the reflective mirrors of vertical LEDs are placed between the p-GaN and Si carrier, which functions not only as a mirror layer but also as a p-contact layer. Hence, both reflectivity and resistivity should be concerned when making the mirror layer.

As we discussed above, Ag is a promising reflection material with high reflectance within visible spectrum range. However, the work function of Ag is 4.3 eV, which is too low compare with p-GaN (7.1 eV). The ohmic contact between Ag and p-GaN is doomed to be bad due to the largely unmatched work function. For combining high reflectivity and low resistivity, the bilayer structure of Ni/Ag reflector is developed, as illustrated in Figure 8.8b. With a few angstroms thick of Ni layer deposited on p-GaN surface, ohmic contact can be easily formed. The Ag layer laid above Ni layer is acting as the reflection layer.

The thickness of Ni layer should be properly controlled. If the Ni layer is too thin, the ohmic contact is hard to be formed, resulting in high resistivity; and if it is too thick, the Ni layer will absorb the light and reduce reflectivity. The Ni layer is often deposited by electron beam evaporation, and the required Ni thickness is a few angstroms, which is very hard to be precisely controlled.

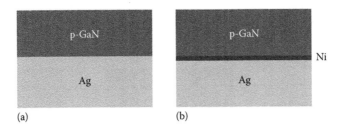

FIGURE 8.8 Structure of reflective mirror with (a) single Ag layer and (b) Ni/Ag bilayer.

A modified technique so called "Ni sacrifice" is introduced to improve the practicality of making high reflectivity and low resistivity reflector.[34] The "sacrifice" can be understood through the procedures of deposition-annealing-removing Ni layer. First, a thick enough Ni layer is deposited on the p-GaN surface and followed by annealing in N_2 ambient to form ohmic contact; then the original Ni layer is removed by wet etching; and finally, Ag reflector is deposited on p-GaN. The trick of sacrificing Ni is in the annealing and removing steps: a small amount of Ni atoms will be diffused into p-GaN and form ohmic contact during annealing process, and the extra Ni atoms are washed off to ensure a high reflectivity of Ag.

8.2.3 Film Transfer

The film transferring technique of GaN film grown on Si substrate involves wafer bonding and substrate removal processes.[2] Before removing the original Si substrate, the process of attaching the wafer to a new carrier substrate is called wafer bonding. Bonding media and carrier materials are the prerequisites for wafer bonding. The candidate of carrier material should be good in electrical conduction, aiming for a low operation voltage; thermal conductivity of carrier material should also be good as it plays the role of conducting heat away from the device. Si is an outstanding carrier material with acceptable electrical conductivity, good thermal conductivity, high chemical stability, and matched thermal expansion coefficient. Compare with Si, metals have better electrical and thermal properties, which attracts many researches on metal carriers. The metal carrier can be attached to the GaN film via wafer bonding or electrical plating. The thermal expansion coefficient difference among metal carrier, GaN film, and Si substrate makes the strain status very complicated during the film transferring process. Multiple layers of metals with matched thermal expansion coefficients and proper structure design can help to compensate the strain induced by the carrier and fabricate LED device with a metal carrier. But Si substrate is more common in the industry.

For the bonding media, adhesivity is the first priority; the other concerns are melting point, conductivity, and chemical stability. Various materials can be selected as the bonding media. Without considering production cost, Au is a promising bonding media with excellent properties in the aspects of adhesivity, thermal and electrical conductivity and chemical stability. Indeed Au was widely used as the bonding media in semiconductor manufacturing. It is known that Au is costly, instead of pure Au, Au-Sn alloy with Au weight percentage from 10%–90% is used as the bonding media to reduce the production cost. The Au-Sn alloy brings another advantage: by utilizing alloy composition to the eutectic point of Au-Sn system, bonding temperature can be dramatically reduced. Although Au-Sn alloy can reduce production cost to some extent, the price of is still too high. A diffusion bonding technique is developed by using base metals as bonding media. The principle of diffusion bonding is based on diffusion of miscible metals; if two pieces of metals with a smooth surface are laid together and sustained compressive force is applied at high temperature, their atoms will diffuse into each other and form a continuous interface.

Detailed procedures of diffusion bonding are illustrated in Figure 8.9. After Ag reflector is prepared, three layers of metals are deposited above the reflector sequentially via electron beam deposition, namely, diffusion barrier layer, bonding layer, and diffusion layer, respectively. The barrier layer above

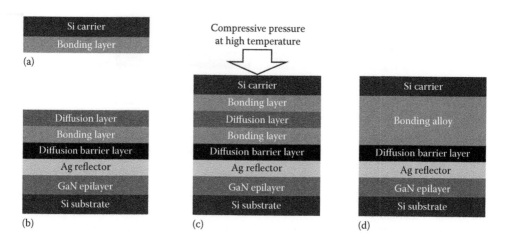

FIGURE 8.9 Schematic illustration of diffusion bonding with (a) a carrier covered by a bonding layer and (b) the epilayer covered by a reflector, a diffusion boding barrier, a bonding layer and a diffusion bonding barrier in sequence. (c) By applying a stress to compress the two together at a high temperature, and (d) the diffusion bonding layer will diffuse into the bonding layer on both sides to form a continuous alloy phase.

Ag reflector is to protect the reflector from damaging by the atoms diffusion from the bonding layer. The bonding layer, normally in the form of high melting point metals like Cu, Mo or Ag, is acting as a solvent; and the diffusion layer, normally in the form of low melting point metals like In or Sn, is acting as a solute. A new Si carrier with the surface covered by bonding layer is attached to as-prepared wafer. Under compressive stress at a proper high temperature, the atoms in the diffusion layer will diffuse into the bonding layers and finally form a continuous alloy phase.

After the bonding process completes, the original Si substrate is removed by wet etching. Benefited from the mature Si-based technologies of IC industry, wet etching of Si substrate is much easier than that of GaN. Both anisotropic and isotropic etching mode can be used. Anisotropic wet etching of Si is designed to create shapes like corners, grooves or trenches. High degree of anisotropic can be achieved. A strong alkaline solution such as aqueous KOH or NaOH is commonly used to selectively etch Si in certain crystal orientations faster than others. {110} and {100} orientation can be etched very fast, while {111} orientation mostly cannot be etched. As GaN LED on Si substrate is mostly grown on (111) plane, anisotropic wet etching is not a good choice for Si substrate removal. In the Si substrate removal process, isotropic etching of Si with aqueous HNO_3/HF is often used. The etching mechanism includes two steps: oxidize Si to SiO_2 by HNO_3 first, and then dissolve SO_2 by HF. Related reactions are stated below:

$$Si + 4HNO_3 \rightarrow SiO_2 + 4NO_2 + 2H_2O$$

$$SiO_2 + 6HF \rightarrow H_2SiF_6 + 2H_2O$$

The etching rate of Si with aqueous HNO_3/HF is determined not only by the concentration of solutes but also by the ratio of HNO_3/HF. When HF is excessive, HNO_3 concentration limits the rate of Si oxidization, temperature becomes an important factor that will largely affect the oxidization rate, and therefore, the reaction becomes more temperature dependent. In the extreme case when HNO_3 is absent, Si cannot be etched without oxidant but SiO_2 still can be etched, this can be used in selectively etching of SiO_2 and Si. When HNO_3 is excessive, the diffusion rate of HF to Si surface limits the total reaction rate. Sometimes, ethylic acid is added as a diluent to make the etching of Si more controllable.

With surface covered by Au thin film, new Si carrier can be protected during etching, and original Si substrate will be removed. The GaN film is then carried by the new Si substrate and the film transfer process complete. The position of n- and p-GaN is upside down with p-GaN on the button and n-GaN exposed to the air.

8.2.4 Surface Roughening

It is known that the refractive index of GaN is relatively large (~2.5), leading to a small total reflection angle. When light is emitted from the active region and transmitted to the interface between GaN and the air, only a small portion of light can escape from the surface; the majority is reflected back and oscillates inside the film and is absorbed gradually. Unlike lateral LEDs with light emitting from five faces, vertical LEDs have only one light-emitting face. Light is very difficult to be extracted from the device. Fortunately, the polarity of transferred GaN film provides an opportunity for enhancing light extraction efficiency via surface roughening.[35]

The polarity of GaN grown on sapphire and Si are normally Ga-polar, which is resistive to most chemicals at room temperature. But after the film transferring process of vertical LEDs, the polarity of GaN is reversed to N-polar. N-polar GaN can be selectively etched by alkaline. A related reaction is stated below[36]:

$$2GaN + 3H_2O \xrightarrow{OH^-} 2Ga_2O_3 + 2NH_3$$

In N-polar GaN, Ga atoms can be attacked by the hydroxyl (OH^-) group, and react with H_2O to form Ga_2O_3. But why is Ga-polar GaN resistive to alkaline? The answer is attributed to the difference in bond structure.[37] As illustrated in Figure 8.10, in N-polar GaN, each nitrogen atom has one dangling bond perpendicular to the surface, and hydroxyl groups are free to contact with Ga atoms in the lattice and react with them, forming hexagonal pyramid shaped etching surface. While in Ga-polar GaN, each nitrogen atom stretches out three dangling bonds which cover the Ga atoms. The hydroxyl groups are repelled away from Ga atoms by the negative charged dangling bonds; thus, Ga-polar GaN is resistive to etching by alkaline.

The surface morphology of a wet etched N-polar surface is shown in Figure 8.11; hexagonal shaped pyramids are formed after etching. The facets of the hexagonal pyramids are of {10$\bar{1}$0} orientation. It can be observed that the sizes of the pyramids are not very uniform and empty spaces are found between some pyramids, and this may be attributed to the nonuniform defects distribution on the surface. A presputtering with argon plasma on the surface can help to generated uniform distributed defects which lead to a uniform surface morphology. The density of pyramids can be adjusted by the presputtering process, and the size of pyramids depends on the etching rate and etching time.

When a beam of light enters a hexagonal pyramid, it has a high probability to escape from the pyramid to the air after a few rounds of reflections. Thus, surface roughening is an effective way of improving light extraction. In the vertical LEDs, the light extraction efficiency can be enhanced by 30%–50% percentage if surface roughening is applied.

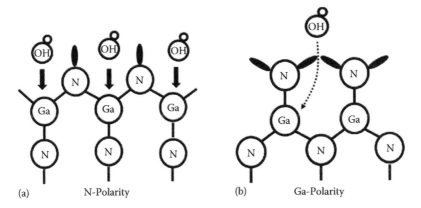

(a)   N-Polarity   (b)   Ga-Polarity

FIGURE 8.10 GaN etching of different polarities (a) N-polar GaN can be etched by alkaline and (b) dangling bonds of N atoms prevent Ga atoms from attacking by alkalizes in Ga-polar GaN.

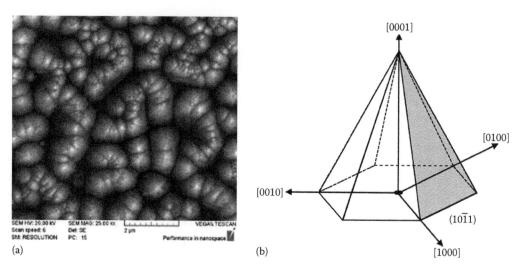

FIGURE 8.11   (a) SEM image of roughened N-polar surface by wet etching in KOH solution and (b) the related crystal orientation.

8.2.5 N-Polar n-Contact

In lateral LEDs, the ohmic contact to Ga-polar n-GaN can be easily formed by Al- or Ti- based metals. While in vertical LEDs, the polarity of n-GaN surface changed to N-polar after film transferring, resulting in high resistive ohmic contact with low stability. The reason can be explained by the formation of V_{Ga}-O_N complex on the surface of N-polar GaN with following reactions involved[38]:

$$O + V_N \rightarrow O_N$$

$$V_{Ga} + O_N \rightarrow V_{Ga} - O_N$$

In GaN lattice, formation of V_{Ga}-O_N not only consumes V_N, but also acts as acceptor and compensates donor like defects V_N, resulting in low electron concentration and leading to a poor ohmic contact and high device voltage.

A surface treatment prior to n-contact deposition with Ar plasma can help to improve the behavior of ohmic contact to N-polar n-GaN. With Ar plasma treatment, V_N concentration can be dramatically increased, and the existence of V_{Ga}-O_N will not largely affect the electron concentration on the surface. Therefore, ohmic contact to N-polar n-GaN with high stability can be easily formed with the assistance of Ar plasma treatment.[39]

8.2.6 Device Passivation

Reliability is an important index to the performance of a LED device. Without protection, a LED may be suffered from surface recombination and diffusion of foreign atoms, resulting in low stability and short life time. Passivation is an essential technique for commercialized LEDs to improve the reliability of the LEDs.

For a vertical LED without passivation, as illustrated in Figure 8.12a: surface recombination occurs when electrons and holes travel via the surface of the device, reduces the radiative recombination efficiency[40]; foreign ions such as Na^+ may diffuse into the active region, causes current leakage; and H atoms may diffuse into p-GaN,[41] resulting in degradation of p-GaN. For hindering these events, the surface of GaN can be covered by dielectric materials such as SiN_x or SiO_2. In general, passivation on the

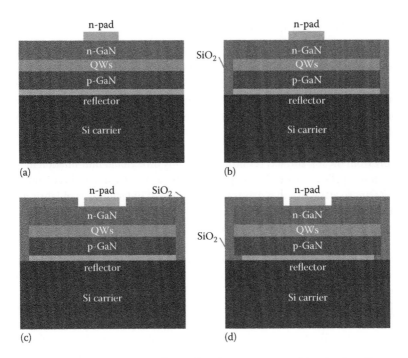

FIGURE 8.12 Schematic structure of a vertical LED (a) without passivation, (b) with sidewalls passivated, (c) with entire surface passivated, and (d) with double side passivated.

sidewalls, as figured in Figure 8.12, is enough to protect the device. However, passivation of the entire surface is more practical, as the SiO_2 layer covers the entire surface when it is deposited via PECVD. The quality and the thickness of the passivation layer should be properly controlled to avoid light extraction impediment.[42]

A modified passivation method is introduced to further improve the reliability of vertical LED device. As illustrated in Figure 8.13, beside the n-surface and the sidewalls, p-surface is also partially passivated[43] near the edge. With such a design, current is forced to flow away from the surface of the sidewall, further preventing recombination of carriers at the interface of GaN/SiO_2. And leakage current can be dramatically reduced.

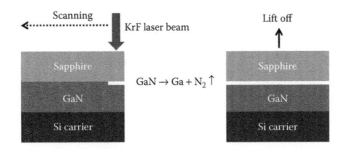

FIGURE 8.13 Schematic of laser lift-off process.

8.2.7 Vertical LEDs with Other Substrates

Compared with lateral structure, thin film vertical structure brings many attractive advantages: with two contacts laid on the bottom and top surface, respectively, the current flows directly across the junction area without turning direction, which can avoid current crowding; the carrier with high thermal conductance can help to reduce junction temperature effectively; these two issues are the essential factors for operation of high power LEDs. Moreover, with the simple design of reflector and surface roughening, light extraction efficiency can be easily enhanced to over 80%. Furthermore, in a vertical structure LED, as the epitaxial film is very thin, sidewall emissions can be negligible, and the light can be extracted only from the top surface, resulting in a highly directional illumination mode. Hence, the vertical LEDs can be applied in directional lighting fields such as flashlight, headlamp, and projector. These advantages are so attractive that drives the development of techniques of fabrication LEDs with SiC and sapphire substrates to the vertical structure.

For LED film grown on SiC substrate, it can directly proceed to vertical structure if the SiC substrate is conductive. However, if the conductive SiC substrate is of low transparency, the light emitted from active region may be absorbed by the SiC substrate, leading to low extraction efficiency. Besides, the buffer layer for GaN growth with high resistivity increases the serials resistance of the device, resulting in high operation voltage. Therefore, fabrication of vertical LEDs with conductive SiC substrate is not very practical; the technique of fabricating vertical LEDs on SiC substrate base on substrate removal is developed. It is known that as SiC has noble chemical properties and high mechanical strength, the substrate removal process is very difficult. A common approach is the thinning of the SiC substrate to a proper degree by mechanical grinding first, and then removing the remaining SiC by plasma dry etching. The device structure and other fabrication processes are similar to that of the vertical LEDs on Si substrate.

For LED film grown on sapphire substrate, the key technology of fabrication vertical structure LED is laser lift-off (LLO).[44] The main idea of LLO is to use high-energy laser beam to decompose GaN at the interface between sapphire and GaN. As illustrated in Figure 8.13, the GaN wafer is attached to a new Si carrier to form a sapphire/GaN/Si structure, and the backside of sapphire smoothly polished to reduce surface scattering of the laser beam. KrF excimer laser with $\lambda = 248$ nm is commonly used as the laser source, as its photon energy is higher than the bandgap energy of GaN and lower than the bandgap energy of sapphire. The laser beam is injected through the back side of sapphire and irradiated at the interface of sapphire/GaN, causes the GaN at the interface to decompose to liquid Ga and gaseous nitrogen. After scanning the laser beam in a matrix covers the entire wafer area, the sapphire substrate can be easily separated from the GaN film. Vertical structure LEDs can be fabricated similar to the processes for LED film grown on Si substrate.

The LLO technique realizes vertical structure LED on sapphire substrate, but many problems still trouble the manufacturing process. The biggest problem is that the incident laser may destruct the epitaxial film of GaN, resulting in low device reliability and low manufacturing yield. For minimizing the harm from laser beam, proper fluence should be controlled. It can be calculated that the minimum laser fluence for GaN decomposition is 0.3 J/cm².[45] As a matter of experience, approximately half of the laser fluence is absorbed by sapphire substrate or reflected by the interface of sapphire/GaN. Thus the required laser fluence is 0.6 J/cm².

8.3 Flip-Chip LED Structure and the Process

Base on the structure of lateral LEDs, a modified LED structure called flip-chip LED (FCLED) is developed in order to improve the device performance. The basic fabrication process of FCLED is similar to that of conventional lateral LEDs. Instead of introducing the fabrication process, the feature of flip-chip structure is focused in this section. As illustrated in Figure 8.14, the n-pad and p-pad of a flip chip are laid on the same side of the LED, which is similar to lateral LEDs, thus the current flowing mode of FCLEDs is the same as that of lateral LEDs. However, there are big differences between FCLEDs and

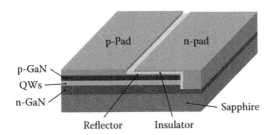

FIGURE 8.14   Schematic structure of a flip-chip LED.

lateral LEDs in mounting mode and light extraction mode. The lateral LEDs are mounted to submounts with the sapphire surface contacting the submount; while the FCLEDs are mounted to submounts with the contact pads contacting the submount. With reflective p-contact on the surface of p-GaN, majority of the light is extracted from the top surface of sapphire in a FCLED. Whereas for lateral LEDs, the reflector is on the backside of sapphire and the light is mainly extracted from the top surface of GaN.

The flip-chip structure brings many advantages compare with conventional lateral structure; the greatest feature is that it solves the problem of poor thermal conduction. As mentioned before, the poor thermal conductivity of sapphire substrate restricts the operation condition for lateral LEDs. For FCLEDs, heat is much easier to be conducted away through the metallic bonding pad, instead of the thick sapphire substrate with low thermal conductivity, which allows the FCLED chip working at high temperature and high input power. Second, with the entire p-surface covered by p-contact, better current spreading can be achieved, thus the FCLEDs can be work under high current density. And moreover, the n-pad and p-pad can be directly soldered on the submount with no wire bonding required, which provides ease for device packaging.

Summary

This section describes the main steps of processing in the viewpoint of three major type LED structure. The fabrication process of optoelectronic devices is a systematic work. Besides those featured techniques introduced above, some general techniques such as wafer cleaning, photolithography, and electronic beam evaporation are also important. Limitations of space prevent us from covering the entire range of the topic. There may be many choices of structures and techniques during the fabrication process, one should consider not only the efficiency and reliability of the device but also the production cost, the manufacturing yield and the repeatability of the process.

References

1. Nakamura, S., T. Mukai, and M. Senoh. 1991. High-power GaN Pn junction blue-light-emitting diodes. *Japanese Journal of Applied Physics* 30 (12A): L1998–L2001.
2. Xiong, C., F. Jiang, W. Fang, L. Wang, H. Liu, and C. Mo. 2006. Different properties of GaN-based LED grown on Si (111) and transferred onto new substrate. *Science in China Series E* 49 (3): 313–321.
3. Wierer, J. J., D. A. Steigerwald, M. R. Krames, J. J. Oshea, M. J. Ludowise, G. Christenson, Y. C. Shen et al. 2001. High-power algainn flip-chip light-emitting diodes. *Applied Physics Letters* 78 (22): 3379–3381.
4. Stocker, D. A., E. F. Schubert, and J. M. Redwing. 1998. Crystallographic wet chemical etching of GaN. *Applied Physics Letters* 73 (18): 2654.
5. Zhuang, D, and J. H. Edgar. 2005. Wet etching of GaN, AlN, and Sic: A review. *Materials Science and Engineering: R: Reports* 48 (1): 1–46.
6. Youtsey, C., G. Bulman, and I. Adesida. 1998. Dopant-selective photoenhanced wet etching of GaN. *Journal of Electronic Materials* 27 (4): 282–287.

7. Youtsey, C., L. T. Romano, and I. Adesida. 1998. Gallium nitride whiskers formed by selective photoenhanced wet etching of dislocations. *Applied Physics Letters* 73 (6): 797–799.

8. Visconti, P., K. M. Jones, M. A. Reshchikov, R. Cingolani, H. Morkoç, and R.J. Molnar. 2000. Dislocation density in GaN determined by photoelectrochemical and hot-wet etching. *Applied Physics Letters* 77 (22): 3532–3534.

9. Huang, D., P. Visconti, K.M. Jones, M. A. Reshchikov, F. Yun, A. A. Baski, T. King, and H. Morkoç. 2001. Dependence of GaN polarity on the parameters of the buffer layer grown by molecular beam epitaxy. *Applied Physics Letters* 78 (26): 4145–4147.

10. Pearton, S. J., C. R. Abernathy, F. Ren, J. R. Lothian, P. W. Wisk, A. Katz, and C. Constantine. 1993. Dry etching of thin-film Inn, AlN and GaN. *Semiconductor Science and Technology* 8 (2): 310.

11. Lin, M. E., Z.F. Fan, Z. Ma, L. H. Allen, and H. Morkoc. 1994. Reactive ion etching of GaN using Bcl3. *Applied Physics Letters* 64 (7): 887–888.

12. Lee, H., D. B. Oberman, and J. S. Harris Jr. 1995. Reactive ion etching of GaN using Chf3/Ar and C2clf5/Ar Plasmas. *Applied Physics Letters* 67 (12): 1754–1756.

13. Shul, R. J., G. B. McClellan, S. A. Casalnuovo, D. J. Rieger, S. J. Pearton, C. Constantine, C. Barratt et al. 1996. Inductively coupled plasma etching of GaN. *Applied Physics Letters* 69 (8): 1119–1121.

14. Foresi, J. S. and T. D. Moustakas. 1993. Metal contacts to gallium nitride. *Applied Physics Letters* 62 (22): 2859–2861.

15. Luther, B. P., S. E. Mohney, T. N. Jackson, M. A. Khan, Q. Chen, and J. W. Yang. 1997. Investigation of the mechanism for ohmic contact formation in Al and Ti/Al contacts to N-Type GaN. *Applied Physics Letters* 70 (1): 57–59.

16. Ruvimov, S., Z. Liliental-Weber, J. Washburn, K. J. Duxstad, E. E. Haller, Z-F. Fan, S. N. Mohammad et al. 1996. Microstructure of Ti/Al and Ti/Al/Ni/Au ohmic contacts for N-Gan. *Applied Physics Letters* 69 (11): 1556–1558.

17. Wang, D.-F., F. Shiwei, C. Lu, A. Motayed, M. Jah, S N. Mohammad, K. A Jones, and L. Salamanca-Riba. 2001. Low-resistance Ti/Al/Ti/Au multilayer ohmic contact to N-Gan. *Journal of Applied Physics* 89 (11): 6214–6217.

18. Motayed, A., R. Bathe, M. C. Wood, O. S. Diouf, R. D. Vispute, and S. N. Mohammad. 2003. Electrical, thermal, and microstructural characteristics of Ti/Al/Ti/Au Multilayer ohmic contacts to N-Type GaN. *Journal of Applied Physics* 93 (2): 1087–1094.

19. Liu, Q. Z. and S. S. Lau. 1998. A review of the metal–GaN contact technology. *Solid-State Electronics* 42 (5): 677–691.

20. Jang, J.-S., I.-S. Chang, H.-K. Kim, T.-Y. Seong, S. Lee, and S.-J. Park. 1999. Low-resistance Pt/Ni/Au ohmic contacts to P-type GaN. *Applied Physics Letters* 74 (1): 70.

21. Margalith, T., O. Buchinsky, D. A. Cohen, A. C. Abare, M. Hansen, S. P. DenBaars, and L. A. Coldren. 1999. Indium tin oxide contacts to gallium nitride optoelectronic devices. *Applied Physics Letters* 74 (26): 3930–3932.

22. Kim, D. W., Y. J. Sung, J. W. Park, and G. Y. Yeom. 2001. A study of transparent indium tin oxide (Ito) contact to P-Gan. *Thin Solid Films* 398: 87–92.

23. Hibbard, D. L., S. P. Jung, C. Wang, D. Ullery, Y. S. Zhao, H. P. Lee, W. So, and H. Liu. 2003. Low resistance high reflectance contacts to P-Gan using oxidized Ni/Au and Al or Ag. *Applied Physics Letters* 83 (2): 311–313.

24. Sheu, J. K., Y.-K. Su, G.-C. Chi, P. L. Koh, M. J. Jou, C. M. Chang, C. C. Liu, and W. C. Hung. 1999. High-transparency Ni/Au ohmic contact to P-Type GaN. *Applied Physics Letters* 74 (16): 2340–2342.

25. Qiao, D., L. S. Yu, S. S. Lau, J. Y. Lin, H. X. Jiang, and T. E. Haynes. 2000. A study of the Au/Ni ohmic contact on P-Gan. *Journal of Applied Physics* 88 (7): 4196–4200.

26. Horng, R.-H., D.-S. Wuu, Y.-C. Lien, and W.-H. Lan. 2001. Low-resistance and high-transparency Ni/Indium Tin oxide ohmic contacts to P-Type GaN. *Applied Physics Letters* 79 (18): 2925–2927.

27. Chen, C. H., S. J. Chang, Y. K. Su, G. C. Chi, J. K. Sheu, and J. F. Chen. 2002. High-efficiency InGaN-GaN Mqw green light-emitting diodes with cart and Dbr structures. *IEEE Journal of Selected Topics in Quantum Electronics* 8 (2): 284–288.

28. Hsu, Y. P., S. J. Chang, Y. K. Su, C. S. Chang, S. C. Shei, Y. C. Lin, C. H. Kuo, L. W. Wu, and S. C. Chen. 2003. InGaN/Gan light-emitting diodes with a reflector at the backside of sapphire substrates. *Journal of Electronic Materials* 32 (5): 403–406.

29. De, M. P., J. M. Bethoux, H. P. D. Schenk, M. Vaille, E. Feltin, B. Beaumont, M. Leroux, S. Dalmasso, and P. Gibart. 2002. Vertical cavity InGaN LEDs grown by movpe. *Physica Status Solidi* 192 (2): 335–340.

30. Krames, M. R., J. Bhat, D. Collins, N. F. Gardner, W. Götz, C. H. Lowery, M. Ludowise et al. 2002. High-power III-nitride emitters for solid-state lighting. *Physica Status Solidi* 192 (192): 237–245.

31. Park, S. H., J. Kim, H. Jeon, T. Tan, S. N. Lee, S. Chae, Y. Park et al. 2003. Room-temperature GaN vertical-cavity surface-emitting laser operation in an extended cavity scheme. *Applied Physics Letters* 83 (11): 2121–2123.

32. Xiong, C., F. Jiang, W. Fang, L. Wang, C. Mo, and H. Liu. 2007. The characteristics of GaN-based blue led on Si substrate. *Journal of Luminescence* 122: 185–187.

33. Jiang, F., L. Wang, and W. Fang. 2007. Semiconductor Light-Emitting Device and Method for Making Same. US patent. US7919784B1, Filed September 26, 2006 and Issued April 5, 2007.

34. Wang, G., X. Tao, F. Feng, C. Xiong, J. Liu, M. Zhang, and F. Jiang. 2011. Effects of Ni-assisted annealing on P-type contact resistivity of GaN-based led films grown on Si (111) substrates. *Acta Physica Sinica* 60 (7): 078503.

35. Fujii, T., Y. Gao, R. Sharma, E. L. Hu, S. P. DenBaars, and S. Nakamura. 2004. Increase in the extraction efficiency of GaN-based light-emitting diodes via surface roughening. *Applied Physics Letters* 84 (6): 855–857.

36. Zhou, Y., T. Yingwen, R. Jianping, and J. Fengyi. 2009. Improvement for extraction efficiency of vertical GaN-based led on Si substrate by photo-enhanced wet etching. *Acta Optica Sinica* 29 (1): 252–255.

37. Li, D., M Sumiya, S Fuke, D. Yang, D. Que, Y Suzuki, and Y Fukuda. 2001. Selective etching of GaN polar surface in potassium hydroxide solution studied by X-ray photoelectron spectroscopy. *Journal of Applied Physics* 90 (8): 4219–4223.

38. Song, J.-O, D.-S. Leem, S.-H. Kim, and T.-Y. Seong. 2003. Formation of vanadium-based ohmic contacts to N-GaN. *Korean Journal of Materials Research* 13 (9): 567–571.

39. Liu, J., F. Feng, Y. Zhou, J. Zhang, and F. Jiang. 2011. Stability of Al/Ti/Au contacts to N-polar N-GaN of GaN based vertical light emitting diode on silicon substrate. *Applied Physics Letters* 99 (11): 111112.

40. Martinez, G. L., M. R. Curiel, B. J. Skromme, and R. J. Molnar. 2000. Surface recombination and sulfide passivation of GaN. *Journal of Electronic Materials* 29 (29): 325–331.

41. Meneghini, M., L. R. Trevisanello, U. Zehnder, T. Zahner, U. Strauss, G. Meneghesso, and E. Zanoni. 2007. High-temperature degradation of GaN leds related to passivation. *IEEE Transactions on Electron Devices* 53 (12): 2981–2987.

42. Liu, J., and Qiu, C. 2010. Research of passivation and anti reflecting layer on GaN based blue led on silicon substrate. *Acta Optica Sinica* 30 (10): 2978–2982.

43. Jiang, F., J. Liu, and L. Wang. 2008. Semiconductor light-emitting device with double-sided passivation. US Patent. US 2011/0001120 A1, Filed March 25, 2008 and Issued May 13, 2008.

44. Chu, C.-F., F.-I. Lai, J.-T. Chu, C.-C. Yu, C.-F. Lin, H.-C. Kuo, and S. C. Wang. 2004. Study of GaN light-emitting diodes fabricated by laser lift-off technique. *Journal of Applied Physics* 95 (8): 3916–3922.

45. Chu, C.-F., C. K. Lee, C. C. Yu, Y. K. Wang, J. Y. Tasi, C.-R. Yang, and S. C. Wang. 2001. High etching rate of GaN films by Krf Excimer laser. *Materials Science and Engineering: B* 8 (1): 42–44.

Power Electronics

Principles and Properties of Nitride-Based Electronic Devices

An-Jye Tzou,
Chun-Hsun Lee,
Shin-Yi Ho, Hao-
Chung (Henry) Kuo,
and Jian-Jang Huang

9.1 Overview of Semiconductor Power Applications

Power electronic device plays a critical role in the generation, storage, conversion, and distribution of the electrical energy. Because the global energy shortage is, from time to time, an international crisis and economic threat, people have the demand on more advanced power semiconductor to provide efficient control of the power system. In the past several decades, Si has been widely used in power electronics, ranging from several volts to thousand voltage operations. It has the advantage of low cost and high reliability. However, the development of Si power electronics has reached its theoretical limit. The semiconductor power industry has developed very efficient topologies of power circuits and approaches for power management by fully utilizing the electrical and material characteristics of Si power devices. To further bring up the performance of power systems, alternative semiconductor materials with higher reliability at high switching speeds, current levels, and breakdown field are explored. In the past several years, with the mature nitride semiconductor in optical and electrical devices, gallium nitride (GaN) is set to displace Si power devices as III-nitride compound materials with better power handling and higher switching speed. The conversion efficiency is increased, allowing the implementation of essential future "cleantech" innovations, where power, weight, and volumetric efficiency are the key requirements.

9.2 Overview of GaN Power Applications

Wide bandgap (WBG) semiconductor power devices not only provide outstanding performances but also promise their functions in harsh environments, where the silicon power device is prohibited. Table 9.1 shows the comparisons of the fundamental material properties of GaN, Si, GaAs, SiC, and InP, respectively [1–3]. The physical properties of GaN exhibit WBG (E_g) around 3.4 eV, high breakdown

TABLE 9.1 Properties of WBG Semiconductor Materials in Power Electronics

Property	GaN	Si	GaAs	SiC	InP
Electron mobility μ (cm²/V-s)	2000	1300	5000	260	5400
Dielectric constant	9.5	11.4	13.1	9.7	12.5
Bandgap E_g (eV)	3.4	1.1	1.4	2.9	1.35
Breakdown electric field E_c (kV/cm)	3300	300	400	2500	500
Saturated electron drift velocity v_{sat} (×10⁷ cm/s)	2.5	1	1	2	1
Thermal Conductivity (W/cm-K)	1.3	1.5	0.46	4.9	0.7

electric field (E_c) of 3.3 MV/cm, high electron saturation velocity (v_{sat}) of 2.5 × 10⁷ cm/s, and high density (~1 × 10¹³ cm⁻²) of the two-dimensional electron gas (2DEG) with high mobility (μ) of around 2,000 cm²/V-s [2]. Those advantages promote that GaN is a considerable compound semiconductor material for high-frequency and high-power devices regarding the areas of optoelectronics, especially in power switching transistors.

One of the main advantages of GaN is the heterojunction between AlGaN/GaN. The *n*-channel was intrinsically generated by AlGaN/GaN heterostructure, leading to high 2DEG carrier concentrations (~1 × 10¹³ cm⁻²) with high electron mobility (2,000 cm²/V-s) through the polarization engineering without impurity doping, which is so-called "modulation doping" [3]. The high sheet concentration and electron mobility perform a conductive channel, which is responsible for lower on-resistance (R_{on}). The lower R_{on} of GaN-based heterostructure device promises higher current under the same bias as compared with Si-based transistors, which enables lower cost at the same device area. In addition, the high electric breakdown field of GaN is attributed to its WBG of 3.4 eV at room temperature. The high band-gap allows the power device operates under high bias, which is an essential criterion for high power switching application. The WBG of GaN also leads to ideal operating temperatures above 300°C, even to 700°C is possible. Furthermore, the better thermal conductivity of GaN describes the efficiently extract the dissipated power from the device, which also should be considered in power switching applications. Materials with lower thermal conductivity typically lead to device degradation at elevated temperature, even poor lifetime, and short endurance. GaN promote a thermal conductivity of 1.3 W/cm-K, which is comparable to Si (1.5 W/cm-K). The good thermal conductivity is a key parameter for the choosing material in the present electronics industry. Those outstanding material properties clearly indicate why GaN is a candidate for the next generation power device.

The above unique properties of GaN play a crucial role in the regulation and distribution of power and energy in different social areas. GaN-based power devices can sustain higher operation voltage and current. They can also be operated at a higher temperature and faster-switching speed at a lower power loss. Therefore, they have the potential to be applied in motor control, power distribution systems, electric vehicles, and avionics. Some of the prominent applications for GaN-based high-power devices are listed in Figure 9.1.

Beyond the material properties, several figures of merit (FOMs) can be proposed to benchmark different semiconductor materials. The higher value of FOM shows better performance for the material. The FOMs usually combine with relevant material properties and some of the merits with respect to the applications. For power devices, FOMs indicate that the high-power and high-frequency applications into one number, as compared with the strengths of the other competing materials. The first FOM is Johnson's FOM (JFOM), which gives an idea of material suitability for high-power application at high frequency. The JFOM can be calculated by using the Equation 9.1 [4]:

$$\text{JFOM} = \left(\frac{E_c v_{sat}}{2\pi} \right)^2 \tag{9.1}$$

where:

E_c is the breakdown electric field and

v_{sat} is the saturation electron velocity

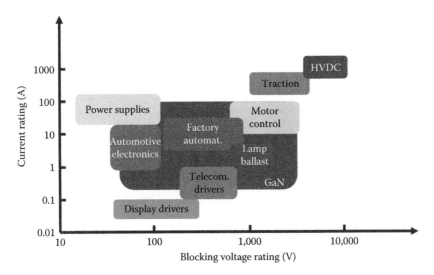

FIGURE 9.1 Applications of GaN-based power switching transistors.

The higher bandgap material will present their advantages in higher E_c, and heterostructure electronic device will present a higher v_{sat}, which means that the WBG material with 2DEG will be much suitable for high-power application at high frequency by JFOM.

Baliga's figure of merit (BFOM) presents the ultimate performance of field-effect transistors at low-frequency power-switching applications [5]:

$$BFOM = \varepsilon_r \mu E_g^3 \tag{9.2}$$

The BFOM is based on the assumption that the power losses are resulted in the power dissipation by current flow through the on-resistance of the power FET. Therefore, the value of BFOM indicates that the system operates at the lower frequency where the conduction losses are dominant. The BFOM can be related to an experimental FOM involving the on-resistance of the device and its critical electric field, as

$$FOM = \frac{V_{BR}^2}{R_{on}} \equiv \varepsilon_r \mu \frac{E_c^3}{4} \tag{9.3}$$

The BFOM defines the intrinsic limit of a power semiconductor. Figure 9.2 shows the specific on-resistance calculated in Si, SiC, and GaN power semiconductors. Most of the data published for the Si MOSFETs are clearly close to the theoretical limit. The early GaN commercial devices clearly show that AlGaN/GaN HEMTs are capable of delivering a breakdown voltage and on-resistance beyond the material limits of Si and SiC semiconductors used today for high-power switching applications.

Baliga's high-frequency FOM (BHFOM) benchmarks these devices at high-frequency, where switching losses dominate the value [5]

$$BHFOM = \frac{\mu E_c^2 V_G^{0.5}}{2 V_{BR}^{1.5}} \tag{9.4}$$

where V_G is the gate drive voltage and V_{BR} is the breakdown voltage. The BHFOM clearly indicates that the device with extremely high breakdown voltage cannot be applied in high-frequency operation.

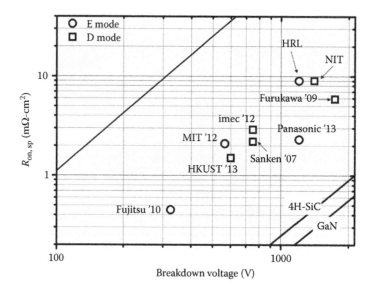

FIGURE 9.2 Comparisons of Specific on-resistance of GaN, SiC, and Si switching power devices.

Other FOMs have been proposed for more specific cases. The Keyes FOM (KFOM) considers thermal limitations due to transistor switching [5]:

$$\text{KFOM} = \chi \left(\frac{c v_{\text{sat}}}{4 \pi \varepsilon_r} \right)^{0.5} \tag{9.5}$$

where:

χ is thermal conductivity
c is the speed of light

Finally, a Combined FOM (CFOM) was developed to simultaneously account for high-frequency, high-power, and high-temperature performance [6,7]:

$$\text{CFOM} = \chi \varepsilon_r \mu v_{\text{sat}} E_c^2 \tag{9.6}$$

Table 9.2 compares several FOMs of the candidate for the next generation materials. The high-power and high-frequency performance of these materials are listed and shown in comparison. This table shows that GaN is an excellent candidate for high-frequency power applications. For GaN, the JFOM is about 270 to 480 times that of Si, about 135 to 240 times that of GaAs and about 1.5 times that of SiC [5]. The BFOM of GaN is about 17 to 34 times that of Si, about 1.5 to 2.5 times that of GaAs and about 3 times that of SiC [5]. It is clear that GaN provides much better high-power/high-frequency performance

TABLE 9.2 The FOMs for WBG Semiconductor Materials in Power Electronics

Property	GaN	Si	GaAs	SiC
Johnson's FOM (*JFOM*)	270–480	1.0	2	324–400
Keyes FOM (*KFOM*)	1.4	1.0	0.4	4.5–4.8
Baliga's FOM (*BFOM*)	17–34	1.0	13	6–12
Baliga's High-Frequency FOM (*BHFOM*)	86–172	1.0	10	57–76
Combined FOM (*CFOM*)	108–290	1.0	4	275–310
T_{max} (°C)	700	300	300	600

possibilities than in Si and GaAs. SiC shows better high-power performance, but the saturation velocity and on-resistance limit the high-frequency activities.

Therefore, WBG semiconductors provide better performance characteristics for high-frequency and high-power applications. The use of WBG materials can be considered and meet the requirements of modern power electronics. Among the WBG semiconductors, GaN shows a strong potential in terms of wafer-size availability (6-inch or more), technology and market prospects.

9.3 Device Physics of GaN/AlGaN HEMTs

A high electron mobility transistor (HEMT), also known as heterostructure field-effect transistor (HFET) or modulation-doped FET (MODFET), is a field-effect transistor incorporating a junction between two materials with different bandgaps without doped region as an inversion channel (as compared with Metal–oxide–semiconductor FET, MOSFET). HEMTs are the most mature structure for III–V-based transistors, which rely on the use of heterojunctions for their operation. The heterojunctions are formed between semiconductors of different alloy compositions and bandgaps, such as AlGaAs and GaAs. The pseudomorphic HEMT (pHEMT) was constructed in a lattice-matched AlGaAs/GaAs system [8], which has been widely studied and used in RF, microwave, and millimeter-wave applications.

In 1993, Khan et al. demonstrated the first AlGaN/GaN MODFET (with n-doped barriers), with a transconductance of 23 mS/mm and electron mobility of 563 cm²/V-s at room temperature [9]. Due to the combined effect of spontaneous polarization and piezoelectric polarization, electrons are confined in the GaN side of the heterojunction and are called two-dimensional electron gas (2DEG), as shown in Figure 9.3. They also reported the first microwave results with current gain cut-off frequency (f_T) of 11 GHz and maximum frequency of oscillation (f_{max}) of 14 GHz [10]. In the early stages, the GaN-based MODFETs exhibited very low transconductance and relatively poor frequency response. However, the state-of-the-art in GaN-based HEMTs has been widely discussed and developed by far, with many improvements in the material quality and process technology. In recent years, GaN-on-Si HEMTs dominate the DC power device application due to its lower cost and large area wafer size. But the RF applications still require GaN-on-SiC for the purpose of high quality of GaN-based material.

III-nitride compound semiconductors can be found in three crystal structures: wurtzite, zincblende, and rock salt, respectively. The wurtzite structure is the thermally stable phase for GaN at room temperature and atmosphere. The wurtzite GaN consists of two interpenetrating hexagonal close-packed (HCP) lattices, c_0 and a_0, where c_0 is the height of a hexagonal cell. The basis of a unit cell for the wurtzite GaN belongs to four atoms, two nitrogen and two gallium atoms, as shown in Figure 9.4. It is characterized by two lattice constants, a_0 (3.18 Å) and c_0 (5.18 Å) [11]. In general, the III-nitride compound always grows along with the c-axis on the c-plane sapphire substrate.

The chemical bond of GaN is tetrahedrally bonded to four atoms (Ga, N). All atoms on the same plane at each side of a bond are the same. Hence, the wurtzite GaN crystal shows two distinct faces, commonly

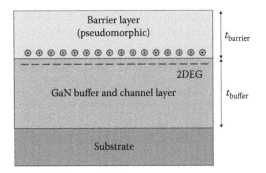

FIGURE 9.3 Schematic device structure of a GaN-based HEMT.

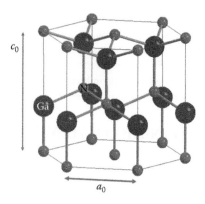

FIGURE 9.4   Wurtzite crystal structure of GaN.

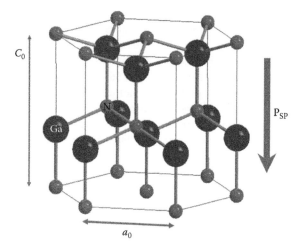

FIGURE 9.5   Atomic arrangement of Ga-face GaN wurtzite structure. Spontaneous polarization vector, P_{sp} is also shown.

known as Ga-face and N-face, which correspond to the [0001] and [000-1] of Miller index with (0001) and (000-1) faces as shown in Figure 9.5 [12].

In an ideal wurtzite crystal, the c_0 to a_0 ratio equal to $\sqrt{(8/3)} = 1.633$. Because of the different metal cations, the c_0 to a_0 ratios of AlN, GaN, and InN are different. Table 9.3 shows a summary of the lattice parameters for wurtzite III-nitrides at 300 K [13]. It is clear to observe that the GaN is more close to the ideal wurtzite structure, followed by InN and AlN. The fact implies that the degree of non-ideality is a significant factor for determining the strength of polarization in III-nitrides, which will be discussed later.

Due to the c_0 to a_0 ratios of AlN, GaN, and InN are not equal to the ideal value, and there is an absence of internal electric fields. The total polarization (P) of a GaN or AlGaN layer is the sum of spontaneous polarization (P_{SP}) and piezoelectric polarization (P_{PE}). The spontaneous polarization arises due to the

TABLE 9.3   Lattice Parameters of Wurtzite III-Nitrides at 300 K

Parameter	Ideal	AlN	GaN	InN
a_0 (Å)	–	3.112	3.189	3.54
c_0 (Å)	–	4.982	5.185	5.705
c_0/a_0 (expected)	–	1.6010	1.6259	1.6116
c_0/a_0 (calculated)	1.633	1.6190	1.6336	1.6270

asymmetry of the wurtzite crystal, whereas piezoelectric polarization results from the lattice mismatch between two materials. The lattice constants of III-nitrides play a significant role in determining spontaneous polarization, which explains the differences in the polarization of GaN and AlN [12]. The spontaneous polarization was raised by increasing the non-ideality of the GaN and AlN. The ternary alloy, such as AlGaN grown on GaN shows different polarization due to the lattice constant was modified by Al content. The spontaneous polarization along the c-axis is given by $P_{SP} = P_{SP} \cdot \hat{z}$, where the piezoelectric polarization is given by the piezoelectric coefficients e_{33} and e_{31} as:

$$P_{PE} = e_{33}\,\varepsilon_z + e_{31}(\varepsilon_x + \varepsilon_y) \tag{9.7}$$

and $\varepsilon_z = (c - c_0/c_0)$, which presents the strain of a III-nitride grown on another III-nitride compound along the c-axis. The in-plane strain is assumed to be isotropic and is given by $\varepsilon_x = \varepsilon_y = (a - a_0/a_0)$, where a_0 and c_0 are the equilibrium values of the lattice constant. The relationship of the lattice constants between two wurtzite III-nitride compounds is given by Equation 9.8.

$$\frac{c - c_0}{c_0} = 2\,\frac{C_{13}}{C_{33}}\,\frac{a - a_0}{a_0} \tag{9.8}$$

where C_{13} and C_{33} are the elastic constants.

By combining Equations 9.7 and 9.8, the net polarization along the c-axis can be calculated as:

$$P_{PE} = 2\,\frac{a - a_0}{a_0}\,(e_{31} - e_{33})\frac{C_{13}}{C_{33}} \tag{9.9}$$

Since $(e_{31} - e_{33})(C_{13} / C_{33}) < 0$ for AlGaN grown on GaN (no matter in any Al composition), the piezoelectric polarization is negative for tensile and positive for compressive strain, respectively [13]. The spontaneous polarization for Ga-face AlGaN grown on GaN is negative because of the direction along the vector from substrate to the surface, as shown in Figure 9.6 [12]. In contrast, spontaneous polarization direction vector is away from the substrate for N-face III-nitrides. For both Ga-face and N-face materials, the piezoelectric and spontaneous polarizations are parallel in the case of tensile strain and anti-parallel for compressively strained barrier layers which are also shown in Figure 9.6.

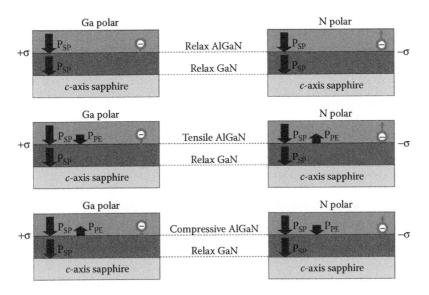

FIGURE 9.6 Directions of spontaneous and piezoelectric polarization in Ga and N-face AlGaN/GaN heterostructures.

The polarization-induced charge density in space is given by:

$$\rho_p = \nabla \cdot P \tag{9.10}$$

The polarization sheet charge density (σ) at an abrupt interface of an AlGaN/GaN or GaN/AlGaN heterostructure is given by Ref. [12,14]:

$$\sigma = P(\text{bottom}) - P(\text{top})$$

$$\sigma = [P_{SP}(\text{bottom}) + P_{PE}(\text{bottom})] - [P_{SP}(\text{top}) + P_{PE}(\text{top})]$$

$$= [P_{PE}(\text{bottom}) - P_{PE}(\text{top})] + [P_{SP}(\text{bottom}) - P_{SP}(\text{top})] \tag{9.11}$$

$$= \sigma(P_{PE}) + \sigma(P_{SP})$$

However, if the polarization induced sheet charge density is positive ($+\sigma$), free electrons will try to compensate for the charges induced by polarization. These charges will accumulate at the interface between the heterojunction to create a two-dimensional electron gas (2DEG). In the case of Ga-face structures with AlGaN grown on relaxed GaN, the polarization-induced sheet charge density is positive, as shown in Figure 9.5. If the heterostructure is grown pseudomorphically, the piezoelectric polarization from the tensile strain will be raised, leading to an increase in the sheet charge density ($+\sigma$) and the electron confinement in the 2DEG. In order to calculate the amount of the polarization-induced sheet charge density (σ) at the AlGaN/GaN interface, which is dependent to the Al-content x as the $Al_xGa_{1-x}N$, the following relations from the physical properties of GaN and AlN are used [12,14]:

Lattice constant (unit: m)

$$a(x) = (-0.77x + 3.189) \cdot 10^{-10} \tag{9.12}$$

Elastic constants (unit: GPa)

$$C_{13}(x) = (5x + 103)$$
$$\tag{9.13}$$
$$C_{33}(x) = (-32x + 405)$$

Piezoelectric constants (unit: C/m²)

$$e_{31}(x) = (-0.11x - 0.49)$$
$$\tag{9.14}$$
$$e_{33}(x) = (0.73x + 0.73)$$

Spontaneous polarization (unit: C/m²)

$$P_{SP}(x) = (-0.52x - 0.029) \tag{9.15}$$

The total amount of polarization induced sheet charge density for an undoped $Al_xGa_{1-x}N$/GaN heterostructure can be calculated, which is given by Ambacher et al. [12,14]

$$|\sigma(x)| = |P_{PE}(Al_xGa_{1-x}N) + P_{SP}(Al_xGa_{1-x}N) - P_{SP}(GaN)|$$

$$= \left| a \frac{a(0) - a(x)}{a(x)} \left[e_{31}(x) - e_{33}(x) \frac{C_{13}(x)}{C_{33}(x)} + P_{SP}(x) - P_{SP}(0) \right] \right| \tag{9.16}$$

The magnitude of the polarization-induced sheet charge density (σ) for both Ga-face or N-face heterostructures is always the same but the sign changes (Anti-parallel directions). But we should notice that the Ga-face AlGaN/GaN was generally employed for the electronic devices without special purpose.

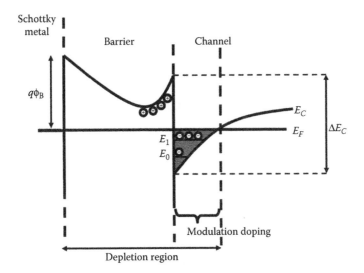

FIGURE 9.7 The energy band alignment of an AlGaN/GaN HEMT structure.

Figure 9.7 shows the energy band diagram of an AlGaN/GaN structure for a HEMT system. The bandgap offset at the AlGaN/GaN interface (ΔE_C), forms a quantum well at the interface under AlGaN barrier layer, as shown in Figure 9.6. The electrons are confined in a two-dimensional space and perform as an electron channel with ultrafast electron mobility. The inherent polarization effect contributes charge from barrier (AlGaN) to the channel above GaN, leading to a high electron density, so-called "modulation doping." The modulation doping promises AlGaN/GaN HEMT for a high-frequency and high-power applications.

Since the modulation doping in AlGaN/GaN HEMT for 2DEG, the density of the 2DEG is usually around 1×10^{13} cm^{-2} with mobility around 2,000 cm^2/V-s at room temperature [14,15]. The 2DEG sheet carrier concentration is given by [12,14]:

$$n_s(x) = \frac{+\sigma(x)}{e} - \left[\frac{\varepsilon_0 \varepsilon(x)}{de^2} \right] \left[e\varphi_b(x) + E_F(x) - \Delta E_C(x) \right] \tag{9.17}$$

where:
d is the thickness of the Al$_x$Ga$_{1-x}$N barrier
$e\varphi_b$ is the Schottky barrier height of the gate contact
E_F is the Fermi level
E_C is the conduction band offset at the AlGaN/GaN interface

For determining the sheet carrier concentration from the polarization induced sheet charge from the Equation 9.17, the following approximations were employed to calculate [14]:

Dielectric constant:

$$\varepsilon(x) = -0.3x + 10.4 \tag{9.18}$$

Schottky barrier height (Unit: eV):

$$e\varphi_b = 1.3x + 0.84 \tag{9.19}$$

Fermi energy:

$$E_F = E_0(x) + \frac{\pi \hbar^2}{m^*(x)} n_s(x) \tag{9.20}$$

where the ground sub-band level of 2DEG is given by [15]

$$E_0(x) = \left[\frac{9\pi\hbar e^2}{8\varepsilon_0 \sqrt{8m^*(x)}} \frac{n_s(x)}{\varepsilon(x)} \right]^{2/3}$$

(9.21)

with the effective electron mass, $m^*(x)$ is $0.22m_e$ for GaN.

Band offset

$$\Delta E_C = 0.7\left[E_g(x) - E_g(0) \right]$$

(9.22)

where the bandgap of $Al_xGa_{1-x}N$ is given by

$$E_g(x) = xE_g(AlN) + (1-x)E_g(GaN) - x(1-x)b$$

$$= xE_g(6.2) + (1-x)E_g(3.4) - x(1-x)1.0$$

(9.23)

From Equation 9.17, it can be seen that the sheet concentration is dependent on the Al composition and the thickness of the $Al_xGa_{1-x}N$ barrier layer. To further understand how the 2DEG sheet carrier concentration can be controlled by the $Al_xGa_{1-x}N$ barrier layer, Figure 9.8 shows a theoretical calculation of the 2DEG sheet carrier density against different alloy compositions for various AlGaN barrier thicknesses [12], respectively. For example, a constant barrier thickness of 30 nm AlGaN is considered, the 2DEG sheet carrier concentration (n_s) will be approximately 0.85×10^{13}, 1.5×10^{13}, 2×10^{13} cm^{-2} for $Al_xGa_{1-x}N$ compositions of $x = 0.2, 0.3$, and 0.4, respectively. If the thickness of an AlGaN barrier of Al composition of 25% is increased from 10 nm to 20 nm and 30 nm, the 2DEG sheet carrier concentration will be increased from approximately 0.75×10^{13} to 1.1×10^{13}, and 1.2×10^{13} cm^{-2}. The polarization induced 2DEG sheet carrier density of N-face AlGaN/GaN heterostructure will be slightly lower than in Ga-face structure, which is related to the smaller piezoelectric constants of GaN in comparison to AlGaN [12].

The Al composition was generally employed by 20% to 25% to AlGaN/GaN HEMT as well as approximately 20 nm of AlGaN barrier thickness [2,16,17]. Due to the presence of 2DEG at the AlGaN/GaN interface allows a conduction channel without gate bias; hence, the AlGaN/GaN HEMTs generally belong to depletion-mode transistors. There is no 2DEG sheet carrier density below a certain $Al_xGa_{1-x}N$ barrier layer thickness; 3 nm for $x = 0.35$, is known as the critical barrier thickness. This limitation resulted in poor confinement of the polarization induced electron sheet charge because the conduction band offset becomes too small. In addition, the thick barrier layer or high Al composition are expected to obtain higher polarization induced sheet carrier density, but the fact is contrary. The high density of structural defects and rough interface will lead to a limitation for the 2DEG mobility,

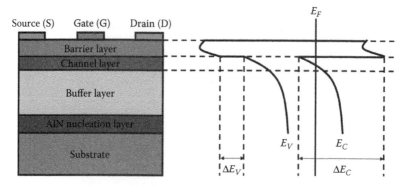

FIGURE 9.8 A basic D-mode AlGaN/GaN HEMT with corresponding energy band diagram.

which has unfavorable effect for the HEMTs [18–20]. Therefore, it is crucial to optimize the layer structure of AlGaN/GaN heterostructure as we fabricate a GaN-based HEMT during MOCVD growth.

9.4 Operation of the AlGaN/GaN HEMT

A typical AlGaN/GaN HEMT was shown in Figure 9.8 consists of an AlGaN barrier layer around 20 nm thick grown on a GaN channel layer. The device structures are generally grown on *c*-plane sapphire, semi-insulating silicon carbide (SiC), or Si (111) substrates. Due to the thermal expansion coefficient mismatch and lattice mismatch between substrate and GaN, the wafer bowing, strain-induce defect, and threading dislocations will result in degraded device quality [21,22]. Therefore, the nucleation layer (AlN) structure and the thick buffer layer (generally thicker than 4 μm) are necessarily introduced before the high-quality GaN channel layer is grown.

A typical device structure of a depletion-mode (D-mode) HEMT shows normally-on operation. The D-mode HEMT requires a negative bias in order to deplete the channel layer under the gate electrode, and perform the cut-off stage. The band diagram for the D-mode AlGaN/GaN HEMT structure is also shown in Figure 9.8. The gate electrode is a Schottky contact, which can be used to switch the device. The source/drain electrode belongs to ohmic contacts with low contact resistance, providing high current operation without power loss. The three-terminal device constructs the device structure and leads to high performance in DC or RF applications.

Figure 9.9 illustrates the output characteristics of a D-mode HEMT. The three-terminal HEMT is in the common source (S) configuration, as illustrated in the inset of Figure 9.8. The gate (G) is the input, where the drain (D) is the output. The input signal into gate will switch the device from off-state to on-state. Switching off a D-mode HEMT requires a negative gate bias to deplete the channel of 2DEG, which blocks the channel so that there is no current flow; this is known as the pinch-off state. The expression which relates the number of carriers in the channel is controlled by the gate bias

$$n_S = \frac{\varepsilon_{\text{AlGaN}}}{q(d_{\text{AlGaN}} + \Delta d)}(V_{\text{GS}} - V_{\text{TH}}) \tag{9.24}$$

where:

n_s is the charge density of the 2DEG

d_{AlGaN} is the thickness of the AlGaN barrier layer

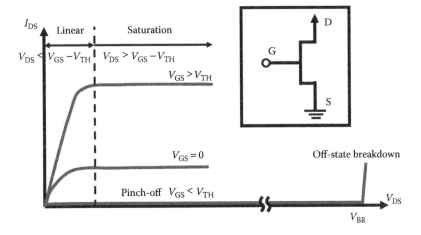

FIGURE 9.9   Typical I_{DS}-V_{DS} (drain-source current-voltage) characteristics of a D-mode AlGaN/GaN HEMT. The inset shows a common source configuration for a transistor.

Δd is the effective distance of the 2DEG from the interface
V_{GS} is the gate bias
V_{TH} is the threshold voltage

When $V_{GS} = V_{TH}$, the term $(V_{GS}-V_{TH})$ will be equal to zero, leading to no sheet carriers and the device shall be kept in off-state. When $V_{GS} > V_{TH}$, the channel starts to accumulate electrons and then current flows under drain bias. Here, the drain bias $V_{DS} < V_{GS} - V_{TH}$, which is called linear region shows that the device is operating as a linear curve, where the current will be linearly increased with applied drain electric field regarding the channel electron velocity is proportional to the applied electric field. The current flowing between the source and drain at this region can be given by

$$I_{DS} = q n_S v_{eff} W_G \tag{9.25}$$

where:
v_{eff} is the effective electron velocity in the 2DEG channel
W_G is the gate width

The velocity of electrons in the channel depends on their mobility, and the electric field applied and is given by the relationship [22]

$$v = \mu_n E \tag{9.26}$$

where:
μ_n is the electron mobility
E is the applied electric field

The mobility of electrons in the 2DEG of an AlGaN/GaN HEMT is affected by scattering through defects and dislocations due to the semiconductor crystal, impurity, and alloy disorder scattering [23]. According to Equation 9.26, the electron velocity increases with the applied electric field; hence, the drain current shows linearly increasing at low electric field when $V_{DS} < V_{GS} - V_{TH}$. Following the increasing of applied electric field as the drain bias to $V_{DS} > V_{GS} - V_{TH}$, the electron velocity becomes independent to the applied electric field and finally saturates. Unfortunately, there is no accurate function that can describe the activity in this region. So far, only the two-step approximate method can be nearly fitted to the experimental data.

Afterward, the lateral bias beneath the gate depletion region begins to pinch the channel off at the gate edge, as the drain bias is continuously increasing when $V_{DS} > V_{GS}$. This pinch point will limit the number of electrons which can flow from the source to drain due to the continuously constricted channel width, leading to a saturated drain current. In this region, the drain current keeps a constant no matter even if we increase the drain bias any further, and is called the saturation region. The saturate drain current can be presented by

$$n_S = \frac{\varepsilon_{AlGaN} v_{sat} W_G}{q(d_{AlGaN} + \Delta d)}(V_{GS} - V_{TH}) \tag{9.27}$$

where v_{sat} is the saturated electron velocity.

Furthermore, the off-state breakdown voltage (V_{BR}) is commonly defined as the drain current that reaches to 1 mA/mm at off-state with gate-bias lower than V_{TH}. The off-state drain current (as an off-state leakage) shows a strong shooting peak after the bias over V_{BR}, generally, corresponds to device failure without recovery. The detailed mechanism will be discussed later.

9.5 Enhancement Mode AlGaN/GaN HEMTs

One of the technology challenges for incorporating GaN HEMTs into power electronics circuits is their "normally-on" (D-mode) operation. For conventional AlGaN/GaN HEMT structures, a large amount of two-dimensional electron gas (2DEG) exists because of the strong built-in polarization electric field in the Ga-face *c*-plane epi-structure. The 2DEG in the channel cannot be easily depleted by the Schottky gate contact at zero gate bias. GaN HEMTs are in the on-state when its V_{GS} (gate-source voltage) is at 0 V and in the off-state when the V_{GS} is below 0 V. For most switching power supplies, the pulse width modulation (PWM) scheme switches the device OFF when the gate voltage is set to 0 V. Thus, under an OFF state, there is still current flow across the GaN HEMT channel. The intrinsic normally-on property makes GaN HEMTs difficult to build an offline converter.

A more complex circuit design is required when using the depletion-mode transistors. There have been many technologies proposed to raise the conduction band energy level underneath the gate contact so that a positive threshold voltage is obtained to ensure enhancement-mode (E-mode) operation. Examples include using the gate recess structure [24,25], the fluorine treatment [26,27], the p-type cap layer [27–33], the piezo-neutralization layer [34], the nonpolar *a*-plane channel [35], and the Metal–oxide–semiconductor field-effect transistor (MOSFET) structure [36]. Figure 9.10 compares several popular technologies to achieve E-mode operation. Among the aforementioned methods, literatures on HEMTs using a p-type cap layer have reported the threshold voltage (V_{TH}) ranging from 1 to 3 V with the applied gate voltage larger than 5 V [27–33].

Other than the development of direct E-mode structure, a hybrid circuit provides an alternative solution. As shown in Figure 9.11, an E-mode low-voltage Si MOSFET was connected to D-mode high-voltage GaN HEMT in series while the gate of the GaN HEMT was connected to the source of the Si MOSFET. This hybrid configuration, also named cascade structure, demonstrates an effective E-mode power device. It is safe in case of faulty gate control. Moreover, the structure is compatible with existing Si gate drivers, as well as the freedom to optimize of GaN HEMTs without complicating the input circuits. Even though internally it is separated into two devices, the final packaged chip is a three-terminal device. The midpoint of the Cascode is embedded, as is the upper gate to lower source connection.

	Nonpolar plane	F-doped gate	Recessed gate	p-type gate
Structure	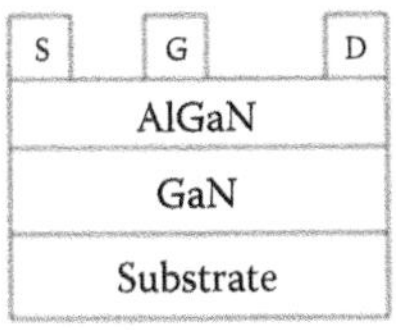	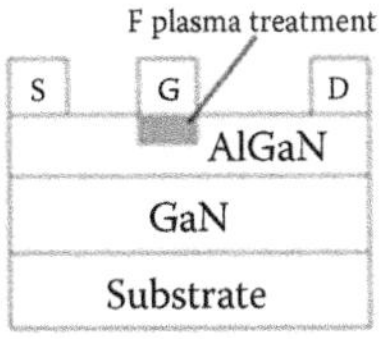	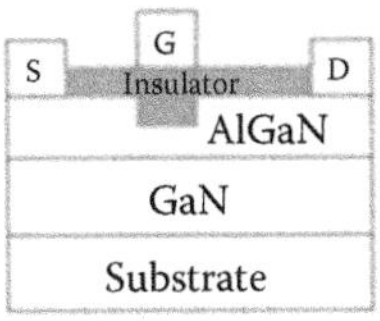	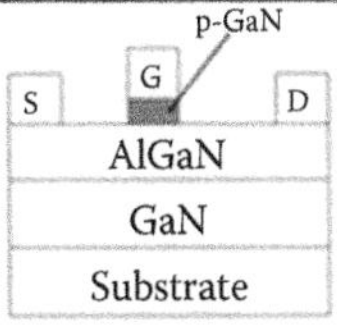
Advantage	➤ Simple process ➤ Simple structure	➤ Low I_{leak} ➤ Controllability of V_{TH}	➤ Low I_{leak} ➤ Large I_{max}	➤ Controllability of V_{TH} ➤ Good reliability
Challenge	➤ Epitaxial grow ➤ No polarization Induced charge ➤ Increase I_{max}	➤ Stability of doped fluorine ➤ Increase V_{GS_max}	➤ Recess control ➤ Stability of Insul./Semi. Surface	➤ p-GaN etching control

FIGURE 9.10 Comparisons of technologies for achieving E-mode operation.

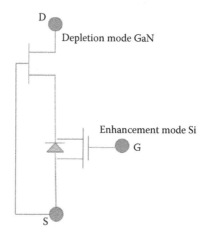

FIGURE 9.11 The cascode normally off GaN HEMT integrated with an enhancement mode Si MOS device.

9.6 Effects of Device Fabrication on the Device Performance

9.6.1 Current Collapse

In device operation, current collapse is one of the most important issues needs to be considered [37–42]. The on-resistant (R_{on}) will be raised during device switching, which is attributed to the trapping effect in AlGaN/GaN HEMTs. Figure 9.12a shows the I_{DS}-V_{DS} (drain-source current-voltage) characteristics with different drain voltage of a AlGaN/GaN HEMT. The dynamic R_{on} is increased with an increasing drain bias. The electrons may leak from the gate and fill trap states on the surface and then forming a virtual gate, as shown in Figure 9.12b The virtual gate modulates the depletion region, partially depleting the electrons of channel subsequently. Therefore, the trapping effect is responsible for the reduction of the current, which results in a collapsed current. This effect can be recovered with time but it is not appropriate for power electronic applications.

The trapping effect means electrons are captured by trap states from III-nitride epi-layers. These traps are mostly related to crystalline imperfection during material growth and device processing: (i) lattice-mismatch-induced defect/threading dislocation in buffer layers within foreign substrates such as Si, SiC, and sapphire; (ii) growth condition (V/III ratio) induced surface states; (iii) unintentional doping impurities (*i.e.*, C), and compensation doping (*i.e.*, Fe, Be); and (iv) thermal, physical, and chemical damage from device fabrication process. The trapping states lead to degraded performance under high-power operation, in such of the DC/RF performance of AlGaN/GaN HEMTs. The possible sources of traps are presented in Figure 9.13.

Therefore, it is essential to improve current collapse phenomenon by using some modifications to minimize trapping effects through the control of surface/interface states. It has been reported that the surface

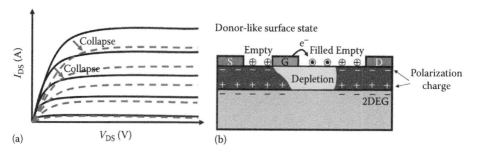

FIGURE 9.12 (a) Devices performance operated w/ and w/o current collapse. (b) Illustration of the current collapse phenomenon in a GaN HEMT structure.

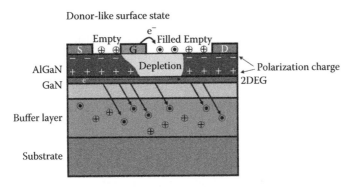

FIGURE 9.13   The possible sources of traps in AlGaN/GaN HEMT structure.

states in GaN-based materials are electrically active states at threading dislocations and nitrogen vacancies and oxygen impurities as shallow surface donors [43–46]. In addition, many researchers have investigated the origin, location, and density of traps and reported various physical models of trapping mechanisms [41]. Many researches have reported how to suppress the current collapse. These modifications are surface passivation [47–49], a thin AlN barrier layer in AlGaN/GaN interface [50], application of a field plate [51], surface treatments with chemicals or plasma [52], and postgate annealing [53].

9.6.2 Effects of Passivation Layer

AlGaN/GaN HEMTs typically feature a short distance (e.g., ~20 nm) from the polarized surface to 2DEG channel, and could easily suffer from surface-state-induced adverse effects such as current collapse, particularly at high-drain-voltage switching conditions. An effective way is surface passivation with the dielectric layer, such as Si_3N_4 and SiO_2 [48,49]. Some groups have reported that surface passivation can reduce electron trapping states to prevent charge trapping-induced current collapse [48]. In general, we can increase the drain current regarding the surface passivation, which means that the 2DEG density is increased. After the surface passivation, the net surface charge and net charge from AlGaN/GaN interface become less negative and less positive, respectively. Therefore, the depleted 2DEG channel can be improved. The current collapse can be improved in terms of surface passivation, which is attributed to the increased positive charge at dielectric/AlGaN interface [48,49].

Therefore, we can summarize the effect of passivation. First, the passivation layer buries surface trap state, leading to the surface state as infeasible to capture the electrons from the gate and then preventing the virtual gate that depletes the 2DEG channel. Second, the passivation layer reduces the electric field at gate edge, resulting in more donor-like surface states by keeping the surface traps empty. The field plate is also employed to suppress the electric field at the gate edge to drain side, resulting in an increased breakdown voltage and also reducing the electron trapping on the AlGaN surface [51].

The previous study reports the surface passivation studies of SiO_2, Si_3N_4, and SiON on AlGaN/GaN HEMTs [54]. The unpassivated HEMTs are also compared with the passivated HEMTs. Due to the occurrence of deep traps, the SiO_2 passivated HEMTs show slow switching speed with high breakdown voltage. The SiO_2, Si_3N_4, and SiON dielectrics with a thickness of 100 nm were deposited at 300°C by plasma enhanced chemical vapor deposition (PECVD). Refractive indices of the as-deposited SiO_2, Si_3N_4, and SiON are 1.47, 1.88, and 1.58, respectively. The X-ray photon spectroscopy (XPS) revealed that the SiO_2 and Si_3N_4 dielectrics were slightly silicon rich and SiON was slightly nitrogen rich.

After the PECVD grown SiO_2, Si_3N_4, and SiON dielectrics passivated, an increased drain current has been observed after the passivation from SiO_2, Si_3N_4, and SiON dielectrics. A small increased gate leakage was observed on SiON passivated HEMTs. Low drain current collapse with low gate leakage was observed on SiON passivated AlGaN/GaN HEMTs. The SiON passivation is a promising candidate for AlGaN/GaN HEMTs because of its better breakdown voltage with small drain current collapse.

9.6.3 Effects of High-κ Insulator Passivation

As mentioned above, we can remove the surface states by using insulator passivation. The typical SiO_2 and Si_3N_4 dielectric layer are generally grown by PECVD or LPCVD. The material selection for the surface passivation is a key issue for improving the current collapse. In the last section, we compared the SiO_2, Si_3N_4, and SiON passivated and unpassivated AlGaN/GaN HEMTs. The SiO_2 passivated HEMT shows the highest breakdown voltage, but the serious gate leakage was performed due to the high density of deep-level traps in SiO_2. Another choice is high-κ insulator with atomic layer deposition (ALD). The ALD-grown aluminum oxide (Al_2O_3) layer significantly improves the power performance of the no-field plate GaN HEMT [55]. The unique physical properties of Al_2O_3, in particular, the WBG, high breakdown field, and high thermal stability make it extremely attractive as a passivation material for GaN HEMTs. The ALD process promising ultrathin, uniform, conformal, and pin-hole free films makes it highly desirable as a manufacturing technology. The ALD Al_2O_3-passivated GaN HEMTs have demonstrated significant improvement in DC, pulsed-IV, and RF power performance over the device with Si_3N_4-passivated HEMT. These results have been attributed to the interface quality between Al_2O_3 and GaN. The ALD Al_2O_3-passivated GaN HEMT shows a high manufacturability and serves as a high-performance candidate for the power device applications.

Furthermore, K. J. Chen et al. reported another effective method to suppress the surface states [56]. It has been well known that the surface states are created by dangling bonds, threading dislocation, and ions absorbed from the ambient environment [57]. As the oxide-based high-κ dielectrics such as Al_2O_3 [58,59] and HfO_2 [60] have been employed to passivate GaN-based HEMT by using ALD, and also promising the gate dielectric layer in GaN-based metal–insulator–semiconductor HEMTs (MIS-HEMTs). According to some research results, it has been revealed that high densities of interface states can be formed at the between oxide interface -based high-κ dielectrics and GaN-based materials [61,62]. The high density of interface states was formatted from Ga–O bonds [63]. Therefore, the current collapse cannot be fully improved as we employed the oxide-based high-κ dielectric layer for passivation. Thus, nitride-based high-κ dielectrics such as AlN is a better choice for GaN-based HEMT passivation with low interface states. Moreover, the *in situ* remote plasma is a powerful tool for decreasing the Ga-O dangling bonds in plasma-enhanced ALD (PEALD). K. J. Chen et al. reported an *in situ* low-damage plasma pretreatment (RPP) that could remove the surface native oxide with minimum surface damage [56]. The *in situ* surface treatment by NH_3 plasma [64] shows an improved result than in SiN_x-passivated AlGaN/GaN HEMTs. The RPP process combined with ALD-grown AlN can obtain an atomically sharp interface, resulting in significant reduction of the current collapse and the dynamic on-resistance (dynamic R_{on}) in AlGaN/GaN HEMTs during device switching. This result promotes the ALD process to be a key technology to achieve high-performance GaN HEMTs.

9.6.4 Breakdown Voltage Improvement

The GaN-on-Si HEMTs are available at a relatively low cost and can be used with conventional integrated circuit process tools for fabricating GaN-on-Si power devices. However, devices with low-resistivity substrates tend to leak current from the channel to substrate, resulting in a reduction of V_{BR} [65]. The vertical leakage of GaN-on-Si HEMTs may be effectively eliminated by reducing the density of threading dislocations (TDs) in the GaN epilayers [66] or by increasing the resistivity of the buffer layers [65]. The possible leakage paths are shown in Figure 9.14.

Several approaches of high resistivity buffer layer have been reported, such as thicker GaN buffer layer [67], substrate removal [68], Silicon-on-Insulator substrate (SOI) [69], and using high resistivity of carbon-doped buffer layers [70]. The carbon-doped semi-insulating (S.I.) GaN buffer layer could be comprehensively obtained by tetrabromomethane (CBr_4) precursor [71] or manipulation of the growth pressure, V/III ratio, and growth temperature under the metalorganic chemical vapor deposition (MOCVD) growth procedure [72]. The carbon-doped S.I. GaN buffer layer has been demonstrated to

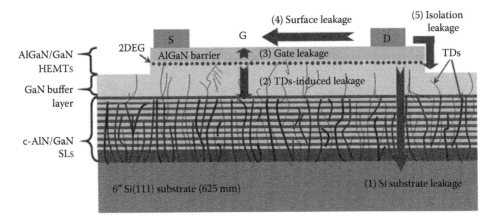

FIGURE 9.14   The leakage paths of GaN-on-Si HEMTs.

enhance the BV toward power switching and microwave applications [73,74]. However, some previous studies [74,75] reported that the virtual gate associated with surface traps had been verified to act as the dominant mechanism causing the current collapse (CC) in AlGaN/GaN HEMTs. Nevertheless, carbon-doped S.I. GaN shows poor epitaxial quality due to its lower growth temperature and growth pressure [76], which is also responsible for CC through charge trapping in deep levels within the S.I. buffer layer of GaN-on-Si HEMTs [77,78]. In general, the deep acceptor and deep donor traps coordinate into the GaN buffer layers, which will accompany with charging/discharging interactions during the high electric field was applied [74,79]. Such interactions are associated with the reduction of V_{BR} and also correlated to decrease the drain current [74,79]. The CC phenomenon results in low reliable device performance, especially in power switching applications. Therefore, it is essential to overcome current collapse issue via reduction of buffer trap density and threading dislocations.

H.C. Kuo et al. reported the growth conditions of carbon-doped GaN spacer layer via CBr$_4$ precursor doping [80]. Furthermore, carbon-doped related CC was also investigated and discussed, which we purposed to optimize a GaN-on-Si HEMT with high V_{BR} but low CC in a reliable performance. The GaN-on-Si HEMT with low carbon-doped AlGaN back barrier shows a comparable BV, but the CC could be eliminated, which is a possible candidate structure for high-power applications (Figures 9.15 and 9.16).

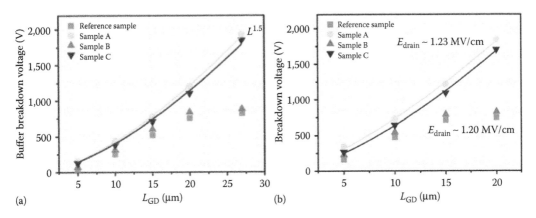

FIGURE 9.15   (a) Two-terminal buffer V_{BR} versus L_{SD} for HEMT samples. (b) Three-terminal V_{BR} versus L_{GD} for HEMT samples. The V_{BR} was captured for leakage currents up to 10 μA/mm with different distances between pads. (From Tzou, A. J. et al., *Electronics*, 5, 28, 2016. With permission.)

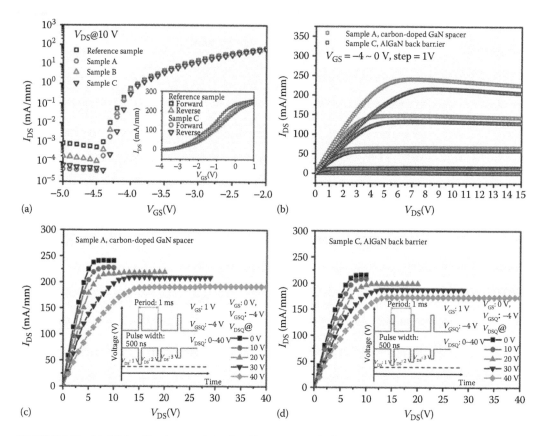

FIGURE 9.16 (a) DC I_{DS}–V_{GS} characteristics for all specimens. The hysteresis I_{DS}–V_{GS} curves for the Reference sample and Sample C are shown in the inset. (b) DC I_{DS}–V_{GS} characteristics for Sample A and Sample C. Pulsed I_{DS}–V_{GS} characteristics for (c) Sample A and (d) Sample C. Pulsed I_{DS}–V_{GS} characteristics were extracted from a quiescent gate bias (V_{GSQ}) of –5 V to an on-state at 0 V in 500 ns and a separation of 1 ms. Afterward, the quiescent drain bias (V_{DSQ}) was swept from 0 to 40 V (in 10-V increments). (From Tzou, A. J. et al., *Electronics*, 5, 28, 2016. With permission.)

9.7 Conclusion

Research on GaN devices has been going on for many years. The unique properties of GaN for high current and high voltage have attached great attention to power system engineers. Even though the market penetration of GaN electronics to power systems is still small, given the mature of technological development and continuing cost-down of the fabrication cost, GaN HEMT switches and rectifiers will poise the potential for next-generation high-efficiency switching power supply.

References

1. B. Ozpineci, L. M. Tolbert, S. K. Islam, and M. Chinthavali, Comparison of wide bandgap semiconductors for power applications, In *European Conference on Power Electronics and Applications*, Toulouse, France, pp. 2–4, 2003.
2. U. Mishra, P. Parikh, and Y.-F. Wu, AlGaN/GaN HEMTs-an overview of device operation and applications, *Proc. IEEE*, 90, 6, 1022–1031, June 2002.
3. R. Kemerley, H. Wallace, and M. Yoder, Impact of wide bandgap microwave devices on DoD systems, *Proc. IEEE*, 90, 6, 1059–1064, June 2002.

4. F. A. Marino, N. Faralli, D. K. Ferry, S. M. Goodnick, and M. Saraniti, Figures of merit in high-frequency and high-power GaN HEMTs, *J. Phys: Confr. Ser.*, 193, 1, 012040, 2009.

5. B. J. Baliga, Power semiconductor device figure of merit for high-frequency applications, *IEEE Electron Dev. Lett.*, 10, 10, 455–457, October 1989.

6. G. B. P. Devices, Cost-effective revolutionary performance, *Power Electron. Europe*, 7, 2008.

7. M. Shur and R. Davis, *GaN-Based Materials and Devices: Growth, Fabrication, Characterization and Performance.* Singapore: World Scientific, 2004.

8. L. Nguyen, L. Larson, and U. Mishra, Ultra-high speed modulation-doped fielde effect transistors: A Tutorial Review, *Proc. IEEE*, 80, 4, 494–518, April 1992.

9. J. N. K. M. Asif Khan, A. Bhattarai and D. T. Olson, High electron mobility transistor based on a GaN/AlxGa1-xN heterojunction, *Appl. Phys. Lett.*, 63, 9, 1214–1216, August 1993.

10. M. A. Khan, J. Kuznia, D. T. Olson, W. Scha, J. Burm, and M. Shur, Microwave performance of a 0.25 mm gate AlGaN/GaN heterostructure field effect transistor, *Appl. Physi. Lett.*, 65, 9, 1121–1123, August 1994.

11. A. Trampert, O. Brandt, and K. Ploog, Crystal structure of group {III} nitrides, In *Gallium Nitride (GaN), Series: Semiconductors and Semimetals*, Pankove, J. I. and Moustakas, T. D. (eds.). Amsterdam, the Netherlands, Elsevier, Vol. 50, pp. 167–192, 1997.

12. O. Ambacher et al., Two-dimensional electron gases induced by spontaneous and piezoelectric polarization charges in N- and Ga-face AlGaN/GaN heterostructures, *J. Appl. Phys.*, 85, 6, 3222–3233, March 1999.

13. F. Bernardini, V. Fiorentini, and D. Vanderbilt, Spontaneous polarization and piezoelectric constants of III–V Nitrides, *Phys. Rev. B*, 56, R10024–R10027, October 1997.

14. A. Bykhovski, B. Gelmont, and M. Shur, The influence of the strain-induced electric field on the charge distribution in GaN-AlN-GaN structure, *J. Appl. Phys.*, 74, 11, 1993.

15. M. Shur, G. Simin, S. Rumyantsev, R. Jain, and R. Gaska, Insulated gate nitride-based field effect transistors, In *Fundamentals of III–V Semiconductor MOSFETs*, Oktyabrsky, S. And Ye, P. (eds.). US, Springer, pp. 379–422, 2010.

16. Y.-J. Lee, Y.-C. Yao, C.-Y. Huang, T.-Y. Lin, L.-L. Cheng, C.-Y. Liu, M.-T. Wang, and J.-M. Hwang, High breakdown voltage in AlGaN/GaN HEMTs using AlGaN/GaN/AlGaN quantum-well electron-blocking layers, *Nanoscale Res. Lett.*, 9, 1, 433, 2014.

17. J. Ibbetson, P. Fini, K. Ness, S. Denbaars, J. Speck, and U. Mishra, Polarization effects, surface states, and the source of electrons in AlGaN/GaN heterostructureeld eect transistors, *Appl. Phys. Lett.*, 77, 250, July 2000.

18. P. Xu, Y. Jiang, Y. Chen, Z. Ma, X. Wang, Z. Deng, Y. Li, H. Jia, W. Wang, and H. Chen, Analyses of 2-DEG characteristics in GaN HEMT with AlN/GaN super-lattice as barrier layer grown by MOCVD, *Nanoscale Res. Lett.*, 7, 1, 1–6, 2012.

19. C. Wang, X. Wang, G. Hu, J.Wang, and J. Li, Inuence of Al content on electrical and structural properties of Si-doped AlxGa1-xN/GaN HEMT structures, *Phys. Stat. Sol. (C)*, 3, 3, 486–489, 2006.

20. N. Goyal and T. A. Fjeldly, Effects of strain relaxation on bare surface barrier height and two-dimensional electron gas in AlxGa1-xN/GaN heterostructures, *J. Appl. Phys.*, 113, 1, 014505, January 2013.

21. Y. Ohno and M. Kuzuhara, Application of GaN-based heterojunction FETs for advanced wireless communication, *IEEE Trans. Electron Dev.*, 48, 3, 517–523, March 2001.

22. R. F. Pierret et al., *Semiconductor Device Fundamentals.* Reading, MA: Oxford University Press, 2013.

23. M. Germain, M. Leys, S. Boeykens, S. Degroote, W. Wang, D. Schreurs, W. Ruythooren, K.-H. Choi, B. V. Daele, G. V. Tendeloo, and G. Borghs, High electron mobility in AlGaN/GaN HEMT grown on sapphire: Strain modification by means of AlN interlayers, In *MRS Proceedings*, Cambridge, Cambridge University Press, Vol. 798, pp. Y10–Y22, 2003.

24. T. Oka and T. Nozawa, AlGaN/GaN recessed MIS-gate HFET with high-threshold-voltage normally-off operation for power electronics applications, *IEEE Electron Devic. Lett.*, 29, 668–670, July 2008.

25. Y.-C. Lee, C.-Y. Wang, T.-T. Kao, and S.-C. Shen, Threshold voltage control of recessed-gate III-N HFETs using an electrode-less wet etching technique, In *CS MANTECH Conference*, Boston, MA, April 2012.

26. W. Chen, K.-Y. Wong, and K. J. Chen, Monolithic integration of lateral field-effect rectifier with normally-off HEMT for GaN-on-Si switch-mode power supply converters, In *International Electron Devices Meeting (IEDM)*, pp. 1–4, December 2008.

27. Y. Cai, Y. Zhou, K. J. Chen, and K. M. Lau, High-performance enhancement-mode AlGaN/GaN HEMTs using fluoride-based plasma treatment, *IEEE Electron Dev. Lett.*, 26, 435–437, July 2005.

28. Y. Uemoto, M. Hikita, H. Ueno, H. Matsuo, H. Ishida, M. Yanagihara, T. Ueda, T. Tanaka, and D. Ueda, Gate injection transistor (GIT)—A normally-off AlGaN/GaN power transistor using conductivity modulation, *IEEE T. Electron Dev.*, 54, 3393–3399, December 2007.

29. O. Hilt, A. Knauer, F. Brunner, E. Bahat-Treidel, and J. Wurfl, Normally-off AlGaN/GaN HFET with *p*-type GaN Gate and AlGaN buffer, In *International Symposium on Power Semiconductor*, pp. 347–350, June 2010.

30. O. Hilt, F. Brunner, E. Cho, A. Knauer, E. Bahat-Treidel, and J. Wurfl, Normally-off high-voltage p-GaN gate GaN HFET with carbon-doped buffer, In *International Symposium on Power Semiconductor*, pp. 239–242, May 2011.

31. S. L. Selvaraj, K. Nagai, and T. Egawa, MOCVD grown normally-OFF type AlGaN/GaN HEMTs on 4 inch Si using p-InGaN cap layer with high breakdown, In *Device Research Conference*, pp. 135–136, June 2010.

32. I. Hwang et al., 1.6 kV, 2.9 mΩ cm2 normally-off p-GaN HEMT device, In *International Symposium on Power Semiconductor*, pp. 41–44, June 2012.

33. H. Chonan, T. Ide, X. Q. Shen, and M. Shimizu, Effect of hole injection in AlGaN/GaN HEMT with GIT structure by numerical simulation, *Phys. Status Solidi C*, 9, 847–850, January 2012.

34. K. Ota, K. Endo, Y. Okamoto, Y. Ando, H. Miyamoto, and H. Shimawaki, A normally-off GaN FET with high threshold voltage uniformity using a novel piezo neutralization technique, In *International Electron Devices Meeting* (IEDM), pp. 1–4, Dec. 2009.

35. M. Ishida, M. Kuroda, T. Ueda, and T. Tanaka, Nonpolar AlGaN/GaN HFETs with a normally off operation, *Semicond. Sci. Tech.*, 27, 024019, January 2012.

36. Y. Chang, W. Chang, H. Chiu, L. Tung, C. Lee, K. Shiu, M. Hong, J. Kwo, J. Hong, and C. Tsai, Inversion-channel GaN MOSFET using atomic-layer-deposited Al2O3 as gate dielectric, *Appl. Phys. Lett.*, 93, 053504–053504-3, August 2008.

37. U. K. Mishra, L. Shen, T. E. Kazior, and W. Yi-Feng, GaN-based RF power devices and amplifiers, *Proc. IEEE*, 96, 287–305, 2008.

38. M. A. Khan, Q. Chen, C. J. Sun, J. W. Yang, M. Blasingame, M. S. Shur, and H. Park, Enhancement and depletion mode GaN/AlGaN heterostructure field effect transistors, *Appl. Phys. Lett.*, 68, 514, 1996.

39. X. Hu, G. Simin, J. Yang, M. A. Khan, R. Gaska, and M. S. Shur, Enhancement mode AlGaN/GaN HFET with selectively grown pn junction gate, *Electron. Lett.*, 36, 8, 753–754, 2000.

40. M. A. Khan, M. S. Shur, Q. C. Chen, and J. N. Kuznia, Current/voltage characteristic collapse in AlGaN/GaN heterostructure insulated gate field effect transistors at high drain bias, *Electron. Lett.*, 30, 2175–2176, 1994.

41. P. B. Klein and S. C. Binari, Photoionization spectroscopy of deep defects responsible for current collapse in nitride-based field effect transistors, *J. Phys.: Cond. Matt.*, 15, R1641–R1667, 2003.

42. N. Q. Zhang, B. Moran, S. P. DenBaars, U. K. Mishra, X. W. Wang, and T. P. Ma, Effects of surface traps on breakdown voltage and switching speed of GaN power switching HEMTs, In *IEDM Technical Digest*, pp. 25.5.1–25.5.4, 2001.

43. P. J. Hansen, Y. E. Strausser, A. N. Erickson, E. J. Tarsa, P. Kozodoy, E. G.Brazel, J. P. Ibbetson, U. Mishra, V. Narayanamurti, S. P. DenBaars, and J. S. Speck, Scanning capacitance microscopy imaging of threading dislocations in GaN films grown on (0001) sapphire by metal organic chemical vapor deposition, *Appl. Phys. Lett.*, 72, 2247–2249, 1998.

44. H. J. Im, Y. Ding, J. P. Pelz, B. Heying, and J. S. Speck, Characterization of individual threading dislocations in GaN using ballistic electron emission microscopy, *Phys. Rev. Lett.*, 87, 106802, September 2001.

45. J. W. P. Hsu, M. J. Manfra, D. V. Lang, S. Richter, S. N. G. Chu, A. M. Sergent, R. N. Kleiman, L. N. Pfeiffer, and R. J. Molnar, Inhomogeneous spatial distribution of reverse bias leakage in GaN Schottky diodes, *Appl. Phys. Lett.*, 78, 1685–1687, 2001.

46. D. M. Schaadt, E. J. Miller, E. T. Yu, and J. M. Redwing, Lateral variations in threshold voltage of an AlxGa1-xN/GaN heterostructure field-effect transistor measured by scanning capacitance spectroscopy, *Appl. Phys. Lett.*, 78, 88–90, 2001.

47. Y. Ohno, T. Nakao, S. Kishimoto, K. Maezawa, and T. Mizutani, Effects of surface passivation on breakdown of AlGaN/GaN high-electron-mobility transistors, *Appl. Phys. Lett.*, 84, 2184–2186, 2004.

48. B. Luo et al., Influence of MgO and Sc2O3 passivation on AlGaN/GaN high-electron mobility transistors, *Appl. Phys. Lett.*, 80, 1661–1663, 2002.

49. S.Arulkumaran, T. Egawa, H. Ishikawa, T. Jimbo, and Y. Sano, Surface passivation effects on AlGaN/GaN high-electron-mobility transistors with SiO2, Si3N4, and silicon oxynitride, *Appl. Phys. Lett.*, 84, 613–615, 2004.

50. J. S. Lee, J. W. Kim, J. H. Lee, C. S. Kim, J. E. Oh, and M. W. Shin, Reduction of current collapse in AlGaN/GaN HFETs using AlN interfacial layer, *Electron. Lett.*, 39, 750–752, 2003.

51. Y. F. Wu, A. Saxler, M. Moore, R. P. Smith, S. Sheppard, P. M. Chavarkar, T. Wisleder, U. K. Mishra, and P. Parikh, 30-W/mm GaNHEMTs by field plate optimization, *IEEE Electron Dev. Lett.*, 25, 117–119, March 2004.

52. A. P. Edwards, J. A. Mittereder, S. C. Binari, D. S. Katzer, D. F. Storm, and J. A. Roussos, Improved reliability of AlGaN-GaN HEMTs using an NH3/plasma treatment prior to SiN passivation, *IEEE Electron Dev. Lett.*, 26, 225–227, 2005.

53. N. Miura, T. Nanjo, M. Suita, T. Oishi, Y. Abe, T. Ozeki, H.Ishikawa, T. Egawa, and T. Jimbo, Thermal annealing effects on Ni/Au based Schottky contacts on n-GaN and AlGaN/GaN with insertion of high work function metal, *Sol.-State Electron.*, 48, 689–695, May 2004.

54. S. Arulkumaran, T. Egaw, H. Ishikawa, and T. Jimbo, and Y. Sano, Surface passivation effects on AlGaN/GaN high-electron-mobility transistors with SiO2, Si3N4, and silicon oxynitride, *Appl. Phys. Lett.*, 84, 613–615, 2004.

55. D. Xu, K. Chu, J. Diaz, W. Zhu, R. Roy, L. Mt. Pleasant, K. Nichols, P.-C. Chao, M. Xu, and P. D. Ye, 0.2-μm AlGaN/GaN high electron-mobility transistors with atomic layer deposition Al_2O_3 passivation, *IEEE Electron Dev. Lett.*, 34, 744–746, 2013.

56. S. Huang, Q. Jiang, S. Yang, C. Zhou, and K. J. Chen, Effective passivation of AlGaN/GaN HEMTs by ALD-grown AlN thin film, *IEEE Electron Dev. Lett.*, 33, 516–518, 2012.

57. B. M. Green, K. K. Chu, E. Martin Chumbes, J. A. Smart, J. R. Shealy, and L. F. Eastman, The effect of surface passivation on the microwave characteristics of undoped AlGaN/GaN HEMT's, *IEEE Electron Dev. Lett.*, 21, 268–270, 2000.

58. T. Hashizume, S. Ootomo, and H. Hasegawa, Suppression of current collapse in insulated gate AlGaN/GaN heterostructure field-effect transistors using ultrathin Al2O3 dielectric, *Appl. Phys. Lett.*, 83, 2952–2954, 2003.

59. D. H. Kim, V. Kumar, G. Chen, A. M. Dabiran, A. M. Wowchak, A. Osinsky, and I. Adesida, ALD Al_2O_3 passivated MBE-grown AlGaN/GaN HEMTs on 6H-SiC, *Electron. Lett.*, 43, 129–130, 2007.

60. J. Shi, L. F. Eastman, X. Xin, and M. Pophristic, High performance AlGaN/GaN power switch with HfO2 insulation, *Appl. Phys. Lett.*, 95, 042103, July 2009.

61. S. Huang, S. Yang, J. Roberts, and K. J. Chen, Characterization of Vth-instability in Al2O3/GaN/ AlGaN/GaN MIS-HEMTs by quasi-static C–V measurement, In *International Conference on Nitride Semiconductors*, Glasgow, U.K., 2011.

62. C. Mizue, Y. Hori, M. Miczek, and T. Hashizume, Capacitance–voltage characteristics of Al_2O_3/ AlGaN/GaN structures and state density distribution at Al_2O_3/AlGaN interface, *Jpn. J. Appl. Phys.*, 50, 021001, February 2011.

63. K. Mishra, P. Schmidt, S. Laubach, and K. Johnson, Localization of oxygen donor states in gallium nitride from first-principles calculations, *Phys. Rev. B*, 76, 035 127, July 2007.

64. A. P. Edwards, J. A. Mittereder, S. C. Binari, D. S. Katzer, D. F. Storm, and J. A. Roussos, Improved reliability of AlGaN–GaN HEMTs using an NH3 plasma treatment prior to SiN passivation, *IEEE Electron Dev. Lett.*, 26, 225–227, 2005.

65. C. H. Seager, A. F.Wright, J. Yu, W. Götz, Role of carbon in GaN, *J. Appl. Phys.*, 92, 6553–6560, 2002.

66. Q. F. Shan, D. S. Meyaard, Q. Dai, J. Cho, E. F. Schubert, J. K. Son, C. Sone, Transport-mechanism analysis of the reverse leakage current in GaInN light-emitting diodes, *Appl. Phys. Lett.*, 99, 253506, 2011.

67. S. L. Selvaraj, A. Watanabe, A. Wakejima, T. Egawa, 1.4-kV Breakdown voltage for AlGaN/GaN High-electron-mobility transistors on silicon substrate, *IEEE Electron Dev. Lett.*, 33, 1375–1377, 2012.

68. P. Srivastava et al., Record breakdown voltage (2200 V) of GaN DHFETs on Si with 2-m buffer thickness by local substrate removal, *IEEE Electron Dev. Lett.*, 32, 30–32, 2011.

69. Q. Jiang, C. Liu, Y. Lu, and K. J. Chen, Mechanism of PEALD-grown AlN passivation for AlGaN/ GaN HEMTs: Compensation of interface traps by polarization charges, *IEEE Electron Dev. Lett.*, 34, 357–359, 2013.

70. E. B. Treidel, F. Brunner, O. Hilt, E. Cho, J. Würfl, and G. Tränkle, AlGaN/GaN/GaN:C back-barrier HFETs with breakdown voltage of over 1 kV and low Ron×A, *IEEE Trans. Electron Dev.*, 57, 3050–3058, 2010.

71. R. Xuan, W. H. Kuo, C. W. Hu, S. F. Lin, J. F. Chen, Enhancing threshold voltage of AlGaN/GaN high electron mobility transistors by nano rod structure: From depletion mode to enhancement mode, *Appl. Phys. Lett.*, 101, 112105, 2012.

72. D. D.Koleske, A. E. Wickenden, R. L. Henry, and M. E. Twigg, Influence of MOVPE growth conditions on carbon and silicon concentrations in GaN, *J. Cryst. Growth*, 242, 55–69, 2002.

73. C. Poblenz, P. Waltereit, S. Rajan, S. Heikman, U. K. Mishra, and J. S. Speck, Effect of carbon doping on buffer leakage in AlGaN/GaN high electron mobility transistors, *J. Vac. Sci. Technol. B.*, 22, 1145–1149, 2004.

74. C. Zhou, Q. Jiang, Sen. Huang, and K. J. Chen, Vertical leakage/breakdown mechanisms in AlGaN/GaN-on-Si devices, *IEEE Electron Dev. Lett.*, 33, 1132–1134, 2012.

75. Z. Q Fang, B. Claflin, D. C. Look, D. S. Green, and R. Vetury, Deep traps in AlGaN/GaN heterostructures studied by deep level transient spectroscopy: Effect of carbon concentration in GaN buffer layers, *J. Appl. Phys.*, 108, 063706, 2010.

76. J. Selvaraj, S. L. Selvaraj, and T. Egawa, Effect of GaN buffer layer growth pressure on the device characteristics of AlGaN/GaN high-electron-mobility transistors on Si, *Jpn. J. Appl. Phys.*, 48, 121002, 2009.

77. M. J. Uren, J. Möreke, and M. Kuball, Buffer design to minimize current collapse in GaN/AlGaN HFETs, *IEEE Trans. Electron Dev.*, 59, 3327–3333, 2012.

78. M. J. Uren, K. J. Nash, R. S. Balmer, T. Martin, E. Morvan, N. Caillas, S. L. Delage, D. Ducatteau, B. Grimbert, and J. C. De Jaeger, Punch through in short-channel AlGaN/GaN HFETs, *IEEE Trans. Electron Dev.*, 53, 395–398, 2006.

79. L. Zhang, L. F. Lester, A. G. Baca, R. J. Shul, P. C. Chang, C. G. Willison, U. K. Mishra, S. P. Denbaars, and J. C. Zolper, Epitaxially-grown GaN junction field effect transistors, *IEEE Trans. Electron Dev.*, 47, 507–511, 2000.

80. A. J. Tzou, D. H. Hsieh, S. H. Chen, Y. K. Liao, Z. Y. Li, C. Y. Chang, and H. C. Kuo, An investigation of carbon-doping-induced current collapse in GaN-on-Si high electron mobility transistors, *Electronics*, 5, 28, 2016.

10

Power Conversion and the Role of GaN

Srabanti Chowdhury

Power conversion underlines almost every application that requires electrical power to function. While the complexity and the stages of the power conversion system behind these applications can be widely different, the basic building blocks are similar. Buck or boost to step down or step up the voltage levels, an inverter to transform a DC signal to AC signal, a power factor controller (PFC) to reduce the reactive power from the output, and rectifiers to convert AC to DC are few such building blocks. Solid-state electronic devices, in the form of diodes and transistors, are an integral part of these power converters and their role in power conversion is rapidly increasing to refine an existing or define a new functionality. Silicon (Si) has been the backbone of the semiconductor industry and enabled power electronics much like computational electronics. In the form of metal–oxide–semiconductor field-effect transistor (MOSFET) and its various adaptations, particularly CoolMOSFET [1,2], insulated gate bipolar transistor (IGBT) [1], bipolar junction transistor (BJT) and diodes, Si has served the power electronics industry extremely well so far.

However, the demand to operate at higher power density to reduce the chip area, higher frequency to shrink the size of the passive components and ultimately reduce the overall copper loss, and higher operating temperature to significantly reduce cooling needs calls for wide bandgap (WBG) semiconductors.

10.1 WBG Semiconductors and Their Advantages

WBG semiconductors due to of their large bandgap (typically >3eV) can sustain a higher temperature of operation compared to Si, preventing a thermal runaway. The critical electric field (E_c), also a function of the bandgap, is greater than Si, which allows higher doping density (resulting in higher conductivity) in these materials, ultimately permitting higher power density. All these attractive properties qualify WBG semiconductors to be the material of choice for next generation power electronics. Figure 10.1 shows a

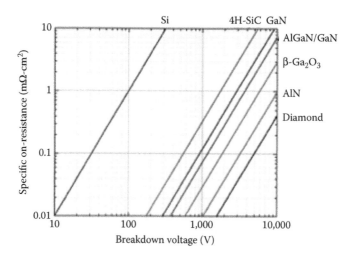

FIGURE 10.1 DC FOM: Specific ON resistance versus breakdown voltage compared for various semiconductors using their material properties. Note that these comparison assumes a vertical geometry and uniform electric field set up by a p-n junction that holds the voltage.

comparison of various WBG semiconductors featuring their DC figure of merit (FOM). Breakdown voltage and On-resistance (R_{on}) (used as the two axes of Figure 10.1) are the two prevalent metrics used for qualifying a semiconductor as a power conversion switch or a power device. For a given breakdown voltage, the semiconductor offering a lower R_{on} is attractive since R_{on} has a direct impact on conduction loss. While candidates like aluminum nitride (AlN), diamond and gallium oxide (Ga$_2$O$_3$) look promising in the list, the two leading contenders of the present time are silicon carbide (SiC) and gallium nitride (GaN). Both have shown exceptional progress over the last decade, demonstrating their merit as was predicted theoretically. In addition to forming a power electronic device material, SiC serves as a substrate for GaN LEDs. Its role as an LED substrate helped in the maturing of the SiC substrate technology, critical for all device development. Commercialization of GaN devices was initiated by the development of LEDs and lasers, closely followed by devices for radio frequency (RF) applications and currently by power electronics. Presently, both SiC and GaN technologies have entered the market, supported by their relevant application space and undergoing various kinds of assessments that will eventually play a big role in shaping the future of the sustainable market for each of them.

10.2 GaN: Material Properties Relevant to Power Electronics

The material properties of GaN have been thoroughly discussed in various literatures [3,4]. The purpose of presenting the topic here is to stimulate the discussion from a power electronics point of view.

GaN presents a unique feature where it can be alloyed suitably with Gr III elements like Al and In, to modify its bandgap and other associated material properties, as suggested by Figure 10.2. The critical electric field in bulk GaN material predicted to be around 3MV/cm increases in Al$_x$Ga$_y$N consistent with the mole fraction (x) of the Al in the layer. The electron mobility in bulk GaN has recently been reported to reach 1160 cm^2/V-s [5] a number much higher than those achievable in Si or SiC. When a thin Al$_x$Ga$_y$N layer is grown heteroepitaxially on GaN, two-dimensional electron gas (2DEG) is formed due to the difference in the interfacial polarization (piezoelectric plus spontaneous) charges, used as a channel in high electron mobility transistors (HEMTs). Unlike the conventional AlGaAs/GaAs HEMTs, the 2DEG in AlGaN/GaN is formed without doping the wider bandgap cap layer. The advantage of a polarization-induced 2DEG is realized by its very high room temperature Hall mobility, reaching up to 2,000 cm^2/V-s or often higher, repetitively. The charge density of the 2DEG, a function of the polarization charge in the

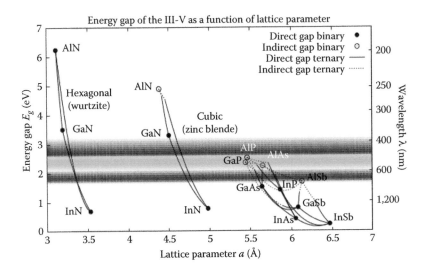

FIGURE 10.2 Bandgap versus lattice parameters of III-V. Wurzite III-nitrides presents a band gap between 0.7 and 6.2 eV. Similar to Opto-electronics, power electronics exploits the range of bandgap that III-nitrides offer. While AlGaN /GaN combination have been heavily explored and used to make devices, research is on-going to wider bandgap materials like high composition Al(GaN) materials and up to AlN. (From Caro Bayo, M. Á. Theory of elasticity and electric polarization effects in the group-III nitrides, PhD Thesis, University College Cork, 2013.)

AlGaN layer and therefore its mole fraction and the AlGaN thickness, is typically measured between $9E12-1E13 cm^{-2}$ (for a ~28% Al, ~25 nm-thick- AlGaN). The product of high mobility with high charge density allows high current capability in these devices. High current, high voltage and high mobility of electrons simultaneously achieved through the AlGaN/GaN heterostructure make III-Nitride HEMTs appropriate for RF applications. It is interesting to note that the power electronic application benefits from the same set of abovementioned properties of GaN that make it attractive for the RF application [6], albeit power electronic devices operate at much lower frequencies.

One remarkable advantage of GaN and other nitrides is the access to engineer band gap, which plays a significant role in power electronics over any other wide band gap semiconductor. As shown in Figure 10.2 using appropriate composition of Al in AlGaN one can expand the energy bandgap window significantly. In addition to bandgap tailoring, III-nitrides allows impurity free polarization-based doping of the material, both uniformly and delta doped, owing to it spontaneous and piezoelectric polarization properties at each unit cell.

Let us now focus on the functionality of a switch. A power electronic device operates in the Class D or E or higher modes in a power converter and commonly referred as a switch. An *ideal* switch behaves in the following way:

1. When "On," the voltage drop across the terminals should be zero, regardless of the amount of current flow, that is, it should behave as a short circuit.
2. When "Off," no current should flow across the terminal, and a very high voltage can be supported across the terminals, that is, it should behave as an open circuit.
3. The transition from On to Off state and vice versa should happen in an infinitesimally small amount of time.

Due to these assumptions, an ideal switch is lossless. Like an ideal transformer, an ideal switch is imaginary but definitely serves as a scale to benchmark other semiconductor switches. The two characteristic parameters formulated for benchmarking any power electronic device are V_{br} and R_{on}. Both R_{on} and V_{br} can be correlated to their material parameters and have formed the basis of different FOM [7–9]

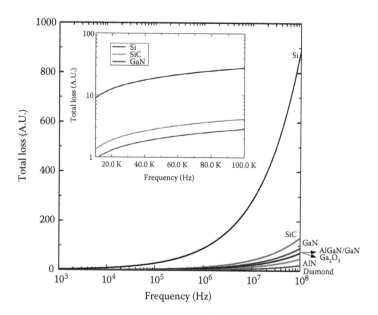

FIGURE 10.3 Huang's Switching FOM: Total loss of a switch made of various WBG semiconductor. The plot is generated using the analysis discussed by Alex Huang [9], where the total loss is minimized with respect to the device area.

calculations. Figure 10.1 showed the DC FOM plot of various semiconductors based on their performance projected by the E_c and μ in these materials. It is critical to remind the readers that the comparison is made based on vertical geometry and under the assumption of a p-n junction reverse biased diode holding the voltage. Please note that the AlGaN/GaN plot in the Figure 10.1 is not exact since the electric field in the channel is not uniform, and carries peaks around the corners of gates and other field plates. However, the plot is the closest representation of a HEMT which shows that the high electron mobility in the channel reduces R_{on} significantly.

Higher current (proportional to charge in the channel and mobility) and higher critical electric field when achieved simultaneously ensures faster switching speed and higher power density, thereby making GaN a very attractive solution. Figure 10.3 uses Huang's FOM approach of estimating the switching FOM and compares all the WBG materials. Clearly, the free carrier mobility allows us to compare these semiconductors, particularly when comparing them by their high-frequency performances.

10.3 Lateral GaN: HEMTs for Medium Power Application

10.3.1 AlGaN/GaN HEMT: Key Design Advantages

Utilizing the horizontal nature of the 2DEG AlGaN/GaN HEMTs (Figure 10.4) are designed as power switches where the voltage is held laterally between the gate and the drain terminals. Due to the inbuilt polarization charges defining a triangular quantum well at the AlGaN and GaN interface populated with electrons, AlGaN/GaN HEMTs are normally-on devices where a negative potential needs to be applied to the gate to turn the device off. Power electronic application requires normally-off devices, as shown in Figure 10.5 [10], which are achieved through various gating structures (p-AlGaN, p-GaN, trenched gate, fluorine treatment) or a cascode [11–16] connection with a normally off (Si) FET. Figure 10.6 shows a cascode configuration used in a package to provide a normally-off solution.

Under the off-state operation in a HEMT, the electrons in the 2DEG are depleted exposing the positive polarization charges at the AlGaN/GaN interface. These positive charges emanate electric fields that

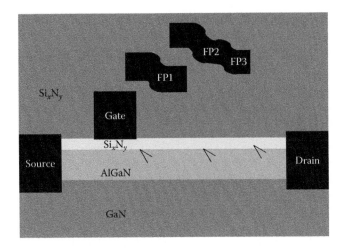

FIGURE 10.4 A lateral AlGaN/GaN HEMT. Field plates are employed to reduce peak electric field (shown by the short solid lines crowding at each metal corner) that can lead to premature breakdown and current collapse under switching operation.

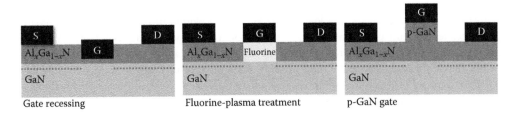

FIGURE 10.5 Various device fabrication approaches to make normally-off AlGaN/GaN HEMT.

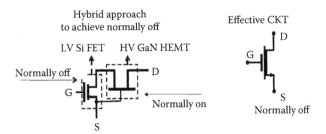

FIGURE 10.6 Cascode normally off GaN HEMT allows for standard silicon-based drive circuits while delivering the high-voltage performance of GaN devices.

terminate on the gate electrode causing a peak electric field greater than the E_c at the drain edge of the gate. Field plates are designed appropriately to mitigate these peak electric field points and distribute them uniformly to enhance the breakdown voltages. The field plates are connected to the gate or source, depending on the architecture employed to achieve normally-off operation. For a single chip normally-off architecture, source connected field plates are used, while two-chip cascode solution prefers gate connected field plates. Besides enhancing the breakdown in the device, field plates help suppress the dispersion in the HEMTs related to the surface states. The surface of AlGaN/GaN HEMT carries donor like states that can trap electrons assisted by the peak electric fields and gate leakage current during the off-operation of the switching cycle, leading to an increased channel resistance under the on-operation.

This phenomenon, also known as dispersion causes an increase in the R_{on} during the switching operation. The R_{on} measured during the switching operation and termed as dynamic R_{on} ($R_{on.dy}$), is a critical parameter that measures the conduction loss ($I^2R_{on.dy}$) per switching cycle, and is, therefore, used to benchmark different power electronic devices.

10.3.2 A Discussion on the State-of-the Art HEMT Designs for Medium Power Conversion

Since the early results on high-voltage AlGaN/GaN HEMTs were reported [17], there has been a tremendous development in the HEMT technology pioneered predominantly by industries. The first JEDEC (the standard used for Si-industry) qualified HEMT 600V GaN-on-Si power HEMT were announced in early 2013 by Transphorm Inc.

Most of the HEMTs for power electronics are typically grown on 4"–6" Si substrates [16,18,19]. These epi films are developed to have a low defect density (2×10^9–1×10^{10}) cm^{-2} with a high 2DEG mobility and high charge density such that both a high breakdown field and a low channel resistance are achieved. The device structure consists of a resistive (AlGaN)-based buffer layer capable of sustaining the rated voltage and the design of which is such that the final wafer has less than 20-micron bow. The path adopted by Transphorm and IR (now a part of Infineon Technologies) to achieve normally-off operation is the cascoded configuration. To quantify performance, let us go through an example of 0.1Ω-600 V-40pF HEMT specs. A typical drain current characteristic of a depletion mode GaN-based HEMTs with insulated gates cascoded with a low-voltage Si FET are plotted in Figure 10.7. At the maximum V_D, the room temperature drain leakage is as low as 10–50 nA and gate leakage is 3–15 nA. As shown in Figure 10.8, the off-state leakages are very small. At 150°C, the typical drain leakage is between 1 and 10 µA while the gate leakage is about 0.1–1 µA at 600 V. GaN-on-Si results were compared to GaN-on-SiC and the performance of GaN-on-Si was no different than GaN-on-SiC, establishing not only the success of the GaN-on-Si technology but also setting up an example against the age old myth about role of dislocation on device leakage since the dislocation density realized for GaN-on-SiC is an order of magnitude lower than that of GaN-on-Si.

The plastic-packaged GaN HEMTs are designed with a threshold of +2.1 V typical at 1mA drain current. The $R_{on.dy}$ shows a maximum increase of 20% compared to the DC R_{on}, which suggests excellent design of field plates and quality of the dielectric layers used in the device. The pulsed drain current is 70 A at a $V_G = 8$ V and $V_D = 10$ V. The continuous (CW) drain current in a typical HEMT can be between

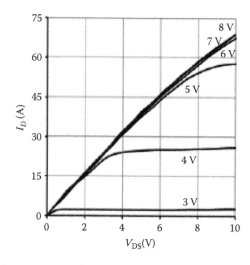

FIGURE 10.7 *I–V* curves of a 600 V HEMT from Transphorm (product TPH3006).

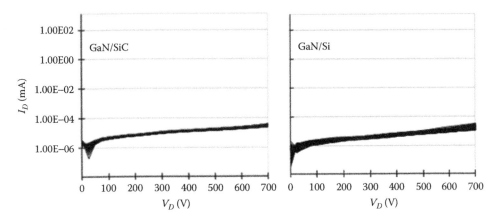

FIGURE 10.8 GaN/Si and GaN/SiC HEMT drain leakage comparison data shows minimal impact of dislocations/ GaN-Si material on achieving comparable and lower leakage than GaN on SiC devices. The gate leakage is even smaller owing to a robust dielectric layer under the gate.

14 and 25 A at a case temperature of 25°C. For higher current ratings, HEMTs are paralleled to enhance the total current capability. The packaging is done very conservatively to reduce bond wire length as much as possible. The bond wires introduce parasitic inductance destabilizing the device performance under switching operation.

The advantage of the high-frequency switches is discussed with two important power electronic applications—motor drive and PV inverter. The high pulse width modulation (PWM) frequency allows integration of compact output filters resulting in a 3-phase pure-sine-wave inverter for PV or motor drive application (Figure 10.9a). An actual motor operation test at 100-kHz PWM showed a significant electromechanical efficiency boost by the GaN inverter by 8%, 4%, and 2% at low, mid, and high load, respectively, compared to a state-of-the-art IGBT inverter at 16 kHz [20].

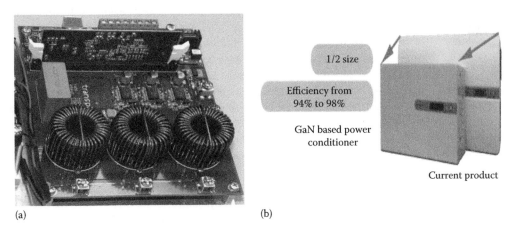

FIGURE 10.9 (a) Role of high frequency: inclusion of GaN devices allowed designing a compact filter on a 5″ × 6″ board. The compact filter is fitted on the same board as the inverter to filter out the higher harmonics of the PWM signal. The high frequency allows driving the circuit with smaller inductors and capacitors leading to very small footprint of the electrical component. (b) The figure shows a PV inverter using GaN devices. The reduction of the volume of the inverter (by 40% compared to Si-based design) was a result of reduction of the heat sinks and the passive components. High temperature operation of GaN allowed the system to operate at higher temperature without compromising the overall efficiency of the system. In other words, the heat sink requirement can be reduced significantly by switching to GaN was proved by this joint effort of Yaskawa and Transphorm in 2015.

Yaskawa Electric Corporation has recently demonstrated a PV-inverter using Transphorm's GaN devices. Higher efficiency at higher frequency and higher temperature resulted in reduction of the volume of the unit by 40% as shown in Figure 10.9b [21].

In power electronic applications the potential benefit of running a system at higher frequency and higher temperature is often discounted since Si-based devices are not capable of performing under such conditions as clear from Figure 10.3. GaN-based switches are, therefore, necessary to introduce new functionalities in the system that was never possible with Si.

The two examples explained here show the impact that a switch brings to the end application, resulting in endless possibilities in technologies.

10.4 Vertical GaN: CAVETs and Other Designs for High Power Application

Despite its capability to handle up to 1 KV or slightly higher voltage as a "switch," the lateral geometry starts to look unattractive as the power level goes beyond 15–20 kW. This is due to the increased distance between the gate and the drain considerably increasing the chip area and the output capacitance of the device. Vertical GaN power devices [16] appear to be the obvious solution for high-power applications, following the examples of Si and SiC power devices. Although the exact limits of power level for efficient switching operation at which vertical devices become preferred over the lateral GaN technology are yet to be determined, it is undeniable that for applications >20 kW the lateral topology becomes increasingly unattractive both in cost and manufacturability. Another significant drawback of the lateral HEMTs is its limitations to handle current. When applications demand over 40–50 A of current, switches with vertical geometry start looking as the only viable option.

Vertical GaN devices are becoming practical solutions with the availability of highquality, freestanding GaN substrates. While GaN's electric field is at least 10 times higher than Si, it is almost comparable to SiC's (similar bandgap) critical electric field. The advantage of vertical GaN device over SiC counterparts can be attributed to (1) the recently reported high bulk mobility in bulk GaN outperforming both Si and SiC [5] (2) the possibility of using polarization doping in these structures appropriately, for example, AlGaN/GaN/AlN-based channel to boost the mobility in that region, as illustrated in a current aperture vertical electron transistor (CAVET). A CAVET is a vertical device, which utilizes the high mobility AlGaN/GaN lateral channel to form part of its conduction path. One can enhance the mobility in the channel by using a thin AlN interfacial layer as used in HEMTs.

The first CAVET reported in 2000 by Ben Yaacov et al. [22] was designed for RF application. These early CAVETs were grown on sapphire since bulk GaN wafers were not available of suitable size or quality to fabricate these devices. Later in 2009 the device was redesigned and reported as a high-voltage switch for power electronics application and demonstrated very promising normally-on and normally-off device performance (Chowdhury et al. [23]). The first report on CAVETs fabricated on bulk GaN substrates provided by Toyota Motor Corporation, Japan, was published in 2012, where the breakdown voltage was measured up to 280 V on a 3 μm drift region and R_{on} was less than 2.2 mΩcm^2 (Chowdhury et al. [24])

Kanechika et al. [25] reported a vertical insulated gate HFET, similar to the CAVET, on bulk GaN substrates in 2007 with $R_{on} = 2.6$ mΩcm^2. Otake et al. [26] reported Vertical MOSFETs on bulk GaN with a threshold voltage of 3.7 V and R_{on} of 9.3 mΩcm^2 also in 2007.

Nei et al. (Avogy) reported vertical GaN transistors offering greater than 2.3 A of saturation current, breakdown voltages of 1.5 kV, area differential specific on-resistance of 2.2 mΩcm^2 [27]. In 2015 Toyoda Gosei reported vertical transistors blocking over 1.2 kV. A list of all the major developments in vertical GaN is presented in Table 10.1

In this chapter, in the vertical device, technology is discussed with the help of a CAVET. Common to all vertical transistors there are some critical design rules, which are discussed next.

TABLE 10.1 A Summary of the Significant Results in Vertical GaN Research

Year	Title	Group	V_{br} (V)	R_{on} (mΩcm²)	V_{th} (V)
February 2004	AlGaN/GaN current aperture vertical electron transistors with regrown channels	Ben Yaacov/UCSB	65	–	–6
May 2007	A vertical insulated gate AlGaN/GaN heterojunction field-effect transistor	Kanechika, Toyota Motor Corporation	–	2.6	–16
January 2008	Vertical GaN-based trench gate metal oxide semiconductor field-effect transistors on GaN bulk substrates	Otake, ROHM Co.	–	9.3	+3.7
June 2008	Enhancement and depletion mode AlGaN/GaN CAVET with Mg-Ion-Implanted GaN as current blocking layer	Chowdhury, UCSB	Low	2 (D-mode), 4 (E-mode)	–2, +0.6
June 2010	Dispersion-free AlGaN/GaN CAVET with lowR_{on} achieved with plasma MBE regrown channel with Mg-ion-implanted current blocking layer	Chowdhury, UCSB	200	2.5	–7
October 2010	Vertical heterojunction field-effect transistors utilizing re-grown AlGaN/GaN two-dimensional electron gas channels on GaN substrates	Okada, Sumitomo Electric Industries	672	7.6	–1.1
September 2013	1000 V vertical JFET using bulk GaN	Q. Diduck, Avogy Inc	1000	4.8	+1
January 2014	Vertical GaN-based trench metal oxide semiconductor field-effect transistors on a free-standing GaN substrate with blocking voltage of 1.6 kV	Oka, TOYODA GOSEI Co	775 (no FP) 1605 (FP)	12.1	+7
June 2014	Low ON-resistance and high current GaN Vertical Electron Transistors with buried p-GaN layers	Yelluri, UCSB	–	0.37	–9
September 2014	1.5-kV and 2.2-mΩ-cm² Vertical GaN Transistors on Bulk-GaN Substrates	Nei, Avogy	1500	2.2	0.5
April 2015	1.8 mΩ-cm² vertical GaN-based trench metal–oxide–semiconductor field-effect transistors on a free-standing GaN substrate for 1.2-kV-class operation	Oka, TOYODA GOSEI Co	1250	1.8	3.5
December 2016	1.7 kV/1.0 mΩcm² Normally-off vertical GaN transistor on GaN substrate with regrown p-GaN/AlGaN/GaN Semipolar Gate Structure	Shibata, Panasonic Corporation	1700	1	2.5

10.4.1 Design and Operation of a CAVET

CAVET (see Figure 10.10) has sources and gate on the top and the drain at the bottom. The electrons controlled by the gate, or more precisely by the effective gate of length L_{go}, flows through the bulk of the material into the drain through an aperture between current blocking layers (CBL). The CBLs have been typically achieved by either p-type doping of GaN layer or by isolation implantation. The lateral high mobility electron channel achieved by AlGaN/GaN layer is merged to a thick vertical GaN drift region in order to achieve low R_{on} and high breakdown voltage. Since vertical devices sustain the blocking voltage in the vertical direction into the bulk material of the device, the chip area is smaller for a given operating

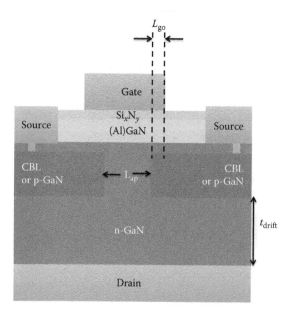

FIGURE 10.10 Schematic layout of a CAVET with all the critical components and dimensions marked. Note that the top layer (Al)GaN can be replaced by n-GaN layer to make it into a MIS-CAVET. All the gating technology discussed in Figure 10.5 can be applied to CAVET to render it e-mode or normally-off.

current compared to the lateral design. In CAVETs the high electric field region, unlike in the lateral HEMTs, are buried in the bulk of the material implying no current collapse due to surface traps. Vertical device, therefore, should not require extensive field plates, although appropriate field termination inside the device structure is important. The CBLs are designed and placed appropriately to prevent electrons from flowing vertically into the drain from any other region in the device other than the aperture. The CBL also behave as a field plate equivalent providing image charges for the space charge in the depleted drift region. The conductivity of the aperture is an important design parameter and is designed to be higher than the channel conductance so that the voltage drop occurs only across the AlGaN/GaN channel in the on-state of operation. Increasing the aperture conductivity by increasing the doping concentration in the aperture is also undesirable because it leads to a premature breakdown at the aperture of the device.

The results (see Figure 10.11) obtained from early CAVETs, reported in 2010 with Mg-implanted CBL using MBE regrowth, showed promising R_{on} (2.2mΩcm^2) and V_{BD} (200 V) [24]. The I_{don}/I_{doff} of the CAVET was ~5 × 10^4. On-off current ratio in a CAVET is determined primarily by the gate-drain overlap (L_{go}), which serves as the effective gate length of the device. R_{on} decreases with the increase of L_{ap} since the resistance in the aperture decreases being inversely proportional to L_{ap}. Saturation in R_{on} is observed when the channel and drift region offers the dominant part of the resistance. Using an increased aperture length beyond the value that gives the saturated value of R_{on} is not desirable as it increases chip area. A blocking field of 1 MV/cm was observed in these CAVETs. Panasonic Corporation has recently presented 1.7 kV/ 1.0 mΩcm^2 normally-off vertical GaN transistor on GaN substrate with regrown p-GaN/AlGaN/GaN semipolar gate structure [28], making an excellent case for vertical devices for power electronic applications, such as automotive inverters and generators. Low dispersion and high blocking electric field beyond what the best lateral designs can provide make CAVETs and its variants suitable for high power switching.

In a CAVET the intrinsic current flow occurs in two dimensions; electrons first flow horizontally through the 2DEG and then move vertically through the aperture region. This is quite different from the HEMT or the bipolar transistor, where current flow is confined to one dimension. It is, therefore, critical to develop an accurate model in order to identify which parameters primarily determine the device characteristics.

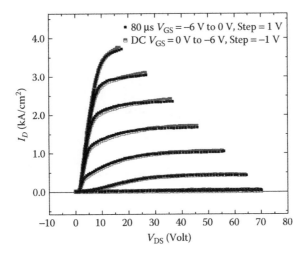

FIGURE 10.11 The peak electric field in a CAVET is buried under the gate into the bulk GaN. This alleviates surface state-related dispersion, which plagues lateral HEMT designs, necessitating field plate designs in HEMTs. CBL provides surface for the space charges to image, hence managing the electric field. Dispersion less CAVET was measured at 80µs pulse widths were reported in 2010. In these devices the CBL was achieved through Mg implantation.

Electrons (current flows opposite to the flow of electrons) first travels horizontally through the 2DEG, until it reaches the gate. The gate only modulates the electrons in the 2DEG, so pinch-off occurs in the horizontal direction inside the 2DEG underneath the gate, just like in a standard FET.

Electrons, which pass the pinch-off point in the channel, continue to travel horizontally at their saturated velocity v_{sat} until they arrive at the aperture, travel downward through the aperture, and are collected at the drain. It is critical that the conductivity of the material inside the aperture as well as in the drain region be much larger than that of the 2DEG so that the entire voltage drop between the source and drain occurs in the 2DEG. This condition ensures that the total current passing through the device is entirely determined by the conductivity of the 2DEG. If this condition is not met, then a significant amount of the applied source-drain voltage is supported across the aperture. In this case, until V_{DS} is very large, the 2DEG does not pinch off, and the current does not reach its saturation value. This is analogous to quasi-saturation in a bipolar transistor, which can occur at large injection currents when the ohmic drop $I_c. R_c$ across the collector drift region becomes comparable to the total base-collector voltage V_{CB}. In addition, the conductivity of the 2DEG must be much higher than that of the adjacent bulk GaN directly below the 2DEG to ensure current flow through the 2DEG rather than through the bulk GaN.

Extensive simulation-based studies were conducted during the development of a CAVET to optimize the device dimensions for making it suitable for high power application. CAVET has a complex geometry, and hence the design rules are not simple. However, there are five very important parameters as identified in Figure 10.10. Each of their roles is visited in this section.

10.4.2 N_{ap} and L_{ap} (Doping and Length of the Aperture)

For the proper functioning of the device, the aperture needs to be more conductive than the 2DEG. The product of the aperture length and the aperture doping, which gives the conductance of the region has to be high. Now increasing the aperture doping is not very desirable because that increases the chances of the breakdown occurring in the aperture. So the only parameter that can increase is L_{ap}. Two CAVETs were modeled, one with a higher aperture resistance than the other. The aperture resistance was varied using a combination of L_{ap} and N_{ap}.

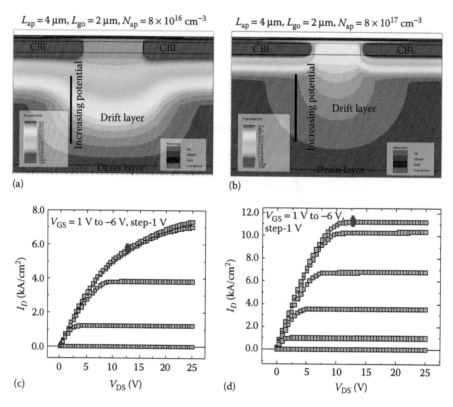

FIGURE 10.12 (a) A CAVET simulated with Silvaco ATLAS with a resistive aperture (b) CAVET in (a) with increased conductivity of the aperture. CAVET with resistive aperture leads to slow saturation as seen from the *I–V* curve in (c) As the resistance of the aperture region is decreased, the current saturates as shown in (d). The voltage distribution is shown in the potential contour, which is shown in the simulated snap shots above. The red dots in the plots indicate the bias condition at which the above simulation snapshots were taken.

From the simulation (Silvaco ATLAS) result shown in Figure 10.12a and b, it can be seen that the device with higher aperture resistance does not show saturation of the current. Or in other words since a major part of the applied V_{DS} is now absorbed in the aperture, a much higher voltage needs to be applied to gain saturation. Device shown in Figure 10.12b with lower aperture resistance easily saturates.

The results indicate that for lower aperture doping which is necessary for high breakdown voltage design, wider apertures are needed.

10.4.3 t_{drift} (Drift Region Thickness)

The CAVET is designed to support most of the applied voltage in the drift or lightly doped (n-) layer. This is the very principle based on which the vertical devices work. While the channel architecture can vary between different flavors of vertical transistors, the expected properties of the drift region and therefore its design is common to all. Majority of the voltage drop under the off-state operation occurs in the drift region, as shown in Figure 10.13. Therefore, the drift region has to be lightly doped to deplete under the off-state drain bias. However, increasing thickness and lower doping of the region although essential for higher blocking voltages increases the R_{on} under the on-state. The effect is more pronounced for higher voltage-classes of vertical devices, where drift region is made thicker to sustain the high-voltage switching. Again in these simulations CBL is modeled as a Mg-doped

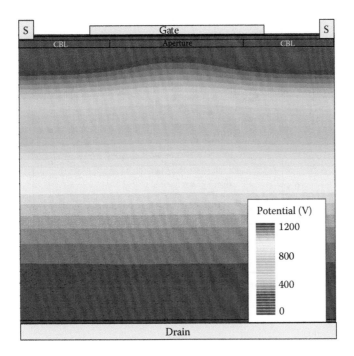

FIGURE 10.13 Potential contour in a CAVET under the off-state operation showing the majority of the voltage dropping across the drift region. (Courtesy of Dong Ji.)

GaN layer. From the potential contours, one can see that most of the applied voltage is supported by the drift region. As the applied voltage is increased, the depletion region extends and the potential contours are pushed into the aperture. The peak field region is buried in the bulk material and occurs at the lower edge of the CBL.

What will play an important role in vertical transistor performances is the mobility of the drift region. The improvement of the bulk GaN material over the past decade leads us to predict that the differentiator of GaN technology with other competing technologies like SiC (which presents similar advantages due to its material properties) is indeed the mobility of the material. Increase in mobility of the bulk GaN that forms the drift region will lower the R_{on} without paying off any significant penalty. The drift region mobility is therefore a key component and impacts the performance and the roadmap just like the critical electric field. A simple simulation done in Silvaco ATLAS to illustrate how low R_{on} can go based on drift region mobility is shown in Figure 10.14 which also plots the R_{on} as a function of aperture length (proportional to the chip area).

10.4.4 The Channel Thickness and Effective Gate Length (L_{go})

Besides ensuring proper gate control and maintaining a good transconductance (g_m) in the device, channel thickness and effective gate length play another important role in these devices best understood from the leakage analysis of the device. Three main leakage paths exist and are shown in Figure 10.15.

10.4.4.1 Through the CBL

If the CBL does not provide a high enough barrier to the current, current can flow from source to drain through it. This is an undesirable path for the current and can be of large significance in the device performance.

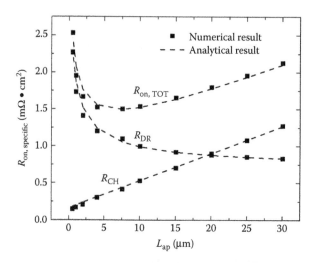

Mobility (drift region) in cm²/Vs	$V_{br}^{*}(V)$	R_{on} mΩ.cm¹ (Lateral channel JFET)
1100	1240	1.4
900	1260	1.7
700	1280	2.2
500	1300	3.1

FIGURE 10.14 R_{on} of CAVET is comprised of the channel and the drift region resistance. The plot captures both of those components as a function of L_{ap}. Total R_{on} as a function of the electron mobility in the drift region is listed. Bulk GaN mobility will play a key role in distinguishing GaN vertical devices from SiC counterparts (simulation done by Dong Ji).

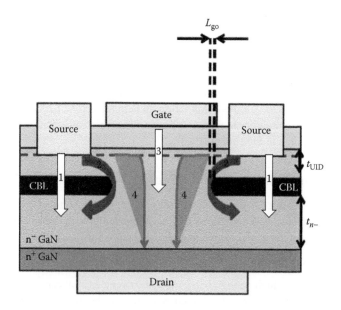

FIGURE 10.15 Three critical leakage paths in a CAVET (1) through CBL (2) unmodulated electrons, and (3) Gate leakage. The leakage currents are labeled as 1, 2 and 3 in the figure. Current that is *controlled* is labeled as 4.

10.4.4.2 Unmodulated Electrons

Under normal functioning, all the electrons from the source should flow through the 2DEG to the gate horizontally and then vertically through the aperture to the drain.

However, if the confinement of the electrons in the channel is not sufficient, then some electrons can find an easy path from the source to the drain via the conductive aperture, bypassing the gate. Such electrons add to the leakage current and should be cut off from flowing. L_{go}, the length of the gate that overlaps the aperture (which is analogous to a std. HEMT gate length) and t_{UID} are the two knobs that control this path.

If L_{go} is small and or t_{UID} is large, the effective distance from the source to the aperture is decreased and the chance of electrons flowing to the drain, without being modulated by the gate, increases. So L_{go} has to be large enough to enhance gate control injected by the source. On the other hand, t_{UID} needs to be of minimum thickness.

However, if t_{UID} is too small, the electrons in the 2DEG would be close to the CBL. This increases the chance of the electrons getting captured (especially if the CBL has traps) causing dispersion when the gate signal is pulsed.

This sets a lower limit in t_{UID} determined by the dispersion criterion.

10.4.4.3 Through the Gate

The other leakage takes place through the gate and can be addressed by suitable gate dielectric

In summary, vertical devices have developed into full-blown technology primarily due to the availability of free standing GaN. While scaling of wafer size and material will be important to the advancement of the technology into manufacturing, recent results on devices look very promising to encourage more R&D on every front.

10.5 Future of GaN and Key Challenges

The future of GaN power electronics looks very positive. 600 V GaN HEMTs have already been JEDEC qualified and shown superior performance at system level when compared to state-of-the-art IGBTs in inverters driving induction and permanent magnet motors and PV inverters. The development of GaN-based vertical transistors to address the power level beyond that of the lateral GaN is crucial for sustaining the GaN-based power electronics market. Bulk GaN technology have made excellent progress over the last 5 years, thanks to the solid state lighting market, which continues to stimulate the development of vertical GaN devices for power conversion. With lateral and vertical GaN devices, one can address the entire spectrum of power electronic application, starting from hundreds of watts to few megawatts creating a sustainable market for GaN-based power electronics.

However, there are key challenges that remain to be solved for the researchers and technologists to ensure the permanence of GaN. Two most important technical challenges, along with a market argument are discussed as an endnote to this chapter

- Single chip reliable E-mode operation: While companies like Panasonic and EPC have announced products that are single chip E-mode devices, the reliability is still under evaluation and has not been fully mastered. Reliable single chip E-mode devices with threshold voltage >3 V are definitely something that the commercial converter market is hoping for.
- Integration: Gate drivers using GaN has been a difficult proposition due to lack of a hole channel devices, unlike the CMOS technology in Si. While there is work on going to address complementary logic using GaN [29], not much progress can be reported. This leaves us with the option of integrating with Si. GaN-on-Si although has matured very fast on [111] orientation, it has not been so successful on [100] traditionally used in CMOS fabrication. Integration with Si will be a key to the path forward and requires major attention.
- Market acceptance: Lateral and vertical GaN are currently considered complementary in nature of their application space. It is essential to have both of these technologies succeed to establish a sustainable

market in GaN. For lateral and vertical GaN to penetrate the power electronics market, solutions need to be evaluated at the system level. Cost argument at the device level is not the right metric for comparing GaN with Si. The system cost will drive the technology along with performances at high temperature and higher frequencies. Driving systems' efficiently at higher frequencies and higher temperature allows the reduction of form factor and potentially reducing system cost. It also allows increasing the functionality significantly enabling better integration with smarter technologies. At this point of time, consistent design wins into applications are critical in both lateral and vertical GaN domain to ensure acceptance in the next generation power electronics market.

References

1. B. Jayant Baliga, *Fundamentals of Power Semiconductor Devices*. Springer: New York, 2008.
2. A. Kadavelugu, V. Baliga, S. Bhattacharya, M. Das and A. Agarwal, Zero voltage switching performance of 1200V SiC MOSFET, 1200V silicon IGBT and 900V CoolMOS MOSFET, *2011 IEEE Energy Conversion Congress and Exposition (ECCE)*, Phoenix, AZ, pp. 1819–1826, 2011.
3. S. J. Pearton and C. Kuo, GaN and related materials for device applications, *MRS Bulletin*, 22 (2), 17–21, 1997.
4. S. J. Pearton, J. C. Zolper, R. J. Shul, and F. Ren, GaN: Processing, defects and devices, *J. Appl. Phys.* 86, 1–78 and references therein, 1999.
5. P. Kruszewski, J. Jasinski, T. Sochacki, M. Bockowski, R. Jachymek, P. Prystawko, M. Zajac, R. Kucharski, and M. Leszczynski, Vertical schottky diodes grown on low-dislocation density bulk GaN substrate, *The International Workshop on Nitride Semiconductor*, Wroclaw, Poland, 2014.
6. U. K. Mishra, P. Parikh, and Y. F. Wu, AlGaN/GaN HEMTs-An overview of device operation and application, *Proc. IEEE*, 90(6), 1022–1031, 2002.
7. B. J. Baliga, Semiconductors for high voltage, vertical channel FET's, *J. Appl. Phys.*, 53, 1759–1764, 1982.
8. B. J. Baliga, Power semiconductor device figure-of-merit for high frequency applications, *IEEE Electron Dev. Lett.*, 10, 455–457, 1989.
9. A. Q. Huang, New unipolar switching power device figure of merits, *IEEE Electron Dev. Lett.*, 25(5), 298–301, May 2004.
10. C. S. Suh, A. Chini, Y. Fu, C. Poblenz, J. S. Speck and U. K. Mishra, p-GaN/AlGaN/GaN Enhancement-Mode HEMTs, *64th Device Research Conference*, State College, PA, pp. 163–164, 2006.
11. R. Chu, A. Corrion, M. Chen, L. Ray, D. Wong, D. Zehnder, B. Hughes, and K. Boutros, 1200-V normally Off GaN-on-Si field-effecransistors with low dynamic on–resistance, *IEEE Electron Dev. Lett.*, 32(5), 632–634, May 2011.
12. T. Ueda, T. Tanaka, and D. Ueda, Gate injection transistor (GIT)—A normally-off AlGaN/GaN power transistor using conductivity modulation, *IEEE Trans. Electron Dev.*, 54(12), 3393–3399, December 2007.
13. M. Kanamura, T. Ohki, T. Kikkawa, K. Imanishi,,T. Imada, A. Yamada, and N. Hara, Enhancement-mode GaN MIS-HEMTs with n-GaN/i-AlN/n-GaN triple cap layer and high-k gate dielectrics, *IEEE Electron Dev. Lett.*, 31(3), 189–191, March 2010.
14. Y. Cai, Y. Zhou, K. J. Chen, and K. M. Lau, High-performance enhancement-mode AlGaN/GaN HEMTs using fluoride-based plasma treatment, *IEEE Electron Dev. Lett.*, 26, (7), 435–437, July 2005.
15. Y. Niiyama, H. Kambayashi, S. Ootomo S, T. Nomura, S. Yoshida and T P. Chow, Over 2 A operation at 250°C of GaN metal-oxide-semiconductor field effect transistors on sapphire substrates, *Jpn. J. Appl. Phys.* 47(9), 7128–7130, September 2008.

16. S. Chowdhury and U. K Mishra, Lateral and vertical transistors using the AlGaN/GaN hetero-structures, *IEEE Trans. Electron Dev.*, 60(10), 3060–3066, October 2013.

17. Y. Dora, A. Chakraborty, L. McCarthy, S. Keller, S. P. Denbaars, and U. K. Mishra, High break-down voltage achieved on AlGaN/GaN HEMTs with integrated slant field plates, *IEEE Electron Dev. Lett.*, 27(9), 713–715, September 2006.

18. S. L. Selvaraj, T. Suzue, and T. Egawa, Breakdown enhancement of AlGaN/GaN HEMTs on 4-in silicon by improving the GaN quality on thick buffer layers, *IEEE Electron Dev. Lett.*, 30(6), 587–589, June 2009.

19. B. Lu and T. Palacios, High breakdown (>1500 V) AlGaN/GaN HEMTs by substrate-transfer tech-nology, *IEEE Electron Dev. Lett.*, 31(9), 951–953, September 2010.

20. K. Shirabe, M. Swamy, J. Kang, M. Hisatsune, Y. Wu, D. Kebort and J. Honea, Efficiency compari-son between Si-IGBT based drive and GaN based drive" *IEEE Trans. Indus. Appl.*, 50(1), 566–572, January 2014.

21. Yaskawa Annual Report 2013, http://www.yaskawa.co.jp/ir/ir_document/annualreport/2013/en/ar2013e.pdf

22. I. Ben-Yaacov, Y.-K. Seck, U. K. Mishra, and S. P. Denbaars, AlGaN/GaN current aperture vertical electron transistors with regrown channels, *J. Appl. Phys.*, 95(4), 2073–2078, Febuary 2004.

23. S. Chowdhury, B. L Swenson, and U. K. Mishra, Enhancement and depletionMode AlGaN/GaN CAVET with Mg-Ion-implanted GaN as current blocking layer, *IEEE Electron Dev. Lett.*, 29(6), 543–545, June 2008.

24. S. Chowdhury, M. H. Wong, B. L. Swenson, and U. K. Mishra, CAVET on bulk GaN substrates achieved with MBE-Regrown AlGaN/GaN layers to suppress dispersion, *IEEE Electron Dev. Lett.*, 33(1), 41–43, January 2012.

25. M. Kanechika, M. Sugimoto, N. Soeima, H. Ueda, O. Ishiguro, M. Kodama, E. Hayashi, K. Itoh, T. Uesugi, and T. Kachi, A vertical insulated gate AlGaN/GaN hetrojunction field effect transisitor, *Jpn. J.Appl. Phys.*, 46(21), L503–L505, June 2007.

26. H. Otake, K. Chikamatsu, A. Yamaguchi, T. Fujishima, and H. Ohta, Vertical GaN-based trench gate metal oxide semiconductor field-effect transistors on GaN bulk substrates, *Appl. Phys. Express*, 1, 011105-1–011105-3, January 2008.

27. H. Nie, Q. Diduck, B. Alvarez, A. Edwards, B. Kayes, M. Zhang, D. Bour, Dave, I. Kizilyalli, 1.5kV and 2.2mΩcm² Vertical GaN Transistors on Bulk-GaN Substrates, *IEEE Electron Dev. Lett.*, 35(9), 939–941, September 2014.

28. D. Shibata, R. Kajitani, M. Ogawa, K. Tanaka, S. Tamura, T. Hatsuda, M. Ishida, and T. Ueda, 1.7 kV/1.0 mΩcm2 normally-off vertical GaN transistor on GaN substrate with regrown p-GaN/AlGaN/GaN semipolar gate structure, *IEEE International Electron Devices Meeting*, San Francisco, CA, 2016.

29. R. Chu, Y. Cao, M. Chen, R. Li, and D. Zehnder, An experimental demonstration of GaN CMOS technology, *IEEE Electron Dev. Lett.*, 37(3), 269–271, March 2016.

30. M. Á. Caro Bayo, Theory of elasticity and electric polarization effects in the group-III nitrides. PhD Thesis, University College Cork, 2013.

11

Recent Progress in GaN-on-Si HEMT

Kevin J. Chen
and Shu Yang

11.1 Introduction: Challenges in GaN-on-Si HEMT

With the benefits of low-loss conduction, high-frequency switching, and high-temperature operation, wide bandgap GaN heterojunction power devices are attractive candidates for next-generation energy-efficient and compact power conversion systems [1–5] including household appliances, photovoltaic inverters, data centers, motor drives, and so on. Compared with mainstream Si power devices, GaN heterojunction power devices are capable of delivering substantially lower ON resistance (R_{ON}) for the same voltage rating [6]. Such superiority originates from the fundamental material properties including high breakdown strength of GaN (10 times higher than Si), and high density (~10^{13} cm^{-2}) of two-dimensional electron gas (2DEG) with high electron mobility (~2000 cm^2/Vs) in GaN-based heterostructures. Lower R_{ON} leads to reduced conduction loss and higher power conversion efficiency, while high switching frequency brings the benefit of smaller passives and compact systems. Both the higher conversion efficiency and high-temperature operating capability result in reduced burden of thermal management, and enable simpler/low-cost cooling solutions.

From commercialization point of view, growing group III-nitride epitaxial structure on Si substrate is a compelling and cost-competitive choice to implement GaN-based lateral power devices. The benefits include not only low-cost and highly scalable silicon substrates, but also the capability of manufacturing the GaN power chips using fully-depreciated 6- and 8-inch Si fabrication lines. It also opens up the possibility of heterogeneous integration of GaN and Si electronics.

Despite tremendous progress in material growth, device fabrication, and circuit demonstration, GaN-on-Si power transistors are confronted with several major challenges as shown in Figure 11.1, including (1) normally-off operation and (2) device stability and reliability associated with various traps

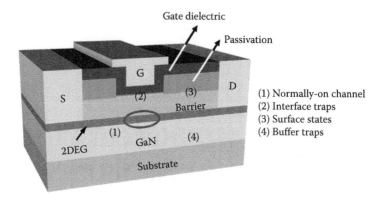

FIGURE 11.1 Illustration of the major challenges in GaN-on-Si power transistors using Metal–insulator–semiconductor high-electron-mobility transistor (MIS-HEMT) as an example.

(e.g., interface traps at the gate-dielectric/III-nitride interface, surface states in the gate-to-drain access region, and bulk traps in the III-nitride (typically GaN) buffer layer). These challenges need to be addressed by developing new processing technologies and structure designs based on solid understanding of the behaviors of interface/bulk traps and their impact on device characteristics.

11.1.1 Challenges in the Gate Region

For high-voltage power electronics applications, normally-off or enhancement-mode (E-mode) power transistors with positive threshold voltage (V_{TH}) is highly desirable, because of their inherent fail-safe characteristic and simpler gate drive circuitry. However, as a result of the strong polarization effect in Ga-face AlGaN/GaN heterostructure [7], the 2DEG channel is naturally a normally-on channel that requires a negative gate voltage to pinch off.

To realize normally-off operation, two approaches are currently under development: (1) a two-chip hybrid cascode configuration including a high-voltage normally-on GaN transistor and a low-voltage normally-off Si MOSFET and (2) a single-chip configuration with a normally-off GaN transistor.

A high-performance E-mode GaN power transistor should achieve normally-off operation while simultaneously maintaining a low ON resistance (R_{ON}) to minimize the conduction loss. R_{ON} of an E-mode GaN transistor consists of (1) channel resistance (R_{ch}) underneath the gate electrode that can be modulated by gate bias, (2) gate-to-source/drain access region resistance (R_{ac}), and (3) source/drain contact resistance. In addition to creating a normally-off channel in the gate region, it is essential to make full use of the high conductivity of the normally-on 2DEG in the access region so as to minimize R_{ac}. Therefore, a desirable E-mode GaN power transistor should feature an E-mode channel and D-mode access regions [8].

E-mode GaN transistor technologies can be classified into two categories, depending on whether the starting epitaxial wafer is D-mode or E-mode. The commonly used non-patterned epitaxial growth indiscriminately creates uniform V_{TH} throughout the entire wafer. Therefore, post-epitaxy V_{TH} modulation techniques are required to locally convert the D-mode gate region to E-mode one if the as-grown wafer is D-mode, or alternatively, convert the E-mode access region to D-mode one if the as-grown wafer is E-mode.

As shown in Figure 11.2, several categories of E-mode GaN transistor technologies have been developed, including p-type GaN [9] or AlGaN cap layer [2,10], fluorine ion implantation [11,12], and partially [13] or fully [14,15] recessed barrier.

E-mode GaN transistor technologies except for p-cap approach, as well as D-mode ones for cascade configuration, prefer insulated gate over Schottky gate, as the former group can provide suppressed gate

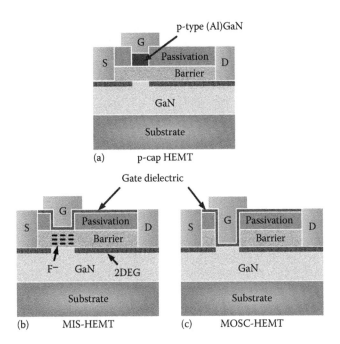

FIGURE 11.2 Typical E-mode GaN transistors using (a) p-type doped (Al)GaN cap layer, (b) fluorine ion implantation, and (c) barrier recess technologies. In (b) and (c), insulated-gate MIS-HEMT and recessed-gate MOS-channel-HEMT (MOSC-HEMT) are used for illustration.

leakage, enlarged gate swing, and more scalable V_{TH}. However, in metal–insulator–semiconductor high-electron-mobility transistor (MIS-HEMT) and recessed-gate MOS-channel-HEMT (MOSC-HEMT), interface traps at the gate-dielectric/III-nitride interface and bulk traps inside the gate dielectric (including the near-interface bulk traps, or border traps) present a grand challenge to the devices' stability and reliability [16–30].

Therefore, gate stack engineering, targeting at modulating V_{TH} and reducing interface/bulk traps in the gate region, is of critical importance to achieving high-performance and high-reliability E-mode GaN power transistors.

11.1.2 Challenges in the Access Region and Buffer Layer

Trapping effects in the access region and III-nitride (typically GaN) buffer layer (Figure 11.1) could also adversely influence the stability and dynamic performance (e.g., dynamic R_{ON}) of GaN-on-Si power devices, due to the following mechanisms: (1) At OFF-state with high drain bias, the large negative V_{GD} could cause electrons to be injected from the gate electrode and trapped at the polarized III-nitride surface in the gate-to-drain access region [31]. Meanwhile, in the vertical direction, the large positive drain-to-substrate voltage could cause electrons to be injected from Si substrate and trapped in the GaN buffer layer [32,33]. (2) At high power state during hard-switching transition, hot electrons can be generated in the channel, then injected to and trapped at the III-nitride surface or in the GaN buffer layer [34,35]. When devices are switched from OFF-state to ON-state, some of the surface states and buffer traps cannot release electrons immediately due to relatively long emission time constants (τ_e), while the 2DEG in the gate-to-drain access region could remain partially depleted, resulting in increased dynamic R_{ON} (or current collapse) and conduction loss. To alleviate the surface-induced current collapse, passivation techniques with dielectrics deposited on the gate-to-drain access region are necessary to modify or compensate surface states [36–42].

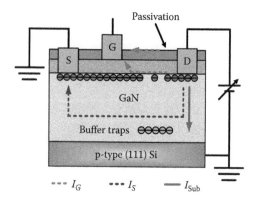

FIGURE 11.3 Illustration of the effect of GaN buffer traps on the electrical performance of GaN-on-Si power transistors. The arrows indicate three primary OFF-state leakage paths: gate leakage (I_G), drain-to-source leakage through buffer layer (I_S), and vertical drain-to-substrate leakage (I_{Sub}). It is also illustrated that the negatively charged buffer traps could partially deplete the 2DEG channel and cause the buffer-induced current collapse when devices are switched from OFF-state to ON-state. (From Yang, S. et al., *Appl. Phys. Lett.*, 104, 013504-1–013504-4, 2014. With permission.)

In general, there exists plenty of bulk traps in the GaN buffer layer grown on Si substrate. The GaN buffer traps not only affect dynamic performance, but also could have significant impact on the breakdown voltage of the GaN-on-Si power devices [32,43–50]. As illustrated in Figure 11.3 [51], the OFF-state leakage of GaN-on-Si lateral power transistor primarily consists of three paths, the drain-to-gate leakage, the drain-to-source leakage through the buffer layer, and the drain-to-substrate vertical leakage. The breakdown voltage of the GaN-on-Si power devices is ultimately limited by the GaN-to-Si vertical leakage. In industry-standard GaN-on-Si epitaxial wafers, the conductive Si substrate allows the high drain-to-substrate voltage at OFF-state to drop across the buffer layer. The high electric field could induce the ionization/deionization of GaN buffer traps, which will in turn affect the vertical leakage/breakdown [32]. Meanwhile, the ionization of acceptor buffer traps could generate negative space charges and partially deplete the 2DEG in the access region when devices are switched to ON-state, resulting in buffer-induced current collapse. Therefore, in order to optimize the GaN-on-Si platform for high breakdown voltage and suppressed current collapse, it is of utmost significance to understand the mechanisms of vertical leakage/breakdown, and optimize the buffer structure design.

In this chapter, the primary challenges and recent progress in GaN-on-Si power transistors are discussed and summarized. In Section 11.2, gate stack technologies toward high-performance and high-reliability normally-off (E-mode) GaN power transistors are reviewed, from the perspectives of interface engineering, barrier-layer engineering, as well as gate dielectric technology. Surface passivation techniques are discussed in Section 11.3. Section 11.4 focuses on III-nitride buffer layer-related issues, including mechanisms of buffer traps in influencing vertical breakdown and dynamic performance, and the optimization techniques of buffer layer. A brief summary is given in Section 11.5.

11.2 Gate Stack Technology

As introduced in Section 11.1.1, insulated-gate GaN transistors (e.g., MIS-HEMTs and MOSC-HEMTs) are highly preferred over the conventional Schottky-gate HEMTs for high-voltage power switching applications. However, the insertion of gate dielectric brings about a grand challenge to devices' stability and reliability that arises from interface traps at the gate-dielectric/III-nitride interface and bulk traps inside the gate dielectric [16–26]. As shown in Figure 11.4, when overdriving GaN power transistors to minimize R_{ON}, the overdrive voltage ($V_{G\_ON}-V_{TH}$) drops across the gate dielectric and barrier layer (if any), while interface traps and near-interface bulk traps (i.e., border traps) could be filled with electrons.

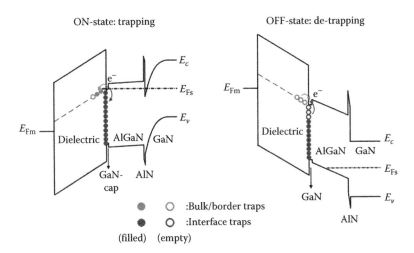

FIGURE 11.4 Illustration of the charging/discharging of interface and border traps in the gate-dielectric/ GaN-cap/AlGaN/GaN gate stack at positive (left) and negative (right) gate bias, respectively. (From Yang, S. et al., *Semicond. Sci. Technol.*, 31, 024001-1–024001-10, 2016. With permission.)

At lower V_G (e.g., OFF-state), interface and border traps could partially release electrons. These charging and discharging processes induce time-/temperature-dependent V_{TH} instability [18–20,23,30,52] and long-term reliability issues [27–29]. In particular, when V_{TH} is positively shifted, R_{ON} will increase due to a reduced overdrive voltage ($V_{G\_ON}-V_{TH}$) [53].

On the other hand, barrier-layer engineering, such as barrier recess and implanting fluorine ions into the barrier layer (Figure 11.2), can effectively modulate V_{TH} and realize E-mode GaN power transistors.

In this section, approaches toward high-performance and high-reliability GaN-on-Si power transistors will be discussed, including interface engineering, barrier engineering, and gate dielectric technology.

11.2.1 Interface Optimization

Interface engineering techniques have been developed to improve the interface quality and reduce interface traps in insulated-gate GaN transistors, including plasma nitridation, thermal nitridation, plasma oxidation, and annealing [22,55–59].

A mono-crystalline interfacial-layer is highly desirable for passivating dangling bonds and achieving low trap density at the interface [60]. *In situ* remote plasma nitridation implemented in plasma-enhanced atomic layer deposition (PEALD) system can create mono-crystalline AlN nitridation interfacial-layer (NIL) [56]. After removing the low-quality native oxide by *in situ* NH_3/Ar remote plasma, the pre-cleaned GaN surface was subsequently exposed to N_2 remote plasma, leading to a nitrogen terminated surface with Ga dangling bonds passivated and possible nitrogen vacancies compensated (Figure 11.5a). After the *in situ* deposition of gate dielectric Al_2O_3, a mono-crystalline interfacial layer containing Al-N bonds can be formed at the interface, as verified by X-ray photoelectron spectroscopy and transmission electron microscopy [56].

In addition to PEALD-AlN, PEALD-SiN_x [25,57] and MOCVD-SiN_x [22,59] have also been used to realize surface nitridation. Both amorphous [59] and mono-crystalline [22] interfacial-layers between MOCVD-SiN_x and III-nitride have been observed, while the crystalline structure of PEALD-SiN_x deposited on III-nitride at a lower temperature is not clear. Figure 11.5b shows the schematic gate stack structure with SiN_x interfacial layers.

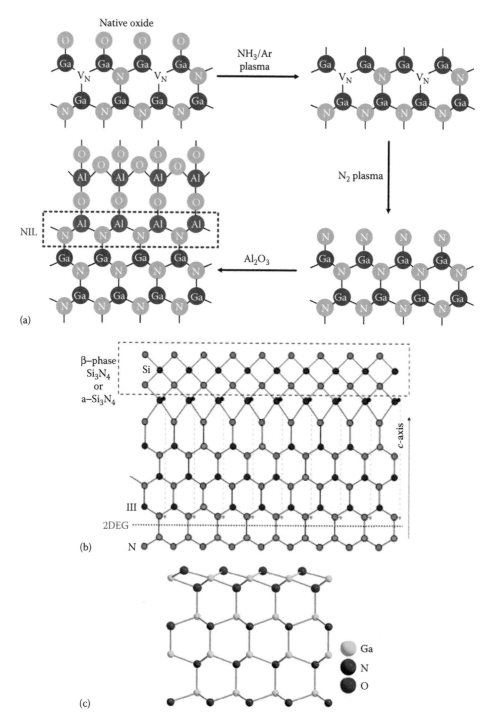

FIGURE 11.5 Schematics of the gate stack structures using various interface engineering techniques: (a) AlN interfacial-layer. (From Yang, S. et al., *IEEE Electron Device Lett.*, 34, 1497–1499, 2013. With permission.) (b) SiN$_x$ interfacial-layer. (From Moens, P. et al., *IEEE Int. Symp. on Power Semiconductor Devices and IC's (ISPSD)*, 374–377, 2014. With permission.) (c) 2 monolayer oxide on III-nitride surface. (From Qin, X. et al., *Appl. Phys. Lett.*, 105, 141604-1–141604-5, 2014; Miao, M. et al., *J. Appl. Phys.*, 107, 123713-1–123713-11, 2010. With permission.)

On the other hand, Miao et al. [61] proposed a favorable 2 monolayer (ML) crystalline oxide on III-nitride based on first-principle calculation, as shown in Figure 11.5c. In practice, Qin et al. found that the crystalline structure of the oxide depends on the oxidation approach (plasma treatment or annealing) and temperature [58]. Mono-crystalline oxide interfacial-layer was also observed after a high-temperature (850°C) annealing process [55]. The most challenging task is to avoid crystallization of the bulk oxide at high temperatures. For example, a typical 20-nm ALD-Al_2O_3 grown on GaN tends to crystalize and become leaky once temperature exceeds 800°C [62].

The mono-crystalline AlN NIL technique has been implemented in GaN MIS-HEMT and recessed-gate MOSC-HEMT, delivering low-trap-density interface [52,56,63,64]. As shown in Figure 11.6, the MIS-HEMT with Al_2O_3 (AlN NIL) gate dielectric shows a steeper subthreshold swing and smaller hysteresis, compared with the control device without NIL.

On the other hand, the conventional GaN-based MIS-HEMT is D-mode because of the polarization-induced 2DEG in the III-nitride hetero-structure, while E-mode GaN transistors are highly preferred for power switching applications (Section 11.1.1). Recessed-gate MOSC-HEMT (Figure 11.2c) is one of the attractive approaches to realize E-mode GaN transistors [14,15,63,65]. In MOSC-HEMT, the polarized III-nitride barrier layer in the gate region is fully recessed, and the III-nitride hetero-structure is still maintained in the access region. In this way, the E-mode gate and D-mode access regions can simultaneously enable normally-off operation and low R_{ON} (Section 11.1.1). To mitigate the dry etching-induced issues (e.g., surface damage, poor uniformity, etc.), digital recess processes featuring multiple cycles of plasma oxidation and wet etching have been developed [63,65].

The incorporation of AlN NIL into MOSC-HEMT not only creates high-quality interface and border region [52], but also boosts the field-effect mobility (μ_{FE}) in the recessed gate region and subsequently reduces R_{ON} (Figure 11.7) [63].

11.2.2 Barrier-Layer Optimization

Even at higher junction temperature, wide bandgap GaN power devices are capable of delivering high conversion efficiency and high switching speed [66,67]. As they are destined to operate within a wide temperature range, it is important to investigate the temperature-dependent stability and performance of GaN transistors [16,67–71].

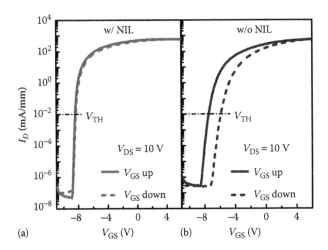

FIGURE 11.6 Comparison of transfer characteristics of Al_2O_3 MIS-HEMTs (a) with and (b) without NIL. A relatively high sweeping rate of 700 mV/s was used to reduce the measurement-induced recovery effect. (From Yang, S. et al., *IEEE Electron Device Lett.*, 34, 1497–1499, 2013. With permission.)

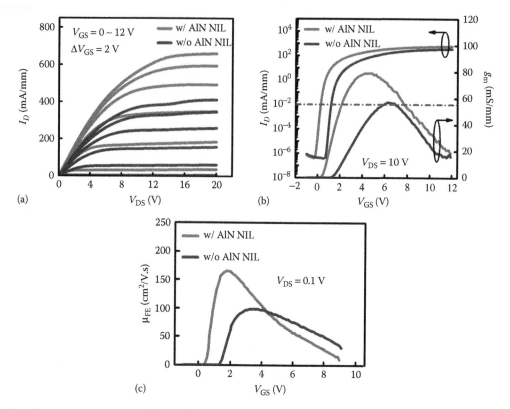

FIGURE 11.7 (a) Output and (b) transfer characteristics of Al_2O_3 MOSC-HEMTs with and without AlN NIL. (c) Extracted field-effect mobility as a function of V_{GS}. (From Liu, S. et al., *IEEE Electron Device Lett.*, 35, 7, 723–725, 2014. With permission.)

It is revealed that recessed-gate MOSC-HEMT with AlN NIL (Section 11.2.1) is thermally stable (Figure 11.8a), whereas MIS-HEMT with the same interface engineering technique exhibits evident thermally induced V_{TH} shift [70]. The mechanism responsible for such distinction in V_{TH}-thermal-stability is illustrated in Figure 11.8b. In *buried-channel* MIS-HEMT, the critical gate-dielectric/III-nitride interface is spatially separated from the channel by a polarized III-nitride barrier layer. The strong polarization field raises the energy band at the interface, such that the relatively deep interface traps can emerge above the Fermi level at pinch-off. These traps are "frozen state" [16] at lower temperature due to long τ_e, but can emit electrons at higher temperature when τ_e is exponentially shortened (Equation 11.1). Such thermally-sensitive emission process of interface traps within a wide energy range could result in thermally induced V_{TH} instability [72]. By contrast, in the recessed-gate MOSC-HEMT, deep interface traps in the gate region are always buried below the Fermi level and maintain fixed charging states at pinch-off regardless of temperature. Even at pinch-off with larger V_{DS} (i.e., more negative V_{GD}), the depletion region would laterally extend toward the drain instead of raising up the energy band in the gate region, leading to thermally stable V_{TH}.

$$\tau_e = \frac{1}{\upsilon_{th}\sigma_n N_C}\exp\left(\frac{E_C - E_T}{kT}\right) \tag{11.1}$$

where υ_{th}, σ_n, N_C, E_C–E_T, k, T are the electron thermal velocity, electron capture cross-section, effective density of states in the conduction band, activation energy of trap level, Boltzmann constant, and temperature, respectively.

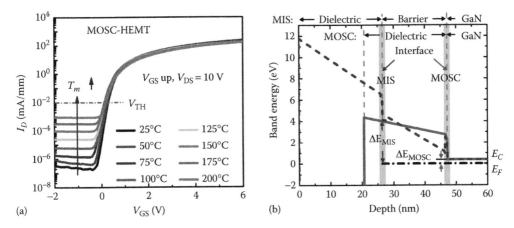

FIGURE 11.8 (a) Multi-temperature DC transfer characteristics of Al_2O_3(AlN NIL)/GaN MOSC-HEMT with measurement temperature (T_m) increasing from 25°C to 200°C. (b) Comparison of energy band diagrams between MIS-HEMT (dashed line) and MOSC-HEMT (solid line) at pinch-off under quasi-steady state. (From Yang, S. et al., *IEEE International Electron Devices Meeting (IEDM)*, 17.2.1–17.2.4, 2014. With permission.)

Despite its superior V_{TH}-thermal-stability, MOSC-HEMT with fully recessed barrier in the gate region exhibits relatively high R_{ON} due to lower channel electron mobility. In order to realize normally-off GaN power transistors with better performance/thermal-stability tradeoff, it is necessary to optimize the barrier-layer structure. The abovementioned mechanisms provide a guideline for the optimization strategies. To enhance V_{TH}-thermal-stability, the energy range of interface trap levels above Fermi level at pinch-off need to be narrowed down, which can be achieved through (1) barrier recessing or (2) incorporating negative charges (e.g., fluorine ions, or F^-) into the barrier to partially cancel out the polarization field (Figure 11.9). Meanwhile, the negatively charged fluorine ions facilitate the realization of E-mode operation without the need to completely remove the barrier, and thus, high channel mobility and low R_{ON} can be maintained.

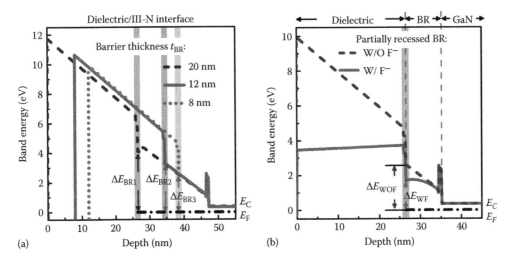

FIGURE 11.9 Simulated energy band diagrams of the MIS-HEMTs that can provide guidelines for barrier-layer engineering: (a) with varying barrier thickness (t_{BR}); (b) with and without F^- while the barrier layer is partially recessed to 8 nm. The negatively charged F^- can reduce the upward band-bending in AlGaN barrier, leading to $\Delta E_{WF} < \Delta E_{WOF}$ at pinch-off. (From Yang, S. et al., *IEEE International Electron Devices Meeting (IEDM)*, 17.2.1–17.2.4, 2014. With permission.)

Fluorine ion implantation (Figure 11.2b) is one of the widely utilized techniques to realize E-mode operation (Section 11.1.1). Because of the strong electronegativity, fluorine atom can capture one electron and become negatively charged in III-nitride system, leading to depleted 2DEG channel and effective V_{TH} modulation [8,11,12,73–75]. Diffusion process and thermal stability of fluorine ions in III-nitride system, as well as the stability/reliability of the as-fabricated E-mode GaN transistors, have been intensively studied by means of molecular dynamics simulation, physical, and electrical characterizations [74,76–78].

According to the barrier-layer design guideline, an optimized E-mode MIS-HEMT with partially recessed and fluorine implanted barrier has been developed (Figure 11.10a) [71]. As shown in Figure 11.10b and c, the optimized E-mode MIS-HEMT is capable of delivering the desirable normally-off operation up to 200°C, excellent thermal stability, and low R_{ON} comparable with conventional D-mode MIS-HEMT.

11.2.3 Gate Dielectric Technology

Device stability can be improved by interface and barrier-layer optimization techniques, whereas the long-term reliability and lifetime is more relevant to the bulk quality of gate dielectric. In addition to dielectrics deposited at relatively low temperature (e.g., by ALD, plasma-enhanced chemical vapor deposition, or PECVD, etc.), high-temperature grown gate dielectric films by low-pressure chemical vapor deposition (LPCVD) and metalorganic chemical vapor deposition (MOCVD) have been incorporated

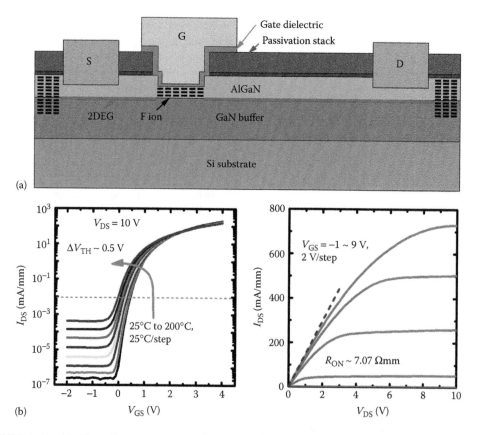

FIGURE 11.10 (a) Schematic cross section of the optimized E-mode MIS-HEMT with partially recessed and fluorine-implanted barrier. (b) Multi-temperature transfer and (c) room-temperature output characteristics of the optimized MIS-HEMT. (From Liu, C. et al., *IEEE Electron Device Lett.*, 36, 318–320, 2015. With permission.)

TABLE 11.1 Summary of Recently Reported TDDB Results of GaN MIS-HEMTs

Gate Dielectric	t (nm)	V_G (V)	t_{BD} @ 63% (s)	β	T (°C)	Reference
ALD-Al_2O_3	8	3.5, 3.7, 4	~10^3, 2×10^2, 30	2.87	25	[80]
ALD-Al_2O_3	15	10	9.8×10^3	~1.2	25	[81]
MOCVD-SiN_x	–	~10	1.0×10^2	~1	125	[22]
LPCVD-SiN_x	20	20, 22, 24	2.4×10^3, 7.2×10^2, 228	2.4	25	[29]
		20	2.4×10^3, 4.2×10^2, 70		25, 125, 200	

Source: Hua, M. et al., *IEEE Trans. Electron Devices*, 62, 3215–3222, 2015.

into GaN power transistors [22,24,27,29,59]. The LPCVD-SiN_x shows superior properties, including low oxygen contamination (compared with PECVD-SiN_x), low leakage, and high breakdown strength [24,79]. Most importantly, GaN desorption at high temperature in LPCVD system is negligible, and the crystalline structure of III-nitride is still well ordered after LPCVD-SiN_x deposition [24].

Table 11.1 summarizes recently reported time-dependent dielectric breakdown (TDDB) characteristics of GaN MIS-HEMTs [22,29,80,81]. LPCVD-SiN_x MIS-HEMT exhibits long time-to-breakdown (t_{BD}) even at higher electric field, verifying that LPCVD-SiN_x can enable much longer lifetime at modest electric field and can significantly enhance device reliability.

11.3 Surface Passivation Technology

As discussed in Section 11.1.2, at OFF-state with high drain bias or during hard switching, electrons captured by surface states in the access region as well as buffer traps could result in dynamic R_{ON} degradation (or current collapse) [31,33–35,41,42]. To alleviate the surface-induced current collapse, passivation techniques with dielectrics deposited on the gate-to-drain access region are necessary to modify or compensate surface states [36–42].

SiN_x passivation is widely used to suppress current collapse in both GaN microwave and power transistors by way of surface state modification [37,42,82–85]. In general, SiN_x passivation still needs to be combined with field plate technique in order to achieve low dynamic R_{ON} at high voltage [37].

Compared with the commonly used SiN_x, AlN has a wider bandgap, higher thermal conductivity, higher breakdown strength, and less lattice mismatch to GaN, and thus, is a compelling choice for GaN-on-Si power device passivation (Table 11.2) [86–88]. PEALD-epitaxial AlN (including AlN/SiN_x stack) passivation techniques have been developed [38–40,89–91]. Without the implementation of field plate, AlN passivation is capable of suppressing current collapse while simultaneously maintaining low OFF-state leakage up to 600 V [39,90].

The PEALD-epitaxial AlN exhibits atomically sharp interface and well-ordered atomic arrangement on polarized III-nitride, suggesting that the epitaxial AlN thin film grown by PEALD is also polarized. The large amount of positive polarization charges at the AlN/III-nitride interface and provide effective and robust compensation to any surface/interface states that may degrade R_{ON} [40,89]. Even in the worst-case scenario, that is, all the surface/interface states in the access region are filled with electrons,

TABLE 11.2 Comparison of Material Properties between SiN_x and AlN

Passivation Dielectric	Bandgap (eV)	Thermal Cond. (mWK^{-1}m^{-1})	Breakdown Strength (MV/cm)	Lattice Mismatch to GaN
SiN_x	5.1	1	3	−7.9%
AlN	6.2	200	6	−2.4%

Sources: Hwang, J.H. et al., *Solid-State Electronics*, 48, 363–366, 2004; Takizawa, T. et al., *J. Electron. Mater.*, 37, 628–634, 2008; Tsurumi, N. et al., *IEEE Trans. Electron Devices*, 57, 980–985, 2010.

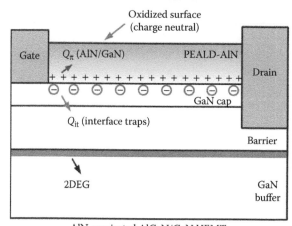

FIGURE 11.11 Schematic cross section of the PEALD-AlN-passivated AlGaN/GaN HEMT showing the charge compensation mechanism. (From Huang, S. et al., *IEEE Electron Device Lett.*, 34, 193–195, 2013. With permission.)

all the surface/interface states can still be sufficiently compensated as long as polarization charge density (Q_π) is higher than trap density (Q_{it}), and thus, surface-induced current collapse can be well suppressed. The mechanism of AlN passivation technique is different from that of the widely used SiN_x, and can reduce dynamic R_{ON} in a fundamentally distinct way. In the AlN-passivated GaN power transistors, the depletion region in the access region at high-voltage OFF-state would be a pure field effect, while 2DEG in the access region would recover instantaneously without any trap-related transients after devices are switched back to ON-state, leading to excellent dynamic performance during high-frequency operation [89] (Figure 11.11).

11.4 III-Nitride Buffer Layer in GaN Power Devices

III-nitride (typically GaN) buffer layer featuring high resistivity and current blocking capability is a necessity to realize high-voltage GaN power switching transistors. Compared with other substrate materials, Si substrate is a favorable choice to implement GaN power devices (Section 11.1). However, a large amount of bulk traps exist in GaN buffer layer grown on Si substrate originating from (1) high-density (10^9–10^{10} cm^{-2}) of crystalline defects and dislocations due to lattice and thermal expansion mismatch between GaN and Si [45,48], (2) residual impurities such as silicon or oxygen introduced during epitaxial growth [46], and (3) intentionally incorporated compensation dopants such as carbon [43] or iron [92]. Crystalline defects [93] that require further optimization of growth techniques to reduce. The intentional dopants (carbon or iron), acting as acceptor traps in wurtzite GaN, can compensate the background donor doping in order to obtain high resistivity and low leakage current in the buffer layer [94]. Compared with iron doping, carbon doping induces less memory effect during epitaxial growth and provides higher breakdown strength [44,47], and thus, has become the most attractive approach to obtain semi-insulating GaN buffer layer for high-voltage power electronics applications. Carbon-doped GaN buffer can successfully deliver 650-V voltage rating GaN-on-Si power devices with low off-state leakage currents [44,49].

It should be noted that unlike GaN microwave devices implemented on semi-insulating substrate, the conductive Si substrate in industry-standard GaN-on-Si high-voltage power devices allows the high drain-to-substrate voltage (occurring at OFF-state) to drop across the buffer layer. The GaN-to-Si vertical leakage ultimately limits the breakdown voltage of the lateral GaN-on-Si power devices. Zhou et al. [32] revealed that the GaN-to-Si vertical leakage/breakdown is closely associated with the interaction between high electric field and dynamics of donor/acceptor GaN buffer traps, following the space-charge-limited current (SCLC) model [95]. In the meantime, the ionization of deep-level acceptor traps (such as carbon-related traps) could produce negative space charges and cannot recover immediately when devices are switched back to ON-state due to slow emission process, resulting in partially depleted 2DEG channel and buffer-induced current collapse. In addition to SCLC, Moens et al. [96] also identified the presence of Poole–Frenkel conduction mechanism that leads to better dynamic performance at higher voltage.

In order to achieve high breakdown voltage while simultaneously maintain low dynamic R_{ON}, buffer structure and composition optimization are of great significance. Chevtchenko et al. [46] found that there is a tradeoff between breakdown voltage and dynamic performance within a certain voltage range, while both parameters show strong dependency on carbon doping concentration and thickness of unintentionally doped (UID) GaN channel on top of the carbon-doped GaN buffer. Buffer structure with higher carbon doping concentration and thinner UID GaN channel on top of the carbon-doped GaN buffer leads to higher breakdown voltage but worse dynamic performance, due to severe trapping effects near the channel. On the other hand, lower carbon concentration and thicker UID GaN channel can mitigate current collapse by suppressing trapping effects near the channel, however at the expense of reduced breakdown strength of the buffer layer, as potential barrier is lowered and punch-through is more easily to occur. It is demonstrated that the tradeoff between high voltage and dynamic performance can be optimized by placing an AlGaN back-barrier between the channel and carbon-doped buffer, so that the 2DEG channel are less prone to high-field induced trapping during high-voltage switching [44]. In this approach, the poorer thermal conductivity of AlGaN than GaN should be noted.

11.5 Conclusion

In this chapter, we have described and summarized the primary challenges and recent progresses toward high-performance and high-reliability GaN-on-Si lateral heterojunction power devices that are promising candidate for high-voltage power electronics. To overcome the challenges in the gate region and realize the desirable normally-off operation while enhancing device stability/reliability, key technologies of interface engineering, barrier-layer engineering, and gate dielectric have been reviewed and discussed in Section 11.2. To tackle the challenges in the access region and buffer layer in order to achieve high breakdown voltage during suppress current collapse, the influences of surface/buffer traps on device characteristics are discussed, and advanced surface passivation as well as buffer layer optimization techniques have been reviewed in Sections 11.3 and 11.4. The elucidation of the correlation between key technologies implemented in different regions (gate and access regions, buffer layer, etc.) and device performance/reliability can provide a comprehensive insight into further development of GaN-on-Si power devices.

References

1. B. J. Baliga, Gallium nitride devices for power electronic applications, *Semicond. Sci. Technol.*, 28(7): 074011-1–074011-8, 2013.
2. M. Ishida, T. Ueda, T. Tanaka, and D. Ueda, GaN on Si technologies for power switching devices, *IEEE Trans. Electron Dev.*, 60(10): 3053–3059, 2013.
3. K. Shenai, Future prospects of widebandgap (WBG) semiconductor power switching devices, *IEEE Trans. Electron Devices*, 62(2): 248–257, 2015.
4. J. Millan, P. Godignon, X. Perpina, A. Perez-Tomas, and J. Rebollo, A survey of wide bandgap power semiconductor devices, *IEEE Trans. Power Electron.*, 29(5): 2155–2163, 2014.
5. A. Lidow, GaN transistors: Giving new life to Moore's Law, in *Proceedings of the International Symposium on Power Semiconductor Devices and IC's (ISPSD)*, June 2015, pp. 1–6. New York: IEEE.

6. W. Saito, I. Omura, T. Ogura, and H. Ohashi, Theoretical limit estimation of lateral wide band-gap semiconductor power-switching device, *Solid State Electron.*, 48(9): 1555–1562, 2004.

7. O. Ambacher, J. Smart, J. R. Shealy, N. G. Weimann, K. Chu, M. Murphy, W. J. Schaff et al., Two-dimensional electron gases induced by spontaneous and piezoelectric polarization charges in N- and Ga-face AlGaN/GaN heterostructures, *J. Appl. Phys.*, 85(6): 3222–3233, 1999.

8. K. J. Chen and C. H. Zhou, Enhancement-mode AlGaN/GaN HEMT and MIS-HEMT technology, *Phys. Stat. Soli. A*, 208(2): 434–438, 2011.

9. A. Lidow, J. Strydom, M. de Rooij, A. Ferencz, and R. V. White, Driving eGaN FETs in high performance power conversion systems, in K. Shenai, R. Garg, R. Ma, M. Dudley, A. Khan, Eds. *Gallium Nitride and Silicon Carbide Power Technologies*, vol. 41(8), pp. 113–125, Pennington, NJ: The Electrochemical Society, 2011.

10. T. Morita, M. Yanagihara, H. Ishida, M. Hikita, K. Kaibara, H. Matsuo, Y. Uemoto, T. Ueda, T. Tanaka, and D. Ueda, 650 V 3.1 mΩcm² GaN-based monolithic bidirectional switch using normally-off gate injection transistor, in *Int. Electron Devices Meeting (IEDM) Tech. Dig.*, December 2007, pp. 865–868.

11. Y. Cai, Y. Zhou, K. J. Chen, and K. M. Lau, High-performance enhancement-mode AlGaN/GaN HEMTs using fluoride-based plasma treatment, *IEEE Electron Device Lett.*, 26(7): 435–437, 2005.

12. R. Wang, Y. Cai, C. Tang, K. M. Lau, and K. J. Chen, Enhancement-mode Si_3N_4/AlGaN/GaN MISHFETs, *IEEE Electron Device Lett.*, 27(10): 793–795, 2006.

13. W. Saito, Y. Takada, M. Kuraguchi, K. Tsuda, and I. Omura, Recessed-gate structure approach toward normally off high-voltage AlGaN/GaN HEMT for power electronics applications, *IEEE Trans. Electron Devices*, 53(2): 356–362, 2006.

14. T. Oka and T. Nozawa, AlGaN/GaN recessed MIS-Gate HFET with high-threshold-voltage normally-off operation for power electronics applications, *IEEE Electron Device Lett.*, 29(7): 668–670, 2008.

15. W. Huang, Z. Li, T. P. Chow, Y. Niiyama, T. Nomura, and S. Yoshida, Enhancement-mode GaN hybrid MOS-HEMTs with $R_{ON, sp}$ of 20 mΩcm², in *Proceedings of the International Symposium on Power Semiconductor Devices and IC's (ISPSD)*, June 2008, pp. 295–298. New York: IEEE.

16. M. Miczek, C. Mizue, T. Hashizume, and B. Adamowicz, Effects of interface states and temperature on the *C–V* behavior of metal/insulator/AlGaN/GaN heterostructure capacitors, *J. Appl. Phys.*, 103(10): 104510-1–104510-11, 2008.

17. S. Huang, S. Yang, J. Roberts, and K. J. Chen, Threshold voltage instability in Al_2O_3/GaN/AlGaN/ GaN metal-insulator-semiconductor high-electron mobility transistors. *Jpn. J. Appl. Phys.*, 50: 110202-1–110202-3, 2011.

18. P. Lagger, C. Ostermaier, G. Pobegen, and D. Pogany, Towards understanding the origin of threshold voltage instability of AlGaN/GaN MIS-HEMTs, in *Int. Electron Devices Meeting (IEDM) Tech. Dig.*, December 2012, pp. 13.1.1–13.1.4. New York: IEEE.

19. S. Yang, Z. Tang, K.-Y. Wong, Y.-S. Lin, Y. Lu, S. Huang, and K. J. Chen, Mapping of interface traps in high-performance Al_2O_3/AlGaN/GaN MIS-heterostructures using frequency- and temperature-dependent *C-V* techniques, in *Int. Electron Devices Meeting (IEDM) Tech. Dig.*, December 2013, pp. 6.3.1–6.3.4. New York: IEEE.

20. P. Lagger, M. Reiner, D. Pogany, and C. Ostermaier, Comprehensive study of the complex dynamics of forward bias-induced threshold voltage drifts in GaN based MIS-HEMTs by stress/recovery experiments, *IEEE Trans. Electron Devices*, 61(4): 1022–1030, 2014.

21. G. P. Lansbergen, K.-Y. Wong, Y.-S. Lin, J.-L. Yu, F.-J. Yang, C.-L. Tsai, and A. S. Oates, Threshold voltage drift (PBTI) in GaN D-mode MISHEMTs: Characterization of fast trapping components, in *Proc. Int. Rel. Phys. Symp. (IRPS)*, June 2014, pp. 6C.4.1–6C.4.6. New York: IEEE.

22. P. Moens, C. Liu, A. Banerjee, P. Vanmeerbeek, P. Coppens, H. Ziad, A. Constant et al., An industrial process for 650 V rated GaN-on-Si power devices using *in-situ* SiN as a gate dielectric, in *Proceedings of the International Symposium on Power Semiconductor Devices and IC's* (ISPSD), June 2014, pp. 374–377. New York: IEEE.

23. N. Ramanan, B. Lee, and V. Misra, A novel methodology using pulsed-*IV* for interface or border traps characterization on AlGaN/GaN MOSHFETs, in *Proceedings of the International Symposium on Power Semiconductor Devices and IC's* (ISPSD), June 2014, pp. 366–369. New York: IEEE.

24. M. Hua, C. Liu, S. Yang, S. Liu, K. Fu, Z. Dong, Y. Cai, B. Zhang, and K. J. Chen, GaN-based metal-insulator-semiconductor high-electron-mobility transistors using low-pressure chemical vapor deposition SiN$_x$ as gate dielectric, *IEEE Electron Device Lett.*, 36(5): 448–450, 2015.

25. T. L. Wu, D. Marcon, B. De Jaeger, M. Van Hove, B. Bakeroot, S. Stoffels, G. Groeseneken, S. Decoutere, and R. Roelofs, Time dependent dielectric breakdown (TDDB) evaluation of PE-ALD SiN gate dielectrics on AlGaN/GaN recessed gate D-mode MIS-HEMTs and E-mode MIS-FETs, in *Proc. Int. Rel. Phys. Symp.* (IRPS), June 2015, pp. 6C.4.1–6C.4.6. New York: IEEE.

26. W. Choi, H. Ryu, N. Jeon, M. Lee, N. H. Lee, K. S. Seo, and H. Y. Cha, Impacts of conduction band offset and border traps on Vth instability of gate recessed normally-off GaN MIS-HEMTs, in *Proc. Int. Symp. on Power Semiconductor Devices and IC's* (ISPSD), June 2014, pp. 370–373. New York: IEEE.

27. T.-L. Wu, D. Marcon, M. B. Zahid, M. Van Hove, S. Decoutere, and G. Groeseneken, Comprehensive investigation of on-state stress on D-mode AlGaN/GaN MIS-HEMTs, in *Proc. Int. Rel. Phys. Symp.* (IRPS), June 2013, pp. 3C.5.1–3C.5.7. New York: IEEE.

28. J. Wu, X. Lu, S. Ye, J. Park, and D. Streit, Electrical characterization and reliability analysis of Al2O3/AlGaN/GaN MISH structure, in *Proc. Int. Rel. Phys. Symp.* (IRPS), June 2014, pp. CD.6.1–CD.6.5. New York: IEEE.

29. M. Hua, C. Liu, S. Yang, S. Liu, K. Fu, Z. Dong, Y. Cai, B. Zhang, and K. J. Chen, Characterization of leakage and reliability of SiN$_x$ gate dielectric by low-pressure chemical vapor deposition for GaN-based MIS-HEMTs, *IEEE Trans. Electron Dev.*, 62(10): 3215–3222, 2015.

30. J. H. Bae, I. Hwang, J. M. Shin, H. I. Kwon, C. H. Park, J. Ha, and J. Lee, Characterization of traps and trap-related effects in recessed-gate normally-off AlGaN/GaN-based MOSHEMT, in *Int. Electron Devices Meeting* (IEDM) *Tech. Dig.*, December 2012, pp. 13.2.1–13.2.4. New York: IEEE.

31. R. Vetury, N. Zhang, S. Keller, and U. K. Mishra, The impact of surface states on the DC and RF characteristics of AlGaN/GaN HFETs, *IEEE Trans. Electron Dev.*, 48(3): 560–566, 2001.

32. C. Zhou, Q. Jiang, S. Huang, and K. J. Chen, Vertical leakage/breakdown mechanisms in AlGaN/GaN-on-Si devices, *IEEE Electron Device Lett.*, 33(8): 1132–1134, 2012.

33. M. Meneghini, P. Vanmeerbeek, R. Silvestri, S. Dalcanale, A. Banerjee, D. Bisi, E. Zanoni, G. Meneghesso, and P. Moens, Temperature-dependent dynamic R_{ON} in GaN-based MIS-HEMTs: Role of surface traps and buffer leakage, *IEEE Trans. Electron Dev.*, 62(3): 782–787, 2015.

34. J. Joh and J. A. del Alamo, A current-transient methodology for trap analysis for GaN high electron mobility transistors, *IEEE Trans. Electron Dev.*, 58(1): 132–140, 2011.

35. D. Jin and J. A. del Alamo, Mechanisms responsible for dynamic ON-resistance in GaN high-voltage HEMTs, in *Proceedings of the International Symposium on Power Semiconductor Devices and IC's* (ISPSD), June 2012, pp. 333–336. New York: IEEE.

36. G. Koley, V. Tilak, L. F. Eastman, and M. G. Spencer, Slow transients observed in AlGaN/GaN HFETs: Effects of SiN$_x$ passivation and UV illumination, *IEEE Trans. Electron Dev.*, 50(4): 886–893, 2003.

37. R. Chu, A. Corrion, M. Chen, R. Li, D. Wong, D. Zehnder, B. Hughes, and K. Boutros, 1200-V normally off GaN-on-Si field-effect transistors with low dynamic ON-resistance, *IEEE Electron Dev. Lett.*, 32(5): 632–634, 2011.

38. S. Huang, Q. Jiang, S. Yang, C. Zhou, and K. J. Chen, Effective passivation of AlGaN/GaN HEMTs by ALD-grown AlN thin film, *IEEE Electron Dev. Lett.*, 33(4): 516–518, 2012.

39. Z. Tang, Q. Jiang, Y. Lu, S. Huang, S. Yang, X. Tang, and K. J. Chen, 600-V normally off SiN_x/AlGaN/GaN MIS-HEMT with large gate swing and low current collapse, *IEEE Electron Dev. Lett.*, 34(11): 1373–1375, 2013.

40. S. Huang, Q. Jiang, S. Yang, Z. Tang, and K. J. Chen, Mechanism of PEALD-grown AlN passivation for AlGaN/GaN HEMTs: Compensation of interface traps by polarization charges, *IEEE Electron Dev. Lett.*, 34(2): 193–195, 2013.

41. G. Longobardi, F. Udrea, S. Sque, G. A. M. Hurkx, J. Croon, E. Napoli, and J. Sonsky, Impact of donor traps on the 2DEG and electrical behavior of AlGaN/GaN MISFETs, *IEEE Electron Dev. Lett.*, 35(1): 27–29, 2014.

42. M. F. Romero, A. Jimenez, F. G. P. Flores, S. Martin-Horcajo, F. Calle, and E. Munoz, Impact of N_2 plasma power discharge on AlGaN/GaN HEMT performance, *IEEE Trans. Electron Dev.*, 59(2): 374–379, 2012.

43. P. B. Klein, S. C. Binari, K. Ikossi, A. E. Wickenden, D. D. Koleske, and R. L. Henry, Current collapse and the role of carbon in AlGaN/GaN high electron mobility transistors grown by metalorganic vapor-phase epitaxy, *Appl. Phys. Lett.*, 79(11): 3527–3529, 2001.

44. J. Wurfl, O. Hilt, E. Bahat-Treidel, R. Zhytnytska, P. Kotara, F. Brunner, O. Krueger, and M. Weyers, Techniques towards GaN power transistors with improved high voltage dynamic switching properties, in *Int. Electron Devices Meeting (IEDM) Tech. Dig.*, December 2013, pp. 6.1.1–6.1.4. New York: IEEE.

45. S. C. Binari, P. B. Klein, and T. E. Kazior, Trapping effects in GaN and SiC microwave FETs, *Proc. IEEE*, 90(6): 1048–1058, 2002.

46. S. A. Chevtchenko, E. Cho, F. Brunner, E. Bahat-Treidel, and J. Wurfl, Off-state breakdown and dispersion optimization in AlGaN/GaN heterojunction field-effect transistors utilizing carbon doped buffer, *Appl. Phys. Lett.*, 100(22): 223502-1–223502-3, 2012.

47. M. J. Uren, J. Moreke, and M. Kuball, Buffer design to minimize current collapse in GaN/AlGaN HFETs, *IEEE Trans. Electron Dev.*, 59(12): 3327–3333, 2012.

48. M. J. Uren, M. Casar, M. A. Gajda, and M. Kuball, Buffer transport mechanisms in intentionally carbon doped GaN heterojunction field effect transistors, *Appl. Phys. Lett.*, 104(26): 263505-1–263505-4, 2014.

49. P. Moens, P. Vanmeerbeek, A. Banerjee, J. Guo, C. Liu, P. Coppens, A. Salih et al., On the impact of carbon-doping on the dynamic Ron and off-state leakage current of 650V GaN power devices, in *Proceedings of the International Symposium on Power Semiconductor Devices and IC's (ISPSD)*, June 2015, pp. 37–40. New York: IEEE.

50. B. Lu, E. L. Piner, and T. Palacios, Breakdown mechanism in AlGaN/GaN HEMTs on Si substrate, in *Device Research Conference (DRC)*, June 2010, pp. 193–194. New York: IEEE.

51. S. Yang, C. Zhou, Q. Jiang, J. Lu, B. Huang, and K. J. Chen, Investigation of buffer traps in AlGaN/GaN-on-Si devices by thermally stimulated current spectroscopy and back-gating measurement, *Appl. Phys. Lett.*, 104(1): 013504-1–013504-4, 2014.

52. S. Liu, S. Yang, Z. Tang, Q. Jiang, C. Liu, M. Wang, B. Shen, and K. J. Chen, Interface/border trap characterization of Al_2O_3/AlN/GaN metal-oxide-semiconductor structures with an AlN interfacial layer, *Appl. Phys. Lett.*, 106(5): 051605-1–051605-4, 2015.

53. S. Yang, Y. Lu, H. Wang, S. Liu, C. Liu, and K. J. Chen, Dynamic gate stress-induced V_{TH} shift and its impact on dynamic R_{ON} in GaN MIS-HEMTs, *IEEE Trans. Electron Dev.*, 37(2): 157–160, 2015.

54. S. Yang, S. Liu, C. Liu, M. Hua, and K. J. Chen, Gate stack engineering for GaN lateral power transistors, *Semicond. Sci. Technol.*, 31(2) 024001-1–024001-10, 2016.

55. S. Yang, S. Huang, M. Schnee, Q. T. Zhao, J. Schubert, and K. J. Chen, Fabrication and characterization of enhancement-mode high-κ $LaLuO_3$-AlGaN/GaN MIS-HEMTs, *IEEE Trans. Electron Dev.*, 60(10): 3040–3046, 2013.

56. S. Yang, Z. Tang, K.-Y. Wong, Y.-S. Lin, C. Liu, Y. Lu, S. Huang, and K. J. Chen, High-quality interface in Al$_2$O$_3$/GaN/AlGaN/GaN MIS structures with *in situ* pre-gate plasma nitridation, *IEEE Electron Dev. Lett.*, 34(12): 1497–1499, 2013.

57. W. Choi, H. Ryu, N. Jeon, M. Lee, H. Y. Cha, and K. S. Seo, Improvement of V_{th} instability in normally-off GaN MIS-HEMTs employing PEALD-SiN$_x$ as an interfacial layer, *IEEE Electron Dev. Lett.*, 35(1): 30–32, 2014.

58. X. Qin, H. Dong, J. Kim, and R. M. Wallace, A crystalline oxide passivation for Al$_2$O$_3$/AlGaN/ GaN, *Appl. Phys. Lett.*, 105(14): 141604-1–141604-5, 2014.

59. M. Van Hove, S. Boulay, S. R. Bahl, S. Stoffels, X. W. Kang, D. Wellekens, K. Geens, A. Delabie, and S. Decoutere, CMOS process-compatible high-power low-leakage AlGaN/GaN MISHEMT on silicon, *IEEE Electron Dev. Lett.*, 33(5): 667–669, 2012.

60. X. Wang, L. Dong, J. Zhang, Y. Liu, P. D. Ye, and R. G. Gordon, Heteroepitaxy of La$_2$O$_3$ and La$_{2-x}$Y$_x$O$_3$ on GaAs (111)A by atomic layer deposition: Achieving low interface trap density, *Nano Lett.*, 13(2): 594–599, 2013.

61. M. Miao, J. Weber, and C. G. Van de Walle, Oxidation and the origin of the two-dimensional electron gas in AlGaN/GaN heterostructures, *J. Appl. Phys.*, 107(12): 123713-1–123713-11, 2010.

62. Y. Hori, C. Mizue, and T. Hashizume, Process conditions for improvement of electrical properties of Al$_2$O$_3$/n-GaN structures prepared by atomic layer deposition, *Jpn. J. Appl. Phys.*, 49(8R): 080201-1–080201-3, 2010.

63. S. Liu, S. Yang, Z. Tang, Q. Jiang, C. Liu, M. Wang, and K. J. Chen, Al$_2$O$_3$/AlN/GaN MOS-channel-HEMTs with an AlN interfacial layer, *IEEE Electron Dev. Lett.*, 35(7): 723–725, 2014.

64. K. J. Chen, S. Yang, Z. Tang, S. Huang, Y. Lu, Q. Jiang, S. Liu, C. Liu, and B. Li, Surface nitridation for improved dielectric/III-nitride interfaces in GaN MIS-HEMTs, *Phys. Stat. Soli. A*, 212(5): 1059–1065, 2015.

65. Y. Wang, M. Wang, B. Xie, C. P. Wen, J. Wang, Y. Hao, W. Wu, K. J. Chen, and B. Shen, High-performance normally-off Al$_2$O$_3$/GaN MOSFET using a wet etching-based gate recess technique, *IEEE Electron Dev. Lett.*, 34(11): 1370–1372, 2013.

66. P. G. Neudeck, R. S. Okojie, and L. Y. Chen, High-temperature electronics – A role for wide bandgap semiconductors? *Proc. IEEE*, 90(6): 1065–1076, 2002.

67. R. Chu, D. Brown, D. Zehnder, X. Chen, A. Williams, R. Li, M. Chen, S. Newell, and K. Boutros, Normally-off GaN-on-Si metal-insulator-semiconductor field-effect transistor with 600-V blocking capability at 200°C, in *Proceedings of the International Symposium on Power Semiconductor Devices and IC's (ISPSD)*, June 2012, pp. 237–239.

68. P. Kordoš, D. Donoval, M. Florovič, J. Kováč, and D. Gregušová, Investigation of trap effects in AlGaN/GaN field-effect transistors by temperature dependent threshold voltage analysis. *Appl. Phys. Lett.*, 92(15): 152113-1–152113-1, 2008.

69. M. Tapajna, K. Cico, J. Kuzmik, D. Pogany, G. Pozzovivo, G. Strasser, J. F. Carlin, N. Grandjean, and K. Frohlich, Thermally induced voltage shift in capacitance-voltage characteristics and its relation to oxide/semiconductor interface states in Ni/Al$_2$O$_3$/InAlN/GaN heterostructures, *Semicond. Sci. Technol.*, 24(3): 035008-1–035008-3, 2009.

70. S. Yang, S. Liu, C. Liu, Z. Tang, Y. Lu, and K. J. Chen, Thermally induced threshold voltage instability of III-Nitride MIS-HEMTs and MOSC-HEMTs: Underlying mechanisms and optimization schemes, in *Int. Electron Devices Meeting (IEDM) Tech. Dig.*, December 2014, pp. 17.2.1–17.2.4. New York: IEEE.

71. C. Liu, S. Yang, S. Liu, Z. Tang, H. Wang, Q. Jiang, and K. J. Chen, Thermally stable enhancement-mode GaN metal-insulator-semiconductor high-electron-mobility transistor with partially recessed fluorine-implanted barrier, *IEEE Electron Dev. Lett.*, 36(4): 318–320, 2015.

72. S. Yang, S. Liu, C. Liu, Y. Lu, and K. J. Chen, Mechanisms of thermally induced threshold voltage instability in GaN-based heterojunction transistors, *Appl. Phys. Lett.*, 105(22): 223508-1–223508-4, 2014.

73. Y. Cai, Y. Zhou, K. M. Lau, and K. J. Chen, Control of threshold voltage of AlGaN/GaN HEMTs by fluoride-based plasma treatment: From depletion mode to enhancement mode, *IEEE Trans. Electron Dev.*, 53(9): 2207–2215, 2006.

74. L. Yuan, M. Wang, and K. J. Chen, Atomistic modeling of fluorine implantation and diffusion in III-nitride semiconductors, in *Int. Electron Devices Meeting (IEDM) Tech. Dig.*, December 2008, pp. 543–546. New York: IEEE.

75. K. J. Chen, L. Yuan, M. Wang, H. Chen, S. Huang, Q. Zhou, C. Zhou, B. Li, and J. N. Wang, Physics of fluorine plasma ion implantation for GaN normally-off HEMT technology, in *Int. Electron Devices Meeting (IEDM) Tech. Dig.*, December 2011, pp. 19.4.1–19.4.4. New York: IEEE.

76. C. Yi, R. Wang, W. Huang, C. Tang, K. M. Lau, and K. J. Chen, Reliability of enhancement-mode AlGaN/GaN HEMTs fabricated by fluorine plasma treatment, in *Int. Electron Devices Meeting (IEDM) Tech. Dig.*, December 2007, pp. 389–392. New York: IEEE.

77. M. Wang, L. Yuan, K. J. Chen, F. Xu, and B. Shen, Diffusion mechanism and the thermal stability of fluorine ions in GaN after ion implantation, *J. Appl. Phys.*, 105(8): 083519-1–083519-4, 2009.

78. C. Ma, H. Chen, C. Zhou, S. Huang, L. Yuan, J. Roberts, and K. J. Chen, Reliability of enhancement-mode AlGaN/GaN HEMTs under ON-state gate overdrive, in *Int. Electron Devices Meeting (IEDM) Tech. Dig.*, December 2010, pp. 20.4.1–20.4.4. New York: IEEE.

79. A. Stoffel, A. Kovacs, W. Kronast, and B. Muller, LPCVD against PECVD for micromechanical applications, *J. Micromech. Microeng.*, 6(1): 1–13, 1996.

80. J. Wu, X. Lu, S. Ye, J. Park, and D. Streit, Electrical characterization and reliability analysis of Al_2O_3/AlGaN/GaN MISH structure, in *Proc. Int. Rel. Phys. Symp. (IRPS)*, June 2014, pp. CD.6.1–CD.6.5. New York: IEEE.

81. S. Huang, X. Liu, K. Wei, G. Liu, X. Wang, B. Sun, X. Yang, B. Shen, C. Liu, S. Liu, M. Hua, S. Yang, and K. J. Chen, O_3-sourced atomic layer deposition of high quality Al_2O_3 gate dielectric for normally-off GaN metal-insulator-semiconductor high-electron-mobility transistors, *Appl. Phys. Lett.*, 106(3): 033507-1–033507-5, 2015.

82. H. Kim, R. M. Thompson, V. Tilak, T. R. Prunty, J. R. Shealy, and L. F. Eastman, Effects of SiN passivation and high-electric field on AlGaN-GaN HFET degradation, *IEEE Electron Dev. Lett.*, 24(7): 421–423, 2003.

83. A. P. Edwards, J. A. Mittereder, S. C. Binari, D. S. Katzer, D. F. Storm, and J. A. Roussos, Improved reliability of AlGaN-GaN HEMTs using an NH_3 plasma treatment prior to SiN passivation, *IEEE Electron Dev. Lett.*, 26(4): 225–227, 2005.

84. T. Kikkawa, M. Nagahara, N. Okamoto, Y. Tateno, Y. Yamaguchi, N. Hara, K. Joshin, and P. M. Asbeck, Surface-charge controlled AlGaN/GaN-power HFET without current collapse and gm dispersion, in *Int. Electron Devices Meeting (IEDM) Tech. Dig.*, December 2001, pp. 25.4.1–25.4.4. New York: IEEE.

85. G. Koley, V. Tilak, L. F. Eastman, and M. G. Spencer, Slow transients observed in AlGaN/GaN HFETs: Effects of SiN_x passivation and UV illumination, *IEEE Trans. Electron Dev.*, 50(4): 886–893, 2003.

86. J. H. Hwang, W. J. Schaff, B. M. Green, H. Y. Cha, and L. F. Eastman, Effects of a molecular beam epitaxy grown AlN passivation layer on AlGaN/GaN heterojunction field effect transistors, *Solid-State Electron.*, 48(2): 363–366, 2004.

87. T. Takizawa, S. Nakazawa, and T. Ueda, Crystalline SiN_x ultrathin films grown on AlGaN/GaN using *in situ* metalorganic chemical vapor deposition, *J. Electron. Mater.*, 37(5): 628–634, 2008.

88. N. Tsurumi, H. Ueno, T. Murata, H. Ishida, Y. Uemoto, T. Ueda, K. Inoue, and T. Tanaka, AlN passivation over AlGaN/GaN HFETs for surface heat spreading, *IEEE Trans. Electron Dev.*, 57(5): 980–985, 2010.

89. Z. Tang, S. Huang, X. Tang, B. Li, and K. J. Chen, Influence of AlN passivation on dynamic ON-resistance and electric field distribution in high-voltage AlGaN/GaN-on-Si HEMTs, *IEEE Trans. Electron Dev.*, 61(8): 2785–2792, 2014.

90. Z. Tang, Q. Jiang, S. Huang, Y. Lu, S. Yang, C. Liu, X. Tang, S. Liu, B. Li, and K. J. Chen, Monolithically integrated 600-V E/D-mMode SiN$_x$/AlGaN/GaN MIS-HEMTs and their applications in low-standby-power start-up circuit for switched-mode power supplies, in *Int. Electron Devices Meeting (IEDM)*, December 2013, pp. 6.4.1–6.4.4. New York: IEEE.

91. A. D. Koehler, N. Nepal, T. J. Anderson, M. J. Tadjer, K. D. Hobart, C. R. Eddy, and F. J. Kub, Atomic layer epitaxy AlN for enhanced AlGaN/GaN HEMT passivation, *IEEE Electron Dev. Lett.*, 34(9): 1115–1117, 2013.

92. Y. C. Choi, M. Pophristic, H. Y. Cha, B. Peres, M. G. Spencer, and L. F. Eastman, The effect of an Fe-doped GaN buffer on OFF-state breakdown characteristics in AlGaN/GaN HEMTs on Si substrate, *IEEE Trans. Electron Dev.*, 53(12): 2926–2931, 2006.

93. A. Hierro, D. Kwon, S. A. Ringel, M. Hansen, J. S. Speck, U. K. Mishra, and S. P. DenBaars, Optically and thermally detected deep levels in n-type Schottky and p(+)-n GaN diodes, *Appl. Phys. Lett.*, 76(21): 3064–3066, 2000.

94. J. L. Lyons, A. Janotti, and C. G. Van de Walle, Carbon impurities and the yellow luminescence in GaN, *Appl. Phys. Lett.*, 97(15): 152108-1–152108-3, 2010.

95. M. A. Lampert, Simplified theory of space-charge-limited currents in an insulator with traps, *Phys. Rev.*, 103(6): 1648–1656, 1956.

96. P. Moens, A. Banerjee, M. J. Uren, M. Meneghini, S. Karboyan, I. Chatterjee, P. Vanmeerbeek et al., Impact of buffer leakage on intrinsic reliabiliity of 650 V AlGaN/GaN HEMTs, in *Int. Electron Devices Meeting (IEDM) Tech. Dig.*, December 2015, pp. 35.2.1–35.2.4. New York: IEEE.

12

Reliability in III-Nitride Devices

Davide Bisi,
Isabella Rossetto,
Matteo Meneghini,
Gaudenzio
Meneghesso,
and Enrico Zanoni

12.1 Introduction

III-nitrides are very promising compound semiconductors which attracted strong attention for both microwave and power switching applications. The electrical properties, namely the high carrier mobility and current density, the outstanding breakdown field and the possibility of operating at very high temperature, make this material very suitable for the fabrication of the next generation of microwave, millimeter-wave, and power switching devices.

In order to achieve the successful implementation of high-performance and high-reliability devices and the introduction of the future telecommunication and power conversion electronic systems, a deep understanding of the parasitic effects and failure mechanisms, as well as of methods for their suppression, is required. The main examples of parasitic phenomena which limit the device performance are charge trapping, promoting dynamic on-resistance increase, and threshold voltage instabilities; leakage currents, which induce premature breakdown and contributes to conduction/switching losses; time-dependent degradation and time-dependent breakdown mechanisms, which hamper the long-term reliability.

In this chapter, we will discuss the main characterization methodologies for the analysis of parasitic effects, the latest results concerning the understanding of their origin and the implications of the above-mentioned aspects for the state-of-the-art technologies.

Section 12.2 introduces the study of deep-level effects in GaN high electron mobility transistors (HEMTs); a detailed evaluation of the characterization techniques is presented, including double-pulse measurements, drain current transient analysis, and gate frequency sweep (GFS) method; we then discuss a method for charge trapping localization; the role of compensation doping and hot electrons on the trapping mechanisms and the impact of trapping-induced leakage current on time-dependent breakdown.

Section 12.3 discusses the stability of dielectrics adopted in GaN HEMT and metal-insulator-semiconductor (MIS) HEMT structures, including bias temperature threshold voltage instabilities and time-dependent dielectric breakdown (TDDB).

Section 12.4 concludes the chapter by providing an overview of the main breakdown mechanisms in state-of-art III-N devices.

This chapter does not discuss the issues related to the stability of the Schottky gate junction, reviewed in [1], including converse piezoelectric effects [2] and GaN electrochemical oxidation [3], and the reliability of the ohmic contacts [4].

12.2 Deep-Level Effects in GaN HEMTs

Defects are originated by the deviation in the crystalline structure of the perfect periodic arrangement of atoms in the ideal lattice. They include point-defects (vacancies, interstitials, and antisites), extended-defects (screw, edge, and mixed dislocations) and interface defects. From the electrical point-of-view, their major effect is the introduction of allowed energy-levels within the forbidden energy-gap (Figure 12.1). Depending on their position within the gap, they can be categorized in *shallow-levels*, responsible for doping, and *deep-levels*, responsible for charge-trapping and/or nonradiative recombination effects. Their impact on the device operations is related to their electrical configuration, their concentration, and to their location within the device structure.

12.2.1 Current Collapse, Virtual-Gate Effects, and Dynamic R_{ON} Increase

In GaN-based HEMTs, one of the most evident effects of deep levels is charge-trapping and the so-called current collapse, that is, the dynamic decrease in drain current experienced by the device when operated

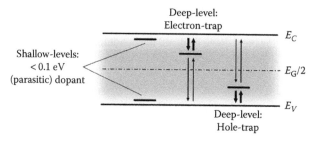

FIGURE 12.1 Depending on their position within the gap, defect-states and impurities can be categorized in *shallow-levels*, mainly responsible for parasitic doping effects, and *deep-levels*, mainly responsible for charge-trapping and/or non-radiative recombination effects. If deep-levels are located in the upper-half of the energy-gap, they have higher probability to capture and emit electrons from and to the conduction-band, behaving as electron-traps. On the other hand, if they are located in the lower-half of the energy-gap, they have higher probability to capture and emit holes from and to the valence-band, behaving as hole-traps.

with large gate-drain voltage swings. This effect is strictly related to charge trapping at deep levels, which can be present in the as-grown material, related to the processing conditions, and/or induced by electrical stress.

Charge trapping can be promoted by several causes and located in multiple locations within the device, including surface states [5,6], traps in the GaN buffer ascribed to intentional and unintentional impurities and defects, for example, carbon [7–11] and iron [12,13], traps below the gate region ascribed to defect states generated by processing steps [14], traps in the gate dielectrics in MIS-gate HEMTs [15–18].

The main effects of charge-trapping are instabilities of the threshold voltage (V_{TH}), decrease in the transconductance (g_m) and an increase in the dynamic on-resistance (R_{ON}).

Current collapse affects the performance of GaN-HEMTs designed for microwave and millimeter wave applications and for power switching applications. In microwave devices, current collapse leads to the reduction of the output power and the degradation of the RF performance [5,19–21]. In power devices, current collapse leads to the increase in the on-resistance and the increase in the conductive losses [22–24].

Deep levels can be theoretically predicted by means of first-principle calculations [25], and they can be experimentally characterized by means of deep-level transient spectroscopy (DLTS) techniques [26–28].

In the following, we review the main characterization techniques for the investigation of charge-trapping effects, the impact on device performance, the identification and the localization of deep levels, and the analysis of charge-trapping mechanisms observed in the current generation of AlGaN/GaN HEMTs.

12.2.2 Characterizing Traps in GaN HEMTs: Double-Pulse Testing and Drain Current Transient Spectroscopy

The analysis of the current collapse effects and the properties of the deep levels located in a HEMT structure is of fundamental importance in the following aspects: (1) the identification of the defect states and their localization within the HEMT structure is an important feedback for device design, crystal growth and device fabrication improvement and (2) the experimental characterization of the device performance as a function of bias and temperature can be used for accurate physics-based device modeling, including parasitic effects and time-dependent mechanisms.

In this chapter, we discuss three key measurement techniques employed for the analysis of charge-trapping phenomena: (1) the double-pulse current-voltage (*I–V*) characterization, useful to quantify the effects of charge-trapping on the dynamic parameters of devices under test (in terms of threshold-voltage shift, transconductance degradation, and on-resistance increase); (2) the drain current transient spectroscopy (DCTS), useful to identify the involved deep trap states; and (3) the GFS measurement technique, useful to investigate the dynamic behavior or the devices close to the real switching conditions.

12.2.2.1 Double-Pulsed I-V Measurements

A quick and reliable characterization of the current collapse can be obtained through the execution of pulsed drain current versus drain or gate voltage (I_D-V_D or I_D-V_G) measurements, which allow one to investigate the changes in drain current (I_{DSS}), on-resistance (R_{ON}), transconductance (g_m), and threshold voltage (V_{TH}) induced by the capture of carriers at trap states within the HEMT structure.

A schematic diagram of the double-pulse measurement is depicted in Figure 12.2. Gate and drain contact are pulsed from a quiescent bias point ($V_{G,\,Q}$;$V_{D,\,Q}$) to a measurement bias point ($V_{G,\,M}$;$V_{D,\,M}$) [29]. By varying the measuring bias point, the full I_D-V_D and I_D-V_G domains are acquired. By varying the quiescent bias point, it is possible to inhibit or promote charge-trapping effects. The duration of the "quiescent" phase should be sufficiently long to allow charge trapping to occur; on the contrary, duration of the "measurement" phase must be short enough to prevent charge detrapping. If these conditions are satisfied, pulsed *I–V* characteristics may be used for evaluating "trapped" or "collapsed" device behavior. Typically, with the quiescent bias point ($V_{G,\,Q}$;$V_{D,\,Q}$) = (0 V;0 V), we acquire the reference I_D-V_D and I_D-V_D

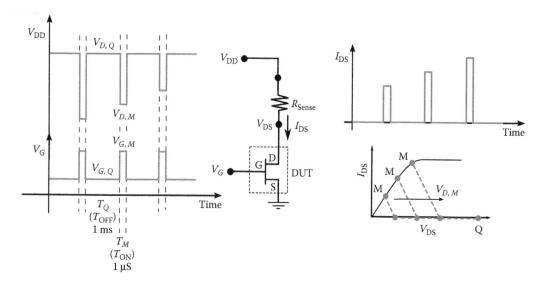

FIGURE 12.2 Double-pulse I-V characterization. Gate and drain contact are pulsed from a quiescent-bias point $(V_{G,Q}; V_{D,Q})$ to a measurement bias point $(V_{G,M}; V_{D,M})$, and the drain-current (I_{DS}) is acquired as the potential drop across a sense resistor.

characteristics, not affected by current collapse. Then, with quiescent bias points in high-field off-state $(V_{G,Q}; V_{D,Q}) = (V_{TH}-2\,\mathrm{V}; V_{DD})$ or semi-on-state $(V_{G,Q}; V_{D,Q}) = (V_{TH} + 0.5\,\mathrm{V}; V_{DD})$, we promote the charge-trapping effects and monitor the related degradation of electrical performance. The pulse width, that is, the measuring-bias period, is typically 1 µs (T_{ON}), whereas the pulse period ranges typically from 100 µs to 1 ms (T_{OFF}).

Representative double-pulsed I_D–V_D and I_D–V_G measurements performed on a test HEMT structure exposed to multiple off-state quiescent bias points are reported in Figure 12.3. The monitored parameters that can be affected by charge-trapping effects include the saturation drain current (I_{DSS}), the on-resistance (R_{ON}), the threshold voltage (V_{TH}), the transconductance peak $(g_{m,\,peak})$, and the transconductance droop $(g_{m,\,droop})$ evaluated at high gate voltage $(V_{TH} + 3/4\,\mathrm{V})$.

By investigating how these parameters degrade when the device is exposed to off-state quiescent-bias one can obtain a preliminary localization of defect states within the complex HEMT structure. Meneghini et al. demonstrated that traps located in the region under the gate or in the access regions can be separately identified [14].

Samples with a large density of trapped charge located in the intrinsic device, that is, in the region below the gate contact, predominantly exhibit a dynamic threshold voltage shift (ΔV_{TH}), as predicted in [30].

On the other hand, the presence of trapped charge in the ungated source and drain access regions would increase the resistivity of the extrinsic device, promoting the dynamic decrease of the transconductance, with no significant modification of V_{TH}.

In the following, we show three representative case-studies to illustrate in more detail, the methodology reported so far.

The first case-study (case-study A) is represented by a GaN-based HEMT grown by metalorganic chemical vapor deposition (MOCVD) on a silicon carbide substrate. Device structure consists of a 20 nm $Al_{0.36}Ga_{0.64}N$ barrier layer, a 0.7 nm AlN layer, a 10 nm GaN channel layer, a 1 µm thick $Al_{0.04}Ga_{0.96}N$ back-barrier layer and a 300 nm $Al_{0.09}Ga_{0.91}N$ layer. Gate length and gate width are 1 and 150 µm, respectively, gate–drain distance (L_{GD}) is equal to 2 µm, while gate–source distance (L_{GS}) is 1 µm. The gate of this device was formed by etching 5 nm through the AlGaN barrier by fluorine-based plasma etching (gate recess) and by depositing 250 nm of indium tin oxide (ITO, a transparent conductive oxide). The device was passivated with a silicon nitride layer.

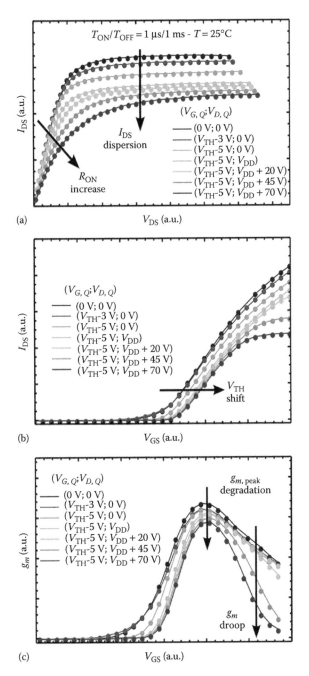

FIGURE 12.3 Double-pulsed (a) I_D–V_D and (b) I_D–V_G acquired on a test HEMT structure exposed to multiple off-state quiescent biases with increasing drain-voltage. (c) g_m–V_G derived from I_D–V_G.

Figure 12.4 shows the pulsed g_m versus V_{GS} characteristics acquired by exposing the device to different quiescent bias points, namely $(V_{G-qb}; V_{D-qb}) = (0\ V; 0\ V)$, $(-4\ V; 0\ V)$, and $(-4\ V; 10\ V)$. Results indicate that a -4 V gate quiescent bias is sufficient to induce a significant positive shift of the threshold voltage, with respect to the curve measured starting from rest conditions $(0\ V; 0\ V)$. This effect indicates that traps are present underneath the gate contact at the interface between the ITO and the AlGaN layer: when charged, these traps located below the gate contact change the threshold voltage of the device,

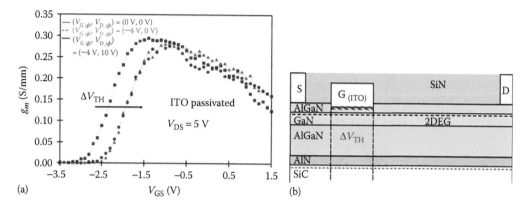

(a)   (b)

FIGURE 12.4 (a) Pulsed g_m versus V_{GS} acquired on a test HEMT structure employing a 5-nm recessed ITO gate (case-study A). When exposed to negative quiescent gate bias, the device experiences a positive threshold voltage shift, suggesting the presence of traps underneath the gate contact (b), at the interface between the AlGaN and the ITO. (From Meneghini, M. et al., *IEEE Trans. Electron Devices*, 58(9), 2996–3003, 2011. © 2011 IEEE. With permission.)

without affecting the transconductance peak. As reported in [14], the adoption of a different gate structure, namely, a Ni/Au/Ni metal stack, was then sufficient to suppress the trap density below the gate contact and to prevent the related threshold voltage shift.

The second case-study (case-study B) is represented by a Schottky-gated HEMT grown on a 3" SiC substrate by means of metalorganic chemical vapor deposition (MOCVD) and devoted to microwave applications. The device structure consists of a 1.8 μm GaN buffer layer, with a 22-nm thick AlGaN (18% Al) barrier. Gate-source spacing, gate-drain spacing, gate length and gate width are respectively 1 μm, 3.5 μm, 0.5 μm and 100 μm. The devices have a 1.3 μm source-connected field plate.

Figure 12.5 shows the I_D–V_D, I_D–V_G, and derived g_m–V_G acquired exposing the device at different off-state quiescent bias points, up to $V_{D,Q} = 50$ V. For the $(V_{GQ}, V_{DQ}) = (-6\ \text{V};50\ \text{V})$ quiescent bias point,

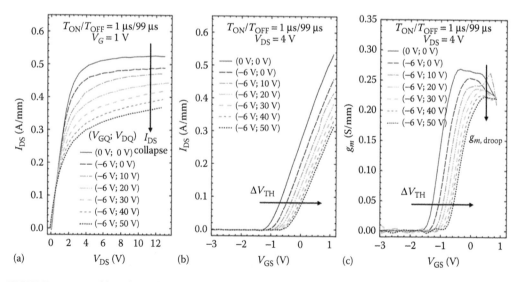

(a)   (b)   (c)

FIGURE 12.5 Double-pulsed (a) I_D–V_D, (b) I_D–V_G, and (c) g_m–V_G acquired on a test HEMT structure devoted to microwave operations (case-study B) exposed to off-state quiescent-biases: the current collapse is caused by a remarkable positive V_{TH} shift and a $g_{m,peak}$ reduction, suggesting charge trapping within the epitaxial layers in the region below the gate contact and the access regions. The employed pulse width/period ratio is 1 μs/100 μs.

we measured a 40% decrease in I_{DSS}, associated with a remarkable V_{TH} positive shift (+700 mV), slight R_{ON} increase, and 20% g_m peak reduction. Results indicate that, since the current collapse is related to both a positive V_{TH} positive shift and transconductance droop, charge-trapping mechanism(s) happen(s) both below the gate contact and in the access regions. Results from DCTS (see Section 12.2.2.2) revealed in fact the presence of deep levels within the (Al)GaN epitaxial layers, which are able to trap charge both below the gate contact and in the access regions, possibly due to the high-field of this device with relatively scaled geometries for RF applications.

The case-study C is represented by a test HEMT structure designed for power switching application. The device structure is grown by means of MOCVD on a 4" SiC substrate, and consists of a 1.8 μm-thick GaN buffer layer, with a 17 nm-thick AlGaN (22% Al) barrier. With respect to the case-study B, it features an up-scaled geometry in order to accommodate higher operating voltages: gate-source spacing, gate-drain spacing, and gate length are 2, 20, 2 μm, respectively. The devices have a 4 μm gate-connected field plate (Γ-gate, see the schematic structure in Figure 12.6). The field plate is used for lowering the maximum value of the longitudinal electric field in the gate-drain region and to the smoothening of its profile.

Figure 12.6 shows the I_D-V_D, I_D-V_G, and derived g_m-V_G acquired exposing the device at different off-state quiescent bias points, up to $V_{D,Q}$ = 200 V. When subjected to high drain voltage quiescent bias points, the device experiences a dramatic increase in the dynamic R_{ON}. From 0 V to 100 V, the dynamic on-resistance smoothly increases from 33.1 to 45.8 Ω·mm; from 100 V, it experiences a much steeper worsening, reaching 102 Ω·mm at $V_{D,Q}$ = 200 V. As can be noticed from I_D-V_G and g_m-V_G in

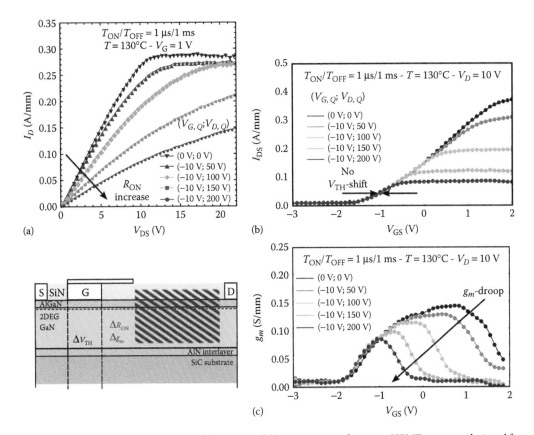

(a) (b) (c)

FIGURE 12.6 Double-pulsed (a) I_D-V_D (b) I_D-V_G, and (c) g_m-V_G acquired on a test HEMT structure designed for power applications (case-study C) exposed to multiple off-state quiescent biases, with increasing drain-voltage up to 200 V. No V_{TH}-shift and remarkable R_{ON}-increase suggest that charge-trapping mechanisms are localized in the gate-drain access region.

Figure 12.6b and c, the R_{ON}-increase is manifested in the drop of the transconductance, with no appreciable V_{TH} shift. Contrarily to the case-studies A and B, these results indicate that, since the current collapse is mainly related to R_{ON}-increase and g_m-droop, the charge-trapping mechanism that causes the majority of current collapse occurs within the gate-drain access region between the field-plate edge and the drain contact, that is, where the electric field in off-state bias is highest. As electrons need to achieve sufficient energy to reach deep levels of the GaN buffer, AlGaN barrier or SiN passivation, most trapping occurs within the gate-drain region, where the electric field is maximum (Figure 12.6).

12.2.2.2 Drain Current Transient Spectroscopy

In order to identify the deep levels responsible for the current collapse observed during the double-pulse measurements, DLTS techniques are employed.

Capacitance-mode DLTS (C-DLTS) and deep-level optical spectroscopy (DLOS) have been widely employed to identify the deep levels [31–34]; in most of the cases, the measurements were carried out on large-area Schottky diodes, since these techniques could be applied to small periphery transistors only with difficulty due to the small gate capacitance. Unfortunately, the distribution of the electric field is significantly different in diodes and actual transistors; therefore, the results obtained by analyzing diodes provide information on the properties of the deep levels, but do not give a clear indication of where those levels are located in a HEMT structure, and on how they can influence the dynamic behavior of the actual devices.

Conversely, the effective description of the properties of the traps relevant for GaN-based transistors operation can be obtained by studying the time-resolved step-response of drain current induced by exposure to a trapping voltage, either through current-mode DLTS (I-DLTS), where the DLTS signal is reconstructed by sweeping the temperature and sampling the current at multiple fixed time boxcar windows [32,31,35] or DCTS, where the transient signal is acquired continuously over several decades of time at multiple constant temperatures [27,36–39]: depending on the measurement conditions, the latter method can be faster, since it does not require to carry out a full temperature scan, but only to measure the current transients at some (5–10) temperature levels.

A schematic representation of drain current transient measurement is depicted in Figure 12.7. First, the device under test is subjected to a trapping phase ($V_{G,F}$;$V_{D,F}$, typically in the off-state or semi-on-state, for a period of 100 s). Then, it is biased in a low-field, low-power on-state ($V_{G,M}$;$V_{D,M}$, typically in the linear region, or in the saturation region close to the knee voltage) to enable the acquisition of the drain current response and the analysis of the related time-constant spectrum.

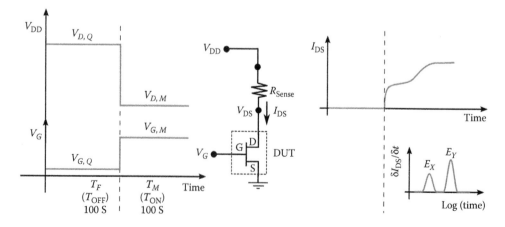

FIGURE 12.7 Drain current transient spectroscopy. Device under test is subjected to a stress/trapping phase ($V_{G,F}$; V_{DF}, typically in the off-state or semi-on-state for a period of 100 s); then, it is biased in a low-field, low-power on-state ($V_{G,M}$; V_{DM}, typically in linear region, or right after the knee voltage) to permit the acquisition of the drain-current response and the analysis of the related time-constant spectrum.

Several methods can be used for the extrapolation of the time constants of the charge emission processes: the multiexponential fitting [27,40,41], the analysis of the derivative of the current transients (the peak of the derivative is located in correspondence of the time constant of the emission process) [89,39], and the stretched exponential fitting [36] have been adopted by various research groups. The choice of the method can modify the results of the investigation since it can significantly change the shape and slope of the Arrhenius plot.

We implemented and compared three different methods for the evaluation of time constants: (1) the fitting of transient data by a polynomial function, and the subsequent extrapolation of time constants from the peaks of the $\partial I_{DS}/\partial \log(t)$ (derivative) curves [39], (2) the fitting of the data by a function represented by the sum of 100 exponentials with fixed time constants and variable amplitude coefficients [27], and (3) the fitting of the data by the stretched multiexponential function

$$I_{DS}(t) = I_{DS,\text{final}} - \sum_{i}^{N} A_i e^{-\left(\frac{t}{\tau_i}\right)^{\beta_i}}$$

where the fitting parameters A_i, τ_i and β_i are respectively the amplitude, the typical time constant and the nonexponential stretching factor of the N detected charge emission ($A_i > 0$) or capture ($A_i < 0$) processes (N typically ranges between 2 and 4 depending on the analyzed samples). The comparative results of the different fitting algorithms are summarized in Figure 12.8. As can be noticed, the three methods provide different behaviors: in the ideal case of almost purely exponential transient (trap E2), the three methods return similar fitting parameters, hence leading to the same Arrhenius plot (see level E2 in Figure 12.8); conversely, in the presence of stretched nonexponential processes (e.g., trap E5) the output of both polynomial and fixed-tau multiexponentials fitting methods are affected by spectral dispersion, which manifest itself as the distortion of the differential current peaks (for the polynomial fit case) or the splitting of spectral lines (for the multi-exp. fit). As a consequence, the time constants extrapolated by means of polynomial fitting and fixed-tau multiexponential can significantly differ from those obtained through stretched exponential fitting: this difference may lead to a wrong extrapolation of the Arrhenius plots (level E5 in Figure 12.8).

Literature ascribed the presence of strongly nonexponential transients to the following causes: (1) current transients originate from the superposition of different deep-level emission processes [42]; (2) current transients are related to the presence of high density defect states, as in the case of extended line defects, that lead to trapped electron delocalization and formation of mini bands [43,44]; (3) current

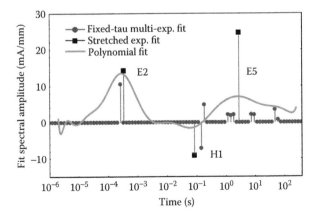

FIGURE 12.8 Comparison among the three data fitting algorithms: the sum of 100 exponentials with fixed time constants and variable amplitude coefficients; the sum of few stretched exponentials with variable time constants and amplitude coefficients, and a polynomial function with the subsequent extrapolation of time constants from the peaks of the $\partial I_{DS}/\partial \log(t)$.

transients originate from the presence of defect states in alloy materials, such as $Al_xGa_{(1-x)}N$, which lead to deep-level splitting due to band discontinuity [45]; (4) the charge emission parameters vary with the nonuniform distribution of the electric field within the complex HEMT structure [43]; and (5) trapping/detrapping kinetics are not governed by thermal emission, but by other mechanisms, such as hopping or trap-assisted-tunneling in the case of surface traps [6,46].

12.2.2.3 Deep-Level Characteristics

For a better understanding of the properties of the deep levels responsible for the current collapse, it is necessary to extrapolate the Arrhenius plots of the traps and their signature in terms of activation energy (E_A) and capture cross section (σ_c). These data are obtained by carrying out the current transient investigation at several temperature levels. Figure 12.9 reports the results of DCTS performed over a test HEMT structure designed for microwave applications (case study A) in [28]. Three different deep-level signatures have been identified. The signatures of E2, E5, and H1 correspond to apparent activation energies and apparent capture cross sections of 0.62 eV and 8.7×10^{-15} cm² for E2; 1.10 eV and 3.9×10^{-12} cm² for E5; and 0.91 eV and 3.3×10^{-13} cm² for H1.

Neglecting the self-heating of the devices during the analysis of the current transients may lead to a wrong extrapolation of the Arrhenius plots. In order to take into account the self-heating of the device during the measurement and to avoid results misinterpretation, the actual device temperature has been estimated taking into account the dissipated power during the measurement phase and the thermal resistance of the device. The latter is measured by means of the pulsed-current method described by Joh et al. [47]. Further details on the importance of temperature correction are extensively reported by Chini et al. in [48].

In Figures 12.10 and 12.11, we report the comparison between the signatures of the deep levels detected within this work and previous literature references.

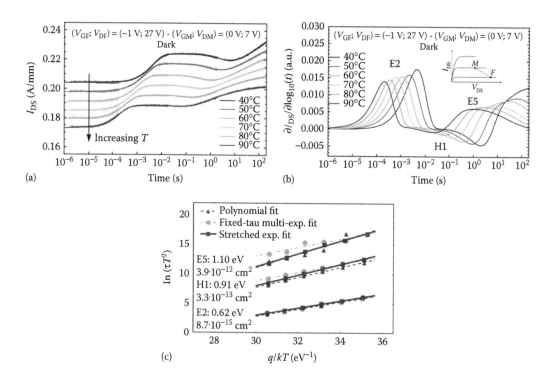

FIGURE 12.9 (a) Drain current transient acquired at multiple temperatures and (b) related time-constant spectra. (c) Arrhenius plot with apparent activation energies (E_A) and capture cross-sections (σ_c) extrapolated by stretched exponential fit of E2- E5- and H1-related de-trapping processes. Comparison with multiple fitting methods is also reported.

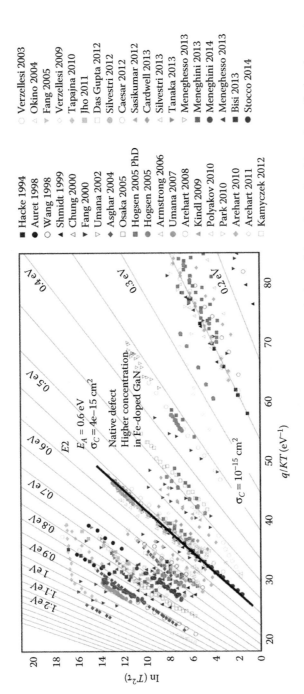

FIGURE 12.10 Superposition of deep-level signatures reported in the literature on GaN-based devices and those obtained in this work, ascribed to electron traps.

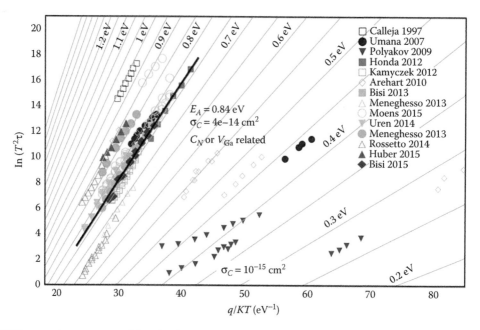

FIGURE 12.11 Superposition of deep-level signatures reported in the literature on GaN-based devices and those obtained in this work, ascribed to hole-traps.

As can be noticed, the signature of trap E2 ($E_A = 0.62$ eV; $\sigma_c = 8.7 \times 10^{-15}$ cm^2) reveals good matching with the electron-trap E2/E_C-0.60 eV, which is broadly detected in several works regarding both n-type GaN Schottky-diode and HEMT structures [12,27,49–63]. This level has been generally ascribed to GaN defects, even though its origin is still controversial: Nitrogen anti-site native point defect [49,53] and foreign impurities and dopants (C [58], Si [64], Mg-H complex [51], Fe [12]) have been also reported.

Level E5 ($E_A = 1.10$ eV; $\sigma_c = 3.9 \times 10^{-12}$ cm^2) reveals similar signature of deep levels associated with GaN dislocations [33,34,52,65], which would be consistent with the logarithmic filling time dependence reported in [28].

Level H1 ($E_A = 0.91$ eV; $\sigma_c = 3.3 \times 10^{-13}$ cm^2) displays similar signature with V_{GA}-related deep acceptor states reported in [56,66–68] (in this case the 0.91 eV apparent activation energy represents the trap-level energy from the valence band maximum). The level E_V + 0.9 eV has been also reported more recently in [69] as possible charge-state transition of carbon impurities in nitrogen substitutional position (C_N).

Finally, very low thermal activation and very slow detrapping typical time constant have been reported in literature [39,40] and could be explained by the following: (1) electron transport mechanism dominated by RC discharging processes, (2) superficial shallow states filled and emptied by hopping conduction mechanisms, or (3) deep-level charge detrapping by tunneling mechanisms.

12.2.2.4 Gate Frequency Sweep

Double-pulsed I-V characterization and drain current DLTS (I-DLTS) can provide significant information on the influence of charge-trapping on the device behavior. These techniques are essential tools for a detailed investigation of the nature of traps and related defects. Nevertheless, these methodologies provide only little indication of the device behavior in the real operating switching conditions. De Santi et al. [70] suggested a method to emulate the device dynamic response to traps when operating in real conditions. The GFS is a custom test method that allows to evaluate the impact of the trap states in a real operating condition and to define the main trap properties (e.g., activation energy, and capture cross section [70]) under these conditions.

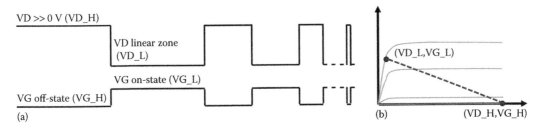

VD >> 0 V (VD_H)

VD linear zone
(VD_L)

VG on-state (VG_L)

VG off-state (VG_H)

(a)

(b)

(VD_L,VG_L)

(VD_H,VG_H)

FIGURE 12.12 (a) Schematic representation of the signal (square wave) applied to the gate and to the drain terminal and (b) example of the filling and detrapping bias conditions during gate frequency sweep measurements.

The GFS method consists in the application of a signal which emulates the power switching operating condition by imposing a square wave with a duty cycle of 50% to both the gate and the drain terminals. The frequencies considered should allow the device to reach the steady-state in the higher frequency. A principle of operation is described in Figure 12.12 and can be divided into two main phases: (1) in phase I, the device is kept in a trapping/filling condition, according to the device main application. In the case shown in Figure 12.12, the transistor is kept in off-state. (2) In phase II, the device is kept in a detrapping/recovery condition. In the case considered in Figure 12.12, the device is kept in on-state in the linear zone. During the second phase, the current waveform is measured. The current waveform is then acquired for several frequencies in order to study the trapping/detrapping processes and to evaluate the relation between the capture and the emission time constants. As carrier trapping is usually faster than detrapping, fewer and fewer electrons are released from deep levels as the frequency increases, thus inducing a decrease of the drain current.

The above-mentioned three techniques, namely double pulse, drain current transient, and GFS, mainly differ for the bias condition applied and the trapping/detrapping time used (Figure 12.13). Conversely to the GFS, which emulates the operating bias condition, the double pulse and the drain current transient method may use an overrate bias point in order to provide more information concerning the mechanism and the kinetics of the trap states under analysis. Figures 12.13 and 12.14 show the different trapping/detrapping times considered in these techniques: (1) DCT uses a duty cycle of 50% and a constant filling time. During the DCT, the recovery phase is monitored within a defined time range (usually 1 μs–100 s) after having applied a filling pulse with a duration equal to the upper value of the recovery time range; (2) GFS, similarly to the DCT, uses a duty cycle equal to 50%. The filling and

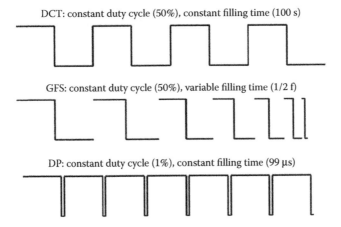

DCT: constant duty cycle (50%), constant filling time (100 s)

GFS: constant duty cycle (50%), variable filling time (1/2 f)

DP: constant duty cycle (1%), constant filling time (99 μs)

FIGURE 12.13 Schematic representation of pulse waveforms with different duty cycle, employed to emulate multiple characterization methodologies (e.g., double-pulse and drain-current transient).

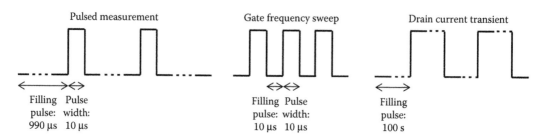

FIGURE 12.14 Filling times applied during the three techniques used, namely double pulse, drain current transient, and gate frequency sweep, and corresponding to a frequency of 50 kHz.

detrapping times vary according to the range of frequencies considered. (3) The pulsed measurements, like "double-pulse," use a constant filling and detrapping time. A lower duty cycle (usually 0.1%–1%) is used. The measurement pulse duration is intentionally kept short in order not to induce significant detrapping during the pulse.

The authors in [70] validated GFS by testing gate injection transistors (GITs) submitted to off-state bias condition: Figure 12.15b reports the drain current waveforms acquired after that the steady–state is reached as a function of the applied frequency (from 10 Hz to 1 MHz). In the case under analysis, the drain current decreases with the frequency, indicating that the detrapping (recovery) process is slower than the trapping process. This effect is mitigated at high temperature (Figure 12.15a), because the temperature can be responsible for higher emission rate of the trapped state, hence favoring the detrapping process.

It is worth mentioning that the above-mentioned three techniques may lead to discrepant results since they are aimed at the investigation of different aspects concerning the trapping condition. An example is discussed in Figure 12.16.

In the case under analysis pulsed measurements (Figure 12.16b) and drain current transients (Figure 12.16a) may enhance the trapping effects with respect to the GFS technique. The discrepancy observed can be mainly ascribed to the different filling (trapping) times used (see Figure 12.14).

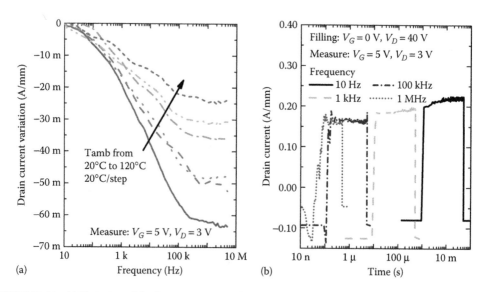

FIGURE 12.15 (a) Variation of the drain current as a function of frequency, acquired with multiple chuck temperatures. (b) Example of drain current waveforms measured during the phase II at several frequencies applied.

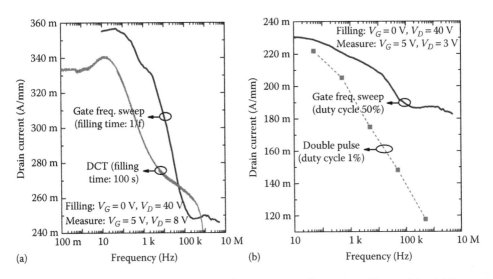

FIGURE 12.16 Comparison among three different techniques evaluated in consistent bias condition. (a) Comparison between the drain current dynamic variation evaluated with the gate frequency sweep and the drain current transient analysis. (b) Comparison between the drain current dynamic decrease measured with the double pulse and the gate frequency sweep method.

Figure 12.16b shows the comparison of the drain current decrease (due to trap states) measured with pulsed measurements and GFS. With the aim of reproducing the same conditions used during GFS the pulsed measurements were performed at different pulse widths and constant duty cycles. The equivalent frequency of the pulsed measurement has been chosen as half the reciprocal of the pulse width. Due to the different duty cycles used, during GFS, the device is kept in a trapping condition for a shorter time leading to less severe trapping conditions.

Analogous considerations can be done for the DCT (Figure 12.16a). During the DCT the recovery is monitored for a time ranging from 1 μs to 100 s; in order to maintain a duty cycle of 50%, the device is biased continuously with an equal filling and recovery time (100 s). The longer trapping time considered during the DCT, therefore, promotes the enhancement of the trapping effects.

12.2.3 Charge Trapping Location

In order to locate the deep levels responsible for current collapse within the HEMT structure and to investigate the behavior of the device under different bias conditions, double-pulse measurements and DLTS can be performed by employing different quiescent bias points.

- *Off-state*: In the off-state, the access region of the device is exposed to lateral and vertical electric field, hence promoting possible charge trapping either in the epitaxial layers and/or at the surface.
- *Back-gating*: In back-gating, the entire device is exposed to the vertical field between the substrate and the top contacts (gate, drain, source), hence promoting possible charge-trapping in the channel and buffer layers; since no lateral field is applied, surface trapping should be negligible.
- *Semi-on-state*: In the semi-on-state, the combination of high drain voltage and relatively high drain current promotes hot electrons which can overcome the 2DEG confinement, being injected and trapped into epitaxial defect-state or even on the device surface or within passivation.
- *On-state*: In the on-state regimes, with positively biased gate, high electric-field affect mostly the gate-stack, hence promoting charge-trapping in the AlGaN barrier or gate dielectric layers (in the case of MIS-gated HEMTs) leading to metastable V_{TH} instabilities.

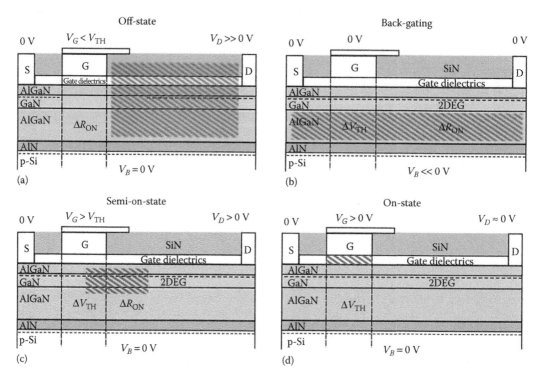

FIGURE 12.17 (a) In the off-state, the access region of the device is exposed to lateral and vertical electric field, hence promoting possible charge trapping either in the epitaxial layers and/or at the surface. (b) In back-gating, the entire device is exposed to vertical field between the substrate and the top contacts (gate, drain, source), hence promoting possible charge-trapping in the channel and buffer layers; since no lateral field is applied, surface trapping should be negligible. (c) In the semi-on-state, the combination of high drain voltage and relatively high drain current promotes hot-electrons which can overcome the 2DEG confinement, being injected and trapped into epitaxial defect-state. (d) In the on-state regimes, with positively biased gate, high electric-field affect mostly the gate-stack, hence promoting charge-trapping in the AlGaN barrier or gate dielectric layers (in case of MIS-gated HEMTs) leading to metastable V_{TH} instabilities.

In the following, we report case studies demonstrating how the characterization technique reported in this chapter has been useful to identify charge-trapping issues related to carbon doping, hot electrons, and gate dielectrics (Figure 12.17).

12.2.4 Trapping in the GaN Buffer: The Role of Carbon

Carbon is an effective impurity used to compensate the unintentional doping of III-nitrides materials, achieving highly insulating buffer layers for GaN-based electronic devices.

In MOCVD, the carbon impurity commonly exists in the materials due to the use of metal-organic precursors. The incorporation of carbon in the MOCVD III-N layers can be effectively controlled by tuning the growth conditions, including pressure, temperature, III/V ratio, growth rate, and Al content [71–76]. With no segregation or memory effect in the growth system and the epitaxial stack, carbon allows a precise control of the doping profile and a precise control of GaN film resistivity [73].

In GaN HEMTs, the use of carbon-doped buffer allows the reduction of lateral and vertical buffer leakage current [76–79], the increase of the breakdown voltage [74,78,80], and the suppression of the short-channel effects [20]. As a drawback, the deep levels introduced by carbon promote charge-trapping, current collapse, and the degradation of dynamic performance (degradation of the transconductance (g_m) and of the on-resistance (R_{ON})) [7–10,81,82].

Deep levels ascribed to carbon doping have been predicted by first principle calculations and experimentally observed by means of DLTS techniques, both in diode and transistor structures.

Carbon in GaN has an amphoteric nature [83,84]. Depending on the Fermi level, carbon can be incorporated either in nitrogen substitutional position (C_N), behaving as acceptor-state, in gallium substitutional (C_{Ga}), or interstitial position (C_i), behaving as donor-state. Two scenarios have been proposed to explain the semi-insulating behavior of carbon-doped GaN. The first is the self-compensation of acceptor species (C_N) and donor species (e.g., C_{Ga}, or C_i), incorporating in GaN with similar formation energies [83] (Figure 12.18a). The second, proposed more recently, report that carbon in nitrogen substitutional position (C_N) could behave as deep acceptor with the (0/−) transition level at $E_V + 0.9$ eV [85,84] (Figure 12.18b); if incorporated with sufficient concentration, carbon would compensate the unintentionally n-type doping, pin the Fermi level at $E_V + 0.9$ eV, and achieve semi-insulating behavior.

Both donor-like and acceptor-like deep levels associated with carbon impurities have been detected in experimental research works, with DLOS and DLTS. Deep levels have been reported at $E_C - 0.11$, 1.35, 2.05, 3.0, and 3.28 eV and $E_V + 0.86$, 0.9 eV [86–90]. The detection of deep levels at $E_V + 0.9$ eV provides experimental evidence of the possible existence of the deep-acceptor-state C_N at $E_V + 0.9$ eV theorized in [85].

In AlGaN/GaN HEMTs, deep-level signatures with apparent activation energies ranging from 0.71 eV to 0.94 eV (Arrhenius plot in Figure 12.19) were identified to be the primary cause of current collapse and dynamic R_{ON}-increase [7–11]. Rossetto et al. [8] and Huber et al. [9] explicitly demonstrated increasing trap density with increasing carbon-doping concentration.

From the experiment published in [91], time-resolved stress/recovery R_{ON} transients were acquired by means of sampled transient measurements. Figure 12.20 depicts the stress/recovery transients acquired when applying an off-state stress bias at $(V_G;V_{DS};V_B) = (-8 \text{ V};25 \text{ V};0 \text{ V})$. It can be noticed that temperature influences both the rate of the R_{ON}-recovery and the rate of the R_{ON}-increase. The activation energy of R_{ON}-increase and R_{ON}-recovery are 0.90 eV and 0.95 eV, respectively. This discovery is at the base of the positive temperature dependent current collapse observed in [91].

To confirm that the observed current collapse was related to charge trapping in the buffer and not at the surface, auxiliary back-gating measurements were performed. During the stress phase, a negative potential was applied to the substrate terminal ($V_B = -25$ V) and no potential difference was applied between source, gate and drain terminals ($V_G = V_D = V_S = 0$ V). Under back-gating, surface trapping is supposed to be negligible because no bias is applied between gate and drain terminals, and the formed 2DEG would screen the superficial layers from the field-effect induced by back–gating.

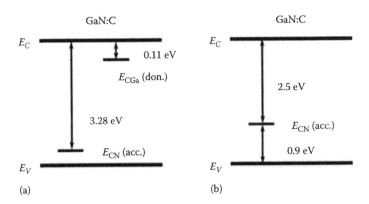

FIGURE 12.18 Schematic representation of the energy levels assumed in the C-doped GaN layer according to the (a) self-compensation model (From Wright, A. F., *J. Appl. Phys.*, 92, 2575, 2002. With permission.) and (b) the dominant deep-acceptor model (From Lyons, J. L. et al., *Phys. Rev. B*, 89, 035204, 2014. With permission.)

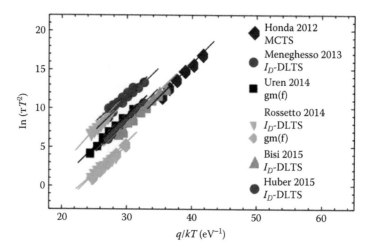

FIGURE 12.19 Deep-level signature with apparent activation energies ranging from 0.71 eV to 0.94 eV identified to be the primary cause of current-collapse and dynamic R_{ON}-increase in high electron mobility transistors equipped with carbon-doped buffer.

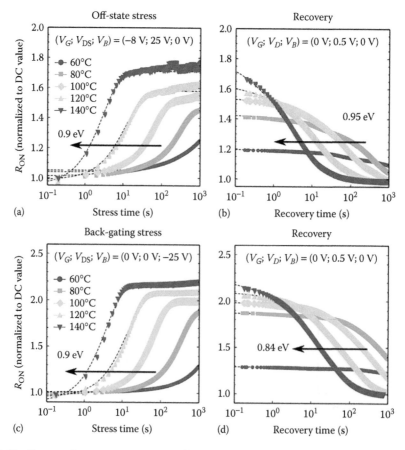

FIGURE 12.20 Stress and recovery transients performed in (a–b) Off-state and (c–d) back-gating stress. Both charge-trapping and charge-detrapping are thermally activated. Off-state-stress and backgating-stress lead to similar stress/recovery transients, suggesting unique charge-trapping mechanism localized in the buffer.

Results reveal that back-gating stress leads to similar stress/recovery transients (Figure 12.20c and d), with similar kinetics, similar timings, and similar thermal activation than those acquired during off-state stress (Figure 12.20a and b). This indicates that the observed charge-trapping process is located in the buffer region and is the dominant cause promoting current dispersion during off-state regimes.

As experimentally reported in [92] and [93] the dynamic R_{ON}-increase is proportional to the gate-drain length (L_{GD}). This suggests that the charge-trapping leading to the dynamic R_{ON}-increase related to buffer trapping is not located at the gate edge, but it is distributed over the entire gate-drain access region.

The R_{ON}-increase ascribed to carbon-related deep levels could be explained by two different mechanisms. The first one considers the involvement of electron traps, and electron injection into traps caused by leakage current [23,94,95].

The second hypothesis considers the involvement of deep acceptors and their slow ionization and neutralization processes [96,97]. During high-voltage off-state conditions, the gate-drain access region is exposed to high drain potential, and the depletion width would extend within the GaN buffer. The slow response of the deep acceptors to change their charge state and the low conductivity of the buffer would lead to strong delays to reach the electrostatic steady-state during on-off-on transitions. The strong temperature dependence of the R_{ON}-increase would account for the ionization of the deep-acceptor state (0.9 eV) by the thermal emission of trapped holes, whereas the strong temperature dependence of the R_{ON}-recovery would account for the availability of free holes required to neutralize the deep acceptors and restore the original charge balance.

The solutions proposed to prevent and mitigate the carbon-related dynamic R_{ON} increase involve the optimization of both the epitaxial structure and of the device architecture. Carbon-related dynamic R_{ON} increase could be suppressed by optimizing the carbon concentration, by using a u.i.d. GaN channel layer, by increasing the vertical spacing between the carbon-doped buffer and the 2DEG [74], and by implementing a superlattice buffer [98].

A different approach is to promote the introduction of free holes in the buffer to avoid the ionization of deep-acceptor states. This solution can be implemented (1) by ensuring vertical current paths for leakage current [11], (2) by using an additional p-GaN region beside the drain electrode [99], and (3) by using a photonic-ohmic drain and the photon pumping of deep traps [100].

12.2.5 Hot-Electron Induced Trapping

The OFF/ON switching approach is a key aspect of the design of efficient power converters. Depending on the system specifications (switching loss, frequency, electromagnetic interference, cost, and complexity) either hard- or resonant soft-switching architectures may be preferred [101]. For a successful integration of GaN electronics power systems, the implications of OFF/ON switching must be meticulously investigated not only at the circuit level [102] but also at the device level [103]. In fact, during the OFF/ON transitions, the hard-switching load-line trajectories expose the transistor to critical current voltage conditions, with the combination of very high voltage and relatively high current. This may lead to the parasitic generation of highly energetic electrons (hot electrons) which can promote not only long-term reliability issues [104,105], but also the instantaneous drop of dynamic performance, caused by enhanced charge-trapping effects [106].

By means of numerical simulations, Braga et al. demonstrated that with moderated drain voltages (slightly higher than $V_{DS, SAT}$) an appreciable density of electrons gain enough energy to overcome the 2DEG confinement, to be deflected and getting trapped at crystallographic defect states [107].

Experimental results have been reported by Wang et al. [108] and Meneghini et al. [109]. In [109], the effects of hot electrons on the device performance have been studied by means of pulsed I_D–V_D and I_D–V_G curves acquired by exploring semi-on-state quiescent bias points. As can be noticed from data in Figure 12.21, in addition to the R_{ON}-increase promoted by high drain voltage in the off-state, a further worsening in the R_{ON} is detected in the semi-on-state, when $V_{G, Q} > V_{TH}$ ($I_{DS} > 1$ A/mm).

Coherently with the proposed SEMI-ON charge-trapping mechanism, a 400 mV positive V_{TH} shift has been observed in pulsed I_D–V_G characteristics (not shown here), indicating that electrons, accelerated by

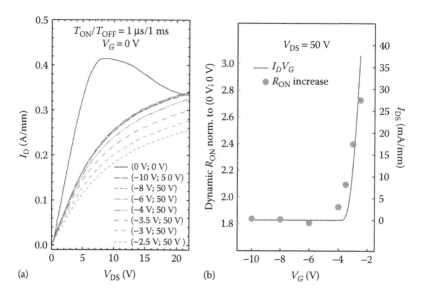

FIGURE 12.21 (a) Pulsed I_D–V_D characteristics acquired with constant quiescent V_{DS}, and multiple quiescent V_G. Dynamic R_{ON} increase worsen as $V_G > V_{TH}$. (b) Good correlation between I_{DS} and dynamic R_{ON} increase suggesting the influence of hot-electrons-related trapping mechanism. (Reprinted with permission from Meneghini, M. et al., *Appl. Phys. Lett.*, 104, 143505, 2014. Copyright 2014, American Institute of Physics.)

the electric field in the proximity of the gate-edge, get trapped not only in the gate-drain access region but also underneath the gate contact.

To identify the trap states responsible for hot electrons-related charge-trapping mechanisms, DCTS was performed by employing both off-state and semi-on-state stress.

From results depicted in Figure 12.22, it can be noticed that in the semi-on-state an additional trap signal (labeled E3) emerges, indicating the presence of an additional trap mechanism. By performing DCTS at multiple temperatures (Figure 12.22c), the activation energy of the different detrapping processes have been extrapolated. The apparent activation energies and capture cross sections of trap state E3 and E6 are 0.60 eV and 1.2×10^{-17} cm², and 0.96 eV and 5.8×10^{-15} cm², respectively (Figure 12.22d). The analysis demonstrates that hot electrons can access trap sites which cannot be reached by gate-injected carriers in reverse-bias conditions.

12.2.6 Trapping-Induced Drain-Source Time Dependent Breakdown

Charge-trapping can be responsible not only for the collapse of the dynamic properties of the device but also for reliability issues, including time-dependent degradation and time-dependent breakdown. Time-dependent drain-source breakdown induced by trapping was experimentally demonstrated in GaN-based HEMTs [110,111]. Constant voltage stress in the off-state showed a time-dependent increase of the subthreshold drain-source leakage current, which did not affect the gate leakage current (Figure 12.23). This phenomenon is also observed by I-V sweep measurements of breakdown carried out in the current-controlled mode with a gate voltage close to the threshold voltage, as a consequence of the increase of a punch-through effect.

At the beginning of the stress, the main contribution of the drain current is provided by the gate-drain leakage due to surface conduction or to the leakage of the reverse-biased Schottky junction. For longer stress times, source current is found to change its sign from positive to negative, indicating that the current is flowing out of the source. The drain-to-source current (and the noise to it superimposed) is furthermore found to gradually increase with the stress time, as a consequence of the generation/accumulation

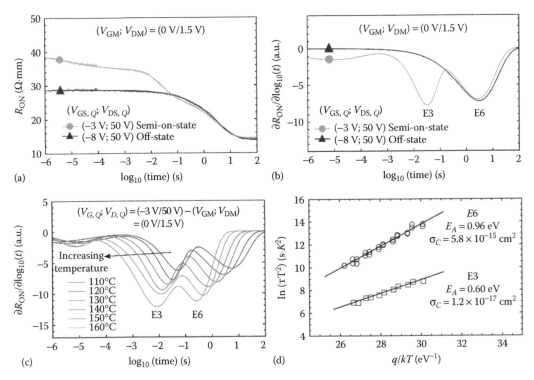

FIGURE 12.22 (a) Time-resolved R_{ON}-recovery transients and (b) related derivative employed for deep-level spectroscopy acquired after 100s off-state and semi-on-state stress. Coherently with pulsed measurements in Figure 12.4, an additional trap signal (labelled E3) appears after semi-on-stress, evidence of hot-electrons-related trapping mechanism. (c) and (d) Thermal activation signatures of trap-state E3 and E6. (Reprinted with permission from Meneghini, M. et al., *Appl. Phys. Lett.*, 104, 143505, 2014. Copyright 2014, American Institute of Physics.)

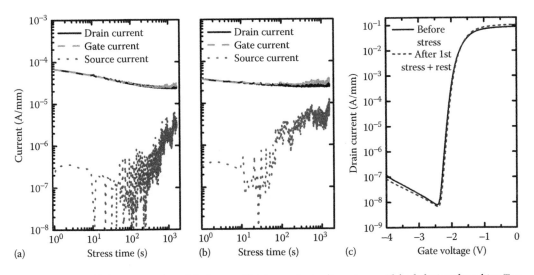

FIGURE 12.23 (a) Current monitored during an off-state constant voltage stress with high drain voltage bias. Test indicates (a) a recoverable mechanism for short stress times and (b) permanent degradation at long stress times. (c) The recovery after 1h of rest is demonstrated.

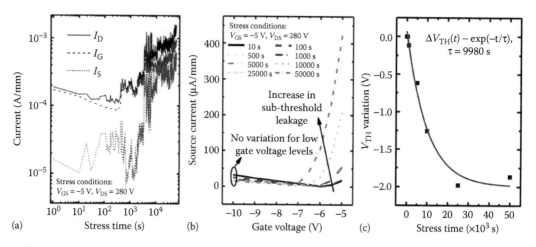

FIGURE 12.24 (a) Current monitored during an off-state constant voltage stress with high drain voltage bias, (b) source current versus gate voltage (I_S–V_G) curves monitored at defined steps during the stress, (c) corresponding threshold voltage shift induced by the accumulation of positive charge under the gate. (From Meneghini, M., et al., *IEEE Trans. Electron Devices*, 61, 1987, 2014. © 2014 IEEE. With permission.)

of positive charge in the region under the gate due to the high electric field. This assumption is confirmed by the increase of the source current (I_S–V_G curve) measured in the sub-threshold region at defined steps during the stress (Figure 12.24) and by the correlation between the measured increase of the I_S–V_G curve and the leftward shift of the pinch-off voltage detected (Figure 12.24). The positive gate charge may originate from holes for impact ionization and/or band transition processes promoted by defect states; the possible recovery of the phenomenon for short stress times suggests that the generation of positive charge is not due to donor defects at the AlGaN/GaN interface.

Emission microscopy clarifies that HEMTs submitted to off-state stress emit a weak electroluminescence in the proximity of localized regions with a nonhomogeneous intensity. The dependence between the emission microscopy and the drain current indicates that the increased (source) current flows through preferential leakage paths and with a nonuniform behavior along the gate (Figure 12.25).

FIGURE 12.25 (a) False-color emission microscopy detected and (b) corresponding damage detected after the failure. (From Meneghini, M. et al., *IEEE Trans. Electron Devices*, 61, 1987, 2014. © 2014 IEEE. With permission.)

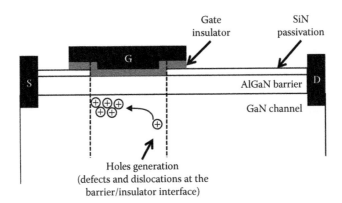

FIGURE 12.26 Representative schematic depicting the hole generation mechanism which contributes to the trapping-induced drain-to-source breakdown.

In the first stage of the stress, the above-discussed phenomenon is recoverable and induces a self-limited shift of the threshold voltage. For longer stress times, the increase of source-drain current can lead to a permanent degradation, as a consequence of the thermal runaway, in correspondence of preferential leakage paths, and/or to the achievement of a critical density of positive charge, that can locally increase the electric field and contribute to the failure. It is worth noticing that no variation of the gate leakage current is observed even after the failure.

Analogous considerations were demonstrated by Bahl et al in MISHEMT structures [112]. The authors of this paper demonstrate that at high drain voltage levels, a time-dependent breakdown may occur on the source-side of the gate as a consequence of hole injection (at high V_{DS}) in the gate insulator (Figure 12.26). The hole injection is mainly ascribed to defects at the AlGaN/Al$_2$O$_3$ interface; the impact ionization is suggested not to play a significant role due to the increase of the degradation with the temperature. Optical beam induced resistance change (OBIRCH) analysis confirms the hypothesis. The degraded devices, conversely to the fresh ones, demonstrate failure spots which are similar to the ones previously discussed in the analysis of the stress-induced leakage current (SILC) phenomenon due to the dielectric wear-out [113]. It is worth noticing that, since device failure occurs in the gate insulator, the breakdown voltage depends on the gate dielectric thickness and that the time to failure follows a Weibull distribution, consistently with the percolation theory subsequently described in Section 12.3.2.

12.2.7 Charge-Injection Mechanisms: Trap Effects in p-Gate E-Mode Devices

The application of the GaN devices to power conversion imposes several requirements which can be summarized as: (1) e-mode operation for safety reasons, (2) low on-resistance values, (3) a high gate bias swing, (4) low leakage current values, and (5) high robustness toward both the OFF- and the ON-states. With respect to this performance, excellent characteristics were demonstrated by p-gate GaN-based HEMTs employing a p-GaN or a p-AlGaN gate layer. E-mode switching transistors with $\approx$70 mΩ/600 V were demonstrated with both the structures [99,114].

An initial explanation of the operating principles in these structures was discussed in detail by Uemoto et al. [115,116]. The authors demonstrated a device (called GIT) with the normally off operation and high current, which can withstand high positive gate voltage levels and low gate current.

In GITs the e-mode operation is enabled by the presence of a p-AlGaN layer between the i-AlGaN barrier layer and the gate contact, which lifts up the potential at the channel, as shown in the band diagram (Figure 12.27). The basic operation of the device can be described by three main operation modes: (1) at a null gate

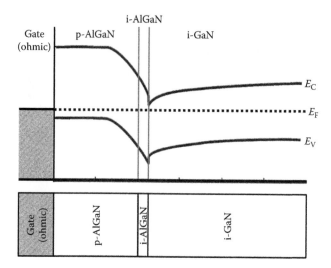

FIGURE 12.27 Band diagram of a gate-injection transistor at null gate bias level.

voltage the conduction band is beyond the Fermi level, leading to a fully depleted channel; (2) at gate voltages lower than the built-in voltage of the p-n junction the device works as a field-effect transistors; (3) high gate bias levels results in the hole injection from the p-AlGaN layer to the channel. These latter, in order to maintain the neutrality at the channel, allow the accumulation of an equal number of electrons, which flow to the drain due to the higher mobility, thus leading to a "conductivity modulation" which results in an increase of the drain current and a low gate leakage at corresponding bias levels. It is worth noticing that the electron flow from the channel to the gate contact is suppressed by the AlGaN/GaN hetero-junction. The "conduction modulation" mechanism and the consequent increase of the drain current are confirmed by the presence of a second peak in the transconductance, whose corresponding light emission faces a peak at the optical band gap of the GaN (365 nm) [117].

The use of a Mg-doped GaN layer has been proposed as an alternative to the p-AlGaN layer [118,119] demonstrating analogous performance in terms of the threshold voltage and V_{br} versus R_{ON} behavior.

12.2.7.1 Surface and Bulk Traps in GIT Devices

According to the requirements imposed by the application in the power conversion field, the GaN-based devices with a p-type gate must fulfill the achievement of a low on-state resistance, which must be guaranteed even after the application of very high drain voltages in the off-state or, dynamically, during switching. The reduction of dynamic losses is indeed a key issue for efficient power switching devices. Preliminary experimental [116] investigations demonstrated that the p-gate e-mode devices can achieve significant suppression of surface trap effects without the use of field plates and/or of improvements in the geometry. Negligible variation in the dynamic on-state resistance was initially shown up to a drain voltage of 60 V [116].

Although several solutions have been proposed so far to reduce the trapping effects (e.g., improvement of the surface and/or passivation, use of field plates structures), the dynamic increase of the on-resistance as a consequence of parasitic effects still represents an important technical issue for GaN-based devices aimed at high operating voltages ($V_D > 600$ V), and it has been the subject of fruitful discussion in the literature.

The increase of the dynamic on-resistance is due to a temporary charge trapping induced by the switching from the off-state to the on-state. Under this condition a negative charge trapping may take place, especially on the drain-side edge of the gate corresponding to the electric field peak, leading to a slowdown of the electron transport in the channel region and, thus, to a significant reduction of the conduction mechanisms and of the system efficiency. Trapping effects seem to be mainly induced by the deep level states located in the buffer, although in principle also traps in the passivation and/or in the AlGaN barrier layer may contribute. Since the charging/discharging process from the deep levels does not occur immediately after the switching, returning to the charge equilibrium after the switching requires a non-negligible amount of time, depending on the properties of the deep levels and their energetic position [120].

Therefore, in the technology evolution toward high-voltage switching devices, the design of the buffer has to be optimized with respect to the lateral/vertical breakdown and the dynamic switching properties. A trade-off is indeed established between the high-voltage robustness of the buffer and the dynamic on-resistance increase due to the parasitic effects. Würfl et al. showed that a good trade-off in the performance can be obtained by incorporating Fe in the buffer or by combining C-doped buffer structures with AlGaN back-barrier designs [121].

The trade-off between the breakdown voltage and the dynamic on-resistance is depicted in Figure 12.28 [120]. A buffer carbon doping compensation provides a strong increase of the gate-drain robustness, although it contributes to a significant increase of the parasitic effects in terms of the on-resistance dynamic change. The opposite condition is obtained by means of the use of Fe doping compensation or an AlGaN back-barrier structure. The lower dynamic increase of the on-resistance is counterbalanced by the low breakdown strength value reached (40–50 V/μm). An interesting solution was developed by Würfl et al. [120], on the basis of device electric field simulations. The suggested structure is characterized by an AlGaN barrier, placed close to the channel, and a carbon-doping buffer compensation in the portion of the buffer layer defined by a low change of the electric field during switching condition. This combination led to a breakdown strength of 80 V/μm and a R_{on}_dynamic increase of only 10% at 65 V [120].

The above-mentioned technological approaches were furthermore discussed by studying the improvements on the switching transients induced by the use of a back-barrier structure. The devices under test were submitted to an off-state voltage equal to 500 V [120]. The dynamic increase of the on-resistance during switching transients was discussed for devices with the AlGaN back-barrier and a medium-scale carbon-doped buffer (2×10^{18}/cm³), which demonstrate an increase limited to a factor of 2.5, and for

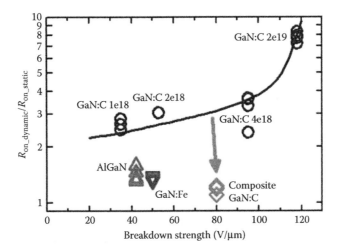

FIGURE 12.28 Trade-off between dynamic on-state resistance increase and device breakdown strength in dependence on different technologies tests. (From Würfl, J. et al., *IEEE International Electron Devices Meeting*, 6.1.1–6.1.4, 2013. © 2013 IEEE. With permission.)

devices with no AlGaN back-barrier and a highly carbon-doped GaN buffer layer (2×10^{19}/cm^3). These latter demonstrate an increase equal to a factor of 880 immediately after turning on the device and of 220 after 10 μs.

State-of-the-art devices show that a proper design of the buffer can lead to an on-resistance dynamic variation lower than a factor of 2.5 at $V_D = 650$ V [114]. In the devices under analysis, the nonnegligible influence of the substrate used was furthermore discussed. Switching transients of GaN-on-SiC test transistors demonstrate a monotonic increase of the dynamic on-resistance with the drain voltage level applied, reaching a variation lower than 2 at $V_D = 500$ V. Switching transients of GaN-on-Si test transistors showed similar performance at $V_D = 500$ V, although a nonmonotonic behavior of the dynamic on-resistance with the drain voltage applied is observed, facing a maximum increase higher than 2 at $V_D = 300$ V.

12.2.7.2 Time- and Field-Dependent Trapping in GIT

Previous analyses reported in the literature suggest that, even if GIT transistors do not suffer from trapping effects when exposed to high drain voltage levels, continuous exposure to high drain voltage may induce a significant increase in the on-resistance. This aspect is of fundamental importance if we consider the high drain voltage levels applied at the operating condition and must be carefully considered. Evidence of time- and field-dependent trapping mechanisms in GIT transistors was investigated in detail by Meneghini et al. [187] The authors demonstrate, by means of combined electrical and optical investigation, that long-term exposure to high drain bias in on-state condition may induce to a significant and recoverable increase of the on-resistance Figure 12.29. It is worth noticing that experimental results indicated that the above-mentioned trapping effects are promoted only by the operation in the on-state condition (Figure 12.30).

The authors of the paper suggest that the (recoverable) increase of the on-resistance is ascribed to charge trapping in the gate-drain region. This hypothesis is supported by emission microscopy analysis. When submitted to high drain bias the electroluminescence (EL) peak shifts toward the drain contact, suggesting that the profile of the electric field changes due to the accumulation of negative charges in the gate-drain region. The increase of negative charges in the gate-drain region contributes to the decrease of the electric field (and thus of the EL signal) at the gate edge and to the increase of the drain EL signal.

The on-resistance variation is fully recoverable with an exponential dependence on time. The recovery kinetics can be furthermore accelerated by the ambient temperature with and activation energy of 0.47 eV.

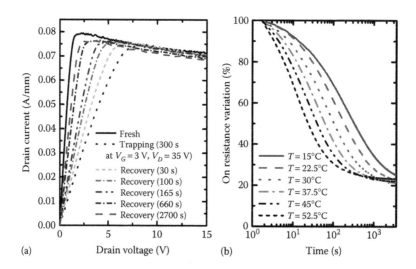

FIGURE 12.29 (a) DC current-voltage curves measured on a fresh device (solid line), after 300 s trapping period (dashed line) and subsequent recovery periods. (b) Recovery kinetics evaluated during the recovery phase at different ambient temperature levels.

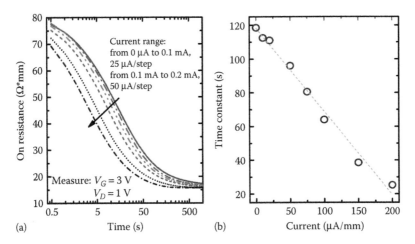

FIGURE 12.30 Variation (a) of the on-resistance recovery and (b) of the time constants with different levels of gate current injected.

Figure 12.30 demonstrates that the recovery can be significantly enhanced by the injection of holes from the p-type gate layer. This aspect is confirmed by figure 12.30, which shows that the recovery kinetics can be accelerated by injecting a different (higher) hole current level from the gate, which demonstrates the dependence of the recovery time constants on the injected gate current. The contribution of the injected holes to the detrapping kinetics can be explained as follows: the injection of holes may recall the trapped electrons to the gate region, promoting the recovery process.

12.2.7.3 Hybrid Drain-Embedded GIT

A different and innovative approach to suppress the current collapse and the parasitic effects at high drain voltage levels was discussed by Kaneko et al. [99]. The authors suggest a novel structure to improve the device performance with no change in the design of the buffer layer and/or buffer doping compensation. The current collapse at high drain voltage levels in a GaN-based transistor is reduced by means of the introduction of a p-GaN layer below the drain of a GaN-based device with a p-GaN gate. The proposed device, also called hybrid drain-embedded GIT—HD-GIT—demonstrated no current collapse and stable system operation up to a drain voltage of 850 V.

The Figure 12.31 shows a schematic cross section of a conventional GIT and a HD-GIT. The main differences consist of the introduction of a p-GaN layer under the drain contact and the recess of the gate structure. At high drain voltage levels during off-state operation, the holes are injected from the drain terminal to the i-AlGaN and to the i-GaN layer. The release of trapped electrons during the switching, consequent to the holes injection, significantly contribute to the suppression of the current collapse (and thus, increase of the dynamic on-resistance).

The authors demonstrated that in the case under analysis [99] the HD-GIT has a similar robustness than the conventional GIT. Nevertheless, the HD-GIT is found to be current collapse free up to 850 V, while in a conventional GIT the dynamic on-resistance increases for drain voltages higher than 600 V. The HD-GIT demonstrates stable conditions even after switching operation of 300 hours in terms of the threshold voltage, on-state resistance, and off-state leakage current. Significant reliability toward reverse bias high temperature test was also demonstrated.

An improvement to this technology was discussed by Okita et al. [122]. In this paper, the authors demonstrate a new technology, called "through recessed and regrowth gate." According to this process, the AlGaN barrier is initially fully removed by the gate region and epitaxially regrowth. This solution significantly improves the standard deviation of the measured threshold voltage and static on-resistance value, without inducing any variation in the dynamic and/or fast switching properties.

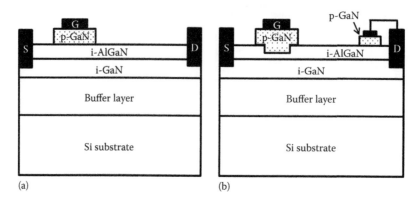

FIGURE 12.31   (a) Cross section of a conventional GIT and (b) of the HD-GIT structure.

A noticeable enhancement of the switching properties is thus achieved without compromising the attenuation of the trapping effects, already achieved and demonstrated in the HD-GIT, up to a drain voltage of 850 V.

12.3 Stability of Dielectrics in GaN MISHEMTs and HEMTs

The requirements imposed by the applications in power switching converters already described in Section 12.2.7 can be fulfilled by the use of a metal-insulator-semiconductor high-electron-mobility transistor (MISHEMT) structure. The insertion of a dielectric layer under the gate contact can reduce the leakage current and increase both the off- and on-state breakdown voltage, allowing (with the use of a proper recess) the implementation of normally off devices [123,124].

Nevertheless, the insertion of a dielectric under the gate may enable the worsening of the device performance, especially in terms of trapping effects in the dielectric or at the dielectric/III-Nitride interface, and of the robustness toward forward gate bias. Indeed, the MISHEMT structures, especially when submitted to a forward gate bias, are found to be extremely sensitive to time-dependent degradation mechanisms and susceptible of TDDB. The above-mentioned aspects, which are of fundamental importance consistently to the final application of the devices, are described in detail in Sections 12.3.1 and 12.3.2. In Section 12.3.2 the TDDB is further developed, in order to investigate the origin of the failure and to predict the time to failure of other GaN-based devices such as HEMTs with a p-GaN gate and d-mode AlGaN/GaN HEMTs submitted to the additional off-state condition.

12.3.1 Bias Temperature Threshold Voltage Instability in MISHEMTs

The insertion of a thin dielectric under the gate in MISHEMT structure is found to significantly improve the performance of the devices. However, the performance reached in terms of robustness and parasitic current is limited by the significant increase of trapping effects, mainly originating from trap states located in the dielectric or at the dielectric/III-Nitride interface. The presence of charge trapping under the gate may induce detrimental instabilities in the threshold voltage (V_{TH}). This worsening of the performance is accelerated at higher ambient temperatures and significantly enhanced by the application of a positive/negative voltage, leading to the so-called "positive and negative bias temperature instabilities" (PBTI and NBTI, respectively).

This issue can be limited by the use of several dielectrics and/or different deposition methods in order to improve the intrinsic defectiveness of the dielectric layer and/or of the dielectric/AlGaN interface [16,125,126].

Table 12.1 reports a summary of bias temperature instabilities discussed in the literature. A detailed explanation is provided in Sections 12.3.1.1 through 12.3.1.3.

TABLE 12.1 Summary of the Main Discussed Threshold Voltage Instabilities

Authors	Typology of Instability	Dielectric Used	Summary of the Discussed Results
Lagger et al. [125]	PBTI	• SiN, SiO$_2$, HfO$_2$, Al$_2$O$_3$, HfSiO$_2$. • Several barrier and dielectric thickness	• The threshold voltage instability is promoted by a high gate bias, which enables the creation of a second channel at the dielectric interface and the trapping of electrons in the dielectric or in the dielectric/AlGaN interface. • Dependence of the V_{TH} shift on the dielectric used and on the thickness of the barrier and dielectric layer.
Lagger et al. [127]	PBTI	• ALD Al$_2$O$_3$	• V_{TH} shift with negligible change in the transconductance peak. • Time constants follow a broad distribution with no dependence on temperature.
Lagger et al. [128]	PBTI	• SiO$_2$	• The broad distribution of the time constants is described and explained considering several hypotheses.
Mizue et al. [129]	PBTI, NBTI	• ALD Al$_2$O$_3$	• The shape of the C-V curve is influenced by the accumulation of electrons in the GaN/AlGaN and in the AlGaN/dielectric interface. • The density of states at the interface influences the C-V slope at high V_G values. • Negligible NBTI at room temperature.
Wu et al. [18]	PBTI	• e-mode MISHEMTs • ALD Al$_2$O$_3$, PEALD SiN	• Power low dependence of V_{TH} shift on stress time. • A defect band model is proposed. • The narrower/wide distribution of the defect levels and their position with respect to the conduction band influences the instability and the corresponding accessibility of the trap states.
Rossetto et al.[126]	PBTI	• PEALD SiN	• Correlation between the V_{TH} shift and the injected forward gate current.
Meneghesso et al. [16]	PBTI	• PEALD SiN, RTCVD SiN, ALD Al$_2$O$_3$	• Correlation between the V_{TH} shift and the injected forward gate current. • Dependence of the V_{TH} shift on the dielectric and deposition used.
Wu et al. [130]	PBTI	• PEALD SiN, RTCVD SiN,	• PEALD deposition reduces the V_{TH} shift.
Bisi et al. [131]	PBTI	• *In situ* Al$_2$O$_3$ MOSHEMTs	• The trap conduction mechanism depends on the field regime. • Low-field: Gate-current variation and the flat-band voltage shift are ascribed to the charge/discharge of oxide-traps at the interface (tunneling mechanism). • High-field: increase of the parasitic leakage current and of the flat-band shift due to an additional transport mechanism, presumably due to traps far from the interface.
Meneghini et al. [132]	NBTI	• PEALD SiN	• High temperature induces significant V_{TH} shift even at low voltage with an activation energy of 0.37 eV. • V_{TH} shift recovers showing a log trend with time and a non-negligible dependence on temperature. • Correlation between R_{ON} and V_{TH} variation. • The negative bias enables the trap states depletion thermally, through tunneling and/or defect-assisted conduction.
Guo et al. [133]	NBTI	• SiO$_2$/Al$_2$O$_3$ MOSHEMTs	• Correlation between V_{TH} shift and subthreshold swing. • At low stress regime the V_{TH} recoverable shift is due to electron detrapping from pre-existing traps. At mid-stress regime the V_{TH} shift is due to field-induced electron trapping. At high-stress regime a nonrecoverable V_{TH} shift is observed.

12.3.1.1 Positive Bias Temperature Instability

Positive bias temperature instability was extensively studied in GaN-based MISHEMTs. Several approaches were described in order to investigate the origin of the instability and to limit the effects on the device performance.

A detailed investigation of the mechanisms which contribute to the instability was presented by Lagger et al. [125,128]. The dependence of the threshold voltage instability on the gate bias applied is explained by distinguishing among three different regimes, schematically shown in Figure 12.32. According to the band diagram at thermal equilibrium (a) there is no electron flow between the channel and the interface. (b) At positive (low) gate bias levels, there may be electron flow via the barrier toward the III-N interface, consistently with the change of the initial potential barrier. (c) The threshold voltage instability is promoted by the third regime, also called spillover regime, where a high gate bias level is applied. In this condition a second channel is created at the dielectric interface, and no change is observed in the voltage drop $(V_D–V_B)$, thus providing electrons which may get trapped in defect states located in the dielectric or the dielectric/AlGaN interface. The trapped electrons are responsible for the increase of the ΔN_{it} (namely the number of trapped electron density at the interface). Experimental results confirm that a positive gate bias induces a threshold voltage shift with no change in the transconductance peak [127]; furthermore, the drain current transient analysis establishes that the corresponding time constants follow a broad distribution with a negligible dependence on the applied temperature.

Several hypotheses for the broad distribution of the capture/emission time constants were suggested and discussed by Lagger et al. [125]: (1) broad and homogeneous distribution of individual defects at the dielectric/AlGaN interface according to Shockley–Read–Hall (2) Spatial distribution of border traps in the dielectric. The broad distribution can be due to a uniform distribution of tunneling distances to

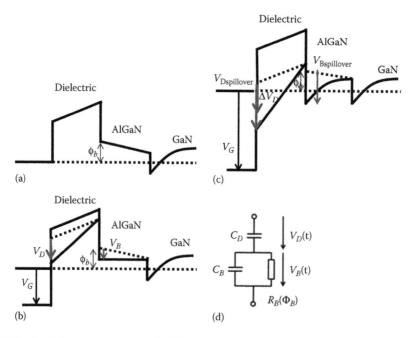

FIGURE 12.32 Band diagram variation with different positive gate bias levels. (a) At the thermal equilibrium there is no net electron flow between the channel and the interface. (b) At positive gate bias levels, according to the change of the initial potential barrier, an electron flow may be observed via the barrier toward the III-N interface. (c) At high positive gate bias levels (called spillover condition in [125]) a second channel is created at the dielectric interface. The electrons provided by the second channel may get trapped in defect states. (d) Equivalent circuit model for the gate-stack under forward gate bias conditions.

the individual dielectric defects, if we consider the contributions to the time constant in a tunneling process from the dielectric/AlGaN interface to the bulk defects in the dielectric; (3) distribution of activation energies due to the effects of the lattice relaxation effects (disordered materials used as dielectric); (4) presence of multistate defects; (5) according to the discussion provided in [128] the gate stack in forward gate bias conditions can be modelled as the equivalent circuit reported in Figure 12.32d. The effective trap time constant is therefore influenced by the barrier potential dependent time constant of the GaN/AlGaN barrier layer which acts as a rate limiter. On the basis of previous studies [134], the existence of discrete leakage paths in the barrier layer can lead to a distributed resistance and, thus, contribute to the broadening of the corresponding time constants.

The explanation is consistent with the early discussion provided by Mizue et al. [129]. By means of capacitance–voltage measurements and simulations, Mizue et al. demonstrated the following results: (1) a C-V curve with two steps is observed, which correspond to the accumulation of electrons at the AlGaN/GaN and Al_2O_3/AlGaN interfaces, respectively; (2) the simulations suggest that the C-V slope, experimentally measured with a high gate forward bias, is influenced by the density of states at the interface; and (3) device submitted to high reverse gate bias at room temperature demonstrate negligible threshold instabilities (and thus negligible electron capture mechanisms at the interface (or in the dielectric)). Experimental results are supported by the simulation of the band diagram in a consistent bias point.

An explanation concerning the origin of the PBTI mechanism and differences among different dielectrics used was provided by Wu et al. [18], by means of a detailed comparison between e-mode MISHEMT structures characterized by SiN (PEALD deposition) or Al_2O_3 (ALD deposition) dielectric under the gate. Although, according to the capacitance–voltage measurements, a higher D_{it} is calculated in the devices with a Al_2O_3 gate insulator, the SiN-based devices are severely more affected by threshold voltage instabilities. Both the materials indicated a nonlinear (power law) dependence of the threshold voltage on stress time. The results were accurately interpreted according to a defect band model assuming a uniform distribution of defects inside the gate dielectric and a thermodynamic equilibrium. According to experimental measurements, the trap levels associated to the PEALD SiN devices are defined by a wide distribution of defect levels centered below the GaN conduction band; the trap states related to the Al_2O_3 devices are due to a narrower distribution which is far from the GaN conduction band. Following the model considered, a low dependence of ΔV_{TH} on gate voltage applied corresponds to a higher distribution of defects around the Fermi level, where the electrons can be trapped even at low voltage values. Conversely, although a higher defect density is observed (e.g., in the case of Al_2O_3), a higher ΔV_{TH} dependence on gate voltage suggests a distribution of defects narrower and far from the Fermi level, resulting in defects which are much less accessible at low gate voltage levels (Figure 12.33).

Rossetto et al. [126] and Meneghesso et al. [16] demonstrated that, independently from the dielectric and/or the deposition method used, the threshold voltage instability is well correlated not only to

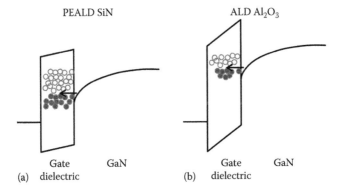

FIGURE 12.33 Defect distribution inside the gate dielectric on the basis of the model explained in [18] for (a) PEALD SiN and (b) ALD Al_2O_3.

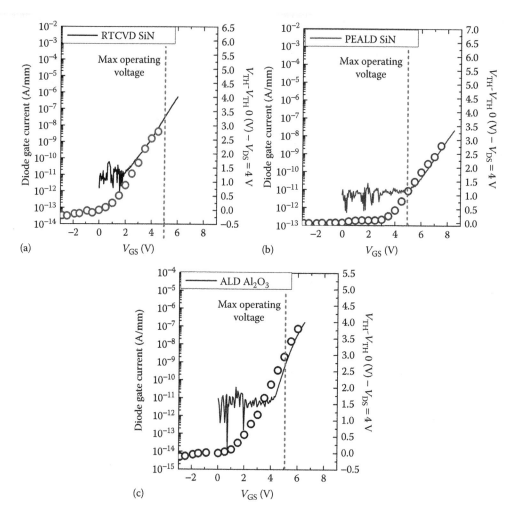

FIGURE 12.34 Correlation between the threshold voltage dynamic shift (void circles) and the gate leakage forward current (solid line) for partially recessed-gate AlGaN/GaN MISHEMTs with (a) 15-nm rapid-thermal-chemical-vapor-deposition (RTCVD) SiN, (b) 15-nm plasma-enhanced-atomic-layer-deposition (PEALD) SiN, and (c) 15-nm PEALD Al$_2$O$_3$. (Reprinted from *Microelectronics Reliab*, 58, Meneghesso, G. et al., Trapping and reliability issues in GaN-based MIS HEMTs with partially recessed gate, 151, Copyright 2015, with permission from Elsevier.)

the forward gate bias but also to the forward gate leakage current. Consistently with the origin of the trapping mechanism described by Lagger et al. [125], the threshold voltage instability is found to be promoted by the intensity of the injection of electrons in the gate insulator when a forward gate bias is applied. An example of correlation is shown in Figure 12.34.

The different trapping mechanisms induced by a low-field and a high-field regime were studied by Bisi et al. [131] in *in situ* Al$_2$O$_3$ metal–oxide–semiconductor capacitors. During injection in the low-field regime, a low gate current flows and decreases following a power-law; the gate-current variation and the flat-band voltage shift are ascribed to the charge/discharge of oxide traps at the interface due to tunneling mechanisms (Figure 12.35a and c). This hypothesis is confirmed by the power law which describes the gate current stress/relaxation kinetics. The high-field regime induces an increase of the parasitic leakage current and of the flat-band shift (Figure 12.35b and c). The corresponding enhancement of charge trapping at the interface is characterized by slow recovery transients. Under this condition, the

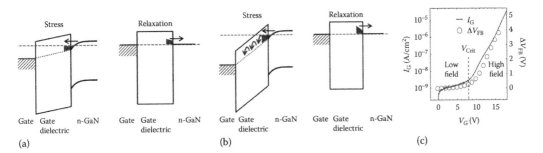

FIGURE 12.35 (a) Correlation between the gate current increase and the threshold voltage shift evaluated for low- and high- field regimes. Schematic depicting the mechanism occurring during the stress and the relaxation phase under (b) low and (c) high-field regime. (Reprinted with permission from Bisi, D. et al., *Appl. Phys. Lett.*, 108, 112104, 2016. Copyright 2016, American Institute of Physics.)

onset of additional transport mechanisms leads to a further charge-trapping effect, presumably due to traps far for the interface (in the bulk and/or in the oxide). It is worth noticing that the threshold voltage shift can be ascribed to the injection of electrons in pre-existing traps and/or to the formation of negatively charged oxide defects.

12.3.1.2 Negative Bias Temperature Instability

Mizue et al. [129] demonstrated, on the basis of the experimental capacitance–voltage measurements and of the simulated band diagrams, that the MISHEMT structures are stable and face negligible trapping effects when submitted to reverse gate bias. Nevertheless, this conclusion is based on measurements performed at room temperature and do not exclude the presence of threshold voltage instabilities when a higher temperature is applied.

The stability of the threshold voltage in GaN-based MIS HEMTs submitted to reverse bias was evaluated by Meneghini et al. [132]. In the case under analysis, MIS HEMT structures have a SiN dielectric deposited under the gate with a technique based on plasma-enhanced-atomic-layer-deposition; consistent results were observed in devices with different dielectric materials and deposition techniques. The authors of this paper demonstrate, by means of trapping/recovery transients, the existence of an NBTI phenomenon even at low-stress voltages for sufficiently high temperatures. According to the chosen bias condition (null drain bias), the contribution of the buffer-related process is limited and the trap states detected should be located near the AlGaN/dielectric interface. Although a negligible threshold voltage shift (< 0.5 V) was found at room temperature after 5000s, a negative shift of more than 4 V was observed at high temperature (150°C) in consistent bias conditions (Figure 12.36). The threshold voltage shift is found to be accelerated by the temperature with an activation energy of 0.37 eV (Figure 12.37), and it fully recovers following a trend logarithmic with time having a nonnegligible dependence on the temperature (Figure 12.36). Analogous considerations can be done for the variation of the on-resistance in analogous trapping/recovery condition, which demonstrates a strong dependence on the temperature and a high correlation with the instability of the threshold voltage (Figure 12.36). This correlation, therefore, suggests that the same physical process is responsible for the variation of the two parameters.

The authors of the paper interpreted the experimental results on the basis of the mechanism described in (Figure 12.37). The NBTI is supposed to originate because of a distribution of defects at the insulator/AlGaN interface and/or the insulator layer (the stress voltage used is indeed too low to modulate the traps in the III-N buffer). At the equilibrium, the traps are negatively charged if below the Fermi level and neutral if above. When a negative bias is applied, the trap states are depleted thermally, through tunneling and/or defect-assisted conduction, resulting in a net positive

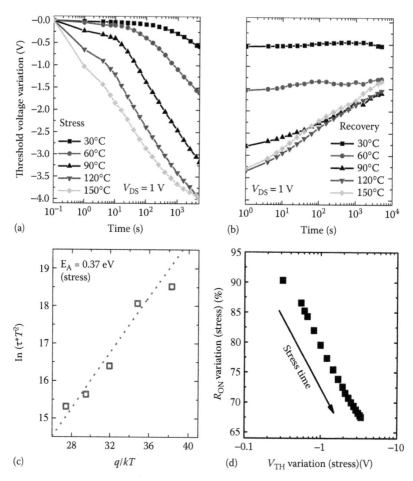

FIGURE 12.36 (a) Variation of the threshold voltage measured (a) during the stress and (b) during the recovery at several ambient temperatures ranging from 30°C to 150°C and bias condition of $V_{GS} = -10$ V, $V_{DS} = 0$ V. (c) Correlation between the shift of the threshold voltage and the variation of the on-resistance reported for an ambient temperature equal to 90°C. (d) Corresponding activation energy of the trap states responsible for the shift of the threshold voltage. (From Meneghini, M. et al., *Electron Device Lett.*, 37, 474–477, 2016. With permission.)

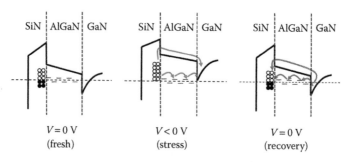

FIGURE 12.37 Schematic representation of the mechanism which contributes to the shift of the threshold voltage. (From Meneghini, M., Rossetto, I., Bisi, D., Ruzzarin, M. et al., *Electron Device Lett.*, 37, 474–477, 2016. With permission.)

charge that induces a negative threshold voltage shift and a decrease of the on-resistance. It is worth noticing that the stretched exponential behavior is consistent with the broad distribution of the energy levels. During the recovery the voltage applied is null, the defects are below the Fermi level and must be refilled.

A detailed analysis of negative bias temperature instability was furthermore provided by Guo et al. [133]. The devices under analysis are GaN-based MOSHEMT with a recessed gate structure and a gate stack defined by a composite SiO_2/Al_2O_3 dielectric. The results discussed are consistent with previous works reported in the literature concerning the NBTI of Si MOSFETs, GaAs-based MOSFET, GaN-based MISHEMTs, and MOSHEMTs.

The threshold voltage instability is found to be correlated to the subthreshold swing, and it can be ascribed to three main processes depending on the regime considered. Low stress induces a recoverable negative threshold voltage which increases with time and a negligible subthreshold swing. The power law behavior of the threshold voltage variation and the weak temperature dependence suggest that the shift is the result of electron detrapping from pre-existing traps (in the oxide or/at the oxide/interface border) and that the process may occur through tunneling. Mid-stress regime induces an initial negative threshold voltage shift which becomes positive with longer stress times; a strong correlation between the threshold voltage and subthreshold swing is observed. The above-discussed variations follow a long-time trend with negligible dependence on temperature. It is worth noticing that, although the direction of the threshold voltage shift is opposite to the one mainly reported in NBTI experiments, it is consistent with a further analysis of an e-mode GaN-based MOSFET structure [135]. Capacitance–voltage curves furthermore indicate a temporary *charge buildup* around the threshold which recovers after detrapping. These (recoverable) processes could arise from field-induced electron trapping in the GaN channel under both the edges of the gate on the source and drain side. The authors [133] explain this mechanism with a process defined as Zener trapping (Figure 12.38); in this condition, the electron trapping takes place in the channel under a high reverse electric field. Under high stress a permanent degradation may occur: the threshold voltage shift, consistently with the variation of the subthreshold swing and the softening of the capacitance–voltage curve, suggests the formation of interface states (probably ascribed to broken H bonds).

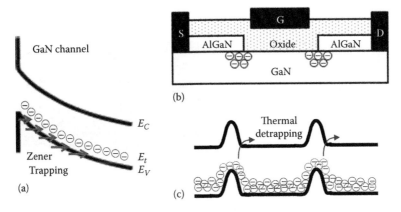

FIGURE 12.38 (a) Band diagram along Y-axis illustrating charge-trapping in the GaN MOSFET channel during reverse gate-bias, (b) cross-section and (c) band diagram along X-axis of the GaN MOSFET channel. Authors suggest charge-trapping in the GaN channel during reverse gate-bias is localized at the gate edges. (From Alex, G. and Jesús A. del Alamo, *IEEE International*, 4A–1, 2016. © 2016 IEEE. With permission.)

12.3.1.3 Materials and Deposition Techniques to Limit the Trapping Effects

The deleterious influence of the PBTI and NBTI mechanisms in GaN-based MISHEMTs leads to the necessity of studying several materials and deposition techniques in order to reduce the threshold voltage instability. Lagger et al. [125] demonstrated that the design of the gate-dielectric stack and the dielectric material used have a strong impact on the trapping effects. The authors explain how the variation of the dielectric and of the barrier thickness change the gate voltage required for the creation of a second channel, and consequently have a significant impact on threshold voltage instabilities.

Several materials were considered in the early studies: SiO_2 demonstrate good performance thanks to its large bandgap despite the low permittivity. Improvements were furthermore discussed in MISHEMT structures employing SiO_2 bilayers (SiO_2/Al_2O_3) [136]. Although the lower bandgap and the lower permittivity may represent an issue, an interesting solution was found in SiN/GaN structures, especially due to the possibility of *in situ* deposition; a significant reduction of threshold voltage instabilities was demonstrated by bilayer gate insulators (SiN/Al_2O_3) [137]. A strong effort was spent to optimize Al_2O_3, in

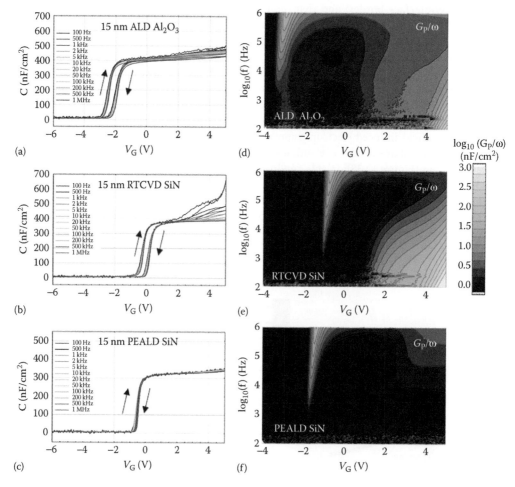

FIGURE 12.39 (a–c) Capacitance–voltage-frequency measurements and (d–f) normalized-conductance as a function of voltage and frequency carried out on partially recessed-gate AlGaN/GaN MISHEMTs with 15-nm plasma-enhanced-atomic-layer-deposition (PEALD) Al_2O_3 (see [a] and [d]); 15-nm rapid-thermal-chemical-vapor-deposition (RTCVD) SiN (see [b] and [e]); and 15-nm PEALD SiN (see [c] and [f]). (Reprinted from *Microelectronics Reliab*, 58, Meneghesso, G. et al., Trapping and reliability issues in GaN-based MIS HEMTs with partially recessed gate, 151, Copyright 2015, with permission from Elsevier.)

consideration of its large bandgap, high permittivity, and high breakdown field. Nevertheless, the high density of fixed charges, whose origin still remains unclear, requires further investigation [138].

Growth at different temperatures was found [139] to have a significant influence on the positive or negative threshold voltage shift (in the case of positive shift due to the electron injection into the oxide or to the increased negative charge) and to influence the slope of the capacitance–voltage curve (due to interface states). It is furthermore found that post-deposition annealing in forming gas reduces the threshold voltage instabilities, mainly under reverse bias, as a consequence of the decrease of the corresponding fixed charges and of the field in the oxide.

A significant effort was moreover spent in order to study the improvement of the trapping effects achievable by means of a different deposition technique. Plasma-enhanced atomic layer deposition was extensively tested for SiN gate insulators. Rossetto et al. [126] and Meneghesso et al. [16] discussed the impact of different deposition techniques on MISHEMTs characterized by a SiN layer.

C–V measurements from depletion to accumulation detect threshold voltage shift and demonstrate that devices experience a frequency-dependent capacitance dispersion which is influenced by the dielectric and by the deposition technique used (Figure 12.40). The trapping mechanisms can be ascribed to (1) the influence of the AlGaN layer on the frequency response to the gate bias [140], and/or (2) the charge (de)trapping at border-traps by means of tunneling mechanisms [141], and/or (3) parasitic conduction mechanisms including leakage currents [142]. The effect of the trap states seems to be significantly reduced in devices with a PEALD SiN gate insulator.

The dependence of the trapping effects on the deposition technique and/or the dielectric used was confirmed by the analysis with pulsed measurements. A strong correlation between the threshold voltage shift and the forward gate leakage current flowing was furthermore demonstrated.

The reduction in the threshold voltage instability provided by a PEALD deposition technique, with respect to the RTCVD, was furthermore confirmed by Wu et al. [130]

The threshold voltage instability has an even more severe impact in the e-mode devices as a lower distance separates the dielectric from the channel. As a consequence stronger electron scattering mechanisms may happen since the trap states are closer to the channel (Figure 12.40).

12.3.2 Time-Dependent Dielectric Breakdown in GaN MISHEMTs and HEMTs

The reduction of the parasitic currents and the enhancement of the robustness under forward gate bias is of fundamental importance in devices for power switching converters. The use of a p-(Al)GaN layer or the insertion of a dielectric under the gate (with the use of a proper recess) effectively fulfills these requirements and enables e-mode operation. Nevertheless, both the above-mentioned structures are found to be extremely sensitive to TDDB especially when submitted to forward gate bias. This typology of failure must be carefully investigated since it mirrors the operating condition of the devices targeted for power switching applications.

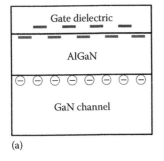
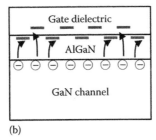
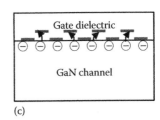

(a)　　　　　　　　　　(b)　　　　　　　　　　(c)

FIGURE 12.40 Schematic cross-section of the gate module for (a) AlGaN/GaN MISHEMT, (b) partially recessed-gate AlGaN/GaN MISHEMT, and (c) fully recessed GaN MOSC-HEMT. Charge-trapping at the dielectric-traps and dielectric/(Al)GaN interface is influenced by presence and by the thickness of the AlGaN barrier.

The time-dependent dielectric breakdown has been extensively studied in thin oxide films, as a consequence of its importance in CMOS-integrated circuits. A fruitful discussion concerning the main properties and origin of the TDDB mechanism in oxide films was provided by Degraeve et al. [143]. In order to describe the oxide wear-out the authors of that paper provided a detailed model based on a percolation concept which links the previous theories concerning the oxide breakdown, namely the anode injection model [144] (injected electrons generate holes at the anode that can tunnel back into the oxide until a critical hole fluence is reached) and the electron trap generation model [145] (the breakdown is reached at a critical density of electron traps generated during stress). A schematic representation of the described process is depicted in Figure 12.41. Traps are located (even in a fresh device) in a random position inside the oxide. Each trap is imagined as a sphere with a fixed radius which enables a conduction path when it overlaps to a neighboring sphere. When the oxide is submitted to stress, traps are generated until a conducting path is created from one interface to another, thus leading to the breakdown condition.

In the model suggested the TDDB can be described by a Weibull function; the Weibull cumulative distribution function shown is $F(t) = 1 - \exp(-t/\eta)\beta$, where β is the shape parameter and η is the scale parameter, corresponding to the time at which the 63.2% of samples failed. The corresponding Weibull statistics is characterized by the following properties: the shape factor decreases with the oxide thickness, since in thinnest oxide a smaller number of traps influences the breakdown determining a larger statistical spread; the breakdown distribution scales with the area of the oxide, which moreover is found to have a strong impact on the breakdown charge.

According to the above-mentioned theory, the shape factor becomes an essential parameter to define the typology of breakdown mechanism detected. Based on the Weibull shape parameter the breakdown can be considered as intrinsic or extrinsic. With high shape factor β, intrinsic breakdown occurs, mainly ascribed to an oxide wear-out failure. Low shape factor β corresponds to early failures. In this case, the failure can be ascribed to the specific technological steps used to grow the oxide, and the breakdown is characterized by a fixed field-independent percentage of failures.

Furthermore, the shape factor is relevant for the determination of the number of defects forming the percolation path, according to the formula: $\beta = Mn$, where n is the trap generation rate and M is the number of defects which form the conductive path.

Recent studies in the literature demonstrate that the theory of TDDB described in [143] can be applied to GaN-based devices in order to define the properties of the breakdown and to predict a mean time to failure. Preliminary reports demonstrate TDDB mechanism in different typologies of GaN-based devices, for example, GaN-based diodes [146], d-mode and e-mode GaN-based transistors with Schottky [147,148], MIS [16,124,149] and p-type structures [150,151]. The following analysis discusses the TDDB phenomenon in Schottky-based GaN HEMTs submitted to off-state constant voltage stress, MISHEMT

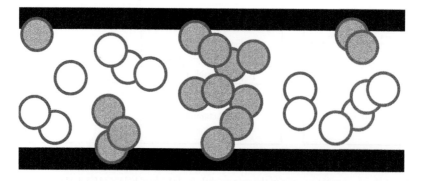

FIGURE 12.41 Schematic representation of the model suggested by Degraeve et al. [143] to describe the TDDB phenomenon in oxide according to the percolation concept.

(d-mode and e-mode) submitted to on-state and forward gate voltage stress, and e-mode GaN-based HEMTs with a p-GaN gate.

12.3.2.1 TDDB in GaN HEMTs Stressed in Off-State Condition

The evidence of a TDDB breakdown mechanism in d-mode devices submitted to off-state constant voltage stress is discussed by Moens et al. [147]. In the case under analysis, the 2DEG is depleted, as a consequence of the reverse bias and of the high drain voltage applied. Under this condition, the GaN buffer can be considered as behaving as a defective dielectric. TDDB mechanism and the corresponding Weibull distribution are confirmed for 10A-rated devices tested at 200°C for drain voltage values higher than 900 V, in order to test both the lateral and the vertical robustness (Figure 12.42). According to the high shape factor of the TDDB, four different models were used in order to extrapolate the corresponding distribution at the operating voltage ($V_D = 600$ V).

The analysis of the long-term reliability and the presence of a TDDB mechanism on GaN-based devices submitted to off-state stress is furthermore discussed by Meneghini et al. [148]. The comparison of the (off-state) breakdown measurements and the constant voltage tests demonstrates the evidence of a time-dependent failure. The catastrophic failure occurs, for longer stress times, at a drain voltage value lower than the measured breakdown voltage. The time to failure is found to be field-dependent (Figure 12.43), as demonstrated by its dependence on the voltage level applied, and strongly dependent on the device-to-device (and process) variability. The TTF can be described by a Weibull distribution characterized by a low shape factor β, as suggested by the high spread in the TTF (Figure 12.43). The calculated shape factor, equal to 0.5, indicates that the failure rate decreases with time and that early failure occurs.

The presence of a time-dependent process which can be described by a Weibull distribution is explained by the hypotheses formulated concerning the origin of the failure mechanism. The TDDB process, together with the high activation energy detected (1.73 eV), suggests the failure of the dielectric (passivation layer) under the gate overhead when the device is submitted to a high electric field (and/or to a high temperature). It is worth noticing that the early failure detected is consistent with the possible variability in the dielectric deposition. The hypothesis concerning the failure of the dielectric under the gatehead was confirmed by means of simulations of the electric field and improvements provided in the second generation of devices in order to reduce the electric field at the edge of the gate.

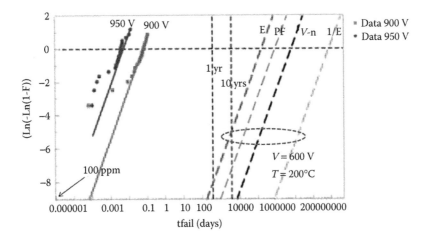

FIGURE 12.42 TDDB mechanism in GaN-based HEMTs submitted to off-state constant voltage test at high drain voltage levels. (From Moens, P. et al., *45th European Solid State Device Research Conference (ESSDERC)*, 64–67, 2015. © 2015 IEEE. With permission.)

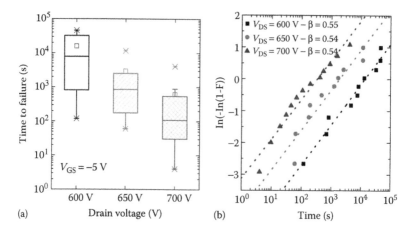

FIGURE 12.43   (a) Dependence of the time to failure on the drain bias level applied. (b) Weibull plot defining the time to failure in devices submitted to an off-state constant voltage test. The low shape factor calculated indicates the presence of an early failure.

12.3.2.2 TDDB in GaN MISHEMTs Submitted to On-State Stress

MISHEMT structures are demonstrated to be an interesting solution to satisfy the requirements imposed by the applications in power switching converters. The insertion of a thin dielectric under the gate and the use of a proper recess enable the implementation of e-mode devices with low leakage current and improved breakdown voltage. In the final application (power switching converters), e-mode MISHEMTs are submitted to both off-state and on-state (forward gate bias) condition. Therefore these devices must fulfill a high gate bias swing and a good robustness in both the above-mentioned operating conditions.

When a MISHEMT structure is submitted to a gate forward bias, the breakdown of the dielectric under the gate can be modeled according to the theory used for oxides submitted to a positive bias stress. Therefore, the analysis of the TDDB has been extensively used to study the origin of the breakdown mechanisms in MISHEMTs submitted to a forward gate bias and/or to on-state stress.

Wu et al. [124], on the basis of TDDB analysis, provided a fruitful explanation of the failure mechanism, which may occur on devices characterized by a 15 nm thick *in situ* SiN/Al$_2$O$_3$ bilayer and submitted to an on-state stress. Constant voltage tests performed at high ambient temperature demonstrate that the failure is time-dependent and Weibull distributed with a high shape factor. It is worth noticing that the initial formation of a leakage path is suggested by the increase of the noise superimposed to the gate current. The small variability and the high shape factor indicate an intrinsic breakdown mechanism suggesting a possible failure of the dielectric. According to the percolation model [143], the trap cross section, the dielectric thickness, and the shape factor allow the calculation of the minimum number of traps required to create a conduction path which contributes to the catastrophic failure.

The percolation model adopted to describe the TDDB suggests that the time to failure can be significantly enhanced by increasing the dielectric layer thickness due to the reduction of the possibility of forming a conductive path. The number of traps required to form a conductive path (Figure 12.41) indeed increases with the thickness of the insulator layer. The percolation model furthermore explains that the TDDB can be described by a Weibull distribution whose distribution scales with the area considered. The hypotheses were confirmed by experimental results [130,124]: by increasing the insulator layer from 25 to 35 nm the estimated operating voltage for a 20-year robustness at $T = 150°C$ is enhanced from 4.9 to 13 V (exponential law).

The results were confirmed by means of a further detailed discussion on devices with different dielectric thicknesses, namely 15 and 25 nm, and recess depth ranging from 4.4 to 21.7 nm [124]. This latter results in e-mode devices as a consequence of the recess into the channel. It is worth noticing that in the d-mode devices a nonnull AlGaN barrier is left under the gate dielectric, thus indicating that the generated defects corresponding to (soft) breakdown may, in principle, be located also in the AlGaN barrier layer. In all the cases a high shape factor is measured, suggesting the presence of an intrinsic breakdown mechanism; this latter is confirmed by the scaling of the time to failure (and of the corresponding scale parameter) with the device area. The shape factor is found to be strongly influenced by both the dielectric thickness and the remaining AlGaN barrier layer: the first aspect leads to a higher shape factor; conversely the reduction of the remaining AlGaN layer contributes to a reduction of the Weibull shape factor, thus indicating an increase of the spread in the time to failure. The time to failure is found to be enhanced by the dielectric thickness (as previously discussed) and in the case of e-mode structures, that is, with no remaining AlGaN barrier layer under the gate dielectric. The higher robustness of e-mode structures can be explained by the necessity for the electrons to be formed in the channel before charging the dielectric; the higher threshold voltage, thus, implies higher gate voltages for the catastrophic failure.

It is worth noticing that the failure in a fully recessed gate MISHEMTs (characterized by the absence of a remaining AlGaN layer) should indicate that the percolation path is formed in the gate dielectric. As a consequence, the devices with an equal dielectric thickness and no remaining AlGaN layer should need the same amount of traps to form a conductive path, and thus, be characterized by a similar Weibull distribution in terms of shape factor. A possible difference suggests that a further phenomenon can influence the TDDB mechanism in e-mode devices with no remaining AlGaN barrier layer. The percolation path can be formed around the gate corner, which is characterized by a thinner gate dielectric, thus resulting in a different (lower) shape factor. A representation of the occurring failure is shown in Figure 12.44. [124].

Several studies report a TDDB analysis to define the quality of the dielectric and of the deposition technique used and to estimate the corresponding mean time to failure. Meneghesso et al. [16] reported a comparison of the TDDB evaluated in devices with a different gate insulator, namely SiN and Al_2O_3, and a different deposition technique. The devices with a SiN gate insulator show an increase of the gate current before the hard breakdown, suggesting a higher increase of the trap centers in the insulator and/or an increase of the electric field at high forward gate bias levels. The deposition technique, in the case considered plasma-enhanced atomic layer deposition, is found to provide significant improvements (with respect to rapid thermal chemical vapor deposition) in terms of enhancement in the time to failure and in the shape factor, leading to a superior reliability. The comparison is shown in Figures 12.45 and 12.46.

Recent works demonstrate that the TDDB and the corresponding lifetime prediction can be overestimated as a consequence of charging-induced dynamic stress relaxation (CiDSR) effects. Evidence of

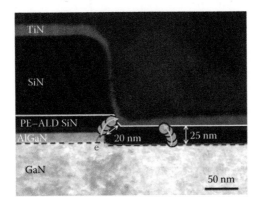

FIGURE 12.44 TEM picture indicating the location of possible percolation paths in e-mode MISHEMT structures. (From Tian-Li, W. et al., *IEEE International Reliability Physics Symposium*, 6C.4.1–6C.4.6, 2015. © 2015 IEEE. With permission.)

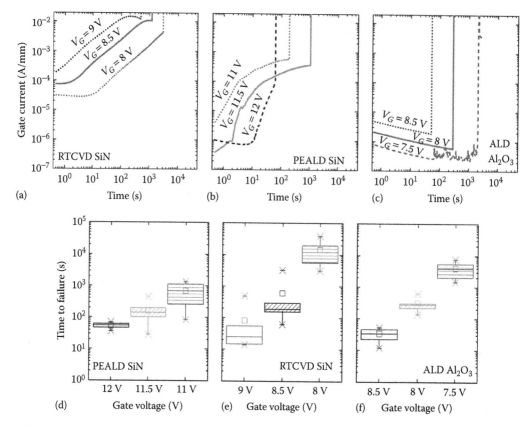

FIGURE 12.45 Time-dependent degradation of GaN-based MIS-HEMTs submitted to constant voltage stress; three different typologies of dielectric or deposition techniques are shown, namely 15 nm (a) RTCVD SiN, (b) 15 nm PEALD SiN, (c) 15 nm ALD Al₂O₃. (c–f) Average TTF measured under different gate voltages on the three wafers.

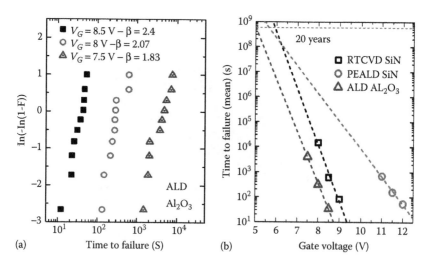

FIGURE 12.46 (a) Weibull distribution of time-to-failure on GaN-based MIS-HEMTs under forward gate stress and (b) lifetime extrapolation based on the results of forward-bias stress tests, namely on (circle) PEALD SiN, (square) RTCVD SiN, (triangle) ALD Al₂O₃.

CiDSR effect was reported by Okada et al. [152]. The authors discuss anomalous TDDB statistics [152], mainly defined by a nonlinear Weibull plot and an overestimated dependence on field/voltage applied during the stress. The analysis of the TDDB behavior in devices with a different dielectric thickness revealed that inaccuracies can be introduced by carrier charging to initial and stress-generated defects. Under constant voltage stress, the effective stress field is dynamically changed by the carrier charging, thus inducing a prolongation of the lifetime. As a consequence, the estimated lifetime is influenced by the defect generation rate (according to the theory of the TDDB) and by the carrier charging to defects which, as in the case study discussed in [152], undermines the validity of the Weibull statistics.

12.3.2.3 TDDB in HEMTs with a p-GaN Gate Submitted to Forward Gate Bias

The necessity of a large gate bias swing and of the development of e-mode devices required for high power application leads to a strong interest in the possible failure mechanism accelerated when the devices are submitted to a forward gate bias. Concerning the GaN HEMTs with a p-GaN gate, three main hypotheses were discussed in recent works.

1. Rossetto et al. [150] investigated the origin of the failure mechanism under gate forward bias stress by means of dc characterization, 2D simulations, and emission microscopy analysis. The authors suggest that the electric field in the AlGaN, conversely to the electric field in the p-GaN and the SiN, may not play a significant role in the failure.

   The electric field in the AlGaN layer is found to decrease with the increase of the gate forward bias applied. The peak of the electric field across the AlGaN, in the case considered in [150], is found to decrease from 1.9 MV/cm (with $V_G = 0$ V) down to 0.32 MV/cm (with $V_G = 9$ V). The results, therefore, indicate that the field across the AlGaN is much lower during the stress than under equilibrium, confirming the negligible role of the electric field in the AlGaN in the failure process.

   The 2D simulations indicate that under gate forward bias a high electric field is calculated in the p-GaN layer and the SiN passivation. Although the electric field in the passivation is much lower than the one reported for the breakdown of the SiN, nonuniform field distribution can be present, leading to a local increase in the electric field in the SiN and generating weak spots, thus promoting the catastrophic failure. The operation under positive bias induces a bending of the band diagram and an increase of the electric field across the p-GaN (Figure 12.47).

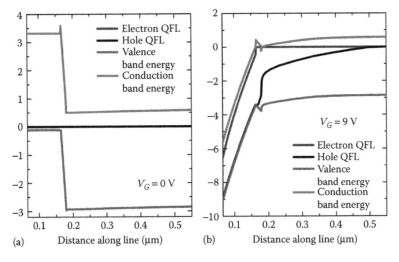

FIGURE 12.47 Band diagram of a GaN HEMT with a p-GaN gate (a) under equilibrium ($V_G = 0$ V) and (b) under gate bias overstress ($V_G = 9$ V).

The increase of the electric field, helped by the possible nonuniformities and the high gate current flowing, can promote the time-dependent generation of defects in the material and can lead to the catastrophic failure with long stress times.

The generation of leakage paths under high gate forward bias is confirmed by emission microscopy. The increase of the gate current and of the noise to it superimposed corresponds to the generation of hot spots detected and to the increase of the electroluminescence signal. According to previous results reported in the literature [153] the increase in the hot spot intensity with the gate current demonstrates the formation of degradation/leakage paths with high-stress times.

Spectral analysis performed on an already damaged device shows a peak at $\lambda = 550$ nm (2.25 eV), which can be ascribed to the yellow luminescence (YL) of GaN. According to the literature the corresponding EL signal results from the contribution of trap states [68,69,154] related to gallium vacancies [68] and/or carbon-related defects [69]. Furthermore, the EL signal is found to increase with the wavelength, as a consequence of bremsstrahlung (BS) radiation related to hot electrons, possibly generated in the p-type material as a consequence of the avalanche multiplication.

The evaluation of the failure mechanism demonstrates that the failure in p-GaN devices submitted to a forward gate bias stress is due to a time-dependent mechanism. The corresponding times to failure can be described by a Weibull distribution. Rossetto et al. [150] show that a time-dependent failure is observed in devices submitted to a constant voltage test with a time to failure which is found to be dependent on the gate voltage applied (Figure 12.48). The time to failure is found to be highly correlated to the initial gate leakage current, thus

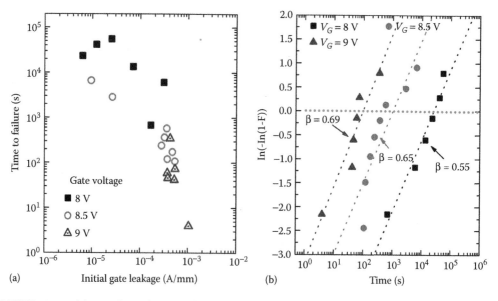

FIGURE 12.48 (a) Correlation between the time to failure and the initial gate leakage current measured in consistent bias point, indicating the influence of the initial defectiveness on the failure mechanism. (b) Weibull plot of the time to failure measured in devices with a p-GaN gate submitted to a forward gate bias stress.

demonstrating that the initial defectiveness may influence the degradation process. The statistical analysis confirms that the time to failure (in p-GaN devices submitted to a forward gate voltage stress) can be described by a Weibull distribution with a shape factor lower than one, thus suggesting that, in the case considered, an extrinsic breakdown mechanism may occur.

It is worth noting that a time to failure distribution with a Weibull shape factor higher than 1 was demonstrated in the literature for GaN-based devices with a p-GaN gate submitted to a forward gate bias stress. This aspect proves that, in principle, the catastrophic failure of these devices can be due to intrinsic or extrinsic breakdown mechanisms, depending on the technology of tested devices.

Analogous results were shown by Tapajna et al. [151] on similar devices tested at a bias levels corresponding to the increase of the noise superimposed on the gate current. This latter aspect was already detected in reverse-bias Schottky gated HEMTs and ascribed to a defect percolation process within the AlGaN layer; as demonstrated by Rossetto et al. [150], in p-GaN HEMTs the situation is different since the electric field in the AlGaN decreases with forward gate bias levels.

2. Tapajna et al. [151,155] studied the soft breakdown mechanism in devices with a p-GaN formed via Ni/Au metallization. The authors suggest that the failure can be ascribed to a formation of a conductive path due to trap generation and/or a percolation mechanism, promoted by the high electric field and the high gate current flowing. The role of the trap generation is discussed by analyzing the degradation mechanism in devices submitted to a constant forward gate bias. Three main stress regimes can be identified: an initial recoverable change in the gate current, presumably ascribed to traps; a nonrecoverable increase of the gate current preceded by the increase of the noise superimposed to the curves; and a strong increase of the noise superimposed before hard breakdown. The gradual increase of the gate current and the presence of current fluctuations even before the breakdown mirrors the increase of the trapping effects and the formation of conductive paths within the stress [148]. Tapajna et al. [151] demonstrated that the increase of the ambient temperature induces a decrease of the shape factor and a reduction of the time to failure with activation energy of 0.1 eV. The authors suggest that the degradation may be due to a p-GaN layer near the interface via a donor-like trap degradation which causes a reduction of the electric field in the AlGaN and an increase of the localized leakage paths.

3. Wu et al. [156] studied the origin of the failure mechanism in devices with a TiN gate metal evaporated on top of the p-GaN to form a Schottky contact. The author suggests that the breakdown is due to an avalanche process as a consequence of the electron/hole pair generation.

The forward gate breakdown mechanism was evaluated at different ambient temperatures, indicating a positive temperature dependence equal to $5 \cdot 10^{-3}$ °C/V. According to previous data reported for Si CMOS technology [157] and AlGaN/GaN HEMTs stressed in off-state condition [158], the positive temperature dependence can be explained by an avalanche breakdown mechanism. Under high forward gate bias, the Schottky metal/p-GaN diode is in reverse bias. In this condition, the electrons in the channel can be emitted over the AlGaN layer and be injected into the p-GaN; the high electric field can provide them enough energy to promote the avalanche breakdown.

The hypothesis of the avalanche breakdown is furthermore supported by the avalanche electroluminescence detected before the gate hard breakdown under high gate bias condition (Figure 12.49). The EL signal created is mainly ascribed to the recombination of the generated electron-holes pairs.

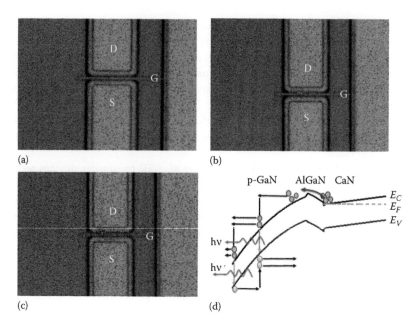

FIGURE 12.49 EMMI measurement in p-GaN gate AlGaN/GaN HEMTs at $V_G = 9$ V (a), $V_G = 9.5$ V (b), and $V_G = 9.8$ V (c). Schematic of electron-hole recombination (d). (From Wu, T.-L. et al., *IEEE Electron Device Lett.*, 36, 1001–1003, 2015. © 2015 IEEE. With permission.)

12.4 Breakdown Mechanism

Breakdown is the key mechanism limiting maximum voltage ratings, device robustness, and long-term reliability. Breakdown causes the loss of the off-state blocking properties, the increase of leakage currents, and the catastrophic burn-out of the device.

Hereafter, breakdown will be defined as the condition when drain-source, drain-gate, or drain-substrate leakage current reaches a certain level (usually around 1 mA/mm) during off-state. Causes for leakage current and breakdown events include defect generation and defect-assisted conduction [159], band-to-band tunneling, and impact-ionization [160]. Depending on the involved mechanisms, breakdown can be field-, current-, time- and temperature-dependent. Breakdown events can be nondestructive (as in the case of controlled avalanche breakdown [161]) or destructive (as in the case of percolative path formation or of thermal runaway).

In AlGaN/GaN high-electron mobility transistor, breakdown can occur at multiple locations, namely:

- Lateral drain-source breakdown of the (Al)GaN buffer caused by subthreshold leakage current and punch-through effects [7,162].
- Drain-gate breakdown at the drain-side gate edge caused in the off-state by high electric field between the gate terminal and the channel in the access region [1].
- Lateral drain-gate breakdown of the passivation layers [163].
- Vertical drain-substrate breakdown of the (Al)GaN buffer layers [147,164].
- Vertical breakdown of the gate dielectric layer of MIS-gated devices, caused by the vertical electric field between the gate and the 2DEG. Critical forward-biased gate operations [16,124,165].

In order to characterize the breakdown event and to identify the breakdown mechanisms, several measurement techniques have been proposed in the literature; these include current- and voltage-controlled sweep measurements [7], drain current injection technique [166], constant-current and constant-voltage stress tests [148], electroluminescence microscopy and spectroscopy [167].

During *I–V* sweep measurements, drain, gate, source and substrate currents are acquired in order to identify the main contribution to the leakage and to identify the leakage paths which is contributing to the breakdown [168]. Usually, the current-controlled *I–V* sweep is preferred over voltage-controlled *I–V* sweep in order to prevent the catastrophic burn-out of the device and to allow the physical and failure analysis after the breakdown event.

Current-controlled and voltage-controlled sweep measurements can be performed at multiple gate bias levels in order to discern between lateral and vertical breakdown mechanisms, identify the breakdown mechanisms, which depend on the gate voltage (such as subthreshold leakage and punch-through), and the breakdown mechanisms which do not depend on the gate voltage (e.g., those ascribed to the vertical drain to substrate potential).

To investigate the time-dependence of breakdown mechanisms—very important for the long-term device reliability—constant-current and constant-voltage stress test are also performed [147].

Combining the electrical characterization with the electroluminescence microscopy and spectroscopy is important to gather further information on the nature and the location of breakdown events. When exposed to high field, highly energetic carriers can undergo intra- and/or intervalley-scattering mechanisms, inducing electroluminescence signal. Electroluminescence can be localized at hotspots, hence revealing the presence of percolative paths at pre-existent or stress-induced defects (see Figure 12.53a, [169,170]) or the areas of maximum electric-field, for example at sharp nonoptimized device corners (see Figure 12.53b, [92]). On the other hand, EL signal can be uniformly distributed over the entire active area, revealing intrinsic mechanisms such as lateral punch-through [167] or vertical avalanche breakdown (see Figure 12.59, [171]).

12.4.1 Lateral Source-Drain Breakdown

One of the main causes of the lateral breakdown is the drain-source subthreshold leakage (Figure 12.50). The increase of the drain-source leakage in the breakdown can be modulated by varying the gate bias level (Figure 12.51). As discussed by Uren et al. [81,162], this mechanism is more evident in devices with a small gate length (short channel devices) and/or in devices with noncompensated unintentionally doped GaN buffer layers. Source-drain leakage current can be ascribed to the nonoptimized depletion of the channel which allows the flow of the current deep into the GaN buffer. This latter can be ascribed to

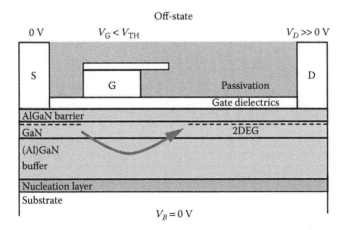

FIGURE 12.50 Lateral drain-source breakdown of the (Al) GaN buffer in AlGaN/GaN HEMTs caused by subthreshold leakage current and punch-trough effects.

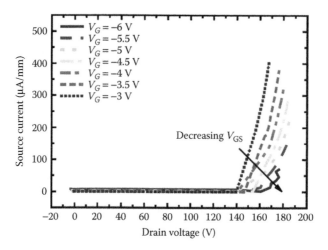

FIGURE 12.51   Variation of drain–source (sub-threshold) leakage with varying gate voltage. (Reprinted with permission from Meneghesso, G. et al., *Jpn. J. Appl. Phys.*, 53, 100211, 2014. Copyright 2014, The Japan Society of Applied Physics.)

punch-through source-drain current and to the drain induced barrier lowering phenomenon. As a major implication, the improper depletion of the buffer layer and the premature source-drain breakdown prevent the breakdown voltage to scale with the gate-drain spacing [7]. Multiple technological solutions can be used to prevent these parasitic effects, including the use of a double-heterostructure with an AlGaN back-barrier to improve the carrier confinement [120,168], and the compensation of unintentional n-type doping of (Al)GaN buffer layers by introducing deep-acceptor species (e.g., carbon and/or iron) [172].

12.4.2 Lateral Gate-Drain Breakdown

During off-state, the drain side gate edge is exposed to maximum electric-field peaks and represents a critical element for the overall device breakdown capabilities (Figure 12.52). In Schottky-gated HEMTs, the degradation of the gate edge has been extensively investigated in the literature. The step-like

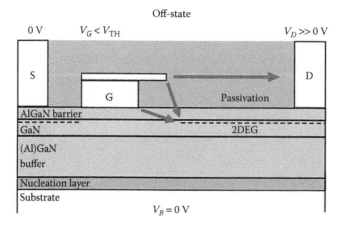

FIGURE 12.52   Drain-gate breakdown at the drain-side gate edge in AlGaN/GaN HEMTs, caused in the off-state by high electric-field between the gate terminal and the channel in the access region, and lateral drain-gate breakdown of the passivation layers.

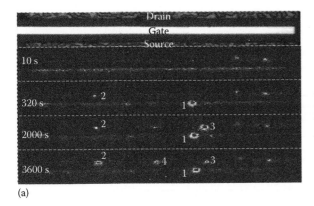 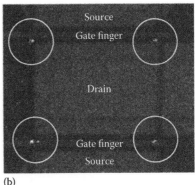

(a)　　　　　　　　　　　　　　　　　(b)

FIGURE 12.53 (a) Time-resolved EL measurements of a Schottky-gated AlGaN/GaN HEMT under reverse-bias stress, showing formation of randomly localized EL hot-spots caused by the degradation of the gate-edge and the formation of percolative paths. (Reprinted with permission from Meneghini, M. et al., *Appl. Phys. Lett.*, 100, 033505, 2012. Copyright 2012, American Institute of Physics.) (b) Electroluminescence microscopy of an MIS-gated AlGaN/GaN HEMT under off-state step stress, just before the occurrence of catastrophic gate-drain breakdown. EL spots are localized at the sharp corners of the drain contact, where the catastrophic breakdown eventually took place. (From Meneghini, M. et al., *IEEE Trans. Power Electron.*, 29, 2199, 2014. © 2014 IEEE. With permission.)

increase of the gate–drain leakage and the formation of electroluminescence hot-spots can be ascribed to the degradation of the gate Schottky junction [1] and/or to the worsening of the insulating properties of the dielectric below the gate head [148]. The degradation of the gate Schottky junction could be ascribed to inverse-piezoelectric effect [2], to a defect generation/percolation process [170], or to electrochemical reaction at the surface of the transistors [3]. These processes may result in the creation of defect-related leakage paths next to the gate which may eventually lead to the catastrophic breakdown of the device. Technological solutions to prevent gate edge gate-drain breakdown include the adoption of the gate-connected and source-connected field plates and the optimization of the epitaxial structure to improve the potential distribution and to suppress the electric-field peak at the drain-side gate edge (Figure 12.53).

Lateral breakdown can happen not only within the III-Nitride epitaxial structure but also at the surface and/or at the SiN passivation layer(s). Several mechanisms can increase the gate-drain leakage current, including surface conduction by hopping at the surface states and the breakdown because of power dissipation and thermal runaway [173]. Surface conduction can be mitigated by means of improved SiN passivation layers.

12.4.3 Vertical Drain-Substrate Breakdown

The presence of a vertical breakdown can be identified by the increase of the drain-substrate current with increasing the drain voltage (Figure 12.54). The increase of the substrate current indicates the contribution of the drain-substrate leakage (vertical breakdown), and it can be limited by improving the epitaxy and the substrate. The vertical breakdown is of pivotal importance in lateral AlGaN/GaN HEMTs grown on silicon substrate, where the full drain-to-substrate potential drops across the few μm-thick III-Nitride epitaxial layers [164].

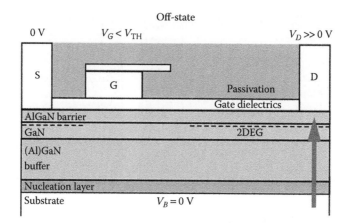

FIGURE 12.54 Vertical drain-substrate breakdown of the (Al) GaN buffer layers in AlGaN/GaN HEMTs.

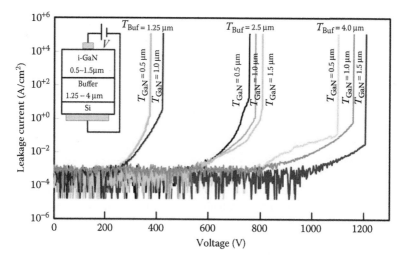

FIGURE 12.55 Vertical I–V breakdown characteristics measured across a two-terminal structure to test the breakdown of the GaN-on-silicon epitaxy. (From Rowena, I. B. et al., *IEEE Electron Device Lett.*, 32, 1534, 2011. © 2011 IEEE. With permission.)

Several technological solutions have been implemented in order to reduce the drain-to-substrate leakage, including (1) the use of an AlGaN back-barrier layer limiting leakage current, as a consequence of the higher band-gap of AlGaN with respect to GaN [174]; (2) the increase of the GaN buffer thickness, reducing the electric-field and promoting the decrease of the dislocation density as experimentally demonstrated [164] (Figure 12.55); (3) the use of Ar-implantation in the substrate which was found to improve the breakdown in both Si and SiC substrates without affecting the dynamic properties of the transistor [175,176]; (4) the removal of the Si substrate (Figure 12.56) under the gate-drain region, in order to suppress the parasitic vertical current without inducing degradation of the on-resistance and of the gate leakage current [177,178]; and (5) the creation of p+ region at the edge of the gate by implanting acceptors atoms in the silicon wafer by Umeda et al. [179], to prevent the leakage of electrons generated between the silicon substrate and the III-Nitride epitaxy.

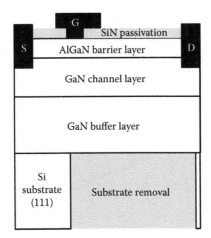

FIGURE 12.56 Schematic cross section of an AlGaN/GaN HEMT with silicon substrate removal.

12.4.4 Avalanche Breakdown

For power switching and rectifying applications, the ruggedness in the breakdown is a critical requirement. A nondestructive breakdown could be achieved if the dominant breakdown mechanism is an avalanche and if the breakdown regime holds for a short-time ($< 1\,\mu s$) to avoid the thermal runaway.

Avalanche breakdown is caused by the generation of electron-hole pairs and the current multiplication due to impact ionization. Impact ionization happens when free electrons traveling through a high field region undergo scattering events and transfer excess energy to bonded electrons in the valence band, lifting them into the conduction band, creating new electron-hole pairs, and increasing the free carrier density; generated holes can at their turn contribute to impact ionization. Impact-ionization coefficient in gallium nitride has been theoretically predicted by means of Monte Carlo calculation [180] and experimentally reported in AlGaN/GaN heterostructure field effect transistors in [181] and [182] by means of gate/drain current measurements and the application of the Hui method [183].

It is worth noticing that avalanche breakdown voltage has a positive temperature coefficient because temperature decreases the electron mean free path and the impact-ionization coefficient; hence, increasing the electric field needed to trigger impact ionization and avalanche breakdown.

Numerical prediction of impact ionization and avalanche breakdown in lateral AlGaN/GaN HEMTs have been reported in [184] and [185]. At high off-state drain bias levels, impact ionization and avalanche breakdown can take place due to electrons injected from the source, the gate or the substrate to the gate-drain access region, where the electric field is maximum.

The holes generated during impact ionization can be collected at the gate or at the source, promoting the build-up of positive charge under the gate and subsequent threshold voltage shift and "kink" effect in *I–V* characteristics, and/or the increase of gate leakage current. In 2002, Brar et al. reported on "kink" effect enhanced at low temperature (down to −50°C), and bell-shaped behavior of gate current for drain voltage greater than 4 V in Ga-polar AlGaN/GaN HEMTs with 0.25 μm Pt/Au T-gate [186].

Impact ionization and avalanche breakdown have been experimentally observed in vertical GaN p-n diodes grown on GaN substrate with relatively low dislocation density between 10^3 cm^{-2} and 10^6 cm^{-2}. Results reported by Kizilyalli et al. in [160] and [161] show (1) the positive temperature dependence of the off-state breakdown voltage (Figure 12.57), (2) the characteristics avalanche waveform under inductive switching conditions (Figure 12.58), and (3) the sustainable avalanche breakdown (ruggedness) of devices exposed to fixed avalanche current up to 10 A for 10^5 times with a frequency of 5 kHz. Uniform electroluminescence signal during avalanche regime caused by the radiative recombination of

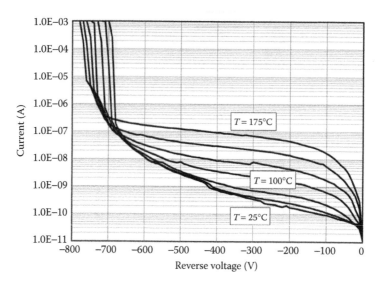

FIGURE 12.57 Temperature dependence of the reverse I–V curves of the GaN p-n diode grown on GaN substrate. Breakdown voltage has a positive temperature dependence, suggesting the involvement of avalanche breakdown. (From Kizilyalli, I. C. et al., *IEEE Trans. Electron Devices*, 60, 3067, 2013. © 2013 IEEE. With permission.)

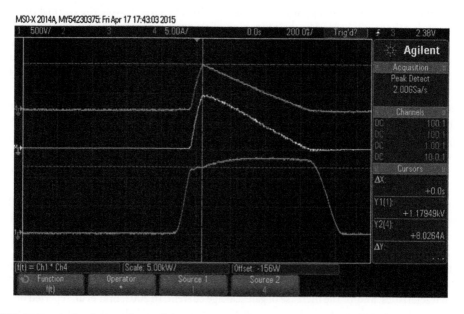

FIGURE 12.58 Avalanche waveform of the GaN p-n diode grown on GaN substrate subjected to inductive switching operation. (Reprinted from *Microelectronics Reliab*, 55, Kizilyalli, I. C. et al., Reliability studies of vertical GaN devices based on bulk GaN substrates, 1654, Copyright 2015, with permission from Elsevier.)

electron–hole pairs created by impact ionization is observed in vertical p-n GaN diodes grown on GaN substrate by Dickerson et al. [171] (Figure 12.59).

The optimized design of the field-plates and of the device epitaxial structure and layout is of key importance for the optimization of the potential distribution, for the suppression of electric field peaks, and for the improvements of overall breakdown voltage capabilities.

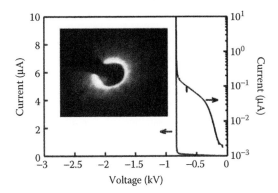

FIGURE 12.59 Electroluminescence of a GaN p-n diode grown on GaN substrate exposed to avalanche breakdown. The EL signal is uniformly distributed, suggesting the radiative recombination of electron-hole pairs generated by impact-ionization. (From Dickerson, J. R. et al., *IEEE Trans. Electron Devices*, 63, 419, 2016. © 2016 IEEE. With permission.)

Summarizing, the proper characterization of the breakdown mechanisms by means of combined electro-optical measurements and dedicated test structures is of key importance to isolate the limiting mechanisms hampering the device life-time and maximum voltage ratings, including intrinsic breakdown mechanisms due to ionization effects, or extrinsic breakdown mechanisms due to material defectiveness or to nonoptimized device design, and to propose effective technological solutions to improve the device performance, the device robustness, and the device reliability.

12.5 Conclusion

The research efforts on the characterization, the modeling and the physics-based understanding of parasitic effects and reliability of III-Nitride devices have demonstrated to be of paramount importance towards the effective technological advancement of wide-band gap semiconductor devices for the next generation of power RF and power switching applications.

In this chapter, we reviewed the main characterization methodologies and the latest results on the state-of-art understanding of parasitic effects and reliability of III-N devices. We reported on the study of deep-level effects in GaN HEMTs, with a detailed evaluation of the characterization techniques, including double-pulse measurements, drain current transient analysis, and GFS method; of the localization of charge trapping mechanisms; and of the role of compensation doping, hot electrons induced trapping and trapping-induced leakage current and time-dependent breakdown. We discussed the stability of dielectrics adopted in GaN HEMT and MIS HEMT structures, including bias temperature threshold voltage instabilities and TDDB. We concluded the chapter by providing an overview of the main breakdown mechanisms in state-of-art III-N devices.

Acknowledgments

This work was supported by the ONR project "GaN HEMT Reliability Physics: from failure mechanisms to testing methods, test structures and acceleration laws" Contract No. N000141410647, the ENIAC project "E2COGaN—Energy Efficient Converters using GaN Power Devices" under Grant 324280, the EDA project "EUGaNIC" Contract No. B1447 IAP1 GP, the ESA/ESTEC project "Preliminary Validation of Space Compatible GaN Foundry Processes" Contract No. 4000106310/12/NL/SFe, and Progetto di Ateneo, University of Padova, Prof. Enrico Zanoni. The authors would like to thank Carlo De Santi, Stefano Dalcanale, Fabiana Rampazzo, Alessandro Barbato, Marco Barbato, Maria Ruzzarin, Alaleh Tajalli, Antonio Stocco, Riccardo Silvestri, Nicola Trivellin, Giovanni Verzellesi, and Alessandro Chini for the fruitful discussions and contributions.

References

1. E. Zanoni, M. Meneghini, A. Chini, D. Marcon, G. Meneghesso, AlGaN/GaN-based HEMTs failure physics and reliability: Mechanisms affecting gate edge and Schottky junction, *IEEE Trans. Electron Dev.*, 60: 3119, 2013.

2. J. Joh, J. A. del Alamo, Critical voltage for electrical degradation of GaN high-electron mobility transistors, *IEEE Electron Device Lett.*, 29: 287, 2008.

3. F. Gao et al., Role of oxygen in the OFF-state degradation of AlGaN/GaN high electron mobility transistors, *Appl. Phys. Lett.*, 99, 223506, 2011.

4. M. Piazza, C. Dua, M. Oualli, E. Morvan, D. Carisetti, F. Wyczisk, Degradation of TiAlNiAu as ohmic contact metal for GaN HEMTs, *Microelectronics Reliab.*, 49: 1222, 2009.

5. R. Vetury, N. Q. Zhang, S. Keller, U. K. Mishra, The Impact of Surface States on the DC and RF Characteristics of AlGaN/GaN HFETs, *IEEE Trans. Electron Devices*, 48(3): 560, 2001.

6. G. Meneghesso, M. Meneghini, D. Bisi, I. Rossetto, A. Cester, U. K Mishra, E. Zanoni, Trapping phenomena in AlGaN/GaN HEMTs: a study based on pulsed and transient measurements, *Semicond. Sci. Technol.*, 28: 074021, 2013.

7. G. Meneghesso, M. Meneghini, D. Bisi, R. Silvestri, A. Zanandrea, O. Hilt, E. Bahat-Treidel, et al., GaN-based power HEMTs: Parasitic, reliability and high field issues, *ECS Transactions*, 58: 187, 2013.

8. I. Rossetto, F. Rampazzo, M. Meneghini, M. Silvestri, C. Dua, P. Gamarra, R. Aubry et al., Influence of different carbon doping on the performance and reliability of InAlN/GaN HEMTs, *Microelectronics Reliab.*, 54: 2248, 2014.

9. M. Huber, M. Silvestri, L. Knuuttila, G. Pozzovivo, A. Andreev, A. Kadashchuk, A. Bonanni, A. Lundskog, Impact of residual carbon impurities and gallium vacancies on trapping effects in AlGaN/GaN metal insulator semiconductor high electron mobility transistors, *Appl. Phys. Lett.*, 107: 032106, 2015.

10. D. Bisi, A. Stocco, I. Rossetto, M. Meneghini, F. Rampazzo, A. Chini, F. Soci et al., Effects of buffer compensation strategies on the electrical performance and RF reliability of AlGaN/GaN HEMTs, *Microelectronics Reliab.*, 55(9): 1662–1666, 2015. doi:10.1016/j.microrel.2015.06.038.

11. M. J. Uren, M. Silvestri, M. Cäsar, G. A. M. Hurkx, J. A. Croon, J. Šonský, M. Kuball, Intentionally Carbon-Doped AlGaN/GaN HEMTs: Necessity for Vertical Leakage Paths, *IEEE Electron Device Lett.*, 35: 327, 2014.

12. M. Silvestri, M. J. Uren, M. Kuball, Iron-induced deep-level acceptor center in GaN/AlGaN high electron mobility transistors: Energy level and cross section, *Appl. Phys. Lett.*, 102: 073501, 2013.

13. M. Meneghini, I. Rossetto, D. Bisi, A. Stocco, A. Chini, A. Pantellini, C. Lanzieri, A. Nanni, G. Meneghesso, E. Zanoni, Buffer traps in Fe-doped AlGaN/GaN HEMTs: Investigation of the physical properties based on pulsed and transient measurements, *IEEE Trans. Electron Dev.*, 61: 4070, 2014.

14. M. Meneghini, N. Ronchi, A. Stocco, G. Meneghesso, U. K. Mishra, Y. Pei, E. Zanoni, Investigation of Trapping and Hot-Electron Effects in GaN HEMTs by Means of a Combined Electrooptical Method, *IEEE Trans. Electron Dev.*, 58(9): 2996–3003, 2011.

15. M. Meneghini, I. Rossetto, D. Bisi, M. Ruzzarin, M. Van Hove, S. Stoffels, T.-L. Wu et al., Negative Bias-Induced Threshold Voltage Instability in GaN-on-Si Power HEMTs, *IEEE Electron Device Lett.*, 37: 474, 2016.

16. G. Meneghesso, M. Meneghini, D. Bisi, I. Rossetto, T.-L. Wu, M. Van Hove, D. Marcon, S. Stoffels, S. Decoutere, E. Zanoni, Trapping and reliability issues in GaN-based MIS HEMTs with partially recessed gate, *Microelectronics Reliab.*, 58: 151, 2016.

17. A. Guo, J. A. del Alamo, Positive-bias temperature instability (PBTI) of GaN MOSFETs, in *IEEE International Reliability Physics Symposium*, 2015, p. 6C.5.1.

18. T.-L. Wu, J. Franco, D. Marcon, B. De Jaeger, B. Bakeroot, S. Stoffels, M. Van Hove, G. Groeseneken, S. Decoutere, Toward understanding positive bias temperature instability in fully recessed-gate GaN MISFETs, *IEEE Trans. Electron Dev.*, 63: 1853, 2016.

19. S. C. Binari, P. B. Klein, T. E. Kazior, Trapping effects in GaN and SiC microwave FETs, *Proceedings of the IEEE*, 90: 1048, 2002.

20. D. Bisi, A. Chini, F. Soci, A. Stocco, M. Meneghini, A. Pantellini, A. Nanni et al., Hot-electron degradation of AlGaN/GaN high-electron mobility transistors during RF operation: Correlation with GaN buffer design, *IEEE Electron Device Lett.*, 36(10): 1011–1014, 2015. doi:10.1109/LED.2015.2474116.

21. A. Benvegnù, S. Laurent, M. Meneghini, D. Barataud, G. Meneghesso, E. Zanoni, R. Quere, On-wafer single-pulse thermal load–pull RF characterization of trapping phenomena in AlGaN/GaN HEMTs, *IEEE Trans. Microw. Theory and Tech.*, 64: 767, 2016.

22. J. Würfl, O. Hilt, E. Bahat-Treidel, R. Zhytnytska, P. Kotara, F. Brunner, O. Krueger, M. Weyers, Techniques towards GaN power transistors with improved high voltage dynamic switching properties, in *IEEE International Electron Device Meeting*, Washington, DC, December 9–11, 2013, p. 6.1.1.

23. P. Moens, P. Vanmeerbeek, A. Banerjee, J. Guo, C. Liu, P. Coppens, A. Salih et al., On the impact of carbon-doping on the dynamic Ron and off-state leakage current of 650V GaN power devices, in *International Symposium on Power Semiconductor Devices & IC's*, Hong-Kong, China, May 10–14, 2015, p. 37.

24. R. Reiner, P. Waltereit, F. Benkhelifa, S. Müller, M. Wespel, R. Quay, M. Schlechtweg, M. Mikulla, O. Ambacher, Benchmarking of large-area GaN-on-Si HFET power devices for highly-efficient, fast-switching converter applications, in *IEEE Compound Semiconductor Integrated Circuit Symposium (CSICS)*, Monterey, CA, October 13–16, 2013, p. 1.

25. C. G. Van de Walle, J. Neugebauer, First-principles calculations for defects and impurities: Applications to III-nitrides, *J. Appl. Phys.*, 95: 3851, 2004.

26. D. V. Lang, Deep-level transient spectroscopy: A new method to characterize traps in semiconductors, *J. Appl. Phys.*, 45: 3023, 1974.

27. J. Joh, J. A. del Alamo, A current-transient methodology for trap analysis for GaN high electron mobility transistors, *IEEE Trans. Electron Dev.*, 58(1): 132–140, 2011.

28. D. Bisi, M. Meneghini, C. De Santi, A. Chini, M. Dammann, P. Brückner, M. Mikulla, G. Meneghesso, E. Zanoni, Deep-level characterization in GaN HEMTs-Part I: Advantages and limitations of drain current transient measurements, *IEEE Trans. Electron Dev.*, 60: 3166, 2013.

29. D. Bisi, A. Stocco, M. Meneghini, F. Rampazzo, A. Cester, G. Meneghesso, E. Zanoni, High-voltage double-pulsed measurement system for GaN-based power HEMTs, in *2014 IEEE International Reliability Physics Symposium*, Waikoloa, HI, June 1–5, 2014, p. CD.11.1.

30. S. P. Kumar, A. Agrawal, S. Kabra, M. Gupta, R. S. Gupta, An analysis for AlGaN/GaN modulation doped field effect transistor using accurate velocity-field dependence for high power microwave frequency applications, *Microelectr. J.*, 37(11): 1339–1346, 2006.

31. A. R. Arehart, Investigation of electrically active defects in GaN, AlGaN, and AlGaN/GaN high electron mobility transistors, PhD dissertation, Ohio State University, Columbus, OH, 2009.

32. A. Sasikumar, A. Arehart, S.A Ringel, S. Kaun, M. H. Wong, U. K. Mishra, J. S. Speck, Direct correlation between specific trap formation and electric stress-induced degradation in MBE-grown AlGaN/GaN HEMTs, *IEEE International Reliability Physics Symposium (IRPS)*, Anaheim, CA, April 15–19, 2012, p. 2C.3.1.

33. L. Stuchlíková, J. Šebok, J. Rybár, M. Petrus, M. Nemec, L. Harmatha, J. Benkovská et al., Investigation of deep energy levels in heterostructures based on GaN by DLTS, *8th International Conference on Advanced Semiconductor Devices & Microsystems (ASDAM)*, Smolenice, Slovakia, October 25–27, 2010, pp. 135–138.

34. Z.-Q. Fang, D. C. Look, D. H. Kim, I. Adesida, Traps in AlGaN/GaN/SiC heterostructures studied by deep level transient spectroscopy, *Appl. Phys. Lett.*, 87(18): 182115, 2005.

35. J.-G. Tartarin, S. Karboyan, F. Olivie, G. Astre, L. Bary, B. Lambert, I-DLTS, electrical lag and low frequency noise measurements of trapping effects in AlGaN/GaN HEMT for reliability studies, *European Microwave Integrated Circuits Conference (EuMIC)*, Manchester, UK, October 10–11, 2011, pp. 438–441.

36. G. Meneghesso, G. Verzellesi, R. Pierobon, F. Rampazzo, A. Chini, U. K. Mishra, C. Canali, E. Zanoni, Surface-related drain current dispersion effects in AlGaN-GaN HEMTs, *IEEE Trans. Electron Dev.*, 51(10): 1554–1561, 2004.

37. M. Faqir, G. Verzellesi, A. Chini, F. Fantini, F. Danesin, G. Meneghesso, E. Zanoni, C. Dua, Mechanisms of RF Current Collapse in AlGaN–GaN High Electron Mobility Transistors, *IEEE Trans. Device Mater. Rel.*, 8(2): 240–247, 2008.

38. M. Caesar, M. Dammann, V. Polyakov, P. Waltereit, W. Bronner, M. Baeumler, R. Quay, M. Mikulla, O. Ambacher, Generation of traps in AlGaN/GaN HEMTs during RF-and DC-stress test, *IEEE International Reliability Physics Symposium (IRPS)*, Anaheim, CA, April 15–19, 2012, pp. CD6.1–CD6.5.

39. M. Ťapajna, R. J. T. Simms, Yi Pei, U. K. Mishra, M. Kuball, Integrated optical and electrical analysis: Identifying location and properties of traps in AlGaN/GaN HEMTs during electrical stress, *IEEE Electron Device Lett.*, 31(7): 662–664, 2010.

40. S. DasGupta, M. Sun, A. Armstrong, R. J. Kaplar, M. J. Marinella, J. B. Stanley, S. Atcitty, T. Palacios, Slow detrapping transients due to gate and drain bias stress in high breakdown voltage AlGaN/GaN HEMTs, *IEEE Trans. Electron Dev.*, 59(8): 2115–2122, 2012.

41. D. Jin, J. A. del Alamo, Methodology for the study of dynamic ON-resistance in high-voltage GaN field-effect transistors, *IEEE Trans. Electron Dev.*, 60(10): 3190, 2013.

42. W. R. Thurber, R. A. Forman, W. E. Phillips, A novel method to detect nonexponential transients in deep level transient spectroscopy, *J. Appl. Phys.*, 53(11): 7397, 1982.

43. O. Mitrofanov, M. Manfra, Mechanisms of gate lag in GaN/AlGaN/GaN high electron mobility transistors, *Superlattices and Microstruct.*, 34(1–2): 33–53, 2003.

44. P. D. Kirchner, W. J. Schaff, G. N. Maracas, L. F. Eastman, T. I. Chappell, C. M. Ransom, The analysis of exponential and nonexponential transients in deep-level transient spectroscopy, *J. Appl. Phys.*, 52(11): 6462, 1981.

45. P. Omling, L. Samuelson, H. G. Grimmeiss, Deep level transient spectroscopy evaluation of nonexponential transients in semiconductor alloys, *J. Appl. Phys.*, 54(9): 5117, 1983.

46. H. Hasegawa, M. Akazawa, Current collapse transient behavior and its mechanism in submicron-gate AlGaN/GaN heterostructure transistors, *J. Vac. Sci. Technol. B*, 27: 2048, 2009.

47. J. Joh, J. A. del Alamo, U. Chowdhury, T.-M. Chou, H.-Q. Tserng, J. L. Jimenez, Measurement of channel temperature in GaN high-electron mobility transistors, *IEEE Trans. Electron Dev.*, 56(12): 2895–2901, 2009.

48. A. Chini, F. Soci, M. Meneghini, G. Meneghesso, E. Zanoni, Deep levels characterization in GaN HEMTs—Part II: Experimental and numerical evaluation, *IEEE Trans. Electron Dev.*, 60: 3176, 2013.

49. P. Hacke, T. Detchprohm, K. Hiramatsu, N. Sawaki, Analysis of deep levels in n-type GaN by transient capacitance methods, *J. Appl. Phys.*, 76: 304, 1994.

50. C. B. Soh, S. J. Chua, H. F. Lim, D. Z. Chi, W. Liu, S. Tripathy, Identification of deep levels in GaN associated with dislocations, *J. Phys.: Condens. Matter*, 16(34): 6305–6315, 2004.

51. P. Hacke, H. Nakayama, T. Detchprohm, K. Hiramatsu, N. Sawaki, Deep levels in the upper band-gap region of lightly Mg-doped GaN, *Appl. Phys. Lett.*, 68(10): 1362–1364, 1996.

52. F. D. Auret, S. A. Goodman, F. K. Koschnick, J-M. Spaeth, B. Beaumont, P. Gibart, Electrical characterization of two deep electron traps introduced in epitaxially grown n-GaN during He-ion irradiation, *Appl. Phys. Lett.*, 73: 3745, 1998.

53. H. M. Chung, W. C. Chuang, Y. C. Pan, C. C. Tsai, M. C. Lee, W. H. Chen, W. K. Chen, C. I. Chiang, C. H. Lin, H. Chang, Electrical characterization of isoelectronic In-doping effects in GaN films grown by metalorganic vapor phase epitaxy, *Appl. Phys. Lett.*, 76(7): 897–899, 2000.

54. Z-Q. Fang, L. Polenta, J. W. Hemsky, D. C. Look, Deep centers in as-grown and electron-irradiated n-GaN, in *International Semiconducting and Insulating Materials Conference SIMC-XI*, 2000, pp. 35–42.

55. H. K. Cho, C. S. Kim, C.-H. Hong, Electron capture behaviors of deep level traps in unintentionally doped and intentionally doped n-type GaN, *J. Appl. Phys.*, 94(3): 1485–1489, 2003.

56. F. D. Auret, W. E. Meyer, L. Wu, M. Hayes, M. J. Legodi, B. Beaumont, P. Gibart, Electrical characterisation of hole traps in n-type GaN, *phys. stat. sol. (a)*, 201(10): 2271–2276, 2004.

57. T. Okino, M. Ochiai, Y. Ohno, S. Kishimoto, K. Maezawa, T. Mizutani, Drain current DLTS of AlGaN-GaN MIS-HEMTs, *IEEE Electron Device Lett.*, 25(8): 523–525, 2004.

58. D. C. Look, Z.-Q. Fang, B. Claflin, Identification of donors, acceptors, and traps in bulk-like HVPE GaN, *J. Cryst. Growth*, 281(1): 143–150, 2005.

59. D. Johnstone, S. Biyikli, S. Dogan, Y. T. Moon, F. Yun, H. Morkoç, Comparison of deep levels in GaN grown by MBE, MOCVD, and HVPE, *Proceedings SPIE 5739, Light-Emitting Diodes: Research, Manufacturing, and Applications IX*, SPIE, San Jose, CA, March, 2005, pp. 7–15.

60. A. R. Arehart, A. Corrion, C. Poblenz, J. S. Speck, U. K. Mishra, S. A. Ringel, Deep level optical and thermal spectroscopy of traps in n-GaN grown by ammonia molecular beam epitaxy, *Appl. Phys. Lett.*, 93(11): 112101, 2008.

61. A. R. Arehart, A. Sasikumar, S. Rajan, G. D. Via, B. Poling, B. Winningham, E. R. Heller et al., Direct observation of 0.57 eV trap-related RF output power reduction in AlGaN/GaN high electron mobility transistors, *Solid-state Elec.*, 80: 19–22, 2013.

62. S. Chen, U. Honda, T. Shibata, T. Matsumura, Y. Tokuda, K. Ishikawa, M. Hori, H. Ueda, T. Uesugi, T. Kachi, As-grown deep-level defects in n-GaN grown by metal–organic chemical vapor deposition on freestanding GaN, *J. Appl. Phys.*, 112(5): 053513, 2012.

63. J. Pernot, P. Muret, Electronic properties of the EC-0.6 eV electron trap in n-type GaN, *J. Appl. Phys.*, 103(2): 023704, 2008.

64. X. D. Chen, Y. Huang, S. Fung, C. D. Beling, C. C. Ling, J. K. Sheu, M. L. Lee, G. C. Chi, S. J. Chang, Deep level defect in Si-implanted GaN n+-p junction, *Appl. Phys. Lett.*, 82: 3671, 2003.

65. Z.-Q. Fang, G. C. Farlow, B. Claflin, D. C. Look, D. S. Green, Effects of electron-irradiation on electrical properties of AlGaN/GaN Schottky barrier diodes, *J. Appl. Phys.*, 105(12): 123704, 2009.

66. C. B. Soh, D. Z. Chi, A. Ramam, H. F. Lim, S. J. Chua, Study of electrically active defects in n-GaN layer, *Mat. Sci. Semicon. Proc.*, 4(6): 595–600, 2001.

67. A. Y. Polyakov, N. B. Smirnov, A. V. Govorkov, A. A. Shlensky, S. J. Pearton, Influence of high-temperature annealing on the properties of Fe doped semi-insulating GaN structures, *J. Appl. Phys.*, 95(10): 5591, 2004.

68. P. Kamyczek, E. Placzek-Popko, Vl. Kolkovsky, S. Grzanka, R. Czernecki, A deep acceptor defect responsible for the yellow luminescence in GaN and AlGaN, *J. Appl. Phys.*, 111(11): 113105, 2012.

69. J. L. Lyons, A. Janotti, C. G. Van de Walle, Carbon impurities and the yellow luminescence in GaN, *Appl. Phys. Lett.*, 97(15): 152108, 2010.

70. C. De Santi, M. Meneghini, H. Ishida, T. Ueda, G. Meneghesso, E. Zanoni, Gate frequency sweep: An effective method to evaluate the dynamic performance of AlGaN/GaN power heterojunction field effect transistors, *Appl. Phys. Lett.*, 105: 073507, 2014.

71. G. Parish, S. Keller, S. P. Denbaars, U. K. Mishra, SIMS investigations into the effect of growth conditions on residual impurity and silicon incorporation in GaN and AlGaN, *J. Electron. Mater.*, 29: 15, 2000.

72. D. D. Koleske, A. E. Wickenden, R. L. Henry, M. E. Twigg, Influence of MOVPE growth conditions on carbon and silicon concentrations in GaN, *J. Cryst. Growth*, 242: 55, 2002.

73. A. E. Wickenden, D. D. Koleske, R. L. Henry, M. E. Twigg, M. Fatemi, Resistivity control in unintentionally doped GaN films grown by MOCVD, *J. Cryst. Growth*, 260: 54, 2004.

74. F. Brunner, E. Bahat-Treidel, M. Cho, C. Netzel, O. Hilt, J. Würfl, M. Weyers, Comparative study of buffer designs for high breakdown voltage AlGaNGaN HFETs, *Phys. Status Solidi C*, 8: 2427, 2011.

75. J.-T. Chen, U. Forsberg, E. Janzén, Impact of residual carbon on two-dimensional electron gas properties in AlxGa1xN/GaN, *Appl. Phys. Lett.*, 102: 193506, 2013.

76. P. Gamarra, C. Lacam, M. Tordjman, J. Splettstosser, B. Schauwecker, M.-A. di Forte-Poisson, Optimisation of a carbon doped buffer layer for AlGaN/GaN HEMT devices, *J. Cryst. Growth*, 414: 232, 2015.

77. C. Poblenz, P. Waltereit, S. Rajan, S. Heikman, U. K. Mishra, J. S. Speck, Effect of carbon doping on buffer leakage in AlGaN/GaN high electron mobility transistors, *J. Vac. Sci. Technol. B*, 22: 1145, 2004.

78. W. Z. Wang, S. L. Selvaraj, K. T. Win, S. B. Dolman, T. Bhat, N. Yakovlev, S. Tripathy, G. Q. Lo, Effect of carbon doping and crystalline quality on the vertical breakdown characteristics of GaN layers grown on 200-mm silicon substrates, *J. Electron. Mater.*, 44: 3272, 2015. doi:10.1007/s11664-015-3832-3.

79. Dong-Seok Kim, C.-H. Won, H.-S. Kang, Y.-J. Kim, Y. T. Kim, I. M. Kang, J.-H. Lee, Growth and characterization of semi-insulating carbon-doped/undoped GaN multiple-layer buffer, *Semicond. Sci. Technol.*, 30: 035010, 2015.

80. E. Bahat-Treidel, F. Brunner, O. Hilt, E. Cho, J. Würfl, G. Tränkle, AlGaN/GaN/GaN:C back-barrier HFETs with breakdown voltage of over 1 kV and low RON × A, *IEEE Trans. Electron Dev.*, 57(11): 3050, 2010.

81. M. J. Uren, J. Möreke, M. Kuball, Buffer design to minimize current collapse in GaN/AlGaN HFETs, *IEEE Trans. Electron Dev.*, 59: 3327, 2012.

82. E. Bahat-Treidel, O. Hilt, F. Brunner, S. Pyka, J. Wurfl, Systematic study of GaN based power transistors' dynamic on-state resistance at elevated temperatures, in *International Symposium on Compound Semiconductors (ISCS)*, 2014.

83. A. F. Wright, Substitutional and interstitial carbon in wurtzite GaN, *J. Appl. Phys.*, 92: 2575, 2002.

84. J. L. Lyons, A. Janotti, C. G. Van de Walle, Effects of carbon on the electrical and optical properties of InN, GaN, and AlN, *Phys. Rev. B*, 89: 035204, 2014.

85. J. L. Lyons, A. Janotti, C. G. Van de Walle, Carbon impurities and the yellow luminescence in GaN, *Appl. Phys. Lett.*, 97: 152108, 2010.

86. A. Armstrong and A. R. Arehart, D. Green, U. K. Mishra, J. S. Speck, S. A. Ringel, Impact of deep levels on the electrical conductivity and luminescence of gallium nitride codoped with carbon and silicon, *J. Appl. Phys.*, 98: 053704, 2005.

87. A. Armstrong, A. R. Arehart, B. Moran, S. P. DenBaars, U. K. Mishra, Impact of carbon on trap states in n-type GaN grown by metalorganic chemical vapor, *Appl. Phys. Lett.*, 84: 374, 2004.

88. A. Armstrong, A. R. Arehart, B. Moran, S. P. DenBaars, U. K. Mishra, J. S. Speck, S. A. Ringel, in *30th International Symposium on Compound Semiconductors*, San Diego, CA, 2003, p. 42.

89. Z.-Q. Fang, B. Claflin, D. C. Look, D. S. Green, R. Vetury, Deep traps in AlGaN/GaN heterostructures studied by deep level transient spectroscopy: Effect of carbon concentration in GaN buffer layers, *J. Appl. Phys.*, 108: 063706, 2010.

90. U. Honda, Y. Yamada, Y. Tokuda, K. Shiojima, Deep levels in n-GaN doped with carbon studied by deep level and minority carrier, *J. J. Appl. Phys.*, 51: 04DF04, 2012.

91. D. Bisi, M. Meneghini, M. Van Hove, D. Marcon, S. Stoffels, T.-L. Wu, S. Decoutere, G. Meneghesso, E. Zanoni, Trapping mechanisms in GaN-based MIS-HEMTs grown on silicon substrate, *Phys. Status Solidi A*, 212: 1122, 2015.

92. M. Meneghini, D. Bisi, D. Marcon, S. Stoffels, M. Van Hove, T-L Wu, S. Decoutere, G. Meneghesso, E. Zanoni, Trapping and reliability assessment in D-Mode GaN-Based MIS-HEMTs for power applications, *IEEE Trans. Power Electron.*, 29(5): 2199, 2014.

93. D. Bisi, M. Meneghini, F. A. Marino, D. Marcon, S. Stoffels, M. Van Hove, S. Decoutere, G. Meneghesso, E. Zanoni, Kinetics of buffer-related RON-increase in GaN-on-silicon MIS-HEMTs, *IEEE Electron Device Lett.*, 35: 1004, 2014.

94. M. Meneghini, P. Vanmeerbeek, R. Silvestri, S. Dalcanale, A. Banerjee, D. Bisi, E. Zanoni, G. Meneghesso, P. Moens, Temperature-dependent dynamic-Ron in GaN-based MIS-HEMTs: Role of surface traps and buffer leakage, *IEEE Trans. Electron Dev.*, 62: 782, 2015.

95. M. H. Kwan, K.-Y. Wong, Y. S. Lin, F. W. Yao, M. W. Tsai, Y.-C. Chang, P. C. Chen et al., CMOS-compatible GaN-on-Si field-effect transistors for high voltage power applications, in *IEEE Electron Device Meeting (IEDM)*, San Francisco, CA, December 15–17, 2014, p. 17.

96. M. J. Uren, M. Caesar, S. Karboyan, P. Moens, P. Vanmeerbeek, M. Kuball, Electric field reduction in C-doped AlGaN/GaN on Si high electron mobility transistors, *IEEE Electron Device Lett.*, 36: 826, 2015.

97. K. Tanaka, T. Morita, H. Umeda, S. Kaneko, M. Kuroda, A. Ikoshi, H. Yamagiwa et al., Suppression of current collapse by hole injection from drain in a normally-off GaN-based hybrid-drain-embedded gate injection transistor, *Appl. Phys. Lett.*, 107: 163502, 2015.

98. S. Stoffels, M. Zhao, R. Venegas, P. Kandaswamy, S. You, T. Novak, Y. Saripalli, M. Van Hove, S. Decoutere, The physical mechanism of dispersion caused by AlGaN/GaN buffers on Si and optimization for low dispersion, in *IEEE International Electron Device Meeting*, Washington, DC, December 7–9, 2015, p. 35–4.

99. S. Kaneko, M. Kuroda, M. Yanagihara, A. Ikoshi, H. Okita, T. Morita, K. Tanaka et al., Current-collapse-free operations up to 850 V by GaN-GIT utilizing hole injection from drain, in *International Symposium on Power Semiconductor Devices & IC's (ISPSD)*, Hong Kong, China, May 10–14, 2015, p. 41.

100. X. Tang, B. Li, Z. Zhang, G. Tang, J. Wei, K. J. Chen, Characterization of static and dynamic behaviors in AlGaN/GaN-on-Si power transistors with photonic-ohmic drain, *IEEE Trans. Electron Dev.*, 63: 2831, 2016.

101. Y.-F. Wu, J. Gritters, L. Shen, R. P. Smith, J. McKay, R. Barr, R. Birkhahn, Performance and robustness of first generation 600-V GaN-on-Si power transistors, in *IEEE Workshop on Wide Bandgap Power Devices and Applications (WiPDA)*, Columbus, OH, October 27–29, 2013, p. 6.

102. X. Huang, Z. Liu, Q. Li, F. C. Lee, Evaluation and application of 600 V GaN HEMT in cascode structure, *IEEE Trans. Power Electronics*, 29: 2453, 2014.

103. J. Joh, N. Tipirneni, S. Pendharkar, S. Krishnan, Current collapse in GaN heterojunction field effect transistors for high-voltage switching applications, in *IEEE Interational Reliability Physics Symposium*, Waikoloa, HI, June 1–5, 2014, p. 6C–5.

104. M. Meneghini, A. Stocco, R. Silvestri, G. Meneghesso, E. Zanoni, Degradation of AlGaN/GaN high electron mobility transistors related to hot electrons, *Appl. Phys. Lett.*, 100: 233508, 2012.

105. Y. Puzyrev, S. Mukherjee, J. Chen, T. Roy, M. Silvestri, R. D. Schrimpf, D. M. Fleetwood, J. Singh, J. M. Hickley, A. Paccagnella, S. T. Pantelides, Gate bias dependence of defect-mediated hot-carrier degradation in GaN HEMTs, *IEEE Trans. Electron Dev.*, 61: 1316, 2014.

106. I. Hwang, J. Kim, S. Chong, H.-S. Choi, S.-K. Hwang, J. Oh, J. K. Shin, U-In Chung, Impact of channel hot electrons on current collapse in AlGaN/GaN HEMTs, *IEEE Electron Device Lett.*, 34(12): 1494, 2013.

107. N. Braga, R. Mickevicius, R. Gaska, X. Hu, M. S. Shur, et al., Simulation of hot electron and quantum effects in AlGaN/GaN heterostructure field effect transistors, *J. Appl. Phys.*, 95: 6409, 2004.

108. M. Wang, D. Yan, C. Zhang, B. Xie, C. P. Wen, J. Wang, Y. Hao, W. Wu, B. Shen, Investigation of surface- and buffer-induced current collapse in GaN high-electron mobility transistors using a soft switched pulsed (I-V) measurement, *IEEE Electron Device Lett.*, 35: 1094, 2014.

109. M. Meneghini, D. Bisi, D. Marcon, S. Stoffels, M. Van Hove, T.-L. Wu, S. Decoutere, G. Meneghesso, E. Zanoni, Trapping in GaN-based metal-insulator-semiconductor transistors: Role of high drain bias and hot electrons, *Appl. Phys. Lett.*, 104: 143505, 2014.

110. M. Meneghini, G. Cibin, M. Bertin, G. A. M. Hurkx, P. Ivo, J. Šonský, J. A. Croon, G. Meneghesso, E. Zanoni, OFF-state degradation of AlGaN/GaN power HEMTs: Experimental demonstration of time-dependent drain-source breakdown, *IEEE Trans. Electron Dev.*, 61: 1987, 2014.

111. E. Zanoni, M. Meneghini, G. Meneghesso, D. Bisi, I. Rossetto, A. Stocco, Reliability and failure physics of GaN HEMT, MISHEMT HEMT and p-gate HEMTs for power switching applications, in *IEEE 3rd Workshop on Wide Bandgap Power Devices and Applications* (*WiPDA*), 2015, Blacksburg, VA. 2015, p. 75.

112. S. R. Bahl, M. Van Hove, X. Kang, D. Marcon, M. Zahid, S. Decoutere, New source-side breakdown mechanism in AlGaN/GaN insulated-Gate HEMTs, in 2013 *25th International Symposium on Power Semiconductor Devices & IC's (ISPSD)*, Kanazawa, Japan, 2013, p. 419.

113. R. Degraeve, T. Kauerauf, M. Cho, M. Zahid, L. A. Ragnarsson, D. P. Brunco, B. Kaczer, Ph. Roussel, S. De Gendt, G. Groeseneken, Degradation and breakdown of 0.9 nm EOT SiO/sub 2/ ALD HfO/sub 2/metal gate stacks under positive constant voltage stress, *IEEE International Electron Devices Meeting, 2005. IEDM Technical Digest*, 2005, pp. 408–411.

114. O. Hilt, R. Zhytnytska, J. Böcker, E. Bahat-Treidel, F. Brunner, A. Knauer, S. Dieckerhoff, J. Würfl, 70 mΩ / 600 V Normally-off GaN transistors on SiC and Si substrates, *Proceedings of the 27th International Symposium on Power Semiconductor Devices & IC's*, IEEE, Hong Kong, China, May 10–14, 2015, pp. 237–240.

115. Y. Uemoto, M. Hikita, H. Ueno, H. Matsuo, H. Ishida, A normally-off AlGaN/GaN transistor with $R_{on}A$=2.6mΩcm2 and BV$_{ds}$=640V using conductivity modulation, *Proceedings International Electron Devices Meeting (IEDM)*, IEEE, San Francisco, CA, December 11–13, 2006, pp. 1–4.

116. Y. Uemoto, M. Hikita, H. Ueno, H. Matsuo, H. Ishida, M. Yanagihara, T. Ueda, Gate injection transistor (GIT)—A normally-off AlGaN/GaN power transistor using conductivity modulation, *IEEE Trans. Electron Devices*, 54(12): 3393–3399, 2007.

117. M. Meneghini, M. Scamperle, M. Pavesi, M. Manfredi, T. Ueda, H. Ishida, T. Tanaka, D. Ueda, G. Meneghesso, E. Zanoni, Electron and hole-related luminescence processes in gate injection transistors, *Appl. Phys. Lett.*, 97: 033506, 2010.

118. O. Hilt, A. Knauer, F. Brunner, E. Bahat-Treidel, J. Würfl, Normally-off AlGaN/GaN HFET with p-type Ga gate and AlGaN, *Proceedings 22nd International Symposium Power Semiconductor Devices*, IEEE, Hiroshima, Japan, June 6–10, 2010, pp. 248–250.

119. O. Hilt, F. Brunner, E. Cho, A. Knauer, E. Bahat-Treidel, J. Würfl, Normally-off high-voltage p-GaN gate GaN HFET with carbon-doped, *Proceedings IEEE 23rd International Symposium Power Semiconductor Devices*, San Diego, CA, May 23–26, 2011, pp. 239–242.

120. J. Würfl, O. Hilt, E. Bahat-Treidel, R. Zhytnytska, P. Kotara, F. Brunner, O. Krueger, M. Weyers, Techniques towards GaN power transistors with improved high voltage dynamic switching properties, *IEEE International Electron Devices Meeting*, Washington, DC, December 9–11, 2013, pp. 6.1.1–6.1.4.

121. J. Würfl, O. Hilt, E. Bahat-Treidel, P. Kurpas, S. A. Chevchenko, O. Bengtsson, E. Ersoy, A. Liero, A. Wentzel, W. Heinrich, N. Badawi, S. Dieckerhoff, Enabling GaN high speed devices: Microwave meets power eElectronics—and vice versa, *Microwave Integrated Circuits Conference (EuMIC)*, IEEE, Nuremberg, Germany, October 6–8, 2013, pp. 176–179.

122. H. Okita, M. Hikita, A. Nishio, T. Sato, K. Matsunaga, H. Matsuo, M. Mannoh, Y. Uemoto, Through recessed and regrowth gate technology for realizing process stability of GaN-GITs, in 2016 *28th International Symposium on Power Semiconductor Devices and ICs (ISPSD)*, Prague, Czech Republic. 2016, p. 23.

123. H. Hahn, F. Benkhelifa, O. Ambacher, F. Brunner, A. Noculak, H. Kalisch, A. Vescan, Threshold voltage engineering in GaN-Based HFETs: A systematic study with the threshold voltage reaching more than 2 V, *IEEE Trans. Electron Dev.*, 62(2): 538–545, 2015.

124. Tian-Li Wu, D. Marcon, B. De Jaeger, M. Van Hove, B. Bakeroot, S. Stoffels, G. Groeseneken, S. Decoutere, R. Roelofs, Time dependent dielectric breakdown (TDDB) evaluation of PE-ALD SiN gate dielectrics on AlGaN/GaN recessed gate D-mode MIS-HEMTs and E-mode MIS-FETs, 2015 *IEEE International Reliability Physics Symposium*, Monterey, CA, April 19–23, 2015, pp. 6C.4.1–6C.4.6.

125. P. Lagger, P. Steinschifter, M. Reiner, M. Stadtmüller, G. Denifl, A. Naumann, J. Müller, L. Wilde, J. Sundqvist, D. Pogany, C. Ostermaier, Role of the dielectric for the charging dynamics of the dielectric/barrier interface in AlGaN/GaN based metal-insulator-semiconductor structures under forward gate bias stress, *Appl. Phys. Lett.*, 105: 033512, 2014.

126. I. Rossetto, M. Meneghini, D. Bisi, A. Barbato, M. Van Hove, D. Marcon, T.-L. Wu, S. Decoutere, G. Meneghesso, E. Zanoni, Impact of gate insulator on the dc and dynamic performance of AlGaN/GaN MIS-HEMTs, *Microelectronics Reliab.*, 55: 1692–1696, 2015.

127. P. Lagger, C. Ostermaier, G. Pobegen and D. Pogany, Towards understanding the origin of threshold voltage instability of AlGaN/GaN MIS-HEMTs, *Electron Devices Meeting (IEDM), 2012IEEE International*, San Francisco, CA, December 10–13, 2012, pp. 13.1.1–13.1.4.

128. P. Lagger, M. Reiner, D. Pogany, C. Ostermaier, Comprehensive study of the complex dynamics of forward bias-induced threshold voltage drifts in GaN Based MIS-HEMTs by stress/recovery experiments, *IEEE Trans. Electron Dev.*, 61(4): 1022–1030, 2014.

129. C. Mizue, Y. Hori, M. Miczek, T. Hashizume, Capacitance–voltage characteristics of Al2O3/AlGaN/GaN structures and state density distribution at Al2O3/AlGaN interface, *Jpn. J. Appl. Phys.*, 50: 021001, 2011.

130. Tian-Li Wu, D. Marcon, B. De Jaeger, M. Van Hove, B. Bakeroot, D. Lin, S. Stoffels, X. Kang, R. Roelofs, G. Groeseneken, S. Decoutere, The impact of the gate dielectric quality in developing Au-free D-mode and E-mode recessed gate AlGaN/GaN transistors on a 200mm Si substrate, *Proceedings of the 27th International Symposium on Power Semiconductor Devices & IC's*, IEEE, Hong Kong, China, May 10–14, 2015, pp. 225–228.

131. D. Bisi, S. H. Chan, X. Liu, R. Yeluri, S. Keller, M. Meneghini, G. Meneghesso, E. Zanoni, U. K. Mishra, On trapping mechanisms at oxide-traps in Al2O3/GaN metal-oxide-semiconductor capacitors, *Appl. Phys. Lett.*, 108: 112104, 2016.

132. M. Meneghini, I. Rossetto, D. Bisi, M. Ruzzarin, M.Van Hove, S. Stoffels, T.-L. Wu, D. Marcon, S. Decoutere, G. Meneghesso, E. Zanoni, Negative bias-induced threshold voltage instability in GaN-on-Si power HEMTs, *Electron Device Lett.*, 37(4): 474–477, 2016.

133. Alex Guo, Jesús A. del Alamo, Negative-Bias Temperature Instability of GaN MOSFETs, in *Reliability Physics Symposium (IRPS), 2016 IEEE International*, Pasadena, CA, April 17–21, 2016, pp. 4A-1-1–4A-1-6.

134. A. Fontserè, A. Pérez-Tomás, M. Placidi, J. Llobet, N. Baron, S. Chenot, Y. Cordier et al., Gate current analysis of AlGaN/GaN on silicon heterojunction transistors at the nanoscale, *Appl. Phys. Lett.*, 101: 093505, 2012.

135. F. Sang, M. Wang, C. Zhang, M. Tao, B. Xie, C. P. Wen, J. Wang, Y. Hao, W. Wu, B. Shen, Investigation of the threshold voltage drift in enhancement mode GaN MOSFET under negative gate bias stress, *Jpn. J. Appl. Phys*, 54(4): 044101, 2015.

136. H. Kambayashi, T. Nomura, H. Ueda, K. Harada, Y. Morozumi, K. Hasebe, A. Teramoto, S. Sugawa, T. Ohmi, High Quality SiO2 / Al2O3gate stack for GaN MOSFET, *Jpn. J. Appl. Phys.*, 52: 04CF09, 2013.

137. M. Van Hove, X. Kang, S. Stoffels, D. Wellekens, N. Ronchi, R. Venegas, K. Geens, S. Decoutere, Fabrication and performance of Au-Free AlGaN/GaN-on-silicon power devices with Al2O3 and Si3N4/Al2O3 gate dielectrics, *IEEE Trans. Electron Dev.*, 60(10): 3071–3078, 2013.

138. M. Choi, J. L. Lyons, A. Janotti, C. G. Van de Walle, Impact of native defects in high-k dielectric oxides on GaN/oxide metal-oxide-semiconductor devices, *Phys. Status Solidi Basic Res.*, 250(4): 787–791, 2013.

139. R. Yeluri, X. Liu, M. Guidry, O. S. Koksaldi, S. Lal, J. Kim, J. Lu, S. Keller, and U. K. Mishra, Dielectric stress tests and capacitance-voltage analysis to evaluate the effect of post deposition annealing on Al2O3 films deposited on GaN, *Appl. Phys. Lett.*, 105: 222905, 2014.

140. M. Capriotti, P. Lagger, C. Fleury, M. Oposich, O. Bethge, C. Ostermaier, G. Strasser, and D. Pogany, Modeling small-signal response of GaN-based metal-insulator-semiconductor high electron mobility transistor gate stack in spill-over regime: Effect of barrier resistance and interface states, *J. Appl. Phys.*, 117: 024506, 2015.

141. Y. Yuan, L. Wang, B. Yu, B. Shin, J. Ahn, P. C. McIntyre, P. M. Asbeck, M. J. W. Rodwell, and Y. Taur, A distributed model for border traps in Al2O3 - inGaAs MOS devices, *IEEE Electron Device Lett.*, 32(4): 485–487, 2011.

142. K. Tang, A. Negara, T. Kent, R. Droopad, A. C. Kummel, P. C. Mcintyre, Border trap analysis and reduction in AL D-high-k InGaAs gate stacks, *Extended Abstracts of the Compound Semiconductor Week*, St. Barbar, CA, 2015.

143. R. Degraeve, G. Groeseneken, R. Bellens, J. L. Ogier, M. Depas, P. J. Roussel, H. E. Maes, New insights in the relation between electron trap generation and the statistical properties of oxide breakdown, *IEEE Trans. Electron Dev.*, 45(4): 904–911, 1998.

144. I. C. Chen, S. Holland, K. K. Young, C. Chang, C. Hu, Substrate hole current and oxide breakdown, *Appl. Phys. Lett.*, 49(11): 669–671, 1986.

145. E. Avni, J. Shappir, A model for silicon-oxide breakdown under high field and current stress, *J. Appl. Phys.*, 64(2): 743–748, 1988.

146. C. De Santi, M. Meneghini, M. Buffolo, G. Meneghesso, E. Zanoni, Experimental demonstration of time-dependent breakdown in GaN-based light emitting diodes, *IEEE Electron Device Lett.*, 37(5): 611–614. 2016.

147. P. Moens, A. Banerjee, P. Coppens, A. Constant, P. Vanmeerbeek, Z. Li, F. Declercq, L. De Schepper, H. De Vleeschouwer, C. Liu, B. Padmanabhan, W. Jeon, J. Guo, A. Salih, M. Tack, Technology and design of GaN power devices, 2015 *45th European Solid State Device Research Conference (ESSDERC)*, Graz, Austria, September 14–18, 2015, pp. 64–67.

148. M. Meneghini, I. Rossetto, F. Hurkx, J. Sonsky, J. A. Croon, G. Meneghesso, E. Zanoni, Extensive investigation of time-dependent breakdown of GaN-HEMTs submitted to OFF-state stress, *IEEE Trans. Electron Dev.*, 62(8): 2549–2015.

149. T.-L. Wu, D. Marcon, M. B. Zahid, M. Van Hove, S.Decoutere, G. Groeseneken, Comprehensive investigation of on-state stress on D-mode AlGaN/GaN MIS-HEMTs, *IEEE International Reliability Physics Symposium Proceedings*, Anaheim, CA, April 14–18, 2013, pp. 3C.5.1–3C.5.7.

150. I. Rossetto, M. Meneghini, O. Hilt, E. Bahat-Treidel, C. De Santi, S. Dalcanale, J. Wuerfl, E. Zanoni, G. Meneghesso, Time-dependent failure of GaN-on-Si power HEMTs with p-GaN gate, *IEEE Trans. Electron Dev.*, 63(6): 2334–2339, 2016.

151. M. Ťapajna, O. Hilt, E. Bahat-Treidel, J. Würfl, J. Kuzmik, Gate reliability investigation in normally-off p-type-GaN Cap/AlGaN/GaN HEMTs under forward bias stress, *IEEE Electron Device Lett.*, 37(4): 385–388, 2016.

152. K. Okada, K. Kurimoto, M. Suzuki, Anomalous TDDB statistics of gate dielectrics caused by charging-induced dynamic stress relaxation under constant–voltage stress, *IEEE Trans. Electron Dev.*, 63(6): 2268–2274, 2016.

153. M. Meneghini, A. Stocco, M. Bertin, N. Ronchi, A. Chini, D. Marcon, G. Meneghesso, E. Zanoni, Electroluminescence analysis of time-dependent reverse-bias degradation of HEMTs: A complete model, *Electron Devices Meeting (IEDM), 2011 IEEE International*, Washington, DC, December 5–7, 2011, pp. 19.5.1–19.5.4.

154. A. Sedhain, J. Li, J. Y. Lin, H. X. Jiang, Nature of deep center emissions in GaN, *Appl. Phys. Lett.*, 96(15): 151902, 2010.

155. M. Tapajna, O. Hilt, E. Bahat-Treidel, J. Würfl, J.Kuzmík, Investigation of gate-diode degradation in normally-off p-GaN/AlGaN/GaN high-electron-mobility transistors, *Appl. Phys. Lett.*, 107(19): 193506, 2015.

156. T.-L. Wu, D. Marcon, S. You, N. Posthuma, B. Bakeroot, S. Stoffels, M. Van Hove, G. Groeseneken, S. Decoutere, Forward bias gate breakdown mechanism in enhancement-mode p-GaN gate AlGaN/ GaN high-electron mobility transistors, *IEEE Electron Device Lett.*, 36(10): 1001–1003, 2015.

157. S. M. Sze, K. K. Ng, *Physics of Semiconductor Devices*, 3rd edition. Hoboken, NJ: Wiley, 2006.

158. N. Dyakonova, A. Dickens, M. S. Shur, R.Gaska, J. W.Yang, Temperature dependence of impact ionization in AlGaN–GaN heterostructure field effect transistors, *Appl. Phys. Lett.*, 72: 2562–2564, 1998.

159. C. Zhou, Q. Jiang, S. Huang, K. J. Chen, Vertical leakage/breakdown mechanisms in AlGaN/GaN-on-Si devices, *IEEE Electron Device Lett.*, 33: 1132, 2012.

160. I. C. Kizilyalli, A. P. Edwards, H. Nie, D. Disney, D. Bour, High voltage vertical GaN p-n diodes with avalanche capability, *IEEE Trans. Electron Dev.*, 60: 3067, 2013.

161. I. C. Kizilyalli, P. Bui-Quang, D. Disney, H. Bhatia, O. Aktas, Reliability studies of vertical GaN devices based on bulk GaN substrates, *Microelectronics Reliab.*, 55: 1654, 2015.

162. M. J. Uren, K. J. Nash, R. S. Balmer, T. Martin, E. Morvan, N. Caillas, S. L. Delage, D. Ducatteau, B. Grimbert, J. C. De Jaeger, Punch-Through in Short-Channel AlGaN/GaN HFETs, *IEEE Trans. Electron Dev.*, 53(2): 395–398, 2006.

163. F. Monti, I. Imperiale, S. Reggiani, E. Gnani, A. Gnudi, G. Baccarani, L. Nguyen et al., Numerical study of GaN-on-Si HEMT breakdown instability accounting for substrate and packaging interactions, in *27th International Symposium on Power Semiconductor Devices & IC's (ISPSD)*, Kowloon Shangri-La, Hong Kong, p. 381, 2015.

164. I. B. Rowena, S. L. Selvaraj, T. Egawa, Buffer thickness contribution to suppress vertical leakage current with high breakdown field (2.3 MV/cm) for GaN on Si, *IEEE Electron Device Lett.*, 32: 1534, 2011.

165. D. Bisi, S. H. Chan, X. Tahhan, O. S. Koksaldi, S. Keller, M. Meneghini, G. Meneghesso, E. Zanoni, U. K. Mishra, Quality and Reliability of in-situ Al2O3 MOS capacitors for GaN-based power devices, in 2016 *28th International Symposium on Power Semiconductor Devices and ICs (ISPSD)*, Prague, Czech Republic, 2016, p. 119.

166. S. R. Bahl, J. A. del Alamo, A new drain-current injection technique for the measurement of off-state breakdown voltage in FET's, *IEEE Trans. Electron Dev.*, 40: 1558, 1993.

167. M. Meneghini, A. Zanandrea, F. Rampazzo, A. Stocco, M. Bertin, G. Cibin, D. Pogany, E. Zanoni, G. Meneghesso, Electrical and electroluminescence characteristics of AlGaN/GaN high electron mobility transistors operated in sustainable breakdown conditions, *Jpn. J. Appl. Phys.*, 52: 08JN17, 2013.

168. G. Meneghesso, M.Meneghini, E.Zanoni, Breakdown mechanisms in AlGaN/GaN HEMTs: An overview, *Jpn. J. Appl. Phys.*, 53: 100211, 2014.

169. E. Zanoni, F. Danesin, M. Meneghini, A. Cetronio, C. Lanzieri, M. Peroni, G. Meneghesso, Localized damage in AlGaN/GaN HEMTs induced by reverse-bias testing, *IEEE Electron Device Lett.*, 30: 427, 2009.

170. M. Meneghini et al., Time-dependent degradation of AlGaN/GaN high electron mobility transistors under reverse bias, *Appl. Phys. Lett.*, 100: 033505, 2012.

171. J. R. Dickerson, A. A. Allerman, B. N. Bryant, A. J. Fischer, M. P. King, M. W. Moseley, A. M. Armstrong, R. J. Kaplar, I. C. Kizilyalli, O. Aktas, J. J. Wierer, Vertical GaN power diodes with a bilayer edge termination, *IEEE Trans. Electron Dev.*, 63: 419, 2106.

172. D. Bisi, A. Stocco, I. Rossetto, M. Meneghini, F. Rampazzo, A. Chini, F. Soci et al., Effects of buffer compensation strategies on the electrical performance and RF reliability of AlGaN/GaN HEMTs, *Microelectronics Reliab*, 55: 1662, 2015.

173. W. S. Tan, P. A. Houston, P. J. Parbrook, D. A. Wood, G. Hill, C. R., Gate leakage effects and breakdown voltage in metalorganic vapor phase epitaxy AlGaN/GaN heterostructure field-effect transistors, *Appl. Phys. Lett.*, 80(17): 3207–3209, 2002.
174. D. Visalli, M. Van Hove, J. Derluyn, S. Degroote, M. Leys, K. Cheng, M., AlGaN/GaN/AlGaN double heterostructures on silicon substrates for high breakdown voltage field-effect transistors with low on-resistance, *Jpn. J. Appl. Phys.*, 48: 04C101, 2009.
175. O. Hilt, P. Kotara, F. Brunner, A. Knauer, R. Zhytnytska, J. Würfl, Improved vertical isolation for normally-off high voltage GaN-HFETs on n-SiC substrates, *IEEE Trans. Electron Dev.*, 60: 3084, 2013.
176. M. Ťapajna, L. Válik, P. Kotara, R. Zhytnytska, F. Brunner, O. Hilt, E. Bahat-Treidel, J. Würfl, and J. Kuzmík, Impact of the buffer structure on trapping characteristics of normally-off p-GaN/AlGaN/GaN HEMTs for power switching applications, *Advanced Semiconductor Devices & Microsystems (ASDAM), 2014 10th International Conference on*, IEEE, Smolenice, Slovakia, October 20–22, 2014, pp. 1–4.
177. B. Lu, T. Palacios, High breakdown AlGaN/GaN HEMTs by substrate-transfer technology, *IEEE Electron Device Lett.*, 31: 951, 2010.
178. N. Herbecq, I. Roch-Jeune, N. Rolland, D. Visalli, J. Derluyn, S. Degroote, M. Germain, F. Medjdoub, 1900 V, 1.6 mΩ cm2 AlN/GaN-on-Si power devices realized by local substrate removal, *Appl. Phys. Express*, 7: 034103, 2010.
179. H. Umeda, A. Suzuki, Y. Anda, M. Ishida, T. Ueda, T. Tanaka, D. Ueda, Blocking-voltage boosting technology for GaN transistors by widening depletion layer in Si substrates, *IEDM Tech. Dig.*, pp. 20–5, 2010.
180. J. Kolnik, I. H. Oguzman, K. F. Brennan, R. Wang, P. P. Ruden, Monte Carlo calculation of electron initiated impact ionization in bulk zinc-blende and wurtzite GaN, *J. Appl. Phys.*, 81: 726, 1997.
181. K. Kunihiro, K. Kasahara, Y. Takahashi, Y. Ohno, Experimental evaluation of impact ionization coefficients in GaN, *IEEE Electron Device Lett.*, 20: 608, 1999.
182. N. Dyakonova, A. Dickens, M. S. Shur, R. Gaska, J. W. Yang, Temperature dependence of impact ionization in AlGaN–GaN heterostructure field effect transistors, *Appl. Phys. Lett.*, 72: 2562, 1998.
183. K. Hui, C. Hu, P. George, P. K. Ko, Impact ionization in GaAs MESFET's, *IEEE Electron Device Lett.*, 11: 113, 1990.
184. H. Hanawa, H. Onodera, A. Nakajima, K. Horio, Numerical analysis of breakdown voltage enhancement in AlGaN/GaN HEMTs with a High-k passivation layer, *IEEE Trans. Electron Dev.*, 6: 769, 2014.
185. Y. Ohno, T. Nakao, S. Kishimoto, K. Maezawa, T. Mizutani, Effects of surface passivation on breakdown of AlGaN/GaN high-electron-mobility transistors, *Appl. Phys. Lett.*, 84: 2184, 2004.
186. B. Brar, K. Boutros, R. E. DeWames, V. Tilak, R. Shealy, L. Eastman, Impact ionization in high performance AlGaN/GaN HEMTs, in *Proceedings IEEE Lester Eastman Conference on High*, Newark, DE, August 6–8, 2002, p. 487.
187. M. Meneghini, C. De Santi, T. Ueda, T. Tanaka, D. Ueda, E. Zanoni, G. Meneghesso, Time- and field-dependent trapping in GaN-based enhancement-mode transistors with p-gate, *IEEE Electron Device Lett.*, 33(3) : 375–377, 2012.

IV

Light Emitters

13

Internal Quantum Efficiency for III-Nitride–Based Blue Light-Emitting Diodes

Zi-Hui Zhang,
Yonghui Zhang,
Hilmi Volkan Demir,
and Xiao Wei Sun

13.1 Introduction

The III-nitride light-emitting diodes (LEDs) have popularly penetrated into the market of the visible light communication, lighting, sensing, and display illumination [1,2]. It is well known that the III-nitride LED-based white-lighting source is able to yield the efficacy much higher than those traditional light sources such as the incandescent light bulbs by more than 15 times, which therefore, leads to a yearly reduction of the CO_2 emission by 1900 Mt [1], and makes significant contribution to relieve the global warming effect. Additionally, compared to the mercury-based fluorescent light tubes, the III-nitride solid-state lighting source produces no pollutions. As a result, the III-nitride–based solid-state lighting has attracted intense global interest and is regarded as the ultimate lighting solution in this century.

The technology of III-nitride LEDs has been developed for more than three decades since Maruska et al. pioneered the world's first single-crystalline GaN material [3,4]. However, the efficiency of the GaN LED at that time was low due to the poor crystal quality and the absence of the p-GaN layer. It was found that GaN can be of p-type conductivity through Mg doping method. The Mg dopants can be further activated by either thermal annealing in the N_2 ambient [5] or low energy electron beam irradiation [6]. Moreover, the adoption of the GaN [7] or the AlN buffer layer [8] enables the excellent crystalline quality for the subsequently grown GaN epilayers. Although the main obstacles which hinder the development of GaN LEDs have been solved, more work is still needed to continuously improve the efficiency of GaN LEDs.

In this chapter, we will first discuss the issues that strongly influence the LED internal quantum efficiency (IQE) by using the well-known ABC model in Section 13.2. Then we will review the most recently published results that discuss the approaches ever adopted to improve the IQE. The different ways to reduce the electron leakage are analyzed and compared in Section 13.3. The designs to promote the hole injection are addressed in Section 13.4. The polarization effect in the [0001] oriented quantum wells is discussed in Section 13.5. Section 13.6 proposes the design methods to improve the current spreading for the LEDs grown on the insulating substrates (e.g., sapphire). Section 13.7 reviews the ways to reduce the Shockley–Read–Hall (SRH) recombination rate. The approaches to reduce the Auger recombination rate are illustrated in Section 13.8. Lastly, we make our conclusion and suggest the future outlooks in Section 13.9.

13.2 Key Factors Which Affect the Quantum Efficiency

The IQE for GaN LEDs is substantially influenced by a few factors, which will be reviewed and discussed subsequently. The architectural energy band of an InGaN/GaN LED is sketched in Figure 13.1. Several events take place during the carrier transport that can be linked to the well-known ABC model [9]. The ABC model can be formulated in Equation 13.1,

$$\eta_{EQE} = \eta_{IQE} \cdot \eta_{LEE} = \frac{\eta_{inj} \cdot Bn^2}{An + Bn^2 + Cn^3} \eta_{LEE} \tag{13.1}$$

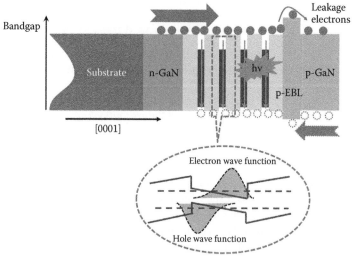

FIGURE 13.1 Schematic energy band diagram for the InGaN/GaN LED architecture. Here, p-EBL denotes the p-type electron blocking layer.

in which η_{EQE}, η_{IQE}, η_{inj}, η_{LEE}, *A*, *B*, *C*, and *n* represent the external quantum efficiency (EQE), IQE, carrier injection efficiency, light extract efficiency, defect-induced SRH recombination coefficient, bimolecular radiative recombination coefficient, Auger recombination coefficient and carrier density, respectively. The carrier injection efficiency (η_{inj}) includes the electron injection efficiency and the hole injection efficiency. Note, this chapter has concentrated on the novel designs to improve the IQE, while the reviews on enhancing the light extraction efficiency have been reported by Geng et al. [10] and Zhmakin [11].

According to Figure 13.1, the electrons are supplied by the n-GaN layer, while the holes are injected from the p-GaN layer. It is well known that electrons and holes are not synchronized in the transport due to the fact that electrons are more mobile than holes. In addition, the low doping efficiency for the p-GaN layer also strongly reduces the hole injection efficiency, which in turn, further results in the electron leakage into the p-GaN layer [12]. Detailed discussions on variously reported approaches for improving the carrier injection will be conducted in Sections 13.1 and 13.2.

For [0001] oriented InGaN/GaN LEDs, the radiative recombination rate is mostly impacted by the polarization induced electric field, which spatially separates the electron and hole wave functions (see Figure 13.1) in the quantum well region, known as the quantum-confined Stark effect (QCSE) [3]. Therefore, the key approach to increasing the radiative recombination rate is to screen the polarization effect in the quantum wells [13], which will be discussed in detail subsequently.

Additionally, the radiative recombination rate is also affected by the current crowding effect [3], which significantly makes the current more nonhomogeneously distributed in the quantum well planes, for example, the current substantially crowds under the p-electrode. Meanwhile, a high local current density also results in serious Joule self-heating effect and increased thermal resistance [14]. It is reported that the current crowding effect increases the ideality factor for the LED device [15], while a lager ideality factor is the signature of the nonradiative recombination. Therefore, suppressing the current crowding effect is also useful to reduce the local carrier accumulation and the corresponding nonradiative recombination.

The IQE is also affected by the defect-related recombination mechanism, denoted as the *An*. Until now, there is no consensus on whether the defect-related recombination is responsible for the efficiency droop, which is defined as the decrease of the quantum efficiency for a LED device as the injection current increases [9]. The report by Schubert et al. shows that the defect-related recombination has a strong impact on the maximum quantum efficiency but does not lead to the efficiency droop [16]. The LED device with a low dislocation density has a pronounced efficiency peak while showing a fast decrease of the quantum efficiency as the injection current further increases. However, the LED device with a high dislocation density exhibits a low peak efficiency accompanied by a small efficiency decrease with the ascending injection current, and in the meanwhile, Shao et al. report that the current where the peak efficiency takes place is also shifted to a higher value once the dislocation density is increased [17]. Nevertheless, the defect-related recombination may also contribute to the efficiency droop if the SRH recombination coefficient scales up with the injection current, and this is well modeled by the density-activated defect recombination (DADR) model [18,19], which takes place when the carrier concentration is high enough to facilitate the carrier delocalization. The DADR model well interprets that the InGaN/GaN quantum well with more InN composition processes a more severe efficiency droop. The methods to suppress the defect-related recombination will be discussed in Section 13.7.

The carriers are also consumed nonradiatively in the quantum wells by the Auger recombination [9]. The Auger recombination rate scales up to the cubic power of the carrier density as can be seen in Equation 13.1, and therefore, the Auger recombination significantly reduces the quantum efficiency especially when the LED device is biased at a high current level. Recently, it has been further confirmed that the efficiency droop for InGaN/GaN LEDs is mainly caused by the Auger recombination [20]. However, it is still disputable if the Auger recombination is the only mechanism for causing the efficiency droop [21]. The Auger recombination coefficient has been characterized by different groups. Through combining the ABC model and the photoluminescence (PL) measurement, Shen and coworkers report that the Auger recombination coefficient ranges from 1.4×10^{-30} to 2.0×10^{-30} cm^6s^{-1} for thick $In_xGa_{1-x}N/GaN$ ($x \sim 9\%$–15%) double heterostructure (DH) [22]. A room-temperature Auger recombination coefficient of $1.8 \pm 0.2 \times 10^{-31}$ cm^6/s in the bandgap range of 2.5–3.1 eV is contributed by Brendel et al. [23], who extract the Auger recombination coefficient through measuring the optical gain spectra for an InGaN/GaN quantum well laser structure. Recently, the Auger recombination for multiple quantum wells is found to strongly depend on the volume of the active region [24]. By considering the polarization effect in the [0001] oriented quantum wells, the effective recombination volume becomes smaller, and as a result, the Auger recombination coefficient is $1.4(\pm 0.2) \times 10^{-31}$ cm^6s^{-1}. However, the Auger recombination coefficient will increase to $5.1(\pm0.2) \times 10^{-30}$ cm^6s^{-1} if the real recombination volume is used during the data extraction. The Auger recombination can also be extracted by using the modulation bandwidth studies, which was conducted by Green and coworkers [25] and Auger recombination coefficient is measured to be $1.0(\pm 0.3) \times 10^{-29}$ cm^6s^{-1} for the 450 nm InGaN/GaN LEDs. However, a collection of Auger recombination coefficient for various LED structures has been summarized by Cho et al. [26] and Piprek et al. [9,27].

From the above discussion, one can see that there are many factors hindering the IQE for InGaN/GaN LEDs. This chapter will mainly review the approaches and also propose the design principles for different InGaN/GaN LED structures to enhance the LED performance and reduce the efficiency droop.

13.3 Designs to Increase the Electron Injection Efficiency for Nitride Light-Emitting Diodes

Electron injection is one of the key factors that matter the InGaN/GaN LED quantum efficiency [9,26,28]. On the one hand, to enhance the electron injection, one can engineer the p-EBL so that the electrons have less chance of escaping from the active region, and on the other hand, one can increase the electron capture rate by designing novel quantum well/quantum barrier structures. In addition, the electron energy can be decreased before being injected into the active region so that the quantum wells are able to trap the electrons more efficiently and prevent them from directly overflying into the p-GaN region. Lastly, an increased hole injection efficiency is very useful in making better use of the electrons for radiative recombination. Nevertheless, the hole injection efficiency will be discussed in Section 13.4.

13.3.1 p-EBL Designs for Improving Electron Blocking Effect

Currently, a p-AlGaN electron blocking layer (EBL) that is lattice-mismatched to the GaN quantum barrier is widely adopted in standard [0001] oriented LED structures. However, the GaN/p-EBL interface possesses strong polarization effect that leads to the electron accumulation and reduces the effective conduction band barrier height, which in turn facilitates the electron leakage [29]. The polarization inverted p-EBL (i.e., [000-1] orientation) can eliminate the electron accumulation [30]. However, this structure induces post-bonding difficulties, thus making it less reliable in reality. Some groups suggest inserting a p-InGaN before growing the p-AlGaN EBL to reduce the electron leakage [31–35], but the origin on the reduced electron leakage was not explained. Here, we propose that the reduced electron leakage by the p-InGaN insertion layer can be well explained by the polarization inversion effect. As has been mentioned, for the [0001] oriented GaN/p-AlGaN structure, there is very strong polarization appearing at the interface and giving rise to the positive polarization interface charges, which significantly bends the conduction band of the GaN quantum barrier downwards. However, if the GaN quantum barrier is replaced by a $GaN/In_xGa_{1-x}N$ heterojunction, then the interface between GaN layer and InGaN layer is polarized by yielding negative charges, and the conduction band of the GaN layer will be bent in the way favoring the electron blocking.

Kim et al. propose a lattice-matched InAlN p-EBL so that the GaN/InAlN heterojunction is free from any piezo-polarization effect, and this helps to confine the electrons better and then improve the IQE in the green regime [36]. InAlN as the p-EBL also proves to be effective in reducing the electron

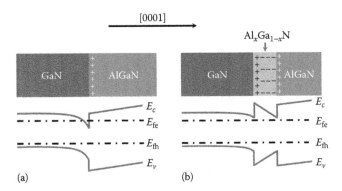

FIGURE 13.2 The schematic energy band diagrams of (a) conventional GaN/p-EBL architecture, in which the p-$Al_{0.20}Ga_{0.80}$N EBL is 20 nm thick, and (b) proposed polarization self-screened GaN/p-EBL architecture for InGaN/GaN LEDs, in which the AlN composition follows a linear grading from 20% to 0.0% within 10 nm thickness and the rest p-$Al_{0.20}Ga_{0.80}$N EBL is 10 nm. (Reprinted with permission from Zhang, Z.-H. et al., *Appl. Phys. Lett.*, 104, 251108, Copyright 2014, American Institute of Physics.)

spillover level for blue InGaN/GaN LEDs as reported by Choi et al. [37,38]. Despite the success of growing the InAlN material, achieving the epitaxial InAlN compound is still more challenging than growing the AlGaN layer.

Recently, we have proposed a design to suppress the polarization discontinuity by using the polarization self-screening effect [39] (the details of the polarization self-screening effect will be demonstrated in Section 13.5). By applying the polarization self-screening effect to the p-EBL, the polarization discontinuity and the electron leakage can be reduced [40]. Compared to those InAlN and InAlGaN p-EBLs, the polarization self-screened p-EBL can be easily grown by MOCVD technology. The schematic layer diagrams of the p-EBL for the studied devices are shown in Figure 13.2a and b, respectively. Figure 13.2a illustrates a conventional AlGaN p-EBL. The investigation into Figure 13.2a shows that the positive polarization induced interface charges at the GaN/AlGaN interface strongly cause the electron accumulation and bend the conduction band of the GaN region. According to the Reference [41], we obtain the formula of $\Phi_b = \Delta E_C - kT \cdot \ln(n_{LB/EBL}/N_C)$, and it shows that the effective conduction band barrier height (Φ_B) is coaffected by ΔE_c, T, $n_{LB/EBL}$ and N_c which denote the conduction band offset between the GaN last barrier (LB) and the p-EBL, the carrier temperature, the electron concentration at the LB/EBL interface and the density of states for electrons, respectively. Therefore, one way to increase Φ_B is to decrease $n_{LB/EBL}$, and this can be achieved by the p-EBL architecture presented in Figure 13.2b, where the 20nm-thick p-EBL is grown by the MOCVD system in such a way that a 10nm-thick Al_xGa_{1-x}N layer has the AlN compositionally graded from 20.0% to 0.0% along the [0001] growth direction and then is followed by a 10nm-thick $Al_{0.20}Ga_{0.80}$N layer. As a result, the negative polarization induced bulk charges (ρ_B^{Pol}) are produced, which can screen the polarization induced positive interface charges (σ_B^{Pol}). Therefore, fewer electrons will be accumulated at the GaN/p- Al_xGa_{1-x}N interface, which enables a reduction of $n_{LB/EBL}$ and then increases Φ_B, leading to a better electron confinement. We have assumed a 40% polarization level due to the strain release by generating dislocations during the epitaxial growth process [42]. The σ_B^{Pol} between the GaN last barrier and the p-$Al_{0.20}Ga_{0.80}$N EBL is 0.36×10^{17} m^{-2}, while the ρ_B^{Pol} in the Al_xGa_{1-x}N region is found to be -3.74×10^{24} m^{-3}. The details for calculating the σ_B^{Pol} and ρ_B^{Pol} can be found in Reference [40].

We define $\Delta\phi$, Φ_{b1}, and Φ_{b2} as the bending level of the last quantum barrier, the effective conduction band barrier height between the last quantum barrier and the p-EBL, as well as the effective conduction band barrier height of the rest p-EBL region, respectively. Φ_{b3} and Φ_{b4} denote the effective valance band barrier height between the last quantum barrier and the p-EBL and the effective valance band barrier height of the rest p-EBL region, respectively. The investigation of Figure 13.3a and b demonstrates that $\Delta\phi$ is smaller in the polarization self-screened p-EBL architecture, which means the last quantum barrier is more effective in confining the electrons. Meanwhile the values of Φ_{b1} and Φ_{b2} in

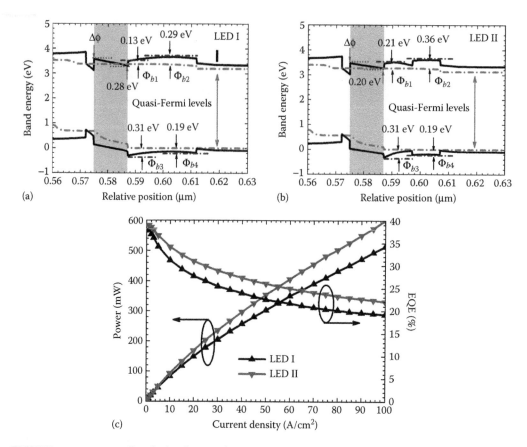

FIGURE 13.3 Numerically calculated energy band diagrams in the vicinity of the p-EBL and the values of $\Delta\phi$, Φ_{b1}, Φ_{b2}, Φ_{b3}, and Φ_{b4} for (a) the LED device with bulk-type p-AlGaN EBL, (b) the LED device with the polarization self-screened p-EBL, and (c) experimentally measured EQE and optical power for the two LEDs. (Reprinted with permission from Zhang, Z.-H. et al., *Appl. Phys. Lett.*, 104, 251108, Copyright 2014, American Institute of Physics.)

Figure 13.3b are larger than those in Figure 13.3a, and this represents a reduced electron leakage level for the p-EBL. On the other hand, the two devices have identical Φ_{b3} and Φ_{b4}, which translate to the unaffected hole injection by the polarization self-screened p-EBL structure. Hence, both the quantum efficiency and the efficiency droop for the LED device with the polarization self-screened p-EBL have been improved according to Figure 13.3c, such that the quantum efficiency for LED II is enhanced by 16.9% at 100 A/cm² when compared to LED I, and the efficiency droop at 100 A/cm² for LED I and LED II is 49.3% and 42.3%, respectively.

Recently, chirped multiple quantum-barrier (MQB) type p-EBL is proposed and proves to be useful in reducing the electron leakage and improving the IQE for the InGaN/GaN LEDs [43–45]. By properly tuning superlattice-type MQB thickness, the MQB structure creates the forbidden energy bands above the natural conduction band edge, and this cuts off the resonant tunneling for electrons, hence facilitating the electron reflectivity and reducing the electron leakage rate. However, according to the report by Piprek et al. [46], the origin of the improved quantum efficiency by the chirped MQB type p-EBL is attributed to the enhanced hole injection. The enhanced hole injection efficiency arises from the strong hole accumulation at the AlGaN/p-GaN interface, which is caused by the negative polarization induced charges. Piprek et al. believe that the electron leakage is also the consequence of the poor hole injection. The same conclusion is also made by us [47]. However, discussions on the hole injection will be conducted subsequently.

13.3.2 Active Region Designs to Better Confine Electrons

Another approach for increasing the electron injection efficiency is to modify the quantum wells and quantum barriers so that the conduction band offset can be increased and the electron thermionic escape can be suppressed. Currently, GaN is used as the quantum barrier material, and the electron injection can be enhanced if AlGaN with an optimized AlN composition is employed. The numerically simulated results by Chang et al. demonstrate the advantage of AlGaN barriers in improving the electron confinement [48]. Zhao et al. report that a high electron current injection efficiency can be realized by using a thin large bandgap barrier (1 ~ 2.5 nm), such as that a thin AlGaN layer is embedded between the InGaN quantum well and the GaN quantum barrier [49,50]. Besides, Liu et al. suggest that a high electron injection efficiency can also be obtained by inserting a thin (1 ~ 2 nm) InAlGaN or InAlN cap layer such as GaN/InAlGaN/InGaN/InAlGaN/GaN and GaN/InAlN/InGaN/InAlN/GaN structures, respectively [51]. The polarization matched AlGaInN barriers are also advisable for achieving a high electron confinement [52,53]. Kuo et al. demonstrate that the GaN/InGaN/GaN type quantum barriers [54] and the InGaN/AlGaN/InGaN quantum barriers can increase the electron injection efficiency and achieve a reduced efficiency droop [55], respectively. However, the mechanism on how the GaN/InGaN/GaN type and InGaN/AlGaN/InGaN type quantum barriers are better in manipulating the electron injection than the conventional GaN quantum barriers are not provided in their works. Besides increasing the conduction band offset through the alloy engineering, by properly increasing the thickness of the last quantum barrier, the electron leakage current can also be decreased [56], because when the length of the last quantum barrier increases, the electron accumulation at the last quantum barrier/p-EBL interface is reduced, while a reduced electron accumulation helps to suppress the electron leakage [40]. However, details for the carrier transport models are not provided in their paper, we believe a thicker last quantum barrier reduces the electron intraband tunneling process, and this, in turn, further alleviates the electron accumulation level at the last quantum barrier/p-EBL interface. For that reason, the electron leakage current level has been substantially decreased thanks to the thicker last quantum barrier. Besides, to avoid the hole blocking effect, the authors for Reference [56] purposely adopt the partially p-doped last quantum barrier when the quantum barrier is thickened.

13.3.3 Designs to "Cool Down" Electrons

The other alternative method to enhance the electron injection is to make electrons "cold". The layer that can make electrons "cold" is named as the electron cooler (EC), electron injector or electron reservoir layer. Otsuji et al. publish their results addressing the effect of the InGaN electron reservoir layer on the electroluminescence efficiency for InGaN/GaN LEDs [57], and they attribute the improved device performance to the enhanced electron capture efficiency by the quantum wells. They tentatively ascribe the increased electron capture efficiency to the variation of the potential field distribution in the quantum wells. Later on, Li et al. conclude that the InGaN insertion layer can reduce the internal electric field in the quantum well through observing blueshift of the wavelength from the cathodoluminescence (CL) spectra and a reduced carrier lifetime derived from the time-resolved CL spectra [58]. More importantly, Li et al. by the electron holography, also measure and show a reduced electrostatic potential for the LED device with the InGaN insertion layer [58]. However, a most recent physical model, developed by Ni et al. [59], Zhang et al. [60], Li et al. [61,62], Avrutin et al. [63], Zhang et al. [64] and Chang et al. [65] suggests that a reduced electron leakage is caused by the phonon-electron scattering happening in the InGaN insertion layer. During the phonon-scattering process, the electrons lose 92 meV in energy, such that the electrons become "cold" and the InGaN insertion layer functions as the EC. The device architectures and the electric field profiles in the InGaN EC layer are shown in Figure 13.4a–c, respectively. The electric field intensity for LEDs I and II is calculated at 20 A/cm², and by following $qV = \int_0^{t_{cooler}} q \times E(y)dy$, the electrons in LED I lose the kinetic energy of 48.10 meV. The kinetic energy of the electrons in LED II is increased by 27.82 meV due to the polarization induced electric field in LED III, and therefore, the electrons will lose the energy of 64.18 meV (92–27.82 meV) during the transport in the InGaN EC layer for LED II.

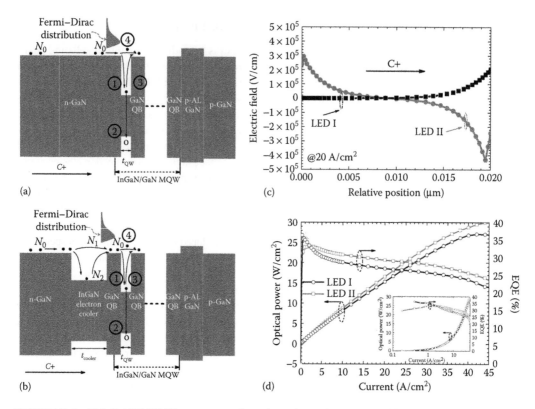

FIGURE 13.4 (a) InGaN/GaN LED structure without the EC layer, that is, LED I; (b) InGaN/GaN LED structure with the EC layer, that is, LED II; (c) calculated electric field profile in the EC layer for LED II, and the electric field profile in the same region is also presented for LED I for comparison; and (d) measured EQE and optical power in terms of the injection current. Four electron/transport processes are also shown in Figure 13.4a and b: ① electrons are captured into the quantum well, ② electrons recombine with holes and at defects, ③ electrons re-escape from the quantum well and ④ electrons directly fly over to a remote position without being captured by the quantum well. (From Zhang, Z.-H. W. et al., *Opt. Express*, 22, A779–A789, 2014. With permission from Optical Society America.)

As a result, LED II still has more chances of capturing electrons into the quantum wells than LED I does, which translates into a reduced efficiency droop of 18.3% from 24.0% and the quantum efficiency enhancement of 7.8% at 35 A/cm², as reported by Zhang et al. [64].

The electron injection efficiency can be improved by adopting the n-AlGaN EBL [66,67]. However, according to our study, the polarization induced electric field in the [0001] oriented n-EBL reduces the electron thermal energy, and hence the quantum wells can better capture electrons [68].

To summarize, this section reviews the methods that have been employed to reduce the electron leakage level and increase the electron injection efficiency into the multiple quantum wells. In order to meet the required targets, the p-EBL has to be engineered either by reducing the lattice mismatch between the p-EBL and the last quantum barrier or by manipulating the polarization charge polarity/polarization charge density at the interface between the p-EBL and the last quantum barrier. In addition, the quantum barrier height can be properly increased through material engineering, for example, a thin AlGaN cap layer. Finally, the electron injection mechanism can be tuned before entering the quantum wells, such that we "cool down" electrons and make them less energetic, which can be obtained by using the InGaN EC layer and the n-AlGaN EBL. More importantly, we also report the impact of the polarization induced electric field in affecting the electron kinetic energy, which requires the thickness and the alloy composition for the InGaN EC layer and the n-AlGaN EBL to be well controlled.

13.4 Designs to Increase the Hole Injection Efficiency for Nitride Light-Emitting Diodes

13.4.1 p-EBL Designs with Reduced Hole Blocking Effect

The electron leakage is also the consequence of the declining hole injection efficiency. The current commonly used InGaN/GaN LED architectures employ the p-type AlGaN EBL, which also hinders the hole injection due to the valence band offset between the AlGaN and the p-GaN layers [69,70]. Therefore, tremendous efforts have been paid to develop novel p-EBL structures, such as the superlattice p-EBL [71–75]. It has been reported by Schubert et al. that the superlattice p-EBL increases the activation of the deep Mg dopants [76–78]. Meanwhile, the staircase p-EBL is also useful in facilitating the hole injection [79–82]. The staircase p-EBL can then be further modified by linearly grading the AlN composition [83–86], which proves to be effective in further reducing the hole blocking effect by the p-EBL.

Recently, some groups report the AlGaN/GaN/AlGaN type p-EBL to increase the hole injection across the p-EBL [41,87,88]. The insertion of the GaN layer helps to reduce the overall valance band barrier height in the p-EBL, and more importantly, a wide GaN insertion layer reduces the AlGaN thickness if the total thickness of the p-EBL is a constant, and this, in turn, triggers the hole intraband tunneling process [87]. Furthermore, we propose the hole injection enhancement through the subband tuning effect, which indicates that the position of the GaN insertion layer is essential in favoring the hole injection [41], as shown in Figure 13.5a. First of all, the GaN insertion layer has to be very thin (1 nm thick GaN is grown in Reference [41]) to produce the subbands in the so-called AlGaN/GaN/AlGaN quantum well, and then the thin GaN insertion layer has to be selected at the position close to the p-GaN side, for example, $x = 3.5$ nm in Figure 13.5a. By doing so, the remaining 3.5 nm thick AlGaN layer guarantees a smooth hole tunneling into the thin GaN insertion layer and leads to a high hole concentration in the thin GaN layer, denoted as $p_{\text{p-GaN}}$. Besides, the

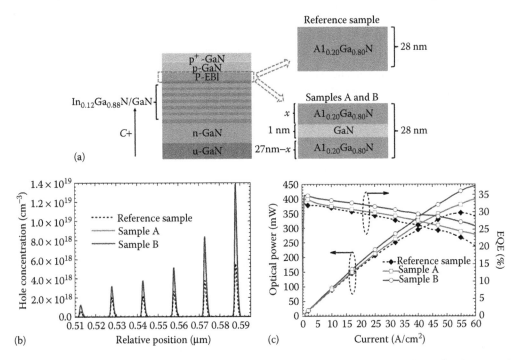

FIGURE 13.5 (a) Schematic diagrams for the reference sample, Sample A ($x = 13.5$ nm), and Sample B ($x = 3.5$ nm); (b) numerically calculated hole concentration in MQW region at 30 A/cm²; and (c) experimentally measured optical output power and EQE for the reference sample, Sample A and Sample B. (From Zhang, Z.-H. et al., *Opt. Lett.*, 39, 2483–2486, 2014. With permission from Optical Society America.)

holes occupy the subbands in the thin GaN layer. Originally the valence band barrier height can be formulated by $\Phi_b = \Delta E_V - kT \ln(p_{p\text{-}GaN}/N_V)$, in which N_V is the effective density of states for holes, k is the Boltzmann constant and ΔE_V is the valence band edge difference between GaN and AlGaN layers, that is, $\Delta E_V = E_{V\_GaN} - E_{V\_AlGaN}$. When holes occupy the subbands in the thin GaN layer, ΔE_V is replaced by ΔE_{Vi}. Here $\Delta E_{Vi} = E_{Vi\_GaN} - E_{V\_AlGaN}$, where E_{Vi_GaN} is the quantized subvalence band edge of the thin GaN layer in the p-AlGaN/GaN/p-AlGaN EBL, and it is easily obtained that ΔE_{Vi} is smaller than ΔE_V. Therefore, employing a thin GaN insertion layer in the p-EBL is useful in promoting the hole injection. More importantly, the position of the thin GaN insertion layer in the p-EBL is crucially important, and we recommend that the GaN insertion layer has to be close to the p-GaN side. Figure 13.5b proves that the subband in the thin GaN insertion layer enables a higher hole concentration in the MQW region. Figure 13.5c presents the measured optical output power and the EQE for the LED samples. Thanks to the subband tuning effect on the hole injection, both samples A and B show improved optical performance than the reference sample with sample B yielding the strongest intensity.

13.4.2 Hole Injection Enhancement with Produced "Hot" Holes

Aside from engineering the p-EBL for increasing the hole injection, we have also proposed and demonstrated a hole accelerator embedded in the p-GaN layer, that is, a polarized p-EBL/p-GaN/p-AlGaN structure, which promises the effectiveness in increasing the kinetic energy of the holes [89], so that the holes have more chance of climbing over the p-EBL. According to the experimental measurement, the efficiency droop has been reduced from 54.2% to 35.9 at 100 A/cm². Note, because the hole accelerator comprises of some heterojunctions, the energy band discontinuity of the heterojunction may cause hole blocking effect, which makes the thickness and the alloy composition in the p-AlGaN layer important. In the meanwhile, the free carriers in the hole accelerator may screen the polarization induced electric field in the p-GaN layer, making the thickness of the p-GaN vitally critical. However, a comprehensive study revealing the impact of the hole accelerator on the hole injection has been conducted by Zhang et al. [90].

13.4.3 Hole Injection Enhancement by Improving the Hole Concentration in the p-GaN Layer

The hole injection is also affected by the doping efficiency of the p-GaN layer. Currently, the thermal annealing process to activate the Mg dopants in the p-GaN layer is indispensable during the LED epitaxial growth [5]. However, the hole concentration is still not competitive to the electron concentration in the n-GaN layer, which can easily achieve the concentration of 5×10^{18} cm^{-3}. According to $\Phi_b = \Delta E_V - kT \ln(p_{p\text{-}GaN}/N_V)$ [41], if the hole concentration ($p_{p\text{-}GaN}$) in the p-GaN layer is low, the valence band barrier height (Φ_b) in the p-EBL will be large. Thus, tremendous efforts ought to be paid to improve the hole concentration in the p-GaN layer. Schubert et al. have reported the AlGaN/GaN superlattice doping [76–78], for which the Mg activation energy can be further reduced if the Al composition is increased, such that the Mg activation energy is 70 meV and 58 meV for $Al_{0.10}Ga_{0.90}N$/GaN superlattice and $Al_{0.20}Ga_{0.80}N$/GaN superlattice, respectively. These values are much smaller than the Mg activation energy (170 meV ~ 200 meV) in the p-GaN layer [78]. This idea has been applied to the InGaN/GaN LED architecture which adopts the AlGaN/GaN superlattice as the p-EBL [72,74].

Besides the superlattice AlGaN/GaN structures, three-dimensional hole gas (3DHG) is also proposed for the N-polar GaN LED by Simon et al. [86], such that the Mg dopants can be more effectively ionized by the polarization induced electric field in the AlGaN layer when the AlN composition is graded. In the meantime, the polarization-doped LED possesses higher optical intensity than the conventional LED. 3DHG can also be generated in the Ga-polar III-nitride semiconductors [85,91–94]. We have tested the 3DHG in an undoped $Al_{0.10\rightarrow0.02}Ga_{0.90\rightarrow0.98}N$/GaN structure which is grown along the [0001] orientation. Furthermore, we also grow and fabricate a blue InGaN/GaN LED device that is p-GaN free and the electrically driven blue emission is obtained [94].

To increase the hole concentration in the p-GaN layer, we have also proposed a hole modulator [47]. The hole modulator is realized by Mg doping the last quantum barrier. The Mg-doped last quantum barrier has ever been reported by Kuo et al. [95]; however, the enhanced hole injection is still unclear until now. The holes donated by the Mg-doped last quantum barrier are depleted and stored in the p-GaN layer, which can increase the overall hole concentration for the p-GaN layer and facilitates the hole injection when the device is biased. Thanks to the increased hole concentration, a reduced electron leakage level is simultaneously obtained, and the EQE is experimentally increased by 25.59% at 80 A/cm^2 for the LEDs in the 450 nm regime.

13.4.4 Increase the Hole Injection by Improving the Hole Transport in Active Region

Meanwhile, insufficient hole injection is also reflected by the nonuniform hole distribution in the multiple quantum wells. A most effective way to reduce the valence band discontinuity between the quantum well and the quantum barrier is to replace the GaN quantum barriers with InGaN quantum barriers [96,97]. Playing with the quantum well thickness [98,99], for example, properly thickening the quantum wells along the [0001] growth orientation is helpful to homogenize the hole density in the quantum wells [98]. Besides engineering the quantum well thickness, a design of properly thinning the quantum barriers is also effective to promote the hole injection [100,101]. Most recently, Piprek has proposed and presented a cascaded active region, which theoretically promises super hole injection efficiency, and leads to an IQE even higher than 100% due to the carrier recycling effect [102,103]. However, the cost of this design lies in the challenge for material growth and the increased forward voltage [104,105]. The hole injection can also be manipulated by material engineering, and Li et al. reveal that a longer hole penetration depth can be enabled by the V-shape pits [106]. Moreover, the hole penetration depth can be enhanced by Mg-doped quantum barriers with a moderate Mg dosage level [107–110]. Kuo et al. initiate the GaN/InGaN-type last quantum barrier to promote the hole injection efficiency [31], and the experimental proofs have also been reported by several groups [32–34]. The GaN/InGaN-type last quantum barrier further evolves into the GaN/InGaN superlattice as the hole reservoir layer [111]. Although the GaN/InGaN-type and the GaN/InGaN-superlattice-type last quantum barriers show the advantage in increasing the hole injection capability both numerically and experimentally, the physical interpretations, nevertheless, are still unclear so far.

13.4.5 Hole Injection Enhancement by Improving the Hole Injection from the p-Electrode into the p-GaN Layer

Lastly, due to the low p-type doping efficiency in the p$^+$-GaN layer, the width of the surface depletion region in the p$^+$-GaN layer is thick, which leads to a low tunneling efficiency between the p$^+$-GaN layer and the p-type contact. In order to promote the hole tunneling efficiency, we have proposed a charge inverter as shown in Figure 13.6b [112]. A charge inverter is a structure that is composed of semiconductor/insulator/metal (SIM) layers. Here, the insulator has to be properly thin to enable the hole tunneling process. In our work, the roughness fluctuation of our p$^+$-GaN layer is less than 1nm, and therefore, we adopt the SiO$_2$ film of 1 nm that is prepared by PECVD at 300°C. The idea of the charge inverter is that, when the LED is forwardly biased, the SIM region has a slight inversion layer, where the electrons accumulate at the p$^+$-GaN/SiO$_2$ interface and shrink the surface depletion region width from 3.5 nm to 2 nm at 100 mA (see Figure 13.6c, chip size is 350×350 μm^2). For that reason, we have obtained the enhanced optical performance according to Figure 13.6d. Note that the declining power enhancement in terms of the injection current in Figure 13.6d is attributed to the electron leakage since more nonequilibrium holes are produced at the p$^+$-GaN/SiO$_2$ interface and attract more electrons to bypass the MQW region. Hence the p-EBL has to be further modified to confine electrons better.

To summarize, we have reviewed the most adopted methods to increase the hole injection efficiency for InGaN/GaN LEDs. One can promote the hole transport by reducing the valence band barrier height of the p-EBL and manipulating the hole transport mechanism, for example, through the subband tuning effect. Besides, the hole injection can also be enhanced by increasing the kinetic energy for holes, which can be achieved by the hole accelerator. In the meanwhile, improving the doping efficiency for the p-GaN

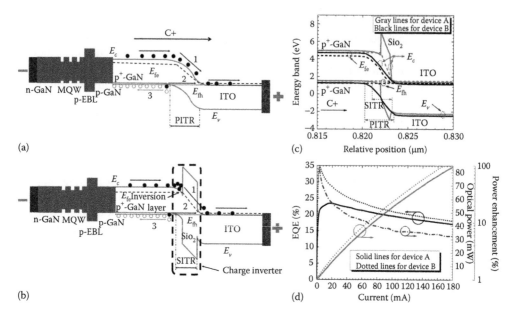

FIGURE 13.6 Schematic energy band diagram for (a) Device A without a charge inverter; (b) Device B with a charge inverter; (c) numerically calculated energy band diagram; and (d) experimentally measured EQE, optical output power and power enhancement in terms of the injection current. Chip size is 350 × 350 μm². (Reprinted with permission from Zhang, Z.-H. et al., *Appl. Phys. Lett.*, 108, 133502, Copyright 2016, American Institute of Physics.)

layer offers another design strategy for enhancing the hole injection efficiency, such as the 3DHG structure and the hole modulator. To increase the hole injection across the p-EBL, it is also essentially important to homogenize the hole distribution within the active region so that the holes can penetrate into those deep quantum wells close to the n-GaN layer. Last but not the least, one has to reduce the surface depletion region width for the p⁺-GaN layer to enable a more efficient hole tunneling from the p-type contact.

13.5 Improving the Quantum Efficiency through Screening the Polarization Effect in the [0001] Oriented Quantum Wells

As has been well known, InGaN/GaN MQWs grown along the [0001] orientation feature a very strong polarization field, which disables the flat band condition for the quantum wells and the quantum barriers. More importantly, the polarization induced electric field spatially separates the electron and the hole wave functions, well known as the QCSE, which consequently reduces the electron-hole recombination [13]. Therefore, in order to increase the radiative recombination rate, it is vital to suppress the polarization effect in the quantum wells.

13.5.1 Screen the Polarization Effect by Engineering the Energy Band for the Active Region

The QCSE can be alleviated by using very thin quantum wells. However, the thin quantum wells will cause poor electron capture efficiency [113]. Currently, the most effective way to screen the polarization effect in the quantum wells is to adopt the nonpolar and semipolar LED architectures [114–121]. Recently, by properly alloying ternary and quaternary quantum barriers, the polarization matched InGaN/InAlN and InGaN/InAlGaN quantum well/quantum barrier structures have been proposed and studied both experimentally and numerically [52,53,122,123]. Another convenient way to suppress the polarization effect in the blue quantum wells is to adopt the staggered quantum wells [124–128].

13.5.2 Screen the Polarization Effect by Si-Doped Quantum Barriers

It is also suggested that the polarization effect in the quantum wells can be more conveniently screened by Si doping the quantum barriers [129], such that the free electrons released by the Si dopants in the quantum barriers are able to screen the polarization induced interface charges in the quantum wells. In addition, the crystal and interfacial quality of the InGaN/GaN architectures can be significantly improved by Si-doping the quantum barriers [130,131] as long as the Si doping concentration is lower than the threshold value [132].

Despite the advantages of the Si-doping feature in the quantum barriers, the Si-doped quantum barriers hinder the hole injection [133]. Therefore, we propose the Si-step-doped quantum barriers in increasing the hole injection efficiency [134]. More importantly, our results show that the doping position in the quantum barrier is extremely important for more effectively screening the polarization effect in the quantum wells, such that the ionized positively charged Si dopants can well compensate the polarization induced negative charges at the quantum well/quantum barrier interface. Similarly, the role of ionized Si-dopants in screening the polarization effect and the recommended doped position for [0001] InGaN/GaN LEDs are then also reported by Kim and coworkers in 2015 [135]. The reduced electric field intensity in the quantum wells has also been obtained from their results. Therefore, the Si-step-doped quantum barriers are very promising for more efficiently screening the polarization effect and improving the device efficiency.

13.5.3 Polarization Self-Screening Effect in Quantum Wells

Inspired by the work [134], we further propose and develop the polarization self-screening effect such that the polarization effect in the quantum wells can be self-screened by the polarization induced bulk charges in the quantum barriers [39]. The device architectures to realize the polarization self-screening effect are depicted in Figure 13.7a and b. The quantum wells are grown along the [0001] orientation, and

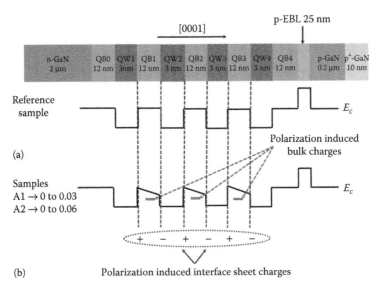

FIGURE 13.7 Schematic device architectures and the conduction band diagrams for (a) reference sample which has $In_{0.15}Ga_{0.85}N$ and GaN as the quantum wells (QWs) and quantum barriers (QBs), respectively and (b) samples A1 and A2, which have InGaN as the quantum barriers, and the InN composition is linearly increased from 0.0% to 3.0% and from 0.0% to 6.0% for samples A1 and A2, respectively. Here, p-EBL denotes the p-type electron blocking layer and the thickness of each layer is shown in the figure. (Reprinted with permission from Zhang, Z.-H. et al., *Appl. Phys. Lett.*, 104, 243501, Copyright 2014, American Institute of Physics.)

TABLE 13.1 Calculated Polarization Induced Interface Charges and Bulk Charges in the MQW Region for the Reference Sample and Samples A1 and A2

	$In_{0.15}Ga_{0.85}N/GaN$	$In_{0.15}Ga_{0.85}N/IN_{0.03}Ga_{0.97}N$	$In_{0.15}Ga_{0.85}N/IN_{0.06}Ga_{0.94}N$
σ_S^{Pol}	0.55×10^{17} m^{-2}	0.44×10^{17} m^{-2}	0.33×10^{17} m^{-2}
	$In_{0.15}Ga_{0.97}N \leftarrow\rightarrow$ GaN within 12 nm	$In_{0.06}Ga_{0.94}N \leftarrow\rightarrow$ GaN within 12 nm	
ρ_S^{Pol}	9.19×10^{23} m^{-3}		1.84×10^{24} m^{-3}

Note: A 40% polarization level is assumed considering the dislocation generation during the epitaxial growth process.

Source: Zhang, Z.-H. et al., *Appl. Phys. Lett.*, 104, 243501, 2014.

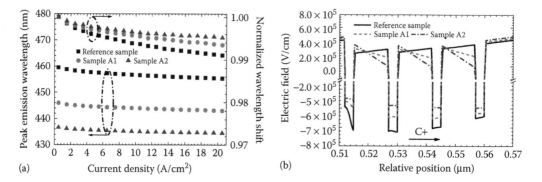

FIGURE 13.8 (a) Experimentally measured peak emission wavelength as a function of the current density and (b) numerically calculated polarization induced electric field profiles in the quantum wells for the reference sample, samples A1 and A2, respectively. The electric field profiles are collected at 0 V. (Reprinted with permission from Zhang, Z.-H. et al., *Appl. Phys. Lett.*, 104, 243501, Copyright 2014, American Institute of Physics.)

this results in the polarization induced sheet charges at the quantum well/quantum barrier interface, as can been seen from Figure 13.7. Meanwhile, once the InN composition in the quantum barriers is linearly increasing along the [0001] orientation, the negative polarization induced bulk charges in the quantum barriers are generated [91,94]. The charge density is calculated and demonstrated in Table 13.1. The details in calculating the polarization induced bulk charges can be found in Reference [39]. Since the grading level in the quantum barriers for sample A2 is larger than that for sample A1, more charges can be produced for sample A2.

In order to demonstrate the self-screening capability, we both experimentally measure the peak wavelength in terms of the injection current density (Figure 13.8a) and numerically calculate the electric field profiles in the quantum well region (Figure 13.8b). The self-screening effect to the polarization in the quantum wells is reflected by the reduced peak emission wavelength when comparing samples A1, A2 and the reference sample. According to Table 13.1, sample A2 has more polarization induced bulk charges than sample A1 does and therefore, sample A2 is better than sample A1 in further reducing the polarization in the quantum wells. Thus, sample A2 has the shortest peak emission length. The theoretical proof is given in Figure 13.8b, agreeing well with the experimental measurement in Figure 13.8a that samples A1 and A2 can better decrease the polarization level in the quantum wells than the reference sample with sample A2 best screening the polarization. It shall be noted that the polarization self-screened MQW structure is free from any hole blocking effect [39].

To summarize, in this subsection, we have reviewed the approaches employed to screen the polarization effect, which often appears in the for example, [0001] oriented III-V nitride quantum wells. More importantly, we have demonstrated the importance of the ionized Si dopants in affecting the polarization induced electric field. We also recommend the preferable Si-doping profiles for the [0001] oriented quantum well. In addition, the ionized dopants can be replaced by the polarization induced bulk charges which are obtained by grading the alloy composition in the quantum barriers. By doing so, we obtain the polarization self-screening effect.

13.6 Current Spreading Layers for Nitride Light-Emitting Diodes

Current crowding induces a high local current density and local heat [136,137], thus causing a substantially nonuniform light distribution and also causes the efficiency droop [3,138,139]. Compared with electrons, holes are even much easier to crowd due to the worse electrical conductivity for the p-GaN layer. The schematic device architecture and the simplified equivalent circuit for the III-nitride–based LED with a lateral current injection scheme are shown in Figure 13.9a and b, respectively [140]. The resistances of the transparent current spreading layer (TCL) and the n-GaN layer are represented by R_{TCL}, $R_{n\text{-}GaN2}$, and $R_{n\text{-}GaN3}$, respectively. The total vertical resistance for each branch is denoted as R_x, and $R_x = R_{p\text{-}GaN} + R_{MQW} + R_{n\text{-}GaN1} + R_{TCL/p\text{-}GaN}$, in which the resistances of p-GaN, MQW, and n-GaN are denoted as $R_{p\text{-}GaN}$, R_{MQW}, and $R_{n\text{-}GaN1}$, respectively, while the interfacial resistance between TCL and p-GaN is $R_{TCL/p\text{-}GaN}$. The total current is divided into I_0, I_1, and I_2. According to Figure 13.9b, the relationship among I_0, I_1, and I_2 is obtained and shown in Equation.13.1 and 13.2. Thus, in order to increase I_1, one can either increase R_x or reduce R_{TCL}. Moreover, a reduced R_{TCL} also benefits I_2. However, the TCL (e.g., ITO) normally is 100 ~ 200 nm, while the n-GaN layer is of 2 ~ 4 μm, and as a result, the area of the cross section through which the current flow for TCL is smaller than that for the n-GaN layer, hence $R_{n\text{-}GaN2} < R_{TCL}$. Note that if the R_{TCL} is tremendously reduced (e.g., thick metal mirror layer for flip-chip LEDs), I_2/I_0 will be larger than 1 that means the current crowds at the edge of the mesa. Nevertheless, according to Equation 13.2, a properly increased R_x (e.g., by a current blocking layer) and a properly reduced R_{TCL} enable an even current distribution. The conclusions we make here are consistent with the report by Guo et al. [141] and Ryu et al. [139].

The common way to increase the R_x is to make certain layers resistive by, for example, selectively ion implanting the p-GaN layer [142,143] or by embedding the insulating layer (e.g., Al_2O_3, SiO_2) between the p-GaN layer and the p-contact [144,145]. Lin et al. selectively cap the p-GaN layer with the Ti metal film while the Ti diffusion can reduce the hole concentration and make the p-GaN layer locally high

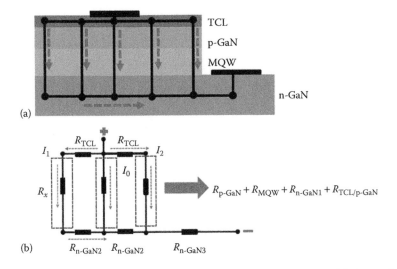

FIGURE 13.9 (a) Schematic device architecture with current flow paths and (b) simplified equivalent circuit for InGaN/GaN LEDs grown on insulating substrates. (From Zhang, Z.-H., Design and Growth of High-Power Gallium Nitride Light-Emitting Diodes, PHD, School of Electrical and Electronic Engineering, Nanyang Technological University, Singapore, 2014.)

resistive [146]. By inductively coupled plasma (ICP) etching, Kuo et al. selectively produce nitrogen vacancies to compensate the holes in the p-GaN layer [147]. Another convenient way to properly increase R_x is to *in situ* grow the barrier junction in the p-GaN layer. Liu et al. propose the short-period i-InGaN/p-GaN superlattice [148,149], and they make use of the energy band discontinuity between the InGaN and GaN layers to modify R_x. Recently, we have developed the (p-GaN/n-GaN/p-GaN)$_x$ current spreading layer [150], and such design avoids the growth difficulty in growing the lattice-mismatched materials. The schematic device architectures for the studied devices are shown in Figure 13.10. In our design, we

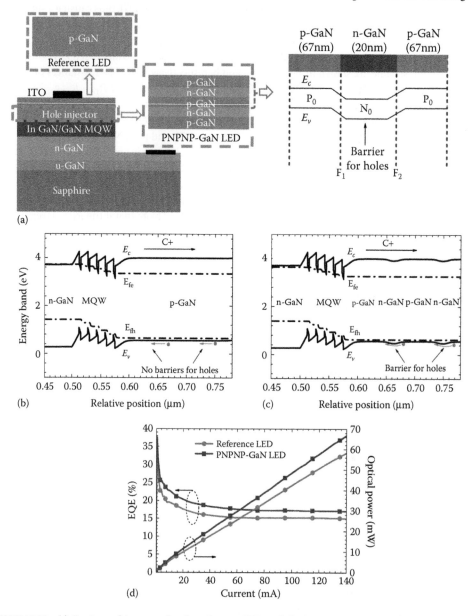

FIGURE 13.10 (a) Device architectures for the reference LED and the PNPNP-GaN LED (the schematic energy band for the PNP-GaN junction is also demonstrated), numerically calculated energy bands for (b) the reference LED and (c) PNPNP-GaN LED, (d) experimentally measured EQE and the optical power in terms of the current for the Reference LED and the PNPNP-GaN LED. (From Zhang, Z.-H. et al., *Opt. Express*, 21, 4958–4969, 2013. With permission from Optical Society America.)

select two p-GaN/n-GaN/p-GaN (PNP-GaN) junctions, that is, p-GaN/n-GaN/p-GaN/n-GaN/p-GaN (PNPNP-GaN). The thickness of the n-GaN layer has to be properly selected so that the n-GaN layer can be completely depleted at the equilibrium state so that each PNP-GaN junction consumes no bias, and this will not increase the forward bias. Meanwhile, Figures 13.10b and c illustrate the calculated energy bands for the reference LED and the PNPNP-GaN LED. The current spreading effect can be improved for the PNPNP-GaN LED thanks to the junction barriers by the PNPNP-GaN structures. As a result, the quantum efficiency is enhanced by 16.98% and 14.37% at 20 mA and 100 mA (see Figure 13.10d), respectively. Note that the device mesa size is 350 μm × 350 μm for the LED devices. According to Equation 13.2, by making the n-GaN layer more resistive, the current can better distribute to the edge of the mesa, and for that purpose, we have also suggested the (n-GaN/p-GaN/n-GaN)$_x$ junction embedded in the n-GaN layer [151]. However, the forward voltage for this proposed device is increased by ~0.6 V at 20 mA for the 350 μm × 350 μm chip, which is due to the nonoptimized p-GaN thickness.

As mentioned, another approach to improving the current spreading is to reduce the R_{TCL}. Therefore, the n$^+$-GaN/p$^+$-GaN tunnel junction is demonstrated to improve the current spreading [152–156]. However, the carrier interband tunneling process requires a stronger electric field in the tunnel junction, which is merely determined by the doping concentration in the n$^+$-GaN/p$^+$-GaN junction. In order to further enhance the carrier interband tunneling efficiency, we propose the polarization tunnel junction, that is, n$^+$-GaN/InGaN/p$^+$-GaN junction [157]. The polarization induced electric field in the InGaN thin layer (3 nm is chosen in our work) follows the same direction as the built-in electric field inside the junction, and this results in a higher electric field intensity in the tunnel junction [158]. Therefore, compared to the n$^+$-GaN/p$^+$-GaN tunnel junction, the n$^+$-GaN/InGaN/p$^+$-GaN tunnel junction reduces the forward bias by ~ 0.7 V (the mesa size is 350 μm × 350 μm). Meanwhile, the polarization tunnel junction improves the hole injection which is effective in improving the EQE.

To summarize, we have proposed a model to design LED devices with the improved current spreading effect. On the one hand, the current crowding underneath the p-electrode can be alleviated by properly reducing the resistance of the current spreading layer (R_{TCL}) and increasing the resistance of the n-GaN layer ($R_{n\text{-}GaN2}$) by, for example, reducing the n-GaN doping concentration. On the other hand, if the R_{TCL} is extremely small, the current tends to grow at the edge of the mesa. However, in any case of different R_{TCL} values, the current can be evenly distributed in the mesa region by optimized the vertical resistance (R_x). Furthermore, we have also reviewed the current technologies adopted to improve the current spreading for LED devices, which can be well explained by our model.

13.7 Effects of Defect Density on Radiative Recombination Rate

To date, there is no consensus on whether the defect-related recombination is responsible for the efficiency droop [9,16–19,159]. However, it is straightforward that, by reducing the defect-related recombination, the LED IQE can be enhanced. An effective way to suppress the impact of defects on the SRH recombination (e.g., DADR) is to introduce the V-defects. Theoretical and experimental studies indicate that V-defects can screen the carriers from the nonradiative recombination centers [160–164]. However, the role of the V-defects in affecting the IQE for InGaN/GaN LEDs is arguable till now [165–167]. Moreover, it is also essential to suppress the defect-related recombination by improving the crystal quality, such as the InGaN/GaN LED epitaxial growth on the nanopatterned substrate [168]. According to the report in the Reference [168] by Li et al., the nanopatterned substrate doubles the IQE for the green InGaN/GaN LED, and this is attributed to the 44% lower dislocation density which is observed through the transmission electron microscopy.

13.8 Role of Auger Recombination in Nitride Light-Emitting Diodes

As has been discussed earlier in this work, the IQE can be significantly promoted if the Auger recombination can be minimized. The Auger recombination scales to the cubic power of the carrier density, which indicates that a suppressed Auger recombination can be obtained by reducing the carrier density in the quantum wells. Chang et al. have numerically concluded that merely increasing the quantum well number makes little impact in reducing the Auger recombination [169]. However, their conclusions make sense if considering the limited hole injection depth into the quantum wells close to the n-GaN side. Hence, Chang et al. propose and demonstrate the polarization matched InGaN/AlGaInN quantum well architectures and the quantum well thickness is properly increased [169], which, on one hand, alleviates the polarization level and flattens the quantum well energy band, and on the other hand, reduces the local carrier density for suppressing the Auger recombination. The report by Chang et al. is consistent with the results by Vaxenburg et al. [170,171].

We have also proposed a gradient InN composition in the quantum wells to suppress the Auger recombination for blue InGaN/GaN LEDs [172]. Two blue InGaN/GaN LEDs are grown by the MOCVD technology, and the peak emission wavelength for the two grown LEDs is ~ 450 nm. Sample A has the conventional quantum well structures such that the $In_{0.15}Ga_{0.85}N$/GaN architecture has the quantum well and the quantum barrier thicknesses of 3 nm and 12 nm, respectively. Nevertheless, the InN composition in sample B is linearly decreased from 15% to 8% within a thickness of 5 nm range to avoid the wavelength variation, that is, $In_{0.15\rightarrow0.08}Ga_{0.85\rightarrow0.92}N$/GaN. The energy band diagrams at 140 A/cm² for the two LED samples are shown in Figure 13.11a and b, respectively, along with which also presents the C1 and HH1 subbands. Clearly, we can see from Figure 13.11b that the conduction band for sample B is flattened and the electrons can be more evenly distributed in each of the quantum wells, which is also reflected in the electron density profiles for samples A and B as illustrated in Figure. 13.11c. The electrons are less locally accumulated in the quantum wells, and this is helpful to suppress the Auger recombination.

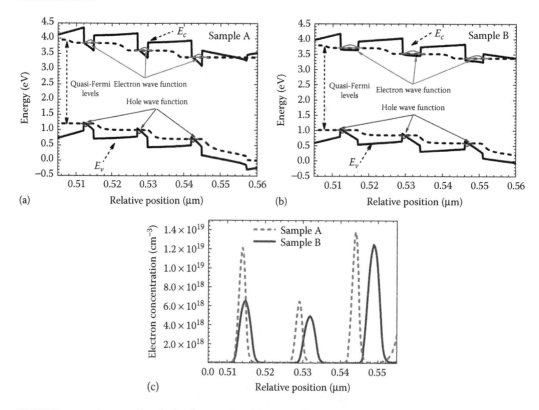

(a)　(b)　(c)

FIGURE 13.11 Numerically calculated energy band diagrams for (a) sample A and (b) sample B and (c) electron density profiles for both samples. Data are collected at 140 A/cm². C1 and HH2 subband wave functions are also included in (a) and (b). E_c and E_v present the conduction band and the valance band, respectively. (Reprinted with permission from Zhang, Z.-H. et al., *Appl. Phys. Lett.*, 105, 033506, Copyright 2014, American Institute of Physics.)

To prove this, we have both numerically calculated and experimentally measured the optical output power and the EQE as a function of the injection current (see Figure 13.12a and c). Investigations into Figures 13.12a and c tell that sample B improves the device performance compared to sample A, and the experimentally measured optical output power for sample B at the current density level of 150 A/cm² is increased by 29.39%. More importantly, according to Figure 13.12b and d, the efficiency droop for sample B is reduced both numerically and experimentally. The experimental droop level decreases from 39.23% to 31.83% at 150 A/cm². The enhanced device performance and the reduced efficiency droop for

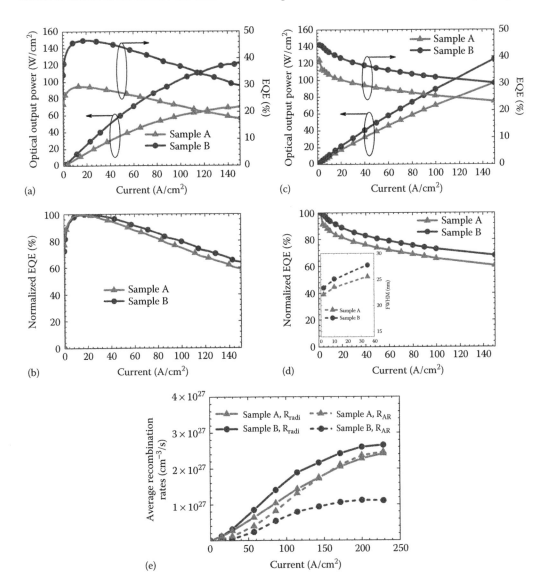

FIGURE 13.12 (a) Numerically calculated optical output power and EQE for samples A and B, (b) numerically calculated and normalized EQE for samples A and B, (c) experimentally measured optical output power and EQE for samples A and B, (d) experimentally measured and normalized EQE for samples A and B, and (e) numerically calculated radiative recombination rate (R_{radi}) and Auger recombination rates (R_{AR}) in terms of the injection current density for samples A and B. The inset figure in (d) shows the full width at half maximum (FWHM) at different current density levels for samples A and B. (Reprinted with permission from Zhang, Z.-H. et al., *Appl. Phys. Lett.*, 105, 033506, Copyright 2014, American Institute of Physics.)

sample B are well attributed to the suppressed Auger recombination in the quantum wells, which have also been calculated and demonstrated in Figure 13.12e, according to which the rising Auger recombination overwhelms the radiative recombination for sample A when the current density exceeds 150 A/cm². However, the Auger recombination is always lower than the radiative recombination for sample B.

To summarize, the effective way to suppress the Auger-recombination-caused carrier loss is to achieve a flat band condition and makes the carriers more uniformly distributed within the quantum wells. In this work, we have reviewed some approaches which realize the flat band condition. However, we also believe the Auger recombination can be suppressed by employing the nonpolar quantum wells. Furthermore, those semipolar quantum wells with a slight [000-1] properties, for example, (11–22) quantum wells reported by Ji et al. [115], are also favorable to obtain the flat band condition and promise the suppressed Auger recombination.

13.9 Conclusion and Future Outlooks

In this chapter, we have reviewed and compared the most recently developed methods to address the issues which hinder the enhancement of the IQE for III-V nitride-based LEDs. We have also proposed and studied alternative design strategies. An increase of the electron injection efficiency can be obtained by, for example, reducing the electron kinetic energy before they enter into the active region and/or suppressing the local electron accumulation level at the interface of the last quantum barrier and the p-EBL. The hole injection efficiency can be promoted, for example, by increasing the hole concentration in the p-GaN layer and/or making the holes more energetic. The hole injection can also be more favored by employing a charge inverter which shortens the surface depletion region width for the p⁺-GaN layer and enables the hole tunneling. Regarding the polarization screening approaches, we have confirmed the effect of the ionized dopants in the quantum barriers on screening the polarization effect in the quantum wells, which makes the doped position in the quantum barriers extremely important. The ionized dopants can then be replaced by the polarization induced bulk charges, that is, polarization self-screening effect. The quantum barriers employing the polarization self-screening feature can even further homogenize the hole distribution in the active region. The current crowding effect in the LED devices can be further alleviated by properly varying the resistances for the p-GaN and the n-GaN layers, respectively. The radiative recombination efficiency is also affected by the SRH recombination that can be weakened by growing LED epi-wafers on the nanopatterned substrate with improved crystal quality. Besides, introducing the intentional V-pits with proper size can also screen the carriers from recombining at the nonradiative recombination centers. As is well known, the Auger recombination consumes a large number of carriers, especially when the device is biased at a high current injection level. Reducing the local carrier density is shown to effectively decrease the Auger recombination. A low local carrier density can be obtained if the quantum well reaches the flat band condition, and this can be realized by grading the InN composition for the polar InGaN quantum wells.

In this work, variously reviewed approaches are effective to improve the IQE for LEDs. However, the origin for the improved device performances has not been well explained for some certain devices, which requires further investigations. Meanwhile, more work will be conducted to evaluate the overall quantum efficiency enhancement when all the proposed technologies are combined into a single LED epi-structure.

In addition, although most of the reviewed work is conducted on the [0001] oriented visible InGaN/GaN LEDs, it is reasonable to extend the proposed device architectures to deep ultraviolet (DUV) LEDs, which have found great potential of being used in skin cure, dissociation of the pollutant materials, and so on, [173] but yet still face great challenges in improving their IQE due to a number of factors, such as the lack of indium in the quantum wells makes the DUV LEDs more sensitive to the threading dislocations, the more resistive AlGaN layers result in a serious current crowding effect, the strong polarization induced electric field in the [0001] oriented quantum wells when [0001] oriented substrates are used, and the low carrier injection efficiency caused by the low doping efficiency [174].

Acknowledgment

This work is supported by National Natural Science Foundation of China (Project No. 51502074), Natural Science of Foundation of Tianjin (Project No. 16JCYBJC16200) and Technology Foundation for Selected Overseas Chinese Scholar by Ministry of Human Resources and Social Security of the People's Republic of China (Project No. CG2016008001).

References

1. S. T. Tan, X. W. Sun, H. V. Demir, and S. P. DenBaars, Advances in the LED materials and architectures for energy-saving solid-state lighting toward "lighting revolution", *IEEE Photonics Journal*, 4: 613–619, 2012.

2. S. Pimputkar, J. S. Speck, S. P. Denbaars, and S. Nakamura, Prospects for LED lighting, *Nature Photonics*, 3: 180–182, 2009.

3. E. F. Schubert, *Light-Emitting Diodes*, 2nd edition. Cambridge, MA: Cambridge University Press, 2006.

4. H. P. Maruska and J. J. Tietjen, The preparation and properties of vapor-deposited single-crystaline GaN, *Applied Physics Letters*, 15: 327–329, 1969.

5. S. Nakamura, T. Mukai, M. Senoh, and N. Iwasa, Thermal annealig effects on p-type Mg-doped GaN films, *Japanese Journal of Applied Physics Part 2-Letters*, 31: L139–L142, 1992.

6. A. Hiroshi, K. Masahiro, H. Kazumasa, and A. Isamu, P-Type conduction in Mg-doped GaN treated with low-energy electron beam irradiation (LEEBI), *Japanese Journal of Applied Physics*, 28: L2112, 1989.

7. S. Nakamura, GaN growth using GaN buffer layer, *Japanese Journal of Applied Physics*, 30: L1705–L1707, 1991.

8. H. Amano, N. Sawaki, I. Akasaki, and Y. Toyoda, Metalorganic vapor phase epitaxial growth of a high quality GaN film using an AlN buffer layer, *Applied Physics Letters*, 48: 353–355, 1986.

9. J. Piprek, Efficiency droop in nitride-based light-emitting diodes, *Physica Status Solidi a-Applications and Materials Science*, 207, 2217–2225, 2010.

10. C. Geng, T. Wei, X. Wang, D. Shen, Z. Hao, and Q. Yan, Enhancement of light output power from LEDs based on monolayer colloidal crystal, *Small*, 10, 1668–1686, 2014.

11. A. I. Zhmakin, Enhancement of light extraction from light emitting diodes, *Physics Reports-Review Section of Physics Letters*, 498: 189–241, 2011.

12. J. Piprek, *Nitride Semiconductor Devices Principles and Simulation*. Weinheim, Germany: WILEY-VCH Verlag GmbH & Co. KGaA, 2007.

13. J.-H. Ryou, P. D. Yoder, J. Liu, Z. Lochner, K. Hyunsoo, S. Choi, H.-J. Kim, and R. D. Dupuis, Control of quantum-confined Stark effect in InGaN-based quantum wells, *IEEE Journal of Selected Topics in Quantum Electronics*, 15: 1080–1091, 2009.

14. S. Huang, B. Fan, Z. Chen, Z. Zheng, H. Luo, Z. Wu, G. Wang, and H. Jiang, Lateral current spreading effect on the efficiency droop in GaN based light-emitting diodes, *Journal of Display Technology*, 9: 266–271, 2013.

15. V. K. Malyutenko and S. S. Bolgov, Effect of current crowding on the ideality factor in MQW InGaN/GaN LEDs on sapphire substrates, *Proceedings of SPIE 7617*, 76171K2010.

16. M. F. Schubert, S. Chhajed, J. K. Kim, E. F. Schubert, D. D. Koleske, M. H. Crawford, S. R. Lee, A. J. Fischer, G. Thaler, and M. A. Banas, Effect of dislocation density on efficiency droop in GaInN/GaN light-emitting diodes, *Applied Physics Letters*, 91: 231114, 2007.

17. X. Shao, H. Lu, D. Chen, Z. Xie, R. Zhang, and Y. Zheng, Efficiency droop behavior of direct current aged GaN-based blue light-emitting diodes, *Applied Physics Letters*,95: 163504, 2009.

18. J. Hader, J. V. Moloney, and S. W. Koch, Temperature-dependence of the internal efficiency droop in GaN-based diodes, *Applied Physics Letters*, 99: 181127, 2011.

19. J. Hader, J. V. Moloney, and S. W. Koch, Density-activated defect recombination as a possible explanation for the efficiency droop in GaN-based diodes, *Applied Physics Letters*, 96: 221106, 2010.

20. J. Iveland, L. Martinelli, J. Peretti, J. S. Speck, and C. Weisbuch, Direct measurement of Auger electrons emitted from a semiconductor light-emitting diode under electrical injection: Identification of the dominant mechanism for efficiency droop, *Physical Review Letters*, 110: 177406, 2013.

21. D. S. Meyaard, G.-B. Lin, J. Cho, E. Fred Schubert, H. Shim, S.-H. Han, M.-H. Kim, C. Sone, and Y. Sun Kim, Identifying the cause of the efficiency droop in GaInN light-emitting diodes by correlating the onset of high injection with the onset of the efficiency droop, *Applied Physics Letters*, 102: 251114, 2013.

22. Y. C. Shen, G. O. Mueller, S. Watanabe, N. F. Gardner, A. Munkholm, and M. R. Krames, Auger recombination in InGaN measured by photoluminescence, *Applied Physics Letters*, 91: 141101, 2007.

23. M. Brendel, A. Kruse, H. Jönen, L. Hoffmann, H. Bremers, U. Rossow, and A. Hangleiter, Auger recombination in GaInN/GaN quantum well laser structures, *Applied Physics Letters*, 99: 031106, 2011.

24. G. Kim, J. H. Kim, E. H. Park, D. Kang, and B.-G. Park, Extraction of recombination coefficients and internal quantum efficiency of GaN-based light emitting diodes considering effective volume of active region, *Optics Express*, 22, 1235–1242, 2014.

25. R. P. Green, J. J. D. McKendry, D. Massoubre, E. Gu, M. D. Dawson, and A. E. Kelly, Modulation bandwidth studies of recombination processes in blue and green InGaN quantum well micro-light-emitting diodes, *Applied Physics Letters*, 102: 091103, 2013.

26. J. Cho, E. F. Schubert, and J. K. Kim, Efficiency droop in light-emitting diodes: Challenges and countermeasures, *Laser & Photonics Reviews*, 7, 408–421, 2013.

27. J. Piprek, F. Römer, and B. Witzigmann, On the uncertainty of the Auger recombination coefficient extracted from InGaN/GaN light-emitting diode efficiency droop measurements, *Applied Physics Letters*, 106: 101101, 2015.

28. G. Verzellesi, D. Saguatti, M. Meneghini, F. Bertazzi, M. Goano, G. Meneghesso, and E. Zanoni, Efficiency droop in InGaN/GaN blue light-emitting diodes: Physical mechanisms and remedies, *Journal of Applied Physics*, 114: 071101, 2013.

29. J. Piprek and Z. M. Simon Li, Sensitivity analysis of electron leakage in III-nitride light-emitting diodes, *Applied Physics Letters*, 102: 131103, 2013.

30. D. S. Meyaard, G.-B. Lin, M. Ma, J. Cho, E. Fred Schubert, S.-H. Han, M.-H. Kim, H. Shim, and Y. Sun Kim, GaInN light-emitting diodes using separate epitaxial growth for the p-type region to attain polarization-inverted electron-blocking layer, reduced electron leakage, and improved hole injection, *Applied Physics Letters*, 103: 201112, 2013.

31. Y.-K. Kuo, Y.-H. Shih, M.-C. Tsai, and J.-Y. Chang, Improvement in electron overflow of near-ultraviolet InGaN LEDs by specific design on last barrier, *IEEE Photonics Technology Letters*, 23: 1630–1632, 2011.

32. R.-M. Lin, S.-F. Yu, S.-J. Chang, T.-H. Chiang, S.-P. Chang, and C.-H. Chen, Inserting a p-InGaN layer before the p-AlGaN electron blocking layer suppresses efficiency droop in InGaN-based light-emitting diodes, *Applied Physics Letters*, 101: 081120, 2012.

33. Z. Liu, J. Ma, X. Yi, E. Guo, L. Wang, J. Wang, N. Lu, J. Li, I. Ferguson, and A. Melton, p-InGaN/AlGaN electron blocking layer for InGaN/GaN blue light-emitting diodes, *Applied Physics Letters*, 101: 261106, 2012.

34. Z. Kyaw, Z.-H. Zhang, W. Liu, S. T. Tan, Z. G. Ju, X. L. Zhang, Y. Ji et al., Simultaneous enhancement of electron overflow reduction and hole injection promotion by tailoring the last quantum barrier in InGaN/GaN light-emitting diodes, *Applied Physics Letters*, 104: 161113, 2014.

35. T. Lu, S. Li, C. Liu, K. Zhang, Y. Xu, J. Tong, L. Wu et al. Advantages of GaN based light-emitting diodes with a p-InGaN hole reservoir layer, *Applied Physics Letters*, 100: 141106, 2012.

36. H. J. Kim, S. Choi, S.-S. Kim, J.-H. Ryou, P. D. Yoder, R. D. Dupuis, A. M. Fischer, K. Sun, and F. A. Ponce, Improvement of quantum efficiency by employing active-layer-friendly lattice-matched InAlN electron blocking layer in green light-emitting diodes, *Applied Physics Letters*, 96: 101102, 2010.

37. S. Choi, M.-H. Ji, J. Kim, H. Jin Kim, M. M. Satter, P. D. Yoder, J.-H. Ryou, R. D. Dupuis, A. M. Fischer, and F. A. Ponce, Efficiency droop due to electron spill-over and limited hole injection in III-nitride visible light-emitting diodes employing lattice-matched InAlN electron blocking layers, *Applied Physics Letters*, 101: 161110, 2012.

38. S. Choi, H. J. Kim, S.-S. Kim, J. Liu, J. Kim, J.-H. Ryou, R. D. Dupuis, A. M. Fischer, and F. A. Ponce, Improvement of peak quantum efficiency and efficiency droop in III-nitride visible light-emitting diodes with an InAlN electron-blocking layer, *Applied Physics Letters*, 96: 221105, 2010.

39. Z.-H. Zhang, W. Liu, Z. Ju, S. Tiam Tan, Y. Ji, Z. Kyaw, X. Zhang, L. Wang, X. W. Sun, and H. V. Demir, Self-screening of the quantum confined Stark effect by the polarization induced bulk charges in the quantum barriers, *Applied Physics Letters*, 104: 243501, 2014.

40. Z.-H. Zhang, W. Liu, Z. Ju, S. Tiam Tan, Y. Ji, X. Zhang, L. Wang, Z. Kyaw, X. W. Sun, and H. V. Demir, Polarization self-screening in [0001] oriented InGaN/GaN light-emitting diodes for improving the electron injection efficiency, *Applied Physics Letters*, 104: 251108, 2014.

41. Z.-H. Zhang, Z. Ju, W. Liu, S. T. Tan, Y. Ji, Z. Kyaw, X. Zhang, N. Hasanov, X. W. Sun, and H. V. Demir, Improving hole injection efficiency by manipulating the hole transport mechanism through p-type electron blocking layer engineering, *Optics Letters*, 39: 2483–2486, 2014.

42. V. Fiorentini, F. Bernardini, and O. Ambacher, Evidence for nonlinear macroscopic polarization in III–V nitride alloy heterostructures, *Applied Physics Letters*, 80: 1204–1206, 2002.

43. S. Chang, Y. Lin, C. Liu, T. Ko, S. Hon, and S. Li, Numerical simulation of GaN-based LEDs with chirped multiquantum barrier structure, *IEEE Journal of Quantum Electronics*, 49: 436–442, 2013.

44. Y. Y. Lin, R. W. Chuang, S. J. Chang, S. G. Li, Z. Y. Jiao, T. K. Ko, S. J. Hon, and C. H. Liu, GaN-based LEDs with a chirped multiquantum barrier structure, *IEEE Photonics Technology Letters*, 24: 1600–1602, 2012.

45. Y. A. Yin, N. Y. Wang, G. H. Fan, and S. T. Li, Advantages of GaN-based light-emitting diodes with polarization-reduced chirped multiquantum barrier, *IEEE Transactions on Electron Devices*, 61: 2849–2853, 2014.

46. J. Piprek and Z. M. S. Li, Origin of InGaN light-emitting diode efficiency improvements using chirped AlGaN multi-quantum barriers, *Applied Physics Letters*, 102: 23510, 2013.

47. Z.-H. Zhang, Z. Kyaw, W. Liu, Y. Ji, L. C. Wang, S. T. Tan, X. W. Sun, and H. V. Demir, A hole modulator for InGaN/GaN light-emitting diodes, *Applied Physics Letters*, 106: 063501, 2015.

48. J.-Y. Chang, M.-C. Tsai, and Y.-K. Kuo, Advantages of blue InGaN light-emitting diodes with AlGaN barriers, *Optics Letters*, 35: 1368–1370, 2010.

49. H. P. Zhao, G. Y. Liu, R. A. Arif, and N. Tansu, Current injection efficiency induced efficiency-droop in InGaN quantum well light-emitting diodes, *Solid-State Electronics*, 54: 1119–1124, 2010.

50. H. P. Zhao, G. Y. Liu, J. Zhang, R. A. Arif, and N. Tansu, Analysis of internal quantum efficiency and current injection efficiency in III-nitride light-emitting diodes, *Journal of Display Technology*, 9: 212–225, 2013.

51. G. Liu, J. Zhang, C. K. Tan, and N. Tansu, Efficiency-droop suppression by using large-band-gap AlGaInN thin barrier layers in InGaN quantum-well light-emitting diodes, *IEEE Photonics Journal*, 5: 2201011–2201011, 2013.

52. J.-Y. Chang and Y.-K. Kuo, Influence of polarization-matched AlGaInN barriers in blue InGaN light-emitting diodes, *Optics Letters*, 37: 1574–1576, 2012.

53. M. F. Schubert, J. Xu, J. K. Kim, E. F. Schubert, M. H. Kim, S. Yoon, S. M. Lee, C. Sone, T. Sakong, and Y. Park, Polarization-matched GaInN/AlGaInN multi-quantum-well light-emitting diodes with reduced efficiency droop, *Applied Physics Letters*, 93: 041102, 2008.

54. Y. K. Kuo, T. H. Wang, J. Y. Chang, and M. C. Tsai, Advantages of InGaN light-emitting diodes with GaN-InGaN-GaN barriers, *Applied Physics Letters*, 99: 91107, 2011.

55. Y. K. Kuo, T. H. Wang, and J. Y. Chang, Advantages of blue InGaN light-emitting diodes with InGaN-AlGaN-InGaN barriers, *Applied Physics Letters*, 100: 31112, 2012.

56. J. R. Chen, T. C. Lu, H. C. Kuo, K. L. Fang, K. F. Huang, C. W. Kuo, C. J. Chang, C. T. Kuo, and S. C. Wang, Study of InGaN-GaN light-emitting diodes with different last barrier thicknesses, *IEEE Photonics Technology Letters*, 22: 860–862, 2010.

57. N. Otsuji, K. Fujiwara, and J. K. Sheu, Electroluminescence efficiency of blue InGaN/GaN quantum-well diodes with and without an n-InGaN electron reservoir layer, *Journal of Applied Physics*, 100: 113105, 2006.

58. T. Li, Q. Y. Wei, A. M. Fischer, J. Y. Huang, Y. U. Huang, F. A. Ponce, J. P. Liu, Z. Lochner, J.-H. Ryou, and R. D. Dupuis, The effect of InGaN underlayers on the electronic and optical properties of InGaN/GaN quantum wells, *Applied Physics Letters*, 102: 041115, 2013.

59. X. Ni, X. Li, J. Lee, S. Liu, V. Avrutin, Ü. Özgür, H. Morkoç, A. Matulionis, T. Paskova, G. Mulholland, and K. R. Evans, InGaN staircase electron injector for reduction of electron overflow in InGaN light emitting diodes, *Applied Physics Letters*, 97: 031110, 2010.

60. F. Zhang, X. Li, S. Hafiz, S. Okur, V. Avrutin, Ü. Özgür, H. Morkoç, and A. Matulionis, The effect of stair case electron injector design on electron overflow in InGaN light emitting diodes, *Applied Physics Letters*, 103: 051122, 2013.

61. X. Li, F. Zhang, S. Okur, V. Avrutin, S. J. Liu, Ü. Özgür, H. Morkoç, S. M. Hong, S. H. Yen, T. S. Hsu, and A. Matulionis, On the quantum efficiency of InGaN light emitting diodes: Effects of active layer design, electron cooler, and electron blocking layer, *Physica Status Solidi (A)*, 208: 2907–2912, 2011.

62. X. Li, S. Okur, F. Zhang, S. A. Hafiz, V. Avrutin, Ü. Özgür, H. Morkoç, and K. Jarašiūnas, Improved quantum efficiency in InGaN light emitting diodes with multi-double-heterostructure active regions, *Applied Physics Letters*, 101: 041115, 2012.

63. V. Avrutin, S. D. A. Hafiz, F. Zhang, Ü. Özgür, H. Morkoç, and A. Matulionis, InGaN light-emitting diodes: Efficiency-limiting processes at highinjection, *Journal of Vacuum Science &Technology A*, 31: 050809, 2013.

64. Z.-H. Zhang, W. Liu, S. T. Tan, Z. Ju, Y. Ji, Z. Kyaw, X. Zhang et al., On the mechanisms of InGaN electron cooler in InGaN/GaN light-emitting diodes, *Optics Express*, 22: A779–A789, 2014.

65. S.-J. Chang and Y.-Y. Lin, GaN-based light-emitting diodes with staircase electron injector structure, *Journal of Display Technology*, 10: 162–166, 2014.

66. S.-H. Yen, M.-C. Tsai, M.-L. Tsai, Y.-J. Shen, T.-C. Hsu, and Y.-K. Kuo, Effect of N-type AlGaN layer on carrier transportation and efficiency droop of blue InGaN light-emitting diodes, *IEEE Photonics Technology Letters*, 21: 975–977, 2009.

67. Y. Ji, Z.-H. Zhang, Z. Kyaw, S. T. Tan, Z. G. Ju, X. L. Zhang, W. Liu, X. W. Sun, and H. V. Demir, Influence of n-type versus p-type AlGaN electron-blocking layer on InGaN/GaN multiple quantum wells light-emitting diodes, *Applied Physics Letters*, 103: 053512, 2013.

68. Z.-H. Zhang, Y. Ji, W. Liu, S. Tiam Tan, Z. Kyaw, Z. Ju, X. Zhang et al., On the origin of the electron blocking effect by an n-type AlGaN electron blocking layer, *Applied Physics Letters*, 104: 073511, 2014.

69. S.-H. Han, D.-Y. Lee, S.-J. Lee, C.-Y. Cho, M.-K. Kwon, S. P. Lee, D. Y. Noh, D.-J. Kim, Y. C. Kim, and S.-J. Park, Effect of electron blocking layer on efficiency droop in InGaN/GaN multiple quantum well light-emitting diodes, *Applied Physics Letters*, 94: 231123, 2009.

70. C. S. Xia, Z. M. Simon Li, and Y. Sheng, On the importance of AlGaN electron blocking layer design for GaN-based light-emitting diodes, *Applied Physics Letters*, 103: 233505, 2013.

71. Y. Y. Zhang and Y. A. Yin, Performance enhancement of blue light-emitting diodes with a special designed AlGaN/GaN superlattice electron-blocking layer, *Applied Physics Letters*, 99: 221103, 2011.

72. S.-J. Lee, S.-H. Han, C.-Y. Cho, S. P. Lee, D. Y. Noh, H.-W. Shim, Y. C. Kim, and S.-J. Park, Improvement of GaN-based light-emitting diodes using p-type AlGaN/GaN superlattices with a graded Al composition, *Journal of Physics D-Applied Physics*, 44: 105101–105105, 2011.

73. J. Kang, H. Li, Z. Li, Z. Liu, P. Ma, X. Yi, and G. Wang, Enhancing the performance of green GaN-based light-emitting diodes with graded superlattice AlGaN/GaN inserting layer, *Applied Physics Letters*, 103: 102104, 2013.

74. J. H. Park, D. Yeong Kim, S. Hwang, D. Meyaard, E. Fred Schubert, Y. Dae Han, J. Won Choi, J. Cho, and J. Kyu Kim, Enhanced overall efficiency of GaInN-based light-emitting diodes with reduced efficiency droop by Al-composition-graded AlGaN/GaN superlattice electron blocking layer, *Applied Physics Letters*, 103: 061104, 2013.

75. S.-J. Lee, S.-H. Han, C.-Y. Cho, S. P. Lee, D. Y. Noh, H.-W. Shim, Y. C. Kim, and S.-J. Park, Improvement of GaN-based light-emitting diodes using p-type AlGaN/GaN superlattices with a graded Al composition, *Journal of Physics D: Applied Physics*, 44: 105101, 2011.

76. E. F. Schubert, W. Grieshaber, and I. D. Goepfert, Enhancement of deep acceptor activation in semiconductors by superlattice doping, *Applied Physics Letters*, 69: 3737–3739, 1996.

77. J. K. Kim, E. L. Waldron, Y.-L. Li, T. Gessmann, E. F. Schubert, H. W. Jang, and J. -L. Lee, P-type conductivity in bulk $Al_xGa_{1-x}N$ and $Al_xGa_{1-x}N/Al_yGa_{1-y}N$ superlattices with average Al mole fraction > 20%, *Applied Physics Letters*, 84: 3310–3312, 2004.

78. I. D. Goepfert, E. F. Schubert, A. Osinsky, P. E. Norris, and N. N. Faleev, Experimental and theoretical study of acceptor activation and transport properties in p-type $Al_xGa_{1-x}N$/GaN superlattices, *Journal of Applied Physics*, 88: 2030–2038, 2000.

79. S.-J. Chang, S.-F. Yu, R.-M. Lin, S. Li, T.-H. Chiang, S.-P. Chang, and C.-H. Chen, InGaN-based light-emitting diodes with an AlGaN staircase electron blocking layer, *IEEE Photonics Technology Letters*, 24: 1737–1740, 2012.

80. Y.-K. Kuo, J.-Y. Chang, and M.-C. Tsai, Enhancement in hole-injection efficiency of blue InGaN light-emitting diodes from reduced polarization by some specific designs for the electron blocking layer, *Optics Letters*, 35: 3285–3287, 2010.

81. B. C. Lin, K. J. Chen, C. H. Wang, C. H. Chiu, Y. P. Lan, C. C. Lin, P. T. Lee, M. H. Shih, Y. K. Kuo, and H. C. Kuo, Hole injection and electron overflow improvement in InGaN/GaN light-emitting diodes by a tapered AlGaN electron blocking layer, *Optics Express*, 22: 463–469, 2014.

82. W. Yang, D. Li, N. Y. Liu, Z. Chen, L. Wang, L. Liu, L. Li, C. H. Wan, W. H. Chen, X. D. Hu, and W. M. Du, Improvement of hole injection and electron overflow by a tapered AlGaN electron blocking layer in InGaN-based blue laser diodes, *Applied Physics Letters*, 100: 031105, 2012.

83. Y.-H. Lu, Y.-K. Fu, S.-J. Huang, Y.-K. Su, R. Xuan, and M. H. Pilkuhn, Efficiency enhancement in ultraviolet light-emitting diodes by manipulating polarization effect in electron blocking layer, *Applied Physics Letters*, 102: 143504, 2013.

84. L. Zhang, K. Ding, N. X. Liu, T. B. Wei, X. L. Ji, P. Ma, J. C. Yan, J. X. Wang, Y. P. Zeng, and J. M. Li, Theoretical study of polarization-doped GaN-based light-emitting diodes, *Applied Physics Letters*, 98: 101110, 2011.

85. L. Zhang, X. C. Wei, N. X. Liu, H. X. Lu, J. P. Zeng, J. X. Wang, Y. P. Zeng, and J. M. Li, Improvement of efficiency of GaN-based polarization-doped light-emitting diodes grown by metalorganic chemical vapor deposition, *Applied Physics Letters*, 98: 241111, 2011.

86. J. Simon, V. Protasenko, C. Lian, H. Xing, and D. Jena, Polarization-induced hole doping in wide-band-gap uniaxial semiconductor heterostructures, *Science*, 327: 60–64, 2010.

87. C. S. Xia, Z. M. S. Li, W. Lu, Z. H. Zhang, Y. Sheng, W. D. Hu, and L. W. Cheng, Efficiency enhancement of blue InGaN/GaN light-emitting diodes with an AlGaN-GaN-AlGaN electron blocking layer, *Journal of Applied Physics*, 111: 094503, 2012.

88. Z. G. Ju, W. Liu, Z.-H. Zhang, T. Swee Tiam, Y. Ji, Z. Kyaw, X. L. Zhang et al., Advantages of the blue InGaN/GaN light-emitting diodes with an AlGaN/GaN/AlGaN quantum well structured electron blocking layer, *ACS Photonics*, 1: 377–381, 2014.

89. Z.-H. Zhang, W. Liu, S. T. Tan, Y. Ji, L. Wang, B. Zhu, Y. Zhang et al., A hole accelerator for InGaN/GaN light-emitting diodes, *Applied Physics Letters*, 105: 153503, 2014.

90. Z.-H. Zhang, Y. Zhang, W. Bi, C. Geng, S. Xu, H. V. Demir, and X. W. Sun, On the hole accelerator for III-nitride light-emitting diodes, *Applied Physics Letters*, 108: 151105, 2016.

91. L. Zhang, K. Ding, J. C. Yan, J. X. Wang, Y. P. Zeng, T. B. Wei, Y. Y. Li, B. J. Sun, R. F. Duan, and J. M. Li, Three-dimensional hole gas induced by polarization in (0001)-oriented metal-face III-nitride structure, *Applied Physics Letters*, 97: 062103, 2010.

92. S. Li, T. Zhang, J. Wu, Y. Yang, Z. Wang, Z. Wu, Z. Chen, and Y. Jiang, Polarization induced hole doping in graded AlxGa1−xN (x = 0.7∼1) layer grown by molecular beam epitaxy, *Applied Physics Letters*, 102: 062108, 2013.

93. S. Li, M. Ware, J. Wu, P. Minor, Z. Wang, Z. Wu, Y. Jiang, and G. J. Salamo, Polarization induced pn-junction without dopant in graded AlGaN coherently strained on GaN, *Applied Physics Letters*, 101: 122103, 2012.

94. Z.-H. Zhang, S. Tiam Tan, Z. Kyaw, W. Liu, Y. Ji, Z. Ju, X. Zhang, X. W. Sun, and H. V. Demir, p-doping-free InGaN/GaN light-emitting diode driven by three-dimensional hole gas, *Applied Physics Letters*, 103: 263501, 2013.

95. Y.-K. Kuo, M.-C. Tsai, S.-H. Yen, T.-C. Hsu, and Y.-J. Shen, Effect of P-Type Last Barrier on Efficiency Droop of Blue InGaN Light-Emitting Diodes, *IEEE Journal of Quantum Electronics*, 46: 1214–1220, 2010.

96. Y.-K. Kuo, J.-Y. Chang, M.-C. Tsai, and S.-H. Yen, Advantages of blue InGaN multiple-quantum well light-emitting diodes with InGaN barriers, *Applied Physics Letters*, 95: 011116, 2009.

97. K. Zhou, M. Ikeda, J. Liu, S. Zhang, D. Li, L. Zhang, J. Cai, H. Wang, H. B. Wang, and H. Yang, Remarkably reduced efficiency droop by using staircase thin InGaN quantum barriers in InGaN based blue light emitting diodes, *Applied Physics Letters*, 105: 173510, 2014.

98. C. H. Wang, S. P. Chang, W. T. Chang, J. C. Li, Y. S. Lu, Z. Y. Li, H. C. Yang, H. C. Kuo, T. C. Lu, and S. C. Wang, Efficiency droop alleviation in InGaN/GaN light-emitting diodes by graded-thickness multiple quantum wells, *Applied Physics Letters*, 97: 181101, 2010.

99. Y.-L. Li, Y.-R. Huang, and Y.-H. Lai, Efficiency droop behaviors of InGaN/GaN multiple-quantum-well light-emitting diodes with varying quantum well thickness, *Applied Physics Letters*, 91: 181113, 2007.

100. X. Ni, Q. Fan, R. Shimada, Ü. Özgür, and H. Morkoç, Reduction of efficiency droop in InGaN light emitting diodes by coupled quantum wells, *Applied Physics Letters*, 93: 171113, 2008.

101. Z. G. Ju, W. Liu, Z.-H. Zhang, S. T. Tan, Y. Ji, Z. B. Kyaw, X. L. Zhang et al., Improved hole distribution in InGaN/GaN light-emitting diodes with graded thickness quantum barriers, *Applied Physics Letters*, 102: 243504, 2013.

102. J. Piprek, Origin of InGaN/GaN light-emitting diode efficiency improvements using tunnel-junction-cascaded active regions, *Applied Physics Letters*, 104: 051118, 2014.

103. J. Piprek, Blue light emitting diode exceeding 100% quantum efficiency, *physica status solidi (RRL)–Rapid Research Letters*, 8: 424–426, 2014.

104. F. Akyol, S. Krishnamoorthy, and S. Rajan, Tunneling-based carrier regeneration in cascaded GaN light emitting diodes to overcome efficiency droop, *Applied Physics Letters*, 103: 081107, 2013.

105. S.-J. Chang, W.-H. Lin, and C.-T. Yu, GaN-Based Multiquantum Well Light-Emitting Diodes With Tunnel-Junction-Cascaded Active Regions, *IEEE Electron Device Letters*, 36: 366–368, 2015.

106. Y. Li, F. Yun, X. Su, S. Liu, W. Ding, and X. Hou, Deep hole injection assisted by large V-shape pits in InGaN/GaN multiple-quantum-wells blue light-emitting diodes, *Journal of Applied Physics*, 116: 123101, 2014.

107. Y. Ji, Z.-H. Zhang, S. T. Tan, Z. G. Ju, Z. Kyaw, N. Hasanov, W. Liu, X. W. Sun, and H. V. Demir, Enhanced hole transport in InGaN/GaN multiple quantum well light-emitting diodes with a p-type doped quantum barrier, *Optics Letters*, 38: 202–204, 2013.

108. F. Zhang, N. Can, S. Hafiz, M. Monavarian, S. Das, V. Avrutin, Ü. Özgür, and H. Morkoç, Improvement of carrier injection symmetry and quantum efficiency in InGaN light-emitting diodes with Mg delta-doped barriers, *Applied Physics Letters*, 106: 181105, 2015.

109. S.-H. Han, C.-Y. Cho, S.-J. Lee, T.-Y. Park, T.-H. Kim, S. H. Park, S. Won Kang, J. Won Kim, Y. C. Kim, and S.-J. Park, Effect of Mg doping in the barrier of InGaN/GaN multiple quantum well on optical power of light-emitting diodes, *Applied Physics Letters*, 96: 051113, 2010.

110. Z.-H. Zhang, S. T. Tan, Y. Ji, W. Liu, Z. Ju, Z. Kyaw, X. W. Sun, and H. V. Demir, A PN-type quantum barrier for InGaN/GaN light emitting diodes, *Optics Express*, 21: 15676–15685, 2013.

111. C. Liu, T. Lu, L. Wu, W. Hailong, Y. Yin, G. Xiao, Y. Zhou, and L. Shuti, Enhanced performance of blue light-emitting diodes with InGaN/GaN superlattice as hole gathering layer, *IEEE Photonics Technology Letters*, 24: 1239–1241, 2012.

112. Z.-H. Zhang, Y. Zhang, W. Bi, C. Geng, S. Xu, H. V. Demir, and X. W. Sun, A charge inverter for III-nitride light-emitting diodes, *Applied Physics Letters*, 108: 133502, 2016.

113. J. P. Liu, J. B. Limb, J.-H. Ryou, D. Yoo, C. A. Horne, R. D. Dupuis, Z. H. Wu et al., Blue light emitting diodes grown on freestanding (11–20) a-plane GaN substrates, *Applied Physics Letters*, 92: 011123, 2008.

114. A. E. Romanov, T. J. Baker, S. Nakamura, J. S. Speck, and E. J. U. Group, Strain-induced polarization in wurtzite III-nitride semipolar layers, *Journal of Applied Physics*, 100: 023522, 2006.

115. Y. Ji, W. Liu, T. Erdem, R. Chen, S. T. Tan, Z.-H. Zhang, Z. Ju et al., Comparative study of field-dependent carrier dynamics and emission kinetics of InGaN/GaN light-emitting diodes grown on (112⁻2) semipolar versus (0001) polar planes, *Applied Physics Letters*, 104: 143506, 2014.

116. M. D. Craven, P. Waltereit, J. S. Speck, and S. P. DenBaars, Well-width dependence of photoluminescence emission from a-plane GaN/AlGaN multiple quantum wells, *Applied Physics Letters*, 84: 496–498, 2004.

117. A. Chitnis, C. Chen, V. Adivarahan, M. Shatalov, E. Kuokstis, V. Mandavilli, J. Yang, and M. A. Khan, Visible light-emitting diodes using a-plane GaN–InGaN multiple quantum wells over r-plane sapphire, *Applied Physics Letters*, 84: 3663–3665, 2004.

118. A. Chakraborty, B. A. Haskell, S. Keller, J. S. Speck, S. P. DenBaars, S. Nakamura, and U. K. Mishra, Nonpolar InGaN/GaN emitters on reduced-defect lateral epitaxially overgrown a-plane GaN with drive-current-independent electroluminescence emission peak, *Applied Physics Letters*, 85: 5143–5145, 2004.

119. A. Chakraborty, B. A. Haskell, S. Keller, J. S. Speck, S. P. Denbaars, S. Nakamura, and U. K. Mishra, Demonstration of nonpolar m-plane InGaN/GaN light-emitting diodes on free-standing m-plane GaN substrates, *Japanese Journal of Applied Physics*, 44: L173–L175, 2005.

120. Y.-D. Lin, A. Chakraborty, S. Brinkley, H. C. Kuo, T. Melo, K. Fujito, J. S. Speck, S. P. DenBaars, and S. Nakamura, Characterization of blue-green m-plane InGaN light emitting diodes, *Applied Physics Letters*, 94: 261108, 2009.

121. T. Detchprohm, M. Zhu, Y. Li, Y. Xia, C. Wetzel, E. A. Preble, L. Liu, T. Paskova, and D. Hanser, Green light emitting diodes on a-plane GaN bulk substrates, *Applied Physics Letters*, 92: 241109, 2008.

122. P. Seoung-Hwan and A. Doyeol, Optical Emission Characteristics of Pseudopolarization-Matched Green AlInGaN/InGaN Quantum Well Structures, *IEEE Journal of Selected Topics in Quantum Electronics*, 19: 1–8, 2013.

123. J. Xu, M. F. Schubert, A. N. Noemaun, D. Zhu, J. K. Kim, E. F. Schubert, M. H. Kim, H. J. Chung, S. Yoon, C. Sone, and Y. Park, Reduction in efficiency droop, forward voltage, ideality factor, and wavelength shift in polarization-matched GaInN/GaInN multi-quantum-well light-emitting diodes, *Applied Physics Letters*, 94: 011113, 2009.

124. R. A. Arif, Y.-K. Ee, and N. Tansu, Polarization engineering via staggered InGaN quantum wells for radiative efficiency enhancement of light emitting diodes, *Applied Physics Letters*, 91: 091110, 2007.

125. C.-T. Liao, M.-C. Tsai, B.-T. Liou, S.-H. Yen, and Y.-K. Kuo, Improvement in output power of a 460 nm InGaN light-emitting diode using staggered quantum well, *Journal of Applied Physics*, 108: 063107, 2010.

126. R. A. Arif, Z. Hongping, Y. K. Ee, and N. Tansu, Spontaneous emission and characteristics of staggered InGaN quantum-well light-emitting diodes, *IEEE Journal of Quantum Electronics*, 44: 573–580, 2008.

127. H. Zhao, G. Liu, X.-H. Li, G. S. Huang, J. D. Poplawsky, S. T. Penn, V. Dierolf, and N. Tansu, Growths of staggered InGaN quantum wells light-emitting diodes emitting at 520–525 nm employing graded growth-temperature profile, *Applied Physics Letters*, 95: 061104, 2009.

128. S.-H. Park, D. Ahn, and J.-W. Kim, High-efficiency staggered 530 nm InGaN/InGaN/GaN quantum-well light-emitting diodes, *Applied Physics Letters*, 94: 041109, 2009.

129. V. Fiorentini, F. Bernardini, F. D. Sala, A. D. Carlo, and P. Lugli, Effects of macroscopic polarization in III-V nitride multiple quantum wells, *Physical Review B*, 60: 8849–8858, 1999.

130. L. W. Wu, S. J. Chang, T. C. Wen, Y. K. Su, J. F. Chen, W. C. Lai, C. H. Kuo, C. H. Chen, and J. K. Sheu, Influence of Si-doping on the characteristics of InGaN-GaN multiple quantum-well blue light emitting diodes, *IEEE Journal of Quantum Electronics*, 38: 446–450, 2002.

131. S.-N. Lee, J. Kim, K.-K. Kim, H. Kim, and H.-K. Kim, Thermal stability of Si-doped InGaN multiple-quantum wells for high efficiency light emitting diodes, *Journal of Applied Physics*, 108: 102813, 2010.

132. L. T. Romano, C. G. Van de Walle, J. W. Ager, W. Götz, and R. S. Kern, Effect of Si doping on strain, cracking, and microstructure in GaN thin films grown by metalorganic chemical vapor deposition, *Journal of Applied Physics*, 87: 7745–7752, 2000.

133. J.-H. Ryou, L. Jae, L. Wonseok, J. Liu, Z. Lochner, Y. Dongwon, and R. D. Dupuis, Effect of silicon doping in the quantum-well barriers on the electrical and optical properties of visible green light-emitting diodes, *IEEE Photonics Technology Letters*, 20: 1769–1771, 2008.

134. Z.-H. Zhang, S. T. Tan, Z. Ju, W. Liu, Y. Ji, Z. Kyaw, Y. Dikme, X. W. Sun, and H. V. Demir, On the effect of step-doped quantum barriers in InGaN/GaN light emitting diodes, *Journal of Display Technology*, 9: 226–233, 2013.

135. D. Y. Kim, G.-B. Lin, S. Hwang, J. H. Park, D. Meyaard, E. F. Schubert, H.-Y. Ryu, and J. K. Kim, Polarization-engineered high-efficiency GaInN light-emitting diodes optimized by genetic algorithm, *IEEE Photonics Journal*, 7: 1–9, 2015.

136. D. Han, J. Shim, D.-S. Shin, E. Nam, and H. Park, Effect of temperature distribution and current crowding on the performance of lateral GaN-based light-emitting diodes, *Physica Status Solidi C*, 7: 2133–2135, 2010.

137. E. Jung, S. Kim, and H. Kim, Effects of temperature on current crowding of GaN-based light-emitting diodes, *IEEE Electron Device Letters*, 34: 277–279, 2013.

138. S. Huang, B. Fan, Z. Chen, Z. Zheng, H. Luo, Z. Wu, G. Wang, and H. Jiang, Lateral current spreading effect on the efficiency droop in GaN based light-emitting diodes, *Journal of Display Technology*, 9: 266–271, 2013.

139. H.-Y. Ryu and J.-I. Shim, Effect of current spreading on the efficiency droop of InGaN light-emitting diodes, *Optics Express*, 19: 2886–2894, 2011.

140. Z.-H. Zhang, Design and Growth of High-Power Gallium Nitride Light-Emitting Diodes, PHD, School of Electrical and Electronic Engineering, Nanyang Technological University, Singapore, 2014.

141. X. Guo and E. F. Schubert, Current crowding in GaN/InGaN light emitting diodes on insulating substrates, *Journal of Applied Physics*, 90: 4191–4195, 2001.

142. Y. W. Cheng, H. H. Chen, M. Y. Ke, C. P. Chen, and J. J. Huang, Effect of selective ion-implanted p-GaN on the junction temperature of GaN-based light emitting diodes, *Optics Communications*, 282: 835–838, 2009.

143. K. H. Lee, K. M. Kang, G. C. Hong, S. H. Kim, W. Y. Sun, and G. M. Yang, Improved light extraction of GaN-based light-emitting diodes by an ion-damaged current blocking layer, *Japanese Journal of Applied Physics*, 51: 6446–6449, 2012.

144. J. H. Son, B. J. Kim, C. J. Ryu, Y. H. Song, H. K. Lee, J. W. Choi, and J.-L. Lee, Enhancement of wall-plug efficiency in vertical InGaN/GaN LEDs by improved current spreading, *Optics Express*, 20: A287–A292, 2012.

145. C.-F. Tsai, Y.-K. Su, and C.-L. Lin, Improvement in the light output power of GaN-based light-emitting diodes by natural-cluster silicon dioxide nanoparticles as the current-blocking Layer, *IEEE Photonics Technology Letters*, 21: 996–998, 2009.

146. R.-M. Lin, Y.-C. Lu, Y.-L. Chou, G.-H. Chen, Y.-H. Lin, and M.-C. Wu, Enhanced characteristics of blue InGaN/GaN light-emitting diodes by using selective activation to modulate the lateral current spreading length, *Applied Physics Letters*, 92: 261105, 2008.

147. T.-W. Kuo, S.-X. Lin, P.-K. Hung, K.-K. Chong, C. I. Hung, and M.-P. Houng, Formation of selective high barrier region by inductively coupled plasma treatment on GaN-based light-emitting diodes, *Japanese Journal of Applied Physics*, 49: 116504, 2010.

148. Y.-J. Liu, C.-H. Yen, L.-Y. Chen, T.-H. Tsai, T.-Y. Tsai, and W.-C. Liu, On a GaN-based light-emitting diode with a p-GaN/i-InGaN superlattice structure, *IEEE Electron Device Letters*, 30: 1149–1151, 2009.

149. Y.-J. Liu, T.-Y. Tsai, C.-H. Yen, L.-Y. Chen, T.-H. Tsai, and W.-C. Liu, Characteristics of a GaN-based light-emitting diode with an inserted p-GaN/i-InGaN superlattice structure, *IEEE Journal of Quantum Electronics*, 46: 492–498, 2010.

150. Z.-H. Zhang, S. T. Tan, W. Liu, Z. Ju, K. Zheng, Z. Kyaw, Y. Ji, N. Hasanov, X. W. Sun, and H. V. Demir, Improved InGaN/GaN light-emitting diodes with a p-GaN/n-GaN/p-GaN/n-GaN/p-GaN current-spreading layer, *Optics Express*, 21: 4958–4969, 2013.

151. Z. Kyaw, Z.-H. Zhang, W. Liu, S. T. Tan, Z. G. Ju, X. L. Zhang, Y. Ji et al. On the effect of N-GaN/P-GaN/N-GaN/P-GaN/N-GaN built-in junctions in the n-GaN layer for InGaN/GaN light-emitting diodes, *Optics Express*, 22: 809–816, 2014.

152. S.-R. Jeon, M. S. Cho, M. A. Yu, and G. M. Yang, GaN-based light-emitting diodes using tunnel junctions, *IEEE Journal of Selected Topics in Quantum Electronics*, 8: 739–743, 2002.

153. S. R. Jeon, Y. H. Song, H. J. Jang, K. S. Kim, G. M. Yang, S. W. Hwang, and S. J. Son, Buried tunnel contact junctions in GaN-based light-emitting diodes, *Physica Status Solidi (A)*, 188: 167–170, 2001.

154. S.-R. Jeon, Y.-H. Song, H.-J. Jang, G. M. Yang, S. W. Hwang, and S. J. Son, Lateral current spreading in GaN-based light-emitting diodes utilizing tunnel contact junctions, *Applied Physics Letters*, 78: 3265–3267, 2001.

155. T. Tetsuya, H. Ghulam, C. Scott, H. Mark, R. P. Schneider, Jr., K. Chris et al. GaN-based light emitting diodes with tunnel junctions, *Japanese Journal of Applied Physics*, 40: L861, 2001.

156. C.-M. Lee, C.-C. Chuo, I. L. Chen, J.-C. Chang, and J.-I. Chyi, High-brightness inverted InGaN-GaN multiple-quantum-well light-emitting diodes without a transparent conductive layer, *IEEE Electron Device Letters*, 24: 156–158, 2003.

157. Z.-H. Zhang, S. T. Tan, Z. Kyaw, Y. Ji, W. Liu, Z. Ju, N. Hasanov, X. W. Sun, and H. V. Demir, InGaN/GaN light-emitting diode with a polarization tunnel junction, *Applied Physics Letters*, 102: 193508, 2013.

158. S. Krishnamoorthy, F. Akyol, P. S. Park, and S. Rajan, Low resistance GaN/InGaN/GaN tunnel junctions, *Applied Physics Letters*, 102: 113503, 2013.

159. Y. P. Zhang, Z.-H. Zhang, W. Liu, S. T. Tan, Z. G. Ju, X. L. Zhang, Y. Ji et al., Nonradiative recombination - critical in choosing quantum well number for InGaN/GaN light-emitting diodes, *Optics Express*, 23: A34–A42, 2015.

160. A. Hangleiter, F. Hitzel, C. Netzel, D. Fuhrmann, U. Rossow, G. Ade, and P. Hinze, Suppression of nonradiative recombination by V-shaped pits in GaInN/GaN quantum wells produces a large increase in the light emission efficiency, *Physical Review Letters*, 95: 127402–127600, 2005.

161. Y.-H. Cho, J.-Y. Kim, J. Kim, M.-B. Shim, S. Hwang, S.-H. Park, Y.-S. Park, and S. Kim, Quantum efficiency affected by localized carrier distribution near the V-defect in GaN based quantum well, *Applied Physics Letters*, 103: 261101, 2013.

162. J. Kim, Y.-H. Cho, D.-S. Ko, X.-S. Li, J.-Y. Won, E. Lee, S.-H. Park, J.-Y. Kim, and S. Kim, Influence of V-pits on the efficiency droop in InGaN/GaN quantum wells, *Optics Express*, 22: A857–A866, 2014.

163. S.-H. Han, D.-Y. Lee, H.-W. Shim, J. Wook Lee, D.-J. Kim, S. Yoon, Y. Sun Kim, and S.-T. Kim, Improvement of efficiency and electrical properties using intentionally formed V-shaped pits in InGaN/GaN multiple quantum well light-emitting diodes, *Applied Physics Letters*, 102: 251123, 2013.

164. K. Kayo, L. Seogwoo, C. Sung Ryong, P. Jinsub, L. Hyojong, H. Jun-Seok, H. Soon-Ku, L. Hyun-Yong, C. Meoung-Whan, and T. Yao, Improvement of light extraction efficiency and reduction of leakage current in GaN-based LED via V-pit formation, *IEEE Photonics Technology Letters*, 24: 449–451, 2012.

165. L. C. Le, D. G. Zhao, D. S. Jiang, L. Li, L. L. Wu, P. Chen, Z. S. Liu et al., Effect of V-defects on the performance deterioration of InGaN/GaN multiple-quantum-well light-emitting diodes with varying barrier layer thickness, *Journal of Applied Physics*, 114: 143706, 2013.

166. J. Kim, Y. Tak, J. Kim, S. Chae, J.-Y. Kim, and Y. Park, Analysis of forward tunneling current in InGaN/GaN multiple quantum well light-emitting diodes grown on Si (111) substrate, *Journal of Applied Physics*, 114: 013101, 2013.

167. L. C. Le, D. G. Zhao, D. S. Jiang, L. Li, L. L. Wu, P. Chen, Z. S. Liu et al. Carriers capturing of V-defect and its effect on leakage current and electroluminescence in InGaN-based light-emitting diodes, *Applied Physics Letters*, 101: 252110, 2012.

168. Y. Li, S. You, M. Zhu, L. Zhao, W. Hou, T. Detchprohm, Y. Taniguchi, N. Tamura, S. Tanaka, and C. Wetzel, Defect-reduced green GaInN/GaN light-emitting diode on nanopatterned sapphire, *Applied Physics Letters*, 98: 151102–151103, 2011.

169. J.-Y. Chang, F.-M. Chen, Y.-K. Kuo, Y.-H. Shih, J.-K. Sheu, W.-C. Lai, and H. Liu, Numerical study of the suppressed efficiency droop in blue InGaN LEDs with polarization-matched configuration, *Optics Letters*, 38: 3158–3161, 2013.

170. R. Vaxenburg, A. Rodina, E. Lifshitz, and A. L. Efros, The role of polarization fields in Auger-induced efficiency droop in nitride-based light-emitting diodes, *Applied Physics Letters*, 103: 221111, 2013.

171. E. Kioupakis, Q. Yan, and C. G. Van de Walle, Interplay of polarization fields and Auger recombination in the efficiency droop of nitride light-emitting diodes, *Applied Physics Letters*, 101: 231107, 2012.

172. Z.-H. Zhang, W. Liu, Z. Ju, S. T. Tan, Y. Ji, Z. Kyaw, X. Zhang, L. Wang, X. W. Sun, and H. V. Demir, InGaN/GaN multiple-quantum-well light-emitting diodes with a grading InN composition suppressing the Auger recombination, *Applied Physics Letters*, 105: 033506, 2014.

173. H. Hirayama, Advances of AlGaN-based high-efficiency deep-UV LEDs, *SPIE*, 7987: 79870G, 2011.

174. A. Khan, K. Balakrishnan, and T. Katona, Ultraviolet light-emitting diodes based on group three nitrides, *Nature Photonics*, 2: 77–84, 2008.

14

White Light-Emitting Diode: Fundamentals, Current Status, and Future Trends

Bingfeng Fan,
Yi Zhuo,
and Gang Wang

14.1 Introduction

Light-emitting diodes (LEDs) are not new topics. The development history of LEDs has experienced more than 100 years since Henry Joseph Round published the first report on electroluminescence in 1907 [1].The first practical visible solid state LED based on GaAsP on GaAs substrates was invented in 1962 by Nick Holonyak and Bevacqua of the General Electric Company [2]. Then the General Electric (GE) Corporation launched the first commercial GaAsP LED emitting in the visible red wavelength range in the early 1960s. The next several years, from the late 1960s to the mid-1970s, red LED application opened up the emerging market in numeric displays. Driven at first by calculators, then by wristwatches, M. George Craford made the first demonstration of a yellow LED in 1972 [3]. New materials and techniques enabled high-brightness (HB) LEDs covering from yellow to red spectrum. Nevertheless, GaP- and GaAs-based materials are difficult to prepare blue band LEDs because of bandgap restriction. There was a lack of blue band in the visible light until the discovery of the GaN-based material. In 1989, Isamu Akasaki and Hiroshi Amano from Nagoya University demonstrated the first true p-type doping and p-type conductivity in GaN. Mg-doping of GaN is the basis for all nitride-based LEDs and laser diodes [4]. In 1992, they reported the first GaN p-n-homojunction LED that emitted light in the ultraviolet (UV) and blue spectral range [5]. While Shuji Nakamura of Nichia Chemical Industries Corporation developed a two-flow organometallic vapor-phase epitaxy (OMVPE) growth system, the first blue InGaN double-heterostructure LEDs were demonstrated with efficiencies up to 10%. Nakamura's series of inventions had made blue LEDs turn to practical [6–10]. In the development history of blue LEDs, Isamu Akasaki and Hiroshi Amano solved the basic scientific problems in the growth of materials especially in the p-type doping in GaN, and Nakamura's contribution had pushed the blue LED technologies to the industry. In view of their contribution, the Nobel Prize in Physics 2014 was awarded jointly to Isamu Akasaki, Hiroshi Amano, and Shuji Nakamura *for the invention of efficient blue light-emitting diodes which has enabled bright and energy-saving white light sources* [11]. A detailed account of their contributions can be referred to the book *The Blue Laser Diode* written by Nakamura and Fasol (1997) [12].

After the invention of blue LED based on GaN materials, Nichia further proposed the phosphor-converted white LEDs [13]. As soon as it emerged, high-power phosphor-converting white LEDs have realized rapid development, due to their promising applicability in solid state lighting. Well recognized as the next generation lighting source, white LEDs have outstanding advantages such as high brightness, low-energy consumption, extremely long life, small volume, quick response, pollution-free, and so on, which the traditional lighting source can never match. For example, conventional light sources have typical luminous efficiencies of 15–100 lm/W, white LEDs have the potential for luminous efficiencies exceeding 300 lm/W. Not only would GaN-based devices be the most efficient visible light sources, but they would also be nontoxic (mercury-, phosphorus- and arsenic-free) and have projected lifetimes longer than 10 years. On the inception of the first white LED, it immediately attracted the great interest of the industry. Several commercial entities have invested in developing white LEDs for illumination applications, typical companies including Lumileds [14], GELcore [15], Nichia Corporation [13], Osram Opto Semiconductors [16], and Cree Lighting Corporation [17]. After years of development, solid-state lighting based on gallium nitride (GaN) semiconductors has made remarkable breakthroughs in efficiency, and LEDs have begun replacing incandescent and fluorescent lights in a number of niche applications.

In this review, we are going to skip the fundamental theory like luminescent mechanism, GaN-based material properties, and process and device, and will present the state-of-the-art concept in GaN-based white LED research and the developments, trends, and challenges. In the first section, we briefly present the generation of white light. In the second section, the fundamental aspects of wavelength-converted white LEDs will be discussed. In the last part, we focus on the evolution and development of LED packaging technology.

14.2 Generation of White Light

In this section, we briefly present the generation of white light to familiarize you with white LED. Since white light is made up with the whole visible light spectrum, there are two ways to generate white light from monochromatic LEDs. One is to use LEDs with wavelength conversion materials, generally achieved by the blue LED exciting the Ce-doped YAG yellow phosphors or the near-UV or UV LEDs exciting the red/green/blue phosphors, and this kind of excitation is called as phosphor-converted LED. The other is to combine multiple LEDs with different spectrum, specifically mixing individual LEDs emitting three primary colors, red, green, and blue (RGB). Figure 14.1 shows the typical method of creating white LED.

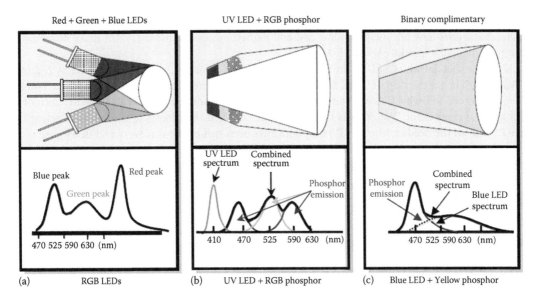

Red + Green + Blue LEDs   UV LED + RGB phosphor   Binary complimentary

(a)   RGB LEDs   (b)   UV LED + RGB phosphor   (c)   Blue LED + Yellow phosphor

FIGURE 14.1 Three methods of generating white light from LEDs: (a) red + green + blue LEDs, (b) UV LED + RGB phosphor, and (c) blue-LED + yellow phosphor. (From Steigerwald, D.A. et al., *IEEE J. Sel. Topics Quantum Electron.*, 8, 310–320, 2002. © 2002 IEEE. With permission.)

14.2.1 Phosphors Excited by Blue-LED Chips

Dichromatic white LEDs formed by yellow phosphors (usually the Ce^{3+}-doped yttrium aluminum garnet, $Y_3Al_5O_{12}:Ce^{3+}$ [YAG:Ce]) and III-nitride blue LEDs are the simplest white LEDs. Figure 14.2 shows a typical emission spectrum of a YAG:Ce phosphor-based white LED. The white LED spectrum consists of blue light band generated by LED with the peak wavelength of 450 to 470 nm, and the yellow band generated by phosphor excited by the blue LED Chip. The YAG phosphors emit yellow light is excited by the blue light and the two lights mix into white light (which can be called as "blue + yellow" white LEDs). According to the principle of colorimetry, we can get a white light with different correlated color temperature (CCT) by adjusting the two emission band ratio. As a result of the use of wavelength-convert material,

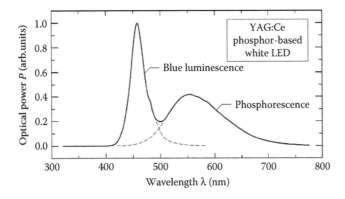

FIGURE 14.2 Emission spectrum of a phosphor-based white LED manufactured by the nichia chemical industries corporation. (From Schubert, E.F., *Light-Emitting Diodes*, Cambridge University Press, Cambridge, MA, 2003. With permission.)

efficiency losses are expected. Compared to the trichromatic and tetrachromatic white LEDs, the dichromatic ones have advantages of low cost and highest luminous efficacy, which make it the most common method to produce white LED.

Most commercial white LEDs is in the manner of light conversion, which is fabricated by coating YAG phosphors on the chip of blue LED. One deficiency for white LEDs that use only YAG:Ce is that they are limited to high CCTs and the relatively lower-color rendering index (CRI), due to a lack of a red spectral component. The other is that Nichia has cornered the patents of white LEDs excited by YAG. Many of the needs for new LED phosphors have been met by the discovery and development of new phosphors over the past 10 years. During this time, the field of LED phosphors has moved from the YAG:Ce phosphors to a variety of silicate, aluminate, nitride, oxynitride, sulfide, and fluoride compositions [20], leading to commercial LEDs that cover a full range of white CCTs. Figure 14.3 presents the developed scheme of phosphors for white LEDs.

14.2.2 Phosphors Excited by UV LED Chips

The disadvantage of the conventional "blue + yellow" white LEDs is the low CRI (R_a). The approach in which phosphors are excited by UV LED chip is similar to that by a blue light source except that the LED output does not contribute to the visible light output. Analogous to the fluorescent lamp, UV or near-UV LEDs exciting red/green/blue phosphors mixture is a solution of white LED with better color rendering. However, due to lower efficiency of UV or near-UV LED chip itself and larger energy losses during wavelength conversion, this method is less efficient than blue-LEDs with yellow phosphor. Yukio Narukawa reported a white LED which consisted of a near-UV (400 nm) LED chip and blue/yellow phosphors (called as "*n*-UV + blue/yellow"), as shown in Figure 14.4a [21]. J. K. Sheu reported the fabrication of white LEDs, which consisted of near-UV LED (400 nm) chips and phosphors (called as "*n*-UV + blue/green/red"), as shown in Figure 14.4b, and their results indicate that such "*n*-UV blue/green/red" white LEDs are much more optically stable than the conventional "blue + yellow" white LEDs [22]. Toshio Nishida demonstrated the high CRI (R_a) potential of the light sources consisting of 350 nm UV LEDs and three-basal-color phosphors (blue/green/red), as shown in Figure 14.4c [23]. Although UV-excited

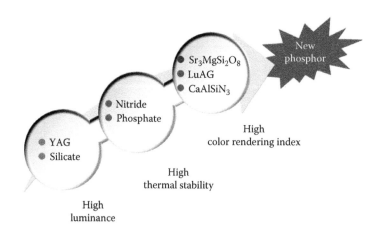

FIGURE 14.3 Developed scheme of phosphors for white LEDs. (From Lin, C.C. and Liu, R.S., *J. Phys. Chem. Lett.* 2, 11, 1268–1277, 2011. With permission.)

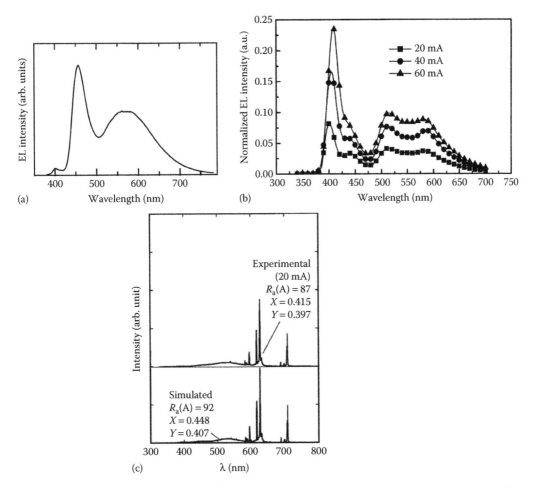

FIGURE 14.4 EL spectra of a white LED in which: (a) a near-UV LED (400 nm) was combined with blue/yellow phosphors. (called as "*n*-UV+blue/yellow") (From Narukawa, Y. et al., *Jpn. J. Appl. Phys.*, 2, 41, 4A, L371–L373, 2002.); (b) a near-UV LED (400 nm) chip was combined with blue/green/red phosphors. (called as "*n*-UV+blue/green/red") (From Sheu, J.K. et al., *IEEE Photon. Technol. Lett.*, 15, 1, 18–20, 2003.); (c) a 350 nm UV LED chip was combined with three-basal-color phosphors. (blue/green/red) (From Nishida, T. et al., *Appl. Phys. Lett.*, 82, 22, 3817–3819, 2003.)

red-green-blue (RGB)-phosphor-converted white LEDs have a higher CRI, UV leakage and attenuation of RGB phosphor under high-intensity UV LED still need to be considered.

14.2.3 Mixed by Red, Green, and Blue LEDs

Mixing red/green/blue LED chips with an electronic circuit that can adjust the relative power ratio of the three LEDs can flexibly generate different color light, and in principle, it can generate the most efficient white light due to the lack of use of the wavelength-convert material. However, change in the emission wavelength of individual LED chip significantly changes the color temperature and makes color stability one of the most important technical problems lying ahead since it is difficult to control the actual color value that changes with temperature, aging, and packaging.

14.2.4 Excited by Quantum Dots

The white light generated by the quantum dot (QD) excitation is a new way, which is put forward in recent years. Quantum dots (QDs) are usually composed of a finite number of atoms of II–VI or III–V elements, and the size of QDs in the three dimensions are all at the nanometer level, generally between 1–10 nm [24]. In a QD, the electrons and holes are in a confined space, which was considered as the zero-dimensional confinement effect, and the electrons and holes energy levels are quantized [25].

QDs have many advantages, for instance, wide absorption spectrum, narrow emission spectrum, long fluorescent life, high-photoluminescence quantum yields, low scattering and good color saturation. But the most remarkable characteristic is that we can control the emission peak position by changing the element composition and particle size [26]. Up to now, scientists have discovered a number of types of QDs, and their emission peak positions have covered the entire visible spectrum [25], even extending to the near infrared region and near UV region [27]. There are a lot of research about the preparation of white LED devices by replacing phosphor with QDs and have achieved good results due to their unique properties. For example, Xi et al. have made a white LED based on Cu:ZnInS/ZnS core/shell quantum dots with CRI up to 96, luminous efficacy of 70–78 lm/W, and color temperature of 3800–5760 K, as shown in Figure 14.5 [28].

Compared with previous phosphor-converted white LEDs, QDs excited white LED have higher CRI and better color stability. Yin et al. found that white LEDs with the encapsulation of yellow phosphors and red QDs exhibited higher CRI and lower sensitivity to temperature than those with the encapsulation of

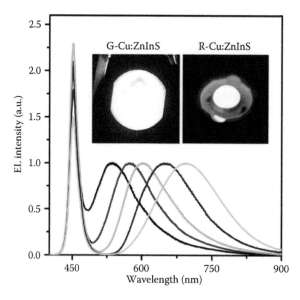

FIGURE 14.5 EL spectra of Cu:ZnInS QD-based LED devices operated at a forward current of 20 mA. Photographs of the 20 mA-driven QD-based LEDs fabricated with G-Cu:ZnInS/ZnS and R-Cu: ZnInS/ZnS QDs are shown in the inset, respectively. (From Yuan, X. et al., *Nanotechnology*, 25, 43, 435202–435202, 2014. With permission.)

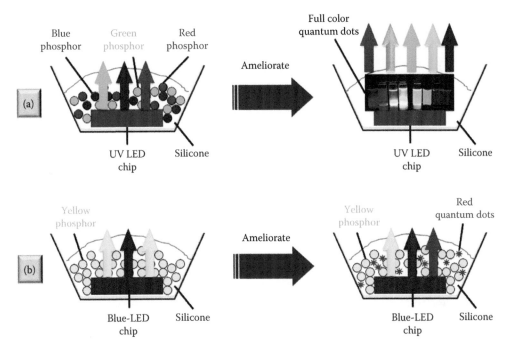

FIGURE 14.6 Structures for generating white light from fluorescent materials, based on (a) UV LED chip and (b) blue-LED chip. (From Lin, C.C. and Liu, R.S., *J. Phys. Chem. Lett.*, 2, 11, 1268–1277, 2011. With permission.)

yellow and red phosphors [29]. Figure 14.6 shows the technology roadmap of white LED upgraded from excited by phosphors to excited by QDs [20].

14.3 Fundamental Aspects of Wavelength-Converted White LEDs

It is necessary to gain some knowledge about colorimetry in order to gain insight into the white LEDs. Readers can refer to some of the professional books about colorimetry to get the basic principles of colorimetry containing color space, color mixing, color analysis, and CIE standard colorimetric system [30]. In this section, we will introduce the fundamental aspects of wavelength-converted white LEDs.

14.3.1 Basic Concepts of White LEDs

14.3.1.1 Luminous Efficacy (Lm/W)

Luminous efficacy is the ratio of the luminous flux emitted by the LED to the electric power consumption. The luminous flux can be obtained by integrating the visual efficiency function $V(\lambda)$ and the distribution function of spectral energy. The electric power consumption is the product of the LED operating voltage and current. In addition to the internal quantum efficiency, luminous efficacy is affected by the phosphor, encapsulation adhesive, and encapsulation lens and so on. Researchers around the world have been working to obtain high-luminous efficiency products for a long time.

The luminous efficacy is defined as:

$$\text{luminous efficacy} = \frac{\Phi_{\text{lum}}}{P} = \frac{\left[683\frac{\text{lm}}{W}\int_\lambda V(\lambda)P(\lambda)d\lambda\right]}{\left[\int_\lambda P(\lambda)d\lambda\right]}$$

where

$P(\lambda)$ is the power spectral density
Φ_{lum} is the luminous flux
$V(\lambda)$ is the visual efficiency function

For the human eye adapted to brightness, the maximum relative spectral sensitivity is at a wavelength of 555 nm. The highest perception of brightness experienced by the human eye is generated with the green light on the 555 nm wavelength. The greatest luminous efficacy that can be theoretically achieved at 555 nm is 683 lm/W. The typical spectrum of a warm white LED achieves a theoretical module luminous efficacy of approximately 320 lm/W. However, since the assumption is that there is a loss-free conversion of physical radiated power into the wavelengths of the spectrum, then the actual realizable module luminous efficacy is much smaller. In future, it may be possible to achieve system luminous efficacy in the range of 200–250 lm/W.

14.3.1.2 Correlated Color Temperature

Color temperature is an important index of white LED color parameters. CCT describes the relative color appearance of a white light source, indicating whether it appears more yellow/gold or bluer, in terms of the range of available shades of white. CCT is given in kelvins (the unit of absolute temperature) and refers to the appearance of a theoretical black body (visualize a chunk of metal) heated to high temperatures. As the black body gets hotter, it turns red, orange, yellow, white, and finally blue. The CCT of a light source is the temperature (in K) at which the heated theoretical black body matches the color of the light source in question. According to the standard file of CIE, the color temperature is divided into three types. Color temperatures higher than 5000 K are called cool colors which make people feel cool, while low-color temperatures, lower than 3000 K, are called warm colors which make people feel warm and steady. Color temperatures between 3000K and 5000K are called the middle colors which make people feel frank and refreshing. Generally, the color temperature of the sunshine is low in the morning and evening, but relatively high at noon.

14.3.1.3 CIE Color Rendering Index

CRI is another important index of white LED color parameters. CRI indicates how well a light source renders colors, on a scale of 0–100, compared to a reference light source. The test procedure established by the International Commission on Illumination (CIE) involves measuring the extent to which a series of eight standardized color samples differ in appearance when illuminated by a given light source, relative to the reference source. The average "shift" in those eight color samples is reported as *Ra* or CRI. In addition to the eight color samples used by convention, some lighting manufacturers report an "R9" score, which indicates how well the light source renders a saturated deep red color.

14.3.2 Wavelength Converter Materials

14.3.2.1 Phosphor

The fluorescent powder has become one of the key materials in the semiconductor lighting technology. Its features directly determine the brightness, CRI, color temperature, and luminous efficiency performance of the phosphor-converted white LEDs.

Currently, it mainly has three kinds of methods to achieve white LEDs using phosphor coating:

1. A blue LED chip is coated with yellow phosphor (YAG) that can be excited by blue light. Then the blue light emitted by the blue LED and the yellow light emitted by yellow phosphors combine into white light. This technology is monopolized by the Japanese Nichia company. But a schematic drawback of this scheme is that the emission spectrum of the phosphor ions Ce^{3+} has no-continuous spectrum characteristic, causing poor color rendering property. It is difficult to meet the requirements of low-color temperature light, and luminous efficiency is not high enough. It needs to be improved by the development of new high-efficiency fluorescent powder.

2. The blue LED chip coated with green and red fluorescent powder can generate white light, which is a composite of blue light emitting by the chip and green and red light emitting by phosphors covered by the chip. It is good to have high color rendering property. However, the fluorescent powders used in this method usually have low-conversion efficiency, especially the red phosphor, which needs to be greatly improved.

3. The violet or UV LED chip, coated with tricolor or multiple color fluorescent powders, emits UVA (370–380 nm) or violet (380–410 nm) light to stimulate phosphors to generate white light. Compared with two previous methods, this method can get the white light with the best color rendering property. However, this method also faces similar problems as with the second method. Currently, on the market, higher conversion efficiency red and green phosphors generally belong to the sulfide system. The stability of such light-emitting phosphors is poor, and their luminous decay is larger than other methods. So the development of phosphors with high-efficiency and low-light decay has become an urgent task.

14.3.2.2 Quantum-Dots

After Alexei Ekimov first discovered the QDs in a glass matrix in the 1980s [31], the research on QDs increase drastically. Now the emission color of QDs can be tuned not only in the visible range (Figure 14.7) [32] but also the near infrared and near UV region [27]. Typical energy level structure is shown in Figure 14.8. Furthermore, QD LED has become a research hotspot in recent years, and electrically pumped QD LEDs are another possible solution for white LEDs due to the large-area and potential for low-cost manufacturing techniques (e.g., roll-to-roll processing).

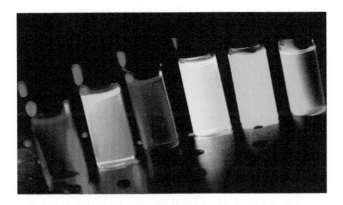

FIGURE 14.7 Composition-tunable emission color of alloyed $Zn_xCd_{1-x}Se$ nanocrystals. (From Regulacio, M.D. and Han, M.Y., *Acc. Chem. Res.* 43, 621–630, 2010. With permission.)

Semiconductor quantum dots

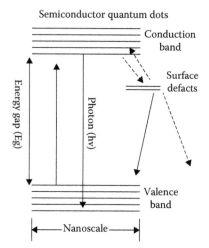

FIGURE 14.8   Quantum dot energy level structure.

As shown in Figure 14.8, QDs have discrete energy level structure. There are three factors affecting the efficiency of QDs: electron-hole recombination, impurity level, and surface defect. Early QDs have a simple structure with a lot of surface defects which results in low-quantum efficiency. After three decades of research and development, some new QD structures have been developed, such as alloy QDs with gradient and core-shell structure. Gradient alloy structure is different from the core-shell structure, where a thin layer semiconductor is grown on the surface of a core semiconductor. Figure 14.9 shows the difference between homogeneous, gradient, and core-shell structure [32]. Core-shell structure modifies the defects of the center core with shell layer, greatly improves the quantum efficiency, and has become the most widely used structure.

QDs have been applied in many fields. (a) QD-LED [33], (b) QD-LED display [34], (c) ion sensor applications [35], (d) solar cell applications [36], and (e) biomedical imaging [37]. We believe that the application of QDs will be more extensive in the future. In the practical application of QDs white LED technology, there remain fundamental science challenges that require further research. The main drawbacks for QDs as down-converters in SSL applications are lifetime and reliability issues. Further research on QD materials is needed to improve their performance at high temperatures and high optical flux densities to allow their on-chip LED application to be compatible with current LED industry manufacturing practices. Besides the material and integration challenges, there are also significant manufacturing challenges for QDs. A more concentrated effort is needed to address reproducibility and scale-up for QD manufacturing.

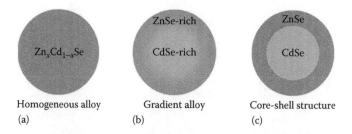

FIGURE 14.9   Schematic representations of spherical nanocrystals having a (a) homogeneous, (b) gradient, and (c) core-shell structure using ZnCdSe as example. (From Regulacio, M.D. and Han, M.Y., *Acc. Chem. Res.* 43, 5, 621–630, 2010. With permission.)

14.4 Development of White LED Packaging

14.4.1 Consideration of White LED Packaging

14.4.1.1 Thermal Management of High Power LED

LEDs, which are regarded as the next generation lighting source, attract more attention in commercial and scientific research fields. To satisfy the requirement of high-luminous flux, the increased electrical currents used to drive the LEDs lead to high-heat flux generated in LED chip. The elevated junction temperature of the LED has been shown to affect the device's light output, light quality, lifetime, and reliability. Thermal management for the solid-state lighting applications is a key issue in LED lighting systems. The temperature as a result of inefficiency becomes the most critical design parameter starting from the LED die with its junction temperature and ends up in the luminary housing with its maximum surface temperatures defined in international standards. The basic system function of "generating light" is getting more and more linked to the question of "dissipating heat". Proper design of a LED system should ensure the continuation of thermal flux from the heat source into the system surrounding atmosphere. Considering the overall efficiency from the mains to the light output about 75% power of a LED lighting system are losses nowadays. Even with future highest efficiency LEDs (150 Lm/W), about 50% of input power needs to be dissipated.

Thermal management is based on the use of thermal resistance networks. A thermal resistance network is analogous to an electrical resistor network using parallel and series resistances to represent heat transfer paths. In place of a voltage potential across a resistor, a temperature potential exists, and instead of current flowing through the electrical circuit, heat flows through the thermal network. Figure 14.10 shows the thermal resistance network for a complete Luxeon LED package mounted on a chip carrier (heat spreader) by solder or bonding adhesive. The heat spreader consists of a high-conductivity material such as copper. Such a network is generally utilized to evaluate the heat transport performances of this system. From Figure 14.10 it can be inferred that these thermal resistances are in a series configuration:

$$R_{system} = R_{cihp} + R_{chip-sub} + R_{submount} + \ldots + R_{heatsink-envir}$$

where R_{system} is the whole system thermal resistance (°C/W); R_{chip} is the thermal resistance of LED chip itself; $R_{chip-sub}$ is the thermal resistance between chip and submount; $R_{submount}$ is the thermal resistance of submount, and so on. The goal of thermal management is to decrease the thermal resistance of the entire LED lighting system. Figure 14.11 shows the evolution of thermal resistance of different packaging forms.

- Chip-level thermal management

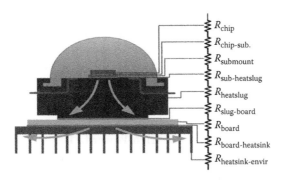

FIGURE 14.10 Typical thermal resistance network of Luxeon LED. (From Pardo, B. et al., *Microelectron. Reliab.*, 53, 1084–1094, 2013. With permission.)

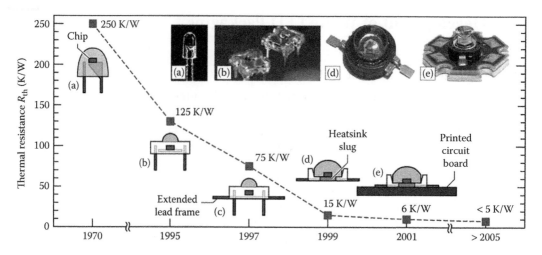

FIGURE 14.11 Thermal resistance of LED package (a) 5 mm, (b) low profile, (c) low profile with extended lead frame, (d) heatsink slug, and (e) heatsink slug mounted on print-circuit board (PCB). Trade names for these packages are "Piranha" (b, Hewlett Packard Corp.), "Barracuda" (d and e, Lumileds Corp.). (From Schubert, E.F., *Light-Emitting Diodes*, 2nd edition, Cambridge University Press, Cambridge, MA, 2006. With permission.)

First, we discuss the source of heat generation in the LED chip. Heat can be generated in the contacts, cladding layers, and the active region. At low-current levels, heat generation in the parasitic resistances of contacts and cladding layers is small due to the I^2R dependence of Joule heating. The dominant heat source at low-current levels is the active region, where heat is created by nonradiative recombination. At high-current levels, the contribution of parasitic becomes increasingly important and can even become dominated. For a more detailed study of the semiconductor physics, the reader is advised to refer to the published handbooks of the topic [19]. It is essential to solve the heat dissipated from the LED and reduce the PN junction temperature.

In the actual thermal design, heat sources were defined in the active layer. The heat was assumed to be generated uniformly within the heat source. The volumetric heat generation Q_{vol} in the active layer of the device was determined from the equation: $Q_{vol} = (1-\eta)UI/V$, where η is the wall-plug efficiency (WPE), U is the LED voltage, I is the current, and V is the active layer volume. The heat generation in the active layer is typically 1W when at 350 mA current, and η is about 17%. The thermal resistance (R_{JC}) between the heat source and the package was determined by $R_{JC} = (T_J - T_C)/P$, where T_J is the junction temperature of the active layer, T_C is the case temperature measured at the case bottom, and P is the applied power, respectively. The junction temperature of the device was determined using the average temperature within an active layer.

The thermal performance was greatly influenced by the type of chip configuration. Figure 14.12 shows the comparison of thermal design concept between the conventional face-up LED and the flip-chip LED. In the flip-chip configuration, the LED is commonly "flip-chip" bonded to a high thermal conductivity submount by the eutectic AuSn solder attachment process. Unlike conventional LED packaging, the heat generated in the nitride-based thin active layer of flip-chip LEDs was directly transferred to the bottom of the die.

14.4.1.2 Heat Dissipation Technology of High-Power LED

With the development of LED technology, LED power continuously improves, but it's also leading to more challenges for the LED heat dissipation. Now, energy conversion efficiencies of the LED is low, about 20%–40% percentage of input energy is converted to light, resulting in a significant improvement of the PN-junction temperature. In the LED working process, this heat has to be conducted away from the junction.

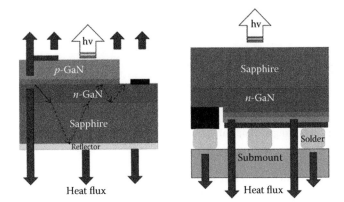

FIGURE 14.12 Comparison of thermal design concept between the conventional face-up LED in which the heat is evacuated from the top and bottom of the die (left) and the flip-chip LED where the thermal transfer occurs at the bottom of the die via submount only (right).

Effective thermal management is essential for the life and reliability of the LED. Based on the research of the LED heat dissipation design, there is more and more heat dissipation technology applied to power LED.

14.4.2 Fabrication of White LED Packaging

14.4.2.1 Phosphor Coating Method

As mentioned above, although white light with high CRI can be generated by mixing individual red, green, and blue LEDs or exciting red, green, and blue phosphors by UV LEDs, high-cost and low-conversion efficiency strictly limit their application. Therefore, the most widely used white light LEDs are Blue LEDs with yellow phosphor till now. As mentioned above, concerns of white LED include luminous efficiency, CCT, and CRI, furthermore angular color uniformity, which is often evaluated by angular-dependent CCT, is also an important one. In addition to the chip and the phosphor, the packaging issues strongly affect the optical properties of white LEDs. The traditional white LED packaging has five main steps. The first one is to attach the die to a substrate by a thermally conductive adhesive. The second one is Gold wire bonding for electrical connection. The third one is phosphor coating and encapsulation. The fourth is optical lens attaching. The last step is singulation and test. Among these five steps of phosphor-converted white LED, phosphor coating method is vital for fabricating white LEDs with high luminous efficiency and angular color uniformity because it controls the shape and arrangement of the phosphor layer. In this section, we will discuss three typical phosphor coating methods.

The generic diagram of the dispense phosphor coating, also called "phosphor-in-cup" [40], shows in Figure 14.13a. Because of its easy fabrication and low cost, this approach is the most common method used in the industry. The yellow-emitting phosphor powders mixing with transparent epoxy resin are uniformly dispersed in a cup reflector with a mounted LED chip. Therefore, phosphor powders are located in the vicinity of the chips. However, it is difficult to precisely control the thickness and shape of the coating layer resulting in an inhomogeneous distribution of phosphor and poor angular color uniformity, which often causes "yellow ring."

In order to overcome the disadvantages of the dispense phosphor coating the conformal phosphor coating (Figure 14.13b) had been proposed [41]. The conformal coating means that a phosphor layer with uniform thickness has the same shape of a LED chip surface to obtain high angular color uniformity. Many methods to realize the conformal phosphor coating have been proposed such as spin coating, electrophoresis, slurry settling, spray coating, and using capillary microchannel [42,43], and

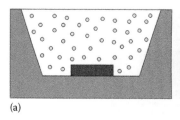 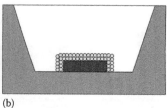 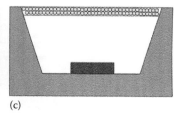

(a)   (b)   (c)

FIGURE 14.13 Phosphor coating method: (a) dispense phosphor coating, (b) conformal coating, and (c) remote phosphor coating.

spray coating is the most widely used method nowadays. However, in the case of the conformal coating phosphor powders are applied directly on the top of a LED chip, much more proximate than that of the dispensed coating. And that brings up two problems. First, the heat from the operating LED chip will increase the working temperature of the phosphor material and seriously decrease their emission efficiency. Second, nearly 60% of the light re-emitted by the phosphor backscatter to the LED chip [44], which are finally reabsorbed by the LED chip, resulting in low-luminous efficiency.

Since the origin of the drawback of the conformal coating comes from the little distance between the phosphor layer and the LED chip, moving the phosphor layer away from the LED chip should be a solution (Figure 14.13c). This phosphor coating method is named the remote phosphor coating. Using remote phosphor coating, higher phosphor conversion efficiency and lower backscattering can be achieved by reducing the probability of re-emitting light being absorbed by the LED chip [40], but still large amount of light gets trapped between the phosphor, the reflector cup, and the LED chip, becoming the primary loss of the remote phosphor [45]. In pursuit of higher light extraction efficiency, phosphor thermal-isolated package [46], scattered photon extraction package [45], ring-remote structure [47], and a duel-layer structure [48] have been proposed. However, the remote phosphor structure still suffers from the concave surface and nonuniform angular-dependent CCT [49]. Therefore, the patterned structure of remote phosphor structure [50] has been developed.

Besides the three main kinds of phosphor structures, some hybrid structures have also been developed such as combining dispense and conformal phosphor structure, [51] and combining remote and conformal phosphor structure [52].

14.4.3 Evolution of White LED Packaging Methods

14.4.3.1 Integrated Packaging Technology

14.4.3.1.1 Chip-on-Board Packaging

Chip-on-board (COB) arrays packaging typically uses a large array of small LED dies that is directly mounted onto a metal-core printed circuit board (MCPCB) or a ceramic substrate, and then connects the LED chips to the circuit on the PCB by wire, and finally, the LEDs are covered with a phosphor mixed silicone. COB packages are typically used in products needing high lumen (up to 14,000 lumens) from compact optical source or extremely high lumen density (e.g., high-bay and low-bay lighting). With a good thermal substrate, these COB arrays can have high enough color uniformity and lumen stability to achieved high-power packages as long as the operating temperature is kept within specification. Compared to typical LED Package-on-Board stack shown in Figure 14.14, COB arrays structures have several more advantages including higher integration, space-saving, high optical power density, high uniformity of light-emitting, high performance of heat dissipation, higher design flexibility, and low packaging cost [53]. The appearance of COB package provides a more reliable and economical method to manufacture the high-power LED surface light source.

COB encapsulation processes are as follows: First, PCB is prepared usually using a metal substrate or a ceramic substrate and then LED chips are mounted to the corresponding position of the PCB. In order

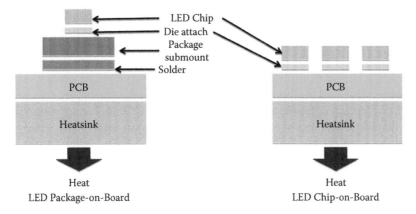

FIGURE 14.14   LED chip-on-board structure versus package-on-board Structure.

to improve light efficiency, the reflective structure can be made on the PCB [54]. Second, the LED chips are electrically connected to the circuit on the PCB by wires. Generally, all LED chip structures (lateral, flip-chip, and vertical) can be utilized in designing LED COB products, and wire bonding process are not needed for flip-chip packaging; Third, a closed surrounding structure is formed using encapsulation on the PCB, and which is called dam-and-fill encapsulation in COB packaging; Finally, the entire LED are covered with phosphor material. The key processes are shown in Figure 14.15.

COB arrays package simplifies the assembly process and the structure of packaging device, reducing the distance of heat diffusion, and thus improves the thermal dissipation performance [55]. However, multichip LED modules and high-power COB modules provide more lumens and wider light-emitting areas, while simultaneously producing more heat. The generation of extra heat results in many problems for LEDs affecting their lifetime, stability, wavelength shift, and luminous efficiency. Thus, heat management

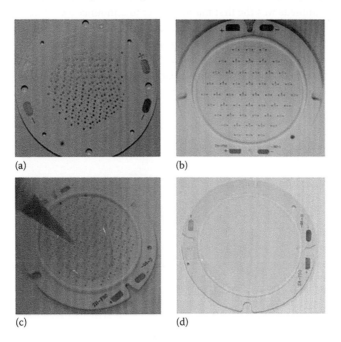

FIGURE 14.15   Key packaging processes in LED COB arrays: (a) die bonding, (b) wire bonding, (c) dam-and-fill encapsulation, and (d) phosphor-and-fill encapsulation. (Courtesy of Foshan Evercore Optoelectronic Technology Co., Ltd.)

is one of the most important topics in high-power LED research currently. In order to improve the thermal performance of the COB package, the metal substrate with low-thermal resistance is generally used to make the PCB; however, an insulating layer is needed between the circuit and metal substrate, which will affect the thermal performance because the thermal resistance of the insulating layer is usually high. Chuluunbaatar, Z. et al. tried to use a nanopore silicon-based (NPSB) substrate to improve the heat dissipation performance [56]. And Jae-Kwan, S. et al. improved the heat dissipation performance of COB LED by applying the technology of low-temperature co-fired ceramic COB packaging, removing the insulating layer [55]. In addition, the size of the LED chips, the chip array density, and phosphor encapsulation methods also have a great effect on the thermal performance of COB LED. According to the study of Ying, S. -P. et al., the heat dissipation performance of a COB LED consisted of many small chips is better than the COB LED consisted of a few large chips when the two kinds of COB LED have the same total light-emitting area [57]. Therefore, the proper chip size and array density can improve the performance of COB LED and reduce costs.

As for the phosphor package, Juntunen, E. and their team's study shows that the encapsulation methods of the remote phosphor can improve the heat dissipation performance, because the excitation of phosphor will produce heat, the heat is away from the LED chip when the phosphor is away from the LED chip [58]. And the remote phosphor method is also proved to perform higher optical efficiency. The optical and thermal performance of COB package remains to be improved, but the advantages of COB package has made it extensively used.

14.4.3.1.2 *Chip-Scale Packaging for LEDs*

Chip-scale packages, which are the mainstay of the semiconductor industry, generally refers to the size of the package that is not more than 120% of the bare chip size. Chip-scale packaging (CSP) is introduced into the LED field from the conventional IC CSP and driven by miniaturization, improved thermals, higher reliability, simplified packaging process, easily integrated, and reduces the material cost of the packaging process. Philips Lumileds shows the miniaturization trend in LED industry as shown in Figure 14.16, and Lumileds was the first manufacturer to announce a CSP LED back in early 2013 at strategies in light (SIL), and now CSP has made inroads into the LED industry. Lumileds, Samsung, and other LED manufacturers also claim to be working on such LED device packages.

Chip-scale packages in LED, also called package-free LEDs or white chips, are traditionally defined as fully functional packages that are of equal size or slightly larger than (<20%) the actual size of the die or in this case the active area of the LED. These packages typical do not require an additional submount or substrate

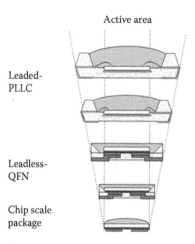

FIGURE 14.16 Miniaturization trend in LED industry. Traditional package migration path for semiconductor devices applied to LEDs shows that the CSP is the eventual end point. (Courtesy of Philips Lumileds.)

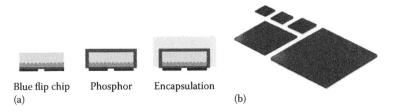

Blue flip chip Phosphor Encapsulation
(a)   (b)

FIGURE 14.17 (a) CSP manufacturing approach (From Shatil Haque, DOE SSL R&D Workshop, Raleigh, NC, February 2016.) and (b) recent example of the scalability of commercial CSPs. (From Samsung media center, May 2016.)

and can be directly attached to Level 2 boards. These chip scale packages are further characterized by having both P and N as bottom contacts and are surface mount devices (SMD) that can make use of standard packaging testing and s assembly equipment rather than wire bonds. Chip-scale packages with their simplified standard spaced bonding pads enable standard testing and pick and place equipment to be used with similar cost and placement accuracy without adding complexity. The typical CSP structure is illustrated in Figure 14.17a. Recent examples of CSP LED products include a variety of sizes to meet the different lighting application needs to replace the mid-power, high-power and small COB packages, as shown Figure 14.17b.

The advantages of chip-scale package can be summarized as [59–61]:

- Better thermal contact to the substrate through metal–metal interface of the bottom epi layer to the heat sink
- Higher current densities, driving higher lumen output per device
- High reliability with no wire bonds and a reduction in an attach process and package
- Higher packaging density due to reduced footprint
- Ease of integration on boards through surface mount technology
- Flexibility of attach-AuSn, SAC, UBM
- good electrical properties
- ease to optical design
- small size
- low cost

Currently, there are usually two ways to implement the chip-scale package:

1. Mount the LED bare chips to a substrate, follow by phosphor encapsulation process, and finally get LED modules with single or multiple LED chips by the crack process. This packaging process is similar to the conventional packaging process, and the craft is mature.
2. Fix the flip-chip LED on a film, with the electrode side down, and uniformly cover the upper surface and the four sides of the LED bare chip with a phosphor layer. This kind of CSP LED is not mounted on a substrate and is directly given to the downstream lighting manufacturers for light applications. It can be freely integrated according to the demand of light applications. This method executes the packaging process in the chip manufacture process, so that it is suitable for mass production, and many LED manufacturers are racing to develop this technology.

The chip-scale package needs to uniformly cover the upper surface and the four sides of the LED bare chip with a phosphor layer, and also need to ensure the small package size, so that the traditional phosphor encapsulation method cannot desire the demand. Usually, the spraying method or electrophoretic deposition method (EPD) method are applied in CSP [62]. These methods can ensure the uniformity of the phosphor, and several chips can be packaged simultaneously for mass production. The reliability of CSP LED is highly related to the welding of the CSP chip and the substrate. As the CSP chips are surrounded by phosphor silicone and the thermal expansion coefficient of the phosphor silicone is bigger than that of the LED chip, some methods must be applied to avoid the bonding's mechanical damage because of the thermal expansion. Usually, bumps are built on the substrate so that there some room between the phosphor silicone and the substrate after welding. Normally, CSP LED emits light from five sides. Except for the side of the welding electrode, the remaining five sides are coated with phosphor. But in order to control the angle of the emitting light and ensure the luminous uniformity, the bottom and side walls can be sprayed with a reflective layer or a light blocking layer to make a single-sided light-emitting device. Overall, the CSP technology simplifies the packaging processes, miniatures the package devices, reduces the packaging costs, and makes it more easy to optically design because of the high integration density.

14.4.3.1.3 Wafer-Level Packaging

Conventional LED manufacturing process comprises epitaxial growth of the device layers on the substrate, processing of the wafer to define individual chips, dicing of the wafer to produce the individual chip, and package. Through the chip process, thousands of LED chips are formed on the epi wafer at the same time but each of them has to be diced off from the wafer and packaged one by one, which means this kind of packaging process has lower productivity and higher cost per LED chip. Therefore, in the past several years, there has been an uprising trend for LED wafer-level packaging (WLP).

IC packaging industries have already proven that WLP can lead to great cost reduction in components at high volume production. Since conventional LED packaging method follows conventional chip-level IC packaging, WLP should also be an answer for future LED packaging. Studies show that wafer-level LED packaging may not only decrease the total cost but also improve its reliability and yield [63]. The basic point of WLP is to package the whole wafer rather than individual chips.

Here come two different approaches. One is chip-to-wafer wafer-level packaging. The LED units are diced off from the epi wafer first and then mounted on a carrier wafer as a package substrate. After subsequent wafer-level packaging process, the carrier wafer is diced into components. Since it still requires dicing the LED wafer, it is also regarded as semi WLP. In general, semi WLP processes require the patterned silicon submount wafer as a carrier. Through silicon vias (TSVs) may be built in for electric interconnection. Both wire bonding and flip-chip technology are feasible connection methods. Two examples corresponding to two different connection methods are shown in Figures 14.18 and 14.19 respectively [64,65].

The fabrication process flow of example A is shown in Figure 14.20. Cavities and TSVs are formed using a wet etching process, after which the metal layer is fabricated using sputtering, and TSVs are filled by electroplating. After die attaching and wire bonding, phosphor and epoxy mixture is squeezed into the cavity [64]. Example B (Figure 14.18) has a little difference with example A since flip-chip technology

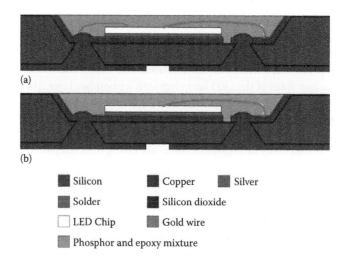

FIGURE 14.18 Schematic cross-sectional view of proposed LED packaging, with (a) lateral LED chip and (b) vertical LED chip. (From Lv, Z. et al. *IEEE Trans. Compon. Packag. Manuf. Technol.*, 3, 1123–1129, 2013. © 2013 IEEE. With permission.)

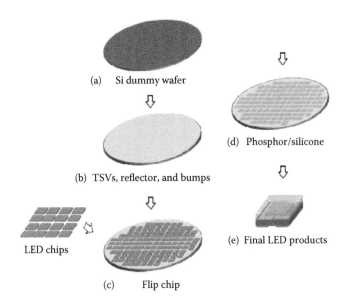

FIGURE 14.19 Concept of wafer-level packaging for flip-chip LEDs: (a) Si dummy wafer as raw material, (b) through silicon vias (TSVs), reflector and bumps are formed on Si wafer, (c) LED chips are mounted on the wafer by flip chip to wafer process, (d) phosphor and silicone is coated on wafer surface, and (e) final packaged LEDs are completed by dicing and testing, and then can be used for different applications. (From Dong C. et al. Silicon based wafer-level packaging for flip-chip LEDs. In *16th International Conference on Electronic Packaging Technology*, 2015.)

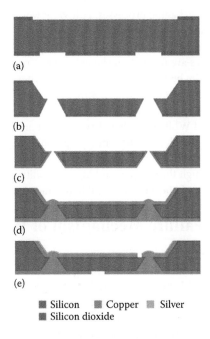

FIGURE 14.20 Fabrication process flow for the silicon substrate. (a) Thermal oxidation and RIE. (b) Wet etching. (c) Thermal oxidation and Cu layer sputtering. (d) Copper plating. (e) Sliver layer sputtering and copper etching. (From Lv, Z. et al. *IEEE Trans. Compon. Packag. Manuf. Technol.*, 3, 1123–1129, 2013. © 2013 IEEE. With permission.)

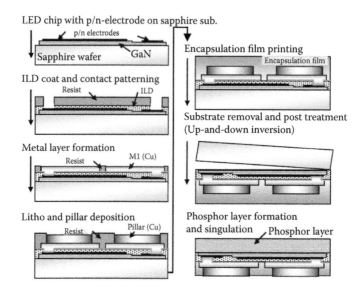

FIGURE 14.21 Fabrication process flow of a full WLP. (From Kojima, A. et al. A fully integrated novel wafer-level LED package (WL2P) technology for extremely low-cost solid state lighting devices. In *IEEE International Interconnect Technology Conference*, 2012. With permission.)

is used. Backside bump is formed first, and then TSVs is fabricated by dry etching. After reflector and bump formation, LED chips are mounted on topside bumps with flux, and then reflow to form a metal connection. After all, phosphor and silicone are coated in sequence [65].

The other one is wafer-to-wafer wafer-level packaging, named Full WLP. LED wafer is diced into components after all wafer-level packaging processes. Full WLP would give a rise in cost if it decreases the density of LED chips on a wafer, so the chip-size package is needed. That is why full WLP is also called wafer-level chip-scale package (WLCSP). An example of full WLP is shown in Figure 14.21. After chip process, p- and n- electrodes are formed on the epi wafer. Copper wires and pillars are electroplated after interlayer dielectric film being coated and patterned. After an encapsulation film is coated to provide mechanical strength and protection, the sapphire substrate is removed by laser lift-off process. Before device singulation, the phosphor coating is done [66].

Considering both semi and full WLP, researchers have focused on several key technologies such silicon submount wafer fabrication [64,67], wafer-level bumping technology [68], wafer bonding technology [69], wafer-level optics [70,71], and wafer-level phosphor coating method [72,73]. As mentioned above, WLP has the merit of low costs, high throughput, and high reliability, and has attracted more and more attention. However, more work should be done for WLP to be practical in LED industries.

14.5 Reliability and Failure Mechanism of White LED

LEDs are expected as the excellent candidates for the next-generation light sources. For the availability and commercial consideration, the reliability and lifetime problems become increasingly prominent. As we can perceive in daily life, the LED bulb can be darker, changed in color in visual and failed. LEDs' performance degradations under stresses, such as electric currents, ESD, high temperature, moisture, mechanical vibration and so on, were frequently observed. The degradation mechanisms under single stresses and coupled stresses (currents & temperature or currents & temperature & moisture) were investigated, but the mechanisms are still open issues.

14.5.1 Thermal and Current Stresses

In working conditions, thermal and current stresses in LEDs are two unavoidable working stresses leading to degradation of luminous efficiencies. There are some reasons for the thermal creating, such as the joule heat of current, light internal reflection in the silicone. Temperature is a key factor determining the lifetime of LEDs. With the on/off daily working, the temperature of the LEDs can be thought as a thermal and current stresses aging process. With the great demand for high power and standard scale integration packaging LEDs, thermal management and high-temperature reliability of the devices are becoming more critical.

14.5.1.1 Thermal Stresses

Some researchers observed that GaN LEDs could fail after stressing at 250°C–300°C for a few hours even without an applied bias. Different semiconductor and packaging materials have various coefficients of thermal expansion (CTE). With the thermal stresses aging process, a partial detachment of the contact materials can take place, such as the semiconductor chip/electrode metal, phosphor/silicon lens. The twice stressing was also observed, which could result in an increase in the series resistance of the LEDs and induces temperature gradients in LEDs. On the other hand, the partial detachment has an influence on the refractive index of the interface. Emission crowding was identified as another main degradation mode. It can be a vicious spiral as a consequence of high temperature.

14.5.1.2 Current Stresses

Electric currents were reported to decrease the light output power of GaN-based blue LED chips. Even very small current densities would degrade the light output power. Electric currents stresses can provide two mechanisms mainly, the heating effect and the degradation in the semiconductor chip.

As we have discussed, the heating effect is always the ohmic heating or joule heating. The heat energy combine with the thermal stresses can be the heat affect, and that is one of the reasons why the current and thermal stresses are always coupled and investigated.

In order to remove joule-heating effects as much as possible, two methods are usually employed. One is to improve the thermal conductivities of devices so as to keep the device temperature as close to the room temperature as possible, and that needs an excellent cooling system. The other is using pulsed currents instead of direct currents that can avoid the heating influence pretty well.

The failure was usually attributed to the increase of nonradiative recombination centers in the active layer. Several mechanisms will lead to the increase of nonradiative recombination rates and defects, including:

1. Migration of defects to the active layer.
2. Point defects related with the diffusion of Mg elements to the active layer along dislocations.
3. Formation of nitrogen vacancies in the active layer.

High-temperature annealing will accelerate atoms "vibration and defects" generation. These defects can be thought as trap centers of electron and holes, and thus assist electrons or holes to hop through the original bandgap. Tunneling currents will increase.

14.6 Concluding Remarks and Perspectives

The development history of white LED is only a short period of more than 20 years since Shuji Nakamura invented GaN blue LED. The technology of white LEDs has not ceased, and white LED has gradually entered a new field of application from the traditional lighting, such as micro-LED, visible light communication (Lifi), UV LED, and so on.

References

1. Round, H. J. 1907. A note on carborundum. *Electrical World.* 49:309.
2. Holonyak, N. and S. F. Bevacqua. 1962. Coherent (visible) light emission from Ga($As_{1-x}P_x$) junctions. *Applied Physics Letters.* 1(4):82–83.
3. Craford, M. G., R. W. Shaw, A. H. Herzog et al. 1972. Radiative recombination mechanisms in GaAsP diodes with and without nitrogen doping. *Journal of Applied Physics.* 43 (10):4075–4083.
4. Amano, H., M. Kito, K. Hiramatsu et al. 1989. P-type conduction in Mg-doped GaN treated with low-energy electron beam irradiation (LEEBI). *Japanese Journal of Applied Physics, Part 2 (Letters).* 28(12):L2112–L2114.
5. Akasaki, I., H. Amano, K. Itoh et al. 1992. GaN-based UV/blue light emitting devices. Gallium Arsenide and Related Compounds 1992. *Proceeding of the Nineteenth International Symposium,* Karuizawa, Japan.
6. Nakamura, S., M. Senoh, and T. Mukai. 1993. p-GaN/n-InGaN/n-GaN double-heterostructure blue-light-emitting diodes. *Japanese Journal of Applied Physics, Part 2 (Letters).* 32 (1A/B):L8–L11.
7. Nakamura, S., M. Senoh, and T. Mukai. 1993. High-power InGaN/GaN double-heterostructure violet light emitting diodes. *Applied Physics Letters.* 62 (19):2390–2392.
8. Nakamura, S., T. Mukai, and M. Senoh. 1994. Candela-class high-brightness InGaN/AlGaN double-heterostructure blue-light-emitting diodes. *Applied Physics Letters.* 64 (13):1687–1689.
9. Nakamura, S., M. Senoh, N. Iwasa et al. 1995. High-brightness InGaN blue, green and yellow light-emitting diodes with quantum well structures. *Japanese Journal of Applied Physics, Part 2 (Letters),* 34(7A):L797–L799.
10. Nakamura, S., M. Senoh, S. I. Nagahama et al. 1996. Room-temperature continuous-wave operation of InGaN multi-quantum-well structure laser diodes. *Applied Physics Letters.* 69(26):4056–4058.
11. The Nobel Prize in Physics 2014. Nobelprize.org. Nobel Media AB 2014, accessed October 13, 2016, https://www.nobelprize.org/nobel_prizes/physics/laureates/2014/.
12. Nakamura, S. and G. Fasol. 1997. *The Blue Laser Diode.* Berlin, Germany: Springer.
13. Nichia's, http://www.nichia.co.jp.
14. Lumileds', http://www.lumileds.com/.
15. GELcore's, http://www.gelighting.com.
16. Osram's, http://www.osram.com.
17. Cree's, http://www.cree.com/.
18. Steigerwald, D. A., J. C. Bhat, D. Collins et al. 2002. Illumination with solid state lighting technology. *IEEE Journal of Selected Topics in Quantum Electronics.* 8:310–320.
19. Schubert, E. F. 2003. *Light-Emitting Diodes.* Cambridge, MA: Cambridge University Press.
20. Lin, C.C. and R. S. Liu. 2011. Advances in phosphors for light-emitting diodes. *Journal of Physical Chemistry Letters.* 2(11):1268–1277.
21. Narukawa, Y., I. Niki, K. Izuno et al. 2002. Phosphor-conversion white light emitting diode using InGaN near-ultraviolet chip. *Japanese Journal of Applied Physics Part 2-Letters.* 41(4A):L371–L373.
22. Sheu, J. K., S. J. Chang, C. H. Kuo et al. 2003. White-light emission from near UV InGaN-GaN LED chip precoated with blue/green/red phosphors. *IEEE Photonics Technology Letters.* 15(1):18–20.
23. Nishida, T., T. Ban, and N. Kobayashi. 2003. High-color-rendering light sources consisting of a 350-nm ultraviolet light-emitting diode and three-basal-color phosphors. *Applied Physics Letters.* 82(22):3817–3819.
24. Ziaudeen, S. A., R. R. Gaddam, P. K. Pallapothu et al. 2013. Supra gap excitation properties of differently confined PbS-nano structured materials studied with opto-impedance spectroscopy. *Journal of Nanophotonics.* 7(1):5952–5960.
25. Vasudevan, D., R. R. Gaddam, A. Trinchi et al. 2015. Core-shell quantum dots: Properties and applications. *Journal of Alloys and Compounds.* 636:395–404.

26. Aboulaich, A., M. Michalska, R. Schneider et al. 2014. Ce-doped YAG nanophosphor and red emitting CuInS₂/ZnS core/shell quantum dots for warm white light-emitting diode with high color rendering index. *ACS Applied Materials & Interfaces.* 6(1):252–258.

27. Sun, H., H. Zhang, J. Ju et al. 2008. One-step synthesis of high-quality gradient CdHgTe nanocrystals: A prerequisite to prepare CdHgTe-polymer bulk composites with intense near-infrared photoluminescence. *Chemistry of Materials.* 20(21):6764–6769.

28. Yuan, X., J. Hua, R. Zeng et al. 2014. Efficient white light emitting diodes based on Cu-doped ZnInS/ZnS core/shell quantum dots. *Nanotechnology.* 25(43):435202–435202.

29. Yin, L., Y. Bai, J. Zhou et al. 2015. The thermal stability performances of the color rendering index of white light emitting diodes with the red quantum dots encapsulation. *Optical Materials.* 42:187–192.

30. Wyszecki, G. and W. S. Stiles. 1982. *Color Science: Concepts and Methods, Quantitative Data and Formulae*, 2nd Edition. New York: John Wiley & Sons.

31. Ekimov, A. I. and A. A Onushchenko. 1981. Quantum size effect in three-dimensional microscopic semiconductor crystals. *JETP Letters.* 34:345–349.

32. Regulacio, M. D. and Ming-Yong Han. 2010. Composition-tunable alloyed semiconductor nanocrystals. *Accounts of Chemical Research.* 43(5):621–630.

33. Lin, H. Y., S. W. Wang, C. C. Lin et al. 2016. Excellent color quality of white-light-emitting diodes by embedding quantum dots in polymers material. *IEEE Journal of Selected Topics in Quantum Electronics.* 22(1):35–41.

34. Yang, Y., Y. Zheng, W. Cao et al. 2015. High-efficiency light-emitting devices based on quantum dots with tailored nanostructures. *Nature Photonics.* 9(4):259–266.

35. Ali, E. M., Y. Zheng, H. H. Yu et al. 2007. Ultrasensitive Pb²⁺ detection by glutathione-capped quantum dots. *Analytical Chemistry.* 79 (24):9452–9458.

36. Luther, J. M., M. Law, M. C. Beard et al. 2008. Schottky solar cells based on colloidal nanocrystal films. *Nano Letters.* 8(10):3488–3492.

37. Qian, H., C. Dong, J. Peng et al. 2007. High-quality and water-soluble near-infrared photoluminescent CdHgTe/CdS quantum dots prepared by adjusting size and composition. *Journal of Physical Chemistry C.* 111(45):16852–16857.

38. Pardo, B., A. Gasse, A. Fargeix et al. 2013. Thermal resistance investigations on new leadframe-based LED packages and boards. *Microelectronics Reliability.* 53:1084–1094.

39. Schubert E. F. 2006, *Light-Emitting Diodes*, 2nd edition, Cambridge, MA: Cambridge University Press.

40. Kim, J. K., H. Luo, E. F. Schubert et al. 2005. Strongly enhanced phosphor efficiency in GaInN white light-emitting diodes using remote phosphor configuration and diffuse reflector cup. *Japanese Journal of Applied Physics Part 2-Letters & Express Letters.* 44(20–23):L649–L651.

41. Krames, M. R., J. B. D. Collins, N. F. Gardner et al. 2002. High-power III-nitride emitters for solid-state lighting. *Physica Status Solidi a-Applied Research.* 192 (2):237–245.

42. Yum, J. H., S. Y. Seo, S. Lee et al. 2001. Comparison of Y₃Al₅O₁₂:Ce₀.₀₅ phosphor coating methods for white-light-emitting diode on gallium nitride. *Proceedings of SPIE 4445 (Solid State Lighting and Displays)*, pp. 60–69.

43. Zheng, H., X. Luo, R. Hu et al. 2012. Conformal phosphor coating using capillary microchannel for controlling color deviation of phosphor-converted white light-emitting diodes. *Optics Express.* 20(5):5092–5098.

44. Narendran, N., Y. Gu, J. P. Freyssinier-Nova et al. 2005. Extracting phosphor-scattered photons to improve white LED efficiency. *Physica Status Solidi a-Applications and Materials Science.* 202(6):R60–R62.

45. Narendran, N. 2005. Improved performance white LED. *Proceedings of the SPIE - The International Society for Optical Engineering.* 5941:594108.

46. Fan, B., H. Wu, Y. Zhao et al. 2007. Study of phosphor thermal-isolated packaging technologies for high-power white light-emitting diodes. *IEEE Photonics Technology Letters.* 19(13–16):1121–1123.
47. Lin, M. T., S. P. Ying, M. Y. Lin et al. 2010. Design of the ring remote phosphor structure for phosphor-converted white-light-emitting diodes. *Japanese Journal of Applied Physics.* 49(7):072101.
48. Chen, K. J., H. C. Chen, M. H. Shih et al. 2013. Enhanced luminous efficiency of WLEDs using a dual-layer structure of the remote phosphor package. *Journal of Lightwave Technology.* 31(12):1941–1945.
49. Chen, K. J., H. C. Chen, C. C. Lin et al. 2013. An investigation of the optical analysis in white light-emitting diodes with conformal and remote phosphor structure. *Journal of Display Technology.* 9(11):915–920.
50. Kuo, H. C., C. W. Hung, H. C. Chen et al. 2011. Patterned structure of remote phosphor for phosphor-converted white LEDs. *Optics Express.* 19(14):A930–A936.
51. Chen, K. J, H. V. Han, B. C. Lin et al. 2013. Improving the angular color uniformity of hybrid phosphor structures in white light-emitting diodes. *IEEE Electron Device Letters.* 34(10):1280–1282.
52. Lin, B. C., J. K. Huang, K. J. Chen et al. 2015. Luminous efficiency enhancement of white light-emitting diodes by using a hybrid phosphor structure. *Journal of Photonics for Energy.* 5(1):057603.
53. Jeon, Eun-chae, Je-Ryung Lee, Tae-Jin Je et al. 2014. Quantitative analysis on air-dispensing parameters for manufacturing dome lenses of chip-on-board LED system. *International Journal of Precision Engineering and Manufacturing.* 15(11):2437–2441.
54. Ha, M. and S. Graham. 2012. Development of a thermal resistance model for chip-on-board packaging of high power LED arrays. *Microelectronics Reliability.* 52(5):836–844.
55. Sim, J. K., K. Ashok, Y. H. Ra et al. 2012. Characteristic enhancement of white LED lamp using low temperature co-fired ceramic-chip on board package. *Current Applied Physics.* 12(2):494–498.
56. Chuluunbaatar, Z., C. Wang, E. S. Kim et al. 2014. Thermal analysis of a nano-pore silicon-based substrate using a YAG phosphor supported COB packaged LED module. *International Journal of Thermal Sciences.* 86:307–313.
57. Ying, S. P. and W. B. Shen. 2015. Thermal analysis of high-power multichip COB light-emitting diodes with different chip sizes. *IEEE Transactions on Electron Devices.* 62(3):896–901.
58. Juntunen, E., O. Tapaninen, A. Sitomaniemi et al. 2013. Effect of phosphor encapsulant on the thermal resistance of a high-power COB LED module. *IEEE Transactions on Components Packaging and Manufacturing Technology.* 3(7):1148–1154.
59. Yuan, C. C. A. 2015. The extended Beer-Lambert theory for ray tracing modeling of LED chip-scaled packaging application with multiple luminescence materials. *Optical Materials.* 50:193–198.
60. Arik, M. and S. Weaver. 2004. Chip scale thermal management of high brightness LED packages. In *Fourth International Conference on Solid State Lighting,* edited by I. T. Ferguson. Dencover, CO. pp. 214–223.
61. Hong, T. H., J. Beleran, K. Y. S. Drake et al. 2011. Packaging approach for integrating 40/45-nm ELK devices into wire bond and flip-chip packages. *IEEE Transactions on Components Packaging and Manufacturing Technology.* 1(12):1923–1933.
62. Wang, X., H. Rao, Q. Lei et al. 2013. An improved electrophoretic deposition method for wafer level white pc-LED array packaging. In *10th China International Forum on Solid State Lighting.*
63. Rao, R. T. 2001. *Fundamentals of Microsystems Packaging.* New York: McGraw-Hill Education.
64. Lv, Z., X. Wang, L. Yang et al. 2013. Study on packaging method using silicon substrate with cavity and TSV for light emitting diodes. *IEEE Transactions on Components Packaging and Manufacturing Technology.* 3(7):1123–1129.
65. Dong C., Z. Li, C. Haijie et al. 2015. Silicon based wafer-level packaging for flip-chip LEDs. In *16th International Conference on Electronic Packaging Technology.*
66. Kojima, A., M. Shimada, Y. Akimoto et al. 2012. A fully integrated novel wafer-level LED package (WL2P) technology for extremely low-cost solid state lighting devices. In *IEEE International Interconnect Technology Conference.*

67. Lv, Z., X. Liu, L. Yang et al. 2012. Silicon substrate with TSV for light emitting diode packaging. In *14th International Conference on Electronic Materials and Packaging*.

68. Wei, T., X. Qiu, J. C. C. Lo et al. Wafer level bumping technology for high voltage LED packaging. Microsystems, Packaging, Assembly and Circuits Technology Conference (IMPACT), 2015 10th International, IEEE, Taiwan, Taipei, 2015, p. 54–57.

69. Lee, S. Y., K. K. Choi, H. H. Jeong et al. 2009. Wafer-level fabrication of GaN-based vertical light-emitting diodes using a multi-functional bonding material system. *Semiconductor Science and Technology*. 24(9):092001.

70. Zhang, R., S. W. R. Lee, and J. C. C. Lo. 2015. Lens forming by stack dispensing for LED wafer level packaging. *IEEE Transactions on Components Packaging and Manufacturing Technology*. 5(1):15–20.

71. Zou, Y., J. Shang, Y. Ji et al. 2014. A novel molding process for wafer level LED packaging using uniform micro glass bubble arrays. In *2014 IEEE 64th Electronic Components and Technology Conference*, 2299–2302.

72. Linsong, Z., R. Haibo, W. Wei et al. 2013. Self-adaptive phosphor coating technology for wafer-level scale chip packaging. *Journal of Semiconductors*. 34(5):96–99.

73. Chen, K., R. Zhang, and S. W. R. Lee. 2010. Integration of phosphor printing and encapsulant dispensing processes for wafer level LED array packaging. In *2010 11th International Conference on Electronic Packaging Technology & High Density Packaging (Icept-Hdp)*. pp. 1386–1392.

15

Current Status and Trends for Green Light-Emitting Diodes

Junxi Wang, Zhe Liu,
and Ning Zhang

15.1 Introduction

InGaN-based green light-emitting diodes (LEDs) are essential components for high efficiency RGB solid-state white lighting to obtain the needed color rendering and high brightness. The first report on green electroluminescence (EL) by semiconductor materials was conducted by Dean P. J. et al. in 1967 when they applied forward voltage to gallium phosphide diodes at the temperature near 300K and observed green near-bandgap EL. Numerous researches on the phosphide for high-efficiency green LEDs were then carried out, but the difficulty to achieve high-quality materials that were required for the 525 nm wavelength limited the application of phosphide in obtaining LEDs in the true green region. Widespread interest in nitride-based materials was revived by Nakamura Shuji [1] who fabricated high-brightness green LEDs by using the AlGaN/InGaN/GaN single quantum well structures and achieved the wavelength of 520 nm with the light output of 3 mW at 20 mA. This breakthrough made it clear that given the improved InGaN/GaN heterostructure design and epitaxial growth process, one could develop high-efficiency green LEDs based on the nitride materials. Further attention was drawn to this possibility with the advent of indium-rich InGaN/GaN multiple quantum wells (MQWs) that were fabricated by Nakamura et al., and the indium-rich InGaN/GaN heterostructure could be obtained via

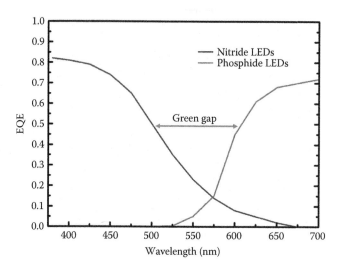

FIGURE 15.1 Efficiency drops of nitride and phosphide LEDs in the green spectral range, which is known as the green gap.

low-temperature growth [2]. The development of commercial metalorganic chemical vapor deposition (MOCVD) equipment and the advanced epitaxial technique have made even more progress for realizing nitride-based green LEDs. However, the external quantum efficiency (QE) in the wavelength ranging from 510 to 550 nm is still lower than that for InGaN-based blue and AlGaInP-based red emitters, well known as "green gap" (shown in Figure 15.1). The challenge to close the "green gap" is to realize the sufficient incorporation of indium atoms into the InGaN quantum wells without degrading the crystal quality. We will then describe the current state of green LEDs and propose some tendency predictions subsequently.

15.2 Indium-Rich InGaN Thin Films

15.2.1 In-Rich InGaN Alloys Growth

Due to the high vapor pressure when growing InN, it is very difficult to obtain In-rich InGaN alloys with high quality for achieving green LEDs. When growing InN, the equilibrium vapor pressure of N_2 is higher than when growing GaN and AlN. The indium composition in InGaN strongly depends on the growth temperature, and the composition increases with decreasing growth temperature under given atmosphere conditions. Therefore, in order to realize In-rich InGaN films with sufficient indium incorporation while maintaining a high crystal quality, higher V/III ratios and lower growth temperature are necessary. In this section, the dependence of indium incorporation efficiency on the growth parameters will be discussed.

During the epitaxial growth of the p-type layer for a completed green LED structure, the underneath InGaN quantum well layers have to suffer from a high temperature that is needed to grow the p-type layer. According to the thermodynamic analysis, In-rich InGaN alloys are under a metastable state, tending to decompose. Studies show that the indium incorporation is strongly related to temperatures. The thermal treatment in the growth process will alter the indium composition and the strain state in the MQWs, thus influencing the optical and electrical characteristics. It has been found that a lower growth temperature will increase the indium incorporation for high-indium InGaN layers and the subsequent high-temperature annealing will decrease the crystal quality, and thus, the optical property.

They demonstrate that the indium composition percentage is raised by 7% and 3% after thermal annealing at the temperature of 500°C–800°C, for InGaN alloys with 33% and 60% In-content [3].

Kim et al. investigated the influences of the growth pressure on the Indium incorporation for InGaN alloys. It was revealed that the pressure increase in the chamber decreases the indium incorporation into InGaN drastically [4]. The enhancement of In composition incorporation is explained by the mass transportation mechanism. It is reported that the In concentration in the InGaN alloys could be decreased by 7.5% when the growth pressure increased by 100 torr, for example, from 150 to 250 torr. They attributed the In-content enhancement to the enhanced mass transportation of precursor gasses through the boundary layer to the surface of the film.

The indium incorporation into InGaN is found to be significantly influenced by the amount and the percentage of hydrogen in the gas mixture. Experiments are performed, and the results show that hydrogen reduction by 100 sccm, for example, from 100 to 0 sccm, can lead to an increased indium incorporation by up to 25% [5]. Furthermore, significant studies are focused on the dependence of indium incorporation on hydrogen (H) irradiation. Okamoto et al. [6] investigated the dependence of the atomic H irradiation in the growth process on the indium incorporation. They proposed that both the increased nitrogen species and the growth mode swapping contributed mainly to the enhancement of indium incorporation.

Chen et al. [7] studied the dependence of V/III ratio on the indium incorporation in theory and suggested that at a given temperature, a higher V/III ratio helps to incorporate more indium atoms into the InGaN layers under a low III-group flux, but further enhancement for the V/III ratio then decreases the indium percentage in the films. Investigations into the effect of the V/III ratio on the crystalline quality for high-indium InGaN were performed by Che et al. [8]. They claimed that the phase separation in high-indium InGaN alloys was prone to occur under metal-rich conditions. In order to keep the composition homogeneity, N-rich and shutter-controlled conditions were needed.

The contribution of the growth interruption (GI) to the indium incorporation has been studied in the InGaN quantum well structures by many research groups [9,10]. The PL measurements show that the InGaN QW quality improves with the GI time increasing, accompanied by the blue-shifting peak wavelength and the decreasing dislocations density. It has been reported that the density of the quantum dot-like regions and well thickness for the MQWs decrease with the increasing interruption time, which is due to the thermal etching effect or the re-evaporation of In atoms. However, there was no direct evidence showing that the interruption time was related to the defect formation. Moreover, the sizes and the number of V-defects did not differ with increasing the interruption time [11].

15.2.2 Dislocations and Phase Separation

The structural defects, such as threading dislocations in the InGaN/GaN heterostructure, have a great influence on the carrier transportation and radiative recombination for the LEDs. Understanding the forming mechanism for dislocations is very important to improve the crystal quality of In-rich InGaN material and can enhance the efficiency of green LEDs. It has been proposed that the misfit strain is relaxed by forming misfit dislocations and stacking faults in indium-rich InGaN layers [12]. They propose that the strain at the In-rich InGaN/GaN interface is partially relaxed through the formation of dislocations and the stacking-order-mismatch-related stacking faults (S. F.) between sub-grains behave as seeds during the forming of the misfit dislocations in the In-rich InGaN layer, shown in Figure 15.2. Moreover, Lei et al. [13] investigated the role of c-screw dislocations on the indium segregation in InGaN alloys and found that the indium atoms were prone to aggregate around the dislocation lines, which then reduces the core energy and forms In-rich clusters.

Park et al. analyzed the phase separation and the corresponding In-rich QDs in the In-rich films that have V-shaped defects. They claimed that the QDs had a size of 2–5 nm and tended to gather near the

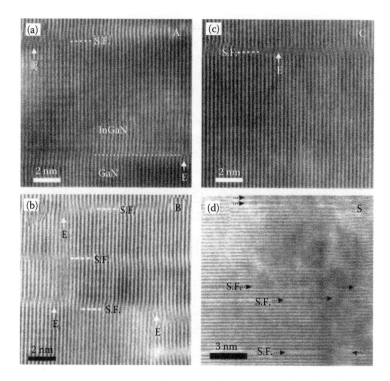

FIGURE 15.2 High resolution transmission electron microscopy images (a–d), viewed along [1-100]$_{GaN}$, show that the stacking faults act as the seeds in the formation of misfit dislocations (symbol E) in the In$_{0.33}$Ga$_{0.67}$N layer. (From Cho, H.K. and Yang, G.M., *J. Cryst. Growth.*, 243, 124–128, 2002. With permission.)

V-pits even in the InGaN layer with a low indium content [14]. With the increasing In content, it was found that the InGaN wells showed expansive characteristics and the lower indium content would lead to a better confinement of dot-like InGaN regions within the quantum wells [15].

The conditions far from thermodynamic equilibrium are needed to suppress the phase separation. It has been revealed that the evolution of the phase separation in In$_x$Ga$_{1-x}$N alloys (x similar to 0.65) is related to the growth rate, and a higher growth rate will produce uniform and single phase InGaN alloys. Increasing the growth rate will improve both the structural and electrical properties of InGaN layer, which is due to the phase separation suppression. Furthermore, the maximum thickness of the InGaN epilayer increases with the increasing growth rate without any phase separation. It is found that the thermal and hydrogen treatment also have effects on indium segregation in InGaN alloys [16]. The use of hydrogen during the GI process is found to be an effective way to suppress the formation of indium clusters in the InGaN well layers. Waltereit et al. reported that under the metal-stable growth conditions, the indium atoms tended to segregate to form In-rich region, resulting in a larger transition energy and a reduced QE. Thus, they proposed N-stable conditions to reduce the In segregation [17].

15.2.3 Semi/Nonpolar InGaN Alloys

The LEDs with InGaN/GaN MQWs grown along the c-axis suffer from the strain-related quantum confine Stark effect, resulting in a reduced radiative recombination and efficiency droop especially when the injection current is high. The growth of InGaN layers along the semi/nonpolar direction can be an alternative way to reduce the polarization fields in the epitaxial film grown along the c-axis.

The crystal orientations have great influence on the indium incorporation in the InGaN layer. Das et al. studied the growth kinetics of nitride materials and found that the incorporation of In is suppressed in InGaN films grown in the (112) direction under a given temperature, incorporation with the (0001) direction, which is shown in Figure 15.3 [18]. When the In compositions are in the range of 0%–40%, the InGaN films have the similar In concentration and growth rate for m-plane, a-plane, and c-plane films [19]. However, a number of studies claimed that the InGaN layers grown along different directions have different indium incorporation efficiency. Lai et al. [20] found that the indium incorporation efficiency into the InGaN layers was estimated to be about 13.9% and 5.1% for c-plane and m-plane InGaN layers, respectively. It was observed that the indium incorporation efficiency in both ($10\bar{2}1$) and ($10\bar{1}0$) directions was enhanced, but the indium incorporation efficiency was reduced for ($11\bar{2}2$) oriented InGaN layer [21]. Dinh et al. proposed that the indium incorporation efficiency (η) on different growth planes had the following behaviors: $\eta(0001) > \eta(11\bar{2}2) > \eta(20\bar{2}1)$ [22]. In order to change the polarization field direction in the MQWs, N-polar InGaN can be grown on substrates with a miscut angle, and the miscut angle can introduce steps favoring the spontaneous two-dimensional nucleation. It is found that increasing the miscut angle within 1.1° would increase the In content. [23] Bai et al. [24] analyzed the dependence of indium on the growth directions. The analysis revealed that the In composition was most increased in InGaN layers with ($11\bar{2}2$), (0001), ($10\bar{1}1$) and ($10\bar{1}0$) planes. In addition, the m-plane QW is observed to have a thicker well width than that of the c-plane QW, which is ascribed to the higher thermal conductivity and the surface off-cut [20].

The effects of substrate temperatures, group III fluxes, and NH_3 flow rate on surface morphologies and indium contents for different polar InGaN films have been conducted by Browne et al. [21]. It is found that the indium compositions remain unchanged with the Ga:In ratio of 1:1 for all planes, while the In contents increase if the growth temperature decreases. When the ammonia flow rate increases, a slower growth rate can have higher indium incorporation efficiency.

In addition, the surface morphology has been studied for samples grown along different polar orientations. The relaxed (0001) oriented InGaN film shows the highest indium incorporation efficiency which is 17.4%–48.7% and an island-like surface morphology, while the morphology for relaxed ($11\bar{2}2$) InGaN layers for which the indium incorporation efficiency is 14.0%–40.0% almost remains unchanged compared to the semipolar ($11\bar{2}2$) GaN templates [22]. The ($20\bar{2}1$)-planed InGaN layer exhibits a 3D morphology, which is ascribed to different growth mechanism between ($10\bar{1}1$) and ($10\bar{1}0$) surfaces. For the InGaN layer with (0001) plane, the relaxation is 100%, while the relaxation is reduced to less

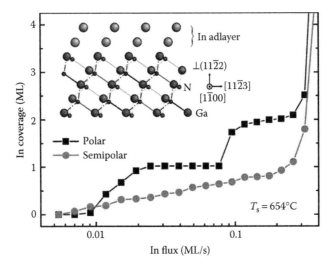

FIGURE 15.3 Indium content in the InGaN layer with different idium flux, measured on a polar (0001) and semipolar ($11\bar{2}2$) GaN layer grown at 654 °C. In the inset, schematic description of the indium adlayer on semipolar GaN ($11\bar{2}2$). (Reprinted with permission from Das, A. et al., *Appl. Phys. Lett.*, 96, 181907 Copyright 2010, American Institute of Physics.)

than 15% for the InGaN layer with (20$\bar{2}$1) plane, and the smaller relaxation is attributed to the lower indium incorporation efficiency for the (20$\bar{2}$1) layers. The surface morphologies of samples grown on the (0001), (11$\bar{2}$2), (20$\bar{2}$1), and (20$\bar{2}\bar{1}$) planes are smoother than those samples grown along with other crystal orientations [22,25]. Furthermore, when the indium content increases, the oxygen and hydrogen incorporations are found to increase for semipolar and nonpolar oriented films [21].

15.3 QE Promotion

In spite of the great progress in growing nitride-based materials, InGaN green LEDs still suffer from the high piezoelectric fields, which leads to a lower radiative efficiency in the In-rich quantum wells. Here, we report various structural designs employed to reduce the polarization fields and enhance the efficiency of the green LEDs.

15.3.1 In-Rich Heterostructure Design and Growth

15.3.1.1 Prestrained Layers

Despite the optimized growth for high-indium InGaN layers, the MQWs of the green LEDs own high stress due to the lattice mismatch between GaN and InGaN layers, giving rise to a high piezoelectric field in the active regions.

It has been well accepted that inserting a prestrained structure before the growing the high-indium InGaN QWs can effectively enlarge the lattice constants of the subsequent grown GaN layer, thus enhancing the emission efficiency and reducing spectral shifts for green LEDs [26–30]. Huang et al. [31] reported that the adoption of InGaN prestrained layer could increase the EL intensity by 182%, and decrease the spectral blueshift by 46% at the injection current level of 50 mA. The LED showed a weaker carrier localization effect if the prestrained growth is conducted.

The effect of the pre-strained structure on the enhancement of electron injection has also been studied [32,33]. Park et al. [27] investigated the inserted supper-lattices layers to enhance the electron injection into the active region. It was found that the radiative recombination rate was increased and the electron overflow was suppressed. Thus, the LED showed that the photoluminescence and optical output power was increased by 73.5% and 42.5%, respectively, indicating that the inserted layer acts as both a stress-relaxing buffer layer and an electron cooler. Xia et al. [30] investigated the thicker InGaN layer on the radiative recombination enhancement for the green LEDs. Interplaying of the impurity-related donor-acceptor pair recombination was observed. The researchers attributed the radiative recombination enhancement for green LEDs to the resonance, and the energy level alignments between the defect and levels reroute excitation in the pre-strained layer. In addition, the correlation between structural properties, growth regimes, and designs for the prestrained structure has also ever been reported [34].

15.3.1.2 High Efficiency MQW Structure

The large polarization field within the MQWs results in the band bending for QWs and the separation between electron and hole wave-functions, leading to a strong QCSE and a reduced internal QE. Several MQW designs have been studied to reduce the polarization field in the active region to improve the radiative efficiency for In-rich InGaN quantum wells, such as staggered InGaN QW [35], graded QWs [36], chirped QWs [37], InGaN-delta InN QWs [38], and so on.

Hybrid structures with better lattice matching can effectively reduce the interfacial strain and maximize the IQE and reduce the efficiency droop for the MQWs. Verma et al. [40] proposed a hybrid LED comprising of p-MgZnO/InGaN/n-MgZnO sandwiched structure. The hybrid LED was found to have an IQE of 93% and a reduced droop with respect to the GaN-based and ZnO-based LEDs. Bayram et al. [39] studied the hybrid green LEDs with the design comprising of n-ZnO/(InGaN/GaN)

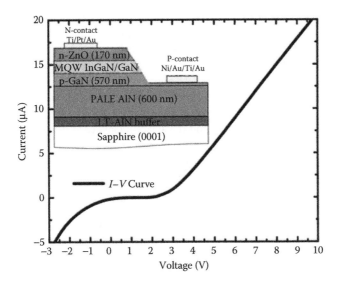

FIGURE 15.4 *I–V* curve of the fabricated LED, which is illustrated in the inset. (From Bayram, C. et al., *Appl. Phys. Lett.*, 93, 081111, 2008. With permission.)

MQWs/p-GaN, exhibiting a turn-on voltage of 2.5 V, a reduced wavelength blue-shift and a narrow EL peak (Figure 15.4). Han et al. [41] demonstrated the LED with strain-compensated type-II InGaN/ZnGeN$_2$/AlGaN QWs. Results revealed that the spontaneous emission rate for the type-II InGaN/ZnGeN$_2$/AlGaN QWs was enhanced by 5.6–6.8 times in comparison with that for the InGaN/GaN QWs due to the between hole confinement in the ZnGeN$_2$ layer.

Doping in the MQWs region can provide ionized dopants and more carriers, resulting in polarization field screening, and thus, better radiative recombination. Zhang et al. [42] reported that the Mg-doping in the barriers can partially screen the polarization field in the MQWs for InGaN-based green LEDs, which was attributed to the fact that the holes generated by the acceptor atoms in the barriers can screen the polarizations. The green LED with Mg dopant concentration of 5×10^{19} cm^{-3} in the barriers exhibits the lowest efficiency droop of 12.4% at a high-injection current. Huang et al. [43] reported that the Mg doping in the range of 3×10^{18} to 5×10^{18} cm^{-3} in the barriers could enhance the output power for semipolar (20–21) LEDs. Simulation and experiment reveal an improved holes injection for LED with Mg doping in the barriers. Zhang et al. [44] investigated the effect of the Mg back-diffusion into the MQWs on the carrier recombination by inserting a low-temperature GaN layer between the MQWs and the AlGaN EBL. It was found that an increased Mg doping depth in the inserted GaN layer could improve the hole injection efficiency and reduce the leakage current, but accompanied by the defect-related nonradiative recombination. Lin et al. [45] investigated the Si doping in barrier layers for green LEDs. It is revealed that, besides screening the electrostatic field, the Si doping in barrier layers with a density of 3.4×10^{16} cm^{-3} can effectively reduce the interface roughness for the InGaN/GaN heterostructure and the In-composition fluctuation. In addition, a suitable Si doping in the barrier layers can improve the current spreading in the lateral direction, resulting in a current crowding alleviation, reducing the local heating and suppressing the efficiency degradation. However, Ryou et al. [46] found that increasing the Si doping concentration in the GaN barriers for the In-rich MQWs would decrease the EL intensity but improve the forward voltage. They attributed the degraded EL intensity to the hole blocking effect since Si doping in the quantum barriers increases the barrier height for holes.

Designs for novel quantum barrier and well structures can effectively reduce the QCSE in the MQWs and enhance the radiative recombination rate. Lee et al. [47] proposed the gradual In-content InGaN QW to suppress the efficiency droop for green LEDs. The In-graded QW LED enhances the IQE by

45.5% and 55.7% at the injection current of 20 and 100 mA, respectively, indicating a considerable reduction of efficiency droop. This efficiency improvement is mainly attributed to gradual InGaN QW that significantly reduces the valence band bending, and thus, enhances the hole injection. Furthermore, the reduced electron leakage also helps to suppress the efficiency degradation for the green LED.

Park et al. [48] investigated the green LEDs with nontrivial semipolar AlInGaN/InGaN quantum well structures. This AlInGaN/InGaN structure owns a lower polarization field, and thus, possess larger optical matrix element. As a result, the optical emission intensity is tripled. Shioda et al. [49] inserted a thin AlGaN interlayer between the InGaN layer and the adjacent GaN layer. Such an interlayer reduced the piezoelectric field and suppressed the efficiency droop for the green LED, which has a 530–580 nm emission wavelength. As a result, a light output power of 12 mW and EQE of 25.4% were achieved at a 20 mA for the green LED with a wavelength of 532 nm. A chirped MQW structure was introduced into green LEDs by Chang et al. [50] and increased the output power by 27% and 15% at 20 and 100 mA, respectively. They attributed the efficiency enhancement to the improved carrier injection into QWs and the improved capability of carrier transport. Li et al. [51] reported green LEDs with low-indium-composition InGaN insertion layer before the InGaN emitting layer. The LED achieved a 28.9% enhancement for the optical output power at 150 mA, which is due to the reduced polarization field in the QW and a corresponding increase I the overlap for electron-hole wave functions. Thick InGaN wells with digitally grown InN/GaN was fabricated by Yu et al. [52]. Such a structure benefits the LED by reducing the localization effect in the QWs, leading to an improved in output power, which is about 23% at 20 A/cm^2. Three-layer staggered InGaN QW LEDs emitting was studied by Zhao et al. [53], and the LED has enhanced the optical output power by 2.0–3.5 times, which was attributed to the better overlap for electron-hole wave functions and then promotes the radiative recombination rate. Yang et al. [54] reported the performance and droop behaviors for green LEDs with various low-temperature-grown GaN cap (LT-cap) layers. It is found that the efficiency degradation was suppressed when LT-cap layer thickness was increased from 0 to 1.5 nm. In order to obtain a green LED without efficiency droop, Sheu and Schiavon [55,56] fabricated a green LED by bonding a green InGaN/GaN MQW structure and a UV LED together, in which UV emission is converted into green light through optical excitation.

15.3.1.3 Electron Blocking Layers

In addition to the higher piezoelectric field, electron overflowing to the p-side at high-injection current is another obstruction suppressing the radiative recombination for green LEDs. In order to stop the electrons from overflowing out of the MQWs, an AlGaN layer, known as the electron blocking layer (EBL) with a higher potential barrier is inserted between the MQWs and the p-type layer. However, the barrier height in the valence of the AlGaN EBL retards the hole injection into the MQWs. In InGaN green LEDs, in order to protect the In-rich InGaN QW, a lower growth temperature for the p-type layer is required and this can lead to a low hole concentration in the p-GaN, further reducing the hole injection into the MQWs. Electron overflowing to the p-side is a big challenge for green LEDs. Therefore, in order to enhance the electron injection efficiency without losing the hole injection, several novel AlGaN EBLs have been studied.

Mao et al. [58] studied the variation of QE at different injection current density levels for green LEDs, for which the p-AlGaN EBL has different Al fractions. Due to the complicated carrier recombination mechanism, with the increasing Al fraction, the QE of LED decreases at lower current densities but increases at higher injection current densities. Kim et al. [57] demonstrated the improved IQE for green LEDs by using lattice-matched InAlN EBLs, and the improved IQE was attributed to the better electron confinement in the MQWs (shown in Figure 15.5). What is more, the LED with the HT InAlN EBL had the highest QE in the measured current ranges. Lin et al. [59] introduced a quaternary InAlGaN/GaN superlattice EBL (SL-EBL) for green LEDs and obtained a low-efficiency droop, such that the LED increased the optical output power by 57% and features an efficiency droop about 30%. Ren and Zhang [60], to weaken the carrier leakage, fabricated an InGaN/GaN green LED with a hybrid EBL (HEBL) which was formed by partially incorporating a small amount of indium in the AlGaN

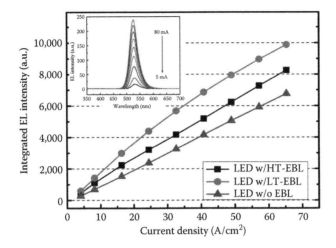

FIGURE 15.5 Light output versus current (*L–I*) curve of green LEDs without EBL, with high temperature (HT) InAlN EBL, and with low-temperature (LT) InAlN EBL; inset shows EL spectra of green LEDs with low-temperature InAlN EBL. (From Kim, H.J. et al., *Appl. Phys. Lett.*, 96, 101102, 2010. With permission.)

conventional EBL. The LED enhanced the optical output power by 80% at 150 mA. According to the simulation, the radiative recombination enhancement was attributed to the higher potential barriers for electrons and lower potential barrier for holes in the new EBL structure, leading to better electron and hole injections. Liu et al. [61] proved that a properly thicker p-AlGaN EBL could reduce the leakage current and improve the efficiency of green LEDs that have large V-pits. As a result, the green LED with an optimized EBL had an output power of 260 mW at 350 mA and an EQE of 31.2% due to the thickness increasing of the energy barrier and few dislocations filtered by the thicker p-AlGaN on the sidewalls of V-pits. Tzou et al. [62] proposed quaternary InAlGaN/GaN superlattice EBL (QSL-EBL) for the green LED with a smoother band diagram and a more uniform carrier distribution in the MQWs, which was attributed to the reduced polarization field in the MQWs. It's found that the light output power for LEDs with QSL-EBL was enhanced by 53%. Kang et al. [63] investigated the green LEDs which had the graded superlattice (GSL) AlGaN/GaN layer. This structure was found be effective in decreasing the effective barrier height for holes and suppressing piezoelectric polarization field near the last barrier, resulting in a better hole injection into the MQWs. As a result, the efficiency is enhanced by 16.4% at 100 A/cm^{-2}.

15.3.1.4 p-Type Layer

Another problem needed to be solved for green LEDs is the growth of p-type layer. To get a better conductivity, Mg-modulation-doped InGaN/GaN superlattice (MD-SLS) designed by Lin et al. [64] was coated on the In-rich MQWs, and this proposed structure enhanced light output power by 100% when compared to the LEDs with a high-temperature grown p-GaN layer. To remove the hydrogen in p-GaN layer and suppress the thermal damage to In-rich MQWs, Kim et al. [65] proposed to anneal the p-GaN layer at a low activation temperature (600°C) by coating a PdZn film on the top of the wafer to remove the hydrogen in the p-GaN. This new annealing method helped to improve the EL intensity by 33% at 20 mA.

15.3.1.5 Localized Surface Plasmon

It has been well known that an improved light emission can be enabled by surface plasmon (SP) thanks to the photon state density of the electron–hole (e–h) pairs. The enhancement can be explained by the following principle: if the energy resonance between e–h pairs and SP is achieved, the energy resonance coupling occurs when the e–h pairs get close to the metal surface. Obviously, the spontaneous recombination rate is expected to increase due to the fact that the recombination rate is proportional to the SP state density.

A method to increase the resonance coupling between excitons in the QDs and LSPs in the Au nanoparticles (NPs) in ZnO films is proposed by inserting a localized surface plasmon (LSP)-enhanced CdSe/ZnS green QD LED with Au NPs in a ZnO electron transport layer [66]. Strong coupling between LSPs in Au NPs and excitons in QDs can be expected, which produces a higher EL intensity due to the excitation of LSP resonance resulting from strong absorption and light scattering by coating QD emissive layer with ZnO solution contains Au NPs. Maximum PL and EL enhancement ratios of 4.47 and 4.54 are observed at 535 and 532 nm, respectively.

Tateishi et al. [67] studied InGaN/GaN QWs with the enhanced efficiency by the SP resonance on aluminum thin films. The enhancement ratio of the green emission reached 80. The authors concluded that the resonance between the excitation light and the SP on the Al surface could improve the excitation efficiency, that is, the light absorption efficiency.

Without sacrificing the p-GaN thickness and effective coupling energy, a method to avoid the exponentially decayed SP field has been introduced by Lu et al. [68]. They improved the IQE and the light extraction efficiency (LEE) for green LEDs by embedding a periodic 2D silver array into the devices. The enhanced IQE and LEE were obtained by showing a PL intensity increment of 280%. Henson et al. [69] used electron beam lithography to fabricate 2D silver nanocylinder array and obtained enhanced emission intensity near green region.

SP effect based on gold NPs was also applied to improve the efficiency of green LEDs (shown in Figure 15.6) [70]. The NPs, which were formed by annealing the Au films, were placed on the p-GaN

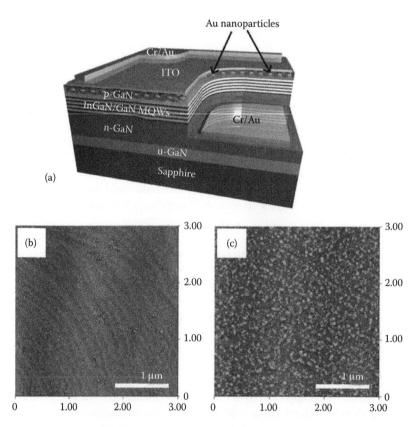

FIGURE 15.6 (a) Schematic of SP-enhanced green LEDs with Au nanoparticles formed on the surface of the P-GaN. AFM images of Au films deposited on a p-GaN layer (b) before and (c) after thermal annealing to form the Au nanoparticles. (Reprinted with permission from Cho, C.Y. et al., *Appl. Phys. Lett.*, 98, 051106, Copyright 2011, American Institute of Physics.)

layer for resonance coupling with excitons in MQWs to improve spontaneous recombination rate. The devices with such design showed an enhanced output power by 86% without electrical characteristics degradation. Kwon et al. [71] found that Au NPs had an absorption peak at 2.3 eV, matching to the photoluminescence emission energy of green MQW and a coupling between the NPs and excitons of In-rich MQWs. With Au NPs, the IQE of the green LEDs was increased by 6.9 times compared to the LEDs without Au particles.

15.3.2 Nanostructure Green LEDs

15.3.2.1 Quantum Dots Green LEDs

It has been well accepted that the self-organized QDs, due to the phase separation of In-rich InGaN, contributed to the high efficiency of green LEDs. The QDs form localized states with deeper levels, which can prevent the carriers from escaping into the neighboring defects. InGaN-based green LEDs with QDs embedded in the QWs through spinodal decomposition were demonstrated by Lu et al. [72]. By decreasing the NH_3 flow rate, the density and average size of the In-rich QDs were increased from 1×10^{11} cm^2 to 2×10^{12} cm^2 and decreased from 8.4 to 2.3 nm, respectively. The LEDs with inserted QDs showed improved light emission efficiency by 10%, which was attributed to a better carrier confinement and a smaller density of defects.

Zhang et al. [73] reduced the green LEDs efficiency droop by 32% by forming $In_{0.25}Ga_{0.75}N$ QDs in the In-rich InGaN quantum wells. The schematic diagram of the InGaN QD LED is shown in Figure 15.7. The temperature dependent photoluminescence measurements revealed a recombination lifetime of 0.57 ns. They attributed these superior optical properties to the smaller piezoelectric field in the QDs.

Lv et al. [74] revealed that the blue shift of EL peak wavelength was negligible in InGaN green QD LEDs. The peak EL wavelength for the QDs LED kept constant at around 527 nm with the increasing current. The negligible wavelength shift was attributed to the suppression of the quantum-confined Stark effect in In-rich QDs due to the strain relaxation [75].

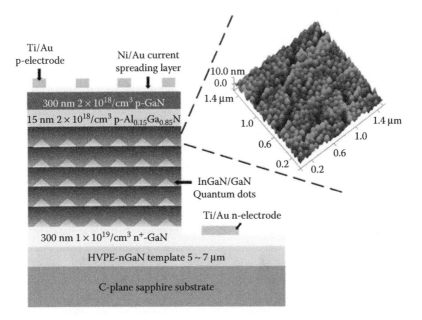

FIGURE 15.7 Schematics of InGaN/GaN quantum dot LED. The inset shows an AFM image of uncapped self-organized $In_{0.25}Ga_{0.75}N$/GaN QDs. (Reprinted with permission from Zhang, M. et al., *Appl. Phys. Lett.*, 97, 011103, Copyright 2010, American Institute of Physics.)

Yu et al. [76] demonstrated green LEDs with coupled InGaN/GaN QW and QDs structure. Studies on tunneling-enhanced carrier transfer showed that carriers could tunnel from shallow QWs to deep QDs nearby. Compared with the conventional single QD layer, the IQE of the QDs was enhanced by more than two times, arriving at 45%.

InGaN QDs inserted at the substrate and MQWs interface were demonstrated to stop the dislocations from propagating into the upper layers [77]. The green LED showed a reduced defect density by a factor of 5, that is, from $5 \times 10^8 cm^2$ to $9.8 \times 10^7 cm^2$ in a GaN layer over the optimized QDs. Such filters reduced the electron and hole trap densities in the upper layer by a factor of 3, and thus, increased the light output power for green LEDs. Additionally, strain-induced QDs were found to effectively suppress the pit/dislocation generation in In-rich layers [78]. The strain modulation on the InGaN layer by using the GaN islands capping is also used to form a surface potential barrier around the dislocation cores, thus blocking carriers from transporting to the neighboring dislocations.

The In-rich self-assembled QDs could be formed by enhancing the phase separation [79]. A way to forming the InGaN QDs by phase separation in the InGaN layer is to increase the surface roughness of the underneath layer. By reducing the QD bandgap, such structure redshifted the emission wavelength for LEDs emitting light from 540 to 610 nm.

15.3.2.2 Nanostructure LEDs

Compared to conventional LEDs with 2D structures, nanostructures offers extraordinary advantages, including reduced dislocation densities, smaller polarization field, and enhanced emission efficiency. Moreover, the nanostructures often come with semipolar and nonpolar planes so that the effect of polarization fields is reduced or eliminated [80,81]. Therefore, optical properties for MQWs grown on these planes are expected to be improved, which is encouraging for lasers and LEDs operated in the visible wavelength [82,83]. It is worth mentioning that Auger recombination and droop effect could be significantly reduced by using defect-free nanowires [84,85]. In 2004, it was reported that LEDs with such design obtained 4.3 times higher light output in contrast to the devices with conventional structures at an injection current of 20 mA[86].

Nanostructure InGaN LEDs can be either fabricated top-down or bottom-up. As far as top-down fabrication method is concerned, the structure is simply obtained by using dry etching technique with the disadvantage of sidewall damaging during the process. This damage will bring in defects and current leakage which would lower the optical output power of the LEDs [87]. On the other hand, the fabrication process of the bottom-up structure is totally different from the previous method by using selective area growth (SAG) technique or simply named mask. In this structure, kinetics of growth on various facets during the epi-growth process can be controlled. The mask can be fabricated by using nanoimprint lithography, which is suitable for mass fabrication [88]. The drawbacks of the method are the fixed diameter or periodic distance of the pattern and cost. Therefore, nanospherical-lens photolithography (NLP) technology is introduced to improve these disadvantages. During a standard NLP process, by UV irradiating the photoresist that is covered with single layer polystyrene spheres (PS) particles, sub-wavelength holes are formed [89]. Therefore, diameter or periodic distance of the patterns becomes adjustable by using PS particles of different sizes [90]. The underlying physics of this core-shell structure was clarified in detail by Reference [91]. InGaN/GaN MQW core-shells that were formed on the GaN NRs suffered from thickness variation and indium compositional gradient due to surface diffusion and vapor-phase diffusion of the selectively grown 3Dstructures, which lead to energy gradient. Emission behaviors of InGaN/GaN core-shell NRs on m-plane can be revealed by the excitation and temperature dependences of PL. On the other hand, with increasing injection current, EL of the InGaN/GaN core-shell-nanorod-embedded 3D LED changed from green to blue. Therefore, an effective approach to finding out the mechanism behind the emission properties of the 3D LEDs will be the analysis of the linewidth, intensity, and emission energy behaviors.

15.3.3 Light Extraction

Photonic crystal possessing a photonic bandgap (PCWG) has been demonstrated to enhance light extraction from the green LED by Kim [92]. The fabrication process is showing in Figure 15.8. It has been observed that the PL intensity enhanced under different PL detection angle and the EL intensity was enhanced by 200%, indicating that the PCWG can improve the light extraction greatly for green LEDs.

To enhance the light extraction, InGaN green LEDs were fabricated using nanoimprint lithography (NIL) to texturize the surface and form a two-dimensional photonic crystal (2DPC) [93,94]. Such a 2DPC structure not only enhanced the light output power but also changed the far-field pattern simultaneously [95]. The LED with a 2DPC achieved an 88.4% enhancement of light output power at 350 mA due to the better light extraction and showed no degradation in the electric characteristics. Hsieh et al. [96] developed a one-shot three-beam laser interference system to fabricate a 2DPC on a green LED to enhance light extraction of LEDs. Submicron-patterns with a diameter about 6 mm were transferred to the surface of the green LED by using the current exposure system, leading to a light extraction efficiency enhanced by a factor of 2.1.

By roughing the p-GaN surface using nanoimprint lithography, Huang et al. [97] demonstrated the light extraction enhancement for green LEDs. The green LED with nano-rough p-GaN surface shows stronger light extraction with a wider view angle, resulting in the enhanced light output power by 48%. Pei et al. [98] reported PL intensity improvement for InGaN/GaN green LEDs grown on wet-etched patterned sapphire substrates. Due to the better light extraction, the PL integral intensity is nearly two times of that on the planar one.

Lin et al. used the fabricated microhole-array pattern to enhance the efficiency of green LEDs through a photoelectrochemical (PEC) process and enhanced the light output power by 65% at 20 mA [99]. Pan et al. [100] proved the enhancement of the light extraction for the green LEDs by nanotexturing the surface of a sapphire substrate to form nanocraters. The LEDs with interface texture showed a 27% improvement in EL intensity.

A hemicylindrical grating structure, which was fabricated by electron-beam lithography and an etching method on the top layer of the diode, can enhance the light output for green LEDs significantly [101]. It was found that the optical transmission for the green LED increased for incidence angles between

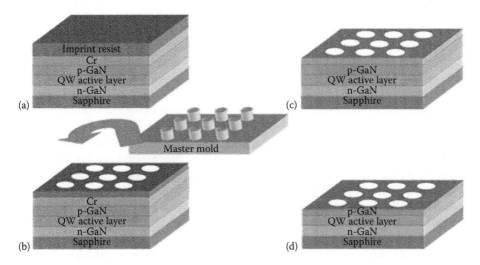

FIGURE 15.8 A schematic of the fabrication flow of PC-LEDs using the nanoimprint lithography and the ICP etching process; (a) sample preparation, (b) the nanoimprint lithography process, (c) residual layer removal and Cr patterning, and (d) the p-GaN ICP etch process. (Reprinted with permission from Kim, J.Y. et al., *Appl. Phys. Lett.*, 91, 181109, Copyright 2007, American Institute of Physics.)

23.58° and 60° and the light output power was increased approximately 4.7 times. Joo et al. [102] demonstrate the enhancement in the light extraction efficiency for green LEDs through periodic aluminum-doped zinc oxide (AZO) subwavelength nanostructure arrays (SNAs). The AZO SNAs are formed on the surface of indium tin oxide electrodes of LEDs by laser interference lithography and a subsequent dry etching after AZO film deposition. The LED had an enhancement of the light output power by 19% at 100 mA and showed no distinct degradation of the electrical characteristics. Lee et al. [103] reported the enhanced light extraction efficiency in green LEDs with gallium oxide hydroxide (GaOOH) rods. The GaOOH rod, with average lengths of 2 μm and a diameter of 50–500 nm, were coated on the surface of indium tin oxide electrodes of LEDs. The light output powers were increased by 24.3% and 26.4% at 20 and 100 mA, respectively.

The usage of ZnO nanostructures prepared by a hydrothermal method to improve the light extraction for green LEDs was studied by Son et al. [104]. The shape of the ZnO nanostructures integrated onto the p-layer had a great influence on the light extraction of green LED. The light output power of LEDs with surface-textured ZnO hemispheres was increased compared to that of the conventional LED, which was attributed to an increase in the multiple scattering, the escape cone, and effective critical angle.

An anodic aluminum oxide (AAO) patterned sapphire substrate was used as a NIL stamp and imprinted onto the surface for the green LED by Jiang et al. [105]. Due to the extraction of more guided modes arising from the fluctuations of the lattice constant and domain orientation of the AAO patterns, light extraction efficiency was significantly improved by 116% as compared to that of the planar LED. Furthermore, the AAO patterns made a uniform broad protrusion in the central area, and some sharp lobes for the green LED.

Green InGaN-based resonant cavity LEDs with dielectric and silver mirrors were fabricated and showed a light output power enhancement of 1.5 times at 600 A/cm² [106]. The mode spectrum exhibits a linewidth of approximately 5.5 nm with a peak wavelength of 525 nm, indicating a quality factor of 100. The LED with mirrors revealed a low thermally induced redshift when the current increased and a stable EL wavelength due to the resonant microcavity effect was also obtained.

15.4 Future

The RGBA (red\green\blue\amber) LED architecture, in which all colors are generated by direct LEDs, can be considered the ultimate embodiment of white light for higher ultimate efficiency, better color rendering, and more flexibility for chromaticity tuning. However, such architecture is dragged by the low efficiencies, and less thermal stability of LEDs at a certain wavelength, especially the realization of efficient LEDs in the green gap is a key technical challenge. Therefore, fundamental research in droop mitigation strategies should benefit green LEDs. Another issue to be resolved is the strain associated lattice mismatches between InGaN (with a high enough indium fraction) and common substrates. An improved understanding of droop and ways to circumvent it in the blue might have broader ramifications to the green as well.

According to the DOE data showing in Table 15.1, IQE and EQE of green LED at 35A/cm² are expected to reach 54% and 46% respectively in 2020. In addition, current droop (relative EQE at 100 A/cm² versus 35A/cm²) of the device will be increased by 10%. As far as power conversion efficiency (PCE) of green LED package is concerned, the value is expected to be 35% in 2020. Last but not least, the thermal stability of green LED will be increased by over 10% [107].

In the meantime, a possible alternative for producing green LEDs is to combine blue LEDs with phosphor-converter. Such proposal requires blue LEDs with very high PCE and phosphor with higher conversion efficiency and less stokes loss [103]. Hence, phosphor-converters with high efficiency and a narrow band would be an important intermediate step to white lighting.

Since the EL measurements show that core/multishell nanowires function as LEDs under forward bias. Most importantly, the devices seem to have a high-QE between 365 nm and 600 nm, and the green LEDs can be used in 3D LEDs with full-color (or white light) emission. Such design has been realized by

TABLE 15.1 Status and Targets of RGBA Emitter by DOE

Metrics	2015 Status	2020 Targets
IQE at 35 A/cm^2	79% (Blue)[1]	90% (Blue)
	39% (Green)	54% (Green)
	75% (Red)	87% (Red)
	13% (Amber)	32% (Amber)
External Quantum Efficiency	69% (Blue)	81% (Blue)
(EQE) at 35 A/cm^2, 25°C	32% (Green)	46% (Green)
	54% (Red)	65% (Red)
	10% (Amber)	24% (Amber)
Power Conversion efficiency	66% (Blue)	80% (Blue)
at 35 A/cm^2	22% (Green)	35% (Green)
	44% (Red)	55% (Red)
	8% (Amber)	20% (Amber)
Current droop–Relative EQE at 100 A/cm^2 versus 35 A/cm^2	85%	95%
Thermal stability–Relative optical flux at 100°C versus 25°C	92% (Blue)	98% (Blue, Green)
	85% (Green)	75% (Red, amber)
	50% (Red)	
	25% (Amber)	

Source: Lee, H.K. et al., *IEEE Photonics Technol. Lett.*, 24, 285–287, 2012.

phosphor-based wavelength conversion and geometrically emissive color mixing [108,109]. Thus, fabrication of flexible displays based on solid-state 3D LEDs is now feasible [110,111].

References

1. S. Nakamura, M. Senoh, N. Iwasa, S.-i. Nagahama, T. Yamada, T. Mukai, (1995). Superbright green InGaN single-quantum-well-structure light-emitting diodes. *Japanese Journal of Applied Physics*. 34(10B):L1332.

2. S. Nakamura, M. Senoh, N. Iwasa, S.-i. Nagahama, (1995). High-brightness InGaN blue, green and yellow light-emitting diodes with quantum well structures. *Japanese Journal of Applied Physics*. 34(7A):L797.

3. T.-Y. Wang, S.-L. Ou, K.-C. Shen, D.-S. Wuu, (2013). Effect of non-vacuum thermal annealing on high indium content InGaN films deposited by pulsed laser deposition. *Optics Express*. 21(6):7337–7342.

4. D.-J. Kim, Y.-T. Moon, K.-M. Song, I.-H. Lee, S.-J. Park, (2001). Effect of growth pressure on indium incorporation during the growth of InGaN by MOCVD. *Journal of Electronic Materials*. 30(2):99–102.

5. E. Piner, M. Behbehani, N. El-Masry, F. McIntosh, J. Roberts, K. Boutros, S. Bedair, (1997). Effect of hydrogen on the indium incorporation in InGaN epitaxial films. *Applied Physics Letters*. 70(4):461–463.

6. Y. Okamoto, K. Takahashi, H. Nakamura, Y. Okada, M. Kawabe, (2000). Effects of atomic hydrogen on the indium incorporation in InGaN Grown by RF-molecular beam epitaxy. *Japanese Journal of Applied Physics*. 39(4B):L343.

7. H. Chen, R.M. Feenstra, J. Northrup, J. Neugebauer, D.W. Greve, (2001). Indium incorporation and surface segregation during InGaN growth by molecular beam epitaxy: Experiment and theory. *Mrs. Internet Journal of Nitride Semiconductor Research* 6:e11.

8. S.-B. Che, T. Shinada, T. Mizuno, X. Wang, Y. Ishitani, A. Yoshikawa, (2006). Effect of precise control of V/III ratio on in-rich InGaN epitaxial growth. *Japanese Journal of Applied Physics*. 45(12L):L1259.

9. S.Y. Kwon, H.J. Kim, H. Na, H.C. Seo, H.J. Kim, Y. Shin, Y.W. Kim, S. Yoon, H.J. Oh, C. Sone, (2003). Effect of growth interruption on In-rich InGaN/GaN single quantum well structures. *Physica Status Solidi (c).* 12(7):2830–2833.

10. C. Du, Z. Ma, J. Zhou, T. Lu, Y. Jiang, P. Zuo, H. Jia, H. Chen, (2014). Enhancing the quantum efficiency of InGaN yellow-green light-emitting diodes by growth interruption. *Applied Physics Letters.* 105(7):071108.

11. M. Cheong, R. Choi, C. Kim, H. Yoon, C. Hong, E. Suh, H. Lee, (2001). Effects of growth interruption on high indium content InGaN/GaN multi quantum wells, *Journal of the Korean Physical Society.* 38(6):701–705.

12. H.K. Cho, G.M. Yang, (2002) Generation of misfit dislocations in high indium content InGaN layer grown on GaN. *Journal of Crystal Growth.* 243(1):124–128.

13. H. Lei, J. Chen, P. Ruterana, (2010). Role of c-screw dislocations on indium segregation in InGaN and InAlN alloys. *Applied Physics Letters.* 96(16):161901.

14. I.-K. Park, M.-K. Kwon, S.-H. Baek, Y.-W. Ok, T.-Y. Seong, S.-J. Park, Y.-S. Kim, Y.-T. Moon, D.-J. Kim, (2005). Enhancement of phase separation in the InGaN layer for self-assembled In-rich quantum dots. *Applied Physics Letters.* 87(6):061906.

15. Y.-S. Lin, K.-J. Ma, C. Hsu, S.-W. Feng, Y.-C. Cheng, C.-C. Liao, C. Yang, C.-C. Chou, C.-M. Lee, J.-I. Chyi, (2000). Dependence of composition fluctuation on indium content in InGaN/GaN multiple quantum wells. *Applied Physics Letters.* 77(19):2988–2990.

16. Y.-T. Moon, D.-J. Kim, K.-M. Song, C.-J. Choi, S.-H. Han, T.-Y. Seong, S.-J. Park, (2001). Effects of thermal and hydrogen treatment on indium segregation in InGaN/GaN multiple quantum wells. *Journal of Applied Physics.* 89(11):6514–6518.

17. P. Waltereit, O. Brandt, K. Ploog, M. Tagliente, L. Tapfer, (2001). Indium surface segregation during growth of (In, Ga) N/GaN multiple quantum wells by plasma-assisted molecular beam epitaxy. *Physica Status Solidi (b),* 228(1):49–53.

18. A. Das, S. Magalhaes, Y. Kotsar, P.K. Kandaswamy, B. Gayral, K. Lorenz, E. Alves, P. Ruterana, E. Monroy, (2010). Indium kinetics during the plasma-assisted molecular beam epitaxy of semipolar (11–22) InGaN layers. *Applied Physics Letters.* 96(18):181907.

19. H. Jönen, U. Rossow, H. Bremers, L. Hoffmann, M. Brendel, A.D. Dräger, S. Metzner, F. Bertram, J. Christen, S. Schwaiger, (2011). Indium incorporation in GaInN/GaN quantum well structures on polar and nonpolar surfaces. *Physica Status Solidi (b).* 248(3):600–604.

20. K. Lai, T. Paskova, V. Wheeler, T. Chung, J. Grenko, M. Johnson, K. Udwary, E. Preble, K. Evans, (2012). Indium incorporation in InGaN/GaN quantum wells grown on m-plane GaN substrate and c-plane sapphire. *Physica Status Solidi (a).* 209(3):559–564.

21. D.A. Browne, E.C. Young, J.R. Lang, C.A. Hurni, J.S. Speck, (2012). Indium and impurity incorporation in InGaN films on polar, nonpolar, and semipolar GaN orientations grown by ammonia molecular beam epitaxy. *Journal of Vacuum Science and Technology.* A 30(4):041513.

22. D.V. Dinh, M. Pristovsek, M. Kneissl, (2016). MOVPE growth and indium incorporation of polar, semipolar (112$\bar{2}$2) and (202$\bar{1}$1) InGaN. *Physica Status Solidi (b).* 253(1):93–98.

23. K. Shojiki, J.-H. Choi, H. Shindo, T. Kimura, T. Tanikawa, T. Hanada, R. Katayama, T. Matsuoka, (2014). Effect of c-plane sapphire substrate miscut angle on indium content of MOVPE-grown N-polar InGaN. *Japanese Journal of Applied Physics.* 53(5S1):05FL07.

24. J. Bai, B. Xu, F.G. Guzman, K. Xing, Y. Gong, Y. Hou, T. Wang, (2015). (11–22) semipolar InGaN emitters from green to amber on overgrown GaN on micro-rod templates. *Applied Physics Letters.* 107(26):5.

25. C.-T. Yu, W.-C. Lai, C.-H. Yen, H.-C. Hsu, S.-J. Chang, (2014). Optoelectrical characteristics of green light-emitting diodes containing thick InGaN wells with digitally grown InN/GaN. *Optics Express.* 22(103):A633–A641.

26. C.F. Huang, T.C. Liu, Y.C. Lu, W.Y. Shiao, Y.S. Chen, J.K. Wang, C.F. Lu, C.C. Yang, (2008). Enhanced efficiency and reduced spectral shift of green light-emitting-diode epitaxial structure with prestrained growth. *Journal of Applied Physics.* 104(12):7.

27. J.Y. Park, J.H. Lee, S. Jung, T. Ji, (2016). InGaN/GaN-based green-light-emitting diodes with an inserted InGaN/GaN-graded superlattice layer. *Physica Status Solidi (a)*. 213(6):1610–1614.
28. W. Lundin, A. Nikolaev, A. Sakharov, E. Zavarin, G. Valkovskiy, M. Yagovkina, S. Usov, N. Kryzhanovskaya, V. Sizov, P. Brunkov, (2011). Single quantum well deep-green LEDs with buried InGaN/GaN short-period superlattice. *Journal of Crystal Growth*. 315(1):267–271.
29. K. Lee, C.-R. Lee, J.H. Lee, T.-H. Chung, M.-Y. Ryu, K.-U. Jeong, J.-Y. Leem, J.S. Kim, (2016). Influences of Si-doped graded short-period superlattice on green InGaN/GaN light-emitting diodes. *Optics Express*. 24(7):7743–7751.
30. Y. Xia, W. Hou, L. Zhao, M. Zhu, T. Detchprohm, C. Wetzel, (2010). Boosting green GaInN/GaN light-emitting diode performance by a GaInN underlying layer. *IEEE Transactions on Electron Devices*. 57(10):2639–2643.
31. C.-F. Huang, T.-C. Liu, Y.-C. Lu, W.-Y. Shiao, Y.-S. Chen, J.-K. Wang, C.-F. Lu, C. Yang, (2008). Enhanced efficiency and reduced spectral shift of green light-emitting-diode epitaxial structure with prestrained growth. *Journal of Applied Physics*. 104(12):123106.
32. J.Y. Park, J.H. Lee, S. Jung, T. Ji, (2016). InGaN/GaN-based green-light-emitting diodes with an inserted InGaN/GaN-graded superlattice layer. *Physica Status Solidi a Applications and Materials Science*. 213(6):1610–1614.
33. W.V. Lundin, A.E. Nikolaev, A.V. Sakharov, E.E. Zavarin, G.A. Valkovskiy, M.A. Yagovkina, S.O. Usov et al., (2011). Single quantum well deep-green LEDs with buried InGaN/GaN short-period superlattice. *Journal of Crystal Growth* 315(1):267–271.
34. K. Lee, C.R. Lee, J.H. Lee, T.H. Chung, M.Y. Ryu, K.U. Jeong, J.Y. Leem, J.S. Kim, (2016). Influences of Si-doped graded short-period superlattice on green InGaN/GaN light-emitting diodes. *Optics Express*. 24(7):7743–7751.
35. H. Zhao, G. Liu, X.-H. Li, R. Arif, G. Huang, J. Poplawsky, S.T. Penn, V. Dierolf, N. Tansu, (2009). Design and characteristics of staggered InGaN quantum-well light-emitting diodes in the green spectral regime. *IET Optoelectronics*. 3(6):283–295.
36. H. Zhao, G. Liu, J. Zhang, J.D. Poplawsky, V. Dierolf, N. Tansu, (2011). Approaches for high internal quantum efficiency green InGaN light-emitting diodes with large overlap quantum wells. *Optics Express*. 19(104):A991–A1007.
37. Y.-A. Chang, Y.-T. Kuo, J.-Y. Chang, Y.-K. Kuo, (2012). Investigation of InGaN green light-emitting diodes with chirped multiple quantum well structures. *Optics Letters*. 37(12):2205–2207.
38. H. Zhao, G. Liu, N. Tansu, (2010). Analysis of InGaN-delta-InN quantum wells for light-emitting diodes. *Applied Physics Letters*. 97(13):131114.
39. C. Bayram, F. Hosseini Teherani, D.J. Rogers, M. Razeghi, (2008) A hybrid green light-emitting diode comprised of n-ZnO/(InGaN/GaN) multi-quantum-wells/p-GaN. *Applied Physics Letters*. 93(8):081111.
40. S. Verma, S.K. Pandey, S.K. Pandey, S. Mukherjee, (2015). Theoretical simulation of Hybrid II-O/III-N green light-emitting diode with MgZnO/InGaN/MgZnO heterojunction. *Materials Science in Semiconductor Processing*. 31:340–350.
41. L. Han, K. Kash, H.P. Zhao, (2014). High efficiency green light-emitting diodes based on InGaN-ZnGeN2 Type-II quantum wells. In: K.P. Streubel, H. Jeon, L.W. Tu, M. Strassburg (Eds.), *Light-Emitting Diodes: Materials, Devices, and Applications for Solid State Lighting Xviii*. Bellingham, Spie-Int Soc Optical Engineering, 2014.
42. N. Zhang, Z. Liu, Z. Si, P. Ren, X.D. Wang, X.X. Feng, P. Dong et al., (2013). Reduction of efficiency droop and modification of polarization fields of InGaN-based green light-emitting diodes via Mg-doping in the barriers. *Chinese Physics Letters*. 30(8):3.
43. C.Y. Huang, Q.M. Yan, Y.J. Zhao, K. Fujito, D. Feezell, C.G. Van de Walle, J.S. Speck, S.P. DenBaars, S. Nakamura, (2011). Influence of Mg-doped barriers on semipolar (20(2)over-bar1) multiple-quantum-well green light-emitting diodes. *Applied Physics Letters*. 99(14):3.

44. J. Zhang, X.J. Zhuo, D.W. Li, L. Yu, K. Li, Y.W. Zhang, J.S. Diao, X.F. Wang, S.T. Li, (2015). Effect of Mg doping in GaN interlayer on the performance of green light-emitting diodes. *IEEE Photonics Technology Letters.* 27(2):117–120.

45. Z.T. Lin, R. Hao, G.Q. Li, S.G. Zhang, (2015). Effect of Si doping in barriers of InGaN/GaN multiple quantum wells on the performance of green light-emitting diodes. *Japanese Journal of Applied Physics.* 54(2):5.

46. J.H. Ryou, J. Limb, W. Lee, J.P. Liu, Z. Lochner, D.W. Yoo, R.D. Dupuis, (2008). Effect of silicon doping in the quantum-well barriers on the electrical and optical properties of visible green light-emitting diodes. *IEEE Photonics Technology Letters.* 20(21–24):1769–1771.

47. Y.J. Lee, C.H. Chen, C.J. Lee, (2010). Reduction in the efficiency-droop effect of InGaN green light-emitting diodes using gradual quantum wells. *IEEE Photonics Technology Letters.* 22(20):1506–1508.

48. S.H. Park, Y.E. Pak, C.Y. Park, D. Mishra, S.H. Yoo, Y.H. Cho, M.B. Shim, S. Kim, (2015). Quaternary AlInGaN/InGaN quantum well on vicinal c-plane substrate for high emission intensity of green wavelengths. *Journal of Applied Physics.* 117(18):6.

49. T. Shioda, H. Yoshida, K. Tachibana, N. Sugiyama, S. Nunoue, (2012). Enhanced light output power of green LEDs employing AlGaN interlayer in InGaN/GaN MQW structure on sapphire (0001) substrate. *Physica Status Solidi a Applications and Materials Science.* 209(3):473–476.

50. Y.A. Chang, Y.T. Kuo, J.Y. Chang, Y.K. Kuo, (2012). Investigation of InGaN green light-emitting diodes with chirped multiple quantum well structures. *Optics Letters.* 37(12):2205–2207.

51. H.J. Li, P.P. Li, J.J. Kang, Z. Li, Y.Y. Zhang, Z.C. Li, J. Li, X.Y. Yi, J.M. Li, G.H. Wang, (2013). Quantum efficiency enhancement of 530 nm InGaN green light-emitting diodes with shallow quantum well. *Applied Physics Express.* 6(5):4.

52. C.T. Yu, W.C. Lai, C.H. Yen, H.C. Hsu, S.J. Chang, (2014). Optoelectrical characteristics of green light-emitting diodes containing thick InGaN wells with digitally grown InN/GaN. *Optics Express.* 22(9):A633–A641.

53. H.P. Zhao, G.Y. Liu, X.H. Li, R.A. Arif, G.S. Huang, J.D. Poplawsky, S.T. Penn, V. Dierolf, N. Tansu, (2009). Design and characteristics of staggered InGaN quantum-well light-emitting diodes in the green spectral regime. *IET Optoelectronics.* 3(6):283–295.

54. J. Yang, D.G. Zhao, D .S. Jiang, P. Chen, J.J. Zhu, Z.S. Liu, L.C. Le et al., (2016). Investigation on the performance and efficiency droop behaviors of InGaN/GaN multiple quantum well green LEDs with various GaN cap layer thicknesses. *Vacuum.* 129:99–104.

55. J.K. Sheu, F.B. Chen, W.Y. Yen, Y.C. Wang, C.N. Liu, Y.H. Yeh, M.L. Lee, (2015). GaN-based photon-recycling green light-emitting diodes with vertical-conduction structure. *Optics Express.* 23(7):A371–A381.

56. D. Schiavon, M. Binder, A. Loeffler, M. Peter, (2013). Optically pumped GaInN/GaN multiple quantum wells for the realization of efficient green light-emitting devices. *Applied Physics Letters.* 102(11):4.

57. H.J. Kim, S. Choi, S.S. Kim, J.H. Ryou, (2010). Improvement of quantum efficiency by employing active-layer-friendly lattice-matched InAlN electron blocking layer in green light-emitting diodes. *Applied Physics Letters.* 96(10):101102.

58. Q.H. Mao, F.Y. Jiang, H.Y. Cheng, C.D. Zheng, (2010). p-AlGaN electron blocking layer with different Al fractions on green InGaN/GaN LEDs grown on Si substrates. *Acta Physica Sinica.* 59(11):8078–8082.

59. D.W. Lin, A.J. Tzou, J.K. Huang, B.C. Lin, (2015). Greatly improved efficiency droop for InGaN-based green light emitting diodes by quaternary content superlattice electron blocking layer. In *Numerical Simulation of Optoelectronic Devices (NUSOD), 2015 International Conference on.* IEEE, pp. 15–16.

60. P. Ren, N. Zhang, Z. Liu, B. Xue, J.M. Li, J.X. Wang, (2015). Promotion of electron confinement and hole injection in GaN-based green light-emitting diodes with a hybrid electron blocking layer. *Journal of Physics D-Applied Physics.* 48(4):7.

61. X.H. Liu, J.L. Liu, Q.H. Mao, X.M. Wu, J.L. Zhang, G.X. Wang, Z.J. Quan, C.L. Mo, F.Y. Jiang, (2016). Effects of p-AlGaN EBL thickness on the performance of InGaN green LEDs with large V-pits. *Semiconductor Science and Technology*. 31(2):6.

62. A.J. Tzou, D.W. Lin, C.R. Yu, Z.Y. Li, Y.K. Liao, B.C. Lin, J.K. Huang, C.C. Lin, T.S. Kao, H.C. Kuo, C.Y. Chang, (2016). High-performance InGaN-based green light-emitting diodes with quaternary InAlGaN/GaN superlattice electron blocking layer. *Optics Express*. 24(11):1387–1395.

63. J.J. Kang, H.J. Li, Z. Li, Z.Q. Liu, P. Ma, X.Y. Yi, G.H. Wang, (2013). Enhancing the performance of green GaN-based light-emitting diodes with graded superlattice AlGaN/GaN inserting layer. *Applied Physics Letters*. 103(10):4.

64. H.C. Lin, G.Y. Lee, H.H. Liu, N.W. Hsu, C.C. Wu, J.I. Chyi, (2009). Polarization-enhanced Mg doping in InGaN/GaN superlattice for green light-emitting diodes. In *Conference on Lasers and Electro-Optics*. IEEE, New York, 2009.

65. J.Y. Kim, M.K. Kwon, S.J. Park, S. Kim, J.W. Kim, Y.C. Kim, (2009). Improving the performance of green LEDs by low-temperature annealing of p-GaN with PdZn. *Electrochemical and Solid State Letters*. 12(5):H185–H187.

66. N.-Y. Kim, S.-H. Hong, J.-W. Kang, N. Myoung, S.-Y. Yim, S. Jung, K. Lee, C.W. Tu, S.-J. Park, (2015). Localized surface plasmon-enhanced green quantum dot light-emitting diodes using gold nanoparticles. *RSC Advances*. 5(25):19624–19629.

67. K. Tateishi, M. Funato, Y. Kawakami, K. Okamoto, K. Tamada, (2015). Highly enhanced green emission from InGaN quantum wells due to surface plasmon resonance on aluminum films. *Applied Physics Letters*. 106(12):5.

68. C.H. Lu, C.C. Lan, Y.L. Lai, Y.L. Li, C.P. Liu, (2011). Enhancement of green emission from InGaN/GaN multiple quantum wells via coupling to surface plasmons in a two-dimensional silver array. *Advanced Functional Materials*. 21(24):4719–4723.

69. J. Henson, E. Dimakis, J. DiMaria, R. Li, S. Minissale, L. Dal Negro, T.D. Moustakas, R. Paiella, (2010). Enhanced near-green light emission from InGaN quantum wells by use of tunable plasmonic resonances in silver nanoparticle arrays. *Optics Express*. 18(20):21322–21329.

70. C.Y. Cho, S.J. Lee, J.H. Song, S.H. Hong, S.M. Lee, Y.H. Cho, S.J. Park, (2011). Enhanced optical output power of green light-emitting diodes by surface plasmon of gold nanoparticles. *Applied Physics Letters* 98(5):051106.

71. M.-K. Kwon, J.-Y. Kim, S.-J. Park, (2013). Enhanced emission efficiency of green InGaN/GaN multiple quantum wells by surface plasmon of Au nanoparticles. *Journal of Crystal Growth*. 370:124–127.

72. C.-H. Lu, Y.-C. Li, Y.-H. Chen, S.-C. Tsai, Y.-L. Lai, Y.-L. Li, C.-P. Liu, (2013). Output power enhancement of InGaN/GaN based green light-emitting diodes with high-density ultra-small In-rich quantum dots. *Journal of Alloys and Compounds*. 555:250–254.

73. M. Zhang, P. Bhattacharya, W. Guo, (2010). InGaN/GaN self-organized quantum dot green light emitting diodes with reduced efficiency droop. *Applied Physics Letters*. 97(1):011103.

74. W. Lv, L. Wang, L. Wang, Y. Xing, D. Yang, Z. Hao, Y. Luo, (2014). InGaN quantum dot green light-emitting diodes with negligible blue shift of electroluminescence peak wavelength. *Applied Physics Express*. 7(2):025203.

75. I.K. Park, M.K. Kwon, J.O. Kim, S.B. Seo, J.Y. Kim, J.H. Lim, S.J. Park, (2007). Green light-emitting diodes with self-assembled In-rich InGaN quantum dots. *Applied Physics Letters* 91(13):3.

76. J.D. Yu, L. Wang, D. Yang, Z.B. Hao, Y. Luo, C.Z. Sun, Y.J. Han, B. Xiong, J. Wang, H.T. Li, (2015). Improving the internal quantum efficiency of green InGaN quantum dots through coupled InGaN/GaN quantum well and quantum dot structure. *Applied Physics Express*. 8(9):4.

77. A. Banerjee, T. Frost, S. Jahangir, P. Bhattacharya, (2014). Green-emitting (lambda=525 nm) InGaN/GaN quantum dot light emitting diodes grown on quantum dot dislocation filters. *IEEE Journal of Quantum Electronics*. 50(4):228–235.

78. G.B. Wang, H. Xiong, Y.X. Lin, Z.L. Fang, J.Y. Kang, Y. Duan, W.Z. Shen, (2012). Green emission from a strain-modulated InGaN active layer. *Chinese Physics Letters*. 29(6):4.

79. I.K. Park, S.J. Park, (2011). Green gap spectral range light-emitting diodes with Self-assembled InGaN quantum dots formed by enhanced phase separation. *Applied Physics Express*. 4(4):3.

80. Z. Li-Xia, Y. Zhi-Guo, S. Bo, Z. Shi-Chao, A. Ping-Bo, Y. Chao, L. Lei, W. Jun-Xi, L. Jin-Min, (2015). Progress and prospects of GaN-based LEDs using nanostructure. *Chinese Physics B*. 24(6):068506.

81. Z.-G. Yu, L.-X. Zhao, X.-C. Wei, X.-J. Sun, P.-B. An, S.-C. Zhu, L. Liu, L.-X. Tian, F. Zhang, H.-X. Lu, (2014). Surface plasmon-enhanced nanoporous GaN-based green light-emitting diodes with Al2O3 passivation layer. *Optics Express*. 22(106):A1596–A1603.

82. Z.-G. Yu, L.-X. Zhao, S.-C. Zhu, X.-C. Wei, X.-J. Sun, L. Liu, J.-X. Wang, J.-M. Li, (2015). Optimization of the nanopore depth to improve the electroluminescence for GaN-based nanoporous green LEDs. *Materials Science in Semiconductor Processing*. 33:76–80.

83. Y. Jin, F. Yang, Q. Li, Z. Zhu, J. Zhu, S. Fan, (2012). Enhanced light extraction from a GaN-based green light-emitting diode with hemicylindrical linear grating structure. *Optics Express*. 20(14):15818–15825.

84. S. Jiang, Y. Feng, Z. Chen, L. Zhang, X. Jiang, Q. Jiao, J. Li, Y. Chen, D. Li, L. Liu, (2016). Study on light extraction from GaN-based green light-emitting diodes using anodic aluminum oxide pattern and nanoimprint lithography. *Scientific Reports*. 6: 21573.

85. T.-W. Yeh, Y.-T. Lin, L.S. Stewart, P.D. Dapkus, R. Sarkissian, J.D. O'Brien, B. Ahn, S.R. Nutt, (2012). InGaN/GaN multiple quantum wells grown on nonpolar facets of vertical GaN nanorod arrays. *Nano Letters*. 12(6):3257–3262.

86. J.-H. Kim, Y.-H. Ko, J.-H. Cho, S.-H. Gong, S.-M. Ko, Y.-H. Cho, (2014). Toward highly radiative white light emitting nanostructures: A new approach to dislocation-eliminated GaN/InGaN core–shell nanostructures with a negligible polarization field. *Nanoscale*. 6(23):14213–14220.

87. M. Tchernycheva, V. Neplokh, H. Zhang, P. Lavenus, L. Rigutti, F. Bayle, F. Julien, A. Babichev, G. Jacopin, L. Largeau, (2015). Core–shell InGaN/GaN nanowire light emitting diodes analyzed by electron beam induced current microscopy and cathodoluminescence mapping. *Nanoscale* 7(27):11692–11701.

88. R. Peng, H. Gang, F. Bing-Lei, X. Bin, Z. Ning, L. Zhe, Z. Li-Xia, W. Jun-Xi, L. Jin-Min, (2016). Selective area growth and characterization of GaN nanorods fabricated by adjusting the hydrogen flow rate and growth temperature with metal organic chemical vapor deposition. *Chinese Physics Letters*. 33(6):68101.

89. W. Guo, M. Zhang, P. Bhattacharya, J. Heo, (2011). Auger recombination in III-nitride nanowires and its effect on nanowire light-emitting diode characteristics. *Nano Letters*. 11(4):1434–1438.

90. B.O. Jung, S.-Y. Bae, S. Lee, S.Y. Kim, J.Y. Lee, Y. Honda, H. Amano, (2016). Emission characteristics of InGaN/GaN core-shell nanorods embedded in a 3D light-emitting diode. *Nanoscale Research Letters*. 11(1):1–10.

91. H.P.T. Nguyen, K. Cui, S. Zhang, M. Djavid, A. Korinek, G.A. Botton, Z. Mi, Controlling electron overflow in phosphor-free InGaN/GaN nanowire white light-emitting diodes. *Nano Letters*. 12(3):1317–1323.

92. J.Y. Kim, M.K. Kwon, K.S. Lee, S.J. Park, S.H. Kim, K.D. Lee, (2007). Enhanced light extraction from GaN-based green light-emitting diode with photonic crystal. *Applied Physics Letters*. 91(18):181109.

93. B.S. Cheng, C.H. Chiu, K.J. Huang, C.F. Lai, H.C. Kuo, C.H. Lin, T.C. Lu, S.C. Wang, C.C. Yu, (2008). Enhanced light extraction of InGaN-based green LEDs by nano-imprinted 2D photonic crystal pattern. *Semiconductor Science and Technology*. 23(5):5.

94. H. Huang, C. Lin, C. Yu, B. Lee, C. Chiu, C. Lai, H. Kuo, K. Leung, T. Lu, S. Wang, (2008). Enhanced light output from a nitride-based power chip of green light-emitting diodes with nano-rough surface using nanoimprint lithography. *Nanotechnology*. 19(18):185301.

95. B. Cheng, C. Chiu, K. Huang, C. Lai, H. Kuo, C. Lin, T. Lu, S. Wang, C. Yu, (2008). Enhanced light extraction of InGaN-based green LEDs by nano-imprinted 2D photonic crystal pattern. *Semiconductor Science and Technology*. 23(5):055002.

96. M.L. Hsieh, K.C. Lo, Y.S. Lan, S.Y. Yang, C.H. Lin, H.M. Liu, H.C. Kuo, (2008). One-shot exposure for patterning two-dimensional photonic crystals to enhance light extraction of InGaN-based green LEDs. *IEEE Photonics Technology Letters*. 20(1–4):141–143.

97. H. Huang, C.H. Lin, C.C. Yu, C.H. Chiu, C.F. Lai, H.C. Kuo, K.M. Leung, T.C. Lu, S.C. Wang, B.D. Lee, (2008). Enhanced light output from a nitride-based power chip of green light-emitting diodes with nano-rough surface using nanoimprint lithography. *Nanotechnology*. 19(18):4.

98. X.J. Pei, L.W. Guo, X.H. Wang, Y. Wang, H.Q. Jia, H. Chen, J.M. Zhou, (2009). Enhanced photo-luminescence of InGaN/GaN green light-emitting diodes grown on patterned sapphire substrate. *Chinese Physics Letters*. 26(2):4.

99. C.F. Lin, C.M. Lin, K.T. Chen, M.S. Lin, J.J. Dai, (2010). Green light-emitting diodes with a photoelectrochemically treated microhole-array pattern. *Electrochemical and Solid State Letters*. 13(3):H90–H93.

100. Y.B. Pan, M.S. Hao, S.L. Qi, H. Fang, G.Y. Zhang, (2010). Effect of interface nanotexture on light extraction of InGaN-based green light emitting diodes. *Chinese Physics Letters*. 27(3):4.

101. Y.H. Jin, F.L. Yang, Q.Q. Li, Z.D. Zhu, J. Zhu, S.S. Fan, (2012). Enhanced light extraction from a GaN-based green light-emitting diode with hemicylindrical linear grating structure. *Optics Express*. 20(14):15818–15825.

102. D.H. Joo, H.K. Lee, J.S. Yu, (2012). Light output extraction enhancement in GaN-based green LEDs with periodic AZO subwavelength nanostructure arrays. *IEEE Photonics Technology Letters*. 24(16):1381–1383.

103. H.K. Lee, M.S. Kim, J.S. Yu, (2012). Enhanced light extraction of GaN-based green light-emitting diodes with GaOOH rods. *IEEE Photonics Technology Letters*. 24(4):285–287.

104. T. Son, K.Y. Jung, J. Park, (2013). Enhancement of the light extraction of GaN-based green light emitting diodes via nanohybrid structures. *Current Applied Physics*. 13(6):1042–1045.

105. S.X. Jiang, Y.L. Feng, Z.Z. Chen, L.S. Zhang, X.Z. Jiang, Q.Q. Jiao, J.Z. Li et al. (2016). Study on light extraction from GaN-based green light-emitting diodes using anodic aluminum oxide pattern and nanoimprint lithography. *Scientific Reports*. 6:10.

106. R.H. Horng, W.K. Wang, S.Y. Huang, D.S. Wuu, (2006). Effect of resonant cavity in wafer-bonded green InGaN LED with dielectric and silver mirrors. *IEEE Photonics Technology Letters*. 18(1–4):457–459.

107. Solid-State Lighting R&D Plan, 2016, U.S. Department of Energy. https://energy.gov/eere/ssl/downloads/solid-state-lighting-2016-rd-plan.

108. Y.J. Hong, C.H. Lee, A. Yoon, M. Kim, H.K. Seong, H.J. Chung, C. Sone, Y.J. Park, G.C. Yi, (2011). Inorganic optoelectronics: Visible-color-tunable light-emitting diodes (Adv. Mater. 29/2011). *Advanced Materials*. 23(29):3224–3224.

109. M. Ebaid, J.-H. Kang, S.-H. Lim, J.-S. Ha, J.K. Lee, Y.-H. Cho, S.-W. Ryu, (2015). Enhanced solar hydrogen generation of high density, high aspect ratio, coaxial InGaN/GaN multi-quantum well nanowires. *Nano Energy*. 12:215–223.

110. X. Dai, A. Messanvi, H. Zhang, C. Durand, J.l. Eymery, C. Bougerol, F.H. Julien, M. Tchernycheva, (2015). Flexible light-emitting diodes based on vertical nitride nanowires. *Nano Letters*. 15(10):6958–6964.

111. N. Guan, X. Dai, A. Messanvi, H. Zhang, J. Yan, E. Gautier, C. Bougerol, F.H. Julien, C. Durand, J. Eymery, (2016). Flexible white light-emitting diodes based on nitride nanowires and nanophosphors. *ACS Photonics*. 3(4):597–603.

16

Ultraviolet Light-Emitting Diodes: Challenges and Countermeasures

Jun Hyuk Park,
Jong Won Lee,
Dong Yeong Kim,
and Jong Kyu Kim

16.1 Introduction

The ultraviolet (UV) light is an electromagnetic wave with a wavelength from 400 to 100 nm. UV spectral region is divided into three sub-regions, based on a convention established during the Second International Congress on Light in 1932. UV light with wavelengths from 315 to 400 nm is classified as UV-A which can penetrate the atmosphere without significant absorption. It damages the deeper layers of the skin or cause cataracts. UV-B ranges from 280 to 315 nm, which is partially absorbed in the atmosphere, and may cause skin cancer. UV light with wavelengths shorter than 280 nm is referred to as UV-C, or deep UV (DUV) which does not reach the earth's surface due to absorption by ozone and oxygen molecules in the stratosphere. UV-C light has the special ability to destroy the DNA and RNA polymerase of microbes including bacteria, viruses, and cancer cells and to treat dermatological diseases with minimum impact on the human body (Gaska et al. 2004; Schubert 2006).

There are a variety of applications for UV-A, UV-B, and UV-C wavelength ranges, respectively. UV-A light can be used for UV curing, photo-lithography, sensing, tanning, and banknote identification. UV-B light is widely employed for medical treatments. UV-C light is very effective to make impossible the reproduction of the microbes including water- and air-borne bacteria, viruses, and other pathogens and render them harmless by damaging the cross-linked DNA. For this reason, UV-C light has been used in air/water/food purification and sterilization to protect humans from serious health threats.

For over 100 years, mercury-vapor (Hg-vapor) lamps have been the only man-made UV light source, especially for UV-C spectral region. However, Hg-vapor lamps have many problems such as bulky and fragile nature, short lifetimes, and time delay getting the lamp to their full power, serious environmental burden arising from Hg usage and disposal, making Hg-vapor lamps simply unsuitable for portable point-of-use applications. These issues are arising strong demands from governments, industry, and consumers to replace the Hg-vapor lamp with an alternative UV light source.

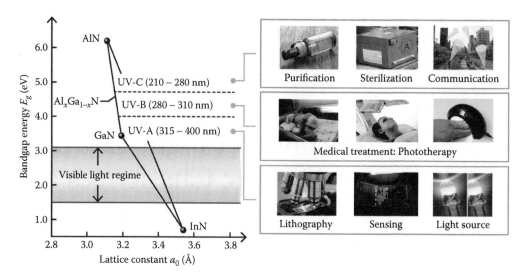

FIGURE 16.1 Band gap energy versus lattice parameter of InN, GaN, AlN and their alloy AlGaInN material system and UV light applications. AlGaInN system can cover wide UV spectra from UV-A to UV-C.

Semiconductor-based light-emitting diodes (LEDs) have lots of economic and technical advantages such as high efficiency, small form factor, low voltage operation, and compatibility with modern microelectronic technologies. Especially, AlGaInN-based LEDs can emit in the short-wavelength region of visible spectrum including green, blue, violet (GaInN) and even in the UV range (AlGaN). Compared to Hg-vapor lamps, AlGaN UV LEDs are compact, fast (instant startup), efficient (low electric power consumption), and environmentally friendly. In addition, since the band gap energy of AlGaN varies from 3.4 eV (GaN) to 6.2 eV (AlN) depending on Al composition, AlGaN UV LEDs can cover a wide range of UV spectral region from UV-A to UV-C, and are thus, emerging UV light sources for a variety of already-existing and future applications, as described in Figure 16.1. The promise of AlGaN UV LEDs can go far beyond just replacing the Hg-vapor lamps in existing applications and include the creation of new market opportunities.

16.2 Challenges in AlGaN UV LEDs

Since AlGaN is a very desirable material system for UV light emitters, vigorous researches have been performed for developing highly-efficient UV LEDs. However, the external quantum efficiency (EQE) for $Al_xGa_{1-x}N$ UV LEDs decreases rapidly as the Al molar fraction x increases, that is, as the wavelength decreases, as shown in Figure 16.2 (Kneissl et al. 2016). Although intensive research on UV LEDs has recently led to remarkable improvements, typical EQE values for UV LEDs especially in the UV-C spectral region are less than 10% and decrease to 10^{-6}% for 210 nm AlN LEDs (Taniyasu et al. 2006).

The reasons for the rapid decrease in EQE of AlGaN UV LEDs with increasing Al molar fraction are related to intrinsic material properties of AlGaN. Figure 16.3a shows a schematic structure of an AlGaN UV LEDs grown on a sapphire substrate consisting of a Si-doped n-type AlGaN, an $Al_xGa_{1-x}N/Al_yGa_{1-y}N$ multiple quantum well (MQW) active region, an AlGaN-based electron blocking layer (EBL), a Mg-doped p-type AlGaN, and a Mg-doped p-type GaN. At first, increasing the Al composition generally leads to a poor crystal quality of AlGaN epitaxial layers, causing poor internal quantum efficiency (IQE). Second, both p-type and n-type dopings in AlGaN becomes difficult due to increase of the ionization energies of acceptor (Mg) and donor (Si) dopants with increasing Al composition, respectively, which causes multiple additional problems; (1) poor electrical efficiency (EE) due to high contact resistance of metal electrodes formed on highly resistive n- and p-type AlGaN; (2) high resistivity of the n-type AlGaN layer causing nonuniform current injection into the active region, referred to as "current

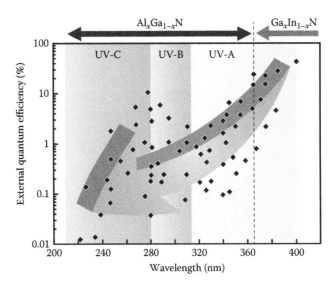

FIGURE 16.2 Reported EQE values of AlGaN UV LEDs as a function of the emission wavelength. EQE of AlGaN UV LEDs decreases rapidly as the emission wavelength decreases. (From Kneissl M., A brief review of III-nitride UV emitters: Technologies and their applications, in *III-Nitride Ultraviolet Emitters Technology and Applications*, Kneissl, M. and Rass, J., eds, Springer International Publishing, Cham, Germany, pp. 1–25, 2016. With permission.)

crowding" (Guo and Schubert 2001); and (3) poor injection efficiency (IE) of holes defined as the fraction of the LED's hole current that is injected into the active region. Although a p-type GaN layer is typically used as a contact and a hole supplying layer to overcome the problems (1) and (3), it causes poor light extraction efficiency (LEE) due to a strong absorption of UV photons. Third, the LEE of AlGaN UV LEDs grown on *c*-plane sapphire substrates is further limited by strong transverse-magnetic (TM)-polarized anisotropic emission due to the unique valence band structure of AlGaN with high Al molar fraction (> 25%). Figure 16.3b describes a power flow chart—consisting of EE, IQE × IE, and LEE which determine the wall-plug efficiency (WPE) defined as the ratio of the optical power out of the LED and the electrical power provided to the LED— of a 280 nm AlGaN DUV LED and a 440 nm GaInN blue LED. Over 70% WPE has been achieved for state-of-the-art GaInN blue LEDs; however, the WPE of typical 280 nm DUV LEDs is still less than 10% due to the material issues of AlGaN causing poor EE, IQE, IE, and LEE. These issues will be described in detail in the following subsections.

16.2.1 Crystalline Quality of AlGaN Epitaxial Layers

Epitaxial growth of a high-quality material is the most important issue for developing efficient optoelectronic devices because defects such as cracks, dislocations, and unintentionally incorporated impurities in materials can act as an energy-loss path by generating defect states inside the band gap or nonradiative recombination centers. High-quality AlGaN growth is particularly difficult due to the large lattice mismatch and thermal expansion mismatch between the AlGaN epitaxial layer and the sapphire substrate, resulting in elastic stress in AlGaN epitaxial layer. The tensile stress in AlGaN epitaxial layer on the sapphire substrate is accumulated as the epitaxial layer thickness increases, and cracks nucleated at the epitaxial layer surface for relaxation (Romanov et al. 2006). Besides crack formation, threading dislocation is another stress relaxation path. Threading dislocation density for AlGaN epitaxial layers on sapphire substrates is typically in the order of 10^{10} cm^{-2} which is much higher than that of GaInN epitaxial layers of 10^8 cm^{-2}. This is caused by the accumulated stress as well as the low surface of Al adatoms, which is responsible for three-dimensional island growth of AlGaN. (Karpov et al. 2002; Imura et al.

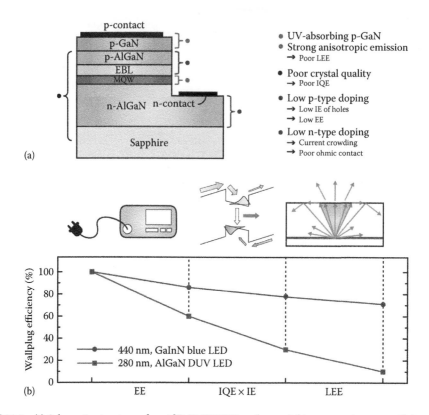

FIGURE 16.3 (a) Schematic structure of an AlGaN UV LED and material issues causing poor efficiency in each layer. (b) Power flow chart of a 440 nm GaInN blue LED and a 280 nm AlGaN DUV LED. The 280 nm AlGaN DUV LED shows much lower WPE compared to the 440 nm GaInN blue LED, which is attributed to the material issues in each layer, as listed in (a).

2007; Khan et al. 2008). Threading dislocations are known to act as nonradiative recombination centers, thus degrading IQE severely. Furthermore, oxygen impurities in AlGaN layer make impurity-related states inside the band gap, causing parasitic emission or phonon generation, which degrades IQE as well (Liao et al. 2011). Such defects including cracks, threading dislocations, and impurities ruin not only the radiative efficiency but also the lifetime, reliability, and breakdown voltage. As a consequence, the growth of high-quality AlGaN epitaxial layer with reduced density of defects is essential to improve the IQE and overall performance of AlGaN UV LEDs.

16.2.2 n-Type and p-Type Doping in AlGaN

Since LED is an optoelectronic device based on p–n junction, which converts electrical power to optical power, both electrical and optical properties are important factors determining the overall performance. Generally, highly doped n- and p-type semiconductor layers are required for desirable electrical and optical properties. N-type conductivity of AlGaN is achieved by incorporating Si atoms that substitute Al or Ga atoms in the lattice, providing loosely bound electrons. However, it is difficult to obtain high n-type conductivity in high-Al-content AlGaN due to large donor ionization energies and self-compensation effect. The ionization energy of the Si donors increases with increasing Al content, ranging from 15 meV for GaN to 62 meV for AlN, which is larger than the thermal energy at room temperature, leading to a highly resistive high-Al-content AlGaN (Neuschl et al. 2013; Park et al. 2016). Insufficient n-type doping in AlGaN exerts a harmful influence on LED performance including resistive ohmic contacts leading to a high operation voltage and poor EE. In addition, resistive n-type AlGaN layer intensifies the localization of the current path, so called "current crowding" effect. Current crowding induces local device heating and degrades the electrical and optical characteristics (Shatalov et al. 2002).

P-type doping in AlGaN has been probably the biggest challenge in the realization of efficient AlGaN DUV LEDs. Although the synthesis of GaN was realized as early as 1932 (Johnson et al. 1932), it was not until 1989 that Amano and Akasaki first achieved p-type doping with Mg for GaN (Amano et al. 1989). However, the Mg acceptor has a very high activation energy of ~170 meV, which is much larger than the thermal energy at room temperature. Moreover, hydrogen atoms present in the deposition system passivate the Mg dopant, which makes high-temperature thermal annealing process necessary for dopants activation (Nakamura et al. 1991). Even though p-type GaN has been obtained, it is still rather difficult to get p-type conductivity in high-Al-content because the Mg dopant activation energy increases with increasing Al composition, up to 630 meV for AlN (Taniyasu et al. 2006). The lack of p-type conductivity in high-Al-content AlGaN causes detrimental problems in developing UV LEDs including (1) extremely poor ohmic contacts for p-type AlGaN, thereby causing a high operating voltage and (2) very poor injection of holes into the active region. In order to mitigate such detrimental problems, p-type GaN layer is typically grown on top of p-type AlGaN in DUV LEDs, as shown in Figure 16.3a, as a contact and hole supplying layer, however, it causes poor LEE due to a strong absorption of UV photons.

On the other hand, the asymmetry in carrier transport related to both concentration and mobility of electrons and holes also becomes severe when Al composition in AlGaN increases. The activation energy of the Mg acceptor is much greater than that of Si donor, which causes a large disparity between electron and hole concentrations. Furthermore, the mobility of holes (~10 cm^2/Vs) is much lower than that of electrons (~400 cm^2/Vs) due to the large effective mass of holes (Bellotti and Bertazzi 2007). These characteristics, taken together, lead to a strong asymmetry in electron and hole transport characteristics.

16.2.3 Extraction of UV Photons

Not all the photons generated in the active region are emitted into the free space. Some photons never escape from the semiconductor due to the light trapping by total internal reflection (TIR) and absorption by metal contacts, epitaxial layers, and substrates. Therefore, extracting photons out of the active region is as important as the generation of photons inside the active region. Various light-extraction-enhancing techniques—such as surface roughening, substrate patterning, incorporating anti-reflective coatings, photonic crystals, and surface plasmonics, removal of absorbing substrates, and so on.—have been developed to show high LEE of GaInN-based visible LEDs over 80% (Krames et al. 2007). However, the LEE of AlGaN UV LEDs, especially high-Al-content AlGaN DUV LEDs, is still lower than 25% despite employing similar LEE-enhancing techniques for GaInN visible LEDs (Shur and Gaska 2010). Such a low LEE of AlGaN UV LEDs mainly originates from two reasons; usage of p-type GaN layer on top of p-type AlGaN and strong anisotropic light emission from the AlGaN active region.

As discussed in the previous section, the lack of p-type conductivity in high-Al-content AlGaN causes poor ohmic contacts on p-type AlGaN and low injection of holes into the active region. In order to alleviate such severe problems, typical DUV LEDs have a p-GaN layer on top of the p-AlGaN as a metal contact formation and hole supplying layer. However, the p-type GaN top layer absorbs almost half of the generated UV light from the active layer. Therefore, severe photon loss inevitably occurs regardless of using conventional LEE-enhancing techniques.

On the other hand, light emission from the high Al-content AlGaN material has strong anisotropic nature due to the unique valence band structure of the AlGaN material (Khan et al. 2005; Shur and Gaska 2010; Kneissl et al. 2016). Emission property is determined by the symmetry of the conduction band and valence band involved in the optical transition (Coldren and Corzine 1995). The Bloch function for the conduction band at $k = 0$ has S-orbital-like isotropic symmetry, while the valence band has

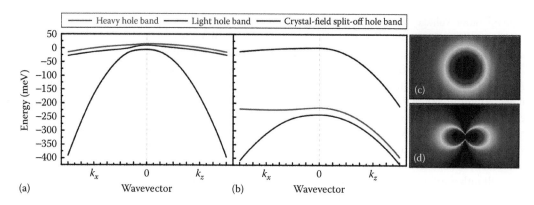

FIGURE 16.4 Valence band structure of *c*-plane (a) GaN and (b) AlN near Γ point calculated by 6 × 6 effective Hamiltonian matrix for the wurtzite semiconductors. Light emission patterns of *c*-plane (c) GaN and (d) AlN, which are isotropic and anisotropic, respectively.

P-orbital-like anisotropic characteristics composed of $|X\rangle$, $|Y\rangle$, and $|Z\rangle$. Generally, both heavy hole band and light hole band possess $|X\rangle$ and $|Y\rangle$ characteristics, and the crystal-field split-off hole band has $|Z\rangle$ characteristic along the *c*-axis (Braunstein and Kane 1962). Figure 16.4 shows the valence band structure of *c*-plane GaN, and AlN near the Γ point, calculated by a 6 × 6 effective-mass Hamiltonian matrix for the wurtzite structure (Chuang and Chang, 1996). The topmost valence band of GaN is a heavy hole band as shown in Figure 16.4a; therefore, light emission from GaN can be understood by the emission from two incoherent dipoles along the *x*- and *y*-axis, respectively, which results in an isotropic emission as shown in Figure 16.4c. Meanwhile, the topmost valence band of AlN is a crystal-field split-off hole band as shown in Figure 16.4b, therefore light emission from AlN can be approximated as the emission from a dipole along the *z*-axis, which results in a strong anisotropic emission as shown in Figure 16.4d (Mathews and Venkatesan 1976; Kim et al. 2015). Consequently, high-Al-content AlGaN having similar valence band structure with AlN emits light propagating mainly along the in-plane direction of AlGaN grown on the *c*-plane sapphire substrate, which can be easily lost by light-trapping and absorption in the LED die. Conventional LEE-enhancing techniques favor extracting isotropic light emission, and are thus very effective for GaInN-based visible LEDs, but turn out to be much less effective for AlGaN DUV LEDs, which calls for a totally new approach overcoming the fundamental LEE limitation in DUV LEDs.

16.3 Countermeasures of the Challenges in AlGaN UV LEDs

In the previous section, the challenges toward highly efficient AlGaN-based DUV LEDs are discussed in terms of the crystal quality, n- and p-type doping, and LEE of AlGaN. Despite the challenges originating from the intrinsic material properties of AlGaN, impressive research efforts have advanced the EQE of AlGaN DUV LEDs from less than 0.1% to about 1 ~ 10% (depending on the emission wavelength) over the past 10 years, which will be discussed in this section.

16.3.1 Growth of High Quality AlGaN Epitaxial Layers

Major challenges in realizing UV LEDs with high IQE include the growth of crack-free and low threading dislocation density AlGaN epitaxial layers. Because of the large lattice mismatch and differences in the thermal expansion coefficient, multiple layers of AlGaN grown on the heterosubstrate, typically sapphire substrate, experience a strain that causes a network of cracks. Many different approaches have been made to adjust the strain in heteroepilayers to avoid crack generation. Insertion of AlGaN/AlN or AlGaN/AlGaN superlattice structures is one of the most useful approaches for strain management.

Superlattice buffers can act as a defect filter to compensate the strain, reduce the dislocation density, and increase the critical thickness. Another widely used method to reduce strain is the formation of low-temperature (LT) AlN or AlGaN nucleation layer (Amano et al. 1986). It is found that the LT interlayer is effective in redefining the in-plane lattice parameter through strain relaxation. The interlayer serves to mediate the elastic tensile mismatch between the adjacent layers and extends the cracking critical thickness (Amano et al. 1999; Kamiyama et al. 2001).

However, it is also reported that the LT nucleation layer is less effective for the growth of high-quality AlGaN than that for GaN (Okada et al. 2007). Unlike trimethylgallium (TMGa) used in GaN growth, trimethylaluminum (TMAl) used in AlGaN growth has a much stronger pre-reaction with ammonia, which affects the quality of AlGaN films. Furthermore, Al atoms do not have enough time to move to energetically favorable sites due to their low surface mobility, resulting in growth of three-dimensional islands instead of smooth two-dimensional AlGaN. In order to overcome these problems, pulsed atomic layer epitaxy or migration-enhanced metal organic chemical vapor deposition (MEMOCVD) has been developed. The growth is initialized by depositing a thin LT-AlN nucleation layer by the conventional low-pressure MOCVD. Then the substrates were heated to grow a high-temperature (HT) AlN film. During the growth of the HT AlN layer, TMAl and ammonia were alternatively supplied into the reactor chamber, with each pulse being several seconds long, as schematically shown in Figure 16.5a. By using MEMOCVD method, the gas phase pre-reaction between the group III (Al) and group V (N) species were greatly suppressed. On the other hand, since the group III and group V atoms are supplied to the reactor chamber at different times, the surface mobility of the adatom is drastically enhanced, which enables them to find energetically favorable sites (Khan et al. 1992; Zhang et al. 2001, 2002).

Besides the MEMOCVD method, the epitaxial lateral overgrowth (ELO) techniques have also been widely accepted to high-quality heteroepitaxial layer growth with low threading dislocation density (Knauer et al. 2013; Nakamura et al. 1998). The principle of ELO technique is that part of the first-step grown AlN with high threading dislocation density is covered with a dielectric mask followed by second growth step. At the beginning of the second growth step, deposition only occurs on the opening areas, not on the dielectric mask. The threading dislocations are prevented from propagating into the second-step grown layer by the dielectric mask, whereas AlN grown above the opening (coherent growth) keeps the same threading dislocation density as the template. Due to the lateral growth, the masked areas are covered by regrown AlN with low dislocation density. Recently, Dong et al. (2013) reported AlN's

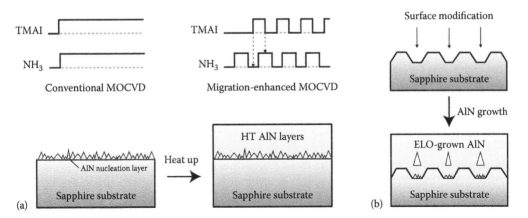

FIGURE 16.5 (a) Process of MEMOCVD High-temperature AlN layer is grown on the low-temperature AlN nucleation layer with alternative supply of TMAl and NH₃ for alleviating low surface mobility of Al atoms. (b) Schematic process of nano-scaled lateral overgrowth. (From Dong P. et al., *Appl. Phys. Lett.*, 102, 241113, 2013. With permission.)

nanoscaled lateral overgrowth which leads to low dislocation densities in AlN and the epitaxial layers above. As shown in Figure 16.5b, the exposed sapphire surface is partially modified by chemical or mechanical damage and the AlN layer growth can be started from the unmodified sapphire region. Each AlN island grows along the lateral direction over the modified areas, and finally, they make full coalescence, resulting in significantly reduced threading dislocation density.

High crystalline quality DUV LED structures with high IQE and reliability can be obtained by using bulk AlN substrate with low threading dislocation density, called homoepitaxy method. The advantage of homoepitaxy is that it reduces the dislocation density of the epilayer by minimizing the lattice and thermal mismatches. The use of bulk AlN substrate allows one to reduce the dislocation density in the epitaxial layers by more than four orders of magnitude down to 10^4–10^5 cm^{-2} (Hu et al. 2003). A conventional growth method of low-dislocation-density bulk AlN crystals is the physical vapor transport (PVT), which uses Al and N$_2$ vapor transport with recrystallization at the temperatures ranging from 1800°C to 2400°C (Dalmau and Sitar 2010). However, significant DUV light absorption caused by Al vacancies and substitutional impurities is observed in the PVT-grown AlN crystals. Recently, Kinoshita et al. (2013) have reported hydride vapor phase epitaxy (HVPE) which can grow thick AlN substrates on the PVT-grown AlN with low threading dislocation density and high transparency in DUV spectral region.

16.3.2 Enhancing Hole Injection Efficiency

A fundamental limitation in achieving high p-type doping in AlGaN causes various critical problems in the realization of highly efficient DUV LEDs including low EE, poor IE of holes causing severe asymmetry in carrier transport, and low LEE by UV-absorbing p-type GaN. Although there is a great need for a major improvement in the p-type doping of AlGaN which can solve the multiple problems, however, the challenge originates from an intrinsic material property (high Mg activation energy). One alternative way to achieve a highly conductive p-type AlGaN is to grow superlattices that have been shown to enhance the free hole concentration in AlGaN by a factor of 10 (Goepfert et al. 2000). The band modulation of superlattices and polarization fields enables a higher acceptor ionization ratio, and thus a higher hole concentration.

Another alternative way for p-type doping in AlGaN, referred as polarization-induced hole doping was proposed (Simon et al. 2010). AlGaInN material system possesses the spontaneous polarization due to non-centrosymmetry of their wurtzite crystal structure. When the crystal is strained, a piezoelectric polarization evolves in proportion to the strain and adds to the spontaneous polarization. Thus, the total polarization for a strained structure is determined by the sum of the piezoelectric polarization and the spontaneous polarization. Such piezoelectric effects can be utilized to manipulate the concentration, spatial distribution, and the mobility of free charge carriers (Schubert 2006). The polarization discontinuity at the abrupt junction, for example, AlGaN/GaN interface, induces bound charges at the interface and this leads to two-dimensional free carriers at the junction by high electric field and severe energy band bending. For the Al-composition-graded AlGaN layer, polarization discontinuity induces three-dimensional bound charges. In the case of AlGaN layer with gradually increasing Al composition along the [000$\bar{1}$] direction, negative bound polarization-induced charges are induced, generating a built-in electric field as shown in the Figure 16.6b and c. Then, free holes are field ionized in order to neutralize the negative polarization-induced charges, preventing energy band bending greater than the band gap of the semiconductor as shown in Figure 16.6d. As a consequence, high-density mobile three-dimensional hole gas is generated. Furthermore, smooth valence band structure of the AlGaN grading layer facilitates vertical hole transmission (Simon et al. 2010). Based on the polarization doping concept, p-type doped high Al content graded Al$_x$Ga$_{1-x}$N ($x = 0.7 \sim 1$) and dopant-free AlGaN-based p-n junction were demonstrated by Li et al. (2012, 2013). Furthermore, Carnevale et al. (2012) demonstrated UV electroluminescence from the polarization-induced p–n diodes with Al composition grading in AlGaN nanowires.

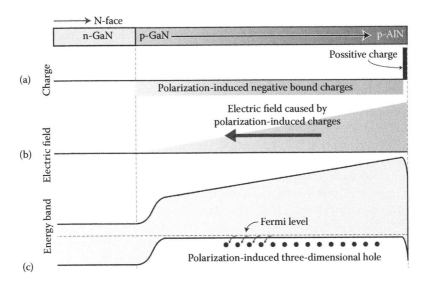

FIGURE 16.6 Schematic illustration showing polarization-induced p-type doping in polar heterostructures. (a) Distribution of the net unbalanced polarization charges. (b) The electric field caused by the unbalanced polarization charge. (c) Energy band structure affected by the electric field. Acceptors are field-ionized in order to neutralize the negative polarization charges, resulting in three-dimensional free hole gas. (From Simon J. et al., *Science* 327, 60–64, 2010. With permission.)

Since p-type doping in AlGaN is limited by intrinsic material property of AlGaN which is hard to overcome in nature, there are approaches to substitute p-type AlGaN layer with UV-transparent p-type semiconductors. Majety et al. (2012) proposed UV LEDs composed of n-type AlGaN and p-type hexagonal BN (*h*BN) instead of p-AlGaN. *h*BN is a member of group III-nitride semiconductor family which is composed of equal numbers of boron and nitrogen atoms with a layered honeycomb lattice similar to graphene. The advantage of *h*BN as a p-type semiconductor later of UV LEDs is a small Mg accepter activation energy of approximately 31 meV and UV transparency due to large band gap energy near 6eV (Dahal et al. 2011). They showed that free hole concentration of Mg-doped p-type *h*BN can reach ~10^{18} cm^{-3} level which is enough for various UV optoelectronics. Majety et al. (2012) demonstrated that the rectifying *I–V* characteristics from the p–n junction diode composed of n-type Al$_{0.62}$Ga$_{0.38}$N and p-type *h*BN. Although there is no report on electroluminescence from the p-type *h*BN included UV LEDs so far, further optimization of p-type *h*BN quality may contribute to the development of highly-efficient nitride-based UV LEDs.

On the other hand, there are several efforts to enhance the injection of holes by modifying the LED epitaxial structures rather than trying to increase the concentration of holes in the p-type AlGaN. There is a strong asymmetry in carrier transport characteristics due to the severe imbalance in concentration and mobility between electron and hole, which causes electron overshooting (or electron leakage). To reduce the electron leakage, AlGaN EBL is typically used between the MQW active region and the p-type region. Highly p-doped AlGaN EBL with high Al composition is desirable because it can effectively suppress the electron leakage while enhancing the hole injection by means of a high potential barrier for electron overflow but a low barrier for holes. However, the realization of such an EBL seems to be limited fundamentally because as Al composition in AlGaN increases (1) the ionization energy of p-type dopant Mg increases and (2) the increase of potential barrier for blocking is limited due to enhanced positive piezoelectric charges at the heterointerface. In order to overcome the challenges, Mg-doped AlGaN/ GaN and AlN/AlGaN superlattices have been introduced for visible and UV LEDs in order to increase the hole concentration and to efficiently block the electrons escaping from the active region (Kim et al. 2004; Hirayama et al. 2010). However, carrier transport along the perpendicular to the superlattice planes is much less than that parallel to the superlattice planes because of multiple potential barriers at the superlattice heterointerface, resulting in inefficient hole injection into the active region. Therefore, super-lattice-EBL with graded Al compositions, called graded superlattice-EBL (GSL-EBL), is adopted to solve anisotropic transport of holes in superlattices (Park et al. 2013). The LEDs with GSL-EBL shows enhanced overall efficiency and reduced droop compared to LEDs with superlattice-EBL and conventional bulk AlGaN EBL due to the enhanced hole injection efficiency.

As discussed in the polarization-induced hole doping part, the material polarization in AlGaInN sys-tem can be utilized to manipulate the concentration and spatial distribution of free charge carriers, and thus enhance the performance of UV LEDs. In order to increase the hole injection efficiency in AlGaN-based DUV LEDs, polarization-engineered tunnel junction (TJ) is adopted by Zhang et al. (2015). This method is particularly useful for AlGaN with initially low hole concentration because nonequilibrium hole injection through TJ can be attainable, resulting in the large numbers of free holes. The schematic design and energy band diagram of a TJ-incorporated UV LED are schematically shown in Figure 16.7. TJ-UV LED consists of a UV LED structure, a GaInN layer acting as TJ and a top n-AlGaN layer rather than p-AlGaN. At the heterointerfaces polarization induced charge density can be quite high (10^{13} cm^{-2}) which can create significantly high electric fields at the AlGaN/GaInN interface. This results in band bending across the GaInN layer over a small distance to align the band edges, thereby increasing the tunneling probability. This TJ layer is reverse biased when the forward bias is applied to the LED, so that electrons in the valence band of p-AlGaN directly tunnels into the empty states in the conduction band of the n-AlGaN, leaving behind holes in the p-type AlGaN as shown in Figure 16.7b. Therefore, the efficient TJ acts as a carrier conversion center enabling the injection of holes into the active region. The additional advantage of such an approach is that the top of the structure is n-type AlGaN, which has low spreading resistance, possibility in forming low-resistance ohmic contacts, and is transparent in nature, resulting in enhanced electrical properties and light extraction (Sarwar et al. 2015).

Oto et al. (2010) proposed quite a different approach to realize a highly efficient DUV emitter based on electron–hole generation by means of electron-beam pumping method. Electron-beam excitation approach uses highly accelerated electrons to generate electron–hole pairs inside the AlGaN/AlN quantum wells grown on the *c*-plane sapphire substrate. At an acceleration voltage of 8 kV, a maximum power efficiency defined as the optical power divided by the electrical power of 40% was reported for 240 nm emission.

16.3.3 Improving Extraction Efficiency

The LEE of AlGaN DUV LEDs is much lower than that of GaInN visible LEDs due to two reasons; (1) usage of UV absorbing p-type GaN layer on top of p-type AlGaN and (2) strong anisotropic light emission from the AlGaN active region. In order to tackle such problems, various efforts have been made.

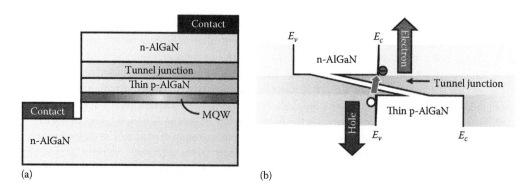

FIGURE 16.7   (a) Schematic epitaxial structure of UV LEDs with tunnel junction. (b) Schematic energy band diagram of the tunnel junction region that enables electron to tunnel from the p-AlGaN to n-AlGaN leaving holes in the p-type region. (From Zhang Y. et al., *Appl. Phys. Lett.*, 106, 141103, 2015. With permission.)

Typical AlGaN UV LEDs have a light-absorbing p-GaN layer thereby abandoning almost half of the DUV emission, which leads to adopt a bottom-emitting flip-chip configuration. In order to maximize the LEE of UV LEDs with flip-chip configuration, a combination of a highly reflective p-type electrode and a UV transparent top layer is required. A reflective p-type electrode can be obtained by using Al-contained electrode rather than conventional Ni/Au electrode widely used for ohmic contacts to p-type GaN because Al has high UV reflectance (over 90%) at DUV region. A highly UV-transparent contact layer can be obtained by eliminating the UV-absorptive p-GaN layer. Maeda and Hirayama (2013) demonstrated that AlGaN DUV LEDs emit at 287 nm with a highly reflective p-contact on the UV transparent layer. The p-GaN contact layer is replaced with high-Al-content p-AlGaN contact layer, and the Ni/Au electrode is replaced with Ni (1 nm)/Al electrode as well. The EQE was enhanced from 2% to 5.5% and 1.7 times of LEE enhancement is measured. However, the proposed LED structure may result in worse electrical properties than conventional structure due to much higher contact resistance on p-type AlGaN than that on p-type GaN.

The large difference in refractive index between air and semiconductor materials causes a low critical angle, and hence, poor light extraction out of the semiconductor. Photons incident out of the critical angle would never escape from the semiconductor with a flat and specular interface with air due to the light trapping by TIR and absorption by metal contacts, epitaxial layers, and substrates, as shown in Figure 16.8a. Breaking the waveguided mode by TIR is probably the most common way to improve the LEE. Therefore, surface roughening or texturing methods that are conventionally used in GaInN visible LEDs have been applied in AlGaN UV LEDs (Kim et al. 2009; Pernot et al. 2010). Roughened (or diffuse) surface can induce a strong light scattering; thus, a few scattering events are sufficient to out-couple the incident light, as schematically shown in Figure 16.8b. Fabricating a textured interface between the AlGaN LED epitaxial layer and sapphire substrate can also break the waves-guided mode, enhancing the LEE. Khizar et al. (2005) fabricated microlenses with a diameter of 12 μm on the sapphire substrate by thermal reflow of photoresist followed by plasma etching and grew AlGaN LED epitaxial layer on top of them. The output powers of the LEDs with and without microlens array were measured and compared with each other. The microlens array formation enhanced the UV LED output power by 55% at 20 mA due to reduced internal reflections of the light (Khizar et al. 2005).

Periodical arrays of micro- and nanostructures composed of III-N materials with different refractive indices can realize two-dimensional (2D) photonic crystals (PCs) that can eliminate in-plane light propagation by forming photonic band gaps (Oder et al. 2004). In other words, by utilizing 2D PCs, the guided-mode in LED epitaxial layer can be controlled. For example, a designed 2D PCs embedded

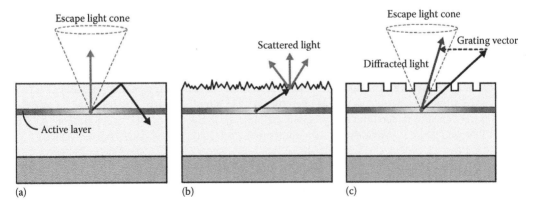

FIGURE 16.8 Schematic illustrations showing light extraction from LEDs with (a) flat surface, (b) roughened surface, and (c) photonic crystal structure. More photons can escape into the free space by employing surface roughening and photonic crystal through breaking the wave-guided mode and making the wavevector of the diffracted light (dark gray arrow) lie inside the escape cone, respectively.

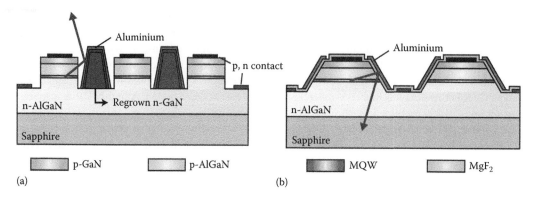

FIGURE 16.9 Schematic structures of (a) top-emitting SEE DUV LEDs that enables sidewall-heading photons to reflect toward top direction by Al-coated mirror and (b) bottom-emitting SEE DUV LEDs, which can extract sidewall-heading photons through the substrates by inclined mirror structure. (From Kim D.Y. et al., *Light: Sci. Appl.*, 4, e263, 2015; Lee J.W. et al., *Sci. Rep.*, 6, 22537, 2016. With permission.)

in LEDs can make laterally-guided photons propagate along the vertical direction for enhancing LEE. When 2D PCs are embedded, the guided modes become bloch modes, and then the tangential component of the wave vector is changed by the grating vector as shown in Figure 16.8c. Consequently, the wavevector of the diffracted light lies inside the escape cone and it can be extracted.

There are several papers that have reported LEE enhancement by PCs in visible LEDs but only a few reports on DUV LEDs because of rigorous fabrication of nanoscale patterning. Shakya et al. (2004) reported 2.5 times light output power enhanced 333 nm UV LEDs by incorporating triangular arrays of the PCs with an optimized lattice constant and hole size. Recently, Inoue et al. (2015) demonstrated a 196% of output power improvement in 265 nm DUV LEDs by utilizing hybrid structure of PCs and subwavelength nanostructures and 90 mW of output power in DC operation as well, which is the highest value ever reported.

Besides the approaches based on surface roughening or texturing, a new LEE-enhancing approach has been proposed to utilize the inherently strong TM polarized light emitted from the AlGaN active region. Figure 16.9a shows schematic side-emission-enhanced (SEE) structures including Al-coated selective-area-grown n-type GaN micro-reflectors, which can help to extract sidewall-heading photons toward the top direction. The SEE DUV LEDs show better electrical properties as well as much enhanced light output power with a strongly upward-directed emission due to an exposed sidewall of the active region and Al-coated microreflectors (Kim et al. 2015).

Lee et al. (2016) proposed improved bottom-emitting SEE DUV LED structure to avoid light absorption at the p-GaN layer as well as to utilize DUV light-emitting to substrate side. They fabricated DUV LED having multiple mesa stripes whose inclined sidewalls are covered by a MgF$_2$/Al omnidirectional mirror as shown in Figure 16.9b. They showed a remarkable improvement in light extraction through the sapphire substrate as well as operating voltage, simultaneously. Furthermore, an analytic model is developed to understand and precisely estimate the extraction of DUV photons from AlGaN DUV LEDs, and hence to provide promising routes to maximize the power conversion efficiency.

16.4 Summary

There is a rapidly-growing demand for efficient UV light sources replacing Hg-vapor lamps for various applications. AlGaN UV LEDs, which can cover a wide range of UV spectral region from UV-A to UV-C, become increasingly attractive due to their remarkable advantages over Hg-vapor lamps, and a variety applications such as purification of air and water, sterilization in food processing, lithography, lighting, defense, and medical applications. However, the efficiency of current AlGaN DUV LEDs decreases rapidly as the Al molar fraction in AlGaN increases, that is, as the emission wavelength decreases. The reasons for the rapid decrease in efficiency are related to material properties of high-Al-content AlGaN including poor crystalline quality of AlGaN heteroepitaxial layers, low n- and p-type doping efficiency, and strong TM-polarized light emission, which, taken together, result in very low EQE, typically less than 5% in the UV-C spectral region. Impressive research efforts to overcome such challenges originating from intrinsic material properties of AlGaN have been made, such as the growth of high crystalline quality AlGaN epitaxial layer, enhancing the concentration and the injection efficiency of holes, and improving light extraction, advancing the EQE of AlGaN DUV LEDs.

References

Amano H, Sawaki N, Akasaki I, Toyoda Y (1986) Metalorganic vapor phase epitaxial growth of a high quality GaN film using an AlN buffer layer. *Appl. Phys. Lett.* 48:353–355.

Amano H, Kito M, Hiramatsu K, Akasaki I (1989) P-type conduction in Mg-doped GaN treated with low-energy electron beam irradiation (LEEBI). *Jpn. J. Appl. Phys.* 28:L2112–L2114.

Amano H et al. (1999) Control of dislocations and stress in AlGaN on sapphire using a low temperature interlayer. *Phys. Stat. Sol. (b)* 216:683.

Bellotti E, Bertazzi F (2007) Transport parameters for electrons and holes. In: *Nitride Semiconductor Devices* (Piprek J, ed), pp. 69–93. Weinheim: Wiley-VCH.

Braunstein R, Kane EO (1962) The balance band structure of the III-V compounds. *J. Phys. Chem. Solids.* 24:1423–1431.

Carnevale SD et al. (2012) Polarization-induced pn diodes in wide-band-gap nanowires with ultraviolet electroluminescence. *Nano Lett.* 12:915–920.

Chuang SL, Chang CS (1996) k·p method for strained wurtzite semiconductors. *Phys. Rev. B* 54:2491–2504.

Coldren LA, Corzine SW (1995) *Diode Lasers and Photonic Integrated Circuits.* New York: John Wiley.

Dahal R et al. (2011) Epitaxially grown semiconducting hexagonal boron nitride as a deep ultraviolet photonic material. *Appl. Phys. Lett.* 98:211110.

Dalmau R, Sitar Z (2010) AlN bulk crystal growth by physical vapor transport. In: *Springer Handbook of Crystal Growth* (Dhanaraj G, Byrappa K, Prasad V, Dudley M, eds), pp. 821–843. Heidelberg, Germany: Springer.

Dong P et al. (2013) 282-nm AlGaN-based deep ultraviolet light-emitting diodes with improved performance on nano-patterned sapphire substrates. *Appl. Phys. Lett.* 102:241113.

Gaska R, Khan MA, Shur MS (2004) III-nitride based UV light emitting diodes. In: *UV Solid-State Light Emitters and Detectors* (Shur MS, Zukauskas A, eds), pp. 59–75. Utrecht, the Netherlands: Springer.

Goepfert ID, Schubert EF, Osinsky A, Norris PE, Faleev NN (2000) Experimental and theoretical study of acceptor activation and transport properties in p-type AlxGa1-xN/GaN superlattices. *J. Appl. Phys.* 88:2030–2038.

Guo X, Schubert EF (2001) Current crowding in GaN/InGaN light emitting diodes. *J. Appl. Phys.* 90:4191–4195.

Hirayama H, Tsukada Y, Maeda T, Kamata N (2010) Marked enhancement in the efficiency of deep-ultraviolet AlGaN light-emitting diodes by using a multiquantum-barrier electron blocking layer. *Appl. Phys. Exp.* 3:031002.

Morkoç H (2008) *Handbook of Nitride Semiconductors and Devices.* Weinheim, Germany: Wiley-VCH.

Hu X et al. (2003) AlGaN/GaN heterostructure field-effect transistors on single crystal bulk AlN. *Appl. Phys. Lett.* 82:1299–1301.

Imura M et al. (2007) Dislocations in AlN epilayers grown on sapphire substrate by high-temperature metal-organic vapor phase epitaxy. *Jpn. J. Appl. Phys.* 46:1458–1462.

Inoue S, Naoki T, Kinoshita T, Obata T, Yanagi H (2015) Light extraction enhancement of 265nm deep-ultraviolet light-emitting diodes with over 90 mW output power via an AlN hybrid nanostructure. *Appl. Phys. Lett.* 106:131104.

Johnson WC, Parson JB, Crew MC (1932) Nitrogen compounds of gallium III. *J. Phys. Chem.* 36:2651–2654.

Kamiyama S et al. (2001) Low-temperature-deposited AlGaN interlayer for improvement of AlGaN/GaN heterostructure. *J. Cryst. Growth* 223:83–91.

Karpov SY, Makarov YN (2002) Dislocation effect on light emission efficiency in gallium nitride. *Appl. Phys. Lett.* 81:4721–4723.

Khan A, Balakrishnan K, Katona T (2008) Ultraviolet light-emitting diodes based on group three nitrides. *Nat. Photonics* 2:77–84.

Khan MA et al. (1992) Low pressure metalorganic chemical vapor deposition of AlN over sapphire substrates. *Appl. Phys. Lett.* 61:2539–2541.

Khan MA, Shatalov M, Maruska HP, Wang HM, Kuokstis E (2005) III-nitride UV devices. *Jpn. J. Appl. Phys.* 44:7191–7206.

Khizar M, Fan ZY, Kim KH, Lin JY, Jiang HX (2005) Nitride deep-ultraviolet light-emitting diodes with microlens array. *Appl. Phys. Lett.* 86:173504.

Kim BJ et al. (2009) Enhancement of light extraction efficiency of ultraviolet light emitting diodes by patterning of SiO_2 nanosphere arrays. *Thin Solid Films* 517:2742–2744.

Kim DY et al. (2015) Overcoming the fundamental light-extraction efficiency limitations of deep ultraviolet light-emitting diodes by utilizing transverse-magnetic-dominant emission. *Light: Sci. Appl.* 4:e263.

Kim JK et al. (2004) p-type conductivity in bulk $Al_xGa_{1-x}N$ and $Al_xGa_{1-x}N/Al_yGa_{1-y}N$ superlattices with average Al mole fraction > 20%. *Appl. Phys. Lett.* 84:3310.

Kim MH et al. (2007) Origin of efficiency droop in GaN-based light-emitting diodes. *Appl. Phys. Lett.* 91:183507.

Kinoshita T et al. (2013) Performance and reliability of deep-ultraviolet light-emitting diodes fabricated on AlN substrates prepared by hydride vapor phase epitaxy. *Appl. Phys. Exp.* 6:092103.

Knauer A et al. (2013) AlGaN layer structures for deep UV emitters on laterally overgrown AlN/Sapphire templates. *Phys. Stat. Soli. A* 210:451–454.

Kneissl M (2016) A brief review of III-nitride UV emitters: Technologies and their applications. In: *III-Nitride Ultraviolet Emitters Technology and Applications* (Kneissl M, Rass J, eds), pp. 1–25. Cham: Springer International Publishing.

Krames MR et al. (2007) Status and future of high-power light-emitting diodes for solid-state lighting. *J. Display Technol.* 3:160–175.

Lee JW et al. (2016) An elegant route to overcome fundamentally-limited light extraction in AlGaN deep-ultraviolet light-emitting diodes: Preferential outcoupling of strong in-plane emission. *Sci. Rep.* 6:22537.

Liao Y, Thomidis C, Kao C, Moustakas TD (2011) AlGaN based deep ultraviolet light emitting diodes with high internal quantum efficiency grown by molecular beam epitaxy. *Appl. Phys. Lett.* 98:081110.

Li S et al. (2012) Polarization induced pn-junction without dopant in graded AlGaN coherently strained on GaN. *Appl. Phys. Lett.* 101:122103.

Li S et al. (2013) Polarization induced hole doping in graded $Al_xGa_{1-x}N$ ($x = 0.7 \sim 1$) layer grown by molecular beam epitaxy. *Appl. Phys. Lett.* 102:062108.

Maeda N, Hirayama H (2013) Realization of high-efficiency deep-UV LEDs using transparent p-AlGaN contact layer. *Phys. Stat. Soli. C* 10:1521–1524.

Majety S et al. (2012) Epitaxial growth and demonstration of hexagonal BN/AlGaN p–n junctions for deep ultraviolet photonics. *Appl. Phys. Lett.* 100:061121.

Mathews PM, Venkatesan KA (1976) *A Textbook of Quantum Mechanics*. New Delhi, India Mcgraw-Hill.

Nakamura S, Mukai T, Senoh M (1991) High-power GaN p–n junction blue-light-emitting diodes. *Jpn. J. Appl. Phys.* 30:L1998–L2001.

Nakamura S, Senoh M, Nagahama SI (1998) Present status of InGaN/GaN/AlGaN-based laser diodes. *J. Cryst. Growth* 189/190:820–825.

Neuschl B et al. (2013) Direct determination of the silicon donor ionization energy in homoepitaxial AlN from photoluminescence two-electron transitions. *Appl. Phys. Lett.* 103:122105.

Oder TN, Kim KH, Lin JY, Jiang HX (2004) III-nitride blue and ultraviolet photonic crystal light emitting diodes. *Appl. Phys. Lett.* 84:466–468.

Okada N et al. (2007) Growth of high-quality and crack free AlN layers on sapphire substrate by multi-growth mode modification. *J. Cryst. Growth* 298:349–353.

Oto T, Banal RG, Kataoka K, Funato M, Kawakami Y (2010) 100 mW deep-ultraviolet emission from aluminium-nitride-based quantum wells pumped by an electron beam. *Nat. Photonics* 4:767–771.

Park JH et al. (2013) Enhanced overall efficiency of GaInN-based light-emitting diodes with reduced efficiency droop by Al-composition-graded AlGaN/GaN superlattice electron blocking layer. *Appl. Phys. Lett.* 103:061104.

Park JH et al. (2016) Variation of the external quantum efficiency with temperature and current density in red, blue, and deep ultraviolet light-emitting diodes. *J. Appl. Phys.* 119:023101.

Pernot C et al. (2010) Improved Efficiency of 255–280 nm AlGaN-based light-emitting diodes. *Appl. Phys. Exp.* 3:061004.

Romanov AE et al. (2006) Cracking of III-nitride layers with strain gradients. *Appl. Phys. Lett.* 89:161922.

Schubert EF (2006) *Light-Emitting Diodes*, 2nd edition. Cambridge, UK: Cambridge University Press.

Sarwar ATMG et al. (2015) Tunnel junction enhanced nanowire ultraviolet light emitting diodes. *Appl. Phys. Lett.* 107:101103.

Shakya J, Kim KH, Lin JY, Jiang HX (2004) Enhanced light extraction in III-nitride ultraviolet photonic crystal light-emitting diodes. *Appl. Phys. Lett.* 85:142–144.

Shatalov M et al. (2002) Lateral current crowding in deep UV light emitting diodes over sapphire substrates. *Jpn. J. Appl. Phys.* 41:5083–5087.

Shur MS, Gaska R (2010) Deep-ultraviolet light-emitting diodes. *IEEE Trans. Electron Dev.* 57:12–25.

Simon J, Protasenko V, Lian C, Xing H, Jena D (2010) Polarization-induced hole doping in wide-band-gap uniaxial semiconductor heterostructures. *Science* 327:60–64.

Taniyasu Y, Kasu M, Makimoto T (2006) An aluminium nitride light-emitting diode with a wavelength of 210 nanometres. *Nature* 441:325–328.

Zhang J et al. (2001) Pulsed atomic layer epitaxy of quaternary AlInGaN layers. *Appl. Phys. Lett.* 79:925–927.

Zhang JP et al. (2002) Pulsed atomic-layer epitaxy of ultrahigh-quality AlxGal-xN structures for deep ultraviolet emissions below 230 nm. *Appl. Phys. Lett.* 81:4392–4394.

Zhang Y et al. (2015) Interband tunneling for hole injection in III-nitride ultraviolet emitters. *Appl. Phys. Lett.* 106:141103.

InGaN/GaN Quantum Dot Visible Lasers

Thomas Frost,
Guan-Lin Su,
John Dallesasse, and
Pallab Bhattacharya

17.1 Introduction

The demonstration of quantum dots (QDs) with emission covering the entire visible spectrum is important for many applications including solid-state lighting [1,2]. Current white light-emitting diodes typically incorporate blue-emitting InGaN/GaN quantum wells and rely on a phosphor to convert some of the blue light to yellow or red. Tuning of the white emission requires the development of new phosphors with the desired emission characteristics. Alternatively, electrically injected devices incorporating blue and red-emitting QDs can be used to directly generate red light, which is tunable by simply changing the indium composition in the dots. Such light-emitting diodes (LEDs) and lasers are also important for display and mobile projector applications, which require blue-, green-, and red-emitting devices [1,2]. InGaN-based QDs may be used for all these wavelengths, negating the need for the use of multiple material systems in these applications. Blue- and green-emitting lasers can be realized with InGaN/GaN-based single or multiple quantum wells [3–10], but red-emitting lasers are typically fabricated using other material systems [11,12] as red-emitting InGaN/GaN quantum well (QW) lasers have yet to be reported.

InGaN/GaN self-assembled QDs (SAQD) have been theoretically predicted and experimentally demonstrated to have superior optical properties compared with InGaN/GaN quantum wells due to their stronger electron-hole (e-h) overlap [19–25]. This results in shorter radiative recombination lifetimes and allows for longer wavelength emission with higher indium composition in the InGaN layer [4–11,19–25]. These QDs form via strain relaxation, and therefore, have reduced piezoelectric polarization. Additionally,

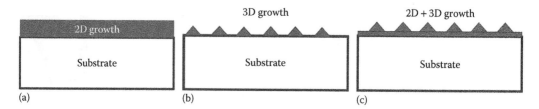

FIGURE 17.1 Schematic of different growth modes: (a) Frank–van der Merwe, (b) Volmer–Weber, and (c) Stranski–Krastanow.

the physical confinement of carriers in space can prevent carrier leakage and escape to dislocations present in the InGaN layer.

InGaN/GaN QDs have been experimentally demonstrated using the Stranski–Krastanow (S–K) growth mode which has been used for the experiments in this chapter [13–20]. Additionally, they have been demonstrated by low-temperature passivation [21], the use of anti-surfactants [22], and post-growth fabrication including quantum size controlled photoelectrochemical (QSC-PEC) etching and site controlled etching [23,24]. The growth of InGaN/GaN QDs has been demonstrated in plasma-assisted molecular beam epitaxy (MBE) [14–20,25], ammonia-based MBE [26], and MOCVD [27].

S–K island growth of self-assembled InGaN/GaN QDs can be achieved when a relatively large lattice mismatch exists between the InGaN layer and the underlying layer (usually GaN). Two other growth modes exist including Frank–van der Merwe (F–M) and Volmer–Weber (V–W) [28,29] depending on the lattice mismatch and the interaction strength of the impinging adatoms on the surface. The three growth modes are shown schematically in Figure 17.1. The V–W growth mode results in the growth of 3-dimensional adatom clusters with the surface and usually occurs when a large lattice mismatch is present (larger than required for S–K growth). This growth mode has recently been shown as the dominant growth mode of high indium content (red-emitting) InGaN "disks" in GaN nanowires [30]. F–M growth results in a layer by layer (planar) growth and can be realized during the growth of lattice matched layers (GaN on GaN or In$_{0.18}$Al$_{0.82}$N on GaN). S–K growth exists between these two extremes and is characterized by an initial 2D growth (typically referred to as a wetting layer) followed by 3D island growth.

17.2 Electronic States in QDs

For facilitating the development of quantum dot-based (QD-based) III-N visible lasers, it is crucial to have a comprehensive model that is able to describe the measured data and extract physical parameters which are not readily available in the literature or directly measured. However, modeling III-N SAQD structures is not trivial. One of the most sophisticated parts in the calculation of electronic states in QD structures is the valence band coupling, which is generally described by the **k·p** method [31–33]. Although realizable, and applying the **k·p** method to three-dimensional structures greatly increases the computational complexity, whether the resultant incremental accuracy is worthwhile needs to be verified. Moreover, it is the geometry of the QD that weights over the number of bands used in the **k·p** method [33]. As a result, instead of using the **k·p** method to calculate the electronic states in the QD system, a single-band effective mass Hamiltonian with corrected hole masses is used in this model.

The hole effective masses in the strained, bulk In$_{0.4}$Ga$_{0.6}$N QD material are corrected using a separate 6-band **k·p** model [34]. Since the strain components in the SAQD structure are now functions of position, spatially averaged values

$$\langle \varepsilon_{ij} \rangle = \frac{\int_{QD} \varepsilon_{ij}(\mathbf{r}) d^3\mathbf{r}}{QD \text{ volume}} \tag{17.1}$$

are used in the formulation of the Hamiltonian.

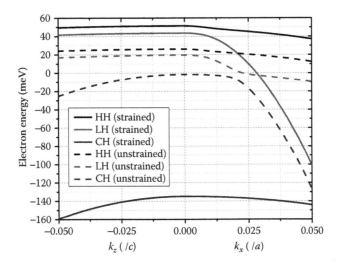

FIGURE 17.2 Valence band structures of unstrained and strained bulk In$_{0.4}$Ga$_{0.6}$N. (From Su, G.L. et al., *Opt. Exp.*, 23, 12850–12865, 2015. With permission of Optical Society of America.)

Figure 17.2 shows the valence band structures of unstrained and strained bulk In$_{0.4}$Ga$_{0.6}$N alloy. While the compressive strain only shifts the position of the heavy hole (HH) band, it drastically changes the energy dispersion of the light hole (LH) and the crystal-field split-off hole (CH) bands. Not only does the large energy separation between the LH and CH band at zone center make the CH band almost irrelevant in the calculation of QD states, but also the LH effective mass in the z direction is almost doubled as listed in Table 17.1. Since the CH band is almost 180 meV away from the LH band at zone center, only the HH and LH bands are considered in the calculation of electronic states and the optical gain spectra.

Using the corrected effective masses, the single-band Hamiltonian can be written as

$$-\frac{\hbar^2}{2}\nabla\cdot(\overrightarrow{\mathbf{m}_i}:\nabla\psi_i)+V(\mathbf{r})\psi_i=E_i\psi_i \tag{17.2}$$

where the subscript *i* refers to an electron, HH, and LH, and the tensor $\overrightarrow{\mathbf{m}_i}$ addresses the anisotropy of the effective masses in III-N materials. The potential operator $V(\mathbf{r})$ is composed of the band offset $V_0(\mathbf{r})$ at the InGaN/GaN heterointerface, the deformation potential $V_{\text{strain}}(\mathbf{r})$ due to strain, and the electrostatic potential $V_{\text{pol}}(\mathbf{r})$ due to the polarization charge distribution.

The band offset at the InGaN/GaN heterointerface is calculated using Anderson's rule and Varshni's parameters [35], assuming the valence band offset (VBO) is 900 meV at room temperature [36].

$$\Delta E_v(T)=\frac{900\text{ meV}}{\Delta E_g(298\text{ K})}\Delta E_g(T)$$

$$\Delta E_c(T)=\Delta E_g(T)-\Delta E_v(T) \tag{17.3}$$

$$E_g(T)=E_g(0\text{ K})-\frac{\alpha T^2}{\beta+T}$$

TABLE 17.1 HH and LH Effective Masses in Bulk In$_{0.4}$Ga$_{0.6}$N Before and After Taking Strain into Consideration

	$m^*_{hh,t}/m_0$	$m^*_{lh,t}/m_0$	$m^*_{hh,z}/m_0$	$m^*_{lh,z}/m_0$
Unstrained	0.213	0.228	1.742	0.884
Strained	0.209	0.217	1.742	1.720

Note: Subscripts *t* and *z* refer to the in- and out-of-*c*-plane directions, and m_0 is the Free Electron Mass

Spatially averaged strain components are again used in the calculation of deformation potential. By taking the advantage of the fact that the spatially averaged shear strain components $\left(\langle \varepsilon_{ij} \rangle,\ i \neq j \right)$ are much smaller than the averaged biaxial strain components $\left(\langle \varepsilon_{ii} \rangle\right)$, the deformation potential can be simplified as

$$V_{\text{strain}}(\mathbf{r}) = \sum_{i=x,y,z} V_i \varepsilon_{ii}(\mathbf{r}) \tag{17.4}$$

where the coefficients V_i's are appropriate deformation potential parameters for either conduction or valence band [37].

The calculation of the polarization-induced electrostatic potential $V_{\text{pol}}(\mathbf{r})$ is relatively straightforward. By using the anisotropic permittivities $\ddot{\varepsilon}$ in III-N materials [38], Poisson's equation gives

$$-\nabla \cdot \left(\ddot{\varepsilon} : \nabla V_{\text{pol}} \right) = \rho_{\text{pol}}(\mathbf{r}) \tag{17.5}$$

Figure 17.3a and b show the wavefunction envelopes of the top three electrons (C) and HH states at different viewing angles, respectively. The hexagonal symmetry in the transverse plane makes the second and third states in either conduction or valence band almost degenerate in energy; thus, the differential gain at C2-HH2 and C2-LH2 transition wavelengths are expected to double. In addition, the polarization field distorts the band edges and makes the electrons stay at the top of the SAQD and the holes stay at the bottom of the wetting layer. The wide separation between the electrons and holes renders poor wavefunction overlaps and hinders bimolecular recombination processes, such as spontaneous and stimulated recombination.

When the carriers enter the system, either through optical or electrical pumping, the spatially separated electron-hole pairs (EHPs) form another electrical dipole in the opposite direction of the

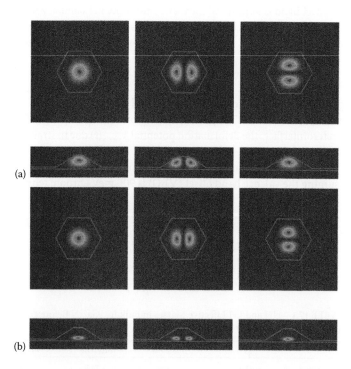

FIGURE 17.3 Wavefunction envelopes of the (a) top three electron and (b) top three HH states, at different viewing angles. The QCSE is visualized by the spatial separation between the electron and HH wavefunction envelopes in the *z* direction. (From Su, G.-L. et al., *Opt. Exp.*, 22, 22716–22729, 2014. With permission of Optical Society of America.)

polarization field. As the pumping level increases, this dipole starts to weaken the QCSE and restores the slanted band edges. The screening effect can be taken into account by adding the screening potential $V_{sc}(\mathbf{r})$. To calculate the screening potential, it is crucial to determine the position of the quasi-Fermi levels. The conduction band quasi-Fermi level is related to the injection via

$$n_{2d} = 2N_{2D} \sum_{\substack{m \in \text{bound} \\ \text{states}}} \int_{-\infty}^{\infty} G_c(E - E_m) f_c(E, T) dE + L_{eff} \int_{E_{cm}}^{\infty} \rho_{3D}(E - E_{cm}) f_c(E, T) dE \qquad (17.6)$$

where N_{2D} is the aerial density of the SAQD, $G(E)$ the inhomogeneous broadening caused by dot size fluctuation, $f_c(E, T)$ the Fermi–Dirac distribution as a function of temperature, $\rho_{3D}(E)$ the three-dimensional density of states (3D-DOS) of the GaN matrix material, and E_{cm} is the conduction minimum of the matrix. It should be noted that the wetting layer states are automatically included in Equation 17.6, as they are also solutions to Equation 17.2 because there is no clear physical boundary between the SAQD structure and the underlying wetting layer in our model. The 3D-DOS of the matrix material is to ensure that the quasi-Fermi levels move continuously with the injection level, as the bound states are discrete in nature. The L_{eff} in Equation 17.6 is the effective height of the QD structure, which is defined as

$$L_{eff} = N_{2D} \times (\text{QD Volume}) \qquad (17.7)$$

in order to accurately convert the aerial density to the volume density. The valence band quasi-Fermi level can be found using a similar expression to Equation 2.6.

The screening charge distribution can be expressed as a function of quasi-Fermi levels:

$$\rho_{sc}(\mathbf{r}) = 2q \sum_{\substack{m \in \text{bound} \\ \text{hole states}}} |\psi_{h,m}(\mathbf{r})|^2 [1 - f_v(E_{h,m}, T)] - 2q \sum_{\substack{n \in \text{bound} \\ \text{electron states}}} |\psi_{e,n}(\mathbf{r})|^2 f_c(E_{e,n}, T) \qquad (17.8)$$

Thus, screening potential can be found using Poisson's equation analogous to Equation 2.3. The electron and hole states are obtained by solving Schrödinger and Poisson's equations self-consistently. Figure 17.4a shows the screened and unscreened energy band diagrams of the SAQD structure, cutting

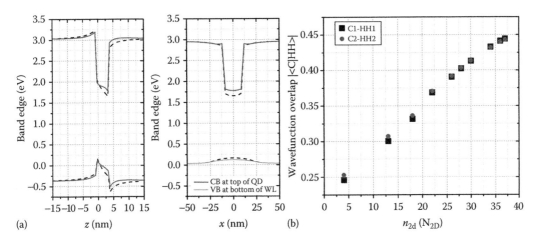

FIGURE 17.4 (a) A comparison between the unscreened and screened band edges of the SAQD at different viewing angles. The injection level is $n_{2d} = 37\ N_{2D}$ and the active region temperature is 120°C. (b) The conduction-valence band wavefunction overlaps (C1-HH1 and C2-HH2) as functions of injection level. (From Su, G.-L. et al., *Opt. Exp.*, 23, 12850–12865, 2015. With permission of Optical Society of America.)

along the z axis, the top of the dot, and the bottom of the wetting layer respectively. While the screening charge formed by EHPs barely changes the potential profile in the transverse direction, it efficiently cancels part of the polarization field in the z direction at high injection levels. The alleviation of the QCSE leads to increased wavefunction overlaps, as shown in Figure 17.4b. The saturation phenomenon at high injection levels is attributed to the large 3D-DOS in the matrix material, which slows down the rate of change of the quasi-Fermi levels.

The screening effect in SAQD structures in the z direction is actually the same as that in quantum wells (QWs). However, it is the extra confinement in the transverse direction that stops the carriers from leaving the quantum structure and recombining at defect centers. The island-like geometry also allows higher indium incorporation and strain relaxation, which ultimately give rise to the formation of indium-rich, much-less-defective active material.

17.2.1 Optical Gain Spectra of the $In_{0.4}Ga_{0.6}N$/GaN SAQD Active Material

As in most of the III-V materials, compressive strain also results in TE-polarized emission in InGaN active materials. Since the effect of CH band has been ruled out in this model, the TE-polarized gain in the $In_{0.4}Ga_{0.6}N$/GaN SAQD is solely considered.

The TE-polarized material gain spectrum can be found using Fermi's golden rule:

$$g^{TE}(\hbar\omega) = \frac{2\pi q^2 N_{2D} \left|\langle S|p_x|X\rangle\right|^2}{n_{r,t}\varepsilon_0 c_0 \omega m_0^2 L_{\text{eff}}} \sum_{\substack{m,n \in \text{bound} \\ \text{EHP states}}} P_{mn}(\hbar\omega) \tag{17.9}$$

where:

q is the elementary charge
$n_{r,t}$ is the refractive index in the transverse direction
ε_0 is the vacuum permittivity
c_0 is the speed of light in vacuum
m_0 is the free electron mass
$\left|\langle S|p_x|X\rangle\right|$ is the momentum matrix element [39]

The optical transition probably between the nth conduction (Cn) and mth valence (Vm) band states is given by

$$P_{mn}(\hbar\omega) = \left|I_{h,m}^{e,n}\right|^2 \left|w_{x,m}\right|^2 \int_{-\infty}^{\infty} G_{cv}(E - E_{h,m}^{e,n})\left[f_c(E_{e,n},T) - f_v(E_{h,m},T)\right] L(E - \hbar\omega)\mathrm{d}E \tag{17.10}$$

where:

$\left|I_{h,m}^{e,n}\right|$ is the wavefunction overlap between the Cn and Vm states
$\left|w_{x,m}\right|$ the in-plane composition of the Vm wavefunction
$E_{h,m}^{e,n}$ the energy difference between the Cn and Vm states
$G_{cv}(E)$ the inhomogeneous broadening function
$L(E - \hbar\omega)$ the homogeneous broadening of the linewidth at energy $\hbar\omega$

Figure 17.5 shows a comparison between the measured Hakki–Paoli gain spectrum right at the threshold and the calculated modal gain spectra. The modal gain is defined as the product of the TE-polarized material gain obtained from Equation 17.9 and the optical confinement factor of the lasing mode. The calculated gain curve follows the experimental data fairly well, except for the portion in the long-wavelength regime. The difference is mainly attributed to the size or indium content fluctuation across different SAQD layers. The active region temperature at the lasing threshold was found to be as high as 120°C; however, as can be seen in the following section, such value corresponds to a thermal impedance

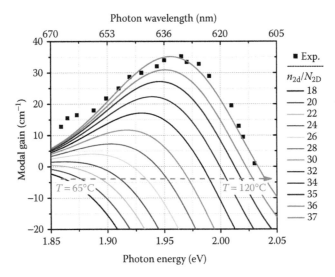

FIGURE 17.5 Calculated (lines) and measured (points) modal gain spectra of an In$_{0.4}$Ga$_{0.6}$N/GaN SAQD-based ridge waveguide laser. The experimental curve was obtained using the Hakki–Paoli method at the lasing threshold. (From Su, G.-L. et al., *Opt. Exp.*, 23, 12850–12865, 2015. With permission of Optical Society of America.)

of 43°C/W, which falls between 30°C/W and 60°C/W, measured from two blue InGaN MQW ridge waveguide lasers grown on GaN and sapphire substrates, respectively [40]. The homogeneous broadening factor used in Equation 2.14 was found to be as around 58 meV, which is relatively large compared to the values in InAs/GaAs SAQDs [41,42]. It is suspected that the dense electronic states resulted from the heavy effective masses in the GaN material system may facilitate phonon scattering events and broaden the gain spectra as the temperature increases. A differential gain of 4.4×10^{-17} cm^2 was derived from our model at the lasing wavelength, 634 nm. This result falls between values of 3.8×10^{-17} cm^2 and 5.3×10^{-17} cm^2, obtained from the threshold current density and small-signal modulation measurements [43,44]. More importantly, the residual strain at In$_{0.4}$Ga$_{0.6}$N/GaN SAQD/matrix interface was found to be around −1.2%, which is only 30% of the lattice mismatch between bulk In$_{0.4}$Ga$_{0.6}$N and GaN. This further confirms the effect of strain relaxation in the SAQD growth process.

17.3 Molecular Beam Epitaxial Growth of InGaN/GaN QDs

Multiple dot layers are typically grown in the active region of lasers to maximize the optical gain and modal confinement factor. It is, therefore, important to investigate the growth mode and relation between the layers. InGaN/GaN QD heterostructures were grown by plasma-assisted MBE (PAMBE) on GaN-on-sapphire substrates, shown schematically in Figure 17.6. An undoped GaN buffer layer of 500 nm thickness is first grown at 710°C with a Ga flux of 2.2×10^{-7} Torr and with 0.66 sccm of ultra-high purity N$_2$ with a plasma source power of 350 W. InGaN/GaN QDs for this study are then grown at 540°C under nitrogen-rich conditions (1.33 sccm/420W N$_2$ plasma power). A variable number of dot layers with varying thickness and GaN barriers of varying thickness are grown, usually with an interruption after the growth of a dot layer. Nominal values of In and Ga fluxes for an In composition of 40% in the QD are 9×10^{-8} and 4×10^{-8} Torr, respectively. The average alloy composition in the QD along the *c*-axis is measured by energy dispersive X-ray spectroscopy on a suitably prepared transmission electron microscopy (TEM) sample, which shows a variation in the alloy composition along the *c*-axis with a

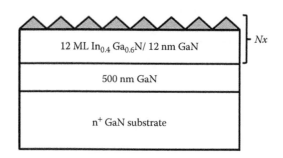

FIGURE 17.6 Quantum dot heterostructure used for characterization of the quantum dot layers. The growth was terminated following the growth of *N* periods of the active region for *N* varying from one to five.

maximum In content of ~40% for red-emitting QDs. The composition measured by X-ray diffraction in a relaxed bulk layer with the same nominal composition is similar.

Single or multiple $In_{0.4}Ga_{0.6}N/GaN$ dot layers were grown under the conditions described above with a GaN barrier thickness of 12 nm and a nominal InGaN thickness of 12 monolayers (ML), as shown schematically in Figure 17.6 Unlike what is observed during the growth of InGaAs quantum dots [45], it is found by atomic force microscopy (AFM) that the first layer of QDs has a smaller dot density (~7 × 10^7 cm^-2), which increases in the second layer and remains relatively constant at 5 × 10^10 cm^-2 in the third and higher layers. AFM images from the first four layers are shown in Figure 17.7. The increase in dot density can clearly be observed in these images with the lowest density in the first layer. These images have been analyzed in terms of the dot sizes and densities. The dot height and density are shown quantitatively in

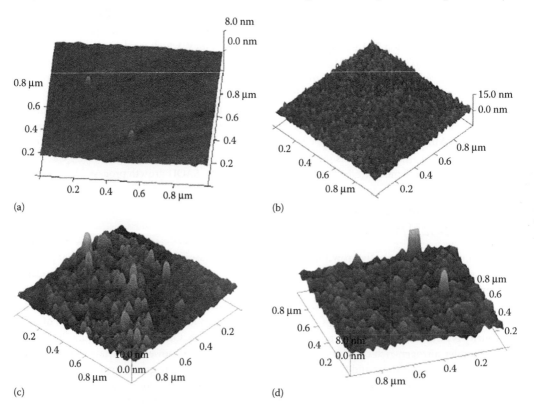

FIGURE 17.7 Atomic force microscopy images of uncapped layers of quantum dots. The heterostructure is shown schematically in Figure 17.6. The values of N are (a) 1, (b) 2, (c) 3, and (d) 4.

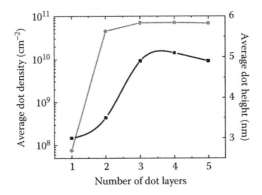

FIGURE 17.8 InGaN island height and aerial density from the AFM imaging on the samples described in Figure 22.6. (Reproduced with permission from Frost, T. et al., High performance red-emitting multiple layer InGaN/GaN quantum dot lasers, *Jpn. J. Appl. Phys.*, 55, 032101, 2016. Copyright 2016 The Japan Society of Applied Physics.)

Figure 17.8. The dot height follows a trend very similar to that seen in the aerial densities with an increase from the first to the third layer, followed by a saturation of the height. Interestingly, the dot base width is roughly constant in all these samples. The average base width under these growth conditions is 37 nm, with a variation of 5 nm. The change in island density from the first QD layer to the second can be understood by considering how In surface segregation impacts the critical thickness for island nucleation in these layers. The resulting composition profile can be understood following the work of Dehaese [46] for a multidot layer structure. The composition increases exponentially toward the intended composition x_{max} during the growth of the QD layer and decays exponentially during the growth of the barrier layer. If the segregation energy is large, it is possible that the composition does not reach x_{max} in the first layer of dots. However, the excess In is available to be incorporated as the film continues to grow, such that x_{max} will be reached in subsequent layers. Similarly, some In incorporation is expected in the barrier layers. In general, the critical thickness is defined as the thickness of the film that grows via a layer-by-layer mode prior to the formation of 3D islands, assuming constant x. When x is not constant, as is the case here, it is more appropriate to consider the critical strain energy for island nucleation, U_{cr}. The strain energy is proportional to f^2h, where f is the misfit strain, and h is the thickness of the film. Because the strain depends linearly on the composition x, the strain energy can be written as $[x(h)]^2h$, where x is now a function of h. The critical thickness is greatest for the first QD layer. Consequently, the first layer of dots requires more deposited material to reach U_{cr} than do subsequent layers. Since each layer is exposed to the growth flux for the same amount of time, the first layer of dots has a shorter amount of time over which the islands grow. The island density is proportional to the product of the island growth time and the nucleation rate. Thus, the first layer will have a lower island density assuming that the nucleation rate is constant.

The relatively low density of QDs on the first layer has several important consequences on heterostructures which incorporate these dots into the active region. First, the very low fill factor will result in a negligible contribution of this first layer to the modal gain of the laser, which is a product of the material gain and the optical confinement factor. The optical confinement factor itself can be considered to be a product of the transverse confinement factor, Γ_z, and the in-plane confinement factor, Γ_{xy}. Due to the relatively large dimensions in the in-plane direction, individual modes are not considered, and instead, Γ_{xy} can be taken as the physical fill factor of the active material in the plane.

While the first layer provides a minimal contribution to the modal gain, it can be used in other applications including in single photon sources which have recently been reported [47], and shown schematically in Figure 17.9. With the low density of QDs in the first layer, a single dot can be optically isolated with an aperture of ~1 μm^2, easily defined by standard photolithography.

It is also observed that growth of the QDs follows a kinetics driven scaling law, in accordance with the observations made by Amar et al. [48]. Under this kinetically driven growth model, adatoms on the

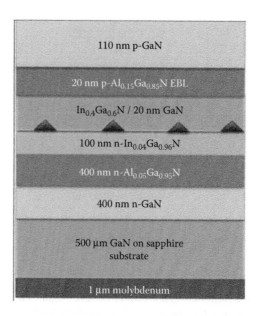

FIGURE 17.9 Schematic of a single photon source using a single layer of the low density InGaN/GaN quantum dots. With an aerial dot density of <10^8 cm^2 and an aperture ~1 μm^2 (which can be defined with standard photolithography) emission from a single dot can be collected.

surface may not reach their thermodynamically favored state and instead are limited by the amount of kinetic energy and mobility they have on the surface to be incorporated into the lattice. Under this model, island formation and stability will be determined by the size of the island. Smaller islands will be relatively unstable with fewer bonds. At some critical island size, given by i, dots will be typically stable. Figure 17.10 shows the lineshape or scaling, function of the size (height) distribution for InGaN/GaN dots grown at 545°C, measured by AFM. The data obtained from the top layer of a 7-layer stack have been analyzed with the scaling function

$$f_i\left(u = \frac{s}{\langle s \rangle}\right) = \frac{N_s(\theta)\langle s \rangle}{\theta} = C_i u^i \exp\left(-i a_i u^{\frac{1}{a_i}}\right) \qquad (17.11)$$

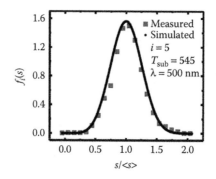

FIGURE 17.10 Distribution of InGaN island heights for typical red-emitting quantum dots. The solid line is the best fit of the scaling function to the measured AFM data. (Reprinted from *J. Cryst. Growth*, 378, Banerjee, A. et al., InGaN/GaN self-organized quantum dot lasers grown by molecular beam epitaxy, 566–570, Copyright 2013, with permission from Elsevier.)

where N_s is the size distribution, s is the island size, θ is the coverage. It is found that the value of i increases from 3 to 5 with increase of substrate temperature from 440°C to 545°C [12]. For comparison, AFM data in relation to several of these scaling functions is shown in Figure 17.10. A lower value of "i" also implies a less uniform distribution of QDs. This may be one limitation on longer wavelength QD devices (beyond red). Sufficient modal gain from a single state may not be reachable unless the uniformity can be improved.

17.4 Optical Properties of InGaN/GaN QDs

Photoluminescence (PL) characteristics of the QDs can be used to investigate the efficiency and some physical characteristics from the dots. The samples were mounted in a closed loop He cryostat and excited nonresonantly by a frequency tripled Ti:Sapphire laser ($h\upsilon = 4.66$ eV). The peak energy in the PL spectra closely follows the Varshni equation [49] with increasing temperature, shown in Figure 17.11 where no clear "S-shaped" behavior is seen [50,51]. In quantum wells, due to potential fluctuations, carriers first recombine in the lowest potential (the region with the highest indium concentration). As the temperature is elevated, the carriers can thermally escape into the lower indium regions leading to an initial increase of the emission energy. As the temperature is further raised, a typical shrinkage of the material bandgap in accordance with the Varshni relation is observed. This "S-shaped" behavior is, therefore, typically associated with indium clustering and nonuniformity. The QDs, lacking this behavior, are likely free of indium clustering.

The temperature-dependent measurements were made as a function of the incident excitation power and the variation of integrated intensity with excitation power is shown in Figure 17.12. In the nitride material system, due to the large polarization field for c-plane growth, it becomes essential to measure η_i at high injection for which the dots (or wells) reach flat band condition, and the polarization field is screened. As reported in the literature [52–54], at low temperatures injected carriers are confined in the localization potential of the QDs or in the potentials due to compositional fluctuations (in quantum wells). With increasing temperature, the carriers acquire sufficient energy to overcome the potential barriers and recombine at nonradiative centers in the barrier and wetting layer regions. Then the ratio of the saturated peak PL intensity at 30 and 300 K at high excitation powers is an approximate measure of the internal quantum efficiency, η_i (at room temperature). The thermionic emission of carriers and recombination in other layers at elevated temperatures may result in an underestimation of η_i. However, by measuring the dots at high excitation where the dots are saturated with carriers, this effect should be

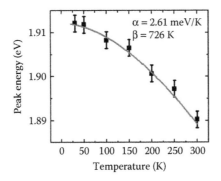

FIGURE 17.11 Variation of peak emission energy from the red-emitting $In_{0.4}Ga_{0.6}N/GaN$ quantum dots as a function of temperature. The line is a fit to the data with the Varshni relation for the given values of α and β. (From Frost, T. et al., *IEEE J. Quantum Electron.*, 49, 923–931, 2013. © 2013 IEEE. With permission.)

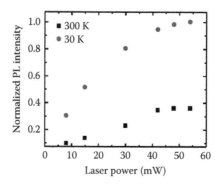

FIGURE 17.12 Normalized integrated photoluminescence intensity as a function of excitation power and temperature. The quantum efficiency of 35.9% is derived by taking the ratio of the intensities under saturation. (From Frost, T. et al., *IEEE J. Quantum Electron.*, 49, 923–931, 2013. © 2013 IEEE. With permission.)

minimized. From the data shown in Figure 17.12a value of $\eta_i = 35.9\%$ is derived. This value of ~36% is typical across red-emitting QD samples with similar growth conditions.

17.5 QD Lasers

Recent demand for visible and ultra-violet LEDs and lasers is immense due to their numerous applications in the fields of solid-state lighting, optical data storage, plastic fiber communication, full-color mobile projectors, heads-up displays, and in quantum cryptography and computing [1,2]. Most research and development into these light sources is being done using nitride-based materials where the emission can be tuned from deep UV in AlN (~6 eV) to the near infrared by using InN (~0.7 eV). Since the first report of a blue-emitting InGaN/GaN QW LED in 1995 [55], much progress has been made in extending the emission to longer wavelengths through the incorporation of more indium in the InGaN QWs [3–10]. While the first blue-emitting laser was demonstrated by Nakamura in 1996 [56], it has since become increasingly difficult to grow and fabricate lasers at longer wavelengths. It was only as recently as 2009 that the first green-emitting laser was demonstrated using InGaN/GaN QWs [57], and red-emission has yet to be shown with such QWs. To fully realize the potential of these nitride-based lasers, it is necessary to further extend the emission wavelength of these devices beyond green into the red which will allow for the production of solid-state projectors and white light sources from a single material system.

The large indium composition and associated strain in the ternary InGaN QWs lead to clustering effects and a large piezoelectric polarization field, especially in c-plane heterostructures, both of which are detrimental in laser performance. The threshold current density of these lasers is generally very large due to reduced electron-hole (e-h) wavefunction overlap in the QWs [38,58–59]. Additionally, a large blue shift of the emission peak with injection is observed due to the quantum-confined Stark effect (QCSE) associated with the polarization field [57]. It has been shown that material inhomogeneities and the piezoelectric field increase in InGaN/GaN QWs with increasing indium content. Further, a wider well width that is needed for emission at longer wavelengths is not an option since the band bending due to a strong polarization field reduces e-h wavefunction overlap significantly. It is for these reasons that red-emitting lasers with InGaN/GaN QWs have not yet been demonstrated. InGaP/InGaAlP double-heterostructure and QW lasers lattice matched to GaAs and emitting in the red wavelength region of 650–670 nm have been reported [11,12]. However, these devices are characterized by very large values (5–10 kA/cm²) and strong temperature dependence (T_0 ~ 50–100 K) of the threshold current density. Both of these characteristics are detrimental in real high-performance applications.

17.5.1 Blue- and Green-Emitting QD Lasers

Green-emitting QD lasers are important for bridging the green gap, and blue- and green-emitting lasers are an alternative to high threshold QW lasers. The laser heterostructures were grown on 500 μm thick c-plane n-GaN bulk substrates (defect density $< 5 \times 10^6 \text{cm}^{-2}$) by plasma-assisted molecular beam epitaxy (PAMBE). The QD laser heterostructure, inclusive of an $Al_{0.15}Ga_{0.85}N$ electron blocking layer, is shown in Figure 17.13. The GaN and AlGaN layers in the laser heterostructure are grown at 740°C and 770°C, and the 8 layers of QDs are grown at 520°C. The inset to Figure 17.13 shows the AFM image of a dot layer, from which a dot density of $5 \times 10^{10} \text{ cm}^{-2}$ is derived. The dot height and diameter are ~3 and 50 nm, respectively. The room temperature PL spectrum of the green QD active region, shown in Figure 17.14, peaks at 529 nm. We have measured excitation dependent luminescence in a InGaN/GaN QD LED emitting at $\lambda = 522$ nm. From the shift in the luminescence peak due to injection, a polarization field of 90 kV/cm is derived, which is much smaller than 2 MV/cm typically measured in equivalent quantum wells [60]. We have also performed time-resolved PL measurements on the QD heterostructures [61]. From this data and the measured internal quantum efficiency of the QDs ~40%, we derive the radiative and nonradiative lifetimes of 1.5 and 1.0 ns, respectively. The greatly reduced radiative lifetime, compared to those in quantum wells, also confirms the low polarization field in the dots.

Broad area edge-emitting lasers were fabricated by mesa etching, p- and n-contact metallization, dielectric deposition and p-contact annealing. Subsequently, the samples were diced into laser bars. Instead of using cleaving to create the mirrors, FIB etching was employed to create smooth and defect-free facets.

Light-current-voltage characteristics were measured at room temperature with a pulsed bias (0.05% duty cycle, 1 μs pulse width) and are shown in Figure 17.15 for a blue-emitting laser and in Figure 17.16a for a green-emitting laser. The laser mount was cooled by a thermoelectric cooler to 5°C to avoid excessive device heating. Even with this, the junction and facet temperatures are estimated to be well above 300 K.

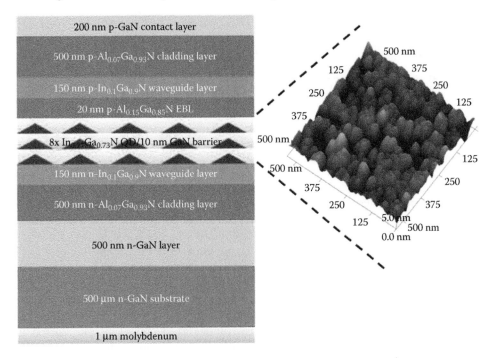

FIGURE 17.13 Schematics of $In_{0.27}Ga_{0.73}N/GaN$ quantum dot laser heterostructure grown by plasma-assisted MBE. Inset shows atomic force microscopy image (500 × 500 nm) of an uncapped self-organized InGaN/GaN quantum dot layer. (Reprinted with permission from Zhang, M. et al., *Appl. Phys. Lett.*, 98, 221104-1–221104-3, Copyright 2011, American Institute of Physics.)

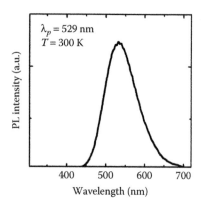

FIGURE 17.14 Room temperature photoluminescence spectrum from 5 layers of $In_{0.27}Ga_{0.73}N/GaN$ quantum dots. (Reprinted with permission from Zhang, M. et al., *Appl. Phys. Lett.*, 98, 221104-1–221104-3, Copyright 2011, American Institute of Physics.)

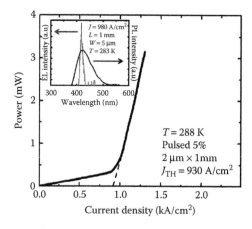

FIGURE 17.15 Light-current characteristics of the blue-emitting quantum dot laser. The spectral characteristics are shown in the inset. (Reprinted with permission from Banerjee, A. et al., *Appl. Phys. Lett.*, 101, 041108-1–041108-4, Copyright 2012, American Institute of Physics.)

The threshold current is 110 mA, which corresponds to a threshold current density of ~1.2 kA/cm². This value is lower than those reported for QW lasers. An output power of 12.5 mW is measured for an injection current of 120 mA. The slope efficiency is determined to be 0.74 W/A from the *L–I* characteristics. The corresponding wall-plug efficiency is 1.1%. It is important to note that lasing is achieved with the FIB-etched facets alone and without any additional deposition of dielectric distributed Bragg reflectors (DBRs). The superior performance characteristics are attributed to the smaller polarization fields and better electron-hole overlap in QDs and smooth laser facets created by FIB etching. A laser made from another heterostructure and emitting at a slightly higher wavelength, in which DBRs were incorporated over the FIB-etched facets, exhibited a threshold current density of 2.0 kA/cm². The spectral output characteristics of the laser were measured below and above threshold and are shown in Figure 17.15 for the blue-emitting laser. The spectral output for an injection current of 120 mA ($J = 1.33$ kA/cm²) under pulsed bias condition is shown in Figure 17.16b for the green-emitting laser. The emission is characterized by a peak at 524 nm and a linewidth of 0.7 nm. In contrast, a broad emission with a linewidth of 54 nm is measured with DC biasing below threshold. The peak of the electroluminescence (EL) spectrum below

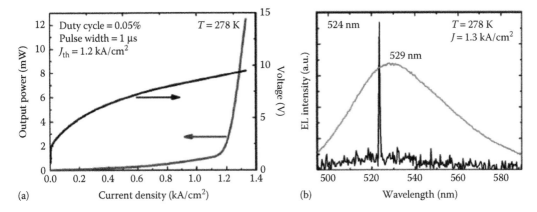

FIGURE 17.16 (a) Light-current-voltage characteristics of InGaN/GaN quantum dot laser measured in the pulsed bias mode at $T = 278$ K. (b) Electroluminescence spectra of $In_{0.27}Ga_{0.73}N$/GaN quantum dot laser below (with DC bias) and above (with pulsed bias) threshold. (Reprinted with permission from Zhang, M. et al., *Appl. Phys. Lett.*, 98, 221104-1–221104-3, Copyright 2011, American Institute of Physics.)

threshold occurs at the same wavelength, 529 nm, as the PL spectrum shown in Figure 17.16b. The slightly wider linewidth of the spontaneous emission, compared to that typically observed for quantum wells, is most probably due to inhomogeneity in QDsize and alloy composition. The blueshift of 5 nm between the lasing and luminescence peaks is smaller than the ~12 nm reported for a InGaN/GaN MQW laser grown on semipolar GaN[105]. This could be due to a lower polarization field in the QDs.

17.5.2 Red-Emitting QD Lasers

Red-emitting InGaN/GaN laser heterostructures, shown schematically in Figure 17.17 were similarly grown on bulk *c*-plane n-GaN substrates. The more commonly used AlGaN waveguide cladding layers are replaced by latticed-matched $In_xAl_{1-x}N$ ($x = 0.18$). The use of this material, which provides a large index step compared with lattice-mismatched AlGaN, in the cladding significantly improves confinement of the optical modes at the longer wavelengths. This leads to reduced cavity loss due to free carrier absorption from the reduced overlap of the optical mode with the heavily doped cladding. Additionally, substrate leakage will also be reduced, further minimizing the cavity loss. The optical confinement provided by $In_{0.18}Al_{0.82}N$ is comparable to that of $Al_{0.46}Ga_{0.54}N$ [62]. The growth of this alloy was done under varying growth conditions to optimize its optical and structural characteristics. Due to the low incorporation of In at high temperatures, epitaxy must be done at a relatively low substrate temperature while ensuring the temperature is sufficiently high to provide surface mobility of Al atoms and prevent a rough surface. The best results were achieved at a substrate temperature of 497°C and with In and Al fluxes of 2.5×10^{-8} and 3.4×10^{-8} Torr, respectively. However, for growth of the p-doped $In_{0.18}Al_{0.82}N$, a lower series resistance was possible at a growth temperature of 480°C. To compensate for the reduced substrate temperature, the In flux was lowered to 2.0×10^8 Torr. It is also important to note that the relatively low growth temperature of the InAlN upper cladding layer reduces In outdiffusion from the InGaN/GaN QDs in the active region, providing an additional advantage over AlGaN-based cladding. To further improve the optical confinement, the GaN barrier layers and waveguide were replaced with $In_{0.08}Ga_{0.92}N$, which also resulted in an increase in the QD internal quantum efficiency to 0.51.

As can be seen in Figure 17.17, 8 $In_{0.42}Ga_{0.58}N$/$In_{0.08}Ga_{0.92}N$ layers have been incorporated in the laser heterostructure to maximize the confinement factor and gain. An additional layer was added to account for the low density of QDs on the first layer, to increase the modal gain. At the same time, adding more dot layers increases the possibility of generating dislocations, increasing the threshold current density

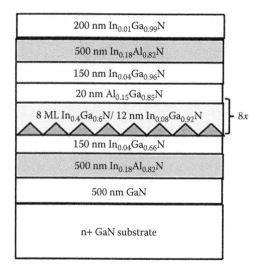

FIGURE 17.17   Red-emitting quantum dot laser heterostructure.

and nonuniform hole injection. The optical confinement factor, calculated by the transfer matrix method, is 0.075 and the QD fill factor is 0.38. The thickness x of the GaN spacer layer between the QDs and the $Al_{0.15}Ga_{0.85}N$ electron blocking layer was varied from 15 to 60 nm. The top p-GaN contact layer has a doping of 5×10^{17} cm^{-3}, and the last 100 nm was grown by metal modulated epitaxy to yield an even higher doping level. The laser diodes have a turn-on voltage of 2.7–3.3 V, a series resistance of ~6 Ω, and a reverse leakage current of 6.6 mA at −5 V. Ridge waveguide edge-emitting lasers of various cavity dimensions were fabricated using standard photolithography, dry etching, and metallization techniques. The ridge is etched down to the waveguide/cladding heterointerface to minimize the cavity loss associated with sidewall roughness. The dielectric thicknesses and facet reflectivity were calculated by transfer matrix method. Measurement of the mirror reflectivity was made on silicon wafers coated with an identical stack of dielectrics. All laser measurements described in the following section were from the output from the low-reflectivity facet. Devices without any facet DBR coating were also characterized. No special device mounting or heating sinking were implemented in these measurements.

17.5.3 Steady State Laser Characteristics

Figure 17.18 shows the output light-current (L–I) characteristics at room temperature and 95°C under continuous wave (cw) bias condition for a device prior to facet coating, and with a GaN spacer thickness $x = 60$ nm. The cavity width and length of this device were 5 μm and 1 mm, respectively. This 60 nm GaN spacer thickness resulted in the best performance and were used for all the measurements in Sections 4.2. The 95°C temperature was chosen in accordance with design specifications for automobile heads-up display applications [63–65]. The threshold current density of this device is $J_{th} = 2.8$ kA/cm^2 at 300 K and 4.8 kA/cm^2 at the higher temperature. The corresponding output slope efficiency decreases from 0.42 W/A ($\eta_d = 0.23$) to 0.03 W/A. This is likely due to thermalization of carriers at the elevated temperature. Device packaging or active cooling techniques may be necessary to improve the high-temperature performance further. The output spectral characteristics at 300 K at an injection of $1.1 J_{th}$ is shown in Figure 17.19. The minimum measured linewidth of the dominant longitudinal mode is 8 Å at a peak emission of 630 nm.

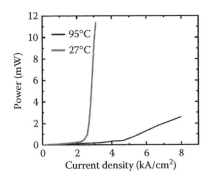

FIGURE 17.18 Light-current characteristics for a 5 × 1 mm laser without high reflectivity facet coating. (Reproduced with permission from Frost, T. et al., High performance red-emitting multiple layer InGaN/GaN quantum dot lasers, *Jpn. J. Appl. Phys.*, 55, 032101, 2016. Copyright 2016 The Japan Society of Applied Physics.)

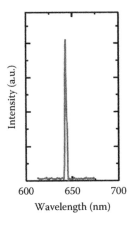

FIGURE 17.19 Electroluminescence spectrum from one of the uncoated facets of the optimized laser. (Reproduced with permission from Frost, T. et al., High performance red-emitting multiple layer InGaN/GaN quantum dot lasers, *Jpn. J. Appl. Phys.*, 55, 032101, 2016. Copyright 2016 The Japan Society of Applied Physics.)

Figure 17.20 shows the *L–I* characteristics of a 10 μm × 1 mm device at 300 K, with DBR facet coatings (0.7 and 0.95), and $x = 60$ nm, and under pulsed (1% duty cycle) biasing conditions. The laser exhibits $J_{th} = 1.7$ kA/cm^2, a slope efficiency of 0.41 W/A, and a wall plug efficiency of 1.6%. It should be noted that while there is still room for improvement, this efficiency is 10x larger than in a similar heterostructure with AlGaN-based cladding. This is due to the increased internal quantum efficiency of the QDs and the increased optical confinement factor in this heterostructure. The maximum output power is 30 mW, putting it in the range needed for projector and heads-up display applications. The temperature dependence of J_{th} under pulsed biasing (1% duty cycle) for the laser of Figure 17.20 is shown in Figure 17.21. The values of T_0 quoted in the figure are obtained by analyzing the data with the relation: $J_{th}(T) = J_{th}(0)\exp(T/T_0)$. These high values of T_0 are extremely encouraging and result from the large band offsets and good carrier confinement in the InGaN/GaN dot heterostructures. The measured $T_0 = 240$ K up to 320 K is comparable with the value reported in the Chapter 3 heterostructure, and the degradation at higher temperatures is likely due to the thermalization of carriers at elevated temperatures. In contrast, red-emitting lasers made with InGaAlP/GaAs heterostructures have $J_{th} \sim 6–8$ kA/cm^2 and $T_0 \sim 60–80$ K [11,12].

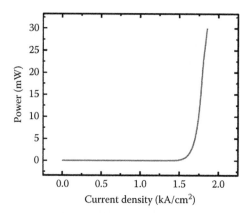

FIGURE 17.20 Light-current characteristics from the low reflectivity of a DBR coated laser at room temperature under pulsed bias. (Reproduced with permission from Frost, T. et al., High performance red-emitting multiple layer InGaN/GaN quantum dot lasers, *Jpn. J. Appl. Phys*, 55, 032101, 2016. Copyright 2016 The Japan Society of Applied Physics.)

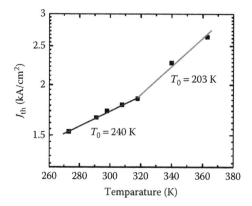

FIGURE 17.21 Temperature dependence of the threshold current density of a laser with DBR coating. (Reproduced with permission from Frost, T. et al., High performance red-emitting multiple layer InGaN/GaN quantum dot lasers, *Jpn. J. Appl. Phys.*, 55, 032101, 2016. Copyright 2016 The Japan Society of Applied Physics.)

17.5.4 Measurement of Differential Gain

Length-dependent characterization of the lasers was performed to measure the laser cavity loss and differential gain. Light-current measurements have been made on lasers of varying cavity lengths and the differential quantum efficiency η_d and J_{th} were recorded for each length. Figure 17.22 shows the variation of η_d^{-1} with cavity length. From this data, a value of $\eta_i = 0.49$ is derived using the relation:

$$\frac{1}{\eta_d} = \alpha_i L \ln \frac{1}{\sqrt{R_1 R_2}} + \frac{1}{\eta_i} \tag{17.12}$$

where:
 η_d is the differential efficiency of the laser
 R_1 and R_2 are the mirror reflectivities
 L is cavity length
 α_i is the cavity loss

The cavity loss is determined to be 8.3 cm^{-1} in these heterostructures, compared with 25 cm^{-1} in the lasers with AlGaN cladding. The reduced cavity loss is due to smaller free carrier absorption and substrate

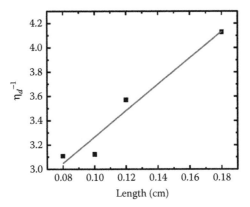

FIGURE 17.22 Variation of inverse differential quantum efficiency with cavity length. (Reproduced with permission from T. Frost et al., High performance red-emitting multiple layer InGaN/GaN quantum dot lasers, *Jpn. J. Appl. Phys.*, 55, 032101, 2016. Copyright 2016 The Japan Society of Applied Physics.)

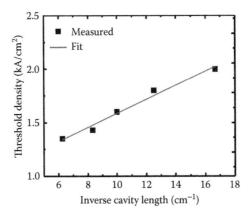

FIGURE 17.23 Variation of threshold current density with inverse cavity length. (Reproduced with permission from T. Frost et al., High performance red-emitting multiple layer InGaN/GaN quantum dot lasers, *Jpn. J. Appl. Phys.*, 55, 032101, 2016. Copyright 2016 The Japan Society of Applied Physics.)

leakage with the presence of the all $In_{0.18}Al_{0.82}N$ cladding. Measured values of J_{th} are plotted against inverse cavity length in Figure 17.23. The differential gain dg/dn is calculated by analyzing this data with the relation the relation [66]:

$$J_{th} = J_{th}^0 + \frac{qd}{\Gamma \eta_i \tau_r \frac{dg}{dn}} \left[\alpha_i + \frac{1}{2L} ln \left(\frac{1}{R_1 R_2} \right) \right] \tag{17.13}$$

where d is the active region thickness calculated as the number of dot layers times the effective dot height, Γ is the product of the optical confinement factor simulated by the transfer matrix method (0.075) and the fill factor (0.38), τ_r is the measured radiative lifetime (2.5 ns), and R_1 and R_2 are 0.7 and 0.95, respectively; the transparency current density J_{th}^0 and dg/dn are fitting parameters for this function. A value of differential gain $dg/dn = 9.0 \times 10^{-17}$ cm² is derived along with a value of $J_{th}^0 = 550$ A/cm². This value of differential gain is ~2× larger than the value found in the AlGaN cladded laser. Such as large value of differential gain is comparable with shorter wavelength blue- and green-emitting QDlasers [13,14]. It is ~5× larger than values reported in shorter wavelength blue-emitting InGaN/GaN quantum-well-based lasers.

17.6 Dynamic Characterization of QD Lasers

17.6.1 Small-Signal Modulation

The small-signal modulation response of 800 μm long ridge waveguide lasers was measured under pulsed bias conditions (5 μs pulses; 0.5% duty cycle) using a sweep oscillator with a bias T, low-noise amplifier, a high-speed silicon detector and a spectrum analyzer. The modulation response is shown in Figure 17.24. The indicated currents refer to the DC bias current. A 10 dBm sinusoidal signal of varying frequency (100 MHz to 3 GHz) is superimposed on the DC bias with the sweep oscillator and bias T. Light from the lasers is collected with a multimode fiber and detected with a Newport D15 40 GHz high-speed detector. The small signal response is amplified and analyzed with an electrical frequency analyzer. The relative change in AC intensity from the lowest measured frequency (100 MHz) are plotted. The measured data have been analyzed with the damped oscillator small signal response model:

$$|M(f)|^2 \propto \frac{1}{(f^2 - f_r^2)^2 + \left(\dfrac{\gamma_d}{2\pi}\right)^2 f^2} \tag{17.14}$$

where:

γ_d is the damping factor

f_r is the resonance frequency of the response

A −3 dB modulation bandwidth of 2.4 GHz was measured at the highest DC injection current of 250 mA and the resonance frequency at this injection level is 1.6 GHz. A higher −3 dB modulation bandwidth may be possible with higher injection currents but this was not possible due to device and facet heating, and no measurements at higher injections were possible. The relatively fast frequencies that these lasers can be modulated at may allow for their use in plastic fiber communication systems up to ~3.9 gigabits per second which have pass bands at this wavelength range. The relatively small −3 dB modulation frequency, in comparison with InGaAs/GaAs QDs, may, however, be due to the relatively large carrier masses in this material system [67].

In addition to demonstrating the potential for these lasers to be used in optical communication systems, small signal analysis can also be used to analyze the gain characteristics of the InGaN QD lasers. The lasers have been fit with the damped oscillator response given in Equation 17.3 and the resultant

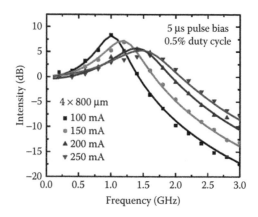

FIGURE 17.24 Small signal modulation response of the In$_{0.4}$Ga$_{0.6}$N/GaN quantum dot laser diodes (points) and fit of the measured response curves (solid lines). (Reprinted from Frost, T. et al., *Appl. Phys. Lett.*, 103, 211111-1–211111-3, Copyright 2013, American Institute of Physics.)

resonant frequencies and damping coefficients are extracted. The differential gain dg/dn is related to the small-signal modulation data using the relation:

$$f_r = \frac{1}{2\pi} \left[\frac{v_g \Gamma (I - I_{th}) \dfrac{dg}{dn} \eta_i}{L w d_{act} q} \right]^{\frac{1}{2}}$$

(17.15)

where:

v_g is the photon group velocity
Γ is the confinement factor
η_i is the QD internal quantum efficiency
L is the cavity length
d_{act} is the thickness of the active region

A value of $\eta_i = 35.9\%$ was obtained from temperature and excitation dependent PL measurements made on the red-emitting QDs. The confinement factor is estimated as the product of the transverse confinement factor Γ_z, where z is the growth diction (0001), and the QD fill factor, Γ_{xy}. Unlike a QW, the in-plane optical confinement factor cannot be considered unity with the discrete and discontinuous QDs. The QD fill factor is used as in-plane confinement factor and is calculated as follows. The pyramidal dots with a base width of 37 nm and height of 5 nm are modeled as equivalent flattened cubes of the same volume and base width with a 3.55 nm effective height. Comparing the volume of an array of these cubes with the volume of the nominal thickness of the deposited InGaN (8 ML), given by the change in reflective high-energy electron diffraction (RHEED) pattern from 2D to 3D growth, results in a fill factor of 0.35. By taking the fill factor into account, the confinement factor is 0.025 for the laser heterostructure. The plot of f_r versus $(I-I_{th})^{1/2}$, obtained from the data of Figure 17.24, is shown in Figure 17.25. The slope of the plot is 3.3 GHz/mA$^{1/2}$, from which a differential gain $dg/dn = 5.3 \times 10^{-17}$ cm^2 is derived from Equation 17.14. The value is comparable to that of shorter wavelength ($\lambda = 430$ nm) strained and strain-compensated InGaN/GaN QW lasers [68,69]. It should be noted that this is comparable with values derived from length dependent characterization of these red InGaN/GaN QD lasers (3.8×10^{-17} cm^2). However, it is considerably smaller than In(Ga)As/GaAs QD lasers [70]. This is largely due to the large carrier effective masses in these materials.

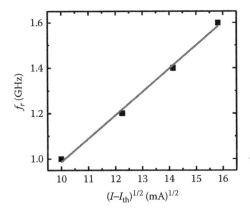

FIGURE 17.25 Variation of the laser resonant frequency with the square root of the injection current above threshold. The plotted resonant frequencies are fit using the damped oscillator model to the measured response curves in Figure 17.24. (Reprinted with permission from Frost, T. et al., *Appl. Phys. Lett.*, 103, 211111-1–211111-3, Copyright 2013, American Institute of Physics.)

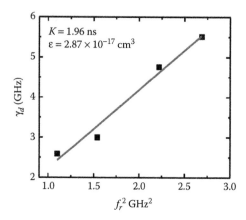

FIGURE 17.26 Variation of the laser damping factor with the square of the resonant frequency. The plotted damping factors are the values fit using the damped oscillator model to the measured response curves in Figure 17.24. (Reprinted with permission from Frost, T. et al., *Appl. Phys. Lett.*, 103, 211111-1–211111-3, Copyright 2013, American Institute of Physics.)

It is known that hole injection is nonuniform in InGaN/GaN multiquantum well LEDs and lasers and a similar situation exists in multiquantum dot devices. Most of the injected holes pile up in the first couple of wells/dots from the injecting p-layer, leading to increased carrier density, and hot-carrier effects, including gain compression. Under gain compression limited modulation response in the devices under study, the damping factor γ_d is related to f_r by the approximate relationship: $\gamma_d = K f_r^2$. The proportionality constant is the K-factor which is a measure of the damping limited bandwidth. A plot of γ_d versus f_r^2 obtained from analysis of the data of Figure 17.24 with Equation 17.13 is illustrated in Figure 17.26. A value of $K = 1.96$ ns is derived from the slope of this plot. The maximum modulation bandwidth of the lasers under this damping model is then given by $f_{-3dB} = 2^{3/2}\pi/K = 4.53$ GHz. These higher values of bandwidth (beyond 2.5 GHz) may be accessible if higher injection currents could be reached, perhaps through device packaging or contact improvement. Under the gain compression limited bandwidth model, a value of the gain compression factor $\varepsilon = 2.87 \times 10^{-17}$ cm^3 is then derived from the approximate relationship:

$$K \cong 4\pi^2 \left(\frac{\epsilon}{v_g \dfrac{dg}{dn}} + \tau_p \right) \tag{17.16}$$

where τ_p is the cavity photon lifetime. This value of ε is comparable to those measured for In(Ga)As/GaAs QD lasers [71].

17.6.2 Large Signal Modulation: Measurement of Auger Recombination Coefficients

Nitride-based LEDs suffer from droop where the light output efficiency continuously and severely decreases with injection after a peak at very low biases. One potential culprit of this phenomenon that has been suggested is Auger recombination [72–74]. While large bandgap III-Nitride-based quantum wells are expected to have relatively small levels of Auger recombination, which is proportional to $(k_B T/E_g)^{3/2}\exp(-E_g/k_B T)$, measured values are typically orders of magnitude larger. It is, therefore, important to investigate if quantum-dot-based devices suffer from similar unexpectedly large Auger recombination coefficients. Previously, luminescence techniques have been used to measure Auger recombination in InGaN materials and heterostructures [75–78]. The Auger coefficient can also be derived from large signal modulation measurements made on lasers [79–80].

When a laser is electrically switched from the off state to a bias state above the threshold, there is a turn-on delay between the electrical pulse and the coherent optical output pulse. For the laser to reach the threshold, the carrier concentration of the QDs, n, must reach its threshold value n_{th}. Below threshold, carriers are injected at a constant rate from the injection current density and lost due to Shockley–Read–Hall (SRH) recombination, spontaneous recombination, or Auger recombination. From the laser rate equations for injected carriers in the active region, the turn-on delay time τ_d can be expressed as

$$\tau_d = qV \int_0^{n_{th}} \frac{1}{I - qVR(n)} \, dn \tag{17.17}$$

where V is the active region volume and $R(n)$ is the total carrier recombination rate given by

$$R(n) = \frac{n}{\tau} = A_{nr}n + R_{sp}n^2 + C_a n^3 \tag{17.18}$$

where

τ is the carrier lifetime
A_{nr} is the Shockley–Read–Hall (SRH) recombination coefficient
R_{sp} is the radiative recombination coefficient

It may also be remembered that

$$J_{th} = \frac{qdn}{\tau} = qdR(n_{th}) = qd(A_{nr}n_{th} + R_{sp}n_{th}^2 + C_a n_{th}^3) \tag{17.19}$$

where d is the thickness of the laser active region. Therefore, accurate measurement of the turn-on delay time and calculation of A_{nr} and R_{sp} allow self-consistent determination of n_{th} and C_a using Equations 17.16 through 17.18.

In the large-signal modulation measurement, the laser is biased with 500 ns pulses (under 1% duty cycle) having a rise time of 100 ps (20%–80%) in switching from current $I = 0$ to $I > I_{th}$. An impedance matching unit is used to reduce reflection and distortion of the pulses. This is similar to the small signal analysis setup with the amplifier and electrical spectrum analyzer replaced with a high-speed oscilloscope. The coherent output light from the laser is collected by a fiber coupled to a high-speed GaAs photodetector and temporally resolved with a 2 GSa/s sampling oscilloscope. The electrical pulsed bias is also routed to the oscilloscope and concurrently measured, as shown in Figure 17.27 at 280 K with an injection of 100 mA. It should be noted that the quoted temperature is the Peltier cooler temperature, and the actual active region temperature may be considerably higher. The relaxation oscillations in the optical pulse are clearly observed. Thus, the turn-on delay τ_d can be measured after properly accounting for the delays in the fiber, rf cable, and the detector. The measured values of τ_d at room temperature for different injection currents from 50 to 200 mA are plotted in Figure 17.28. The delay time decreases with increasing current due to a decrease of carrier lifetime. The solid curve is the calculated using the model described in the previous section. The current-dependent delay times were also measured at different temperatures. Figure 17.29 shows a plot of τ_d at a fixed injection current of 100 mA plotted as a function of temperature. It is evident that the delay time increases with increase of ambient temperature.

In order to analyze the temperature and injection dependent time delay data and to accurately determine the Auger recombination coefficient, it is necessary to calculate R_{sp} and A_{nr}. The value of R_{sp} is calculated using the Fermi's golden rule with an eight-band $\mathbf{k \cdot p}$ description of the bands. The interface strain and polarization field in the dots are taken into account in the model. Thus, at room temperature, $R_{sp} = 1.4 \times 10^{-11}$ cm^3 s^{-1} is derived. This value is very similar to those reported for nitride

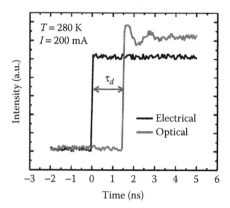

FIGURE 17.27 Measured electrical and optical signals showing the laser diode response to a large signal current pulse driving the laser above threshold. The time delay, τ_d, is indicated in the plot. (Reprinted with permission from Frost, T. et al., *Appl. Phys. Lett.*, 104, 081121-1–081121-4, Copyright 2014, American Institute of Physics.)

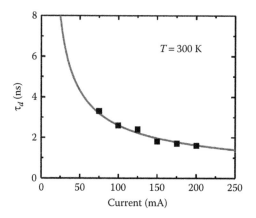

FIGURE 17.28 Variation in the measured values of the delay time with current injection at room temperature. The solid line indicates the calculated delay time with injection. (Reprinted with permission from Frost, T. et al., *Appl. Phys. Lett.*, 104, 081121-1–081121-4, Copyright 2014, American Institute of Physics.)

materials by others [80–81]. Measurement of A_{nr} is carried out by independent transient capacitance measurements made on GaN n^+-p homojunction diodes to determine the presence of deep level traps in PAMBE-grown GaN, assuming that in the laser heterostructure deep traps in the GaN barrier regions between the QD layers and the $In_{0.02}Ga_{0.98}N$ waveguide layer lead to nonradiative recombination. Three electron and two hole trap levels with characteristics listed in Table 17.2 were identified in the GaN layer. Under high injection conditions ($n = p \gg n_i$), $A_{nr} \cong \sigma v_{th} N_T / 2$ where σ and N_T are the trap cross-section and density, respectively. A total value of $A_{nr} = 6.98 \times 10^7$ s^{-1} is calculated taking into account all the trap levels listed in Table 17.2. The value of n_{th} and C_a are then determined by solving Equations 17.16 through 17.18 iteratively and self-consistently for all the injection current. It should be noted that the recombination rate due to SHR recombination ($A_{nr}n$) is relatively small compared with spontaneous radiation recombination ($R_{sp}n^2$) and Auger recombination ($C_a n^3$) are the carrier concentrations typically found in a laser diode (10^{19} cm^3) and the SHR recombination could often be neglected with minimal impact on the model accuracy. From the variation of time delay with temperature, the Auger coefficient is calculated for each temperature using the same value of R_{sp} and trap levels derived

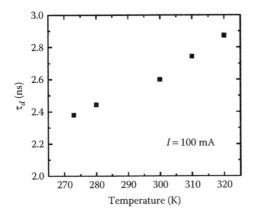

FIGURE 17.29 Variation in the time delay with temperature at a current bias of 100 mA. The small increase in delay with temperature is likely due to electron-hole scattering in the quantum dots. (Reprinted with permission from Frost, T. et al., *Appl. Phys. Lett.*, 104, 081121-1–081121-4, Copyright 2014, American Institute of Physics.)

TABLE 17.2 Characteristics of Deep Level Traps in GaN, Grown by Plasma-Assisted Molecular Beam Epitaxy, Obtained from Transient Capacitance Measurements

Electron Traps			Hole Traps		
$\Delta E(eV)$	$\sigma(cm^2)$	$N_T(cm^{-3})$	$\Delta E(eV)$	$\sigma(cm^2)$	$N_T(cm^{-3})$
0.24	5.154×10^{-16}	2.11×10^{15}	0.387	5.08×10^{-17}	3.62×10^{16}
0.461	2.242×10^{-16}	5.12×10^{15}	0.595	1.136×10^{-16}	6.25×10^{16}
0.674	1.22×10^{-15}	4.63×10^{15}			

above. The values of C_a at different temperatures are plotted in Figure 17.30. At room temperature $C_a = 1.3 \times 10^{-31}$ cm^{6}s^{-1} and $n_{th} = 1.3 \times 10^{19}$ cm^{-3}.

In the temperature range in which the turn-on delay measurements have been made, it is found that τ_d increases with increase of temperature. Similarly, there is a small decrease in the value of C_a, shown in Figure 17.30, with increase of temperature. Interestingly, the trends are identical to those observed for τ_d and C_a in InGaAs/GaAs QDs [82]. The behavior can be explained by invoking electron-hole scattering

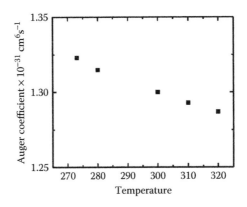

FIGURE 17.30 Variation of the Auger coefficient with temperature from the measured time delays. (Reprinted with permission from Frost, T. et al., *Appl. Phys. Lett.*, 104, 081121-1–081121-4, Copyright 2014, American Institute of Physics.)

in QDs [77,78]. It is assumed that the low energy electron states in the QDs which participate in the Auger process are discrete and that occupation of these states depends on electron-hole scattering and occupation of the hole ground state. The higher energy states into which the third carrier is transferred in the Auger process is in a continuum. In the electron-hole scattering process, which is the dominant mechanism by which hot electrons relax in QDs, electrons in the higher energy states scatter with cold ground state holes and transfer their energy. The holes lose their excess energy and thermalize rapidly via closely spaced hole states by emission of phonons. With increase of temperature, the thermal excitation of holes from the ground state to higher energy states will decrease the rate of electron-hole scattering, and therefore, the population of the electron ground state. As a consequence, the turn-on delay time will increase, and the Auger coefficient will decrease as observed.

Finally, since the Auger coefficient is proportional to $(k_B T/E_g)^{3/2} \exp(-E_g/k_B T)$, it is expected that the coefficient will increase with decrease of bandgap. The value of C_a measured here for $\lambda = 630$ nm ($E_g = 1.97$ eV) is larger than that measured in $In_{0.25}Ga_{0.75}N$/GaN green-emitting ($\lambda \sim 500$ nm) quantum dots-in-nanowire [44] and follows the expected trend of C_a versus bandgap [59].

Auger recombination is a three carrier process in semiconductors in which the excess energy released from the recombination of an electron-hole pair is transferred by Coulombic interactions to a third free carrier (either an electron or hole) deep into its respective band. This carrier can then thermalize back to the ground state by multiphonon emission. The probably of the Auger process, given by the Auger coefficient, C_a, decreases with bandgap, as mentioned in the previous section, proportionally to $(k_B T/E_g)^{3\,2} \exp(-E_g/k_B T)$. It is, therefore, expected that the values of Auger recombination be small in wide bandgap materials, including the III-nitrides an expected value of $C_a \sim 10^{-34}$ cm^{6}s^{-1}.

Calculation of Auger recombination coefficients is typically done under the assumption of defect-free crystalline material. However, reported values of Auger coefficient are typically measured from samples grown on mismatched substrates (typically sapphire or silicon carbide) with defect densities of 10^7–10^8 cm^{-2}. It is likely then that a defect-assisted Auger process is likely the cause of the higher reported values of C_s. QDs, on the other hand, restrict the movement of electrons and holes due to the physical 3-dimensional confinement present in the dots. Defects are then expected to play a smaller role in the Auger process. The measured value of $C_a \sim 10^{-31}$ cm^{6}s^{-1} in the red-emitting QDs is in the range of expected values for a material with a band gap of ~1.9 eV.

In addition to large signal modulation of laser diodes, the Auger recombination coefficient can also be measured through small signal modulation of LEDs (differential carrier lifetime measurements) [83,84], and through the measurement of PL. It should be noted that Auger recombination has also been measured in blue- and green-emitting QDs and nanowires. These materials with their relatively low defect densities have Auger coefficients in the expected range for their respective bandgap.

17.7 Future Prospects and Conclusion

Many of the challenges with InGaN/GaN quantum-well-based lasers can be alleviated through the use of InGaN/GaN QDs. The QDs form by the relaxation of strain, and therefore, have an inherently smaller piezoelectric field. Additionally, the physical 3-dimensional confinement of carriers can reduce non-radiative recombination at defects. This approach has been used in this chapter. InGaN/GaN QDs are demonstrated at emission wavelengths into the red (630 nm) and are incorporated into laser. While these advantages (reduced polarization and nonradiative recombination) are unique to nitride-based QDs, they also have many of the same advantages of QDs demonstrated with other material systems. The incorporation of self-organized QDs in GaAs- and InP-based heterostructures have resulted in superior device performance. These lasers have extremely small threshold current density, wide tunability of the output wavelength, large modulation bandwidth, and near zero chirp and linewidth enhancement factor. III-nitride-based QD lasers can emit across the visible spectrum, and their use allows for use full-color applications.

References

1. O. Nam et al., *Phys. Stat. Sol. (a)*, **201** (12): 2717 (2004).
2. H. Ohta, S. DenBaars, and S. Nakamura, *J. Opt. Soc. Am. B.*, **27** (11): B45 (2010).
3. K. Okamoto, J. Kashiwagi, T. Tanaka, and M. Kubota, *Appl. Phys. Lett.*, **94**: 071105 (2009).
4. U. Strauß et al., *Proc. of SPIE*, **6894**: 689417 (2008).
5. Y. Narukawa, Y. Kawakami, M. Funato, S. Fujita, S. Fujita, and S. Nakamura, *Appl. Phys. Lett.*, **70**: 981 (1997).
6. D. Queren et al., *Appl. Phys. Lett.*, **94**: 081119 (2009).
7. C. Skierbiszewski et al., *Appl. Phys. Lett.*, **86**: 011114 (2005).
8. C. Skierbiszewski et al., *J. Vac. Sci. Technol. B*, **30**: 02B102 (2012).
9. C. Huang et al., *Appl. Phys. Lett.*, **95**: 241115 (2011).
10. D. Bour, C. Chua, Z. Yang, M. Teepe, and N. Johnson, *Phys. Lett.*, **94**: 041124 (2009).
11. M. Ikeda, Y. Mori, H. Sato, K. Kaneko, and N. Watanabe, *Appl. Phys. Lett.*, **47**: 1027 (1985).
12. J. Rennie, Okajima, M. Watanabe, and G. Hatakoshi, *IEEE J. of Quantum. Electron.*, **29** (6): 1857 (1993).
13. A. Banerjee, T. Frost, E. Stark, and P. Bhattacharya, *Appl. Phys. Lett.*, **101**: 041108 (2012).
14. M. Zhang, M., A. Banerjee, C. Lee, J. Hinckley, and P. Bhattacharya, *Appl. Phys. Lett.*, **98**: 221104 (2011).
15. T. Frost, A. Banerjee, K. Sun, S. L. Chuang, and P. Bhattacharya, *IEEE J. Quantum Electron*, **49** (11): 923 (2013).
16. T. Frost, A. Banerjee, and P. Bhattacharya, *Appl. Phys. Lett.*, **103**: 211111 (2013).
17. T. Frost, A. Banerjee, S. Jahangir, and P. Bhattacharya, *Appl. Phys. Lett.*, **104** (8): 081121 (2014).
18. S. Schulz and E. O'Reilly, *Phys. Rev. B*, **82**: 033411 (2010).
19. Y. Wu, Y. Lin, H. Huang, and J. Singh, *J. Appl. Phys.*, **105**: 013117 (2009).
20. B. Damilano, N. Grandjean, S. Dalmasso, and J. Massies, *Appl. Phys. Lett.*, **75**: 3751 (1999).
21. Z. Chen, D. Lu, H. Yuan, P. Han, X. Liu, Y. Li, X. Wang, Y. Lu, and Z. Wang, *J. Cryst. Growth*, **235**: 188 (2002).
22. H. Hirayama, S. Tanaka, and Y. Aoyagi, *Microelectron Eng.*, **49**: 287 (1999).
23. X. Xiao, A. Fischer, G. Wang, P. Lu, D. Koleske, M. Coltrin, J. Wright, S. Liu, I. Brener, G. Subramania, and J. Tsao, *Nano. Lett.*, **14** (10): 5515 (2014).
24. M. Zhang, P. Bhattacharya, and W. Guo, *Appl. Phys. Lett.*, **97**: 011103 (2010).
25. N. Grandjean and J. Massies, *Appl. Phys. Lett.*, **72**: 1078 (1998).
26. S. Dalmasso, B. Damilano, N. Grandjean, J. Massies, M. Leroux, J. L. Reverchon, J. Y. Duboz, *Thin Solid Films*, **380**: 195 (2000).
27. K. Tachibana, T. Someya, Y. Arakawa, *Appl. Phys. Lett.*, **74**: 383 (1999).
28. F. C. Frank and J. H. van der Merwe, *Proc. R Soc Lond A*, **198**: 205 (1949).
29. M. Volmer and A. Weber, *Z. Phys. Chem.*, **119**: 277 (1926).
30. S. Deshpande, T. Frost, L. Yan, S. Jahangir, A. Hazari, X. Liu, J. Millunchick, Z. Mi, and P. Bhattacharya, *Nano Lett.*, **15** (3): 1647 (2015).
31. Y.-R. Wu, Y.-Y. Lin, H.-H. Huang, and J. Singh, *J. Appl. Phys.*, **105**: 013117 (2009).
32. M. Winkelnkemper, A. Schliwa, and D. Bimberg, *Phys. Rev. B*, **74**: 155322 (2006).
33. C. Pryor, *Phys. Rev. B*, **57**: 7190–7195 (1998).
34. S. L. Chuang and C. S. Chang, *Phys. Rev. B*, **54**: 2491–2504 (1996).
35. Y. P. Varshni, *Physica*, **34**: 149–154 (1967).
36. C. -L. Wu, H.-M. Lee, C.-T. Kuo, C.-H. Chen, and S. Gwo, *Appl. Phys. Lett.*, **92**: 162106 (2008).
37. I. Vurgaftman and J. R. Meyer. Band parameters for nitrogen-containing semiconductors. *J. Appl. Phys.* **94**: 3675–3696 (2003).
38. H. Morkoç, *Handbook of Nitride Semiconductors and Devices, GaN-based Optical and Electronic Devices*. Weinheim, Germany: John Wiley & Sons (2008).

39. S. L. Chuang. Optical gain of strained wurtzite GaN quantum-well lasers. *IEEE. J. Quantum Electon.*, **32**: 1791–800 (1996).

40. S. Nakamura, M. Senoh, S.-I. Nagahama, N. Iwasa, T. Yamada, T. Matsushita, H. Kiyoku, Y. Sugimoto, T. Kozaki, H. Umemoto, M. Sano, and K. Chocho. Continuous-wave operation of InGaN/GaN/AlGaN-based laser diodes grown on GaN substrates. *Appl. Phys. Lett.*, **72**: 2014–2016 (1998).

41. O. B. Shchekin and D. G. Deppe. The role of p-type doping and the density of states on the modulation response of quantum dot lasers. *Appl. Phys. Lett.*, **80**: 2758–2760 (2002).

42. M. Sugawara, N. Hatori, H. Ebe, M. Ishida, Y. Arakawa, T. Akiyama, K. Otsubo, and Y. Nakata. Modeling room-temperature lasing spectra of 1.3-µm self-assembled InAs/GaAs quantum-dot lasers: Homogeneous broadening of optical gain under current injection. *J. Appl. Phys.*, **97**: 043523 (2005).

43. T. Frost, A. Banerjee, K. Sun, S. L. Chuang, and P. Bhattacharya. InGaN/GaN quantum dot red (λ = 630 nm) laser. *IEEE J. Quantum Electron.*, **49**: 923–931 (2013).

44. T. Frost, A. Banerjee and P. Bhattacharya. Small-signal modulation and differential gain of red-emitting (λ = 630 nm) InGaN/GaN quantum dot lasers. *Appl. Phys. Lett.*, 103: 211111 (2013).

45. P. Bhattacharya, S. Ghosh, and A. Stiff-Roberts, *Annu. Rev. Mater. Res.*, **34** (1): (2004).

46. O. Dehaese, X. Wallart, and F. Mollot, *Appl. Phys. Lett.*, **66** (1): 52 (1995).

47. S. Deshpande, T. Frost, A Hazari, and P. Bhattacharya, *Appl. Phys. Lett.*, **105**: 141109, (2014).

48. J. G. Amar and F. Family, *Phys. Rev. Lett.*, **74** (11): 2066 (1995).

49. Y. Varshni, *Physica*, **34**: 149 (1967).

50. S. Feng et al., *J. Appl. Phys.*, **92** (8): 4441 (2002).

51. K. Ramaiah et al., *Appl. Phys. Lett.*, **87** (17): 3307 (2004).

52. C. Adelmann et al., *Appl. Phys. Lett.*, **76** (12): 1570 (2000).

53. Z. Gacevic et al., *J. Appl. Phys.*, **109**: 103501 (2011).

54. B. Damilano et al., *Appl. Phys. Lett.*, **75** (24): 3751 (1999).

55. S. Nakamura, M. Senoh, N. Iwasa, and S. Nagahama, *Jpn. J. Appl. Phys.*, **34**: L797–L799 (1995).

56. S. Nakamura, M. Senoh, S. Nagahama, N. Iwasa, T. Yamada, T. Matsushita, Y. Sugimoto and H. Kiyoku, *Appl. Phys. Lett.*, **69**, 4056 (1996).

57. Y. Enya, Y. Yoshizumi, T. Kyono, K. Akita, M. Ueno, M. Adachi, T. Sumitomo, S. Tokuyama, T. Ikegami, K. Katayama, and T. Nakamura, *Appl. Phys. Exp.*, **2**: 082101 (2009).

58. D. Sizov, R. Bhat, Fellow, and C.-E. Zah, *J. Llightwave Tech.*, **30**(5): 679 (2012).

59. J. Piprek, *Phys. Stat. Solidi A*, **207** (10): 2217 (2010).

60. Y. D. Jho, J. S. Yahng, E. Oh, and D. S. Kim, *Phys. Rev. B*, **66**: 035334 (2002).

61. M. Zhang, P. Bhattacharya, and W. Guo, *Appl. Phys. Lett.*, **97**: 011103 (2010).

62. K. Itaya, H. Sugawara, and G. Hatakoshi, *J. Crys. Growth*, **138**: 768 (1994).

63. A. Doshi, S. Y. Cheng, M. Trivedi, *System. Man. and Cybernet.*, **39** (1): 85 (2009).

64. S. Cheng, A. Doshi, and M. Trivedi, *Int. Vehicles. Symposium*, 589 (2007).

65. W. D. Jones, *IEEE Spectrum*, **39** (1), 82 (2002).

66. B. Hakki and T. Paoli, *J. Appl. Phys.*, **46**, 1299 (1975).

67. S. L. Chuang, *Physics of Photonic Devices*, New York: Wiley (2012).

68. H. Zhao, R. Arif, Y. Ee, and N. Tansu, *Opt. Quant. Electron*, **40**: 301 (2008).

69. W. Scheibenzuber, U. T. Schwarz, L. Sulmoni, J. Dorsaz, J.-F. Carlin, and N. Grandjean, *J. Appl. Phys.* **109**: 093106 (2011).

70. N. Kirstaedter, O. G. Schmidt, N. N. Ledentsov, D. Bimberg, V. M. Ustinov, A. Y. Egorov, A. E. Zhukov, M. V. Maximov, P. S. Kop'ev, and Z. I. Alferov. *Appl. Phys. Lett.*, **69**: 1226 (1996).

71. P. Bhattacharya and S. Ghosh, *Appl. Phys. Lett.*, **80** (19): 3482 (2002).

72. Y. C. Shen, G. O. Mueller, S. Watanabe, N. F. Gardner, A. Munkholm, and M. R. Krames, *Appl. Phys. Lett.*, **91**: 141101 (2007).

73. A. David, M. J. Grundmann, J. F. Kaeding, N. F. Gardner, T. G. Mihopoulos, and M. K. Krames, *Appl. Phys. Lett.*, **92**: 053502 (2008).

74. M. Zhang, P. Bhattacharya, J. Singh, and J. Hinckley, *Appl. Phys. Lett.*, **95**: 201108 (2009).

75. W. Guo, M. Zhang, P. Bhattacharya, and J. Heo, *Nano Lett.*, **11** (4): 1434 (2011).

76. K. Itaya, H. Sugawara, and G. Hatakoshi, *J. Crys. Growth*, **138**: 768 (1994).

77. A. Meneghini, N. Trivellin, G. Meneghesso, E. Zanoni, U. Zehnder, and B. Hahn, *J. Appl. Phys.*, **106**: 114508 (2009).

78. A. Laubsch, M. Sabathil, J. Baur, M. Peter, and B. Hahn, *IEEE Trans. Electron. Devices*, **57** (1): 79 (2010).

79. S. Ghosh, P. Bhattacharya, E. Stoner, J. Singh, H. Jiang, S. Nuttinck, and J. Laskar, *Appl. Phys. Lett.*, **79**: 722 (2001).

80. M. Zhang, PhD Thesis, University of Michigan, Growth and Characterization of InGaN/GaN Quantum Well and Quantum Dot Light Emitting Diodes and Lasers, 2011.

81. J. Nie, S. A. Chevtchenko, J. Xie, X. Ni, and H. Morkoç, *Proc. of SPIE*, **6894**: 689424 (2008).

82. E. Kioupakis, P. Rinke, K. Delaney, and C. Van de Walle, *Appl. Phys. Lett.*, **98**: 161107 (2011).

83. A. David and M. J. Grundmann, *Appl. Phys. Lett.*, **97**: 033501 (2010).

84. A. David and M. J. Grundmann, *Appl. Phys. Lett.*, **96**: 103504 (2010).

85. T. Frost, A. Hazari, A. Aiello, M. Z. Baten, L. Yan, J. Mirecki-Millunchick, and P. Bhattacharya. High performance red-emitting multiple layer InGaN/GaN quantum dot lasers. *Jpn. J. Appl. Phys.*, **55**(3): 032101 (February 2016).

18

GaN-Based Surface-Emitting Lasers

Kuo-Bin Hong,
Shen-Che Huang,
Yu-Hsun Chou,
and Tien-Chang Lu

18.1 Introduction to GaN-Based Surface-Emitting Lasers

The debut of semiconductor lasers came in a simple format of Fabry–Pérot (F–P) cavity with two cleaved facets as parallel mirrors. The light is generated in the active region (typically made of multiple quantum wells (MQWs) through the radiative recombination of electron and hole pairs provided by the n- and p-cladding layers, respectively. The n- and p-cladding layers are connected to the external current source to form a p-n diode. These layers are epitaxially grown on the semiconductor substrate by epitaxial equipment, for example, molecular beam epitaxy (MBE) or metalorganic chemical vapor deposition (MOCVD). In order to form an F–P cavity, the substrates have to be cleaved to reveal two end facets so that the light is able to travel back and forth parallel to the junction plane to complete a round-trip oscillation, shown as in Figure 18.1a. After the population inversion and laser threshold are met, the coherent laser light is out-coupling through the edge of the cavity. Therefore, this type of semiconductor laser is termed as the edge-emitting laser (EEL) and has been adopted and commercialized in many optoelectronic systems. However, the laser action of EELs is observed only after the substrates are cleaved. On-wafer testing would be very challenging, and the protection of cleaved facets is also critical. If the laser light emits out of the wafer surface, the cleaved facets are no longer needed, and two-dimensional laser arrays can be implemented. As shown in Figure 18.1b, the emitting direction of laser light is perpendicular to the junction plane. This type of semiconductor laser is called the surface-emitting laser (SEL).

There are several methods to implement SELs. One of the intuitive ways is to make two slanted trenches to redirect the laser light from the EEL to vertical direction, as shown in Figure 18.2a. However, accurate 45° angle of these slanted trenches along with the vertical facets of EELs are required, and that is not a tedious task in terms of the etching process. It has long been known that photonic structures,

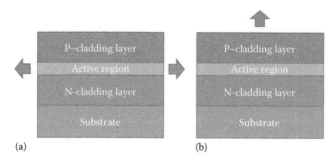

FIGURE 18.1   (a) Schematics of edge-emitting lasers (EELs). (b) Schematics of surface-emitting lasers (SELs).

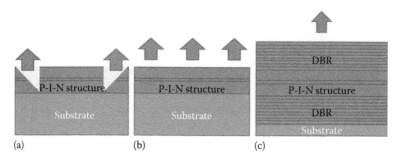

FIGURE 18.2   Typical schematics of surface-emitting lasers: (a) slanted trenches for redirecting laser light toward surface emission, (b) distributed feedback structures for surface emission, and (c) vertical cavity structure with DBRs for surface emission.

such as distributed feedback gratings or photonic crystals are utilized to manage the propagation properties of light. If the distributed feedback grating or photonic crystal (PhC) is incorporated into the EEL structure, the in-plane light shall strongly interact with the photonic structure at the resonant condition to achieve laser action. In addition, these photonic structures can alter the out-coupling direction of light to the vertical direction, as shown in Figure 18.2b. Typically, second-order grating can be regarded as many distributed submicron scale slanted trenches that facilitate surface emission. The SELs incorporated photonic crystasl are called photonic crystal surface-emitting lasers (PCSELs). As a matter of fact, another intuitive way to implement SELs is to change the oscillation direction of the laser cavity to vertical direction as shown in Figure 18.2c. For doing so in a semiconductor, as the cavity length is extremely short, a high-finesse cavity is required to maintain reasonable quality factor to achieve laser action. The high-reflectivity mirrors are commonly achieved by utilizing distributed Bragg reflectors (DBRs). Since the cavity is in the vertical direction, this kind of laser is called vertical-cavity surface-emitting laser (VCSEL).

SELs have several advantages over their counterparts, EELs. For example, the laser performance can be determined directly on the wafer. Probing, mapping, and binning can be significantly simplified in such a wafer-scale testing, which will increase the production throughput and reduce the fabrication cost. In addition, SELs can form a two-dimensional array, which can be applied in high-power emitters and high-density optical transmitters but EELs, on the contrary, can only form one-dimensional emitter array. The output divergence can be well controlled in SELs, and a symmetric output beam with diffraction limit characteristics can be achieved. Therefore, SELs have been used in many applications, such as optical communication, optical interconnect, optical mouse, laser printing, imaging, storage, surveillance, sensing, and medical instruments, to name a few.

GaN material system becomes very attractive because the band gap wavelength can cover large spectra ranging from infrared (IR) to ultraviolet (UV) and has been pervasively applied in blue and green and even UV light-emitting diodes (LEDs). The gradual transformation of solid state lighting using blue

LEDs and yellow phosphors from conventional incandescent light bulbs and fluorescent tubes makes GaN an important III-V material among others. GaN EELs have been demonstrated by Prof. Nakamura back to 1996; and have emerged as important light sources in Blu-ray DVD and automobile headlight lamps (Shuji et al. 1996). However, the demonstration of GaN SELs in terms of VCSELs and PCSELs are relatively late. The immature crystal growth method and difficult fabrication process hinder the development of GaN SELs in spite of many potential advantages and applications. Therefore, this chapter is dedicated to the report of these two kinds of GaN SELs: GaN VCSELs and GaN PCSELs. The operation principle and historical development of these two kinds of lasers will be systematically reviewed. The current status and challenges will also be stated. Conclusion and future prospects will be given at the end of this chapter.

18.2 GaN-Based Vertical Cavity Surface-Emitting Lasers

18.2.1 Development of GaN VCSELs

Like most semiconductor lasers incorporating with an optical gain medium which consisted of a material that absorbs incident radiation of target wavelength, the electrons inside the material can be excited to nonequilibrium of unstable higher energy level under either optical or electrical driven pumping, leading to relaxation processing back from excited states to a lower level via radiative transitions. As the gain is adequate to conquer the loss of internal cavity and mirror reflectors, the state is recognized as the threshold condition, and then coherent light will be emitted. Therefore, the injected radiation can be amplified by stimulated emission with the generation of additional radiative photons. The lasing oscillation has been settled and supported by continuously output pumping power to maintain the positive feedback of population inversion. Figure 18.3 indicates the difference between edge-emitting lasers (EELs) and vertical-cavity surface-emitting lasers (VCSELs).

VCSELs is one of the essential semiconductor light sources because of tremendous inherent features such as circular spot and easily fabricated array development, and several promising applications, inclusive of pico-projector (Freeman et al. 2009), micro-backlight (Inaba et al. 2008), solid state lighting (Wierer and Tsao 2015), and biosensor (Poher et al. 2008). The first VCSEL was presented in 1979 by Prof. K. Iga (Haruhisa et al. 1979). VCSELs possessed the benefits of a circular output beam, low threshold and two-dimensional array capability with respect to EELs type. Moreover, VCSELs have the characteristics of single longitudinal mode due to the short cavity length with few wavelengths order in the direction of light propagation. The longitudinal field is confined by DBRs which sandwich the active region. On the other hand, the transverse field is confined by the selective oxide or ion implanted aperture. In general, the transverse opening of VCSELs to form the current aperture is typically a few microns in diameters. Both techniques of employing the strong suppression of the high-order transverse modes and engineering transverse index guiding profile are realized to support superior confinement performance on the fundamental mode.

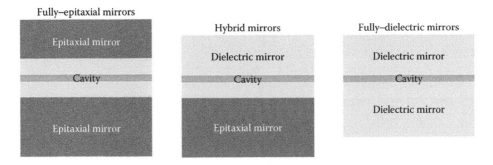

FIGURE 18.3 Schematics of operational difference between edge-emitting lasers (EELs) and vertical-cavity surface-emitting lasers (VCSELs).

To date, IR VCSEL devices have been well developed and commercialized into many applications. Following we briefly describe the development history of nitride-based VCSELs and review the milestones in demonstration of optically and electrically driven nitride-based VCSELs. Development of nitride semiconductors are expanding because of promising potential for practical LEDs and laser diodes among past few decades (Nakamura et al. 1996, 1997; Nakamura 1998). So far in the nitride material system, EELs have been successfully commercialized (Li et al. 1998); however, VCSELs are still limited. As for this nitride-based material system, there are certain difficulties in manufacturing electrically driven nitride-based VCSELs. One is the lack of appropriate substrates as result of higher defect densities. Second is the low resistivity of p-GaN layer by reason of the large activation energy of the Mg dopant. Third is the difficulty in growth of high-quality and high-reflection mirrors by using nitride-based DBRs due to the large lattice mismatch between interfaces. In addition, due to the fact that p-type GaN is of low conductivity, the GaN-based VCSEL devices commonly incorporate thin transparent conducting indium tin oxide (ITO) layer to have capability of the current spreading within the current aperture. The structure with ITO layer still suffers from the nonuniform gain distribution, poor lateral optical confinement and other optical loss which brought about a higher optical loss and threshold gain (Ehsan et al. 2013). Buried AlN current aperture has been further proposed to solve this problem (Cheng et al. 2011). However, introducing the AlN-buried structure had certain problems such as limited gain area, current crowding, and harsh leakage current phenomena over active region.

In spite of the inherent issues in the material quality or other manufacturing, the pleasing milestone makes several research groups dedicated to the nitride-based VCSELs devices. The most key challenge for nitride-based VCSELs is the lattice-matched DBRs. The basic requirement of DBRs needs to be highly reflective. Besides, providing higher refractive index contrast is also essential to the broad stopband spectra and thinner deposited thickness due to the decreased number of pairs. Since there is no denying that the difficulty to obtain high-quality nitride-based DBRs, the substitution of these structural designs can be categorized into three main types as the following description: fully epitaxial, hybrid, and fully dielectric DBRs structures, as shown in Figure 18.4.

18.2.2 GaN VCSELs with Fully Epitaxial DBRs

The past realization of room-temperature continuous-wave (CW) operation of lasing behavior of nitride-based lasers offered the possibility for nitride-based VCSELs, to be applied to high-density optical storage device through the digital versatile disc platform. Since the benefits of VCSELs structure are well known, no one can deny that the tendency of development is to replace the edge-emitting lasers by VCSELs devices.

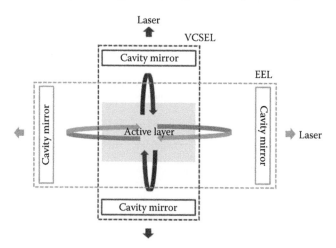

FIGURE 18.4 Schematics of three types of VCSELs with different DBR structures.

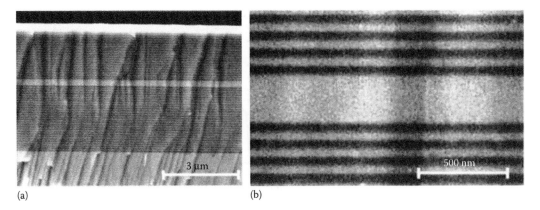

FIGURE 18.5 (a) SEM sketch of a $3\lambda/2$ cavity between two AlInN/GaN DBRs. (b) Enlarged SEM cross-section view of cavity with the abrupt AlInN/GaN interfaces.

As we mentioned before, the nitride-based VCSELs structures can be classified into three major types. The first one is monolithic grown devices, composed of epitaxial growth nitride top and bottom DBRs. Several groups had reported fully epitaxial growth nitride-based VCSELs with $Al_{0.4}Ga_{0.6}N/Al_{0.12}Ga_{0.88}N$ DBRs (structures as shown in Figure 18.5) through MOCVD fabricated technique and observed lasing behavior at 363 nm by using optical pumping (Carlin et al. 2005; Feltin et al. 2005). The all-epitaxial growth and high-quality nitride-based microcavity with lattice-matched AlInN/GaN as the top and bottom reflectors employing MOCVD method has also been illustrated to avoid breaks occurring as a result of the strain of large pairs deposition. The profit of the fully epitaxial device is the controllable cavity thickness which is beneficial to fabricate a microcavity structure. The requirement of epitaxial DBRs needs to be highly reflective, leading to high-quality factor Q value. However, the all-epitaxial nitride case is very difficult to achieve this requirement by means of the related results in the nitride-based material system, which is quite similar to the challenge of long wavelength VCSELs applied to related GaAs or InP systems. As for the wavelength range around 1300 nm, manufacturing VCSELs using GaInAsN/GaAs combination were grown on GaAs substrate but suffered from unstable active materials (Syrbu et al. 2005). In the further design of another telecommunication wavelength at 1550 nm, the VCSELs device was applied with InGaAsP and/or AlGaInAs active region on InP substrate instead of GaAs substrate (Islam et al. 2013). However, it was arduous to find appropriate DBR mirror lattice-matched materials which could offer a superior contrast in refractive index on InP substrate. As for the nitride system, growth difficulty of the lattice-matched material system and small index contrast resulting in a small stopband and more DBR pairs still impede the further development of this kind of VCSEL structure.

18.2.3 GaN VCSELs with Hybrid DBRs

The second VCSEL structure is composed of an epitaxial growth DBR and a dielectric type DBR. The VCSEL with hybrid DBRs can eliminate the growth difficulty and complicated process and keep the feasibility of coplanar contacts with dielectric DBR mesas for the future electrically pumped VCSEL applications. The main requirement for the fabrication of VCSEL with hybrid DBRs is to grow high-reflectivity and high-quality nitride-based DBRs. Since the problem of growing high-reflectivity nitride-based DBRs has been successfully solved by inserting the AlN/GaN Superlattice layers during the growth of AlN/GaN DBRs, it is supposed that the higher opportunity for achieving electrically pumped GaN-based VCSELs is based on the VCSEL structure with hybrid DBRs.

In 2008, T. C. Lu et al. first reported the CW laser operation of electrically pumped nitride-based VCSEL device (Lu et al. 2008c), as shown in Figure 18.6. The VCSEL has a 10-pair InGaN/GaN MQWs active layer

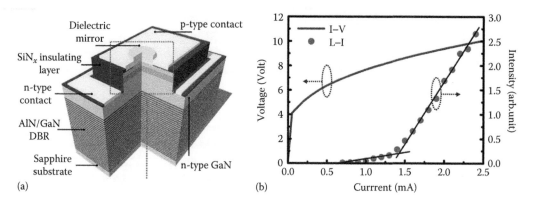

(a)

(b)

FIGURE 18.6 (a) Schematic of VCSEL device with hybrid DBRs containing with dielectric and AlN/GaN DBR structures. (b) The L-I-V characteristics of VCSEL device measured under continuous wave operation at 77 K.

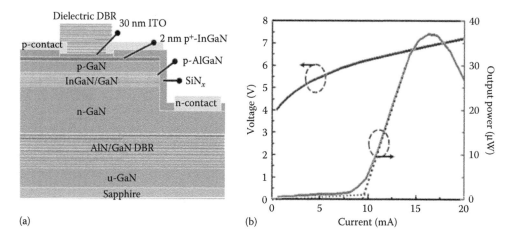

(a)

(b)

FIGURE 18.7 (a) Schematic of VCSEL device with hybrid DBRs structures incorporating with 30 nm indium-tin-oxide layer. (b) The L-I-V characteristics of continuous-wave electrical driven VCSEL device at room-temperature.

with two high-reflectivity mirrors provided by an epitaxial growth AlN/GaN DBR and a Ta_2O_5/SiO_2 dielectric DBR. The laser emitted a blue wavelength at 462 nm with a narrow line width of value 0.15 nm. The laser beam has a divergence angle of value 11.7° with a polarization ratio of value 80%. Moreover, Figure 18.7 indicated the demonstration of the CW laser action on nitride-based VCSELs at room temperature in 2010 (Lu et al. 2010). The laser structure consists of hybrid DBR cavity and 30 nm ITO layer as the transparent conducting layer. In addition, the key design of the structure is to insert an AlGaN electron blocking layer on the top of the InGaN MQW to prevent the carrier flow.

A year later, to achieve superior current and optical confinement and eliminate the absorption loss from ITO layer, B. S. Cheng et al. (2011) has reported the device with a thin AlN layer inserted on top of the InGaN/GaN MQWs as a current blocking layer and an optical confinement layer, as shown in Figure 18.8. The current can be injected more effectively in the device with a buried AlN current aperture. The output emission has a dominant emission peak wavelength at 440 nm with a very narrow line width of 0.52 nm. In 2012, G. Cosendey et al. showed nitride-based blue VCSELs utilizing highly reflective AlInN/GaN DBRs grown on *c*-plane free-standing GaN substrates and demonstrated the lasing behavior operated at room temperature under pulsed electrical injection (Cosendey et al. 2012), as shown in Figure 18.9.

Besides, in order to further investigate the lateral loss of the cavity structure through geometrical changes of current apertures and top mirror design, E. Hashemi et al. (2014) reported investigations

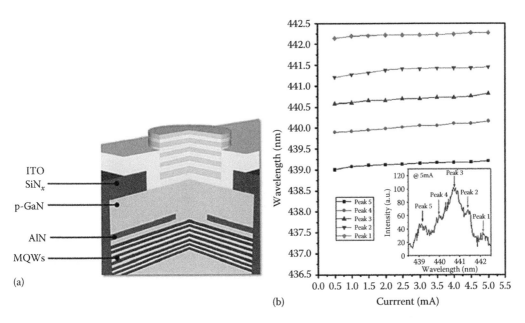

(a)

(b)

FIGURE 18.8 (a) Schematic of VCSEL device with hybrid DBRs structures incorporating with an inserted AlN aperture. (b) The cavity mode wavelength as a function of injected current. The inset showed the spectrum of device operating at 5 mA.

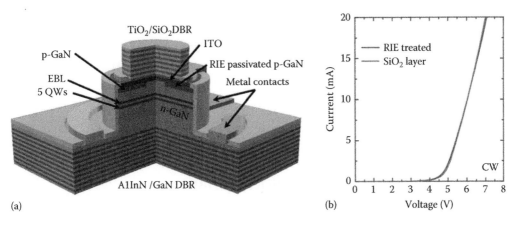

(a)

(b)

FIGURE 18.9 (a) Schematic of VCSEL device with hybrid DBRs structures. (b) The L-I characteristics of VCSEL device operating at room-temperature.

of threshold gain and loss mechanisms via numerical studies as shown in Figure 18.10. At the same year, Y. Y. Lai et al. (2014b) proposed a hybrid structure, as shown in Figure 18.11, combining an ITO layer and an intracavity AlN aperture, which could bring about profits of not only uniform current distribution but also greater lateral optical confinement. The hybrid type design showed single transverse mode operation as a result of suppression of multitransverse modes. Such hybrid design could be employed to realize high-quality electrical driven nitride-based VCSELs for future commercial platforms.

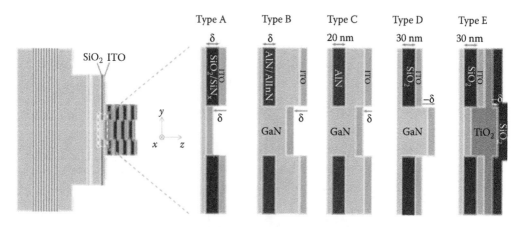

FIGURE 18.10 Schematic analysis of series investigation of current apertures.

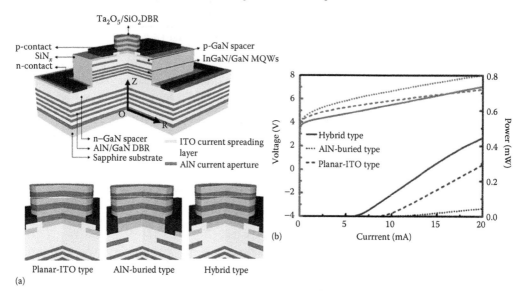

FIGURE 18.11 (a) Schematic of diagram of proposed hybrid type VCSEL and enlarged structures of different compositions. (b) The L-I-V characteristics of VCSEL device under 5 μm current aperture.

18.2.4 GaN VCSELs with Double Dielectric DBRs

The final one is the vertical resonant cavity with whole dielectric top and bottom DBRs. The VCSELs with double dielectric DBRs would be able to possess high-quality Q factors in virtue of the highly reflective DBRs structure, which is relatively easy to make. The large refractive index contrast in dielectric materials facilitate less number of pairs to keep reflectors with properties of high-reflectivity and large-stopband. However, complicated fabrication process through laser lift-off technique of the device with double dielectric DBRs is vital to make thicker the thickness of the GaN cavity. Moreover, it is hard to control the cavity thickness exactly. Such longer cavity length could diminish the microcavity effect and rise up the value of threshold condition.

In 2008, Y. Higuchi et al. reported current-injected CW lasing at room temperature in a nitride-based 7λ-thick VCSEL with two SiO_2/Nb_2O_5 DBRs on a Si substrate as shown in Figure 18.12, fabricated by wafer bonding and laser lift-off techniques (Yu et al. 2008). To improve lasing characteristics, the VCSEL employing a low dislocation density GaN substrate possessed higher output power and longer lifetimes with respect to employing a sapphire substrate in 2009 (Kunimichi et al. 2009), as shown in Figure 18.13.

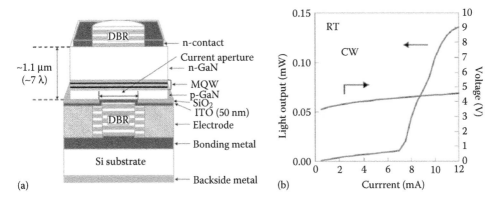

FIGURE 18.12 (a) Schematic of VCSEL device with all-dielectric DBRs structures. (b) The L-I-V characteristics of VCSEL device operating at room-temperature.

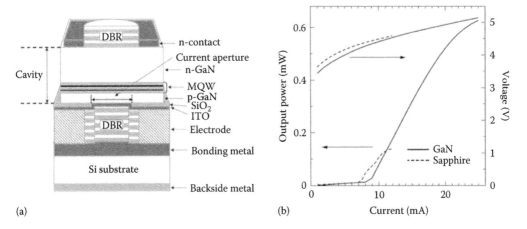

FIGURE 18.13 (a) Schematic of VCSEL device with dielectric DBRs on GaN substrate. (b) The L-I-V characteristics of VCSEL device operating at room-temperature.

T. Onishi et al. (2012) reported on the room temperature CW operation of nitride-based VCSELs with ZrO_2 and SiO_2 film stacks to form dielectric DBR mirrors in 2012, as shown in Figure 18.14. The designed schemes were divided into two parts: one was the device with a long cavity; the other was the structure with a short cavity. The former one is a sturdy design to keep away from thermal-induced peakshift of the gain, further leading up to VCSEL arrays to high-power lasing operation. The latter one might offer a powerful contribution to apply on microcavity effect or high-speed modulation.

Above-mentioned surveys with the room temperature operation were reported on the *c*-plane platform, leading to existing built-in electric field in MQWs due to the polarization-induced charges and lattice-mismatched piezoelectric field. In addition, there were the two following apparent challenges to solve in the nitride-based material system: highly reflective DBR mirrors and precise control of the cavity length. Therefore, C. Holder et al. proposed nonpolar *m*-plane nitride-based VCSEL with top and bottom dielectric DBRs and carried out under pulsed conditions (Casey et al. 2012), as shown in Figure 18.15. These designed structures employed photoelectrochemical (PEC) etching method through undercut etching to $In_{0.12}Ga_{0.88}N$ embedded sacrificial layer to remove the substrate. The proposed scheme brought about the benefit of eliminating the quantum-confined Stark effect (QCSE) phenomena due to the parallel built-in electric field along the MQW interfaces.

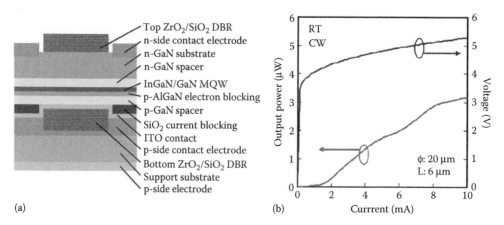

(a)   (b)

FIGURE 18.14 (a) Schematic of VCSEL device with dielectric DBRs on micro-scale polished GaN substrate. (b) The L-I-V characteristics of fabricated 6 μm cavity length VCSEL with 20 μm current aperture operating at room-temperature.

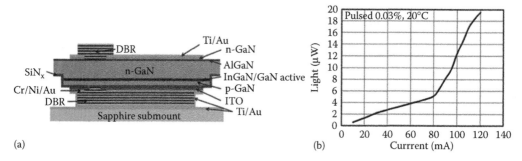

(a)   (b)

FIGURE 18.15 (a) Cross-section schematic of nonpolar VCSEL device with dielectric mirrors via flip-chip design. (b) The L-I characteristics of fabricated VCSEL under pulsed operation at room-temperature.

Similarly, the surface-textured nitride-based device incorporating with high-reflectivity all-dielectric DBRs possessed superior lateral optical confinement via a shallow etched mesa (Lai et al. 2014a), as shown in Figure 18.16. The device with 10-μm-diameter mesa has a higher Q value up to 2,600 and lower threshold energy of 30 nJ due to smaller diffraction loss and scattering loss. Based on the similar conception of superior current and optical confinement performance by employing Al ion implanted aperture and a planar ITO design (Lai et al. 2014b), J. T. Leonard et al. (2015) demonstrated improving nonpolar nitride-based VCSELs with all-dielectric Ta_2O_5/SiO_2 DBRs, as shown in Figure 18.17. To compare with a SiN_x aperture case, this single mode operation of nonpolar VCSELs possessed a decreased threshold current density down to 16 kA/cm².

On the other hand, I. Shouichiro et al. has described on room-temperature CW blue lasing emission at wavelength of 446 nm of nitride-based VCSELs through epitaxial lateral overgrowth (ELO) method, which allowed the precise control of the cavity length and the fabrication of highly reflective dielectric DBRs (Shouichiro et al. 2015), as shown in Figure 18.18. Furthermore, Figure 18.19 shows the follow-up report focusing upon the number issues of quantum wells to achieve higher output power to 1.1 mW under CW operation at lasing wavelength 453.9 nm as well as boron implantation to develop current transfer path instead of inserting a SiO_2 layer between the ITO and p-type GaN layers (Hamaguchi et al. 2016). This achievement seems to solve the formation of cracks and narrow stopband for DBRs. However, scattering loss which originated in the decoupling of the beam profile usually accompanies in the nitride-based VCSELs. Moreover, this type of nitride-based VCSEL also requires a strongly sophisticated process to decline the polished thickness and way for assuring lasing operation.

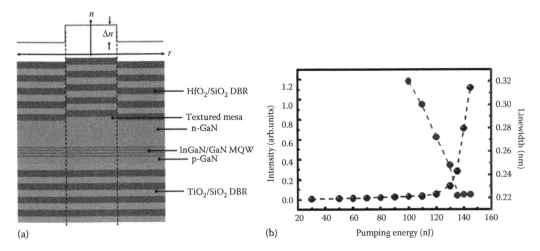

(a)

(b)

FIGURE 18.16 (a) Schematic diagram of textured device and corresponding effective refractive index distribution. (b) Light output intensity and peak line width as a function of pumping energy.

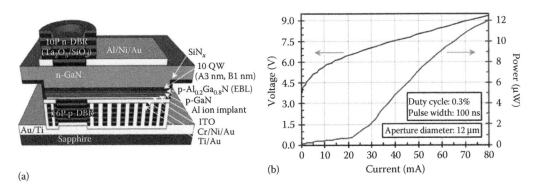

(a)

(b)

FIGURE 18.17 (a) Schematic diagram of flip-chip nonpolar VCSEL with Al aperture by ion implantation. (b) The L-I-V characteristics of VCSEL with 12 μm aperture under pulsed operation.

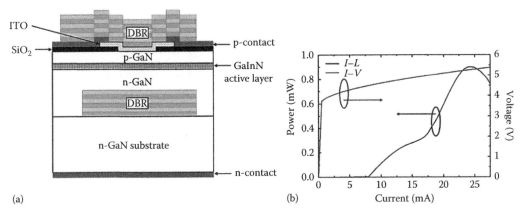

(a)

(b)

FIGURE 18.18 (a) Illustrated schematic of nitride-based VCSEL via epitaxial lateral overgrowth (ELO) method. (b) The continuous-wave L-I-V characteristics of VCSEL at room-temperature.

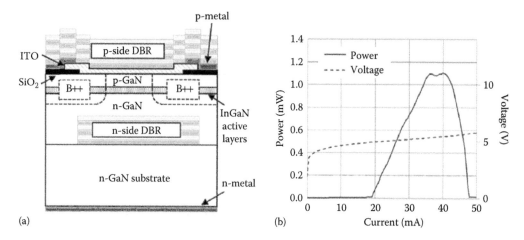

(a) (b)

FIGURE 18.19 (a) Illustrated schematic of ELO VCSEL employing electrically confined by boron implantation. (b) The L-I-V characteristics of VCSEL device with output maximum power 1.1 mW.

18.2.5 Microcavity Exciton Polariton Lasers

Microcavities (MCs) are optical cavities or resonators which have a cavity scale close to or below the wavelength of the light. Common MCs applied VCSEL structures, formed by two DBRs separated by a gain material with only a few wavelengths. An alternative way to confine lights in a smaller space is to apply the photonic crystals, which has a periodic structure in two or three dimensions surrounding an optical defect. Despite the planar structure, cavities with spherical structures which provided total internal reflection condition can produce whispering gallery modes with high-quality factor Q. Exciton-polaritons are quasiparticles, created from strong coupling between excitons and photons in the semiconductor MCs, which behave as composite bosons with very light effective mass and controllable dispersions (Chen et al. 2009a). The corresponding polariton dispersion curves of microcavity are shown in Figure 18.20. Besides, the major polariton relaxation processes in an undoped semiconductor

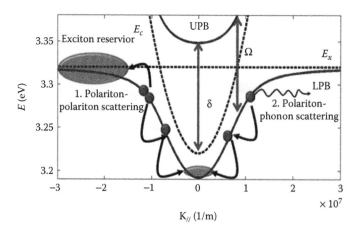

FIGURE 18.20 Schematic of polariton dispersions of a microcavity. δ was the detuning of the microcavity and Ω was the Rabi splitting.

MC were depicted: polariton-polariton scattering and polariton-phonon scattering (Weisbuch et al. 1992; Butté et al. 2006; Shimada et al. 2008; Chen et al. 2009a).

The polaritons could relax to ground state by interacting with other particles such as polaritons and phonons through the excitonic part and condense at ground state through the bosonic part. The polaritons in the condensate will spontaneously emit coherent photons as their lifetime is reached. The laser-like coherent emission is the so-called "polariton laser," which needs no population inversion condition as the traditional lasing action, could lead to realizing the ultralow threshold coherent light sources. The first observation of exciton–polaritons in semiconductor MCs was reported in the GaAs-based material system at low temperatures and soon after realized in the CdTe-based system (Weisbuch et al. 1992). Unfortunately, the small exciton binding energy limits the operation temperature of the strong-coupling phenomenon in GaAs-based or CdTe-based MCs at the cryogenic region, which is hard to perform practical polaritonic devices at room temperature. Since then, the attention of practical polaritonic devices was transferred to the wide band gap material systems, such as GaN and ZnO with much larger exciton binding energies to persist the strong coupling regime (Butté et al. 2006; Shimada et al. 2008).

The polariton condensations and polariton lasers have been observed at room temperature in nitride-based MCs including bulk and GaN/AlGaN MQWs structures due to the exciton binding energies in these systems are comparable to or even larger than the thermal energy (Butté et al. 2006; Christmann et al. 2006, 2008a, 2008b; Christopoulos et al. 2007). In order to achieve the goal of practical laser operation, electrical driven devices under room temperature will be particularly important. Therefore, III-nitride-based MC has been regarded as the most promising material to achieve this goal. The first electrical pumped polariton LED operating at room temperature were confirmed through temperature-dependent and angle-resolved electroluminescence spectra in the nitride-based system (Lu et al. 2011) and the lasing behavior has also been later observed (Bhattacharya et al. 2014), demonstrating a new vertical injected lateral III-nitride cavity to achieve the room-temperature electrical excitation of polariton lasers and observed the double threshold phenomena. These studies would undoubtedly inspire more related applications of polariton lasers. From the prospect, ZnO-based MCs are believed to be another appropriate system for the realization of room-temperature polariton lasers compared to the III-nitride system, since their exciton binding energy and oscillator strength are even larger than that of the GaN system (Christmann et al. 2008b). The polariton lasing actions in ZnO-based MCs from 120 K to room temperature were reported in bulk and nanowire excitonic structures (van Vugt et al. 2006; Guillet et al. 2011). The operating temperature has been increased to 250 K (Helena et al. 2012) and the first room-temperature polariton lasing was realized in a hybrid microcavity with an ultralow threshold (Lu et al. 2012). In recent years, the ZnO polariton laser has been demonstrated to be operated at 353 K with the same cavity structure (Lai et al. 2012).

18.2.6 Summary

We have reviewed and attempted to give the reader a broad overview of current research on the evolution of III-nitride-based VCSELs to date. The motivated development, processing techniques, and concerned examinations were briefly discussed both optical and electrical properties of three aforementioned types of nitride-based VCSEL with DBRs structures: VCSELs with fully epitaxial DBRs, VCSELs with hybrid DBRs and VCSELs with fully dielectric DBRs. These three approaches own their pros and cons, and the improved structures are still under development and await many research groups to dive in. As for the MC exciton-polariton lasers, it is regarded that the physical characteristics and related applications could offer a significant potential and booming opportunity toward commercialized practical use in the near future.

18.3 GaN-Based Photonic Crystal Surface-Emitting Lasers

18.3.1 Fundamental Concept of PhC

PhC is a periodic geometry composed of two or three dielectrics that can be fabricated in one-, two-, or three-dimensional micro/nanostructures. The optical characteristic of the PhC can affect the fundamental behaviors of light (photons) in such periodic structure. This characteristic is similar to the atomic structure influencing the motion of electrons in semiconductors. Electromagnetic waves with different wavelengths propagate in the PhC are addressed modes, connected modes to become bands. And the spacing between bands without the allowing modes at a specific wavelength range is called photonic band gap. Photonic band gap brings unique optical characteristics, such as inhibition of spontaneous emission and high-reflecting mirror, and so on.

The research of PhC dated back to J. W. S. Rayleigh's work in 1888 (Rayleigh 1888). Lord Rayleigh's study showed that such one-dimensional periodic multilayers have a photonic band gap. Over 100 years later, E. Yablonovitch and S. John published two most influential articles on PhC. Yablonovitch expounded that the PhC can alter the density of photon states to achieve effective control of spontaneous emission (Yablonovitch 1987) and an original idea of John indicated that the PhC can be utilized to restrict the motion of light such as localization and propagation (John 1987). Since then, PhCs have attracted considerable attentions as a new optical cavity and are used in various applications of optoelectronic devices. Apart from this, an advantage of PhC allows the PhC devices to be easily integrated with other optical components on semiconductor chips.

Since the PhC micro/nanostructures can affect the propagation of electromagnetic wave in the materials with periodic dielectric contrast to form the photonic band structures, it is important to understand the phenomenon of photonic band gap in order to design the PhC micro/nanostructures. If the electromagnetic wave is monochromatic in the time domain, the time-dependent electromagnetic wave equation can be reduced to the Helmholtz equation by assuming the electromagnetic fields as $E(x, y, z)e^{i\omega t}$ or $H(x, y, z)e^{i\omega t}$. For a 2D PhC system, the electromagnetic wave fields are usually and conveniently decoupled into the transverse electric (TE) and the transverse magnetic (TM) fields. The corresponding two-dimensional electromagnetic wave equations for TE and TM modes can be written as

$$\nabla \times \left(\frac{1}{\varepsilon_r} \nabla \times H_z(x, y) \right) = k_0^2 \mu_r H_z(x, y) \tag{18.1}$$

$$\nabla \times \left(\frac{1}{\mu_r} \nabla \times E_z(x, y) \right) = k_0^2 \varepsilon_r E_z(x, y) \tag{18.2}$$

where $k_0 = 2\pi/\lambda$ is the wavenumber in the vacuum. ε_r and μ_r are relative permittivity and permeability of the dielectric material, respectively. In general, the vast majority of dielectric materials is nonferromagnetic, leading to $\mu_r = 1$. To engineer the band structures of PhC, some computational modeling and simulation methods were implemented and developed, such as plane wave expansion method (PWEM), finite element method (FEM), finite difference time domain method (FDTD), Bloch wave–MoM method (Kastner 1987) and Korringa-Kohn-Rostoker (KKR) method (Korringa 1947; Kohn and Rostoker 1954). These methods are used depending on the exact goal of design and computation resources.

A suitable matrix with periodic inclusions or holes can be made using photolithography etching process. So far, the circular rods or holes arranged to square and hexagonal (also known as triangular) lattices have already been successfully employed. The schematics of square and hexagonal lattices for real and reciprocal spaces are shown in Figure 18.21. Detailed description about the direct and reciprocal lattices are annotated as follows: the gray circles with a radius of r means the lattice points in real space and center-to-center distance called lattice constant, a. The rectangle and parallelogram are the unit cell of square and hexagonal lattices for calculating the photonic band structures. Besides, the

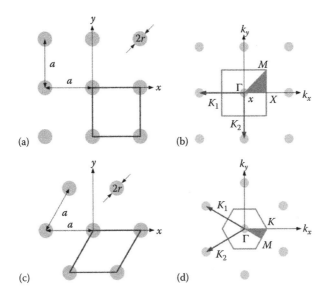

FIGURE 18.21 (a), (c) Real and (b), (d) are the reciprocal spaces for square lattice and hexagonal lattices.

square and hexagon drawn in Figure 18.21b and indicate the first Brillouin zones of square and hexagonal lattices. Γ, X, and M (Γ, K, and M) are the high-symmetry points in reciprocal space for the square (hexagonal) lattice. However, K_1 and K_2 labeled in Figure 18.21b and d denote the reciprocal unit vectors. If the lattice's features are specified, we can solve Equations 18.1 and 18.2 with periodic boundary conditions in particular directions to obtain the band structures. Here, for an illustrated example, a hexagonal lattice consists of two different refractive indices (2.65 for matrix and 1.55 for circular inclusions). The lattice constant and the ratio of r/a are set to 183 nm and 0.25.

The TE mode band diagrams of hexagonal lattice and corresponding mode profiles near the Γ band edges calculated by use of FEM are demonstrated in Figure 18.22. Solid lines in Figure 18.22a are the photonic bands in the Γ-K, K-M, and M-Γ directions. Modes A–F represent the first six band edge modes near the Γ point. We can clearly see that the modes C and D are two degenerated eigenmodes. Similarly,

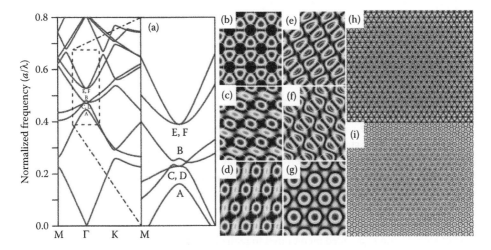

FIGURE 18.22 (a) Photonic TE mode band diagrams of hexagonal lattice, (b)–(g) are corresponding z component of magnetic field of the modes A–F for an infinite periodic lattice, (h) and (i) are the magnetic field profiles of nondegenerated modes A and B for a photonic crystal with finite size.

modes E and F are also degenerate. These six mode profiles of an infinite periodic lattice are well illustrated by the images shown in Figure 18.22b–g. It is worth noting that the mode A is almost entirely confined in the matrix compared to other modes.

On the other hand, the magnetic field profiles $H_z(x,y)$ of nondegenerated modes A and B for a PhC with finite size are also calculated as displayed in Figure 18.22h and i, respectively. The calculated wavelength for modes A and B are 410.8 and 381.1 nm. In addition, the relevant high-quality factors (Q) for these two nondegenerated modes are 4530 and 7455, respectively. In typical laser physics, the transparency condition can be calculated to estimate the threshold gain. The transparency gain can be expressed as $\gamma_{th} = 2\pi n_{av}/\lambda \cdot Q \cdot \Gamma_{xy}\Gamma_z$, where the Γ_{xy} (Γ_z) and n_{av} are the in-plane (out-off plane) optical confinement factor and mode effective index. Based on calculated optical characteristics, we can design the PhC cavity and regulate the emission wavelength. Next section, we will discuss the operation principle and optical characteristics of a photon crystal surface-emitting laser (PCSEL).

18.3.2 Operation Principle and Characteristics of PCSELs

In this section, the classification, optical characteristics, and operation principle of PhC laser will be discussed. Generally, there are two categories of PhC lasers, which has been fabricated and studied over the past few decades. One is the defect type PhC laser that has a point (or line) defect with an optical gain surrounded by a PhC mirror to form an optical resonant cavity. Due to the feature of photonic band gap caused by the PhC nanostructure, light propagated in the PhC would be confined in the defect region and associated lasing actions would be arisen from the defect. This kind of PhC laser is also called as PhC band-gap laser. The PhC band-gap laser typically has a high Q value and a small modal volume. Such defect type PhC micro/nanostructures can achieve a large Purcell effect and low-threshold lasing (Painter et al. 1999; Noda et al. 2000; Park et al. 2004). The brief reviews will summarize specific literature regarding the origin and establishment of the photonic band-gap laser.

In 1999, O. Painter et al. practically demonstrated an optically pumped InGaAsP-based two-dimensional PhC with a single defect (Painter et al. 1999). The schematic diagram of this device, shown in Figure 18.23a, consists of a waveguide with a thickness of $\lambda/2$, and PhC mirrors for vertical and lateral localizations. The scanning electron microscope (SEM) image and simulated lasing mode profile excited by a y-polarized dipole are shown in Figure 18.23b and c. The structure was optically pumped by a semiconductor laser with a wavelength of 830 nm with a focused spot size of about 3 μm. Lasing action was observed at 143 K with pulses of 10 ns. The corresponding lasing peak located at 1504 nm. In 2004, H. G. Park et al. experimentally realized the electrically driven single-cell photonic band gap laser operating at room temperature and the single-mode lasing action was observed at a wavelength of 1519.7 nm (Park et al. 2004). The schematic diagram and SEM image are shown in Figure 18.24.

Above-mentioned special properties such as a small modal volume and a localized lasing mode are important for the development of ultralow threshold lasers and photonic integrated circuits. On the other hand, most of the defect-type PhC lasers with a thin membrane suspended in the air have been realized in InP and GaAs material systems due to the easy removal of underlying sacrificial layers. However, it is rather difficult to fabricate suspended membrane structures in GaN-based micro/nanostructures by selective etching due to their stable chemical bonding properties (Meiera et al. 2006; Arita et al. 2007).

The other type of PhC laser is called PhC band-edge laser, where the lasing action of this kind of PhC laser can be generated without any defined cavity and mirror. Since light at the photonic band edge has a group velocity approaching to zero and constitutes the specific Bragg diffractions to strongly interact with a gain medium, the threshold can be achieved at the band-edge position due to the multidirectional distributed feedback mechanism. In contrast to the PhC band-gap lasers, the PhC band-edge lasers do not require thin suspended membranes to realize vertical optical confinement and the surface-emitting conditions can be acquired by properly selecting the band-edge position above the light cone. For the PhC band-edge lasers, the specific Bragg diffraction could occur at the photonic band edges to achieve surface-emitting condition and laser oscillation in a large lasing area and a low divergence angle

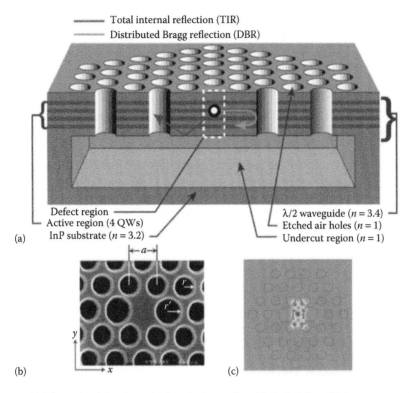

FIGURE 18.23 (a) Schematic of the photonic crystal microcavity with InGaAsP multiple quantum wells. (b) SEM image of defect type photonic crystal, a hole placed in the center of hexagonally arranged pattern is missing. (c) Electric field intensity of defect type PhC for the *y*-dipole mode.

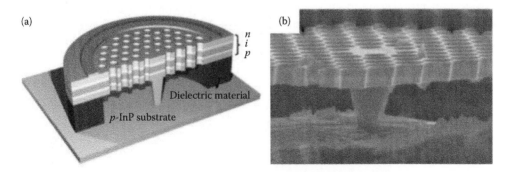

FIGURE 18.24 (a) The schematic of electrically driven single-cell photonic band gap laser. (b) SEM image of single-cell photonic band gap laser, the region around the central post is clearly shown.

by controlling the period and lattice of PhCs which are called PCSELs. This kind of PhC laser could potentially be applied to display and high-power laser. PCSELs has been explored in many different wavelength regions during the past decade (Vurgaftman and Meyer 2003; Kim et al. 2006; Matsubara et al. 2008; Lu et al. 2008b; Sirigu et al. 2008; Kawashima et al. 2010). Following brief reviews will provide a preliminary understanding of PCSELs.

In 1999, M. Imada et al. reported InP-based two-dimensional PhC SEL with a hexagonal lattice operated at room temperature under pulsed current injection (Imada et al. 1999). The lasing action of InP-based PhC was investigated, and PhC is integrated with InGaAsP/InP MQWs by the wafer fusion technique. The designed PhC consists of air circular rods with a radius of 245 mm. The lattice constant was 462 nm and the depth of air rods was designed as 100 nm. The corresponding filling factor was estimated to be 0.2 from the SEM image. They demonstrated the single-mode, large-area, and surface-emitting lasing action with a narrow divergence angle, and analyzed the lasing mechanism that satisfied the Bragg condition, which clearly demonstrated the two-dimensional coherent oscillation based on the nature of the PhC.

In 2004, D. Ohnishi et al. realized the room temperature PhC band-edge laser operated under continuous current injection condition (Ohnishi et al. 2004). This opens a new road toward the large-area single-mode surface-emitting laser, and the relevant schematic diagram is shown in Figure 18.25a. The InP-based PhC contains air holes with depth of about 100 nm which are arranged in square lattice. The lattice constant was 286.25 nm and the optimal air-filling factor was between about 0.1 and 0.15. As shown in Figure 18.25b, the light propagating in Γ-X direction (0°) is considered and the diffracted light travels along the reverse direction (180°) which originated from the Bragg diffraction. The light will also diffract in other Γ-X directions (+90° and −90°). Moreover, light propagating in the four equivalent Γ-X directions are coupled with each other and the light can also be diffracted toward the vertical direction by first-order Bragg diffraction, as shown in Figure 18.25c, for which this device works as a surface-emitting laser. The report revealed that the lasing peak of this device was located at 959.44 nm and output power over 4 mW under CW operation. The line width of the spectrum can be estimated to be 0.35 nm, and the corresponding threshold current was 65 mA. Large-area single mode oscillation with beam divergence angle of about 1.1° was observed. In addition, many interesting and appealing physical investigations about the realizations of watt-class high-power output and dynamic emission control and mode stability for developments of PhC laser devices have been widely discussed (Song et al. 2005; Hirose et al. 2014; Inoue et al. 2014; Liang et al. 2014).

According to the Bragg diffraction theory, the first-order Bragg diffraction with two-dimensional PhC with hexagonal lattice will be introduced. The high-order diffraction mechanism is also shown in this section including K_2 and M_3 PhC modes.

Figure 18.26 shows a photonic band diagram with a PhC triangular lattice. Among of the labels A, B, C, D, and E in the band diagram, each of them presents the different band-edge lasing modes including Γ_1, K_1, M_1, K_2, and M_3 which can control the light propagated in different lasing wavelength and band-edge region. A schematic diagram of the hexagonal PhC in the reciprocal space transferred from the real space is shown in Figure 18.21d. The K_1 and K_2 are the Bragg vectors with the same magnitude, $|K|=2\pi/a$. The parameter a is the PhC lattice constant. Considered the TE modes in the PhC nanostructure,

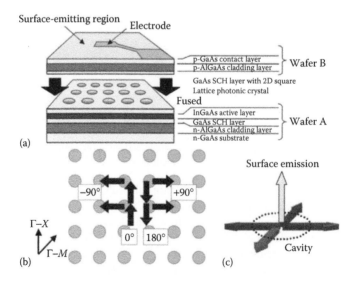

(a)

(b)

(c)

FIGURE 18.25 (a) Schematic diagram of the InP-based PhC band-edge laser with square lattice. (b) A sketch of a square lattice photonic crystal. Thin black arrows denote the Γ-X and Γ-M directions, and bold arrows indicate propagating waves. (c) Dark gray arrows the propagating directions of the coupled waves.

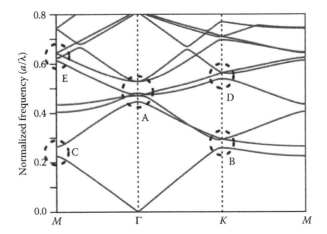

FIGURE 18.26 The band diagram of PhC with hexagonal lattice for lattice constant of 183 nm and the $r/a = 0.25$.

the diffracted light wave from the PhC structure must satisfy the laws of conservation of momentum (Bragg's law) and energy. These two conservation laws can be expressed as follows.

$$k_d = k_i + q_1 K_1 + q_2 K_2, \quad q_{1,2} = 0, \pm 1, \pm 2, \ldots \tag{18.3}$$

$$\omega_d = \omega_i \tag{18.4}$$

where:
k_d is an in-plane wave vector of diffracted light
k_i is the wave vector of incident light
$q_{1,2}$ indicate the order of coupling
ω_d is the frequency of diffracted light wave
ω_i denotes the frequency of the incident light wave

Equation 18.3 represents the linear momentum conservation in the reciprocal space, and Equation 18.4 represents the energy conservation. When both of equations are satisfied, the lasing mode can be determined.

In the calculation, the PhC band-edge lasing behavior would occur at specific points on the Brillouin-zone boundaries including Γ, M, and K which would split and cross. At these PhC lasing band-edge modes, waves propagating in different directions would be coupled and increased the density of state (DOS). Each of these band-edge modes exhibits the different types of wave coupling routes. For example, the coupling at band-edge M_1 (point C) involves two waves, propagating in the forward and backward directions as shown in Figure 18.27c. Depending on different structures, all of them show the similar coupling mechanism but different lasing behavior. Since they can be divided into six equivalent Γ-M directions, the cavity can exist independently in each of the three different directions to form three independent in-plane lasers.

Band-edge K_1 (point B) has a unique coupling characteristic as shown in Figure 18.27b. It forms the triangular shape resonance cavity propagating in three different directions compared with the conventional distributed feedback (DFB) lasers. On the other hand, the point B can also be six Γ-K directions in the structure shown two different lasing cavities in different Γ-K directions coexisted independently.

In Figure 18.27a, as for the Band-edge Γ_1 (point A), the coupling waves in the in-plane directions include six directions: $0°$, $60°$, $120°$, $-60°$, $-120°$, and $180°$. In order to satisfy the first-order Bragg diffraction, the coupled light can emit perpendicular from the sample surface as shown in Figure 18.27d. Therefore, the PhC devices can function as a surface-emitting laser.

Furthermore, higher order Bragg diffractions can function well in the two-dimensional PhC with a hexagonal lattice. For example, at point D, waves satisfied the Bragg's law and Figure 18.28a and b show the in-plane and vertical diffraction of the light wave diffracted in three Γ-K directions to three K points. In the

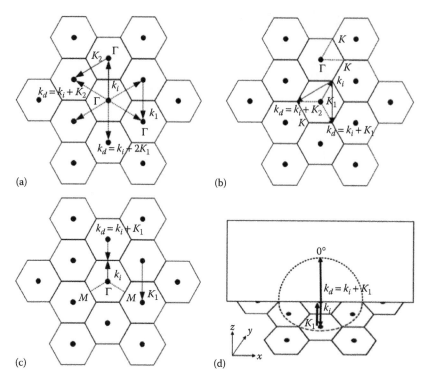

FIGURE 18.27　(a), (b), and (c) are the wave vector diagrams at band-edges Γ_1, K_1, and M_1 in Figure 18.26, k_i and k_d indicate the incident and diffracted light waves. (d) The wave vector diagram at point A in vertical direction.

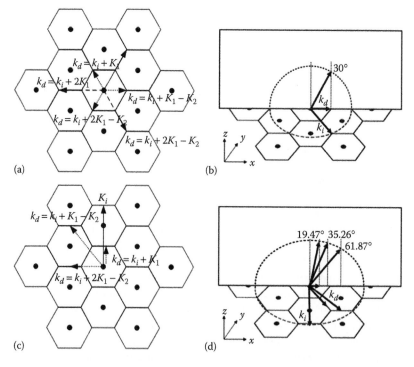

FIGURE 18.28　(a) Wave vector diagram and (b) vertical direction at point D (or band-edge K_2); k_i and k_d indicate incident and diffracted light wave. (c) Wave vector diagram and (d) vertical direction at point E (or band-edge M_3).

wave-vector diagram of one K point, the light wave is diffracted to a tilted angle 30° off from the normal direction to the sample surface as shown in Figure 18.28b. Therefore, the lasing behavior of K_2 mode would emit at this specific angle of about 30°. At point E, Figure 18.28c and drepresent the in-plane and vertical diffraction that the light waves are diffracted in two different Γ-M directions and reach to three M points. Figure 18.28d shows the wave-vector diagram of one M' point where the light wave is diffracted into three independent tilted angles of about 19.47°, 35.26°, and 61.87° off the normal direction to the sample surface.

18.3.3 Development of GaN-Based PCSELs

First lasing action and the mode selection of InP-based PhC with square lattice has been reported at 2001 (Susumu Noda et al. 2001). The following years, several material systems have been applied to promote PhC laser with emission wavelength ranging from visible and near IR (Vurgaftman and Meyer 2003; Kim et al. 2006; Matsubara et al. 2008; Lu et al. 2008a; Sirigu et al. 2008; Kawashima et al. 2010). Moreover, the lasing action of the PCSELs with hexagonal lattice including the organic material has been investigated (Imada et al. 1999; Notomi et al. 2001). Lately, effects of spatial fluctuations in air hole's radius and position on the optical features of PhC nanocavities have been done, and the calculated quality factor was dependent on the spatial fluctuations of air holes (Hagino et al. 2009). Furthermore, developments of three-dimensional coupled-wave theory (CWT) for the optical field calculations of PCSELs with square and hexagonal lattices were carried out (Liang et al. 2012, 2013).

Recently, K. Hirose et al. demonstrated a watt-class high-power 980 nm two-dimensional PCSEL under room temperature CW conditions. The band-edge resonant effect of a PhC fabricated by MOCVD with the regrowth technique gives rise to a high-beam quality of $M_2 \leq 1.1$, narrowing the focus spot by two orders of magnitude compared to VCSELs. Their demonstration shows that surface-emitting lasers can be very promising in high-power applications (Inoue et al. 2014).

As for the development of GaN-based PCSELs, the first electrically driven PCSELs operated at blue-violet wavelength has been reported by Prof. S. Noda's group (Matsubara et al. 2008). A novel fabrication technology called "air holes retained overgrowth" has been utilized to realize current injected PhC laser devices which possess good lasing performances, such as the low threshold current density and narrow beam divergence angle. Shortly afterwards, many reports about the GaN-based PCSELs have been presented (Lu et al. 2008a, 2008b; Chen et al. 2009b, 2010; Wu et al. 2011; Kim et al. 2011; Weng et al. 2011, 2012). In 2009, T. C. Lu et al. fabricated and demonstrated the GaN-based two-dimensional PCSELs with AlN/GaN DBRs. Hexagonal PhC pattern with the lattice constant a ranging from 190 to 300 nm and the circular hole radius r chosen such that r/a is about 0.28. The lasing threshold energy density is about 3.5 mJ/cm^2 under optical pumping at room temperature. One dominant emission wavelength of 424.3 nm with a narrow line width of 1.1 Å above the threshold was observed. The lasing wavelength emitted from PhC lasers with different lattice constants occurs at the calculated band edges provided by the PhC patterns (Lu et al. 2008a). In addition, S. W. Chen et al. studied the GaN-based PCSELs with AlN/GaN DBRs. PhC pattern with the lattice constant a ranging from 190 to 300 nm and the circular hole radius r chosen such that r/a ranged from 0.18 to 0.3. The schematic layer structure of GaN-based PCSELs and SEM image of the PhC structure with hexagonal lattices are shown in Figure 18.29. Clear threshold characteristic under the optical pumping at room temperature was also observed at about 2.7 mJ/cm^2 with lattice constant of 234 nm. Above the threshold, only one dominant peak appears at 401.8 nm with a line width of 0.16 nm. The lasing wavelength emitted from PhC lasers with different lattice constants occurs at the calculated band edges, showing different polarization angles due to the light diffracted in specific directions, corresponding exactly to Γ-, K-, and M-directions in the K space. The PCSEL also showed a spontaneous emission coupling factor β of 5×10^{-3} and a characteristic temperature of 148 K (Chen et al. 2009b).

In 2010, S. W. Chen et al. have further investigated the lasing characteristics of GaN-based two-dimensional PCSELs with different PC lattice constants by using angled resolved spectroscopy (Chen et al. 2010). Due to the Bragg diffraction theory, the normalized frequency of lasing wavelength of PCSELs can be exactly matched with three distinct band-edge frequencies (Γ_1, K_2, and M_3) in the photonic band diagram. The measured ARPL diagram near the (a) Γ_1, (c) K_2, and (e) M_3 band edges are shown in Figure 18.30.

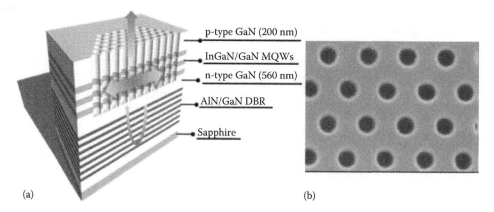

(a)   (b)

FIGURE 18.29 (a) Schematic of GaN-based PCSELs with bottom AlN/GaN DBR. (b) Top-view SEM image of the PhC structure with hexagonal lattices.

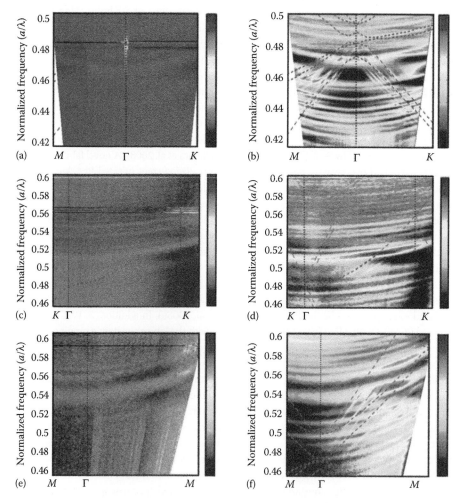

FIGURE 18.30 The measured ARPL diagram near the (a) Γ_1, (c) K_2, and (e) M_3 modes pumped by YVO4 pulse laser, (b) Γ_1, (d) K_2, and (f) M_3 modes pumped by He–Cd laser. The dashed lines represent the calculated photonic band dispersion curves.

Figure 18.31 shows the emission angles and divergence angles of Γ_1, K_2, and M_3 band-edge modes on the normal plane from the sample surface. The lasing emission angles are (0°, 29°, and 59.5°) and the divergence angles of laser beams are (1.2°, 2.5°, and 2.2°) for (Γ_1, K_2, and M_3) band-edge modes, respectively. On the other hand, T. T. Wu et al. (2011) fabricated and investigated the GaN-based PCSELs with different central defects. The threshold energy densities increased from 3.23 to 3.51 mJ/ cm² when the central defect size increased. In addition, lasing wavelengths blue-shifted from 400 to 390 nm for the PCSEL cavities with larger central defects. The tendency of threshold gain and resonance wavelength for PCSELs with different central defects were calculated by the multiple scattering methods and good agreement with the experimental results was carried out. In 2011, D. U. Kim et al. reported the fabrication of a GaN-based membrane-type PhC band-edge laser. Schematic of PhC surface-emitting band-edge laser with honeycomb lattice and SEM images of band-edge laser device are shown in Figure 18.32.

The membrane-type PhC laser claims a smaller PhC active area than previous designs due to strong optical confinement. Threshold energy density of this band-edge laser operated at room temperature under optically pumped condition was 15.5 mJ/cm². Based on polarization angle analysis, the band-edge laser indeed lased at the Γ_1 monopole band-edge mode (Kim et al. 2011). Moreover, P. H. Weng et al.

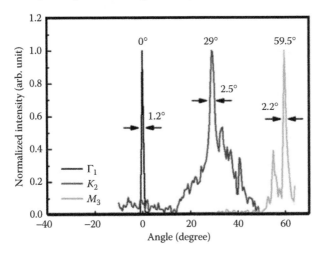

FIGURE 18.31 The emission angles and divergence angles of Γ_1, K_2, and M_3 band-edge modes on the normal plane from the sample surface.

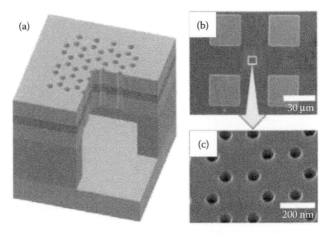

FIGURE 18.32 (a) Schematic of the PhC surface-emitting band-edge laser. SEM images of (b) the band-edge laser device and (c) the honeycomb lattice.

have analyzed threshold gains and lasing modes in GaN-based PCSELs by using the multiple scattering methods (MSM) for hexagonal lattice PhC patterns. The GaN-based PCSELs with different boundary shapes have been fabricated and measured. SEM images of PhC patterns with the circular and hexagonal boundaries of a GaN-based PCSEL are shown in Figure 18.33a and b. As shown in Figure 18.33c, the threshold amplitude gains of modes A–D are calculated as a function of the hole-filling factor by the MSM. However, the resonant mode frequencies calculated by the MSM will approach the band-edge frequencies as the shell number increases. The threshold gains of four resonant modes are varied with the filling factor. Among all the resonant modes, modes A and B have the lowest threshold gain for hole-filling factors of 35% and 30%, while mode C and D have the lowest threshold gain for hole-filling factors of 10% and 15%. This result shows that the hole-filling factor controls the mode selection scheme. The proper hole-filling factor can be selected to fabricate PCSELs operated with a specific mode. Furthermore, the lasing mode at the Γ band edge can be identified by using the angular-resolved spectroscopy and pumping energy densities for the circular boundary, and the hexagonal boundary PCSELs are 2.8 and 3.1 mJ/cm², as shown in Figure 18.33d. Threshold conditions in the GaN-based PCSELs with different boundary shapes are obtained by optical pumping and agree well with simulation results (Weng et al. 2011).

As discussed above, there have been several literatures focusing on the theoretical analysis of the distributed feedback mechanism on PCSELs by using different calculation methods such as PWEM, CWT, MSM, FEM, and FDTD. Although the aforementioned different theoretical methods take advantageous positions in some respects, while analyzing certain laser characteristics, each of them suffers limitations in the calculation of the PCSEL structures. CWT can provide a quick estimation of threshold conditions in PCSELs but the boundary condition needs to be carefully deal with, and the analysis of arbitrary defect or lattice patterns requires further development. FEM and FDTD consume numerous computer memories and calculation time to simulate the finite PC structure. In spite of the three-dimensional calculation, it is a complex way to calculate the threshold gain. On the other hand, PWEM applies only to the infinite PhC patterns instead of the real device such that the boundary

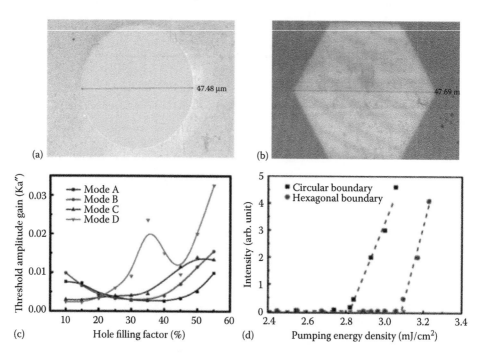

FIGURE 18.33 SEM images of the (a) circular boundary and the (b) hexagonal boundary PhC patterns of a GaN-based PCSEL. (c) Threshold gain of four modes as a function of the hole filling factor. (d) Measured output intensity versus input excitation energy density.

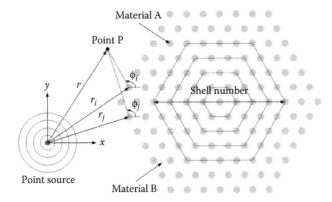

FIGURE 18.34 Representation of the field at point P contributed by scattered fields from various circular photonic structures and a point source used in the MSM.

conditions of the PCSELs are not taken into account. Finally, the MSM shows good capability, such as faster calculation time, considering gain coefficients within PhC materials with an arbitrary defect or lattice patterns. Therefore, we would briefly introduce the multiple scattering method as follows.

The simulation structure is composed of finite two-dimensional PhCs nanostructure with hexagonal lattice patterns and parallel cylinders placed in a uniform background material as shown in Figure 18.34. The complex dielectric constant includes the light amplification in the background material (e.g., GaN) as follows (Weng et al. 2012):

$$\varepsilon_B(\omega) = \varepsilon_b(\omega) - i \frac{2c\sqrt{\varepsilon_b(\omega)}}{\omega} k_b'' \tag{18.5}$$

The constant $\varepsilon_b(\omega)$ represents the dielectric constant varied with the frequency of light, and the k_b'' represents the amplitude gain coefficient of material B. A point source at the origin point emits a monochromatic wave (ω) to cylinders with the finite number, locating at r_i. The scattered field from the jth cylinder can be written as

$$\Phi_s(r, r_j) = \sum_{n=-\infty}^{\infty} i\pi A_n^j H_n(k_B|\mathbf{r} - \mathbf{r}_j|) e^{in\varphi_j} \tag{18.6}$$

where $H_n(k|\mathbf{r}|)$ is the nth-order Hankel function of the first kind, k is the wave number, ϕ is the azimuthal angle, and A_n^j is the nth Bessel coefficient of the jth cylinder.

Similarly, the total incident field around the ith cylinder can be expressed as in the Bessel function

$$\Phi_{in}^i(r) = \sum_{n=-\infty}^{\infty} B_n^i J_n(k_B|\mathbf{r} - \mathbf{r}_i|) e^{in\varphi_i} \tag{18.7}$$

The coefficients A_n^j and B_n^j can be solved by expressing the scattered field for each $i \neq j$, in terms of the modes with respect to the ith cylinder by the addition theorem of the Bessel function. The final field reaching a receiver located at $\mathbf{r}$ is the sum of the direct wave from the source and the scattered fields from all the cylinders. Therefore, the external field of cylinders could be expressed as

$$\Phi_{ext}^i(\mathbf{r}) = \sum_{n=-\infty}^{\infty} \left[B_n^i J_n(k_B|\mathbf{r} - \mathbf{r}_i|) + i\pi A_n^i H_n(k_B|\mathbf{r} - \mathbf{r}_i|) \right] e^{in\varphi_i} \tag{18.8}$$

The above equation expresses a sum of the external field from the incident field and the scattered field outside the ith cylinder. In addition, we assume that the wave inside the ith cylinder can be defined as

$$\Phi_{int}^{i}(\mathbf{r}) = \sum_{n=-\infty}^{\infty} D_n^i J_n(k_A |\mathbf{r} - \mathbf{r}_i|) e^{in\varphi_i} \tag{18.9}$$

where D_n^i is the assumption coefficient of the internal field. Then, the usual boundary conditions are considered. The field and the derivative field should be continuous across the interface between the cylinder and the surrounding material. We can further derive the correlation between A_n^j and B_n^j. Then, the total system matrix of the PhC can be obtained as follows:

$$\Gamma_n^i A_n^i - \sum_{j=1, j \neq i}^{N} \sum_{l=-\infty}^{\infty} G_{l,n}^{i,j} A_l^j = T_n^i \tag{18.10}$$

$$T_n^i = H_{-n}(k_B |\mathbf{r}_i|) e^{-in\varphi_{r_i}} \tag{18.11}$$

$$G_{l,n}^{i,j} = H_{l-n}(k_B |r_i - r_j|) e^{i(l-n)\varphi_{r_i - r_j}}, i \neq j \tag{18.12}$$

Here, A_n^i and Γ_n^i represent the expansion coefficients of scattering field and expansion coefficient of the incident field. Equation 18.10 can be simplified to an eigenvalue problem: $\mathbf{MA} = \mathbf{T}$. The laser oscillation condition that would thus be achieved under $\mathbf{A/T}$ is divergent. Therefore, $\det(\mathbf{M}) = 0$ is the complex determinant equation used to search for the threshold amplitude gain and the frequency of resonant mode. By properly constructing the PhC laser structure, the 2D solution of threshold gain and lasing frequency can be obtained.

18.3.4 Influence of Lattice Types

It is clear that the geometry of PhC lattice would significantly affect the photonic band structure. The PCSELs with commonly seen hexagonal and square lattices have been widely studied and implemented due to their relatively simple patterns. On the other hand, PhCs with honeycomb lattice have prompted many interests due to the flat photonic band near the band-edge positions and fewer PhC air-holes in materials compared with hexagonal and square lattices. These characteristics are beneficial for the realization of low threshold PCSELs. The geometry of PhC lattice would significantly affect both the threshold conditions and the other laser characteristics of PCSELs such as the degree of polarization (DOP) and divergence angle. These laser properties are important for laser projection, display, and solid state lighting applications. In 2015, T. T. Wu et al. fabricated and characterized GaN-based PCSELs with different lattice types including honeycomb, hexagonal, and square lattices (Wu et al. 2015). Laser characteristics including threshold conditions, lasing spectra, polarization, and divergence angle of devices with different lattices have been studied and discussed. Corresponding SEM images, calculated band diagrams, and lasing spectra of GaN-based PhCs with honeycomb, hexagonal and square lattices are illustrated in Figure 18.35a and b.

Figure 18.35d–f present the detailed photonic band structures near Γ_1 band edges of devices with hexagonal, square, and honeycomb lattice, respectively. It can be seen that the corresponding normalized frequencies are calculated to be 0.254, 0.447, and 0.384 for devices with honeycomb, hexagonal, and square lattices, respectively. The lattice constants of honeycomb, hexagonal, and square lattices were designed to be 100, 176, and 150 nm, respectively. In the measurement, the lasing wavelengths for the GaN-based PCSELs with honeycomb, hexagonal, and square lattices were measured to be 393.8, 391.4, and 391.7 nm, respectively. It indicates that GaN-based PCSELs with different lattice type are operated in the similar optical gain region, matching to numerical design. In addition, the measured threshold energy densities and threshold gains calculated by the multiple scattering method are shown in Figure 18.36.

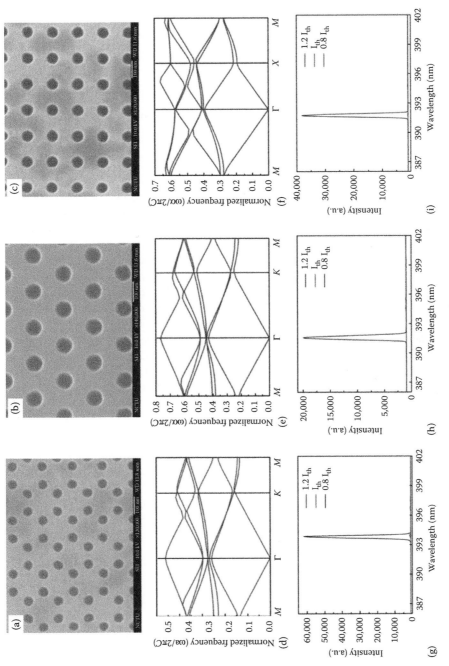

FIGURE 18.35 (a)–(c) SEM images, (d)–(f) calculated band diagram, and (g)–(i) output emission spectrum of the devices with honeycomb, hexagonal, and square lattices, respectively.

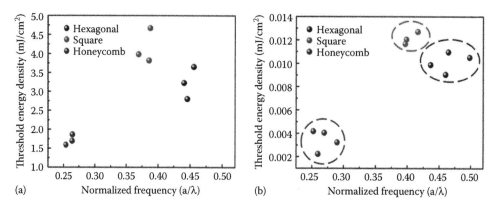

FIGURE 18.36 (a) The measured threshold energy densities of GaN-based PCSELs with different lattice type. (b) Different threshold gain versus normalized frequency calculated by the multiple scattering method.

The threshold energy densities of the device with honeycomb, hexagonal, and square lattice were measured to be approximately 1.6, 2.3, and 3.8 mJ/cm² at room temperature, showing superior characteristics of applying honeycomb as the PhC pattern. Moreover, the experimental results were matched well to the theoretical prediction. Finally, the laser characteristics of devices with three lattice types such as the DOP and divergence angle were analyzed, as drawn in Figure 18.37. On the other hand, DOP of the lasing peak intensity is defined as $(I_{max}-I_{min})/(I_{max}+I_{min})$. Here, I_{max} and I_{min} are the maximum and minimum values of polarization curve. The DOP values of devices with honeycomb, hexagonal, and square lattices were measured to be 86%, 73%, and 70%, respectively. Basically, the DOP value of GaN-based PCSELs would be low since the symmetric feedback mechanism provided by the in-plane k-vectors. On the other hand, Figure 18.37d–f show the divergence angles of each PCSEL on the normal plane from the sample surface. The divergence angles were measured to be 1.3°, 1.7°, and 2.2° for devices with honeycomb, hexagonal, and square lattices. It should be noted that the measured divergence angles for each sample were similar when the measurements were along different in-plane directions.

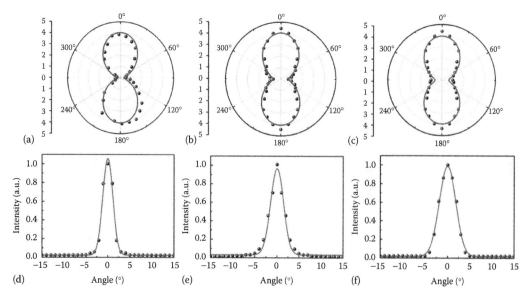

FIGURE 18.37 The measured polar plot of GaN-based PCSELs with (a) hexagonal, (b) square, and (c) honeycomb lattices, respectively. The divergence angles of (d) honeycomb, (e) hexagonal, and (f) square lattices.

These results indicate the lasing emissions of all devices are normal to the device surface with narrow divergence angles.

The overall results concluded that the GaN-based PCSELs with honeycomb lattice showed the lowest divergence angle of 1.3°. GaN-based PCSELs with honeycomb lattice show great laser characteristics and could be potential for high-power applications such as laser display, projection, and solid state illumination systems in the near future.

18.3.5 Photonic Quasi-Crystal Cavities

Except for commonly seen regular PhC, photonic quasi-crystals (PQCs) with the spatial arrangement often situated between the conventional PhC and random scattering patterns have attracted considerable attention in both experimental and numerical researches. PQCs often exhibit particular optical properties because the highly rotational symmetry and missing of long-range translational invariance can lead to the feature of the defect and photonic band gaps for confining optical modes.

In 2015, K. B. Hong et al. designed a novel PhC structure combining with hexagonal lattice and 12-fold PQC supercells (Hong et al., 2015). The goal is to take advantage of PQC to implement the competitive threshold for applications of GaN-based surface-emitting lasers. Laser devices and related lasing characteristics of the GaN-based PQC surface-emitting lasers (PQCSELs) operated at different Γ band-edge modes were fabricated and discussed. The photonic band diagram and mode profiles of GaN-based hexagonal PhC with 12-fold PQC supercells were calculated by finite element method. The ratio of air hole's radius to the lattice constant of PQC supercell and the targeted lasing wavelength were set to 0.185 and 394 nm, respectively. The lattice constant of the hexagonal lattice corresponding to the period between the neighboring PQC supercells is $A = 2[1 + \cos(\pi/6)]a$. Schematic drawing of 12-fold PQC supercell is shown in Figure 18.38a.

Six nondegenerate TE modes located at different Γ band edges were used for investigating the lasing characteristics. The schematic cross section of GaN-based PhC laser and the SEM images of experimental samples were shown in Figure 18.38b–e. The photonic quasicrystal patterns were defined by utilizing electron-beam lithography and inductively coupled plasma etching. Three experimental samples can operate at higher order Γ band-edge modes named S3, S5, and S6, respectively, as shown in Figure 18.38c–e. The measured threshold energy densities and the alculated threshold gains of GaN-based

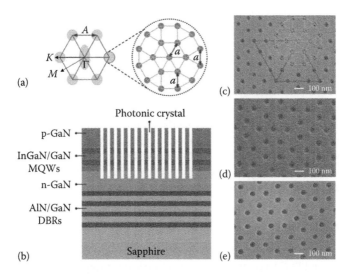

FIGURE 18.38 (a) Schematic drawing of 12-fold PQC supercell, (b) cross section plot of GaN-based PQCSEL, (c)–(e) the SEM images of samples S3, S5, and S6, respectively.

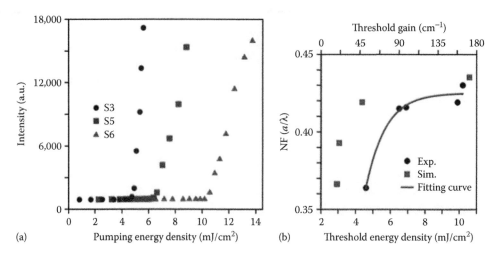

(a)   (b)

FIGURE 18.39   (a) Output emission intensity as a function of the pumping energy at room temperature condition for samples S3, S5, and S6. (b) Measured threshold energy densities and the calculated threshold gains of GaN-based PQCSELs for $r/a = 0.185$.

hexagonal photonic crystals with 12-fold PQC supercells for distinct band-edge modes are illustrated in Figure 18.39. When the pumping energy density was increased, a clear threshold condition could be observed. The threshold pumping energy density of GaN-based PQCSELs for samples S3, S5, and S6 are 4.6, 6.3, and 10.2 mJ/cm², respectively, as shown in Figure 18.39a.

In addition, measured lasing spectra above the threshold condition for samples S3, S5, and S6 showed that main lasing peaks (line width) were measured to be 394.2 nm (0.34 nm), 391.2 nm (0.34 nm) and 392.3 nm (0.25 nm), respectively. The dominated lasing peaks of these three samples are very close to each other. Moreover, the decreasing tendency of threshold energy of PQCSELs was observed while the designed lasing mode moved from the higher Γ band to the lower Γ band, as shown in Figure 18.39b. The relevant normalized frequencies and the tendency of threshold condition for the experimental data showed good agreement with simulations.

In searching for a better design, threshold gains of GaN-based PQCSELs with different ratios of r/a for higher order Γ band-edge modes were calculated as shown in Figure 18.40a. Six modes with corresponding normalized frequencies positioned at different Γ band edges were investigated. Numerical results showed that the S3 mode had the lowest threshold and the threshold gain would be largely varied by tuning the ratio of air hole radius/lattice constant (r/a). Moreover, increasing the value of r/a can largely shrink the threshold gain, especially for S1, S2, and S6 modes. However, Figure 18.40b shows the calculated group velocity (v) of light propagating in the two-dimensional PhC structure. The group velocities near the Γ band edges were normalized by the speed of light (c) in the free space. These curves obviously illustrate the group velocities of light will drop significantly if the wave vector moves closer to the Γ band edges. In addition, the S3 mode has the lowest normalized group velocity of 0.0142 ($10^3 v/c$) which explains exactly why S3 mode owns lower lasing threshold than other modes. Based on experiments and calculations, PQC lasers show the opportunity to realize lower threshold lasers than typical PhC lasers.

18.3.6 Electrically Pumped GaN PCSELs

In previous reports, wafer fusion and the regrowth techniques are common fabrication processes to combine active layers and PhC layers for developing electrically pumped PCSELs. These two processes have their own disadvantages. Wafer fusion technique is known as direct bonding technology, which is the wafer bonding process without using additional intermediate layers. And the regrowth technique

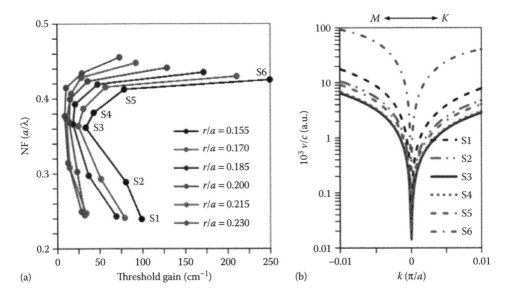

FIGURE 18.40 (a) NF plotted versus the threshold gains of GaN-based PQCSELs for different ratios of r/a. (b) Calculated group velocity near the Γ band-edge for six desired modes for $r/a = 0.185$.

is to grow active layers after the PhC structures are formed on the substrate. Unfortunately, these two techniques have to be carried out at high temperature, which could adversely influence the MQWs' characteristics. Furthermore, as long as the direct bonding or the regrowth technique is applied, the whole fabrication process becomes very complicated and difficult. For instance, the double-sided lithography and chemical-mechanical planarization will be inevitably included, accompanying the quality and yield rate reduction of the samples.

This section will briefly introduce the electrically pumped GaN-based PCSELs by the regrowth technique. It is worth noting that the first electrically driven GaN-based PCSEL operated at blue-violet wavelength has been reported in 2008 (Matsubara et al. 2008). H. Matsubara et al. have developed a fabrication method, named "air holes retained over growth," (AROG) for constructing a two-dimensional GaN/air PhC structure. The schematic of the GaN-based PCSEL, SEM images of the PCSEL with hexagonal lattice and well-defined GaN/air periodic structure are shown in Figure 18.41. The embedded air holes are well preserved because of the special characteristics of GaN growth. A clear threshold current

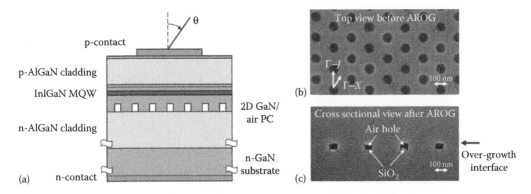

FIGURE 18.41 (a) Schematic of GaN-based PCSEL. (b) SEM image of the hexagonal lattice. (c) Cross-sectional SEM image of periodic GaN/air structure.

condition at 6.7A and corresponding current density of 67 kA/cm² were apparently observed at the pulse driven condition at room temperature. In addition, the spectra of the device operated below and above the threshold current were measured. The lasing peak wavelength was 406.5 nm with a line width of about 0.15 nm. The far-field pattern was measured which showed the beam divergence angle was about 1°. This indicated that the large-area coherent lasing had been achieved.

S. Kawashima et al. (2010) also demonstrated the GaN-based PCSELs with square lattice grown on the sapphire substrate in 2010. The optical microscope (OM) image and schematic cross section of the GaN-based SEL are shown in Figure 18.42a and b. Moreover, the schematic of the PhC with high-aspect ratio air holes arranged in a square lattice. The refractive index distribution of the laser device in the vertical direction and optical power profile calculated by the transfer matrix method are shown in Figure 18.42c and d, respectively. The calculated optical confinement factors of the quantum wells and PhC are 2.9% and 3.0%. In the measurement, the lasing wavelengths of lattice constants of 162.5 and 165 nm are 406 and 410 nm. The lasing action of devices for various sizes of p-contact electrodes are observed under pulsed current operation at room temperature. Measured threshold current densities for p-electrodes with widths of 60, 80, 100, and 120 μm were 28.6, 18.6, 13.2, and 9.7 kA/cm², respectively. Threshold current density depends on the size of the electrode and laser device with a p-electrode of 60 μm² has an opportunity to realize micro SEL array.

As mentioned above, seeking a way to avoid the wafer fusion and the regrowth techniques could be beneficial for developing electrically driven GaN-based PCSELs if the conventional cladding layer can be replaced by other conducting and transparent materials with easy fabrication. Interestingly, indium tin oxide (ITO) exhibits unique optical and electrical characteristics, such as low absorption, high conductivity and low-refractive index that make it a suitable candidate for a cladding material. Recently, B. Cheng et al. (2010) have utilized ITO as a cladding layer for 415 nm InGaN laser diodes. Indeed, replacing the epitaxial nitride cladding layer with ITO has several potential advantages. For green laser diodes, use of ITO in the waveguide structure reduces the epitaxial p-cladding thickness and growth time, which in turn may reduce thermal damage to the active region (Hardy et al. 2013). Therefore, transparent conductive oxides

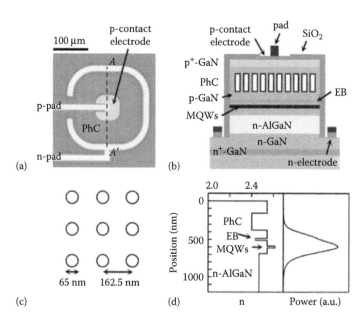

FIGURE 18.42 (a) Optical micrograph of GaN-based distributed feedback surface-emitting laser. (b) Schematic cross section along line *A–A'*. (c) Schematic of PhC with square lattice. (d) Distribution of refractive index and optical power profile of device in the vertical direction.

(TCO) such as ITO, ATO, AZO, GZO, SnO_2, and so on, are likely to become the candidates for implementation of electrically driven GaN-based PCSELs.

18.3.7 Summary

We have reviewed and attempted to give the reader a broad overview of current research on the evolution of GaN-based PCSELs to date. The operation principle, developments, and simulation methods for band diagram and threshold gain calculations of PCSELs were briefly introduced and discussed. Some simulation methods to engineer the band structures of the PhC, were implemented and developed which are named as PWEM, FEM, MSM, FDTD, and CWT. As mentioned above, each of them suffers limitations and drawbacks in the calculation. But the MSM shows good capability, such as faster calculation time, considering gain coefficients within PhC materials with an arbitrary defect or lattice patterns. Although the first electrically driven GaN-based PCSELs under pulsed current operation at room temperature have been realized and demonstrated, related research about GaN PCSELs with current injected approach are still barely seen due to complicated fabricated processes. Nevertheless, the presented results show that the optical characteristics and related applications of GaN PCSELs could offer promising potential and the booming opportunity toward practical use in the near future.

18.4 Conclusion and Prospects

We have reported two kinds of GaN SELs. In GaN-based VCSELs, the historical development has been described from optically pumped devices to electrically pumped one. Due to the difficulties of making GaN-based VCSELs, three kinds of VCSEL structures are introduced, including GaN VCSELs with full epitaxial DBRs, hybrid DBRs, and double dielectric DBRs. This order represents the difficulties in epitaxial growth but the ease in fabrication. Each structure will reflect its own pros and cons. Despite that epitaxially grown GaN-based DBR can be a formidable task, the monolithic DBR shall enjoy huge amount of advantages that 850 nm IR VCSELs have. Therefore, the growth technology, strain engineering, and *in situ* monitoring of GaN DBR layers need to be further developed. That will also contribute to more understanding of GaN materials and to obtain higher crystal quality and to develop more versatile GaN-based devices. On the other hand, a complicated fabrication process is always required in making double dielectric DBR VCSELs, including wafer bonding, thin film and lift-off technology, high precision lapping, and polishing technique, and so on. The development of this technology will help bring mature and highly reliable mass-production flow. Since the VCSEL structure belongs to microcavity and excitons in GaN materials are relatively stable, the microcavity exciton-polariton lasers can be realized with extremely low threshold. The previous demonstration of GaN-based polariton lasers by optical pumping ensures the possibility of such novel coherent light emitters. We believe the demonstration of current injection type GaN-based polariton lasers with VCSEL structures would bring impacts to this field.

On the other hand, we have described the intriguing operation principle of PCSELs and discussed the corresponding laser characteristics, including multiple distributed feedback, single mode operation, and extremely small divergence angle. GaN PCSELs can be regarded as complimentary to GaN VCSELs because GaN PCSELs are capable of achieving high-power operation, which can be the perfect candidate for laser headlight or laser lighting applications. We have discussed the influence of lattice type of PhC. The quasiperiodic PhC was also introduced. Actually, the epitaxial structures and lattice type (or unit cell structure) will significantly influence the PCSEL performance. Since the demonstration of GaN PCSEL came relatively late, the designing rule and modeling tool require systematical development in order to further understanding the potential of GaN PCSELs and to improve the laser performance. Another challenging issue of making GaN PCSELs is to fabricate a large amount of PhC unit cells. Although e-beam lithography has been proved to accurately make these patterns by many research groups, the mass-production and cost issue remain challenging. Nano-imprinting or stepper lithography technology could be possible alternatives, but they need intensive verification.

Finally, we believe the development of GaN SELs, including VCSELs and PCSELs not only represent scientific and technology triumph but also bring more promising applications in the near future.

References

Arita M, Ishida S, Kako S, Iwamoto S, Arakawa Y (2007) AlN air bridge photonic crystal nanocavities demonstrating high quality factor. *Applied Physics Letters* 91:051106.

Bhattacharya P, Frost T, Deshpande S, Baten MZ, Hazari A, Das A (2014) Room temperature electrically injected polariton laser. *Physical Review Letters* 112:236802.

Butté R, Christmann G, Feltin E, Carlin JF, Mosca M, Ilegems M, Grandjean N (2006) Room-temperature polariton luminescence from a bulk GaN microcavity. *Physical Review B* 73:033315.

Carlin J-F, Dorsaz J, Feltin E, Butté R, Grandjean N, Ilegems M, Laügt M (2005) Crack-free fully epitaxial nitride microcavity using highly reflective AlInN/GaN Bragg mirrors. *Applied Physics Letters* 86:031107.

Casey H, James SS, Steven PD, Shuji N, Daniel F (2012) Demonstration of Nonpolar GaN-Based Vertical-Cavity Surface-Emitting Lasers. *Applied Physics Express* 5:092104.

Chen J-R, Lu T-C, Wu Y-C, Lin S-C, Liu W-R, Hsieh W-F, Kuo C-C, Lee C-C (2009a) Large vacuum Rabi splitting in ZnO-based hybrid microcavities observed at room temperature. *Applied Physics Letters* 94:061103.

Chen S-W, Lu T-C, Hou Y-J, Liu T-C, Kuo H-C, and Wang S-C (2010) Lasing characteristics at different band edges in GaN photonic crystal surface emitting lasers. *Applied Physics Letters* 96:071108.

Chen S-W, Lu T-C, Kao T-T (2009b) Study of GaN-based photonic crystal surface-emitting lasers (PCSELs) with AlN/GaN distributed Bragg reflectors. *IEEE Journal of Selected Topics in Quantum Electronics* 15:885–891.

Cheng B, Chua CL, Yang Z, Teepe M, Knollenberg C, Strittmatter A, Johnson N (2010) Nitride laser diodes with nonepitaxial cladding layers. *IEEE Photonics Technology Letters* 22:329–331.

Cheng B-S, Wu Y-L, Lu T-C, Chiu C-H, Chen C-H, Tu P-M, Kuo H-C, Wang S-C, Chang C-Y (2011) High Q microcavity light emitting diodes with buried AlN current apertures. *Applied Physics Letters* 99:041101.

Christmann G, Butté R, Feltin E, Carlin JF, Grandjean N (2006) Impact of inhomogeneous excitonic broadening on the strong exciton-photon coupling in quantum well nitride microcavities. *Physical Review B* 73:153305.

Christmann G, Butté R, Feltin E, Carlin J-F, Grandjean N (2008a) Room temperature polariton lasing in a GaN/AlGaN multiple quantum well microcavity. *Applied Physics Letters* 93:051102.

Christmann G, Butté R, Feltin E, Mouti A, Stadelmann PA, Castiglia A, Carlin J-F, Grandjean N (2008b) Large vacuum Rabi splitting in a multiple quantum well GaN-based microcavity in the strong-coupling regime. *Physical Review B* 77:085310.

Christopoulos S, von Högersthal GBH, Grundy AJD, Lagoudakis PG, Kavokin AV, Baumberg JJ, Christmann G, Butté R, Feltin E, Carlin JF, Grandjean N (2007) Room-temperature polariton lasing in semiconductor microcavities. *Physical Review Letters* 98:126405.

Cosendey G, Castiglia A, Rossbach G, Carlin J-F, Grandjean N (2012) Blue monolithic AlInN-based vertical cavity surface emitting laser diode on free-standing GaN substrate. *Applied Physics Letters* 101:151113.

Ehsan H, Johan G, Jörgen B, Martin S, Gatien C, Nicolas G, Åsa H (2013) Engineering the lateral optical guiding in gallium nitride-based vertical-cavity surface-emitting laser cavities to reach the lowest threshold gain. *Japanese Journal of Applied Physics* 52:08JG04.

Feltin E, Butte R, Carlin JF, Dorsaz J, Grandjean N, Ilegems M (2005) Lattice-matched distributed Bragg reflectors for nitride-based vertical cavity surface emitting lasers. *Electronics Letters* 41:94–95.

Freeman M, Champion M, Madhavan S (2009) Scanned laser pico-projectors: Seeing the big picture (with a small device). *Optic & Photonics News* 20:28–34.

Guillet T, Brimont C, Valvin P, Gil B, Bretagnon T, Médard F, Mihailovic M, Zúñiga-Pérez J, Leroux M, Semond F, Bouchoule S (2011) Laser emission with excitonic gain in a ZnO planar microcavity. *Applied Physics Letters* 98:211105.

Hagino H, Takahashi Y, Tanaka Y, Asano T, Noda S (2009) Effects of fluctuation in air hole radii and positions on optical characteristics in photonic crystal heterostructure nanocavities. *Physical Review B* 798:085112.

Hamaguchi T, Fuutagawa N, Izumi S, Murayama M, Narui H (2016) Milliwatt-class GaN-based blue vertical-cavity surface-emitting lasers fabricated by epitaxial lateral overgrowth. *physica status solidi* (a):n/a-n/a.

Hardy MT, Holder CO, Feezell DF, Nakamura S, Speck JS, Cohen DA, DenBaars SP (2013) Indium-tin-oxide clad blue and true green semipolar InGaN/GaN laser diodes. *Applied Physics Letters* 103:081103.

Haruhisa S, Ken-ichi I, Chiyuki K, Yasuharu S (1979) GaInAsP/InP surface emitting injection lasers. *Japanese Journal of Applied Physics* 18:2329.

Hashemi E, Bengtsson J, Gustavsson J, Stattin M, Cosendey G, Grandjean N, Haglund Å (2014) Analysis of structurally sensitive loss in GaN-based VCSEL cavities and its effect on modal discrimination. *Optics Express* 22:411–426.

Helena F, Chris S, Rüdiger S-G, Gerald W, Marius G (2012) Ballistic propagation of exciton–polariton condensates in a ZnO-based microcavity. *New Journal of Physics* 14:013037.

Hirose K, Liang Y, Kurosaka Y, Watanabe A, Sugiyama T, Noda S (2014) Watt-class high-power, high-beam-quality photonic crystal lasers. *Nature Photonics* 8:406–411.

Hong K-B, Chen C-C, Lu T-C, Wang S-C (2015) Lasing characteristics of GaN-based photonic quasi-crystal surface emitting lasers operated at higher order Γ mode. *IEEE Journal of Selected Topics in Quantum Electronics* 21:4900606.

Imada M, Noda S, Chutinan A, Tokuda T, Murata M, Sasaki G (1999) Coherent two-dimensional lasing action in surface-emitting laser with triangular-lattice photonic crystal structure. *Applied Physics Letters* 75:316.

Inaba Y, Imai K, Fujieda I, Onishi I (2008) P-99: A backlight based on a laser diode and its design considerations. *SID Symposium Digest of Technical Papers* 39:1564–1567.

Inoue T, Zoysa M, Asano T, Noda S (2014) Realization of dynamic thermal emission control. *Nature Materials* 13:928–931.

Islam A, Islam SI, Islam S (2013) Designing an all epitaxial 1,550 nm intra-cavity VCSEL using GaInAsN/AlGaInAs in the active region and AlGaAsSb/AlAsSb in top and bottom DBRs. *Optical and Quantum Electronics* 45:1199–1212.

John S (1987) Strong localization of photons in certain disordered dielectric superlattices. *Physical Review Letters* 58:2486–2489.

Kastner R (1987) On the singularity of the full spectral Green's Dyad. *IEEE Transactions on Antennas and Propagation* 35:1303–1305.

Kawashima S, Kawashima T, Nagatomo Y, Hori Y, Iwase H, Uchida T, Hoshino K, Numata A, Uchida M (2010) GaN surface-emitting laser with two-dimensional photonic crystal acting as distributed-feedback grating and optical cladding. *Applied Physics Letters* 97:251112.

Kim DU, Kim S, Lee J, Jeon SR, Jeon H (2011) Free-standing GaN-based photonic crystal band-edge laser. *IEEE Photonics Technology Letters* 23:1454–1456.

Kim M, Kim CS, Bewley WW, Lindle JR, Canedy CL, Vurgaftman I, Meyer JR (2006) Surface-emitting photonic-crystal distributed-feedback laser for the midinfrared. *Applied Physics Letters* 88:191105.

Kohn W, Rostoker N (1954) Solution of the schrödinger equation in periodic lattices with an application to metallic lithium. *Physical Review* 94:1111.

Korringa J (1947) On the calculation of the energy of a Bloch wave in a metal. *Physica* 13:392–400.

Kunimichi O, Yu H, Kyosuke N, Hiroaki M, Takashi M (2009) Improvement in lasing characteristics of GaN-based vertical-cavity surface-emitting lasers fabricated using a GaN substrate. *Applied Physics Express* 2:052101.

Lai Y-Y, Chou Y-H, Wu Y-S, Lan Y-P, Lu T-C, Wang S-C (2014a) Fabrication and characteristics of a GaN-based microcavity laser with shallow etched mesa. *Applied Physics Express* 7:062101.

Lai Y-Y, Huang S-C, Ho T-L, Lu T-C, Wang S-C (2014b) Numerical analysis on current and optical confinement of III-nitride vertical-cavity surface-emitting lasers. *Optics Express* 22:9789–9797.

Lai Y-Y, Lan Y-P, Lu T-C (2012) High-temperature polariton lasing in a strongly coupled ZnO microcavity. *Applied Physics Express* 5:082801.

Leonard JT, Cohen DA, Yonkee BP, Farrell RM, Margalith T, Lee S, DenBaars SP, Speck JS, Nakamura S (2015) Nonpolar III-nitride vertical-cavity surface-emitting lasers incorporating an ion implanted aperture. *Applied Physics Letters* 107:011102.

Li YM, Yuen W, Li GS, Chang-Hasnain CJ (1998) Top-emitting micromechanical VCSEL with a 31.6-nm tuning range. *IEEE Photonics Technology Letters* 10:18–20.

Liang Y, Okino T, Kitamura K, Peng C, Ishizaki K, Noda S (2014) Mode stability in photonic-crystal surface-emitting lasers with large κ 1DL. *Applied Physics Letters* 104:021102.

Liang Y, Peng C, Ishizaki K, Iwahashi S, Sakai K, Tanaka YK, Noda S (2013) Three-dimensional coupled-wave analysis for triangular-lattice photonic crystal surface emitting lasers with transverse-electric polarization. *Optics Express* 21:565–580.

Liang Y, Peng C, Sakai K, Iwahashi S, Noda S (2012) Three-dimensional coupled-wave analysis for square-lattice photonic crystal surface emitting lasers with transverse-electric polarization: Finite-size effects. *Optics Express* 20:15945–15961.

Lu T-C, Chen J-R, Lin S-C, Huang S-W, Wang S-C, Yamamoto Y (2011) Room temperature current injection polariton light emitting diode with a hybrid microcavity. *Nano Letters* 11:2791–2795.

Lu T-C, Chen S-W, Kao T-T, Liu T-W (2008a) Characteristics of GaN-based photonic crystal surface emitting lasers. *Applied Physics Letters* 93:111111.

Lu T-C, Chen S-W, Lin L-F, Kao T-T, Kao C-C, Yu P, Kuo H-C, Wang S-C, Fan S-H (2008b) GaN two-dimensional surface-emitting photonic crystal lasers with AlN/GaN distributed Bragg reflector. *Applied Physics Letters* 92:011129.

Lu T-C, Chen S-W, Wu T-T, Tu P-M, Chen C-K, Chen C-H, Li Z-Y, Kuo H-C, Wang S-C (2010) Continuous wave operation of current injected GaN vertical cavity surface emitting lasers at room temperature. *Applied Physics Letters* 97:071114.

Lu T-C, Kao C-C, Kuo H-C, Huang G-S, Wang S-C (2008c) CW lasing of current injection blue GaN-based vertical cavity surface emitting laser. *Applied Physics Letters* 92:141102.

Lu T-C, Lai Y-Y, Lan Y-P, Huang S-W, Chen J-R, Wu Y-C, Hsieh W-F, Deng H (2012) Room temperature polariton lasing vs. photon lasing in a ZnO-based hybrid microcavity. *Optics Express* 20:5530–5537.

Matsubara H, Yoshimoto S, Saito H, Jianglin Y, Tanaka Y, Noda S (2008) GaN photonic-crystal surface-emitting laser at blue–violet wavelengths. *Science* 319:445–447.

Meiera C, Hennessy K, Haberer ED, Sharma R, Choi YS, McGroddy K, Keller S, DenBaars SP, Nakamura S, Hu EL (2006) Visible resonant modes in GaN photonic crystal membrane cavities. *Applied Physics Letters* 88:031111.

Nakamura S (1998) The roles of structural imperfections in InGaN-based blue light-emitting diodes and laser diodes. *Science* 281:956–961.

Nakamura S, Senoh M, Nagahama S-I, Iwasa N, Yamada T, Matsushita T, Sugimoto Y, Kiyoku H (1997) Room-temperature continuous-wave operation of InGaN multi-quantum-well-structure laser diodes with a long lifetime. *Applied Physics Letters* 70:868–870.

Nakamura S, Senoh M, Nagahama Si, Iwasa N, Yamada T, Matsushita T, Sugimoto Y, Kiyoku H (1996) Room-temperature continuous-wave operation of InGaN multi-quantum-well structure laser diodes. *Applied Physics Letters* 69:4056–4058.

Noda S, Tomoda K, Yamamoto N, Chutinan A (2000) Full three dimensional photonic bandgap crystals at near-infrared wavelengths. *Science* 289:604–606.

Noda S, Yokoyama M, Imada M, Chutinan A, Mochizuki M (2001) Polarization mode control of two-dimensional photonic crystal laser by unit cell structure design. *Science* 293:1123–1125.

Notomi M, Suzuki H, Tamamura T (2001) Directional lasing oscillation of two-dimensional organic photonic crystal lasers at several photonic bandgaps. *Applied Physics Letters* 78:1325.

Ohnishi D, Okano T, Imada M, Noda S (2004) Room temperature continuous wave operation of a surface-emitting two-dimensional photonic crystal diode laser. *Optics Express* 12:1562–1568.

Onishi T, Imafuji O, Nagamatsu K, Kawaguchi M, Yamanaka K, Takigawa S (2012) Continuous wave operation of GaN vertical cavity surface emitting lasers at room temperature. *IEEE Journal of Quantum Electronics* 48:1107–1112.

Painter O, Lee R-K, Scherer A, Yariv A, O'Brien JD, Dapkus PD, Kim I (1999) Two-dimensional photonic band-gap defect mode laser. *Science* 284:1819–1821.

Park H-G, Kim S-H, Kwon S-H, Ju Y-G, Yang J-K, Baek J-H, Kim S-B, Lee Y-H (2004) Electrically driven single-cell photonic crystal laser. *Science* 305:1444–1447.

Poher V, Grossman N, Kennedy GT, Nikolic K, Zhang HX, Gong Z, Drakakis EM, Gu E, Dawson MD, French PMW, Degenaar P, Neil MAA (2008) Micro-LED arrays: A tool for two-dimensional neuron stimulation. *Journal of Physics D: Applied Physics* 41:094014.

Rayleigh JWS (1888) On the remarkable phenomenon of crystalline reflexion described by Prof. Stokes. *Philosophical Magazine* 26:256–265.

Shimada R, Xie J, Avrutin V, Özgür Ü, Morkoč H (2008) Cavity polaritons in ZnO-based hybrid microcavities. *Applied Physics Letters* 92:011127.

Shouichiro I, Noriyuki F, Tatsushi H, Masahiro M, Masaru K, Hironobu N (2015) Room-temperature continuous-wave operation of GaN-based vertical-cavity surface-emitting lasers fabricated using epitaxial lateral overgrowth. *Applied Physics Express* 8:062702.

Shuji N, Masayuki S, Shin-ichi N, Naruhito I, Takao Y, Toshio M, Hiroyuki K, Yasunobu S (1996) InGaN-based multi-quantum-well-structure laser diodes. *Japanese Journal of Applied Physics* 35:L74.

Sirigu L, Terazzi R, Amanti I, Giovannini M, Faist J, Dunbar A, Houdr´e R (2008) Terahertz quantum cascade lasers based on two-dimensional photonic crystal resonators. *Optics Express* 16:5206–5217.

Song BS, Noda S, Asano T, Akahane Y (2005) Ultra-high-Q photonic double-heterostructure nanocavity. *Nature Materials* 4:207–210.

Syrbu A, Iakovlev V, Suruceanu G, Caliman A, Mereuta A, Mircea A, Berseth C-A, Diechsel E, Boucart J, Rudra A, Kapon E (2005) VCSELs emitting in the 1310-nm waveband for novel optical communication applications. In *Integrated Optoelectronic Devices 2005*. International Society for Optics and Photonics, pp. 167–173.

van Vugt LK, Rühle S, Ravindran P, Gerritsen HC, Kuipers L, Vanmaekelbergh D (2006) Exciton polaritons confined in a ZnO nanowire cavity. *Physical Review Letters* 97:147401.

Vurgaftman I, Meyer J (2003) Design optimization for high-brightness surface-emitting photonic-crystal distributed-feedback lasers. *IEEE Journal of Quantum Electronics* 39:689–700.

Weisbuch C, Nishioka M, Ishikawa A, Arakawa Y (1992) Observation of the coupled exciton-photon mode splitting in a semiconductor quantum microcavity. *Physical Review Letters* 69:3314–3317.

Weng P-H, Wu T-T, Lu T-C (2012) Study of band-edge modes in GaN-based photonic crystal surface-emitting lasers by the multiple-scattering method. *IEEE Journal of Selected Topics in Quantum Electronics* 18:1629–1635.

Weng P-H, Wu T-T, Lu T-C, and Wang S-C (2011) Threshold gain analysis in GaN-based photonic crystal surface emitting lasers. *Optics Letters* 36:1908–1910.

Wierer JJ, Tsao JY (2015) Advantages of III-nitride laser diodes in solid-state lighting. *Physica Status Solidi (A)* 212:980–985.

Wu T-T, Chen C-C, Lu T-C (2015) Effects of lattice types on GaN-based photonic crystal surface-emitting lasers. *IEEE Journal of Selected Topics in Quantum Electronics* 21:1700106.

Wu T-T, Weng P-H, Hou Y-J, Lu T-C (2011) GaN-based photonic crystal surface emitting lasers with central defects. *Applied Physics Letters* 99:221105.

Yablonovitch E (1987) Inhibited Spontaneous Emission in Solid-State Physics and Electronics. *Physical Review Letters* 58:2059–2062.

Yu H, Kunimichi O, Hiroaki M, Takashi M (2008) Room-temperature CW lasing of a GaN-based vertical-cavity surface-emitting laser by current injection. *Applied Physics Express* 1:121102.

V

Emerging
Applications

19

III–V Nitride-Based Photodetection

19.1 Introduction

Since its first introduction, the nitride-based semiconductor has played an important role in the solid-state lighting application (Maruska and Stevenson 1974; Pankove and Hutchby 1976; Akasaki et al. 1989). Many great results have been published and highly efficient light-emitting diodes (LEDs) have been demonstrated (Hiroshi et al. 1990; Shuji et al. 1991; Nakamura et al. 1994). In addition to the light-emitting capability, the nitride-based materials, including aluminum nitride (AlN), gallium nitride (GaN), and indium nitride (InN), can also be very absorptive in different wavelength ranges. Combining with the wide range of band gaps (AlN: 6.2 eV, GaN: 3.39 eV, and InN: 0.77eV), the nitride-based materials can basically cover their photo-response from ultraviolet (UV) into infrared (IR). This wide coverage brings another possible usage for this series of materials: photodetection. Materials in the natural environment are seldom found with such properties. Many applications can be applied, such as photodetectors, solar cells, to name a few. The high internal quantum efficiency provides a good start for nitride-based devices. The chemical and temporal stable properties in this series of materials can be particularly useful when the environment is harsh to the devices, like by the sea or in the desert. From the early stage of the development of nitride materials, the photodetection is always an important option. Walker et al. (1999) demonstrated a GaN-based MSM photodetector with fast response (<10 ns). Liu et al. (2015) proposed and demonstrated an AlGaN/GaN MSM photodetector for UV range. Another very important design in photodetection is the photovoltaic related device. Nitride-based solar cells also draw much attention due to the possible transparent device which can be used in every window, and the wide range collection of the solar spectrum. In years, many groups have shown great progress in this field. As early as 2007, researchers from Georgia Tech provided the evidence of good photovoltaic response and as high as 60% of internal quantum efficiency can be expected from nitride materials (Jani et al. 2007). Attempts to extend the operating wavelength by increasing indium contents were found in Dahal et al. (2009) with 40% of indium incorporated inside the active region of the solar cell however, the resultant photovoltaic IV had very poor fill factor. Matioli et al. (2011) also enhance the external quantum efficiency (EQE) of nitride-based solar cell to 70% by roughening the surface to reduce the Fresnel reflection. Bhuiyan et al. (2012) published a great review on the InGaN-based solar cell performances in 2013. In general, nitride materials are stable under a harsh environment (such as high humidity, high/low temperature, high voltages, etc.) and this characteristic is very useful when the solar panel needs to be deployed outdoors. However, there are several limiting factors that could possibly restrain the future development of nitride-based photodetection devices. The biggest one is the operating wavelength issue. The operating range of the photodetection devices is determined by the active region, and usually, the indium content is the key factor. More indium incorporation is needed to extend the wavelength of operation. But it is difficult to put more than 40% of indium into gallium nitride and still maintain good device quality nowadays. Another one is the cost issue. To have low-cost photodetector or solar cell, the device needs to be capable of large area growth, which is lacking in the nitride-based materials when compared to a 12-inch wafer of silicon. High indium content or purely InN layer as the main active component in the photodetection of nitride-based devices is still a highly sought-after topic for the research. In this chapter, we will review the current progress of the nitride-based photodetection devices in our lab.

19.2 Basic Theory

In this section, we will review basic theory of photodetection related to our nitride-based devices. The photodetection has a long history in optoelectronics. From the beginning of the quantum physics in the twentieth century, in the groundbreaking work by Einstein (1917), the absorption of an incident photon can be predicted if the energy of the photon can overcome the separation of the two quantum levels. By following this idea, the photodetection can be realized in various materials and various formats of devices.

19.2.1 Theory of Photodetection

As the photons impinge upon the semiconductor, if the condition is right (photon energy $h\nu \geq E_g$), the electron-hole pairs can be generated after the absorption of the photons. These carriers, whose concentration can be denoted by δn and δp, can provide photoconductivity in the material. Under constant intensity of photons, excess carrier (δn) can be expressed as $\delta n = G_0 \tau_n$, where G_0 is the generation rate and τ_n is the excess carrier lifetime. If the difference in currents between the dark and illumination conditions is ΔI, its quantity can be: $\Delta I = q(\mu_n + \mu_p)\delta n A V / l$, where μ_n and μ_p are the electron and hole mobilities, respectively. The parameters like A, V, and l are the cross-sectional area, the voltages, and the length of the device, respectively. By probing further, the generation rate G_0 and the responsivity $R_\lambda(A/W)$ can be expressed as:

$$G_0 = \eta \frac{P_{opt}/h\nu}{A} \tag{19.1}$$

$$R_\lambda = \frac{\Delta I}{P_{opt}} = \eta \frac{q(\mu_n + \mu_p)\tau_n V}{h\nu \times \ell^2} \Rightarrow \eta = R_\lambda \frac{h\nu \times \ell^2}{q(\mu_n + \mu_p)\tau_n V} \tag{19.2}$$

where η is the quantum efficiency of the material, P_{opt} is the incident optical power (Chuang 1995, 2009).

In a p-i-n photodiode, this theory can be further extended and considered for the exponential decay of the optical field inside the device (Sze 1981). The quantum efficiency can be expressed as (Sze 1981):

$$\eta = (1-R)\left(1 - \frac{e^{-\alpha W}}{1+\alpha L}\right) \tag{19.3}$$

where R is the surface reflection, W is the depletion width, α is the absorption coefficient of the material, and L is the diffusion length of the minority carrier.

19.2.2 On the Photovoltaic Effect

In addition to photodetection, the photovoltaic effect is also important for the nitride-based devices. In Figure 19.1, a photodiode can show very distinctive IV curves with and without illumination. There are two different regions of operation: photodetection and photovoltaic response. When the voltage of the device is positive, the photovoltaic response is obtained, and the device acts like a solar cell. On the other hand, when the voltage is negative, the device will perform like a regular photodetector. For a p-n junction type of solar cell, the photocurrent can be expressed as (Chuang 1995, 2009):

$$J = J_0\left(e^{qV/kT}-1\right) - J_{ph} = q\left(\frac{D_p p_{n0}}{L_p} + \frac{D_n n_{p0}}{L_n}\right)\left(e^{qV/kT}-1\right) - qG_0(L_p + L_n) \tag{19.4}$$

where p_{n0} and n_{p0} are the thermal equilibrium concentration of minority carriers in n-type and p-type, respectively. From the measured J-V curves, an open-circuit voltage (V_{oc}) and short-circuit current density (J_{sc}) can be obtained, as shown in Figure 19.1. The shape of the J-V (or I-V) curves can affect the overall maximal power of this device. The fill factor (FF) is defined as the ratio between the P_{max} (max power, which is the shaded area in the Figure 19.1) and the product of V_{oc} and J_{sc}.

19.2.3 Specific Nitride-Based Material Concern on Photovoltaic Efficiency

From the photodetection point of view, any photons with a higher than band gap energy are detectable. However, for the solar cell, an optimal band gap is possible for the maximum power conversion efficiency (PCE). To probe this, we need to resort to the detailed balance theory of a solar cell device. In a detailed balance theory, one can idealize the solar cell absorption and carrier generation to find out the

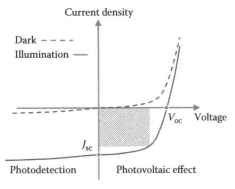

FIGURE 19.1 The photodetection and photovoltaic effects of a diode. The shadow area is the max delivered power P_{max} in photovoltaic mode.

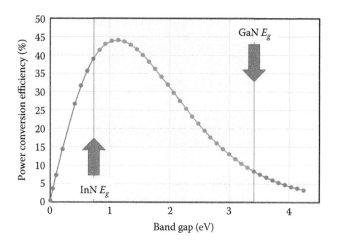

FIGURE 19.2 The ideal solar cell efficiency under detailed balance model. The band gaps of GaN and InN are marked on the same plot.

best band gap for the optimal power conversion efficiency. In the best case scenario, the efficiency of a solar cell can be expressed as (Shockley and Queisser 1961):

$$
\text{PCE} = \frac{x_g \displaystyle\int_{x_g}^{\infty} \frac{x^2}{e^x - 1} dx}{\displaystyle\int_{0}^{\infty} \frac{x^3}{e^x - 1} dx}
\tag{19.5}
$$

where x_g is the band gap energy normalized by the kT_s ($x_g = E_g/(kT_s) = h\nu_g/(kT_s)$) and the T_s is the black body temperature of the Sun. The denominator in Equation 19.5 can be found in the Planck equation and the definite integral can generate a value of $\pi^4/15$. By varying the band-gap energy under the same black body radiation, we are able to calculate the PCE like in Figure 19.2. This calculation does not put any restriction on the device structure and does not consider the diode characteristics at all. Only the band gap of the material is used as the variable (x_g in Equation 19.5) The highest value, which is around 44%, is often called "Shockley–Queisser limit," named after the two authors who wrote the original paper in 1960. This limit, although idealized, illustrates a clear picture for us: the best band gap for the solar cell is around 1.05 eV, and the general blue light emissive InGaN whose band gap falls in the neighborhood of 2.75 eV cannot be favorable for this purpose. To move the material out of this unfavorable zone, one can either change the indium composition to actually narrow its band gap or use other epitaxial methods to lower its effective band gap. In Section 19.3, we will explain our general idea on the material growth to fulfill this requirement.

19.3 Material Growth

Fundamentally, all the devices need good quality materials. The growth of these materials is a task that all the device engineers need to face. In this section, we will discuss the low indium content growth and the pure InN growth in our devices. The low indium content InGaN MQW can be used for transparent photovoltaic devices, while the pure indium nitride can play an important role to further expand the applicable wavelength in the nitride-based devices. The characterization of these materials will be accompanied as the proof of good quality obtained in our system.

19.3.1 Low Indium Content InGaN Growth

The growth of InGaN for LEDs has become a mature technology that changes our daily life. The multiple quantum well design has been proven as an effective method to produce great light emission. However, when the solar cell function is considered, the well-established procedure has to be modified. One major difference is the thickness of the active region. We would prefer thicker InGaN layer in the wafer to enhance the overall performance. However, the large lattice constant difference usually limits the maximum thickness that can be grown without significant reduction in crystal quality. For alleviating this problem, one strategy is to exploit the structure of superlattice or multiple quantum wells (Sheu et al. 2009; Tsai et al. 2015). In the superlattice or multiple-quantum-well design, the materials with different band gaps are grown alternatively to generate good quality material. In this superlattice structure, each layer is in nanometer scale to maintain the stress-lattice balance while increasing the total thickness. The combined effect of this alternative layer arrangement is the formation of absorption band located in between the band gaps of the two constituent materials. By using $In_{0.25}Ga_{075}N$ 40 A and GaN 30 A intermittently, a relatively well-behaved X-ray can be seen at the total thickness of 200 nm. Compared with samples without superlattice but grown with the same thickness of a uniform InGaN, the superlattice structure could provide a better J-V fill factor and much less reverse leakage current. The difference in leakage current can be as high as 5 orders at the $V_{diode} = -1.5$ Volt, which means the superlattice can effectively improve the material quality to enhance the device performances (Sheu et al. 2009).

Another method is to use the graded quantum well (GQW) to extend the absorption range. In the GQW design, the composition of indium can be linearly graded from low to high. While the low indium wells bear much less strain due to lattice mismatch, a much higher indium composition can be inserted in a few quantum wells that can both extend the absorbing photon's wavelength, and put less stress on the overall epitaxial structures. As shown in a previous publication, we can have high indium content of 16% and linearly lower it to 10% (Tsai et al. 2015). Compared with the same number of QW with 13.5% of indium, the GQW design not only alleviates the stress, but also improves the band structure alignment in the device, which is important for current extraction in the solar cell (Tsai et al. 2015). From these two design examples, we could deduce a possible principle in terms of InGaN solar cell design: by reducing the lattice-mismatch strain, either via superlattice or via graded QW, one can insert an absorption layer with higher indium composition without compromising the material quality and maintain the device performance.

19.3.2 InN Growth

The growth of InN poses another difficulty for device applications because of its different conditions between different gas reactants and large lattice mismatch of InN to its substrate (11.15% to GaN) (Bhuiyan et al. 2003; Yu et al. 2005; Park et al. 2012). In the past, molecular beam epitaxy (MBE) was widely applied, and great results in LEDs and solar cells (Hieu Pham Trung et al. 2011; Le et al. 2014) have been demonstrated. In a practical MOCVD system, the growth temperature is often beyond the desorption point of indium nitrides (Bhuiyan et al. 2003). Such limitation restrains the subsequent growth of the capping layer, and thus, affects the overall epitaxial layer design (Huang et al. 2004; Fujii et al. 2010). Another difficulty comes from the high background electron concentration in InN, so the p-type InN needs extra measures to be acquired (Jones et al. 2006; Zhao et al. 2013a, 2013b). People used a specialized method in the MOCVD growth to achieve acceptable material quality, and the pros and cons of these adjustments can be discussed later in the text (Johnson et al. 2004; Ruffenach et al. 2010).

In our study, we tried to improve the InN quality on the surface of the p-type GaN substrate. The p-type GaN substrate can be achieved via proper Mg doping and dopant activation (Hiroshi et al. 1989; Shuji et al. 1992). The natural n-type InN can then form a p–n heterojunction diode for the device application purpose. Meanwhile, the flat thin-film of InN will be difficult to grow due to the large lattice constant mismatch as we mentioned. So it would be better to find out the growth condition for micron scale or even nanometer scale InN growth which can reduce the possible strain developed and maintain the quality of the InN layers (Shen et al. 2006).

For a beginning, the growth temperature of the InN layer has to be reduced. The usual high temperature of the GaN growth can easily make the indium nitride dissociated. The substrate temperature must be set around 450°C–600°C (depending on the growth system) according to the past publication (Ruffenach et al. 2010). In addition to low growth temperature, the III/V gas flow ratio also needs to be modified due to this growth condition. Because the group V element is provided by the ammonia gas, and it needs to be cracked during the growth, the low temperature can lead to inefficient cracking of the molecules, and thus the quantity of the ammonia has to become higher. From the past, the different ratios have been reported, and the dependence on the MOCVD system is obvious. In general, when the V/III ratio is too high, excessive hydrogen will be presented. This situation decreases the Gibbs free energy for InN formation, and thus inhibits the InN deposition. On the other hand, if the ratio is too low, the insufficient ammonia cracking leads to limited N radicals and lower the InN growth rate (Johnson et al. 2004), which can foster the indium droplet formation in turns. Usually the best condition can be found for the ratio 10^5 to 2.6×10^6 (Johnson et al. 2004).

At the beginning, our growth temperature was lowered too much. At 450°C, the InN crystalline was not observed no matter what V/III ratio was tried, as shown in Figure 19.3a–c. From previous literature, the high and low V/III ratio could lead to In droplet formation (Johnson et al. 2004), which can be seen in our SEM pictures in Figure 19.3. The metallic formation of indium cannot be used for device application at all, and thus needs to be excluded. So from this initial test, there exists a proper growth window in terms of temperature: no InN deposition at high temperature, and only indium droplets at low temperature. In Figure 19.4a, the low-temperature (10 K) photoluminescence of these samples were shown.

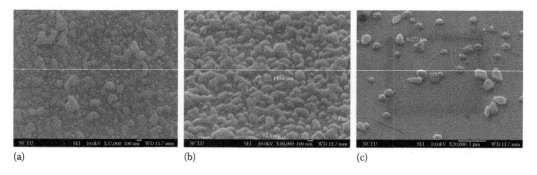

(a)   (b)   (c)

FIGURE 19.3 Surface morphologies of various V/III ratios grown at 450°C (a) V/III = 14241; (b) V/III = 17801; (c) V/III = 71207.

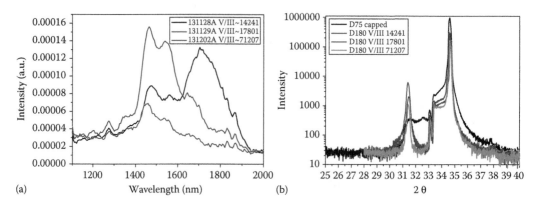

FIGURE 19.4 (a) The PL spectra of the samples; (b) the X-ray diffraction.

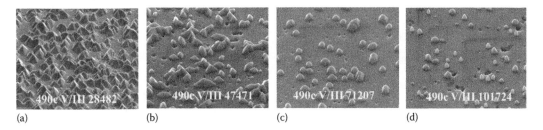

(a)　　　　　　　　(b)　　　　　　　　(c)　　　　　　　　(d)

FIGURE 19.5 Growth temperature: 490°C. Under various V/III ratios. (a) V/III = 28482; (b) V/III = 47471; (c) V/III = 71207; (d) V/III = 101724.

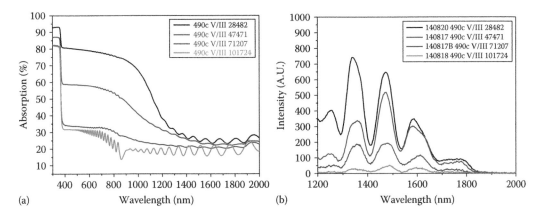

(a)　　　　　　　　　　　　　　　　(b)

FIGURE 19.6 (a) The absorption spectra; (b) the room temperature PL spectra for samples in Figure 19.5.

The weak signal in the IR region indicates low formation or bad quality of InN. From the X-ray diffraction pattern, the InN peak at around 31.3° can be seen in Figure 19.4b, and an extra peak showing possible indium metallic component at 33°, supporting our SEM observation (Yu et al. 2005; Kamimura et al. 2015).

Once the growth temperature was adjusted to 490°C, the morphology of the finished surface began to change. As shown in Figure 19.5, under various V/III ratios, the deposited structures looked entirely different. At ultrahigh V/III ratio (>70,000), the deposited materials on the GaN surface exhibit similar shapes as shown in Figure 19.5c. If one looks closely, the crystalline facets shown in the pyramids in the Figure 19.5a are actually indicators of the better quality of InN. Evidence can be observed from the absorption spectra in IR regime. In Figure 19.6a, the strongest absorption comes from the sample with V/III = 28,182 (in Figure 19.5a). The weakest signal can be obtained from the sample with the highest V/III ratio, which is probably formed by metallic indium droplets or amorphous (In, Ga)N composites (Hsu et al. 2014). A similar trend can be seen when the micro-PL measurement was done with all four samples. As can be seen in Figure 19.6b, the strongest peak intensity comes from V/III = 28,482 sample. From this growth experiments, we could understand that the sharp facet/pyramid-like structure can have much better material quality when we try to judge from the SEM pictures.

From these experiments, a rough growth window of good quality InN can be sketched, as shown in Figure 19.7. As stated previously, exceedingly high or low T_g (growth temperature) can hinder the InN formation and also the possible formation of cubic InN can be observed from the X-ray results. The V/III ratio also controls another knob for material quality. When the V/III ratio becomes exceedingly high (>100,000), the quality of the InN layer is poor in terms of PL strength or IR absorption.

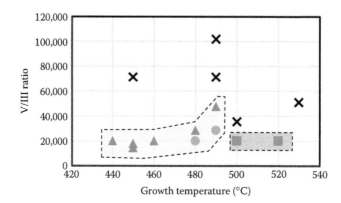

FIGURE 19.7   An acceptable range (left dashed lined area) of growth temperature and V/III ratios for InN grown on the planar n-GaN substrate. The right dashed rectangular window indicates possible formation of cubic InN from the X-ray results.

19.4 Nanostructure Growth Mechanism of InN

From the initial test, we realized that the best choice of the InN growth on GaN could reside in the nanostructures of InN instead of two-dimensional thin film. The thin film growth, although very useful for the device application, will be hindered by the high dislocation density and indium element desorption issues. Meanwhile, the nanoscale structure can be grown with fewer defects, and thus, could produce better materials for the device level application (Ristic et al. 2008). One of the good examples is the InAs quantum dots which were applied to extend the operating wavelengths in GaAs-based optoelectronic devices (Marzin et al. 1994).

Meanwhile, the growth of nanoscale InN on GaN has different mechanisms than the thin film one. At the beginning of the growth, the atoms come to the substrate will tend to sit on the original lattice site. However, once the layers become higher, the built-up stress within the deposited material will tend to break the crystals and form the smaller patches. This change can move the growth from 2D into 3D mode. The Stranski–Krastanov (SK) mode growth in most of the quantum dots can be attributed to this mechanism. Usually, there will be a wetting layer on the substrate.

However, there is another mechanism that can be found in this growth: it is also possible for the atoms to nucleate directly without first depositing as a thin layer. In this case, which is often categorized as Volmer–Weber (V–W) growth mode, the atoms can directly gather into different patches, and the nanoscale islands can later be determined as to whether they can survive or not. In this case, there will be no wetting layer even under the high-resolution TEM. Different surface temperature can determine the critical size of the nanonuclei. The size and density of these nanoscale islands can be determined by the growth temperature and the V/III ratios. The critical size can be larger when the surface temperature increases due to the enhanced diffusion length of the metallic ad-atoms (indium), and the size also grows when the arriving indium atoms increases. However, if the nitrogen concentration is low, the increase of indium could lead to lateral growth (or coalesce) and the two-dimensional film growth is resumed. If the nitrogen is high, the coalescing situation can be stopped due to the preferred incorporation of the indium atoms on the top of the island (Ristic et al. 2008).

When the nanoscale InN island reaches the critical size, the nanocolumn growth mode can proceed. In the nanocolumn growth mode, the indium atoms arrive at the site, either directly or by surface diffusion. Then the metallic atoms climb to the top of the nanoscale island to be incorporated into the crystal. The constant diameter can be explained by high diffusion length of the indium species on the nanocolumn sidewall (Ristic et al. 2008) and the other factor is the very different sticking coefficient of

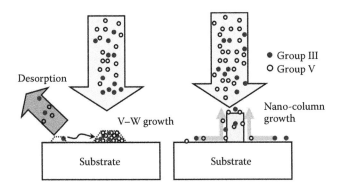

FIGURE 19.8 Two different but connected growth mechanisms for nano-structured InN. (a) Volmer–Weber (V–W) growth and (b) constant diameter nanocolumn growth. (From Ristic, J. et al., *J. Cryst. Growth*, 310, 4035–4045, 2008. With permission.)

the metal element (Indium) between the side wall and the top surface (Ristic et al. 2008). To briefly summarize this section, as shown in Figure 19.8, because of the large strain between the substrate and the deposited materials, usually, there are two connected mechanisms for InN nanostructure: V–W growth and Nanocolumn growth (Ristic et al. 2008). Both of them work together to produce a stable and good quality InN nanoisland on a heterogeneous substrate.

19.5 Device Applications

As mentioned in the introduction of this chapter, various groups have demonstrated great progress in the photodetection and photovoltaic response in the III–V nitride-based materials. In this section, we investigate two different directions: one is how to enhance the performance of the traditional indium gallium nitride solar cell, and the other is to probe the possibility of indium nitride for photodetection. Both of them demonstrated the potential to cope with very different wavelengths of photons (such as UV to IR ranges). The detailed performances and characterization can be shown in the following sections.

19.5.1 Nitride-Based Solar Cell with QD and DBR Enhancement

One of the drawbacks of nitride-based material is that the band gap is often too wide for optimal absorption of the incoming solar photons, as mentioned previously. The colloidal quantum dots (CQDs) and distributed Bragg reflectors (DBRs) can be applied, to compensate the situation partially. The CQDs can extend the absorption range of the generic GaN-based active region, and the DBR structure can reflect the unabsorbed photons back for more possibility of re-absorption by GaN material. Both methods can improve the overall performance of the solar cell. In the following, we will introduce our design and the measurement results.

Figure 19.9 shows the generic device structure. The active region is composed of 14 pairs $In_{0.15}Ga_{0.85}N/$GaN (3 nm/5 nm) undoped MQW sandwiched by a 2 μm thick Si-doped n-GaN layer (n-doping = 2×10^{19} cm^{-3}) and a 200 nm thick Mg-doped p-GaN (p-doping = 2×10^{18} cm^{-3}). The metal contact (Cr/Pt/Au: 50/50/1900 nm) was deposited by electron-beam evaporation. A 11 pairs of HfO_2/SiO_2 was deposited by the sputtering system at room temperature on a glass slide. The reflectivity of the DBR is over 99% between 385 and 460 nm.

After regular semiconductor processes, the spin-coating method was used to apply CdS quantum dots on the top of the device and DBRs were put at the back side of the device. The reflectance spectra and absorption spectra of InGaN MQW solar cell with CdS QDs, the cell with DBR, the cell with both

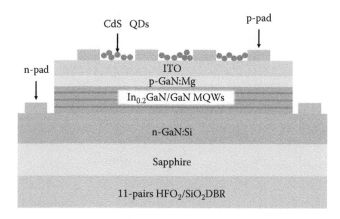

FIGURE 19.9   The schematic diagram of the nitride-based solar cell with DBR and QD enhancement.

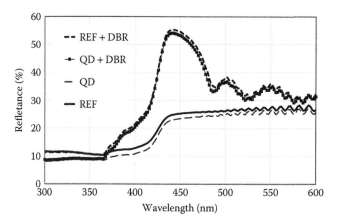

FIGURE 19.10   The measured reflectance spectra of InGaN/GaN MQW solar cell with CdS QDs, the cell with DBR, the cell with both CdS QDs and DBRs, and one bare cell as reference. (From Tsai, Y.-L. et al., *Sol. Energy Mater. Sol. Cells*, 117, 531–536, 2013. With permission.)

CdS QDs and DBRs, and one bare cell as reference are shown in Figure 19.10. The reflectance spectra are significantly different when the DBR layer was placed underneath the device. Strong reflection indicates the good portion of the incident photons that can be turned back to the device if they are not absorbed. On the other hand, the broadband reduction in the reflection can be observed when CQDs are cast on the device. The anti-reflection characteristic of CdS QDs can be caused by the light scattering effect of the nanosphere-like CdS QDs cluster. In addition, these CQD clusters provide a graded refractive index interface, and this results in broadband anti-reflection characteristic for light harvesting (Fendler 1996; Huang et al. 2007). From the measured angular reflectance, this anti-reflection effect brought by CdS QD dispense can be seen up to 60° of the incident angle (Huang et al. 2007; Tsai et al. 2013).

The measured photovoltaic current density-voltage (J-V) of the four types of InGaN/GaN MQW solar cells were performed under a simulated AM1.5G illumination condition and the results are shown in Figure 19.11. The details of measured results are listed in Table 19.1. Compared to the reference cell, the short-circuit current density enhancement of the cell with only CdS QDs and the cell with only DBRs are 5.5% and 16.5%, respectively. With the combination of CdS QDs and DBRs, the enhancement in short circuit current density can be further boosted to 22%. A slight decrement of fill factor can be observed as employing CdS QDs on the cells. This is because that the sprayed CdS QDs over-spilled to

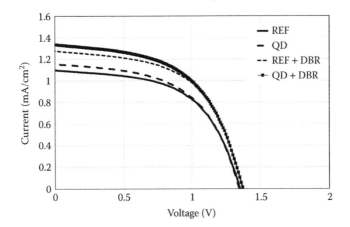

FIGURE 19.11 The measured photovoltaic current density-voltage (J-V) curve of InGaN/GaN MQW solar cell with CdS QDs, the cell with DBR, the cell with both CdS QDs and DBRs, and one bare cell as reference. (From Tsai, Y.-L. et al., *Sol. Energy Mater. Sol. Cells*, 117, 531–536, 2013. With permission.)

TABLE 19.1 Current-Voltage Characteristics of InGaN/GaN MQW Solar Cell with CdS QDs, the Cell with DBR, the Cell with both CdS QDs and DBRs, and one bare cell as reference

	V_{oc} (V)	J_{sc} (mA/cm²)	Fill Factor	PCE (%)
Reference	1.35	1.09	56.03	0.83
QDs	1.34	1.15	54.52	0.85
DBR	1.37	1.27	56.19	0.98
QDs + DBR	1.37	1.33	54.79	1.002

Source: Tsai, Y.-L. et al., *Sol. Energy Mater. Sol. Cells*, 117, 531–536, 2013.

the device's sidewall which led to leakage paths. However, the enhanced power conversion efficiency in the InGaN/GaN MQW solar cells with CdS QDs can still be observed. The cells combine with CdS QDs, and the DBRs shows the highest power conversion efficiency. By comparing with the reference cell, an enhancement of 20.7% can be found.

Figure 19.12 shows the measured EQE of the four types of InGaN/GaN MQW solar cells. The EQE is a measure of photo-generated current at a specific wavelength. From the plot, two regions of enhancement can be identified: short wavelength (<360 nm) and long wavelength (between 360 and 440 nm) ones. In the longer wavelength region of the EQE spectra, which is the majority of photo-generation. The device enhancement comes from the extra reflection brought by DBR. The cross-check finds that the addition of CQD layer in this region does not help the overall performance too much. Meanwhile, CQD's usage can be fully understood when we consider the shorter wavelength region. The increment is although small but effective.

To further understand the influence of CdS QDs on the InGaN/GaN MQW solar cells, detail analysis in EQE is necessary. In general, the enhancement in the long wavelength portion can be attributed to the anti-reflection effect brought by the intermediate refractive index of the CQD layer. A significant peak at a wavelength range of 350–365 nm can be observed in EQE. When we compare this with the DBR only case, this peak is obviously due to the luminescent down-shifting effect of the quantum dots.

Also from Figure 19.13, both of the cases (DBR, DBR+CQDs) show a 1.1 to 1.3-fold enhancement at a wavelength range of 370–440 nm which agrees with the high reflectance of DBRs. The growing trend as the wavelength increases in the enhancement factor can be realized as follows: the absorption length

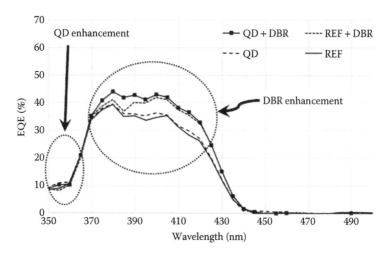

FIGURE 19.12 The measured external quantum efficiency (EQE) of InGaN/GaN MQW solar cell with CdS QDs, the cell with DBR, the cell with both CdS QDs and DBRs, and one bare cell as reference. (From Tsai, Y.-L. et al., *Sol. Energy Mater. Sol. Cells*, 117, 531–536, 2013. With permission.)

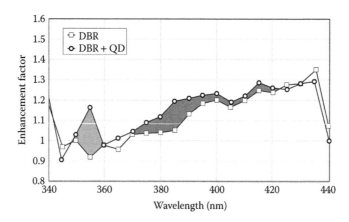

FIGURE 19.13 The enhancement factor of EQE for the cell with DBRs and the cell with both QDs and DBRs. The light-gray area indicates the QD enhanced effect while the dark-gray one is for extra DBR enhancement due to the dispense of QD layer. (From Tsai, Y.-L. et al., *Sol. Energy Mater. Sol. Cells*, 117, 531–536, 2013. With permission.)

of the photons grows as the wavelength increases, and when the DBR is placed at the bottom of the device, the reflected amount of the photons at the longer wavelength is larger than the shorter ones. So the increased EQE, which is directly proportional to the photo-generated current, can become higher in value when the wavelength of the incident photons gets longer.

19.5.2 Indium Nitride Photodetector

Our general device structure for InN-based photodetector can be seen in Figure 19.14. At the beginning, a Schottky-type photodetector and then a heterojunction photodetector was also developed. The substrate is sapphire, and a buffer layer of n-type (Schottky-type) or p-type GaN (heterojunction-type) was grown to become the bottom contact layer. The core/active absorption layer is always InN layer. The low-temperature (LT) GaN was used to have a better contact sometimes, and the ZnO layer can

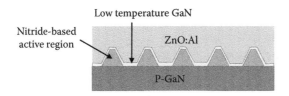

FIGURE 19.14 The cross-section structure of nitride base photodetector with IR capability.

be later deposited as the current conducting layer. After the devices had been finished, a AM1.5G solar spectrum simulator was used to characterize the photocurrent. Since there is GaN inside the device, the photo-current coming from this GaN absorption is inevitable. A long-wavelength pass filter or an IR light source (such as IR LED or tunable laser) can then be used for a more precise measurement in terms of wavelength-dependent quantum efficiency.

19.5.2.1 General Device Fabrication Processes

In the device related epitaxial growth, a Veeco D75 MOCVD was used for nitride-related growth. All the growth was finished on the c-plane sapphire substrates. The low-temperature GaN nucleation layer and undoped GaN layer were grown first to smooth the surface. A 2-μm-thick GaN epilayer (n-type or p-type depends on the application) can be deposited before the indium nitride layer. The indium nitride layer was grown under low temperature (as we discussed in the previous section), and the thickness is usually several tens of nanometers. A low-temperature GaN layer can be later grown on top at the same growth temperature to protect the indium nitride and possibly form a better contact layer to the metal (Huang et al. 2004; Ager et al. 2008; Fujii et al. 2010). After the epitaxial growth, a suitable metal layer, such as Cr/Pt/Au (5 nm/5 nm/190 nm) alloy contact pads, was applied by photolithography and lift-off process steps. The mesa was defined by dry-etch with the size of 14 mil by 14 mil (355 μm by 355 μm). Depending on the experimental purpose, a wafer with and without the LT-GaN capping layer could be grown for comparison.

19.5.2.2 Schottky-Type Photodetector

The easiest structure of a photodetector is to use the natural Schottky barrier formed between conducting metal and InN. Because the InN layer is inherently n-type with a relatively high electron background concentration (can be as high as 5×10^{18} cm^{-3}) (Bhuiyan et al. 2003). The devices were grown at 525°C and the sample showed significant IR response, which can be expected. After the device fabrication had been carried out as per the description in the previous section, they were put under the various tests. Under the regular AM1.5G solar simulator, the device showed great photoresponse. However, this photocurrent generation can be related to GaN material in the device due to the broadband excitation from the solar simulator. Thus, to verify the function of indium nitride, an infrared (IR) source is needed. We first use an external cavity laser with wavelength tunability. Because the IR photons cannot be absorbed by GaN, thus the generated current can only come from InN structure. Both dark current and photocurrent under various laser powers can be recorded. The measured photo-generated currents and laser powers were converted such that the quantum efficiency (defined by [the number of electron-hole pairs]/[the number of the photons]) can be calculated. As seen in Figure 19.15a, a very good linear line can be measured from the photocurrent versus input laser power at 1550 nm of laser wavelength. The slope efficiency, which is 0.443 A/W from the linear extrapolation, can be regarded as the responsivity of the InN structure at this wavelength. The linear extrapolation of the measured photocurrent response in Figure 19.15a exhibit certain "threshold" power needed to initiate the photocurrent extraction. This phenomenon has been documented previously (Munoz Merino et al. 1999; Nakano et al. 2008; Assefa et al. 2010), and we believe that the excess traps and defects between the InN and GaN are the possible reasons to consume the excitation laser photons at the low excitation power level. Figure 19.15b shows

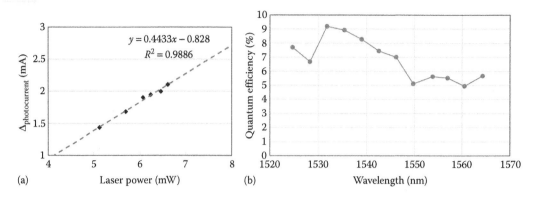

FIGURE 19.15 (a) The reverse biased photo-generated current under various pump laser power. The reverse bias is −0.2 V in the plot, and the linear slope of 0.443 (A/W) can be fitted from the measured data. (b) A quantum efficiency measurement of an InN Schottky type PD by a tunable laser source. (From Hsu, L.-H. et al., *Opt. Mater. Express*, 4, 2565–2573, 2014. With permission.)

the resultant spectral response of quantum efficiency of this InN photodetector at the reverse bias of 0.3 Volts. The tuning range of the DFB laser limits our data acquisition and an EQE as high as 9.2% is recorded.

19.5.2.3 Heterojunction-Type Photodetector

The Schottky type detector is easy to grow and process. However, the required attention on the metal-semiconductor interface might hinder its further development. A heterojunction type of photodetector was proposed and fabricated to overcome this issue.

As shown in Figure 19.14, a multiple-layered structure of the ZnO/LT-GaN/InN photodetectors were grown and fabricated into regular photodetectors. This special arrangement can have a band alignment advantage in addition to the InN IR photodetection benefit. The band alignment between GaN and ZnO is a type-II junction and when the InN layer is inserted, the band diagram similar to the double-heterostructure, which is preferred in most optoelectronic devices, can be expected (Zhu et al. 2008; Zhang et al. 2009). The growth temperature for the InN layer is fixed at 530°C in this case.

For evaluating the photodetection capability of the device, two different types of measurements were made: one is the photovoltaic response, and the other is the EQE of the device. Figure 19.16a demonstrates the J-V characteristics under dark and AM1.5G illumination conditions, although strong photocurrent was measured, the device failed to show any photovoltaic response in the forward bias region, which could be an indication of poor interfacial condition between the InN and GaN layers. Similar to the previous section, the solar simulator provides a wider spectrum of photons, and thus, the absorption in GaN is counted at the same time. Figure 19.16b shows the EQE of the InN photodetector at the reverse bias of 0, 3 and 7 Volts. The efficiencies are peaked at the wavelength of 1770 nm. Both results at −3 and −7 volts show the same uprising trend toward the longer wavelength and the EQE value as high as 3.55% is recorded, while the device showed no photocurrent at all when the reverse bias voltage is zero. Strong photoresponse under solar simulator rises from absorption in both p-GaN and InN layers, but the long wavelength EQE, similar to IR laser in the previous section, can only be absorbed by InN layer. Combining the photocurrent and EQE measurement (Figure 19.16a and b), one can calculate the portion of photocurrents in Figure 19.16a that are brought by InN layer by the following formula:

$$\frac{\text{Photocurrent from InN at reverse bias}}{\text{Total photocurrent measured at reverse bias}} = \frac{q\displaystyle\int_{1200\text{nm}}^{1800\text{nm}} F(\lambda) \times \text{EQE}(\lambda)\,\mathrm{d}\lambda}{I_{\text{photo}}} \qquad (19.6)$$

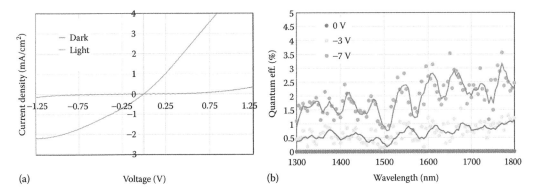

FIGURE 19.16 (a) The plot is the J-V characteristics under dark and simulated AM1.5G illumination. (b) The measured EQE spectra of the InN device for zero bias and reverse bias. The solid curves are the moving average of the data points. (From Hsu, L.-H. et al., *Opt. Express*, 23, 31150–31162, 2015. With permission.)

where I_{photo} is the measured photocurrent, EQE(λ) is the EQE spectrum in Figure 19.16b, and $F(\lambda)$ is the standard AM1.5G solar spectrum. A current, as high as 18.9% of the total photo-current, can be attributed to InN contribution at reverse bias of 0.3 V.

19.6 Conclusion

In this chapter, we demonstrated nitride-based photodetection via different material compositions. Different types of device are applied for photovoltaic and photodetection purposes. The material growth was carried out to evaluate the growth temperature and the ratio of the group V and group III. When properly tuned, the sharp facets on the nanostructure can be observed. This hypothesis can be verified by the PL, absorption, and quantum efficiency measurements, in which the IR response can be obtained. Finally, both the GaN-based solar cell and InN-based photodetectors are both demonstrated, showing the wide range coverage of the nitride material and their potential for photodetection application.

References

Ager JW et al. (2008) Mg-doped InN and InGaN–Photoluminescence, capacitance–voltage and thermo-power measurements. *Phys. Status Solidi B* 245:873–877.

Akasaki I, Amano H, Koide Y, Hiramatsu K, Sawaki N (1989) Effects of ain buffer layer on crystallographic structure and on electrical and optical properties of GaN and Ga1–xAlxN (0 < x ≦ 0.4) films grown on sapphire substrate by MOVPE. *J. Cryst. Growth* 98:209–219.

Assefa S et al. (2010) CMOS-integrated high-speed MSM germanium waveguide photodetector. *Opt. Express* 18:4986–4999.

Bhuiyan AG, Hashimoto A, Yamamoto A (2003) Indium nitride (InN): A review on growth, characterization, and properties. *J. Appl. Phys.* 94:2779.

Bhuiyan AG, Sugita K, Hashimoto A, Yamamoto A (2012) InGaN solar cells: Present state of the art and important challenges. *IEEE J. Photovolt.* 2:276–293.

Chuang SL (1995) *Physics of Optoelectronic Devices*, 1st ed., New York: Wiley.

Chuang SL (2009) *Physics of Photonic Devices*, 2nd ed., New York: Wiley.

Dahal R, Pantha B, Li J, Lin JY, Jiang HX (2009) InGaN/GaN multiple quantum well solar cells with long operating wavelengths. *Appl. Phys. Lett.* 94:063505.

Einstein A (1917) The quantum theory of radiation. *Physikalische Zeitschrift* 18:121.

Fendler JH (1996) Self-assembled nanostructured materials. *Chem. Mater.* 8:1616–1624.

Fujii T et al. (2010) Structural characteristics of GaN/InN heterointerfaces fabricated at low temperatures by pulsed laser deposition. *Appl. Phys. Express* 3:021003.

Hieu Pham Trung N, Yi-Lu C, Ishiang S, Mi Z (2011) InN p-i-n nanowire solar cells on Si. *IEEE J. Sel. Topics Quantum Electron.* 17:1062–1069.

Hiroshi A, Masahiro K, Kazumasa H, Isamu A (1989) P-type conduction in Mg-doped GaN treated with low-energy electron beam irradiation (LEEBI). *Jpn. J. Appl. Phys.* 28:L2112.

Hiroshi A, Tsunemori A, Isamu A (1990) Stimulated emission near ultraviolet at room temperature from a GaN film grown on sapphire by MOVPE using an AlN buffer layer. *Jpn. J. Appl. Phys.* 29:L205.

Hsu L-H et al. (2014) Enhanced photocurrent of a nitride-based photodetector with InN dot-like structures. *Opt. Mater. Express* 4:2565–2573.

Hsu L-H et al. (2015) InN-based heterojunction photodetector with extended infrared response. *Opt. Express* 23:31150–31162.

Huang W, Yoshimoto M, Taguchi K, Harima H, Saraie J (2004) Improved electrical properties of InN by high-temperature annealing with in situcapped SiNx layers. *Jpn. J. Appl. Phys.* 43:L97–L99.

Huang Y-F et al. (2007) Improved broadband and quasi-omnidirectional anti-reflection properties with biomimetic silicon nanostructures. *Nat. Nanotechnol.* 2:770–774.

Jani O, Ferguson I, Honsberg C, Kurtz S (2007) Design and characterization of GaN/InGaN solar cells. *Appl. Phys. Lett.* 91:132117.

Johnson MC, Konsek SL, Zettl A, Bourret-Courchesne ED (2004) Nucleation and growth of InN thin films using conventional and pulsed MOVPE. *J. Cryst. Growth* 272:400–406.

Jones R et al. (2006) Evidence for p-type doping of InN. *Phys. Rev. Lett.* 96:125505.

Kamimura J, Kishino K, Kikuchi A (2015) Growth of very large InN microcrystals by molecular beam epitaxy using epitaxial lateral overgrowth. *J. Appl. Phys.* 117:084314.

Le BH, Zhao S, Tran NH, Mi Z (2014) Electrically injected near-infrared light emission from single InN nanowire p-i-n diode. *Appl. Phys. Lett.* 105:231124.

Liu HY, Hsu WC, Chou BY, Wang YH (2015) Fabrication AlGaN/GaN MIS UV photodetector by H_2O_2 oxidation. *IEEE Photon. Technol. Lett.* 27:101–104.

Maruska HP, Stevenson DA (1974) Mechanism of light production in metal-insulator-semiconductor diodes; GaN:Mg violet light-emitting diodes. *Solid-State Electron.* 17:1171–1179.

Marzin JY, Gérard JM, Izraël A, Barrier D, Bastard G (1994) Photoluminescence of single InAs quantum dots obtained by self-organized growth on GaAs. *Phys. Rev. Lett.* 73:716–719.

Matioli E et al. (2011) High internal and external quantum efficiency InGaN/GaN solar cells. *Appl. Phys. Lett.* 98:021102.

Munoz Merino E et al. (1999) AlGaN-based photodetectors for solar UV applications. *Proc. SPIE 3629, Photodetectors: Materials and Devices IV* pp. 200–210.

Nakamura S, Mukai T, Senoh M (1994) Candela-class high-brightness InGaN/AlGaN double-heterostructure blue-light-emitting diodes. *Appl. Phys. Lett.* 64:1687–1689.

Nakano M et al. (2008) Transparent polymer Schottky contact for a high performance visible-blind ultraviolet photodiode based on ZnO. *Appl. Phys. Lett.* 93:123309.

Pankove JI, Hutchby JA (1976) Photoluminescence of ion-implanted GaN. *J. Appl. Phys.* 47:5387–5390.

Park J, Ryu H, Son T, Yeon S (2012) Epitaxial growth of ZnO/InN core/shell nanostructures for solar cell applications. *Appl. Phys. Express* 5:101201.

Ristic J et al. (2008) On the mechanisms of spontaneous growth of III-nitride nanocolumns by plasma-assisted molecular beam epitaxy. *J. Cryst. Growth* 310:4035–4045.

Ruffenach S, Moret M, Briot O, Gil B (2010) Recent advances in the MOVPE growth of indium nitride. *Phys. Status Solidi A* 207:9–18.

Shen C-H et al. (2006) Near-infrared photoluminescence from vertical InN nanorod arrays grown on silicon: Effects of surface electron accumulation layer. *Appl. Phys. Lett.* 88:253104.

Sheu JK et al. (2009) Demonstration of GaN-based solar cells with GaN/InGaN superlattice absorption layers. *IEEE Electron Device Lett.* 30:225–227.

Shockley W, Queisser HJ (1961) Detailed balance limit of efficiency of p-n junction solar cells. *J. Appl. Phys.* 32:510–519.

Shuji N, Takashi M, Masayuki S (1991) High-power GaN P-N junction blue-light-emitting diodes. *Jpn. J. Appl. Phys.* 30:L1998.

Shuji N, Takashi M, Masayuki S, Naruhito I (1992) Thermal annealing effects on p-type Mg-doped GaN films. *Jpn. J. Appl. Phys.* 31:L139.

Sze SM (1981) *Physics of Semiconductor Devices*, New York: Wiley.

Tsai Y-L et al. (2013) Improving efficiency of InGaN/GaN multiple quantum well solar cells using CdS quantum dots and distributed Bragg reflectors. *Sol. Energy Mater. Sol. Cells* 117:531–536.

Tsai Y-L et al. (2015) Enhanced power conversion efficiency in InGaN-based solar cells via graded composition multiple quantum wells. *Opt. Express* 23:A1434–A1441.

Walker D et al. (1999) High-speed, low-noise metal–semiconductor–metal ultraviolet photodetectors based on GaN. *Appl. Phys. Lett.* 74:762–764.

Yu KM et al. (2005) On the crystalline structure, stoichiometry and band gap of InN thin films. *Appl. Phys. Lett.* 86:071910.

Zhang X-M, Lu M-Y, Zhang Y, Chen L-J, Wang ZL (2009) Fabrication of a high-brightness blue-light-emitting diode using a ZnO-nanowire array grown on p-GaN thin film. *Adv. Mater.* 21:2767–2770.

Zhao S, Liu X, Mi Z (2013a) Photoluminescence properties of Mg-doped InN nanowires. *Appl. Phys. Lett.* 103:203113.

Zhao S et al. (2013b) p-Type InN nanowires. *Nano Lett.* 13:5509–5513.

Zhu H et al. (2008) High spectrum selectivity ultraviolet photodetector fabricated from an n-ZnO/p-GaN heterojunction. *J. Phys. Chem. C* 112:20546–20548.

20

Intersubband Optoelectronics Using III-Nitride Semiconductors

Caroline B. Lim,
Akhil Ajay,
Jonas Lähnemann,
David A. Browne,
and Eva Monroy

20.1 Introduction

The terms "intersubband" (ISB) or "intraband" refer to electronic transitions between confined states in either the conduction band or the valence band of semiconductor heterostructures. The first measurement of ISB absorption in GaAs quantum wells (QWs) was performed by West and Eglash, (West and Eglash 1985) and the results which led to the fabrication of the first QW infrared photodetector (QWIP) (Levine et al. 1987; Levine 1993). In 1994, Faist et al. (1994) presented a major breakthrough in the ISB technology: the quantum cascade laser (QCL). This was the beginning of the tremendous development of the ISB technology, which resulted in commercially available infrared (IR) optoelectronic devices. For an introduction to ISB physics, we refer the readers to Bastard (1988) or Liu and Capasso (2000).

Nowadays, ISB devices based on III-As materials can be tuned from the mid-IR to the THz spectral range. Operation at shorter wavelengths ($<3\,\mu m$) is limited by the available conduction band offset and by material transparency. The III-Nitride semiconductors (GaN, AlN, InN and their alloys), with their wide band gap and a large conduction band offset [≈1.8 eV for GaN/AlN (Binggeli et al. 2001; Cociorva et al. 2002; Tchernycheva et al. 2006b)], are interesting for ISB devices operating in the near-IR spectral range, particularly in the 1.3–1.55 μm wavelength window used for fiber-optic communications (Machhadani et al. 2009; Hofstetter et al. 2010a; Beeler et al. 2013b). GaN is transparent for wavelengths longer than 360 nm (band gap), except for the Reststrahlen band (from 9.6 to 19 μm). Also, absorption in the range of 7.3 to 9 μm has been observed in bulk GaN substrates with carrier concentrations <10^{16} cm^{-3} (Hao et al. 1999; Yang et al. 2005; Welna et al. 2012) and is attributed to the second harmonic of the Reststrahlen band. Although this second band hinders the fabrication of waveguided GaN devices, its effect in planar structures with μm-sized active regions is negligible since the absorption related to two-phonon processes is much smaller than that associated with ISB transitions (Kandaswamy et al. 2009, 2010).

There is also an interest to push the III-Nitride ISB technology toward longer wavelengths, particularly to the THz frequency range. This spectral region has attracted attention due to potential applications in medical diagnostics, security screening, and quality control. GaAs-based QCLs have already demonstrated potential as THz sources in the 1.2–5 THz range; however, their operating temperature and spectral range are limited by the longitudinal optical (LO) phonon emission at 36 meV (8.7 THz). The GaN-based devices have the potential to operate at a higher temperature in the whole THz spectral range due to a large LO-phonon energy of 92 meV (about three times that of GaAs).

20.2 Intersubband Absorption in III-Nitride Heterostructures

20.2.1 Polar GaN/AlGaN QWs

The properties of {0001}-oriented GaN QWs are strongly affected by the presence of spontaneous and piezoelectric polarization (Bernardini et al. 1997). Figure 20.1a presents the band diagram of 1-nm-thick and 2.1-nm-thick GaN QWs in a GaN/AlN multiple quantum well (MQW) structure with 3-nm-thick

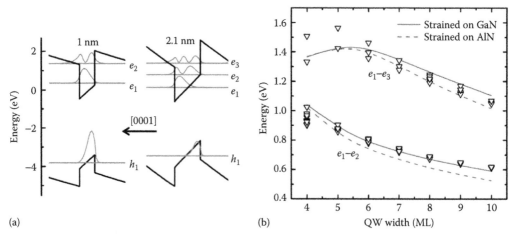

(a)   (b)

FIGURE 20.1 (a) Band diagram of GaN/AlN QWs in an infinite superlattice with 3-nm-thick AlN barriers and 1-nm or 2.1-nm-thick GaN QWs. The structure is considered strained on an AlN substrate. The squared electron wave functions of the ground hole state h_1, the ground electron state, e_1, and the excited electron states e_2 and e_3, are presented. (b) Variation of e_2–e_1 and e_3–e_1 as a function of the QW thickness in GaN/AlN MQW structures with 3-nm-thick barriers. Triangles indicate experimental data and solid and dashed lines correspond to theoretical calculations assuming the structure fully strained on AlN and on GaN substrates, respectively. (Reprinted with permission, from Kandaswamy, P.K. et al., *J. Appl. Phys.*, 104, 093501. Copyright 2008, American Institute of Physics.)

AlN barriers. Calculations were performed using the nextnano[3] 8-band k.p Schrödinger–Poisson solver (Birner et al. 2007) with the material parameters described in ref. (Kandaswamy et al. 2008). In narrow QWs (~1 nm) the energy difference between the ground electron state e_1 and the ground hole state h_1 is mostly determined by the confinement in the QW, whereas for larger QWs (>2 nm) this difference is ruled by the internal electric field, since both levels lie in the triangular part of the QW potential profile. The resulting red shift and reduced oscillator strength of the band-to-band transitions are known as quantum-confined Stark effect (QCSE).

Within the conduction band, the evolution of the e_2–e_1 and e_3–e_1 energy differences with the QW thickness and strain state is presented in Figure 20.1b. The higher e_2–e_1 ISB energy when the MQW is strained on GaN, as compared with MQWs strained on AlN, is related to larger piezoelectric coefficients of AlN (Kandaswamy et al. 2008). Note that certain QW thicknesses can result in a configuration where e_2–e_1 has approximately the same energy as e_3–e_2 (and even as e_4–e_3 for very thick barrier layers), which can lead to the enhancement of nonlinear effects such as two-photon absorption (Hofstetter et al. 2007) or second-harmonic generation (Nevou et al. 2006b).

The IR absorption of AlN/GaN MQWs has been extensively studied by Fourier transform IR spectroscopy (FTIR), using polarizers to discern between the transverse electric (TE) and transverse magnetic (TM) polarized light. Let us remind that ISB transitions can only be induced by light with a nonzero electric field component along the direction of carrier confinement. Figure 20.2 shows the ISB absorption of structures with various quantum well thicknesses (note that one GaN monolayer [1 ML] is approximately 0.25 nm). The samples show a pronounced TM-polarized absorption, attributed to the e_1 to e_2 transition, while no absorption is observed for TE-polarized light within the experimental sensitivity. For large QWs ($\geq$8 ML), the e_1 to e_3 transition is also observed (not shown). This transition is forbidden in symmetric QWs (Yang et al. 1994), but is allowed in nitride QWs because the internal electric field breaks the symmetry of the potential in the well. The experimental values of e_2–e_1 and e_3–e_1 as a function of the QW width are presented in Figure 20.1b, showing a good fit with theoretical calculations.

The full width at half maximum (FWHM) of the absorption remains in the 70–100 meV range for QWs doped at 5×10^{19} cm^{-3}. A record small linewidth of $\approx$40 meV has been measured in nonintentionally doped structures (Guillot et al. 2006a). The spectra present either a Lorentzian shape or are structured with two or three well-defined Lorentzian-shaped peaks, as illustrated in Figure 20.2c and d. These multiple peaks correspond to the expected values of e_2–e_1 in QWs whose thickness is equal to an integer number of GaN monolayers. Regarding the thermal stability of the ISB transition in GaN/AlN MQWs, it has been found that the ISB absorption energy decreases only by $\approx$6 meV at 400°C relative to its room temperature value (Berland et al. 2010).

Using GaN/AlN QWs, the e_2–e_1 energy can be tuned in the 1.0–3.5 μm wavelength range by changing the QW thickness from 1 to 7 nm with AlN barrier thicknesses in the 1.5–5.1 nm range (Iizuka et al. 2002; Kishino et al. 2002; Helman et al. 2003; Tchernycheva et al. 2006b; Liu et al. 2007; Kandaswamy et al. 2008; Andersson et al. 2009; Bayram et al. 2009; Berland et al. 2010; Kaminska et al. 2016). For large QWs larger than $\approx$5 nm, the first two confined electron levels become trapped in the triangular section of the QW, which results in a saturation of the e_2–e_1 value. Therefore, to shift the absorption toward longer wavelengths, it is necessary to reduce the effect of the internal electric field in the QWs. A first approach consisted of using GaN/AlGaN MQWs, thereby reducing the Al mole fraction in the barriers. By changing the geometry and composition, the ISB absorption can be tailored to cover the near-IR range above 1.0 μm and mid-IR region up to 10 μm (Suzuki and Iizuka 1999; Gmachl et al. 2000, 2001b; Ng et al. 2001, 2002; Sherliker et al. 2007; Kandaswamy et al. 2009; Péré-Laperne et al. 2009; Huang et al. 2011; Bayram 2012; Edmunds et al. 2012a; Tian et al. 2012; Chen et al. 2013). A slight red shift of the ISB transition was observed when increasing the compressive strain in the QWs, in agreement with theoretical predictions (Tian et al. 2012).

In order to observe ISB absorption, the first electronic level must be populated. However, high doping levels may affect the transition energy. The ISB absorption energy exhibits a significant blue shift due to many-body effects (Helman et al. 2003; Kandaswamy et al. 2010; Liu et al. 2013), mostly related

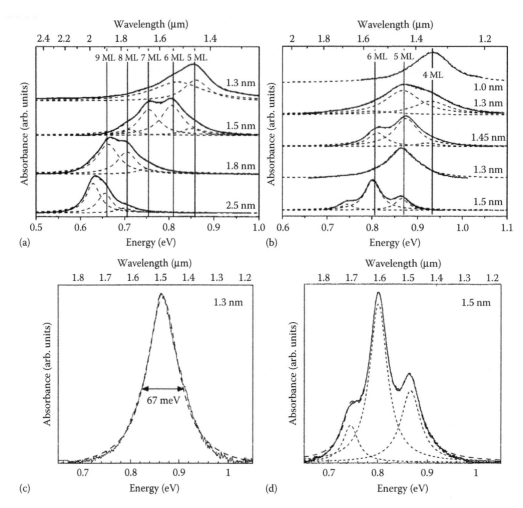

FIGURE 20.2 (a, b) Absorbance spectra of samples consisting of 20 periods of GaN/AlN QWs with 3-nm-thick AlN barriers and various QW thicknesses (indicated in the figure). Vertical lines mark the mean energies of structures. The curves are vertically shifted for clarity. (c) Absorbance of a sample with a 1.3 nm QW (full line) and the corresponding Lorentzian fit (dashed line). (d) Absorbance of a sample with 1.5-nm-thick QWs (solid line), with Lorentzian fitting curves (dotted lines and sum of Lorentzian fits (dashed line). (Reprinted with permission from Tchernycheva, M.L. et al., *Phys. Rev. B*, 73, 125347. Copyright 2006, The American Physical Society.)

to the exchange interaction and the depolarization shift. On the other hand, studies of the effect of the dopant location have shown a dramatic reduction of the ISB absorption FWHM by using a δ-doping technique with Si donors placed at the end of the QW (Edmunds et al. 2012b). This linewidth reduction is attributed to an improvement of the interfacial roughness. It has also been theorized that doping in the wells should provide red-shifted (Wu et al. 2013b) and stronger (Zhuo et al. 2014) ISB absorption as compared to doping in the barriers.

When integrating the III-Nitride nanostructures into complete devices, it is necessary to keep in mind that the magnitude of the carrier distribution depends not only on the Si doping level in the QWs, but also on the presence of nonintentional dopants, and on the carrier redistribution due to the internal electric field. The large polarization discontinuities in the III-N material system can result in a significant (even dominant) contribution to the IR absorption in GaN/AlN superlattices (Kandaswamy et al. 2008).

20.2.2 Far-IR Architectures

Reducing the ISB transition energy below 60 meV (wavelength above 20 μm) requires band engineering to compensate the internal electric field in the QWs. A first proposed architecture consisted of a 3-layer well (step-QW) with a virtually flat potential profile, as illustrated in Figure 20.3 (Machhadani et al. 2010; Beeler et al. 2013a; Wu et al. 2013a; Beeler et al. 2014c). Figure 20.3c shows the conduction band diagram of a step-QW design, in this example, consisting of $Al_{0.1}Ga_{0.9}N/GaN/Al_{0.05}Ga_{0.95}N$ (3 nm/3 nm/10 nm). The design can be broken effectively into two portions; the first is the "barrier," which comprises the high-Al-content $Al_xGa_{1-x}N$ layer and the GaN layer. The second portion is the "well," which is the low-Al-content $Al_xGa_{1-x}N$ layer. The design creates a semi-flat band in the "well" by having the "barrier" balanced at the same average Al percentage, that is, the average polarization in the "barrier" is approximately equal to the average polarization in the "well." Samples following the step-QW design in Figure 20.3 have been synthesized by plasma-assisted molecular beam epitaxy (PAMBE) on GaN templates on float-zone Si (111) to evade problems of substrate absorption (Kandaswamy et al. 2009). The ISB absorption at ≈2 THz (≈70 μm) [illustrated in Figure 20.3d] and at ≈13 THz (≈22 μm) has been reported (Machhadani et al. 2010; Beeler et al. 2014c).

The weakness of the step-QW design lies in the fact that any imbalance in the structure (deviation in the Al mole fraction or thickness of any of the layers) has a drastic effect on the ISB transition energy. The limitations of the step-QW configuration can be surmounted by the insertion of an additional AlGaN layer to separate the GaN layer from the low-Al-content $Al_xGa_{1-x}N$ well

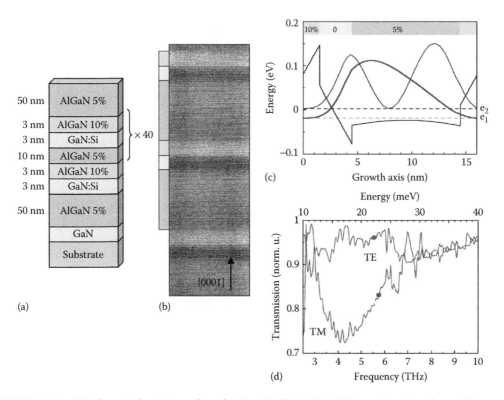

FIGURE 20.3 (a) Schematic description of an $Al_{0.05}Ga_{0.95}N/Al_{0.1}Ga_{0.9}N/GaN$ (10 nm/3 nm/3 nm) step QW structure. (b) High-angle annular dark-field scanning transmission electron microscopy image of the active region in (a). (c) Conduction band profile of a single period of the structure, including the squared wavefunctions of levels e_1 and e_2. (d) Transmission spectra for TM- (square) and TE- (circle) polarized light at $T = 4.7$ K. (Reprinted with permission from Machhadani, H. et al., *Appl. Phys. Lett.*, 97, 191101. Copyright 2010, American Institute of Physics.)

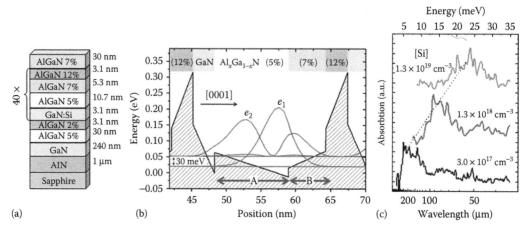

FIGURE 20.4   (a) Schematic description of a pseudo-square QW structure. (b) Band profile of a pseudo-square QW including the squared wavefuctions of e_1 and e_2. The nominal compositions of the layers are indicated above. (c) Spectral absorption of TM-polarized light in samples with different doping levels. (Reprinted with permission from Beeler, M. et al., *Appl. Phys. Lett.*, 105, 131106. Copyright 2014, American Institute of Physics.)

(Beeler et al. 2013a). However, the incorporation of this "separation layer" results in a geometry where the internal electric field is not fully compensated, that is, the QW keeps a triangular potential profile. As a consequence, the 4-layer MQW system is more sensitive to changes in the strain state versus the step-QW design. AlGaN/GaN 40-period MQW structures following this 4-layer MQW design show low-temperature ($T = 5$–10 K) ISB absorption around 27–29 μm (~14 THz). The normalized absorption linewidth for a sample with a doping level [Si] $= 1.5 \times 10^{19}$ cm^{-3} is $\Delta f/f$ ~0.25 (Beeler et al. 2013a), which is a significant improvement in comparison to results in step QWs [$\Delta f/f$ ~0.5 in Reference (Machhadani et al. 2010)].

The 4-layer design uses complex barrier systems in order to achieve the desired robustness. Unfortunately, such barriers inhibit tunneling transport, and therefore, the incorporation of the architecture into ISB devices using the quantum cascade principle. An alternative pseudo-square QW is proposed, consisting of a 4-layer architecture, where the compensation of the polarization-induced internal electric field is obtained by a gradual increase in the polarization field throughout the quantum "trough" generated by three low-Al-content layers (see schematic description in Figure 20.4) (Beeler et al. 2014a). This design has single-layer barriers, which can permit tunneling transport under bias. Experimentally, it is shown that the ISB wavelength can be varied from 150 to 75 μm by changing the size of the "quantum trough," and from 100 to 50 μm by changing the doping level, as illustrated in Figure 20.4c.

20.2.3 Alternative Crystallographic Orientations

The presence of internal electric fields in polar materials increases the design complexity of ISB devices. The use of QWs grown along the nonpolar crystallographic orientations, like the *m*-plane {1–100} or the *a*-plane {11–20} is a simple alternative to obtain heterostructures without internal electric field (Speck and Chichibu 2011). However, nonpolar growth on foreign substrates results in a high density of crystalline defects due to the strong surface anisotropy. Until very recently, high-quality free-standing nonpolar GaN substrates were not available, which has hindered the development of nonpolar ISB technologies, and instead, has supported the growth on semipolar planes (Speck and Chichibu 2011), that is, those (*hkil*) planes with at least two non-zero *h*, *k*, or *i* Miller indices and a nonzero *l* Miller index. Semipolar planes allow a considerable reduction in the internal electric field (Romanov et al. 2006) while presenting a lower in-plane anisotropy than nonpolar surfaces (Lahourcade et al. 2007, 2008). In comparison to polar QWs, semipolar structures exhibit quasi-square potential band profiles with symmetric wavefunctions, due to the reduced electric field, in the range of 0.5–0.6 MV/cm in the QWs for the (11–22) plane (Lahourcade et al. 2008; Machhadani et al. 2013). From the experimental viewpoint, near-IR ISB absorption in the range of 1.5–3.75 μm has been reported on semipolar (11–22)-oriented GaN/AlN MQWs (Lahourcade et al. 2008; Machhadani et al. 2013). The absorption FWHM in the range of 80–110 meV is comparable to the one measured in polar structures (Tchernycheva 2006b; Kandaswamy et al. 2008). However, in semipolar structures, the reduction in the internal electric field results in a red shift of the ISB energy when compared with polar QWs with the same dimensions.

Regarding nonpolar materials, first experimental results obtained by heteroepitaxy of *a*-plane GaN:Si/AlN MQWs on *r*-plane sapphire rendered ISB optical absorption at λ ≈ 2.1 μm with a FWHM = 120 meV (Gmachl and Ng 2003). Nowadays, the availability of high-quality free-standing nonpolar GaN substrates has encouraged research on nonpolar orientations, and especially the growth of *m*-plane MQWs is intensively being studied (Armstrong et al. 2013; Shao et al. 2013a, 2013b). A comparative analysis of *a*- versus *m*-oriented GaN/AlN MQWs, grown on free-standing GaN has shown that the *m*-plane is the most promising nonpolar orientation for ISB application (Lim et al. 2015a, 2016b). Room-temperature near-IR ISB absorption in the 1.5–2.9 μm range was observed in the *m*-plane GaN/AlN MQWs, as illustrated in Figure 20.5. Similar to the semipolar case, the FWHM is comparable to the one measured in polar structures, but the transition energy is red shifted due to the absence of internal electric fields.

By incorporating Ga in the MQWs barriers, heterostructures can be designed for ISB absorption in the mid-IR region, and room temperature ISB absorption in the 4.0–5.8 μm range has been observed recently on *m*-plane GaN/AlGaN MQWs (Kotani et al. 2014; Lim et al. 2015a). The long wavelength limit in this range is set by the substrate absorption associated with the second order of the Reststrahlen band in the GaN substrate.

In spite of these encouraging results, cracks, stacking faults, and Al-rich clusters are present in these near- and mid-IR structures due to the high lattice mismatch (4% along the c axis for GaN/AlN).

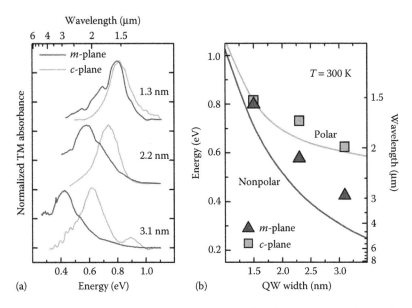

FIGURE 20.5 (a) TM-polarized ISB absorption in *m*-plane and *c*-plane GaN/AlN MQW with ≈3.5 nm AlN barriers and the QW thickness indicated in the figure. (b) ISB energies as a function of the QW width. Solid lines correspond to theoretical simulations assuming that the in-plane lattice parameters of the MQWs correspond to those of a relaxed AlGaN alloy with the average Al composition of the structure. (Reprinted with permission from Lim, C.B. et al., *J. Appl. Phys.* 118, 014309. Copyright 2015, American Institute of Physics.)

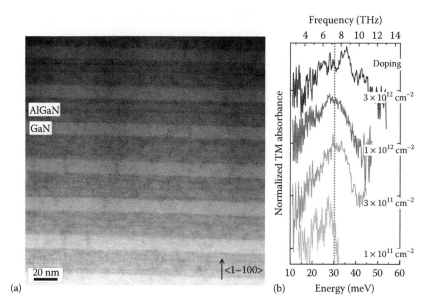

FIGURE 20.6 (a) Cross-section high-angle annular dark-field scanning transmission electron microscopy image of an *m*-GaN/AlGaN MQW viewed along the <11-20> zone axis. (b) Normalized TM-polarized ISB absorption of GaN/Al$_{0.075}$Ga$_{0.925}$N (10 nm/18.5 nm) MQWs with different doping levels (indicated in the figure). The dashed line marks the location of the theoretical ISB transition energy, considering the lowest doping concentration. (Modified with permission from Lim, C.B. et al., *Nanotechnology*, 27, 145201. Copyright 2016, IOP Publishing.)

In contrast, the epitaxy-induced defects (dislocations, stacking faults) are not observed when reducing the Al mole fraction of the QW barriers below 10% (Figure 20.6a), which is feasible when aiming at the THz spectral range. Using *m*-GaN/AlGaN MQWs, low temperature ($T = 5$–10 K) ISB absorption in the 1.5–9 THz range has been demonstrated (Edmunds et al. 2014; Lim et al. 2015a), which confirms that most of the 7–10 THz band forbidden to GaAs-based technologies can be covered by GaN. The effect of QW doping on the ISB transition energy and linewidth was studied on *m*-plane structures in the mid- and far-IR. As illustrated in Figure 20.6b, when increasing the silicon concentration in the wells, the ISB absorption is blue shifted due to the exchange interaction and depolarization shift, and broadened due to scattering by ionized impurities (Edmunds et al. 2014; Kotani et al. 2015; Lim et al. 2016a). The magnitude of the ISB absorption first increases with Si incorporation and then saturates for doping levels around 1×10^{12} cm^{-2}. Also, the blue shift and broadening increase less than theoretically predicted for the samples with higher doping levels. This is explained by the presence of free carriers in the excited electron level due to the increase of the Fermi-level energy.

From the viewpoint of the electron transport, the observation of negative differential resistance in *m*-plane GaN/AlGaN double-barrier heterostructures confirms the feasibility of resonant tunneling in this crystallographic orientation and paves the way to nonpolar ISB devices for THz applications (Bayram et al. 2010).

20.2.4 Quantum Dots

An alternative approach to QW structures for the fabrication of devices is based on optical transitions between bound states in the conduction band of quantum dot (QD) superlattices (Berryman et al. 1997; Phillips et al. 1998). Quantum dot IR photodetectors (QDIPs) are expected to ultimately outperform QWIPs in terms of low dark current, high photoelectric gain, and sensitivity (Ryzhii 1996). Furthermore, under certain conditions, intraband bound-to-continuum transitions in QDs can be nearly independent of the polarization of the excitation (Pan et al. 1998; Chen et al. 2001; Chu et al. 2001).

In the case of III-nitrides, GaN/AlN QD structures can be synthesized by PAMBE through GaN deposition under compressive strain and under N-rich conditions (Daudin et al. 1997; Guillot et al. 2006a). These GaN QDs are hexagonal truncated pyramids with {1–103} facets, generated on top of a 2-ML-thick GaN wetting layer (Chamard et al. 2004). The QD size can be tuned by modifying the deposition time, the growth temperature, or the growth interruption time after deposition of the QDs. By adjusting the growth conditions, QDs with height (diameter) in the range of 1–1.5 nm (10–40 nm), and density between 10^{11} cm^{-2} and 10^{12} cm^{-2} can be synthesized (Guillot et al. 2006b). Silicon dopant atoms can be incorporated without perturbation of the QD morphology.

The Si-doped GaN/AlN QD superlattices exhibit TM-polarized intraband absorption at room temperature, which can be tuned from 1.38 to 1.68 μm as a function of the QD size (Tchernycheva et al. 2005; Guillot et al. 2006b). This absorption line is attributed to transitions from the ground state of the conduction band s, to the first excited electronic state confined along the growth axis, p_z. The broadening of the absorption peak is typically 80–100 meV. The homogeneous linewidth of the s–p_z intraband transition at 1.55 μm in GaN/AlN QDs was assessed by means of nonlinear spectral hole-burning experiments (Nguyen et al. 2010). These measurements demonstrated that electron-electron scattering plays a minor role in the coherence relaxation dynamics since the homogeneous linewidth of 15 meV at 5 K does not depend on the incident pump power. This suggests the predominance of other dephasing mechanisms such as spectral diffusion.

The lateral confinement in the QDs gives rise to additional ISB transitions that can be observed under TE-polarized excitation. The optical signature associated to s–$p_{x,y}$ was first reported by Vardi et al. (2009) who studied near-IR and mid-IR intraband transitions in GaN/AlN QDs using in-plane electronic transport at low temperatures, as illustrated in Figure 20.7. The measured s–$p_{x,y}$ energy (0.1–0.3 eV) was significantly larger than the equivalent transition energy in InGaAs/GaAs QDs. Their analysis shows that the appearance of large s–$p_{x,y}$ energy in GaN/AlN QDs is due to the strong internal electric field in the QDs which results in stronger confinement of the electrons at the top facet of the QD.

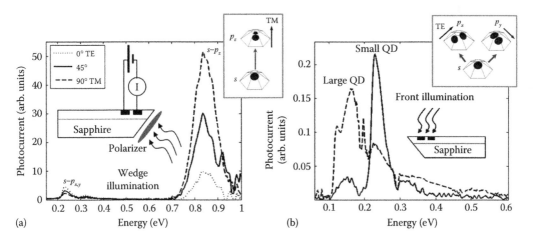

FIGURE 20.7 (a) Polarization-dependent spectral photocurrent of a 20-period GaN/AlN QD sample measured in wedge illumination. Normalized response of both s–p_z peak at near IR and s–$p_{x,y}$ peak at mid-IR; The s–p_z peak is 10 times larger than s–$p_{x,y}$ peak. (b) Mid-IR spectral response of two 20-period GaN/AlN QD samples measured at $T = 12$ K under front illumination and 10 V bias. (Reprinted with permission from Vardi, A. et al., *Phys. Rev. B.*, 80, 155439. Copyright 2009, The American Physical Society.)

20.2.5 Nanowire Heterostructures

Nanowire heterostructures offer a unique situation for devices requiring low defect density in the active region and the combination of materials with large lattice mismatch. The latter should facilitate the integration of binary GaN and AlN compounds for near-IR applications. Furthermore, the three-dimensional strain distribution in nanowire heterostructures results in an attenuation of the internal electric fields (Lähnemann et al. 2011; Songmuang et al. 2011; Beeler et al. 2015), which might extend the accessible wavelength range for simple GaN/AlGaN heterostructures in the mid-IR spectral range.

Intraband transitions in GaN/AlN nanowire heterostructures have been reported in samples grown by PAMBE following the self-assembled approach. Figure 20.8a–c show microscopy images of the samples, and Figure 20.8d contains the schematic sample structure (Beeler et al. 2014b). The GaN nanodisks were doped n-type with germanium (Schörmann et al. 2013; Ajay et al. 2016), and the disk sizes were varied in the 2–8 nm range. A detailed investigation of the interband transitions in these heterostructures using temperature-dependent and time-resolved photoluminescence reveals the efficient quantum confinement in these nanodisks rendering them insensitive to nonradiative carrier losses (Beeler et al. 2015). The intraband energies were measured via FTIR and Figure 20.8e displays the results for a series of samples with GaN/AlN (4 nm/4 nm) nanodisks and different doping levels. The TM-polarized absorption is assigned to the s–p_z intraband transition in the disks. The s–p_z absorption line experiences a blue shift with increasing Ge concentration and a red shift with increasing nanodisk thickness. These intraband transitions are strongly blue shifted due to many-body effects, namely, the exchange interaction and depolarization shift (Beeler et al. 2014b).

A prerequisite for certain types of electrically driven ISB devices is resonant tunneling transport through the quantum barriers. For III-Nitride nanowires, resonant tunneling has been observed in the

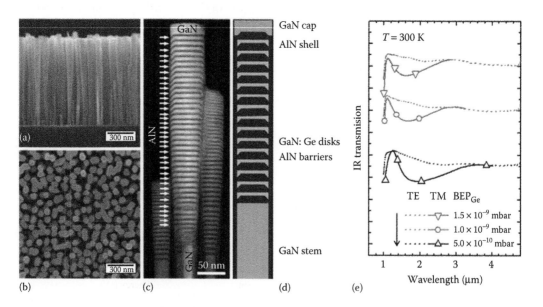

FIGURE 20.8 (a, b) Cross-section and top view scanning electron microscopy and (c) high-angle annular dark field scanning transmission electron microscopy images of self-assembled GaN/AlN nanowire heterostructures. (d) Schematic description of the sample. (e) Room-temperature IR transmission spectra for TE-(dashed) and TM-polarized (solid) light measured for Ge-doped GaN/AlN (4 nm/4 nm) heterostructured nanowires with different doping levels in the GaN NDs. The spectra are vertically shifted for clarity. (Reprinted with permission from Beeler, M. et al., *Nano Lett.*, 14, 1665–1673, 2014. Copyright 2014, American Chemical Society.)

case of an axial double barrier (Songmuang et al. 2010) and for multiple barrier structures (Rigutti et al. 2010), but also in the core-shell geometry (Carnevale et al. 2012).

20.3 All-Optical Switches

The demand for increasing bandwidth in optical communication networks impels the development of all-optical switches at 1.55 µm, which should be capable of sustaining high repetition rates with low switching energy and high modulation depth. Exploiting resonant nonlinearities in semiconductors (Wada 2004), switching can be achieved through the saturation of absorption by an intense control pulse, as originally demonstrated in GaAs structures (Noda et al. 1993). Using GaN/AlGaN, switching can be obtained by saturation of the ISB absorption, so that devices can profit from the ultrafast ISB recovery time, in the 140–400 fs range (Iizuka et al. 2000, 2005; Gmachl et al. 2001a; Heber et al. 2002; Rapaport et al. 2003; Hamazaki et al. 2005), which should allow Tbit/s data rates.

The use of GaN/AlGaN QWs for ISB all-optical switches at telecommunication wavelengths was first proposed by Suzuki et al. (Suzuki and Iizuka 1997; Suzuki et al. 2000). Since then, absorption saturation at ≈1.55 µm with subpicosecond commutation times has been demonstrated by several groups (Rapaport et al. 2003; Iizuka et al. 2004, 2005, 2006a, 2009; Nevou et al. 2009; Li et al. 2007; Sodabanlu et al. 2011). In general, the structures consist of GaN/AlN MQWs embedded in a ridge waveguide. In such a configuration, a critical parameter to minimize transmission losses is the reduction of edge-type dislocations. These defects introduce acceptor centers where electrons can be captured, and therefore, can effectively act as a wire-grid polarizer which leads to selective attenuation of the TM-polarized signal (Iizuka et al. 2006b). Control pulse switching energies of 38 pJ for 10 dB modulation depth (Li et al. 2007) and 25 pJ for 5 dB contrast (Iizuka et al. 2009) have been demonstrated using a waveguide with an AlN cladding below the active GaN/AlN QWs, and GaN or Si_xN_y as the upper cladding layer, respectively. Theoretical calculations predict a reduction of the switching energy by a factor of 30 by replacing the GaN/AlN QWs with properly designed coupled AlN/GaN/AlGaN QWs (Li and Paiella 2006).

Saturation of the intraband absorption in GaN/AlN QDs was first demonstrated by Nevou et al. (2009) in samples containing 200 periods of dots, obtaining saturation for a control power in the range of 15–137 MW/cm² (0.03–0.27 pJ/µm²). In spite of the large signal variation (a consequence of the focusing uncertainty in the sample), even the upper estimate of the saturation intensity for QDs is smaller than the corresponding value for GaN/AlN QWs (9.46 W/µm²) (Li et al. 2007). Based on these results, Monteagudo-Lerma et al. made a comparison of the performance of saturable absorbers consisting of 3 periods of GaN/AlN QWs or QDs inserted in a GaN-on-AlN waveguide structure (Monteagudo-Lerma et al. 2013). In the case of 5-µm-wide QW-based waveguides, a 10 dB change in the transmittance was achieved for input energies of ≈24 pJ with 150 fs pulses. This value was improved by almost a factor of 2 when replacing the QWs with QDs as active elements. The reduction of the waveguide width to 2 µm (monomode waveguide) resulted in a further decrease of the required control pulse energy to ≈8 pJ for 10 dB modulation.

20.4 Electro-Optical Modulators

The electro-optical amplitude and phase modulators allow tuning the amplitude, phase, and/or polarization state of an optical beam as a function of the control voltage. Exploiting ISB transitions in QWs has been proposed as a means to reduce the driving voltage and increase the bandwidth (Holmstrom 2001, 2006; Holmström et al. 2007). Moreover, the oscillator strength of ISB transitions is higher in comparison to band-to-band transitions and should allow improved device miniaturization.

The first electro-absorption ISB modulation experiments on AlN/GaN QWs were based on the electrical depletion of a 5-period AlN/GaN (1.5 nm/1.5 nm) MQW structure deposited on a thick GaN buffer layer (Baumann et al. 2006). The absorption spectrum of such a sample presents two distinct peaks related to ISB transitions in both the AlN/GaN QWs and in the triangular well located at the interface between the GaN buffer layer and the lowest AlN barrier, due to the band bending in the buffer layer generated by the polarization discontinuity. The ratio of these two absorption peaks can be adjusted by applying an external field, which modifies the charge distribution.

For increasing the modulation depth, the interaction of light with the active medium should be enhanced, which can be achieved with a waveguide geometry (Machhadani et al. 2009). Through the use of a 1-μm-thick $Al_{0.5}Ga_{0.5}N$ waveguiding layer on AlN, and with 3 active GaN/AlN QWs operating at 1.55 μm, a modulation depth of 13.5 dB was observed for a −9V/+7V voltage swing (10 dB for 5 V voltage swing).

The intrinsic speed limit can be greatly improved by emptying the active QWs into a local reservoir, instead of transferring carriers over the whole active region. This is the principle of the coupled-QW modulator: the electro-modulation originates from electron tunneling between a wide well (reservoir) and a narrow well separated by an ultrathin (≈1 nm) AlN barrier. Experiments on the GaN QW coupling via Al(Ga)N barriers (Tchernycheva et al. 2006a; Driscoll et al. 2007) have set the basis for the demonstration of room-temperature ISB electro-modulated absorption at telecommunication wavelengths in GaN/AlN coupled QWs with AlGaN contact layers (Nevou et al. 2007a; Kheirodin et al. 2008; Dussaigne et al. 2010). Such devices displayed a −3 dB cut-off frequency (BW_{-3dB}) limited by the device capacitance to 3 GHz for 15 × 15 μm² mesas. According to Hölmstrom, the high-speed performance of such modulators is ultimately determined by the ISB absorption linewidth Γ, since their capacitance depends on the linewidth as $C \sim \Gamma^3$ (Holmstrom 2001, 2006).

All the above-described electro-optical modulators rely on amplitude modulation of the light via ISB absorption. Based on the Kramers-Kronig relations, the ISB absorption should also translate into a variation of the refractive index at wavelengths close to the transition, which can be used for phase modulation. This concept was verified experimentally at mid-IR (≈10 μm) wavelengths using the Stark shift of ISB transitions in GaAs/AlGaAs step QWs (Dupont et al. 1993). The strongly nonlinear susceptibility observed in GaN/AlN QWs (Li et al. 2008; Valdueza-Felip et al. 2008), which might even be enhanced in 3-layer QW designs (Wu et al. 2014), has led to the first theoretical proposals of all-optical cross-phase modulators (Cen et al. 2009).

Using a depletion modulator consisting of 3 GaN/AlN QWs inserted in an $Al_{0.5}Ga_{0.5}N$/AlN ridge waveguide on sapphire, Lupu et al. (2012) reported a variation of the refractive index around ≈1.5 μm as deduced from the shift of the beating interference maxima for different order modes. The change in the refractive index was derived to be $\Delta n = -5 \times 10^{-3}$ as the population evolved from complete depletion to full population of the QWs. This result is in close agreement with the observation of a refractive index variation from -5×10^{-3} to 6×10^{-3} in 100-period Si-doped GaN/AlN (1.5 nm/3 nm) MQWs using a free-space Mach-Zehnder interferometer configuration (Gross et al. 2013). The values of Δn are comparable to those obtained at the same wavelength in phase modulators based on interband transitions in InGaAsP/InP QWs using the QCSE (Zucker et al. 1989), and they are one order of magnitude higher than the index variation obtained in silicon (Soref and Bennett 1987). These results open the way for the realization of ISB Mach-Zehnder interferometer phase modulators in the optical communication wavelength range.

20.5 IR Photodetectors

20.5.1 Quantum Well/Quantum Dot Infrared Photodetectors (QWIPs/QDIPs)

The development of III-Nitride QWIPs has been inspired by their potential applications in optical communications. In a QWIP, the e_2 level in the QWs should be in the proximity of the top of the barrier, to facilitate the extraction of the excited electrons, and the barriers should be wide enough to prevent tunneling. Near-IR photoconductive detectors based on Si-doped GaN/AlN MQWs (Hofstetter et al. 2003; Baumann et al. 2005) and QDs (Doyennette et al. 2005; Vardi et al. 2006, 2009) have been reported. However, photoconductive devices display a low yield and large dark current, originating from structural defects in the highly mismatched GaN/AlN structures. An alternative to circumvent the leakage problem consists of exploiting the device's photovoltaic response, associated to the intrinsic asymmetry of the potential profile in polar GaN wells.

The photovoltaic operation of GaN/AlN QWIPs at telecommunication wavelengths and at room temperature was first studied in detail by Hofstetter et al. (Hofstetter et al. 2006, 2007, 2009; Baumann et al. 2006; Giorgetta et al. 2007). These photovoltaic ISB detectors are based on resonant optical rectification processes (Rosencher and Bois 1991; Hofstetter et al. 2007). In a GaN/AlN superlattice, due to the asymmetry of the potential in the QWs, the excitation of an electron into the upper quantized level is accompanied by a charge displacement in the growth direction, so that an electrical dipole moment is created. For a high electron density and a large number of QWs, these microscopic dipole moments can be detected as an external photovoltage. A strong enhancement of the responsivity has been achieved by using QDs instead of QWs in the active region (Hofstetter 2010b).

A potential application of GaN-based photovoltaic ISB photodetectors is the fabrication of multispectral detectors combining interband and ISB transistors. Following this idea, the monolithic integration of a photoconductive ultraviolet interband (solar-blind) detector based on an AlGaN thin film and a photovoltaic near-IR ISB detector based on an AlN/GaN MQW has been reported (Hofstetter et al. 2008).

Photoconductive QWIPs in the mid- and far-IR spectral regions benefit from a reduction of the lattice mismatch which aids in their fabrication. Devices operating in the 3–5 µm atmospheric window (Rong et al. 2015) and in the THz domain (Sudradjat et al. 2012) have been demonstrated operating at low temperature (5 K and up to 50 K, respectively). The active region of such detectors utilizes a bound-to-quasibound configuration following the step-QW design (Machhadani et al. 2010), as illustrated in Figure 20.9 for the THz QWIP proposed by Sudradjat et al. (2012).

More recently, Pesach et al. have demonstrated InGaN/(Al)GaN QWIPs fabricated on freestanding nonpolar *m*-plane GaN substrates (2013). Devices consisting of 2.5 nm $In_{0.095}Ga_{0.905}N$/56.2 nm $Al_{0.07}Ga_{0.93}N$ and 3 nm $In_{0.1}Ga_{0.9}N$/50 nm GaN superlattices displayed photocurrent peaks at 7.5 and 9.3 µm, respectively, when characterized at low temperature (14 K).

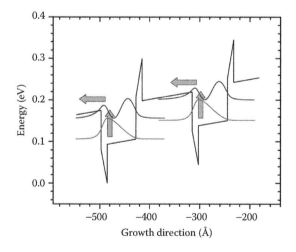

FIGURE 20.9 Conduction-band profile of an AlGaN far-IR QWIP structure, where each ot the repeated units consists of a step $Al_{0.16}Ga_{0.84}N$/$Al_{0.08}Ga_{0.92}N$ barrier and a step GaN/$Al_{0.08}Ga_{0.92}N$ well. The squared envelope functions of the ground state and first excited subbands of each well are also shown, referenced to their respective energy levels. The vertical and horizontal arrows indicate that, respectively, photon absorption and photoelectron escape into the continuum of unbound states over the barriers. (Reprinted with permission from Sudradjat, F.F. et al., *Appl. Phys. Lett.*, 100, 241113, 2012. Copyright 2012, American Institute of Physics).

20.5.2 Quantum Cascade Detectors

Quantum cascade detectors (QCDs) consist of several periods of an active QW coupled to a short-period superlattice, which serves as extractor (Gendron et al. 2004; Giorgetta et al. 2009). In contrast to QWIPs, they operate at zero bias. Under illumination, electrons from the ground state of the active QW, e_1, are excited to the upper state, e_2, and then transferred to the extractor region where they experience multiple relaxations toward the next active QW. When they are operated as photovoltaic detectors, their dark current is extremely low and the capacitance can be reduced by increasing the number of periods, which enables a high-frequency response.

Near-IR GaN/AlGaN QCDs have been reported (Vardi et al. 2008a; Sakr et al. 2010), with the structure illustrated in Figure 20.10. These devices take advantage of the polarization-induced internal electric field to design an efficient AlGaN/AlN (or GaN/AlGaN) electron extractor where the energy levels are separated by approximately the LO-phonon energy ($\approx$90 meV), as shown in Figure 20.10a. Figure 20.10b displays the photovoltaic spectral response of a QCD, compared with the transmittance spectrum. The peak responsivity of these GaN/AlGaN QCDs at room temperature was $\approx$10 mA/W (Vardi et al. 2008a; Sakr et al. 2013). The detectors, containing 40 periods of active region with the size 17×17 μm², exhibit an RC-limited BW_{-3dB} cut-off frequency at 19.7 GHz (Vardi et al. 2008b). However, pump and probe measurements of these devices pointed to an ISB scattering time in the active QW of 0.1 ps and a transit time through the extractor of 1 ps (Vardi et al. 2011). With these data, the intrinsic frequency bandwidth is expected to be above 160 GHz, which is significantly higher than theoretical predictions by Gryshchenko et al. (2012). To improve performance, Sakr et al. (2013) illuminated the side facet of the QCDs perpendicular to the growth axis, reaching a responsivity of at least 9.5 ± 2 mA/W for 10×10 μm² devices at 1.5 μm peak detection wavelength at room temperature with a BW_{-3dB} frequency response of $\approx$40 GHz.

Based on the presence of the internal field, symmetry breaking of the potential in the QWs allows ISB transitions between e_1 and the second excited state, e_3 (Tchernycheva et al. 2006b) This feature was exploited for the fabrication of a two-color GaN-based QCD operating at 1.7 and 1 μm at room temperature (Sakr et al. 2012a).

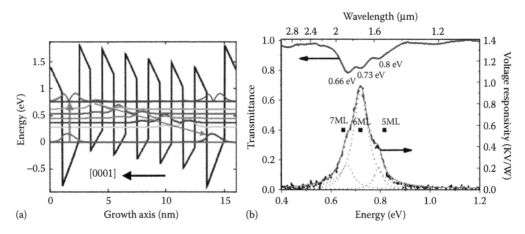

FIGURE 20.10 (a) Conduction band profile and wavefuctions of the confined energy levels in one period of a QCD structure. (b) Room-temperature TM-polarized transmittance (top) and spectral response at zero-bias fitted with three Lorentzians (bottom). Black squares denote the calculated transition energies in a QCD with 5–7 ML active QW. (Reprinted with permission from Vardi, A. et al., *Appl. Phys. Lett.*, 92, 011112, 2008. Copyright 2008, American Institute of Physics.)

To date, PAMBE has been the preferred growth technique for QCDs. Recently, a GaN/Al$_{0.5}$Ga$_{0.5}$N QCD grown by metalorganic vapor phase epitaxy (MOVPE) has been demonstrated with a peak responsivity of 100 μA/W at 4 μm when operated at 140 K (Song et al. 2014).

A simplified QCD design replacing the extractor superlattice by an AlGaN layer has been demonstrated, presenting a peak photoresponse at 1.9 μm (Sakr et al. 2012b). The thickness and composition of the extractor alloy are chosen so that the energy separation between the ground state of the extractor and the ground state of the active QW is close to the LO-phonon energy. The photoresponse of such detectors at normal incidence can be increased by a factor of 30 by using a nanohole array integrated on top of the detector (Pesach et al. 2014). QCDs are also used in combination with photonic crystals and plasmonic antennas to enhance performance (Reininger et al. 2013; Harrer et al. 2014; Wang et al. 2015a, 2015b).

20.6 IR Emission

20.6.1 Light Emission in Superlattices

ISB luminescence is an inefficient process due to the competition with extremely fast nonradiative electron relaxation channels. These competing mechanisms include electron-electron interactions and impurity scattering (tens of picoseconds) and more importantly interactions with LO phonons (subpicosecond). While these processes hinder spontaneous emission, QCLs are still possible because they operate in a regime of population inversion and should benefit from high stimulated gain due to the large oscillator strength of ISB transitions.

Some of the first reports of room-temperature ISB luminescence in the III-Nitride material system were by optical pumping of GaN/AlN MQWs (Nevou et al. 2006a, 2007b; Driscoll et al. 2009). These structures utilized a three-level system where the emission was facilitated via an e_3–e_2 ISB transition tuned to the 2–2.3 μm spectral range. Photoluminescence excitation spectroscopy showed that the emission was only observed for TM-polarized excitation at wavelengths corresponding to the e_1–e_3 ISB transition. Further research also provided mid-IR ISB electroluminescence measurements on chirped AlGaN/GaN MQW structures (Hofstetter et al. 2014). These structures demonstrated a large blue shift of the emission, from 115 meV (FWHM = 38 meV) to 180 meV (FWHM = 58 meV), when increasing the bias from 7 to 14 V.

Room-temperature intraband emission has also been observed in optically pumped GaN/AlN QDs (Nevou et al. 2008). The p_{z-s} intraband luminescence was observed at $\lambda = 1.48$ μm under optical excitation at $\lambda = 1.34$ μm perpendicular to the [0001] growth axis. The population of the p_z state arises from Raman scattering by GaN A_1 LO phonons.

Quantum cascade light emission was also demonstrated in MOVPE-grown GaN/AlGaN superlattices in the mid-IR region as illustrated in Figure 20.11 (Song et al. 2015a). The design employed a diagonal optical transition between upper and lower emitter states and used injector regions to facilitate rapid carrier transport in the superlattice. Furthermore, by estimating a three-dimensional interface roughening, an effective interface grading was included in the calculations, which showed a strong influence on the energy spectrum and the calculated transition levels (Song et al. 2015b). Song et al. included this in their device design and achieved mid-IR emission of $\lambda = 4.9$ μm (FWHM of 110 meV) under pulsed operation at 80 K.

20.6.2 QCL Structures

QCLs rely on optical transitions from electron transitions down a cascade of potential steps where population inversion is maintained via the control of tunneling in a suitably designed semiconductor MQW structure (Faist et al. 1994). Due to the polarization selection rules associated with ISB transitions, these devices are in-plane emitters, with their electric-field vector perpendicular to the plane of the layers.

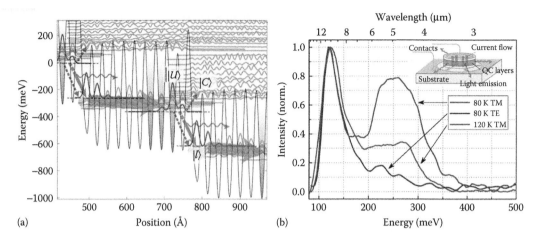

(a)   (b)

FIGURE 20.11 (a) Conduction band structure of the III-Nitride quantum cascade emitter calculated with effective interface grading showing a diagonal optical transition at 4.9 um with an applied bias of 120 kV/cm and Si doping of 1.8×10^{18} cm^{-3} in the shaded wells. The upper ($|u\rangle$) and lower ($|l\rangle$) emitter states, the continuum ($|c\rangle$), and the direction of carrier flow are also indicated (filled horizontal arrows). (b) Emission spectra of the III-Nitride quantum cascade emitter centered around 255 meV with polarized emission observed below 200 meV for pulsed measurements. A clear difference is observed between the TE and TM spectra at 80 K and the emission is still discernible at 120 K. The inset shows a schematic of a mesa device. (Reprinted with permission from Song A.Y. et al., *Appl. Phys. Lett.* 107, 132104, 2015. Copyright 2015, American Institute of Physics.)

An electron injected into the "active" *QWs* first undergoes a radiative ISB transition and is rapidly extracted from the lower laser state by a fast nonradiative transition, which is vital for maintaining the population inversion. Then, the electron tunnels through the "injector region" toward the upper level of the next active QWs. By using several tens or even hundreds of periods of the active region and injector in a series (a cascade), multiple photons per electron are obtained. These complex devices require a precise control of the structure along the multiple periods that compose the active region. Due to the large lattice mismatch of the GaN/AlN system, the fabrication of GaN-based QCLs operating in the near-IR does not appear feasible, despite several theoretical proposals (Jovanović et al. 2003; Hofstetter et al. 2010a; Stattin et al. 2011). However, there is an increasing interest and research effort for the fabrication of GaN QCLs in the far-IR, where the lattice mismatch requirements are lowered, and it should be possible to exploit the large LO-phonon energy of III-nitrides to realize devices operating at room temperature.

Since the first demonstration of a GaAs-based THz QCL in 2001 (Köhler et al. 2002) rapid progress has been made in terms of device performance. To date, QCLs have been demonstrated in the 0.85–5 THz ($\lambda = 60$–350 μm) range (Scalari et al. 2009), with pulsed operation up to 200 K (Fathololoumi et al. 2012), and with pulsed output powers greater than 1 W (Belkin et al. 2008). A continuous-wave operation has been demonstrated at a maximum temperature of 129 K and a maximum power of 136 mW (Williams et al. 2006). Various conduction band designs have been considered, namely the resonant-phonon, the chirped superlattice, bound-to-continuum, and hybrid designs (Williams et al. 2003, 2007a). The two major processes believed to cause degradation of population inversion in THz QCLs at high temperature are:

1. Backfilling of the lower radiative state with electrons from the heavily populated injector, which occurs either by thermal excitation or by reabsorption of nonequilibrium LO-phonons (the hot phonon effect) (Lü and Cao 2006).
2. Thermally activated LO-phonon scattering, as electrons in the upper radiative state acquire sufficient in-plane kinetic energy to emit an LO-phonon and relax nonradiatively to the lower radiative state.

Both of these mechanisms greatly depend on the electron gas temperature, which is 50–100 K higher than the lattice temperature during device operation. The low LO-phonon energy in arsenide compounds (34 meV) constitutes a major bottleneck for operation at higher temperatures. Furthermore, the phonon absorption of the GaAs material system causes an unobtainable spectral region at 8–9 THz (Reststrahlen band).

GaN has an LO phonon energy of 92 meV, much higher than the room temperature thermal energy of 26 meV. Various designs for a GaN-based THz QCL have been proposed (Jovanović et al. 2004; Sun et al. 2005; Vukmirović et al. 2005; Bellotti et al. 2008; Wataru Terashima and Hirayama 2009; Hofstetter et al. 2010a; Yasuda et al. 2011; Chou et al. 2011; Mirzaei et al. 2012), all focusing on a resonant-phonon architecture as shown in Figure 20.12 (Williams et al. 2003). Below, we summarize the efforts of various groups working on this topic, who have introduced design improvements but keep the same underlying concept.

Researchers from the University of Leeds (UK) have engineered one of the first designs for GaN6-based QCLs using a fully consistent scattering rate equation model (Jovanović et al. 2004), and an energy balance method (Harrison et al. 2005). They have created a contour plot outlining the predicted emission wavelength for different well and barrier thickness within a superlattice, taking into account an appropriate strain balancing (Jovanović et al. 2003). They have also proposed a λ = 34 μm QCL design in both the *a* and *c* crystallographic planes (Jovanović et al. 2004).

The group of Paiella and Moustakas at Boston University has proposed a QCL design emitting at 2 THz (λ = 150 μm) (Bellotti et al. 2008). They have also performed a rigorous comparison between GaAs/AlGaAs and GaN/AlGaN THz QCLs emitting at the same wavelength using a microscopic model of carrier dynamics in QCL gain media based on a set of Boltzmann-like equations solved with a Monte Carlo technique (Bellotti et al. 2008, 2009). Results show that the population inversion within GaN lasers is much less dependent on temperature than conventional GaAs designs. Furthermore, they have theoretically studied methods to create lattice-matched QCL structures using quaternary InAlGaN alloys (Shishehchi et al. 2014). From the experimental viewpoint, they have explored tunneling effects in cascade-like superlattices, their temperature dependence, and the effect of bias for multiple device architectures (Sudradjat et al. 2010).

Sun et al. (2013) have modelled a QCL structure based on a 3-well design that depopulates via the LO phonon and emits at 6.77 THz (λ = 44.3 μm), and have proposed the use of a spoof surface plasmon

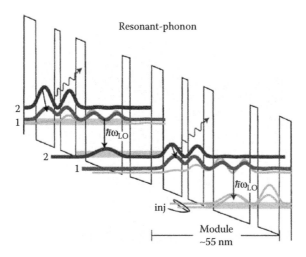

FIGURE 20.12 Conduction band diagram for two periods (Module = 1 period) of a nonpolar resonant phonon QCL design. The squared wavefunctions for the various subband states are plotted. The upper (2) and lower (1) radiative states, and the LO-phonon transition ($\hbar\omega_{LO}$) are labelled. (Reprinted by permission from Macmillan Publishers Ltd. *Nature Photon.*, Williams, B.S., 1, 518. Copyright 2007.)

waveguide instead of a normal surface plasmon waveguide, which should result in an order of magnitude lower losses in the guiding structure.

Mirzaei et al. (2012) have proposed a dual-wavelength QCL to emit at both 33 and 52 μm with similar behavior of the output optical power for both wavelengths. The design is based on the LO-phonon resonance to extract electrons from the lower radiative levels and incorporates a miniband injector, calculated via rate equation analysis to operate properly up to 265 K.

Chou et al. have modeled GaN-based resonant-phonon THz lasers using a transfer matrix method, paying particular attention to the effect of the strain state (Chou et al. 2011). They predict higher THz power in GaN/AlGaN heterostructures as compared to heterostructures incorporating indium (Chou et al. 2012).

Yasuda et al. (2011, 2012) have used the nonequilibrium Green's function to model GaN THz QCL devices, namely a 4-well resonant-phonon InAlGaN/GaN structure on (0001)-oriented GaN, and a 2-well nonpolar GaN/AlGaN structure.

Finally, Terashima and Hirayama (RIKEN, Japan) have presented THz QCL designs based on 4-well resonant-phonon GaN/AlGaN architecture (Wataru Terashima and Hirayama 2009, 2011a, 2011b). The structures have been synthesized by PAMBE and processed in a single-metal plasmon waveguide geometry (Wataru Terashima and Hirayama 2009, 2011a; Hirayama and Terashima 2013). Electroluminescence at 1.37 THz (λ = 219 μm) has been reported in a first structure (Terashima and Hirayama 2011b) grown both on a GaN-on-sapphire template and on bulk GaN. Polarization-dependent electroluminescence at 2.82 THz (λ = 106 μm), slightly tunable by changing the driving voltage in the 20–21 V range, has been reported using a second design (Hirayama and Terashima 2013) grown on an AlN-on-sapphire template. Most recently, Terashima et al. have reported functioning QCLs utilizing a pure 3-level design grown by both PAMBE and MOVPE (Terashima and Hirayama 2015a). The PAMBE laser was grown on a GaN-on-sapphire template consisting of 100 periods of 1.5 nm $Al_{0.20}Ga_{0.80}N$/4 nm GaN/1.5 nm $Al_{0.20}Ga_{0.80}N$/6 nm GaN, as shown in Figure 20.13. Lasing was observed at 5.37 THz (λ = 55.8 μm) for pulsed injection (14.5 V, 1.75 kA/cm²) at a temperature of 5.8 K with a peak optical power observed at 2.8 kA/cm². Low-pressure MOVPE was used to fabricate another QCL with an identical QW structure but with 200 periods and a slightly lower Al concentration ($Al_{0.15}Ga_{0.85}N$) than the PAMBE laser. The reported lasing occurs for pulsed measurements at 5.2 K at a frequency of 6.97 THz (λ = 43.0 μm) with a corresponding threshold current density of 0.75 kA/cm². These recent results are promising for the further development of GaN-based QCLs in the future.

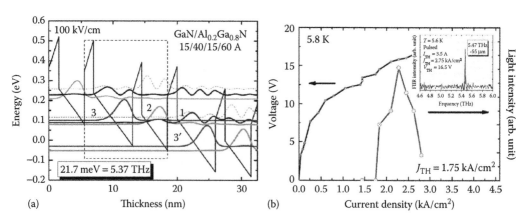

FIGURE 20.13 (a) Conduction band profile and squared wavefunctions of the pure 3 level *c*-plane QCL structure under an external electric field of 100 kV/cm with GaN wells and $Al_{0.20}Ga_{0.80}N$ barriers. (b) The light intensity and voltage versus current density characteristics for the QCL is shown along with a frequency spectrum in the inset. (Reprinted with permission from Terashima, W. and Hirayama, H., *SPIE Proc.*, 9483, 948304. Copyright 2015, Society of Photo-Optical Instrumentation Engineers [SPIE].)

20.7 Conclusion

In this chapter, recent research on III-Nitride ISB optoelectronics was reviewed. In a context of increasing bandwidth demand, the large conduction band offset of the GaN/AlN system and its subpicosecond ISB scattering rate make them very promising materials for high-speed devices operating in the 1.3–1.55 μm telecommunication window. Along these lines, the feasibility of GaN/AlN all-optical switches with subpicosecond commutation times was demonstrated by several groups. Also, room-temperature ISB electro-optical modulators were demonstrated at 1.55 μm. Furthermore, the study of the refractive index variation induced by the ISB absorption opened the way for the realization of ISB Mach-Zehnder interferometer phase modulators in the telecommunication range. GaN-based photovoltaic QWIP and QCDs in the near-IR, and photoconductive QWIP in the 3–5 μm atmospheric window and in the THz domain have been demonstrated. For both all-optical switches and photovoltaic detectors, QDs were demonstrated to be a good alternative to QWs to improve the device performance.

Despite the low efficiency of ISB photoluminescence, near-IR emission was observed in optically pumped GaN/AlN QWs and QDs. Mid-IR ISB electroluminescence was measured on chirped AlGaN/GaN MQWs and in quantum cascade GaN/AlGaN superlattices.

One exciting challenge for ISB optoelectronics is their extension toward the far-IR spectral range. Great effort was made to propose different designs of a GaN-based THz QCL based on the resonant-phonon architecture. From the experimental viewpoint, electroluminescence at 1.37 and 6.97 THz were reported, and lasing was reported at 5.37 and 6.97 THz for pulsed injection at 5 K, opening prospects for the future development of room-temperature III-Nitride-based QCLs.

Acknowledgments

This work is supported by the EU ERC-StG "TeraGaN" (#278428) project.

References

Ajay, A., J. Schörmann, M. Jiménez-Rodriguez, C. B. Lim, F. Walther, M. Rohnke, I. Mouton et al. 2016. Ge doping of GaN beyond the Mott transition. *Journal of Physics D: Applied Physics* 49 (44): 445301. doi:10.1088/0022-3727/49/44/445301.

Andersson, T. G., X. Y. Liu, T. Aggerstam, P. Holmström, S. Lourdudoss, L. Thylen, Y. L. Chen, C. H. Hsieh, and I. Lo. 2009. Macroscopic defects in GaN/AlN multiple quantum well structures grown by MBE on GaN templates. *Microelectronics Journal* 40 (2): 360–362. doi:10.1016/j.mejo.2008.07.065.

Armstrong, A. M., K. Kelchner, S. Nakamura, S. P. DenBaars, and J. S. Speck. 2013. Influence of growth temperature and temperature ramps on deep level defect incorporation in m-plane GaN. *Applied Physics Letters* 103 (23): 232108. doi:10.1063/1.4841575.

Bastard, G. 1988. *Wave Mechanics Applied to Semiconductor Heterostructures*. Les Ulis Cedex, France Les Editions de Physique.

Baumann, E., F. R. Giorgetta, D. Hofstetter, S. Golka, W. Schrenk, G. Strasser, L. Kirste et al. 2006. Near infrared absorption and room temperature photovoltaic response in AlN/GaN superlattices grown by metal-organic vapor-phase epitaxy. *Applied Physics Letters* 89 (4): 41106. doi:10.1063/1.2234847.

Baumann, E., F. R. Giorgetta, D. Hofstetter, H. Lu, X. Chen, W. J. Schaff, L. F. Eastman, S. Golka, W. Schrenk, and G. Strasser. 2005. Intersubband photoconductivity at 1.6 Mm using a strain-compensated AlN/GaN superlattice. *Applied Physics Letters* 87 (19): 191102. doi:10.1063/1.2126130.

Baumann, E, F. R. Giorgetta, D. Hofstetter, S. Leconte, F. Guillot, E. Bellet-Amalric, and E. Monroy. 2006. Electrically adjustable intersubband absorption of a GaN/AlN superlattice grown on a transistorlike structure. *Applied Physics Letters* 89 (10): 101121. doi:10.1063/1.2348759.

Bayram, C. 2012. High-quality AlGaN/GaN superlattices for near- and mid-infrared intersubband transitions. *Journal of Applied Physics* 111 (1): 13514. doi:10.1063/1.3675468.

Bayram, C., N. Péré-laperne, and M. Razeghi. 2009. Effects of well width and growth temperature on optical and structural characteristics of AlN/GaN superlattices grown by metal-organic chemical vapor deposition. *Applied Physics Letters* 95 (20): 201906. doi:10.1063/1.3267101.

Bayram, C., Z. Vashaei, and M. Razeghi. 2010. Reliability in room-temperature negative differential resistance characteristics of low-aluminum content AlGaN/GaN double-barrier resonant tunneling diodes. *Applied Physics Letters* 97 (18): 181109. doi:10.1063/1.3515418.

Beeler, M., C. Bougerol, E. Bellet-Amalric, and E. Monroy. 2013a. Terahertz absorbing AlGaN/GaN multi-quantum-wells: Demonstration of a robust 4-layer design. *Applied Physics Letters* 103 (9): 91108. doi:10.1063/1.4819950.

Beeler, M., C. Bougerol, E. Bellet-Amalric, and E. Monroy . 2014a. Pseudo-square AlGaN/GaN quantum wells for terahertz absorption. *Applied Physics Letters* 105 (13): 131106. doi:10.1063/1.4896768.

Beeler, M., P. Hille, J. Schörmann, J. Teubert, M. de la Mata, J. Arbiol, M. Eickhoff, and E. Monroy. 2014b. Intraband absorption in self-assembled Ge-doped GaN/AlN nanowire heterostructures. *Nano Letters* 14 (3): 1665–1673. doi:10.1021/nl5002247.

Beeler, M., C. B. Lim, P. Hille, J. Bleuse, J. Schörmann, M. de la Mata, J. Arbiol, M. Eickhoff, and E. Monroy. 2015. Long-lived excitons in GaN/AlN nanowire heterostructures. *Physical Review B* 91 (20): 205440. doi:10.1103/PhysRevB.91.205440.

Beeler, M., C. Bougerol, E. Bellet-Amalaric, and E. Monroy. 2014c. THz intersubband transitions in AlGaN/GaN multi-quantum-wells. *Physica Status Solidi (A)* 211 (4): 761–764. doi:10.1002/pssa.201300431.

Beeler, M., E. Trichas, and E. Monroy. 2013b. III-nitride Semiconductors for intersubband optoelectronics: A review. *Semiconductor Science and Technology* 28 (7): 74022. doi:10.1088/0268-1242/28/7/074022.

Belkin, M. A., J. A. Fan, S. Hormoz, F. Capasso, S. P. Khanna, M. Lachab, A. G. Davies, and E. H. Linfield. 2008. Terahertz quantum cascade lasers with copper metal-metal waveguides operating up to 178 K. *Optics Express* 16 (5): 3242–48. doi:10.1364/OE.16.003242.

Bellotti, E., K. Driscoll, T. D. Moustakas, and R. Paiella. 2008. Monte Carlo study of GaN versus GaAs terahertz quantum cascade structures. *Applied Physics Letters* 92 (10): 101112. doi:10.1063/1.2894508.

Bellotti, E., K. Driscoll, T. D. Moustakas, and R. Paiella . 2009. Monte Carlo simulation of terahertz quantum cascade laser structures based on wide-bandgap semiconductors. *Journal of Applied Physics* 105 (11): 113103. doi:10.1063/1.3137203.

Berland, K., M. Stattin, R. Farivar, D. M. S. Sultan, P. Hyldgaard, A. Larsson, S. M. Wang, and T. G. Andersson. 2010. Temperature stability of intersubband transitions in AlN/GaN quantum wells. *Applied Physics Letters* 97 (4): 43507. doi:10.1063/1.3456528.

Bernardini, F., V. Fiorentini, and D. Vanderbilt. 1997. Spontaneous polarization and piezoelectric constants of III-V nitrides. *Physical Review B* 56 (16): R10024–R10027. doi:10.1103/PhysRevB.56.R10024.

Berryman, K. W., S. A. Lyon, and Mordechai Segev. 1997. Mid-infrared photoconductivity in InAs quantum dots. *Applied Physics Letters* 70 (14): 1861–1863. doi:10.1063/1.118714.

Binggeli, N., P. Ferrara, and A. Baldereschi. 2001. Band-offset trends in nitride heterojunctions. *Physical Review B* 63 (24): 245306. doi:10.1103/PhysRevB.63.245306.

Birner, S., T. Zibold, T. Andlauer, T. Kubis, M. Sabathil, A. Trellakis, and P. Vogl. 2007. Nextnano: General purpose 3-D simulations. *IEEE Transactions on Electron Devices* 54 (9): 2137–2142. doi:10.1109/TED.2007.902871.

Carnevale, S. D., C. Marginean, P. J. Phillips, T. F. Kent, A. T. M. G. Sarwar, M. J. Mills, and R. C. Myers. 2012. Coaxial nanowire resonant tunneling diodes from non-polar AlN/GaN on silicon. *Applied Physics Letters* 100 (14): 142115. doi:10.1063/1.3701586.

Cen, L. B., B. Shen, Z. X. Qin, and G. Y. Zhang. 2009. Influence of polarization induced electric fields on the wavelength and the refractive index of intersubband transitions in AlN/GaN coupled double quantum wells. *Journal of Applied Physics* 105 (9): 93109. doi:10.1063/1.3124373.

Chamard, V., T. Schülli, M. Sztucki, T. Metzger, E. Sarigiannidou, J.-L. Rouvière, M. Tolan, and C. Adelmann. 2004. Strain distribution in nitride quantum dot multilayers. *Physical Review B* 69 (12): 125327. doi:10.1103/PhysRevB.69.125327.

Chen, G., Z. L. Li, X. Q. Wang, C. C. Huang, X. Rong, L. W. Sang, F. J. Xu et al. 2013. Effect of polarization on intersubband transition in AlGaN/GaN multiple quantum wells. *Applied Physics Letters* 102 (19): 192109. doi:10.1063/1.4807131.

Chen, Z., O. Baklenov, E. T. Kim, I. Mukhametzhanov, J. Tie, A. Madhukar, Z. Ye, and J. C. Campbell. 2001. Normal incidence InAs/Al[sub x]Ga[sub 1–x]as quantum dot infrared photodetectors with undoped active region. *Journal of Applied Physics* 89 (8): 4558–4563. doi:10.1063/1.1356430.

Chou, H. C., M. Anwar, and T. Manzur. 2012. Active layer design and power calculation of nitride-based THz quantum cascade lasers. In C. Lei and K. D. Choquette (eds.). *Proceedings of SPIE* 8268: 826800. doi:10.1117/12.914477.

Chou, H. C., T. Manzur, and M. Anwar. 2011. Active layer design of THz GaN quantum cascade lasers. *Proceedings of SPIE* 8023: 802309. doi:10.1117/12.888007.

Chu, L., A. Zrenner, M. Bichler, and G. Abstreiter. 2001. Quantum-dot infrared photodetector with lateral carrier transport. *Applied Physics Letters* 79 (14): 2249–2251. doi:10.1063/1.1408269.

Cociorva, D., W. G. Aulbur, and J. W. Wilkins. 2002. Quasiparticle calculations of band offsets at AlN–GaN interfaces. *Solid State Communications* 124 (1–2): 63–66. doi:10.1016/S0038-1098(02)00326-5.

Daudin, B., F. Widmann, G. Feuillet, Y. Samson, M. Arlery, and J. Rouvière. 1997. Stranski-Krastanov growth mode during the molecular beam epitaxy of highly strained GaN. *Physical Review B* 56 (12): R7069–R7072. doi:10.1103/PhysRevB.56.R7069.

Doyennette, L., L. Nevou, M. Tchernycheva, A. Lupu, F. Guillot, E. Monroy, R. Colombelli, and F. H. Julien. 2005. GaN-based quantum dot infrared photodetector operating at 1.38 Mm. *Electronics Letters* 41 (19): 1077–1078. doi:10.1049/el:20052598.

Driscoll, K., A. Bhattacharyya, T. D. Moustakas, R. Paiella, L. Zhou, and D. J. Smith. 2007. Intersubband absorption in AlN/GaN/AlGaN coupled quantum wells. *Applied Physics Letters* 91 (14): 141104. doi:10.1063/1.2794013.

Driscoll, K., Y. Liao, A. Bhattacharyya, L. Zhou, D. J. Smith, T. D. Moustakas, and R. Paiella. 2009. Optically pumped intersubband emission of short-wave infrared radiation with GaN/AlN quantum wells. *Applied Physics Letters* 94 (8): 81120. doi:10.1063/1.3089840.

Dupont, E. B., D. Delacourt, and M. Papuchon. 1993. Mid-infrared phase modulation via stark effect on intersubband transitions in GaAs/GaAlAs quantum wells. *IEEE Journal of Quantum Electronics* 29 (8): 2313–2318. doi:10.1109/3.245560.

Dussaigne, A., S. Nicolay, D. Martin, A. Castiglia, N. Grandjean, L. Nevou, H. Machhadani et al. 2010. Growth of intersubband GaN/AlGaN heterostructures. *Proceedings of SPIE* 7608: 76080H. doi:10.1117/12.847082.

Edmunds, C., J. Shao, M. Shirazi-HD, M. J. Manfra, and O. Malis. 2014. Terahertz intersubband absorption in non-polar m-plane AlGaN/GaN quantum wells. *Applied Physics Letters* 105 (2): 21109. doi:10.1063/1.4890611.

Edmunds, C., L. Tang, D. Li, M. Cervantes, G. Gardner, T. Paskova, M. J. Manfra, and O. Malis. 2012a. Near-infrared absorption in lattice-matched AlInN/GaN and strained AlGaN/GaN heterostructures grown by MBE on low-defect GaN substrates. *Journal of Electronic Materials* 41 (5): 881–886. doi:10.1007/s11664-011-1881-9.

Edmunds, C., L. Tang, J. Shao, D. Li, M. Cervantes, G. Gardner, D. N. Zakharov, M. J. Manfra, and O. Malis. 2012b. Improvement of near-infrared absorption linewidth in AlGaN/GaN superlattices by optimization of delta-doping location. *Applied Physics Letters* 101 (10): 102104. doi:10.1063/1.4751040.

Faist, J., F. Capasso, D. L. Sivco, C. Sirtori, A. L. Hutchinson, and A. Y. Cho. 1994. Quantum cascade laser. *Science* 264 (5158): 553–556. doi:10.1126/science.264.5158.553.

Fathololoumi, S., E. Dupont, C. W. I. Chan, Z. R. Wasilewski, S. R. Laframboise, D. Ban, A. Mátyás, C. Jirauschek, Q. Hu, and H. C. Liu. 2012. Terahertz quantum cascade lasers operating up to ~200 K with optimized oscillator strength and improved injection tunneling. *Optics Express* 20 (4): 3866–3876. doi:10.1364/OE.20.003866.

Gendron, L., M. Carras, A. Huynh, V. Ortiz, C. Koeniguer, and V. Berger. 2004. Quantum cascade photodetector. *Applied Physics Letters* 85 (14): 2824. doi:10.1063/1.1781731.

Giorgetta, F. R., E. Baumann, M. Graf, Q. Yang, C. Manz, K. Kohler, H. E. Beere et al. 2009. Quantum cascade detectors. *IEEE Journal of Quantum Electronics* 45 (8): 1039–1052. doi:10.1109/JQE.2009.2017929.

Giorgetta, F. R., E. Baumann, F. Guillot, E. Monroy, and D. Hofstetter. 2007. High frequency (f = 2.37 GHz) room temperature operation of 1.55 Mm AlN/GaN-based intersubband detector. *Electronics Letters* 43 (3): 185–187. doi:10.1049/el:20073458.

Gmachl, C., S. V. Frolov, H. M. Ng, S. -N. G. Chu, and A. Y. Cho. 2001a. Sub-picosecond electron scattering time for ≃1.55 [Micro Sign]m intersubband transitions in GaN/AlGaN multiple quantum wells. *Electronics Letters* 37 (6): 378. doi:10.1049/el:20010244.

Gmachl, C. and Hock M. Ng. 2003. Intersubband absorption at ~2.1 Mm in A-plane GaN/AlN multiple quantum wells. *Electronics Letters* 39 (6): 567–569. doi:10.1049/el:20030381.

Gmachl, C., H. M. Ng, and A. Y. Cho. 2001b. Intersubband absorption in degenerately doped GaN/Al[sub x]Ga[sub 1–x]N coupled double quantum wells. *Applied Physics Letters* 79 (11): 1590–1592. doi:10.1063/1.1403277.

Gmachl, Claire, Hock M. Ng, S.-N. George Chu, and Alfred Y. Cho. 2000. Intersubband absorption at λ~1.55 Mm in well- and modulation-doped GaN/AlGaN multiple quantum wells with superlattice barriers. *Applied Physics Letters* 77 (23): 3722–3724. doi:10.1063/1.1332108.

Gross, E., A. Nevet, A. Pesach, E. Monroy, S. E. Schacham, M. Orenstein, M. Segev, and G. Bahir. 2013. Measuring the refractive index around intersubband transition resonance in GaN/AlN multi quantum wells. *Optics Express* 21 (3): 3800–3808. doi:10.1364/OE.21.003800.

Gryshchenko, S. V., M. V. Klymenko, O. V. Shulika, I. A. Sukhoivanov, and V. V. Lysak. 2012. Temperature dependence of electron transport in GaN/AlGaN quantum cascade detectors. *Superlattices and Microstructures* 52 (4): 894–900. doi:10.1016/j.spmi.2012.06.013.

Guillot, F., B. Amstatt, E. Bellet-Amalric, E. Monroy, L. Nevou, L. Doyennette, F. H. Julien, and L. S. Dang. 2006a. Effect of Si doping on GaN/AlN multiple-quantum-well structures for intersubband optoelectronics at telecommunication wavelengths. *Superlattices and Microstructures* 40 (4–6): 306–312. doi:10.1016/j.spmi.2006.09.020.

Guillot, F., M. Tchernycheva, L. Nevou, L. Doyennette, E. Monroy, F. H. Julien, L. S. Dang et al. 2006b. Si-doped GaN/AlN quantum dot superlattices for optoelectronics at telecommunication wavelengths. *Physica Status Solidi (A)* 203 (7): 1754–1758. doi:10.1002/pssa.200565129.

Hamazaki, J., H. Kunugita, K. Ema, A. Kikuchi, and K. Kishino. 2005. Intersubband relaxation dynamics in GaN/AlN multiple quantum wells studied by two-color pump-probe experiments. *Physical Review B* 71 (16): 165334. doi:10.1103/PhysRevB.71.165334.

Hao, M., S. Mahanty, R. S. Qhalid Fareed, S. Tottori, K. Nishino, and S. Sakai. 1999. Infrared properties of bulk GaN. *Applied Physics Letters* 74 (19): 2788–2790. doi:10.1063/1.124014.

Harrer, A., B. Schwarz, R. Gansch, P. Reininger, H. Detz, T. Zederbauer, A. M. Andrews, W. Schrenk, and G. Strasser. 2014. Plasmonic lens enhanced mid-infrared quantum cascade detector. *Applied Physics Letters* 105 (17): 171112. doi:10.1063/1.4901043.

Harrison, P., D. Indjin, V. D. Jovanović, A. Mirčetić, Z. Ikonić, R. W. Kelsall, J. McTavish, I. Savić, N. Vukmirović, and V. Milanović. 2005. A physical model of quantum cascade lasers: Application to GaAs, GaN and SiGe devices. *Physica Status Solidi (A)* 202 (6): 980–986. doi:10.1002/pssa.200460713.

Heber, J. D., C. Gmachl, H. M. Ng, and A. Y. Cho. 2002. Comparative study of ultrafast intersubband electron scattering times at ~1.55 Mm wavelength in GaN/AlGaN heterostructures. *Applied Physics Letters* 81 (7): 1237–1239. doi:10.1063/1.1500412.

Helman, A., M. Tchernycheva, A. Lusson, E. Warde, F. H. Julien, K. Moumanis, G. Fishman et al. 2003. Intersubband spectroscopy of doped and undoped GaN/AlN quantum wells grown by molecular-beam epitaxy. *Applied Physics Letters* 83 (25): 5196–5198. doi:10.1063/1.1635985.

Hirayama, H. and W. Terashima. 2013. Recent progress toward realizing GaN-based THz quantum cascade laser. *Proceedings of SPIE* 8993: 89930G–89930G–9. doi:10.1117/12.2039214.

Hofstetter, D., E. Baumann, F. R. Giorgetta, R. Théron, H. Wu, W. J. Schaff, J. Dawlaty et al. 2010a. Intersubband transition-based processes and devices in AlN/GaN-based heterostructures. *Proceedings of the IEEE* 98 (7): 1234–1248. doi:10.1109/JPROC.2009.2035465.

Hofstetter, D., J. Di Francesco, P. K. Kandaswamy, A. Das, S. Valdueza-Felip, and E. Monroy. 2010b. Performance improvement of AlN/GaN-based intersubband detectors by using quantum dots. *IEEE Photonics Technology Letters* 22 (15): 1087–1089. doi:10.1109/LPT.2010.2050057.

Hofstetter, D., E. Baumann, F. R. Giorgetta, M. Graf, M. Maier, F. Guillot, E. Bellet-Amalric, and E. Monroy. 2006. High-quality AlN/GaN-superlattice structures for the fabrication of narrow-band 1.4 Mm photovoltaic intersubband detectors. *Applied Physics Letters* 88 (12): 121112. doi:10.1063/1.2185613.

Hofstetter, D., E. Baumann, F. R. Giorgetta, F. Guillot, S. Leconte, and E. Monroy. 2007. Optically nonlinear effects in intersubband transitions of GaN/AlN-based superlattice structures. *Applied Physics Letters* 91 (13): 131115. doi:10.1063/1.2793190.

Hofstetter, D., E. Baumann, F. R. Giorgetta, R. Théron, H. Wu, W. J. Schaff, J. Dawlaty et al. 2009. Photodetectors based on intersubband transitions using III-nitride superlattice structures. *Journal of Physics: Condensed Matter* 21 (17): 174208. doi:10.1088/0953-8984/21/17/174208.

Hofstetter, D., D. P. Bour, and L. Kirste. 2014. Mid-infrared electro-luminescence and absorption from AlGaN/GaN-based multi-quantum well inter-subband structures. *Applied Physics Letters* 104 (24): 241107. doi:10.1063/1.4883864.

Hofstetter, D., S .S. Schad, H. Wu, W. J. Schaff, and L. F. Eastman. 2003. GaN/AlN-based quantum-well infrared photodetector for 1.55 Mm. *Applied Physics Letters* 83 (3): 572–574. doi:10.1063/1.1594265.

Hofstetter, D., R. Theron, E. Baumann, F. R. Giorgetta, S. Golka, G. Strasser, F. Guillot, and E. Monroy. 2008. Monolithically integrated AlGaN/GaN/AlN-based solar-blind ultraviolet and near-infrared detectors. *Electronics Letters* 44 (16): 986. doi:10.1049/el:20081477.

Holmstrom, P. 2001. High-speed mid-IR modulator using stark shift in step quantum wells. *IEEE Journal of Quantum Electronics* 37 (10): 1273–1282. doi:10.1109/3.952539.

Holmstrom, P. 2006. Electroabsorption modulator using intersubband transitions in GaN–AlGaN–AlN step quantum wells. *IEEE Journal of Quantum Electronics* 42 (8): 810–819. doi:10.1109/JQE.2006.877297.

Holmström, P., X. Y. Liu, H. Uchida, T. Aggerstam, A. Kikuchi, K. Kishino, S. Lourdudoss, T. G. Andersson, and L. Thylén. 2007. Intersubband photonic devices by group-III nitrides. In J. Wang, G. K. Chang, Y. Itaya, and H. Zech (Eds.). *Proceedings of SPIE* 6782: 67821N. doi:10.1117/12.754372.

Huang, C. C., F. J. Xu, X. D. Yan, J. Song, Z. Y. Xu, L. B. Cen, Y. Wang et al. 2011. Intersubband transitions at atmospheric window in Al[sub x]Ga[sub 1–x]N/GaN multiple quantum wells grown on GaN/sapphire templates adopting AlN/GaN superlattices interlayer. *Applied Physics Letters* 98 (13): 132105. doi:10.1063/1.3573798.

Iizuka, N., K. Kaneko, and N. Suzuki. 2004. Sub-picosecond modulation by intersubband transition in ridge waveguide with GaN/AlN quantum wells. *Electronics Letters* 40 (15): 962963. doi:10.1049/el:20045434.

Iizuka, N., K. Kaneko, and N. Suzuki. 2006a. All-optical switch utilizing intersubband transition in GaN quantum wells. *IEEE Journal of Quantum Electronics* 42 (8): 765–771. doi:10.1109/JQE.2006.878189.

Iizuka, N., K. Kaneko, and N. Suzuki. 2002. Near-infrared intersubband absorption in GaN/AlN quantum wells grown by molecular beam epitaxy. *Applied Physics Letters* 81 (10): 1803–1805. doi:10.1063/1.1505116.

Iizuka, N., K. Kaneko, and N. Suzuki. 2005. Sub-picosecond all-optical gate utilizing GaN intersubband transition. *Optics Express* 13 (10): 3835–3840. doi:10.1364/OPEX.13.003835.

Iizuka, N., K. Kaneko, and N. Suzuki. 2006b. Polarization dependent loss in III-nitride optical waveguides for telecommunication devices. *Journal of Applied Physics* 99 (9): 93107. doi:10.1063/1.2195422.

Iizuka, N., K. Kaneko, N. Suzuki, T. Asano, S. Noda, and O. Wada. 2000. Ultrafast intersubband relaxation (≤150 Fs) in AlGaN/GaN multiple quantum wells. *Applied Physics Letters* 77 (5): 648–650. doi:10.1063/1.127073.

Iizuka, N., H. Yoshida, N. Managaki, T. Shimizu, S. Hassanet, C. Cumtornkittikul, M. Sugiyama, and Y. Nakano. 2009. Integration of GaN/AlN all-optical switch with SiN/AlN waveguide utilizing spot-size conversion. *Optics Express* 17 (25): 23247–23253. doi:10.1364/OE.17.023247.

Jovanović, V. D., Z. Ikonić, D. Indjin, P. Harrison, V. Milanović, and R. A. Soref. 2003. Designing strain-balanced GaN/AlGaN quantum well structures: Application to intersubband devices at 1.3 and 1.55 Mm wavelengths. *Journal of Applied Physics* 93 (6): 3194–3197. doi:10.1063/1.1556177.

Jovanović, V. D., D. Indjin, Z. Ikonić, and P. Harrison. 2004. Simulation and design of GaN/AlGaN far-infrared ($\lambda \sim 34$ Mm) quantum-cascade laser. *Applied Physics Letters* 84 (16): 2995–2997. doi:10.1063/1.1707219.

Kaminska, A., P. Strak, J. Borysiuk, K. Sobczak, J. Z. Domagala, M. Beeler, E. Grzanka, K. Sakowski, S. Krukowski, and E. Monroy. 2016. Correlation of optical and structural properties of GaN/AlN multi-quantum wells—Ab initio and experimental study. *Journal of Applied Physics* 119 (1): 15703. doi:10.1063/1.4939595.

Kandaswamy, P. K., F. Guillot, E. Bellet-Amalric, E. Monroy, L. Nevou, M. Tchernycheva, A. Michon et al. 2008. GaN/AlN short-period superlattices for intersubband optoelectronics: A systematic study of their epitaxial growth, design, and performance. *Journal of Applied Physics* 104 (9): 93501. doi:10.1063/1.3003507.

Kandaswamy, P. K., H. Machhadani, C. Bougerol, S. Sakr, M. Tchernycheva, F. H. Julien, and E. Monroy. 2009. Midinfrared intersubband absorption in GaN/AlGaN superlattices on Si(111) templates. *Applied Physics Letters* 95 (14): 141911. doi:10.1063/1.3242345.

Kandaswamy, P. K., H. Machhadani, Y. Kotsar, S. Sakr, A. Das, M. Tchernycheva, L. Rapenne, E. Sarigiannidou, F. H. Julien, and E. Monroy. 2010. Effect of doping on the mid-infrared intersubband absorption in GaN/AlGaN superlattices grown on Si(111) templates. *Applied Physics Letters* 96 (14): 141903. doi:10.1063/1.3379300.

Kheirodin, N., L. Nevou, H. Machhadani, P. Crozat, L. Vivien, M. Tchernycheva, A. Lupu et al. 2008. Electrooptical modulator at telecommunication wavelengths based on GaN/AlN coupled quantum wells. *IEEE Photonics Technology Letters* 20 (9): 724–726. doi:10.1109/LPT.2008.919595.

Kishino, K., A. Kikuchi, H. Kanazawa, and T. Tachibana. 2002. Intersubband transition in (GaN)[sub m]/(AlN) [sub N] superlattices in the wavelength range from 1.08 to 1.61 Mm. *Applied Physics Letters* 81 (7): 1234–1236. doi:10.1063/1.1500432.

Köhler, R., A. Tredicucci, F. Beltram, H. E. Beere, E. H. Linfield, A. G. Davies, D. A. Ritchie, R. C. Iotti, and F. Rossi. 2002. Terahertz semiconductor-heterostructure laser. *Nature* 417 (6885): 156–159. doi:10.1038/417156a.

Kotani, T., M. Arita, and Y. Arakawa. 2014. Observation of mid-infrared intersubband absorption in non-polar m-plane AlGaN/GaN multiple quantum wells. *Applied Physics Letters* 105 (26): 261108. doi:10.1063/1.4905212.

Kotani, T., M. Arita, and Y. Arakawa . 2015. Doping dependent blue shift and linewidth broadening of intersubband absorption in non-polar m-plane AlGaN/GaN multiple quantum wells. *Applied Physics Letters* 107 (11): 112107. doi:10.1063/1.4931096.

Lähnemann, J., O. Brandt, C. Pfüller, T. Flissikowski, U. Jahn, E. Luna, M. Hanke, M. Knelangen, A. Trampert, and H. T. Grahn. 2011. Coexistence of quantum-confined stark effect and localized states in an (In, Ga) N/GaN nanowire heterostructure. *Physical Review B* 84 (15): 155303. doi:10.1103/Phys Rev B.84.155303.

Lahourcade, L., E. Bellet-Amalric, E. Monroy, M. Abouzaid, and P. Ruterana. 2007. Plasma-assisted molecular-beam epitaxy of AlN(11-22) on M sapphire. *Applied Physics Letters* 90 (13): 131909. doi:10.1063/1.2716375.

Lahourcade, L., P. K. Kandaswamy, J. Renard, P. Ruterana, H. Machhadani, M. Tchernycheva, F. H. Julien, B. Gayral, and E. Monroy. 2008. Interband and intersubband optical characterization of semipolar (11–22)-oriented GaN/AlN multiple-quantum-well structures. *Applied Physics Letters* 93 (11): 111906. doi:10.1063/1.2978250.

Levine, B. F. 1993. Quantum-well infrared ohotodetectors. *Journal of Applied Physics* 74 (8): R1–R81. doi:10.1063/1.354252.

Levine, B. F., K. K. Choi, C. G. Bethea, J. Walker, and R. J. Malik. 1987. New 10 Mm infrared detector using intersubband absorption in resonant tunneling GaAlAs superlattices. *Applied Physics Letters* 50 (16): 1092–1094. doi:10.1063/1.97928.

Li, Y., A. Bhattacharyya, C. Thomidis, Y. Liao, T. D. Moustakas, and R. Paiella. 2008. Refractive-index nonlinearities of intersubband transitions in GaN/AlN quantum-well waveguides. *Journal of Applied Physics* 104 (8): 83101. doi:10.1063/1.2996107.

Li, Y., A. Bhattacharyya, C. Thomidis, T. D. Moustakas, and R. Paiella. 2007. Ultrafast all-optical switching with low saturation energy via intersubband transitions in GaN/AlN quantum-well waveguides. *Optics Express* 15 (26): 17922–17927. doi:10.1364/OE.15.017922.

Li, Y, and R Paiella. 2006. Intersubband all-optical switching based on coulomb-induced optical non-linearities in GaN/AlGaN coupled quantum wells. *Semiconductor Science and Technology* 21 (8): 1105–1110. doi:10.1088/0268-1242/21/8/022.

Lim, C. B., A. Ajay, C. Bougerol, B. Haas, J. Schörmann, M. Beeler, J. Lähnemann, M. Eickhoff, and E. Monroy. 2015a. Nonpolar *M* -plane GaN/AlGaN heterostructures with intersubband transitions in the 5–10 THz band. *Nanotechnology* 26 (43): 435201. doi:10.1088/0957-4484/26/43/435201.

Lim, C. B, A. Ajay, C. Bougerol, J. Lähnemann, F. Donatini, J. Schörmann, E. Bellet-Amalric, D. A Browne, M. Jiménez-Rodríguez, and E. Monroy. 2016a. Effect of doping on the far-infrared intersubband transitions in nonpolar m-plane GaN/AlGaN heterostructures. *Nanotechnology* 27 (14): 145201. doi:10.1088/0957-4484/27/14/145201.

Lim, C. B., M. Beeler, A. Ajay, J. Lähnemann, E. Bellet-Amalric, C. Bougerol, and E. Monroy. 2015a. Intersubband transitions in nonpolar GaN/Al(Ga)N heterostructures in the short and mid-wavelength infrared regions. *Journal of Applied Physics* 118: 14309. doi:10.1063/1.4926423.

Lim, Caroline B., Mark Beeler, Akhil Ajay, Jonas Lähnemann, Edith Bellet-Amalric, Catherine Bougerol, Jörg Schörmann, Martin Eickhoff, and Eva Monroy. 2016b. Short-wavelength, mid- and far-infrared intersubband absorption in nonpolar GaN/Al(Ga)N heterostructures. *Japanese Journal of Applied Physics* 55 (5S): 05FG05. doi:10.7567/JJAP.55.05FG05.

Liu, D.. F., J. G. Jiang, Y. Cheng, and J. F. He. 2013. Effect of delta doping on mid-infrared intersubband absorption in AlGaN/GaN step quantum well structures. *Physica E: Low-Dimensional Systems and Nanostructures* 54 (December): 253–256. doi:10.1016/j.physe.2013.06.030.

Liu, H. C, and F. Capasso. 2000. *Intersubband Transitions in Quantum Wells: Physics and Device Applications I.* San Diego, CA: Academic Press.

Liu, X. Y., P. Holmström, P. Jänes, L. Thylén, and T. G. Andersson. 2007. Intersubband absorption at 1.5–3.5 Mm in GaN/AlN multiple quantum wells grown by molecular beam epitaxy on sapphire. *Physica Status Solidi (B)* 244 (8): 2892–2905. doi:10.1002/pssb.200675606.

Lü, J. T. and J. C. Cao. 2006. Monte Carlo simulation of hot phonon effects in resonant-phonon-assisted terahertz quantum-cascade lasers. *Applied Physics Letters* 88 (6): 61119. doi:10.1063/1.2172225.

Lupu, A., M. Tchernycheva, Y. Kotsar, E. Monroy, and F. H. Julien. 2012. Electroabsorption and refractive index modulation induced by intersubband transitions in GaN/AlN multiple quantum wells. *Optics Express* 20 (11): 12541. doi:10.1364/OE.20.012541.

Machhadani, H., M. Beeler, S. Sakr, E. Warde, Y. Kotsar, M. Tchernycheva, M. P. Chauvat et al. 2013. Systematic study of near-infrared intersubband absorption of polar and semipolar GaN/AlN quantum wells. *Journal of Applied Physics* 113 (14): 143109. doi:10.1063/1.4801528.

Machhadani, H., P. Kandaswamy, S. Sakr, A. Vardi, A. Wirtmüller, L. Nevou, F. Guillot et al. 2009. GaN/AlGaN intersubband optoelectronic devices. *New Journal of Physics* 11 (12): 125023. doi:10.1088/1367-2630/11/12/125023.

Machhadani, H., Y. Kotsar, S. Sakr, M. Tchernycheva, R. Colombelli, J. Mangeney, E. Bellet-Amalric, E. Sarigiannidou, E. Monroy, and F. H. Julien. 2010. Terahertz intersubband absorption in GaN/AlGaN step quantum wells. *Applied Physics Letters* 97 (19): 191101. doi:10.1063/1.3515423.

Mirzaei, B., A. Rostami, and H. Baghban. 2012. Terahertz dual-wavelength quantum cascade laser based on GaN active region. *Optics & Laser Technology* 44 (2): 378–383. doi:10.1016/j.optlastec.2011.07.020.

Monteagudo-Lerma, L., S. Valdueza-Felip, F. B. Naranjo, P. Corredera, L. Rapenne, E. Sarigiannidou, G. Strasser, E. Monroy, and M. González-Herráez. 2013. Waveguide saturable absorbers at 155 Mm based on intraband transitions in GaN/AlN QDs. *Optics Express* 21 (23): 27578. doi:10.1364/OE.21.027578.

Nevou, L., F. H. Julien, M. Tchernycheva, F. Guillot, E. Monroy, and E. Sarigiannidou. 2008. Intraband emission at λ≈ 1.48 Mm from GaN/AlN quantum dots at room temperature. *Applied Physics Letters* 92 (16): 161105. doi:10.1063/1.2913756.

Nevou, L., F. H. Julien, R. Colombelli, F. Guillot, and E. Monroy. 2006a. Room-temperature intersubband emission of GaN/AlN quantum wells at λ = 2.3 Mm. *Electronics Letters* 42 (22): 1308–1309. doi:10.1049/el:20062282.

Nevou, L., N. Kheirodin, M. Tchernycheva, L. Meignien, P. Crozat, A. Lupu, E. Warde et al. 2007a. Short-wavelength intersubband electroabsorption modulation based on electron tunneling between GaN/AlN coupled quantum wells. *Applied Physics Letters* 90 (22): 223511. doi:10.1063/1.2745211.

Nevou, L., J. Mangeney, M. Tchernycheva, F. H. Julien, F. Guillot, and E. Monroy. 2009. Ultrafast relaxation and optical saturation of intraband absorption of GaN/AlN auantum dots. *Applied Physics Letters* 94 (13): 132104. doi:10.1063/1.3114424.

Nevou, L., M. Tchernycheva, F. H. Julien, F. Guillot, and E. Monroy. 2007b. Short wavelength (λ = 2.13 Mm) intersubband luminescence from GaN/AlN quantum wells at room temperature. *Applied Physics Letters* 90 (12): 121106. doi:10.1063/1.2715001.

Nevou, L., M. Tchernycheva, F. Julien, M. Raybaut, A. Godard, E. Rosencher, F. Guillot, and E. Monroy. 2006b. Intersubband resonant enhancement of second-harmonic generation in GaN/AlN quantum wells. *Applied Physics Letters* 89 (15): 151101. doi:10.1063/1.2358118.

Ng, H. M., C. Gmachl, J. D. Heber, J. W. P. Hsu, S. N. G. Chu, and A. Y. Cho. 2002. Recent progress in GaN-based superlattices for near-infrared intersubband transitions. *Physica Status Solidi (B)* 234 (3): 817–821. doi:10.1002/1521-3951(200212)234:3<817::AID-PSSB817>3.0.CO;2-4.

Ng, H. M., C. Gmachl, T. Siegrist, S. N. G. Chu, and A. Y. Cho. 2001. Growth and characterization of GaN/AlGaN superlattices for near-infrared intersubband transitions. *Physica Status Solidi (A)* 188 (2): 825–831. doi:10.1002/1521-396X(200112)188:2<825::AID-PSSA825>3.0.CO;2-7.

Nguyen, D. T., W. Wüster, Ph. Roussignol, C. Voisin, G. Cassabois, M. Tchernycheva, F. H. Julien, F. Guillot, and E. Monroy. 2010. Homogeneous linewidth of the intraband transition at 1.55 Mm in GaN/AlN quantum dots. *Applied Physics Letters* 97 (6): 61903. doi:10.1063/1.3476340.

Noda, S., T. Yamashita, M. Ohya, Y. Muromoto, and A. Sasaki. 1993. All-optical modulation for semiconductor lasers by using three energy levels in n-doped quantum wells. *IEEE Journal of Quantum Electronics* 29 (6): 1640–1647. doi:0.1109/3.234416.

Pan, D., E. Towe, and S. Kennerly. 1998. Normal-incidence intersubband (In, Ga)As/GaAs quantum dot infrared photodetectors. *Applied Physics Letters* 73 (14): 1937–1939. doi:10.1063/1.122328.

Péré-Laperne, N., C. Bayram, L. Nguyen-The, R. McClintock, and M. Razeghi. 2009. Tunability of intersubband absorption from 4.5 to 5.3 Mm in a GaN/Al[sub 0.2]Ga[sub 0.8]N superlattices grown by metalorganic chemical vapor deposition. *Applied Physics Letters* 95 (13): 131109. doi:10.1063/1.3242027.

Pesach, A., E. Gross, C.-Y. Huang, Y.-D. Lin, A. Vardi, S. E. Schacham, S. Nakamura, and G. Bahir. 2013. Non-polar m-plane ntersubband based InGaN/(Al)GaN quantum well infrared photodetectors. *Applied Physics Letters* 103 (2): 22110. doi:10.1063/1.4813395.

Pesach, A., S. Sakr, E. Giraud, O. Sorias, L. Gal, M. Tchernycheva, M. Orenstein, N. Grandjean, F. H. Julien, and G. Bahir. 2014. First demonstration of plasmonic GaN quantum cascade detectors with enhanced efficiency at normal incidence. *Optics Express* 22 (17): 21069–21078. doi:10.1364/OE.22.021069.

Phillips, J., K. Kamath, and P. Bhattacharya. 1998. Far-infrared photoconductivity in self-organized InAs quantum dots. *Applied Physics Letters* 72 (16): 2020–2022. doi:10.1063/1.121252.

Rapaport, R., G. Chen, O. Mitrofanov, C. Gmachl, H. M. Ng, and S. N. G. Chu. 2003. Resonant optical nonlinearities from intersubband transitions in GaN/AlN quantum wells. *Applied Physics Letters* 83 (2): 263. doi:10.1063/1.1591247.

Reininger, P., B. Schwarz, A. Harrer, T. Zederbauer, H. Detz, A. Maxwell Andrews, R. Gansch, W. Schrenk, and G. Strasser. 2013. Photonic crystal slab quantum cascade detector. *Applied Physics Letters* 103 (24): 241103. doi:10.1063/1.4846035.

Rigutti, L., G. Jacopin, A. D. Bugallo, M. Tchernycheva, E. Warde, F. H. Julien, R. Songmuang, E. Galopin, L. Largeau, and J. C. Harmand. 2010. Investigation of the electronic transport in GaN nanowires containing GaN/AlN quantum discs. *Nanotechnology* 21 (42): 425206. doi:10.1088/0957-4484/21/42/425206.

Romanov, A. E., T. J. Baker, S. Nakamura, J. S. Speck, and ERATO/JST UCSB Group. 2006. Strain-induced polarization in wurtzite III-nitride semipolar layers. *Journal of Applied Physics* 100 (2): 23522. doi:10.1063/1.2218385.

Rong, X., X. Q. Wang, G. Chen, X. T. Zheng, P. Wang, F. J. Xu, Z. X. Qin et al. 2015. Mid-infrared photoconductive response in AlGaN/GaN step quantum wells. *Scientific Reports* 5: 14386. doi:10.1038/srep14386.

Rosencher, E. and Ph. Bois. 1991. Model system for optical nonlinearities: Asymmetric quantum wells. *Physical Review B* 44 (20): 11315–11327. doi:10.1103/PhysRevB.44.11315.

Ryzhii, V. 1996. The theory of quantum-dot infrared phototransistors. *Semiconductor Science and Technology* 11 (5): 759–765. doi:10.1088/0268-1242/11/5/018.

Sakr, S., P. Crozat, D. Gacemi, Y. Kotsar, A. Pesach, P. Quach, N. Isac et al. 2013. GaN/AlGaN waveguide quantum cascade photodetectors at λ ≈ 1.55 Mm with enhanced responsivity and ~40 GHz frequency bandwidth. *Applied Physics Letters* 102 (1): 11135. doi:10.1063/1.4775374.

Sakr, S., E. Giraud, A. Dussaigne, M. Tchernycheva, N. Grandjean, and F. H. Julien. 2012. Two-color GaN/AlGaN quantum cascade detector at short infrared wavelengths of 1 and 1.7 Mm. *Applied Physics Letters* 100 (18): 181103. doi:10.1063/1.4707904.

Sakr, S., E. Giraud, M. Tchernycheva, N. Isac, P. Quach, E. Warde, N. Grandjean, and F. H. Julien. 2012. A simplified GaN/AlGaN quantum cascade detector with an alloy extractor. *Applied Physics Letters* 101 (25): 251101. doi:10.1063/1.4772501.

Sakr, S., Y. Kotsar, S. Haddadi, M. Tchernycheva, L. Vivien, I. Sarigiannidou, N. Isac, E. Monroy, and F. H. Julien. 2010. GaN-based quantum cascade photodetector with 1.5 Mm peak detection wavelength. *Electronics Letters* 46 (25): 1685–1686. doi:10.1049/el.2010.2181.

Scalari, G., C. Walther, M. Fischer, R. Terazzi, H. Beere, D. Ritchie, and J. Faist. 2009. THz and sub-THz quantum cascade lasers. *Laser & Photonics Review* 3 (1–2): 45–66. doi:10.1002/lpor.200810030.

Schörmann, J., P. Hille, M. Schäfer, J. Müßener, P. Becker, P. J. Klar, M. Kleine-Boymann et al. 2013. Germanium doping of self-assembled GaN nanowires grown by plasma-assisted molecular beam epitaxy. *Journal of Applied Physics* 114 (10): 103505. doi:10.1063/1.4820264.

Shao, Jiayi, Liang Tang, Colin Edmunds, Geoff Gardner, Oana Malis, and Michael Manfra. 2013a. Surface morphology evolution of m-plane (1-100) GaN during molecular beam epitaxy growth: Impact of Ga/N ratio, miscut direction, and growth temperature. *Journal of Applied Physics* 114 (2): 23508. doi:10.1063/1.4813079.

Shao, J., D. N. Zakharov, C. Edmunds, O. Malis, and M. J. Manfra. 2013b. Homogeneous AlGaN/GaN superlattices grown on free-standing (11⁻00) GaN substrates by plasma-assisted molecular beam epitaxy. *Applied Physics Letters* 103 (23): 232103. doi:10.1063/1.4836975.

Sherliker, B., M. Halsall, I. Kasalynas, D. Seliuta, G. Valusis, M. Vengris, M. Barkauskas et al. 2007. Room emperature operation of AlGaN/GaN quantum well infrared photodetectors at a 3–4 Mm wavelength range. *Semiconductor Science and Technology* 22 (11): 1240–1244. doi:10.1088/0268-1242/22/11/010.

Shishehchi, S., R. Paiella, and E. Bellotti. 2014. Numerical simulation of III-nitride lattice-matched structures for quantum cascade lasers. In, edited by Bernd Witzigmann, Marek Osinski, Fritz Henneberger, and Yasuhiko Arakawa, *Proceedings of SPIE* 89800T. doi:10.1117/12.2040709.

Sodabanlu, H., J. S. Yang, T. Tanemura, M. Sugiyama, Y. Shimogaki, and Y. Nakano. 2011. Intersubband absorption saturation in AlN-based waveguide with GaN/AlN multiple quantum wells grown by metalorganic vapor phase epitaxy. *Applied Physics Letters* 99 (15): 151102. doi:10.1063/1.3650929.

Song, A. Y., R. Bhat, A. A. Allerman, J. Wang, T. Y. Huang, C. E. Zah, and C. F. Gmachl. 2015a. Quantum cascade emission in the III-nitride material system designed with effective interface grading. *Applied Physics Letters* 107 (13): 132104. doi:10.1063/1.4932068.

Song, A. Y., R. Bhat, P. Bouzi, C. E. Zah, and C. F. Gmachl. 2015b. Three-dimensional interface roughness in layered semiconductor structures and its effects on intersubband transitions. *Physical Review B* 94: 165307 .

Song, Y., R. Bhat, T. Y. Huang, P. Badami, C. E. Zah, and C. Gmachl. 2014. III-nitride quantum cascade detector grown by metal organic chemical vapor deposition. *Applied Physics Letters* 105 (18): 182104. doi:10.1063/1.4901220.

Songmuang, R., D. Kalita, P. Sinha, M. den Hertog, R. André, T. Ben, D. González, H. Mariette, and E. Monroy. 2011. Strong suppression of internal electric field in GaN/AlGaN multi-layer quantum dots in nanowires. *Applied Physics Letters* 99 (14): 141914. doi:10.1063/1.3646389.

Songmuang, R., G. Katsaros, E. Monroy, P. Spathis, C. Bougerol, M. Mongillo, and S. De Franceschi. 2010. Quantum transport in GaN/AlN double-barrier heterostructure nanowires. *Nano Letters* 10 (9): 3545–3550. doi:10.1021/nl1017578.

Soref, R. and B. Bennett. 1987. Electrooptical effects in silicon. *IEEE Journal of Quantum Electronics* 23 (1): 123–129. doi:10.1109/JQE.1987.1073206.

Speck, J. S. and S. F. Chichibu. 2011. Nonpolar and semipolar group III nitride-based materials. *MRS Bulletin* 34 (5): 304–312. doi:10.1557/mrs2009.91.

Stattin, M., K. Berland, P. Hyldgaard, A. Larsson, and T. G. Andersson. 2011. Waveguides for nitride based quantum cascade lasers. *Physica Status Solidi (C)* 8 (7–8): 2357–2359. doi:10.1002/pssc.201000987.

Sudradjat, F. F., W. Zhang, J. Woodward, H. Durmaz, T. D. Moustakas, and R. Paiella. 2012. Far-infrared intersubband photodetectors based on double-step III-nitride quantum wells. *Applied Physics Letters* 100 (24): 241113. doi:10.1063/1.4729470.

Sudradjat, F., W. Zhang, K. Driscoll, Y. Liao, A. Bhattacharyya, C. Thomidis, L. Zhou, D. J. Smith, T. D. Moustakas, and R. Paiella. 2010. Sequential tunneling transport characteristics of GaN/AlGaN coupled-quantum-well structures. *Journal of Applied Physics* 108 (10): 103704. doi:10.1063/1.3511334.

Sun, G., J. B. Khurgin, and D. P. Tsai. 2013. Spoof plasmon waveguide enabled ultrathin room temperature THz GaN quantum cascade laser: A feasibility study. *Optics Express* 21 (23): 28054. doi:10.1364/OE.21.028054.

Sun, G., R. A. Soref, and J. B. Khurgin. 2005. Active region design of a terahertz GaN/Al0.15Ga0.85N quantum cascade laser. *Superlattices and Microstructures* 37 (2): 107–113. doi:10.1016/j.spmi.2004.09.046.

Suzuki, N. and N. Iizuka. 1997. Feasibility study on ultrafast nonlinear optical properties of 1.55 Mm intersubband transition in AlGaN/GaN quantum wells. *Japanese Journal of Applied Physics* 36 (8A): L1006–L1008. doi:10.1143/JJAP.36.L1006.

Suzuki, N. and N. Iizuka. 1999. Effect of polarization field on intersubband transition in AlGaN/GaN quantum wells. *Japanese Journal of Applied Physics* 38 (4A): L363–L365. doi:10.1143/JJAP.38.L363.

Suzuki, N., N. Iizuka, and K. Kaneko. 2000. Intersubband transition in AlGaN-GaN quantum wells for ultrafast all-optical switching at communication wavelength. Edited by Giancarlo C. Righini, Joseph W. Perry, H. Walter Yao, Ghassan E. Jabbour, Shibin Jiang, Bruce S. Dunn, Geoffrey T. Burnham et al. *Proceedings of SPIE* 3940: 127–138. doi:10.1117/12.381452.

Tchernycheva, M., L. Nevou, L. Doyennette, A. Helman, R. Colombelli, F. H. Julien, E. Guillot, E. Monroy, T. Shibata, and M. Tanaka. 2005. Intraband absorption of doped GaN/AlN quantum dots at telecommunication wavelengths. *Applied Physics Letters* 87 (10): 101912. doi:10.1063/1.2042540.

Tchernycheva, M., L. Nevou, L. Doyennette, F. H. Julien, F. Guillot, E. Monroy, T. Remmele, and M. Albrecht. 2006a. Electron confinement in strongly coupled GaN/AlN quantum wells. *Applied Physics Letters* 88 (15): 153113. doi:10.1063/1.2193057.

Tchernycheva, M., L. Nevou, L. Doyennette, F. Julien, E. Warde, F. Guillot, E. Monroy, E. Bellet-Amalric, T. Remmele, and M. Albrecht. 2006b. Systematic experimental and theoretical investigation of intersubband absorption in GaN/AlN quantum wells. *Physical Review B* 73 (12): 125347. doi:10.1103/PhysRevB.73.125347.

Terashima, W. and H. Hirayama. 2011a. Terahertz intersubband electroluminescence from GaN/AlGaN quantum cascade laser structure on AlGaN template. In *2011 36th International Conference on Infrared, Millimeter and Terahertz Waves (IRMMW-THz)*. doi:10.1109/irmmw-THz.2011.6105201.

Terashima, W. and H. Hirayama. 2009. Design and fabrication of terahertz quantum cascade laser structure based on III-nitride semiconductors. *Physica Status Solidi (C)* 6 (S2): S615–S618. doi:10.1002/pssc.200880772.

Terashima, W. and H. Hirayama. 2011b. Spontaneous emission from GaN/AlGaN terahertz quantum cascade laser grown on GaN substrate. *Physica Status Solidi (C)* 8 (7–8): 2302–2304. doi:10.1002/pssc.201000878.

Terashima, W. and H. Hirayama. 2015a. GaN-based terahertz quantum cascade lasers. *Proceedings of SPIE*, no. 9483: 948304.

Terashima, W. and H. Hirayama. 2015b. Terahertz frequency emission with novel quantum cascade laser designs. *SPIE Newsroom*, doi:10.1117/2.1201507.006058.

Tian, W., W. Y. Yan, Xiong Hui, S. L. Li, Y. Y. Ding, Y. Li, Y. Tian et al. 2012. Tunability of intersubband transition wavelength in the atmospheric window in AlGaN/GaN multi-quantum wells grown on different AlGaN templates by metalorganic chemical vapor deposition. *Journal of Applied Physics* 112 (6): 63526. doi:10.1063/1.4754543.

Valdueza-Felip, S., F. B. Naranjo, M. Gonzalez-Herraez, H. Fernandez, J. Solis, F. Guillot, E. Monroy, L. Nevou, M. Tchernycheva, and F. H. Julien. 2008. Characterization of the resonant third-order nonlinear susceptibility of Si-doped GaN-AlN quantum wells and quantum dots at 1.5 Mm. *IEEE Photonics Technology Letters* 20 (16): 1366–1368. doi:10.1109/LPT.2008.926842.

Vardi, A., N. Akopian, G. Bahir, L. Doyennette, M. Tchernycheva, L. Nevou, F. H. Julien, F. Guillot, and E. Monroy. 2006. Room temperature demonstration of GaN/AlN quantum dot intraband infrared photodetector at fiber-optics communication wavelength. *Applied Physics Letters* 88 (14): 143101. doi:10.1063/1.2186108.

Vardi, A., G. Bahir, F. Guillot, C. Bougerol, E. Monroy, S. E. Schacham, M. Tchernycheva, and F. H. Julien. 2008a. Near infrared quantum cascade detector in GaN/AlGaN/AlN heterostructures. *Applied Physics Letters* 92 (1): 11112. doi:10.1063/1.2830704.

Vardi, A., G. Bahir, S. E. Schacham, P. K. Kandaswamy, and E. Monroy. 2009. Photocurrent spectroscopy of bound-to-bound intraband transitions in GaN/AlN quantum dots. *Physical Review B* 80 (15): 155439. doi:10.1103/PhysRevB.80.155439.

Vardi, A., N. Kheirodin, L. Nevou, H. Machhadani, L. Vivien, P. Crozat, M. Tchernycheva et al. 2008b. High-speed operation of GaN/AlGaN quantum cascade detectors at $\lambda \approx$ 1.55 Mm. *Applied Physics Letters* 93 (19): 193509. doi:10.1063/1.3021376.

Vardi, A., S. Sakr, J. Mangeney, P. K. Kandaswamy, E. Monroy, M. Tchernycheva, S. E. Schacham, F. H. Julien, and G. Bahir. 2011. Femto-second electron transit time characterization in GaN/AlGaN quantum cascade detector at 1.5 Micron. *Applied Physics Letters* 99 (20): 202111. doi:10.1063/1.3660583.

Vukmirović, N., V. D. Jovanović, D. Indjin, Z. Ikonić, P. Harrison, and V. Milanović. 2005. Optically pumped terahertz laser based on intersubband transitions in a GaN/AlGaN double quantum well. *Journal of Applied Physics* 97 (10): 103106. doi:10.1063/1.1900929.

Wada, O. 2004. Femtosecond all-optical devices for ultrafast communication and signal processing. *New Journal of Physics* 6: 183–183. doi:10.1088/1367-2630/6/1/183.

Wang, S., W. Tian, F. Wu, J. Zhang, J. N. Dai, Z. H. Wu, Y. Y. Fang, Y. Tian, and C. Q. Chen. 2015a. Efficient optical coupling in AlGaN/GaN quantum well infrared photodetector via quasi-one-dimensional gold grating. *Optics Express* 23 (7): 8740. doi:10.1364/OE.23.008740.

Wang, S., J. Zhang, F. Wu, W. Tian, J. N. Dai, Y. Tian, and C. Q. Chen. 2015b. Long-range surface plasmon polaritons for efficient optical coupling in AlGaN/GaN quantum well infrared photodetector. *Plasmonics*, 11: 833–838. doi:10.1007/s11468-015-0116-y.

Welna, M., R. Kudrawiec, M. Motyka, R. Kucharski, M. Zając, M. Rudziński, J. Misiewicz, R. Doradziński, and R. Dwiliński. 2012. Transparency of GaN substrates in the mid-infrared spectral range. *Crystal Research and Technology* 47 (3): 347–350. doi:10.1002/crat.201100443.

West, L. C. and S. J. Eglash. 1985. First observation of an extremely large-dipole infrared transition within the conduction band of a GaAs quantum well. *Applied Physics Letters* 46 (12): 1156–1158. doi:10.1063/1.95742.

Williams, B. S. 2007. Terahertz quantum-cascade lasers. *Nature Photonics* 1 (9): 517–525. doi:10.1038/nphoton.2007.166.

Williams, B. S., H. Callebaut, S. Kumar, Q. Hu, and J. L. Reno. 2003. 3.4-THz quantum cascade laser based on longitudinal-optical-phonon scattering for depopulation. *Applied Physics Letters* 82 (7): 1015–1017. doi:10.1063/1.1554479.

Williams, B. S., S. Kumar, Q. Hu, and J. L. Reno. 2006. High-power terahertz quantum-cascade lasers. *Electronics Letters* 42 (2): 89. doi:10.1049/el:20063921.

Wu, F., W. Tian, W. Y. Yan, J. Zhang, S. C. Sun, J. N. Dai, Y. Y. Fang, Z. H. Wu, and C. Q. Chen. 2013a. Terahertz intersubband transition in GaN/AlGaN step quantum well. *Journal of Applied Physics* 113 (15): 154505. doi:10.1063/1.4802496.

Wu, F., W. Tian, J. Zhang, S. Wang, Q. X. Wan, J. N. Dai, Z. H. Wu et al. 2014. Double-resonance enhanced intersubband second-order nonlinear optical susceptibilities in GaN/AlGaN step quantum wells. *Optics Express* 22 (12): 14212. doi:10.1364/OE.22.014212.

Wu, T., Y. Wei-Yi, X. Hui, D. Jian-Nan, F. Yan-Yan, Y. Zhi-Hao, Y. Chen-Hui, and C. Chang-Qin. 2013b. Effects of polarization on intersubband transitions of AlxGa1–xN/GaN multi-quantum wells. *Chinese Physics B* 22 (5): 57302. doi:10.1088/1674-1056/22/5/057302.

Yang, J., G. J. Brown, M. Dutta, and M. A. Stroscio. 2005. Photon absorption in the restrahlen band of thin films of GaN and AlN: Two phonon effects. *Journal of Applied Physics* 98 (4): 43517. doi:10.1063/1.2034648.

Yang, R., J. Xu, and M. Sweeny. 1994. Selection rules of intersubband transitions in conduction-band quantum wells. *Physical Review B* 50 (11): 7474–7482. doi:10.1103/PhysRevB.50.7474.

Yasuda, H., T. Kubis, and K. Hirakawa. 2011. Non-equilibrium green's function calculation for GaN-based terahertz quantum cascade laser structures. In *2011 36th International Conference on Infrared, Millimeter and Terahertz Waves (IRMMW-THz)*. doi:10.1109/irmmw-THz.2011.6105199.

Yasuda, H., I. Hosako, and K. Hirakawa. 2012. Designs of GaN-based terahertz quantum cascade lasers for higher temperature operations. In *2012 Conference on Lasers and Electro-Optics (CLEO)*.

Zhuo, X., J. Ni, J. Li, W. Lin, D. Cai, S. Li, and J. Kang. 2014. Band engineering of GaN/AlN quantum wells by Si dopants. *Journal of Applied Physics* 115 (12): 124305. doi:10.1063/1.4868580.

Zucker, J. E., I. Bar-Joseph, B. I. Miller, U. Koren, and D. S. Chemla. 1989. Quaternary quantum wells for electro-optic intensity and phase modulation at 1.3 and 1.55 Mm. *Applied Physics Letters* 54 (1): 10–12. doi:10.1063/1.100821.

21

Lighting Communications

21.1 Introduction to Visible Light Communication System

21.1.1 LED Lighting and Communication

Light-emitting diode (LED)-based white-lighting is an emerging and booming industry. Over the past few decades, versatile LEDs with visible wavelength ranged from 350 to 750 nm have been widely used as the back lighting source in displays, the lighting component for traffic lights and signs, and the basic element for ubiquitous indoor lighting systems, and so on, because of its high brightness and low power consumption. More than that, the LED-based visible light communication (VLC) system has also been considered as a promising candidate to combine with the white-lighting module for next-generation wireless communication [1–4], which exhibits distinct benefits of electromagnetic interference (EMI) free and free-space data transmission [5,6] when comparing with the traditional radio frequency (RF) wireless and optical fiber wired communication systems. These excellent features also give rise to the flexibility in lighting and communication applications under different environments, such as in flying aircraft, hospitals, [7] and indoor, and so on, which enables to provide higher data rate than the Bluetooth or Wi-Fi-based wireless network. Note that the flash lamp lighting element in mobile phones can also be employed to transmit data in free space [8]. Pang et al. have already proposed an audio communication system that used LEDs to transmit information to televisions, computers, and phones as early as 1999 [9]. As a future prospect, the schematic diagram of the lighting VLC system with LED-based white-light modules for not only visible lighting but also data transmission is illustrated in Figure 21.1. In such systems, the electrical equipment and components involve the functionalities of transmitting and receiving data within the general illumination area. Furthermore, the optical multiple-input multiple-output (MIMO) technology can be employed further to achieve high data rates [10]. As expected, the indoor lighting VLC (also called Light-fidelity, Li-Fi) system for simultaneous white

FIGURE 21.1 The schematic diagram of a lighting communication system in the near future. (From Sweet Home 3D, Copyright © 2005–2015, Emmanuel PUYBARET/eTeks <info@eteks.com>.)

lighting and data transmission would undoubtedly be the future star for designing and configuring the smart home because of its advantages of license-free, high secrecy, EMI immunity, and luminaire integration compatibility [11–18].

In a future vision on potential applications, the VLC can also be employed for traffic communication. The traffic signage or electronic toll collection (ETC) can transmit information to vehicles, which is required by drivers timely. In addition, the car headlights can not only illuminate but also handle the vehicle-to-vehicle information transmission [19]. Indeed, the concept of vehicle information and communication system (VICS) was originally proposed in 1996 [20], which provides information to moving vehicles *via* optical beacons installed at road sides. However, enormous budgets would be needed for allocating beacons on every lane of the road to make such a VICS system hardly to be implemented. For cost-effective consideration, Akanegawa et al. had already used the existing 505 nm diode laser traffic lights to preliminarily demonstrate a 1 Mbit/s traffic information system in 2001 [19]. In 2006, Wook et al. constructed a road-to-vehicle (R2V) communication system by using the LED-based traffic lights with pulse position modulation (PPM) at 1 Mbps [21]. In 2009, Okada et al. also proposed a similar R2V communication system [22] by respectively employing the LED traffic light and photodiode as the transmitter and receiver. In addition, two cameras are used to solve the time-dependent position variation between the transmitter and receiver, and a gyro sensor-based vibrational correction technique was also employed to minimize related vibrations when implementing 1 and 2 Mbit/s data rates for 60 and 40 m free-space distances, respectively. In 2013, Yoo et al. used the LED headlamp combined with a PPM scheme to demonstrate a vehicular VCL system with 10 kbit/s data over 20 m free-space [23]. Later on, Luo et al. presented a mathematical model for car-to-car (C2C) VLC system, showing that the coverage range for communication can be extended up to 20 m at 2 Mbit/s by placing a photodetector in the car at a height of 0.2–0.4 m above the road surface [24].

For indoor lighting VLC, Minh et al. used 16 white LEDs combined with a multiple-resonant equalization technique to implement a 40 Mbit/s VLC in 2008 [25]. Later on, a 100 Mbit/s nonreturn-to-zero (NRZ) VLC using a post-equalized white LED was demonstrated by the same group [26]. In the same year, Vucic et al. used a thin-film high-power phosphorescent white LED and a discrete-multitone modulation to demonstrate a 200+ Mbit/s white-light wireless transmission system [27], and the same group further upgraded the transmission capacity of the white LED to 513 Mbit/s in 2010 [15]. In 2012, by using a commercial phosphorescent white LED combined with a rate-adaptive discrete multitone modulation, Khalid et al. demonstrated a 1 Gbit/s VLC system [28]. In 2012, a 1.1 Gbit/s white LED-based VLC system employing carrier-less amplitude and phase modulation was proposed by Wu et al. [29]. In view of previous works, mixing red, green, and blue (RGB) LEDs [30–32], or combining a blue LED with a yellow phosphor can be considered as two alternative methods for white-light generation [33–35]. As early as 2002, Muthu et al. proposed a white LED source by mixing RGB LEDs [36]. In 2008, Gilewski et al. used an optical feedback circuit to control the lighting color and luminance of RGB LEDs mixed source [37]. In 2010, Wang employed a novel control system to change the lighting color and luminous intensity of RGB LED-based white-light source [38]. In 2015, Hung et al. constructed a RGB LED mixing system to generate white light and different color light [39]. For data transmission, the RGB LEDs can also construct a wavelength division multiplexing (WDM) VLC system. In 2012, Cossu et al. demonstrated 1.5 Gbit/s single-channel and 3.4-Gbit/s WDM transmissions through a 10 cm free-space channel based on RGB LEDs [35]. In 2013, Wang et al. demonstrated a 66 cm bidirectional subcarrier multiplexed and wavelength division multiplexed (SCM-WDM) VLC system with 575 Mbit/s downlink and 225 Mbit/s uplink using RGB LEDs and a phosphor-based LED, respectively [40]. Later on, a WDM VLC system at 3.22 Gbit/s using a single RGB LED with CAP modulation was proposed by Wu et al. [41]. By using the same modulation format, Wang et al. implemented a RGB-LED-based high-speed WDM VLC system at 4.5 Gbit/s in 2015 [42]. Li et al. also proposed a RGB LED-based WDM VLC system with a data rate of 750 Mbps and a transmission distance of 70 cm by using the OFDM signal [43]. In 2015, Chen et al. proposed a hierarchical scheme to detect the rotatable RGB LED array which can be used in the camera of mobile phone [44]. Luo et al. construct an optical camera communication system based on RGB LEDs with a data rate of 150 bit/s [45].

When comparing the white-light source mixed by RGB LEDs, the blue LED illuminated yellow phosphor-based lighting element is more compact and cost-effective. However, it may suffer from a reduced data rate because the passively excited yellow phosphor is only spontaneous emission source with a lifetime much longer than the LED, which could degrade the transmitted data carried by the blue light. Fortunately, this problem can be alleviated by using the discrete multi-tone (DMT) and high QAM-level-based OFDM signals to replace typical data formats [46–50]. However, the bandwidth of commercial blue LEDs is still limited due to its large device area and light emission mechanism induced slow direct modulation speed (~tens to hundreds of MHz) [1,4,26,35], which is difficult to meet the demand for high capacity and long transmission distance. To increase the modulation speed of the single LED device, a large injection current density, and a small device capacitance would be a considerable strategy and is already achieved by the use of micro-LED [51]. However, the increasing of injection current density would lead to a well-known disadvantage of efficiency droop due to the electron overflow [52]. Such issues cannot be perfectly eliminated even though numerous efforts have been proposed, such as ternary or quarternary electron blocking layers (EBLs) [53,54], graded EBL, [55] and superlattice EBL [56], and so on. Not long ago, the micro-LEDs with large driving capacitance and compact device architecture have emerged as an alternative candidate to increase the data bandwidth [51]. In 2014, Chun et al. used a micro-LED combined with a yellow fluorescent copolymer to implement a 1.68 Gbit/s VLC [57], and this is the fastest white-light VLC to the best of our knowledge. Nevertheless, the efficiency droop phenomenon under high current injection level still limits the lighting and modulation performances of micro-LED [52].

To release such a bottleneck that occurred on LED-based VLC, the introduction of visible diode lasers into the lighting communication system is proposed for taking its advantages including broader modulation bandwidth and higher pumping efficiency (with the absence of efficiency droop effect) than currently available LEDs [58–64]. In 2007, Hu et al. used a 650 nm diode laser to demonstrate a 10 Mbps VLC system with transmission distance up to 300 m [65]. In 2012, Lin et al. proposed a 500 Mbps/10 m WDM VLC system by using red and green laser pointers [6]. Recently, Hanson et al. demonstrated a 532 nm VLC at 1 Gbps through a 2 m water pipe [66]. Chen et al. used a red diode laser to demonstrate a bidirectional VLC system with 16-QAM OFDM data at 2.5 Gbps over 20 km single-mode fiber (SMF) and 15 m free-space [67]. The directly-modulated 641 nm laser pointer-based VLC with 4-QAM OFDM data is proposed by Singh et al [68]. Among all visible wavelengths, the gallium nitride (GaN) based blue diode laser is mainly considered as one of the best VLC transmitters [69] because of its cost-effective and high spectral power features. Moreover, it not only enables white lighting after combining with a yellow phosphor, but also promises underwater VLC with reduced absorption and scattering. Recently, Watson et al. demonstrated a 2.5 Gbps nonreturn-to-zero on-off-keying (NRZ-OOK) modulation scheme by using a single GaN-based blue diode laser in a free space link [70]. In 2015, Chi et al. constructed a 450 nm GaN blue diode laser-based 9-Gbps VLC system through a 5 m free-space channel [71]. Lately, Lee et al. used a high-power GaN blue diode laser with 2.6 GHz bandwidth to demonstrate a 4 Gbit/s free-space data transmission link at room temperature [72]. For traffic application, some automobile manufacturers have proposed laser headlight by putting the fluorescent phosphorous plate in front of the blue diode laser for harmless white lighting. The laser headlights can provide more bright white light for long-range illumination with smaller size and longer visual distance as compared to the LED headlights developed for short range illumination. For indoor white-lighting application, Neumann et al. mixed the red, green, blue, and yellow laser beams to obtain a white-lighting source in 2011 [73]. In 2013, Soltic et al. proposed an optimization method on RGB diode laser-based white-lighting system by using delta-function spectra [74]. In 2015, Chun et al. proposed a 15 cm white-lighting link to provide a data rate of up to 6.52 Gbit/s by combining a blue LD with a remote phosphor [75]. Later on, by using 36 parallel information streams with the RGB diode lasers, Tsonev et al. proposed a WDM VLC system, and it can be expected to improve the transmission data rate up to 100 Gbit/s [76]. In the following section of this chapter, the transistor outline (TO38) can packaged GaN blue diode laser-based 9 Gbit/s point-to-point and >5 Gbit/s white-lighting communications over 5 and 0.5 m free-space distances, respectively, are demonstrated. The DC biasing optimization for the GaN blue diode laser on its encoding performance is discussed to maximize its allowable transmission capacity for both cases.

21.2 Phosphor Diffused White Lighting and Visible Light Communication

21.2.1 Building up a GaN Diode Laser-Based White-Lighting System with Phosphor Film

To realize the phosphor diffused diode laser white lighting and VLC module, a TO38 packaged GaN blue diode laser (Opto Semiconductors, PL 450B) without any divergent and diffused element for point-to-point communications, and the same device with a phosphor film-based optical spectral expander and beam diffuser and white-lighting QAM OFDM communications are compared with each other after propagating over a free-space link, as illustrated in Figure 21.2. The blue diode laser with single transverse mode reveals a maximal beam divergence in horizontal and vertical directions of 11° and 25° in far field. The transmitted QAM OFDM data with various QAM-levels and OFDM subcarrier bandwidths was generated by a homemade MATLAB® program, which was uploaded into an arbitrary waveform generator (Tektronix, 70001A) with a sampling rate of 24 GSa/s for upscaling its carrier frequency. After 10-dB-gain amplification with a broadband amplifier (Picosecond Pulse Labs, 5828), the electrical QAM OFDM data was employed to directly modulate the GaN blue diode laser in combination with a DC bias current *via* a bias tee (Mini-Circuit, ZX85-12G-S+). By adding the phosphor film in front of the blue diode laser as a luminescent diffuser, the blue laser beam was spectrally expanded into while light and spatially diverged in free-space for luminescent white-lighting and data transmission simultaneously. After free-space propagation, the blue diode laser beam or diffused while-light spot with carried data was launched into an avalanche photodiode (APD, Hamamatsu, S9073) with a bandwidth of 900 MHz by using a pair of focusing lens. Finally, the received QAM OFDM data after amplification with a 40 dB low-noise amplifier (LNA, Mini-Circuit, ZKL-1R5) was captured by a digital

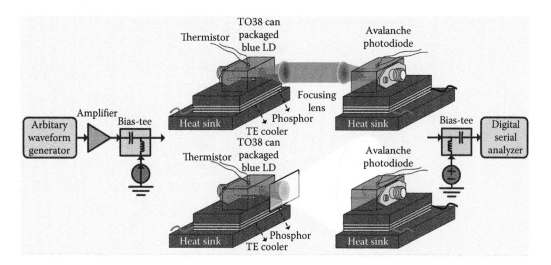

FIGURE 21.2 The experimental setup of TO38 can packaged GaN blue diode laser without and with phosphor film for point-to-point and white-lighting QAM OFDM communications, respectively, over a free-space link.

serial analyzer (Tektronix, DSA71604C) with a sampling rate of 100 GSa/s for further analyzing its constellation plot, error vector magnitude (EVM), signal-to-noise ratio (SNR), and bit error rate (BER).

21.2.2 Diode Laser-Based Point-to-Point Communication

At a room temperature of 24°C, the optical spectrum and power-to-current responses of used GaN blue diode laser are shown in Figure 21.3, which indicates an emitting wavelength of ~450 nm and a threshold current of 35 mA. In addition, a dP/dI slope of >0.6 mW/A observed above threshold represents a differential quantum efficiency of >2.3 × 10^{-4}.

To discuss the transmission performance of the directly modulated GaN blue diode laser, a point-to-point QAM OFDM communication system with a free-space distance of 5 m is firstly demonstrated. Moreover, a 16-QAM OFDM data with an encoding bandwidth of 1 GHz and a raw data rate of 4 Gbit/s is

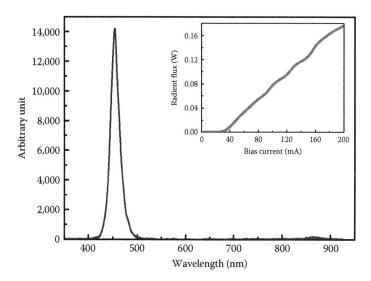

FIGURE 21.3 The optical spectrum and power-to-current curve of used TO38 packaged GaN blue diode laser.

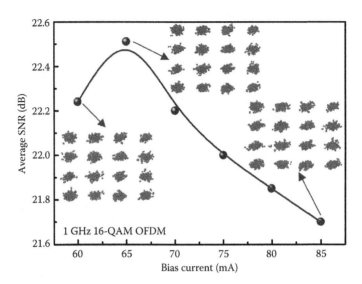

FIGURE 21.4 The average SNRs and related constellation plots of 4-Gbit/s 16-QAM OFDM data carried by the GaN blue diode laser at different bias currents.

employed to directly encode the GaN blue diode laser at different operating conditions for optimization, as shown in Figure 21.4. Whichever the GaN blue diode laser would be operated within 60–85 mA, it can achieve 4 Gbit/s data rate with clear constellation plots after decoding. Increasing the bias current from 60 to 65 mA improves the average SNR of GaN blue diode laser carried 4 Gbit/s data by 0.3 dB, as the suppressed relative intensity noise (RIN) dominates more than the declined frequency response in such a condition. By continuously increasing the bias current up to 85 mA, the overly declined frequency response would inevitably decrease the subcarrier power of transmitted QAM OFDM data carried by high-frequency subcarriers, which then decreases the SNR to >21.5 dB. At an optimized bias current of 65 mA, the lowest EVM of <8.5% and BER of <4 × 10⁻⁸ are observed when comparing with other operating conditions.

To further improve the encoding capacity after optimizing the operating condition, the QAM OFDM data increases its QAM level to 64 and extends its OFDM subcarrier bandwidth to 1.5 GHz so as to increase the transmission capacity to 9 Gbit/s, as shown in Figure 21.5. Although increasing the QAM level from 16 to 64 raises the corresponding FEC criterion from 15.2 to 21.1 dB, the GaN blue diode laser carried 64-QAM OFDM data with a raw data rate as high as 9 Gbit/s still provides a qualified average SNR of 22 dB after receiving. Furthermore, it also exhibits a distinguished constellation plot with an EVM of 5% and a BER <3.8 × 10⁻³. The further improvement on the transmission performance would rely on either improving the coherence of the diode laser or enhancing the bandwidth of the APD in the future.

21.2.3 Diode Laser-Based White-Lighting Communication

To demonstrate the diverged and diffused white lighting, the commonly used phosphor is the commercially available $Y_3Al_5O_{12}:Ce^{3+}$ (YAG) powder, which exhibits an average size of about 13 μm to provide a down-converted luminescent emission peak at wavelength of 550 nm. As a result, the adhered phosphor film provides broadband down-converted luminescence and scattered diffusion to transfer the highly oriented blue laser beam into a widely divergent white-light spot, as shown in Figure 21.6.

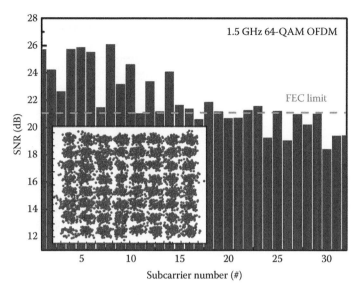

FIGURE 21.5 The subcarrier SNRs and related constellation plot of GaN blue diode laser delivered 9-Gbit/s 64-QAM OFDM data over a 5-m free-space link.

FIGURE 21.6 The images of GaN blue diode laser without and with phosphor film.

To declare the intensity distribution of phosphor diffused white light composed of the blue and yellow components at different angles, the bidirectional transmission distribution functions (BTDF) measurement was carried out. As a result, the Figure 21.7a shows the angle-dependent white-light distributions taken along the long and short axes of the far-field divergent spot from the phosphor diffused diode laser beam, respectively. Both the long- and short-axis distributions show similar Lambertian-like shape, which results from the strong scattering effect caused by the adhered phosphor particle. On the other hand, the angle-dependent yellow light components taken along long and short axes show strong intensity at 0° and Lambertian-like distribution at other degrees. It results from that, before scattering and down-converting by the phosphor particles, the diode laser intensity was highly concentrated at 0°; therefore, the central part of phosphor film would receive stronger diode laser excitation than other regions. To further investigate the white-lighting performance of the GaN blue diode laser with phosphor film, the Commission International de l'Eclairage (CIE) chromaticity coordinate was measured by a color analyzer, as shown in Figure 21.7b, which reveals a chromaticity coordinate (x, y) of (0.34, 0.37)

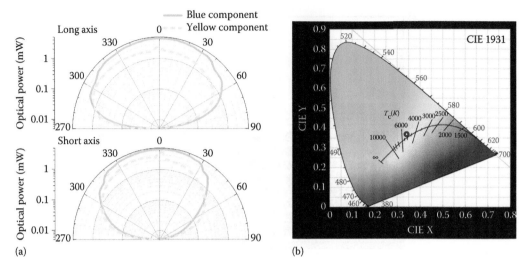

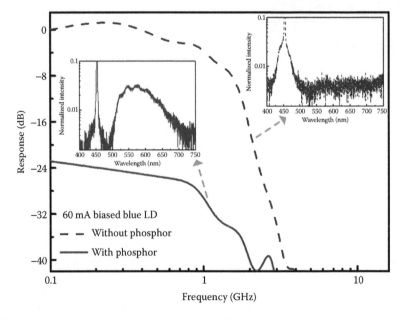

FIGURE 21.7 (a) The phosphor diffused white-light distribution taken along long and short axes from −90° to 90°, which is composed by blue and yellow light components. (b) Chromaticity coordinates result of GaN blue diode laser with phosphor film under CIE 1931.

and a correlated color temperature (CCT) of 5210 K. Note that the measured chromaticity is relatively close to the white-light standard CIE of (0.3333, 0.3333).

Furthermore, the optical spectrum of the blue diode laser without and with phosphor film is shown in the inset of Figure 21.8. In this part, the blue diode laser is biased at 60 mA to strike a balance between the RIN suppression and the throughput intensity-to-frequency declination. Note that the use of phosphor film weakens the blue-light component down to ~10% of its original power such that the yellow-light component can significantly be observed. The Figure 21.8 shows the measured small-signal frequency response of the GaN blue diode laser without and with the phosphor film.

FIGURE 21.8 The optical spectrum and frequency response of the blue diode laser without and with phosphor film.

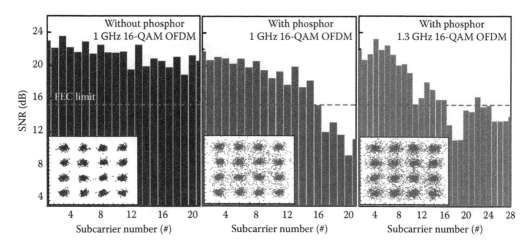

FIGURE 21.9 The constellation plots and related subcarrier SNRs of 4 and >5-Gbit/s 16-QAM OFDM data carried by the diffuser diverged blue diode laser.

With the use of phosphor diffuser for luminescent white lighting, the blue diode laser diverges into the white-light spot and degrades its throughput modulation response by around 25 dB because of the relevant absorption, scattering, reflection, and beam divergence of the blue diode laser beam intensity after passing through the phosphorous diffuser film. In addition, the cut-off frequencies of the TO38 diode laser package and the APD also limit the high-frequency throughout intensity for both cases.

For a fair comparison of the direct modulation performance before and after white-light conversion, the 4 Gbit/s 16-QAM OFDM data is employed to directly encode the diffuser diverged blue diode laser. Without using the phosphor film, the delivered 4 Gbit/s reveals a clear constellation plot with an EVM of <10%, as shown in Figure 21.9a. In contrast, the phosphor-based diffuser slightly blurs the transmitted constellation plot to exhibit an increased EVM of 16%, as shown in Figure 21.9b. That is, the added phosphor helps luminescent lighting at a cost of degrading the average SNR by 4 dB for the transmitted 4 Gbit/s data. Although the BER of the received 4 Gbit/s data is seriously degraded to 2×10^{-3} for the luminescent lighting case, it is still lower than the required BER floor of 3.8×10^{-3} set by the FEC. Furthermore, the OFDM data bandwidth is increased to investigate the upper limitation on transmission capacity of the proposed white-lighting source. Note that a maximal allowable OFDM data bandwidth of >1.25 GHz which represents a raw data rate of >5 Gbit/s is observed, and a FEC qualified BER of 3.6×10^{-3} is also obtained, as shown in Figure 21.9c. Moreover, the constellation plot and related subcarrier SNRs represents an average SNR of 15.2 dB and an EVM of 17.3%.

21.2.4 Color Temperature Tunability of Diode Laser-Based White-Lighting VLC

To realize a more flexible diode laser-based VLC system with color temperature tunability, the effect of phosphor concentration or amount on the white-lighting VLC quality including color temperature and data rate becomes a critical issue. Hence, the phosphor films with different phosphor concentrations are performed to discuss its influence on aforementioned parameters as shown below. The left part of Figure 21.10 shows the images of white-lighting spots from the diode laser after adhering with different phosphor films, which results in CCT values of 27,300, 4,450 and 2,830 K for phosphor film sample 1, 2 and 3, respectively.

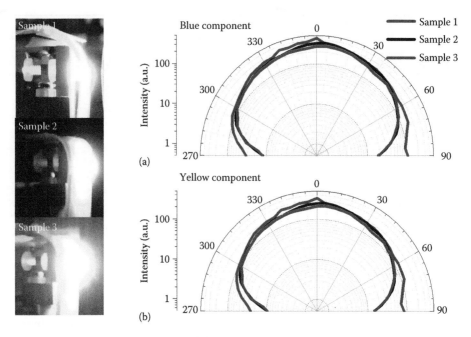

FIGURE 21.10 The white-lighting images of blue diode laser combined with phosphor film with different phosphor concentrations (left) and the angle-dependent white-light intensity distribution (right) of (a) blue and (b) yellow components.

The right part of Figure 21.10 shows the angle-dependent distributions of blue and yellow components within the white light for phosphor film samples 1, 2 and 3, which are detected at a distance of 60 cm away from the phosphor film adhered diode laser. These results indicate similar light field distribution between blue and yellow components, except for the intrinsic Lambertian-like emission character of phosphor particles. When comparing with other degrees, the light field distribution shows stronger power at the central part, especially for the low phosphor concentration (sample 3) case. As the phosphor concentration is increased, both blue and yellow components behave more like the Lambertian distribution, indicating that the light field distribution is mainly affected by the phosphor particles induced diode laser beam diffusion and scattering.

For data transmission, the 16-QAM OFDM data is directly modulated onto the GaN diode laser with different phosphor samples. As a result, the maximal allowable data rates for different samples are ranging from 2 to 6.8 Gbit/s, respectively, as shown in Figure 21.11. In addition, the related EVMs, SNRs, and BERs of $16.9 \pm 0.2\%$, 15.4 ± 0.1 dB and $(3.1 \pm 0.3) \times 10^{-3}$ for different samples are also obtained, respectively. Because of the high sensitivity of OFDM on SNR, decreased diode laser intensity by the absorption and scatter of phosphor particle will inevitably lower the maximum data rate.

As observed, the dynamics of data transmission could somewhat be degraded with the aid of luminescent diffuser; however, it is also the complaint common to all divergent diode laser sources without appropriate beam focusing optics. Nevertheless, these observations have explored a practicability of the phosphor diffuser diverged blue diode laser for simultaneously luminescent white-light lighting and optical wireless communication. Even though the Li-Fi-based wireless streaming is barring some slight softness to the white-light spreading when compared with current LED lighting kit, it already functions good enough, at least with one order of magnitude better-quality data transmission than those ever approached in the LED-based white-lighting system. Although trawling through the access of different

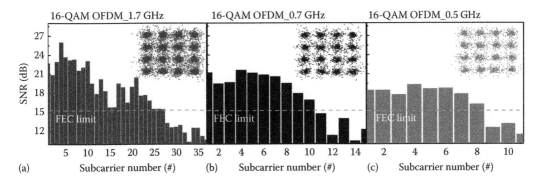

FIGURE 21.11 The BER versus data bandwidth for 16-QAM OFDM data carried by the blue diode laser with (a) sample 1, (b) sample 2, and (c) sample 3 phosphors. The insets in figures are the constellation plots, which correspond to the operation point with the highest data rate.

contents of data-streams has yet to be performed in this work, the extensive capabilities on changing network data formats are easy and intuitive as well.

21.3 Conclusion

As proposed, the blue diode laser-based white-lighting VLC has significantly accelerated its pace for practical application. Numerous research peers have endeavored on related works to make the diode laser-based white-lighting VLC a competitive alternative as compared to currently available LED architectures. The directly modulated blue diode laser-based VLC system with large frequency bandwidth, high encoding rate and dense spectral usage efficiency is indeed the excellent candidate, especially when considering the demand for high-capacity data transmission for tenement, hospital, airplane, mobile phone connectivity, and traffic signs, and so on. This chapter has provided the historical overview on both the LED and the diode laser-based VLC systems with remarkable or milestone demonstrations. Moreover, the GaN blue diode laser-based 9 Gbit/s point-to-point blue-lighting and >5 Gbit/s luminescent white-lighting communications over 5 and 0.5 m free-space distances are also demonstrated, respectively, in which the white-lighting source is implemented by combining the blue diode laser with a compact phosphorous diffuser.

References

1. J. Vučić, C. Kottke, S. Nerreter, K. Habel, A. Büttner, K. D. Langer, and J. W. Walewski. 230 Mbit/s via a wireless visible-light link based on OOK modulation of phosphorescent white LEDs. In *Proceedings of Optical Fiber Communication Conference*, San Diego, CA, p. OThH3, 2010.
2. C.-W. Chow, C.-H. Yeh, Y.-F. Liu, and Y. Liu. Improved modulation speed of LED visible light communication system integrated to main electricity network. *Electronics Letters*, 47(15):867–868, 2011.
3. C.-H. Yeh, Y.-F. Liu, C.-W. Chow, Y. Liu, P. Y. Huang, and H. K. Tsang. Investigation of 4-ASK modulation with digital filtering to increase 20 times of direct modulation speed of white-light LED visible light communication system. *Optics Express*, 20(15):16218–16223, 2012.
4. F. M. Wu, C. T. Lin, C. C. Wei, C. W. Chen, Z. Y. Chen, and H. T. Huang. 3.22-Gb/s WDM visible light communication of a single RGB LED employing carrier-less amplitude and phase modulation. In *Proceedings of Optical Fiber Communication Conference*, Anaheim, CA, p. OTh1G4, 2013.

5. Z. Wang, C. Yu, W. D. Zhong, J. Chen, and W. Chen. Performance of a novel LED lamp arrangement to reduce SNR fluctuation for multi-user visible light communication systems. *Optics Express*, 20(4):4564–4573, 2012.

6. W. Y. Lin, C. Y. Chen, H. H. Lu, C. H. Chang, Y. P. Lin, H. C. Lin, and H. W. Wu. 10m/500Mbps WDM visible light communication systems. *Optics Express*, 20(9):9919–9924, 2012.

7. W. Ding, F. Yang, H. Yang, J. Wang, X. Wang, X. Zhang, and J. Song. A hybrid power line and visible light communication system for indoor hospital applications. *Computers in Industry*, 68(5):170–178, 2015.

8. R. Sagotra and R. Aggarwal. Visible light communication. *International Journal of Computer Trends and Technology (IJCTT)*, 4(4):906–910, 2013.

9. G. Pang, K. L. Ho, T. Kwan, and E. Yang. Visible light communication for audio systems. *IEEE Transactions on Consumer Electronics*, 45(4):1112–1118, 1999.

10. L. Zeng, D. C. O'Brien, H. Le Minh, G. E. Faulkner, K. Lee, D. Jung, Y. Oh, and E. T. Won. High data rate multiple input multiple output (MIMO) optical wireless communications using white LED lighting. *IEEE Journal on Selected Areas in Communications*, 27(9):1654–1662, 2009.

11. Y. Tanaka, T. Komine, S. Haruyama, and M. Nakagawa. Indoor visible light data transmission system utilizing white LED Lights. *IEICE Transactions on Communications*, E86-B(8):2440–2454, 2003.

12. T. Komine and M. Nakagawa. Fundamental analysis for visible-light communication system using LED lights. *IEEE Transactions on Consumer Electronics*, 50(1):100–107, 2004.

13. H. Elgala, R. Mesleh, and H. Haas. Indoor optical wireless communication: potential and state-of-the-art. *IEEE Communications Magazine*, 49(9):56–62, 2011.

14. S. Rajagopal, R. D. Roberts, and S. K. Lim, IEEE 802.15.7 visible light communication: modulation schemes and dimming support. *IEEE Communications Magazine*, 5(3):72–82, 2012.

15. J. Vučić, C. Kottke, S. Nerreter, K. D. Langer, and J. W. Walewski. 513 Mbit/s visible light communications link based on DMT-modulation of a white LED. *Journal of Lightwave Technology*, 28(24):3512–3518, 2010.

16. C.-W. Chow, C.-H. Yeh, Y.-F. Liu, and P. Y. Huang. Mitigation of optical background noise in light-emitting diode (LED) optical wireless communication systems. *IEEE Photonics Journal*, 5(1):7900307, 2013.

17. B. S. Rawat, B. Aggarwal, and D. Passi. LI-FI: A new era of wireless communication data sharing. *Internaional Journal of Scientific & Technology Research (IJSTR)*, 3(10):118–119, 2014.

18. M. Kavehrad. Sustainable energy-efficient wireless applications using light. *IEEE Communications Magazine*, 48(12):66–73, 2010.

19. M. Akanegawa, Y. Tanaka, and M. Nakagawa. Basic study on traffic information system using LED traffic lights. *IEEE Transactions on Intelligent Transportations Systems*, 2(4):193–203, 2001.

20. M. Nishikawa, Y. Katagiri, and M. Hirayama. Vehicle information and communication system. *Oki Technical Review*, 62(157):7–12, 1996.

21. H. B. C. Wook, T. Komine, S. Haruyama, M. Nakagawa. Visible light communication with LED-based traffic lights using 2-dimensional image sensor. *IEICE Transactions on Fundamentals of Electronics, Communications and Computer Sciences*, E89-A(3):654–659, 2006.

22. S. Okada, T. Yendo, T. Yamazato, T. Fujii, M. Tanimoto, and Y. Kimura. On-vehicle receiver for distant visible light road-to-vehicle communication. *2009 IEEE Intelligent Vehicles Symposium*, 1033–1038, 2009.

23. J. H. Yoo, R. Lee, J. K. Oh, H. W. Seo, J. Y. Kim, H. C. Kim, and S. Y. Jung. Demonstration of vehicular visible light communication based on LED headlamp. In *Proceedings 2013 Fifth International Conference on Ubiquitous and Future Networks (ICUFN)*, 465–467, 2013.

24. P. Luo, Z. Ghassemlooy, H. Le Minh, E. Bentley, A. Burton, and X. Tang. Fundamental analysis of a car to car visible light communication system. In *Proceedings 2014 9th International Symposium on Communication Systems, Networks & Digital Signal Processing (CSNDSP)*, 1011–1016, 2014.

25. H. Le Minh, D. O'Brien, G. Faulkner, L. Zeng, K. Lee, D. Jung, and Y. Oh. High-speed visible light communications using multiple-resonant equalization. *IEEE Photonics Technology Letters*, 20(14):1243–1245, 2008.

26. H. Le Minh, D. O'Brien, G. Faulkner, L. Zeng, K. Lee, D. Jung, Y. Oh, and E. T. Won. 100-Mb/s NRZ visible light communications using a postequalized white LED. *IEEE Photonics Technology Letters*, 21(15):1063–1065, 2009.

27. J. Vučić, C. Kottke, S. Nerreter, A. Büttner, K. D. Langer, and J. W. Walewski. White light wireless transmission at 200+ Mb/s net data rate by use of discrete-multitone modulation. *IEEE Photonics Technology Letters*, 21(20):1511–1513, 2009.

28. A. M. Khalid, G. Cossu, R. Corsini, P. Choudhury, and E. Ciaramella. 1-Gb/s transmission over a phosphorescent white LED by using rate-adaptive discrete multitone modulation. *IEEE Photonics Journal*, 4(5):1465–1473, 2012.

29. F. M. Wu, C. T. Lin, C. C. Wei, C. W. Chen, H. T. Huang, and C. H. Ho. 1.1-Gb/s White-LED-based visible light communication employing carrier-less amplitude and phase modulation. *IEEE Photonics Technology Letters*, 24(19):1730–1732, 2012.

30. X. Qu, S. C. Wong, and K. T. Chi. Temperature measurement technique for stabilizing the light output of RGB LED lamps. *IEEE Transactions on Instrumentation and Measurement*, 59(3): 661–670, 2010.

31. J. Hasan and S. S. Ang. A high-efficiency digitally controlled RGB driver for LED pixels. *IEEE Transactions on Industry Applications*, 47(6):2422–2429, 2011.

32. S. K. Ng, K. H. Loo, Y. M. Lai and C. K. Tse. Color control system for RGB LED with application to light sources suffering from prolonged aging. *IEEE Transactions on Industrial Electronics*, 61(4):1788–1798, 2014.

33. V. Jungnickel, J. Vucic, and K. D. Langer. High-speed optical wireless communications. In *Proceedings Optical Fiber Communication Conference*, San Francisco, CA, pp. 9–13, 2010.

34. C. H. Yeh, Y. L. Liu, and C. W. Chow. Real-time white-light phosphor-LED visible light communication (VLC) with compact size. *Optics Express*, 21(22):26192–26197, 2013.

35. G. Cossu, A. M. Khalid, P. Choudhury, R. Corsini, and E. Ciaramella. 3.4 Gbit/s visible optical wireless transmission based on RGB LED. *Optics Express*, 20(26):B501–B5060, 2012.

36. S. Muthu, F. J. P. Schuurmans, and M. D. Pashley. Red, green, and blue LEDs for white light illumination. *IEEE Journal on Selected Topics in Quantum Electronics*, 8(2):333–338, 2002.

37. M. Gilewski and A. Karpiuk. An electronic control of light RGB LEDs. *Przegląd Elektrotechniczny*, 84(8):194–198, 2008.

38. F. C. Wang, C. W. Tang, and B. J. Huang. Multivariable robust control for a red-green-blue LED lighting system. *IEEE Transactions on Power Electronics*, 25(2):417–428, 2010.

39. C. C. Hung, Y. H. Li, and P. H. Yang. Application of the mechanism design to develop the RGB LEDs color mixing. *International Journal of Photoenergy*, 2015:876364, 2015.

40. Y. Wang, Y. Wang, N. Chi, J. Yu, and H. Shang. Demonstration of 575-Mb/s downlink and 225-Mb/s uplink bi-directional SCM-WDM visible light communication using RGB LED and phosphor-based LED. *Optics Express*, 21(1):1203–1208, 2013.

41. F. M. Wu, C. T. Lin, C. C. Wei, C. W. Chen, Z. Y. Chen, H. T. Huang, and S. Chi. Performance comparison of OFDM signal and CAP signal over high capacity RGB-LED-based WDM visible light communication. *IEEE Photonics Journal*, 5(4):7901507, 2013.

42. Y. Wang, X. Huang, L. Tao, J. Shi, and N. Chi. 4.5-Gb/s RGB-LED based WDM visible light communication system employing CAP modulation and RLS based adaptive equalization. *Optics Express*, 23(10):13626–13633, 2015.

43. R. Li, Y. Wang, C. Tang, Y. Wang, H. Shang, and N. Chi. Improving performance of 750-Mb/s visible light communication system using adaptive Nyquist windowing. *Chinese Optics Letters*, 11(8):080605, 2013.

44. S. H. Chen and C. W. Chow. Hierarchical scheme for detecting the rotating MIMO transmission of the in-door RGB-LED visible light wireless communications using mobile-phone camera. *Optics Communications*, 335:189–193, 2015.

45. P. Luo, M. Zhang, Z. Ghassemlooy, H. Le Minh, H. M. Tsai, X. Tang, L. C. Png, and D. Han. Experimental demonstration of RGB LED-based optical camera communications. *IEEE Photonics Journal*, 7(5):7904212, 2015.

46. A. J. Lowery. Design of arrayed-waveguide grating routers for use as optical OFDM demultiplexers. *Optics Express*, 18(13):14129–14143, 2010.

47. R. Mesleh, H. Elgala, and H. Haas. On the performance of different OFDM based optical wireless communication systems. *IEEE/OSA Journal of Optical Communications and Networking*, 3(8):620–628, 2011.

48. J. Y. Sung, C. W. Chow, and C. H. Yeh. Dimming-discrete-multi-tone (DMT) for simultaneous color control and high speed visible light communication. *Optics Express*, 22(7):620–628, 2014.

49. J. Y. Sunga, C. H. Yehb, C. W. Chowa, W. F. Lina, Y. Liu. Orthogonal frequency-division multiplexing access (OFDMA) based wireless visible light communication (VLC) system. *Optics Communications*, 355(15):261–268, 2015.

50. M. S. A. Mossaad, S. Hranilovic, and L. Lampe. Visible light communications using OFDM and multiple LEDs. *IEEE Transactions on Communications*, 63(11):4304–4313, 2015.

51. D. Tsonev, H. Chun, S. Rajbhandari, J. J. D. McKendry, S. Videv, E. Gu, M. Haji et al. A 3-Gb/s single-LED OFDM-based wireless VLC link using a Gallium Nitride μLED. *IEEE Photonics Technology Letters*, 26(7):637–640, 2014.

52. M. H. Kim, M. F. Schubert, Q. Dai, J. K. Kim, E. F. Schubert, J. Piprek, and Y. Park. Origin of efficiency droop in GaN-based light-emitting diodes. *Applied Physics Letters*, 91:183507, 2007.

53. S. Choi, H. J. Kim, S. S. Kim, J. Liu, J. Kim, J. H. Ryou, R. D. Dupuis, A. M. Fischer, and F. A. Ponce. Improvement of peak quantum efficiency and efficiency droop in III-nitride visible light-emitting diodes with an InAlN electron-blocking layer. *Applied Physics Letters*, 96:221105, 2010.

54. Y. K. Kuo, M. C. Tsai, S. H. Yen. Numerical simulation of blue InGaN light-emitting diodes with polarization-matched AlGaInN electron-blocking layer and barrier layer. *Optics Communications*, 282:4252–4255, 2009.

55. C. H. Wang, C. C. Ke, C. Y. Lee, S. P. Chang, W. T. Chang, J. C. Li, Z. Y. Li, H. C. Yang, H. C. Kuo, T. C. Lu, and S. C. Wang. Hole injection and efficiency droop improvement in InGaN/GaN light-emitting diodes by band-engineered electron blocking layer. *Applied Physics Letters*, 97:261103, 2010.

56. Y. Y. Zhang and Y.-A. Yin. Performance enhancement of blue light-emitting diodes with a special designed AlGaN/GaN superlattice electron-blocking layer. *Applied Physics Letters*, 99:221103, 2011.

57. H. Chun, P. Manousiadis, S. Rajbhandari, D. A. Vithanage, G. Faulkner, D. Tsonev, J. J. D. McKendry et al. Visible light communication using a blue GaN μLED and fluorescent polymer color converter. *IEEE Photonics Technology Letters*, 26(20):2035–2038, 2014.

58. J. Piprek. Efficiency droop in nitride-based light-emitting diodes. *Physica Status Solidi A*, 207(10):2217–2225, 2010.

59. M. Atef, R. Swoboda, and H. Zimmermann. Real-time 1.25-Gb/s transmission over 50-m SI-POF using a green laser diode. *IEEE Photonics Technology Letters*, 24(15):1331–1333, 2012.

60. B. Janjua, H. M. Oubei, J. R. R. Durán, T. K. Ng, C. T. Tsai, H. Y. Wang, Y. C. Chi, H. C. Kuo, G. R. Lin, J. H. He, and B. S. Ooi. Going beyond 4 Gbps data rate by employing RGB laser diodes for visible light communication. *Optics Express*, 23(18):23302–23309, 2015.

61. Y. C. Chi, D. H. Hsieh, C. Y. Lin, H. Y. Chen, C. Y. Huang, J. H. He, B. Ooi, S. P. DenBaars, S. Nakamura, H. C. Kuo and G. R. Lin. Phosphorous diffuser diverged blue laser diode for indoor lighting and communication. *Scientific Reports*, 5:18690, 2015.

62. C. Lee, C. Shen, H. M. Oubei, M. Cantore, B. Janjua, T. K. Ng, R. M. Farrell et al. 2 Gbit/s data transmission from an unfiltered laser-based phosphor-converted white lighting communication system. *Optics Express*, 23(23):29779–29787, 2015.

63. H. M. Oubei, J. R. Duran, B. Janjua, H. Y. Wang, C. T. Tsai, Y. C. Chi, T. K. Ng et al. 4.8 Gbit/s 16-QAM-OFDM transmission based on compact 450-nm laser for underwater wireless optical communication. *Optics Express*, 23(18):23302–23309, 2015.

64. J. R. D. Retamal, H. M. Oubei, B. Janjua, Y. C. Chi, H. Y. Wang, C. T. Tsai, T. K. Ng et al. 4-Gbit/s visible light communication link based on 16-QAM OFDM transmission over remote phosphor-film converted white light by using blue laser diode. *Optics Express*, 23(26):33656–33666, 2015.

65. G. Hu, C. Chen, Z., and Q. Chen. Free-space optical communication using visible light. *Journal of Zhejiang University SCIENCE A*, 8(2):186–191, 2007.

66. F. Hanson and S. Radic. High bandwidth underwater optical communication. *Applied Optics*, 47(2):277–283, 2008.

67. C. Y. Chen, P. Y. Wu, H. H. Lu, Y. P. Lin, J. Y. Wen, and F. C. Hu. Bidirectional 16-QAM OFDM in-building network over SMF and free-space VLC transport. *Optics Letters*, 38(13):2345–2347, 2013.

68. S. Singh and R. Bharti. 163m/10Gbps 4QAM-OFDM visible light communication. *International Journal of Engineering and Technical Research (IJETR)*, 2(6):225–228, 2014.

69. R. H. Horng, B. R. Wu, C. H. Tien, S. L. Ou, M. H. Yang, H. C. Kuo, and D. S. Wuu. Performance of GaN-based light-emitting diodes fabricated using GaN epilayers grown on silicon substrates. *Optics Express*, 22(1):A179–A187, 2014.

70. S. Watson, M. Tan, S. P. Najda, P. Perlin, M. Leszczynski, G. Targowski, S. Grzanka, and A. E. Kelly. Visible light communications using a directly modulated 422 nm GaN laser diode. *Optics Letters*, 38(19):3792–3794, 2013.

71. Y. C. Chi, D. H. Hsieh, C. T. Tsai, H. Y. Chen, H. C. Kuo, and G. R. Lin. 450-nm GaN laser diode enables high-speed visible light communication with 9-Gbps QAM-OFDM. *Optics Express*, 23(10):13051–13059, 2015.

72. C. Lee, C. Zhang, M. Cantore, R. M. Farrell, S. H. Oh, T. Margalith, J. S. Speck, S. Nakamura, J. E. Bowers, and S. P. DenBaars. 4 Gbps direct modulation of 450 nm GaN laser for high-speed visible light communication. *Optics Express*, 23(12):16232–16237, 2011.

73. A. Neumann, J. J. Wierer, Jr., W. Davis, Y. Ohno, S. R. J. Brueck, and J. Y. Tsao. Four-color laser white illuminant demonstrating high color-rendering quality. *Optics Express*, 19(S4):A982–A990, 2011.

74. S. Soltic and A. Chalmers. Optimization of laser-based white light illuminants. *Optics Express*, 21(7):8964–8971, 2013.

75. H. Chun, S. Rajbhandari, D. Tsonev, G. Faulkner, H. Haas, and D. O'Brien. Visible light communication using laser diode based remote phosphor technique. In *Proceedings IEEE International Conference on Communication Workshop*, London, UK, pp. 1392–1397, 2015.

76. D. Tsonev, S. Videv, and H. Haas. Towards a 100 Gb/s visible light wireless access network. *Optics Express*, 23(2):1627–1637, 2015.

22

III-Nitride Semiconductor Single Photon Sources

22.1 Introduction

Single photon sources are a critical resources for quantum information science and technologies [1]. Although methods to generate on-demand single photons have been developed for many years on a wide variety of material platforms including atomic vapors, ion traps, crystals, and semiconductors, the quest for a practical single photon source is still ongoing [2]. The challenges include not only finding materials with appropriate properties to enable a precise control of the quantum properties of light, but developing a scalable manufacturing technology at both the device and system levels. In semiconductors, on-demand single photons can be generated from quantum dot (QD) structures with very high brightness and indistinguishability [3–5]. In QDs, the electrons and holes are prohibited from moving freely, resulting in quantized energy states, each of which can be occupied by up to two electrons or holes with opposite spins. If the Coulomb interaction between the electrons and holes are strong enough such that two electrons and two holes form stable exciton and biexciton states, the energy of the biexciton can be spectrally discernible from that of the exciton such that single photon emission from the exciton (or the biexciton) state can be observed and utilized. The stability of excitons is, therefore, important for semiconductor QDs to operate effectively and efficiently as single photon sources. The exciton stability is characterized by its binding energy, which depends on the effective masses of the electrons and holes, the dielectric constant of the QD materials, and the QD potential profile. In bulk III-nitride semiconductors, the exciton binding energy is 28 meV. In QDs, this value can be even larger due to quantum confinement [6]. In comparison, the exciton binding energy in GaAs or InP is only a few meV. Therefore, excitons are much more stable in III-nitride QDs at an elevated temperature. Observation of single photon emission in III-nitride QDs has been made at room temperature by several groups [7,8]. Theoretical prediction of exciton binding energy as large as 1.4 eV has been shown in InN nanowires [9]. The goal of this chapter is to discuss the potentials and challenges for III-nitride QDs as a material of choice for single photon sources for practical applications. The large band gap tunability of the InGaN alloy, between 0.36 and 1.9 μm [10], makes GaN-based single photon sources technologically relevant, both for free-space [11,12] and fiber-based [13,14] applications. We will specifically focus on the aspect of manufacturing scalability which is often overlooked but is critical from a technology point of view.

22.2 Scalable III-Nitride QD Structures

Semiconductor technology is widely recognized as being extremely scalable both from the system complexity and cost perspectives. Moreover, many key functionalities needed for a quantum system including on-demand single photon and entangled photon sources [5,15], quantum memories [16], and quantum gates [17] have all been realized using semiconductor QDs, with performance comparable to, if not better than other technology contenders such as atomic ensembles and nitrogen-vacancy centers in diamonds. Yet, the existing QD-based technology using self-assembly strain-induced QD fabrication processes is not readily scalable. Quantum dots fabricated by these processes often have random locations and exhibit large inhomogeneity in both the static and dynamic properties of QDs [18]. The first critical question to address is, therefore, to develop a scalable fabrication process for III-nitride QDs.

In III-nitride materials, what accompanies the large exciton binding energy is the small exciton Bohr radius (~3 nm). Thus, small QDs are needed to achieve the necessary quantum confinement, if size quantization is used. However, this is not only difficult but can result in a very large inhomogeneous broadening. Hence, to achieve a scalable III-nitride QD structure, we must introduce additional confinement mechanisms. Recently, we have shown that strain-induced quantum confinement can be an effective quantization mechanism [19,20]. If the strain in a compressively-strained III-N nanostructure is relaxed at the surface, the nonuniform strain relaxation profile leads to a hyperbolic secant confinement potential to effectively keep the excitons at the center of the nanostructure, even with a dimension much greater than the exciton Bohr radius. To see an example, we consider a disk-shaped $In_xGa_{1-x}N$ nanostructure with a thickness of 3 nm, a diameter of 30 nm and is sandwiched between the top and bottom by GaN. The nanodisk is otherwise not constrained on the sidewall, and its internal strain is allowed to relax along the perimeter. Assuming the nanodisk is perpendicular to the crystal [0001] direction, the band structure will be modified due to the relaxation of strain via the strong piezoelectric field in InGaN. Figure 22.1a shows the potential profile caused by strain relaxation in an $In_{0.15}Ga_{0.85}N$ nanodisk obtained from two different models: the 1D strain relaxation model classically [20] and the 3D quantum mechanical model using the k·p model. The classical model assumes a monolayer of InGaN sandwiched in GaN. All atoms are connected by "springs." The potential profile $\phi(r)$ can be shown to follow a simple relationship as follows.

$$\phi(r) = B_m(1 - \mathrm{sech}(\kappa r))$$

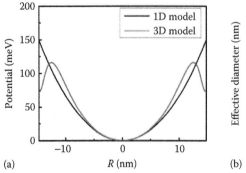
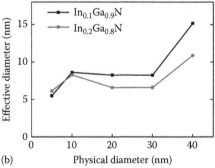

FIGURE 22.1 (a) The strain-induced potential profile in a 30 nm diameter, 3 nm thick $In_{0.15}Ga_{0.85}N$ nanodisk determined both quantum mechanically (3D model) and classically (1D model). (b) The effective QD diameter as a function of the nanodisk diameter in two nanodisks with different indium compositions.

where B_m is the band gap difference between fully strained and fully relaxed InGaN, and $1/\kappa$ is the characteristic length of strain relaxation. $1/\kappa$ can be obtained experimentally by measuring the nanodisk emission wavelength as a function of its diameter [21].

In spite of the simplicity of the 1D classical model, the results agree reasonably well with the quantum mechanical calculations (3D model) except near the edge of the nanodisk (Figure 22.1a). The strain-induced potential profile helps confine both the electrons and holes at the center of the nanodisk even with a nanodisk diameter much greater than the exciton Bohr radius. To see this, we calculated the diameter of nanodisks needed to achieve the same degree of quantum confinement without the piezoelectric field. It is seen in Figure 22.1b that the compressively strained $In_{0.1}Ga_{0.9}N$ nanodisk gives the exciton an equivalent confinement potential as a 7 nm diameter nanodisk without the piezoelectric field. Hence, the strain-induced potential not only relieves the challenging demand of fabricating small QDs but helps increase radiative recombination efficiency of the exciton by increasing the distance to the surface of the QD on which non-radiative recombinations can be significant [22]. Experimentally, single photon emission has been successfully demonstrated in a 3 nm thick $In_{0.15}Ga_{0.85}N$ nanodisk with a diameter up to 35 nm, which is 12 times the exciton Bohr radius in bulk GaN [23].

The nanodisk QD geometry can be easily realized from a scalable top-down fabrication process using lithography and etching as illustrated in Figure 22.2 [24]. Nickel or chrome is used as the etch mask. For electrically pumped devices, a thin layer of Si_3N_4 is first conformally deposited on the nanopillar before the sample is planarized for subsequent deposition of the metal contact [25]. The SiN layer helps electrical insulation between p-GaN and n-GaN. The top-down fabrication process allows the QDs to have precisely controlled locations, dimensions, and shapes. Electron-beam lithography generally can achieve a ± 1.5 nm or better accuracy. This corresponds to a 10 meV or smaller inhomogeneous linewidth in GaN QDs. The top-down process also enables special QD structures that are otherwise difficult to achieve using the bottom-up process. An example is the elliptically-shaped QD, which can be used to generate linearly polarized photons or single photon qubits [26,27]. This will be discussed in the next section.

Although etched QDs have long been considered non-feasible due to their large surface-to-volume ratios, the strain-induced quantum confinement in a III-nitride nanodisk can help keep the exciton away from the surface. Because the surface recombination velocity in GaN is 3–4 orders of magnitude lower than that in III-As materials [28], single photon emission has been observed in an InGaN nanodisk even without any intentional surface passivation or coating (the latter is still necessary if the device reliability is considered to avoid any oxidation of the QD surface). The sample was grown on a *c*-plane sapphire substrate using metal-organic chemical vapor deposition with typical growth conditions used for GaN LEDs. Figure 22.3 shows the locations of electrons and holes inside the nanodisk

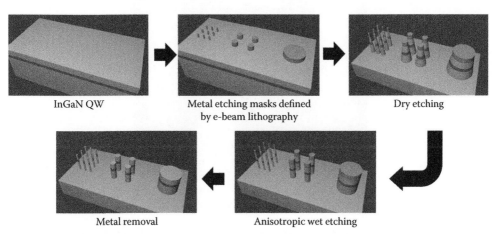

InGaN QW   Metal etching masks defined by e-beam lithography   Dry etching

Metal removal   Anisotropic wet etching

FIGURE 22.2 The illustration of the top-down fabrication process for InGaN QD nanodisks of different diameters sandwiched from top and bottom by GaN starting from a single InGaN quantum well (QW) sample.

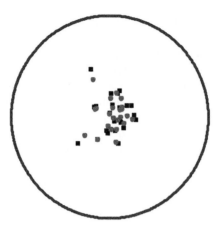

FIGURE 22.3 The distribution of electrons (squares) and holes (circles) in 30 identical 3 nm thick $In_{0.14}Ga_{0.86}N$ nanodisk with a diameter of 30 nm.

calculated from 30 nominally identical structures. The calculations considered the random alloy fluctuation, indium compositional variation along the growth direction and well width fluctuation as well as Coulomb interactions between electrons and holes. The indium compositional fluctuation due to random alloys tends to shift electrons and holes from the center of the QD to the indium-rich regions, which should be randomly distributed across the entire nanodisk. But the strain-induced potential profile helps keep the excitons close to the center of the nanodisks.

22.3 Elliptical QDs as Single Photon Qubit Sources

In this section, we will show one of the key advantages of QDs made from the top-down approach [29]. We will show that single photon qubit sources or single photons with deterministic linear polarization can be realized using elliptically shaped InGaN QDs. Shown in Figure 22.4 are the top view of four groups of elliptical QD nanopillars of different orientations, all fabricated on the same chip with one single patterning process. Although special growth techniques such as selective area epitaxy allow non-circular QDs to

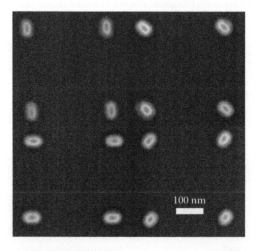

FIGURE 22.4 The scanning electron micrographs of four groups of elliptical QDs oriented along different directions, all fabricated on the same sample. The nominal dimensions of all QDs are 22 nm × 36 nm. (From Teng, C.-H. et al., *Appl. Phys. Lett.*, 107, 191105, 2015. With permission.)

be fabricated by the bottom-up approach, the control is usually not as good and precise as the top-down approach. Moreover, the density of the QDs in the top-down approach is only limited by lithography. This difference can be important in applications that require multiple non-circular QDs to be placed in close proximity.

As the strain field becomes asymmetric in an elliptical shaped QD, the heavy hole, light hole and split-off band are mixed and result in an exciton dipole moment along the direction defined by the long axis of the ellipse. The mixing with the split-off band is important to keep the direction of the dipole moment deterministic in the QD plane (which will be referred to as the x–y direction in the following discussions). The reason is that the z-component of the wavefunction in the split-off band can cancel out the z-component in the light hole band and reduce the uncertainly of light emission in the z direction, which in turn reduces the degree of polarization in the x–y plane.

For studying the emission polarization properties of an elliptical InGaN/GaN QD nanodisk, we performed band structure calculations using an eight-band k·p model [30]. The Coulomb interaction (excitonic effect) was ignored. We were mainly interested in how the direction of the electron-hole dipole moment changes with respect to the shape of the QD, which is not very sensitive to the excitonic effect. Specifically, the three polarization components of the QD emission are proportional to the following three matrix elements:

$$x\text{-polarized:} \left|M_x\right|^2 = \left|\Psi_e \mid p_x \mid \Psi_h\right|^2 \cong \left|S\mid p_x \mid X\right|^2 \left|\varphi_e \mid \varphi_{h,x}\right|^2$$

$$y\text{-polarized:} \left|M_y\right|^2 = \left|\Psi_e \mid p_y \mid \Psi_h\right|^2 \cong \left|S\mid p_y \mid Y\right|^2 \left|\varphi_e \mid \varphi_{h,y}\right|^2$$

$$z\text{-polarized:} \left|M_z\right|^2 = \left|\Psi_e \mid p_z \mid \Psi_h\right|^2 \cong \left|S\mid p_z \mid Z\right|^2 \left|\varphi_e \mid \varphi_{h,z}\right|^2$$

where the z direction is defined along the crystal [0001] direction; p_x, p_y, and p_z are the quantum mechanical momentum operators; $\Psi_e = \varphi_e \cdot S$ and $\Psi_h = \varphi_{h,x} \cdot X + \varphi_{h,y} \cdot Y + \varphi_{h,z} \cdot Z$ are the electron and hole wavefunctions, respectively; S, X, Y, and Z are the Bloch bases; φ_e and φ_h are the envelope functions of the electron and hole states, respectively. The degree of linear polarization (DLP) of the QD emission can be determined by

$$\text{DLP} = \frac{\left|M_y\right|^2 - \left|M_x\right|^2}{\left|M_y\right|^2 + \left|M_x\right|^2}$$

The results are shown in Figure 22.5 for various QD materials. In all materials, the DLP of the QD emission increases with the aspect ratio of the ellipse, which is defined as the ratio of the long axis to the short axis. In QD materials with a split-off band farther apart from the heavy and light holes, characterized by the split-off energy, the DLP is generally lower due to a weaker mixing between the light hole and split-off bands. InGaN exhibits a very small split-off energy compared to other compound semiconductors. Thus, the strong valence band mixing from all three bands leads to a very high DLP in elliptical QDs with a large aspect ratio. Experimentally, this has been confirmed as shown by the data points in Figure 22.5. A DLP as high as 0.99 has been observed in the experiment from elliptical QDs with an aspect ratio approaching 2. Elliptical InGaN/GaN QD nanodisks are, therefore, a potential source for single photon qubits.

The polarization direction of the single photon qubit source can be tuned by changing the QD orientation. As shown in Figure 22.6, the polarization direction tracks the elliptical QD orientation, defined as the direction of the ellipse's long axis, very well. To rule out any crystallographic effect, we measured two different sets of QD orientations: (0°, 45°, 90°, 135°) and (0°, 30°, 60°, 90°). This is to differentiate any effect related to the six-fold crystallographic symmetry with respect to the surface normal of the QD nanodisk. We observed that at all angles, there were no discernible differences in how the direction of the QD polarization followed the QD orientation. The photon antibunching was also confirmed for different orientation QDs as shown in the insets. The g² curves confirm the nature of single photon emission from both 0°- and 90°- oriented elliptical QDs.

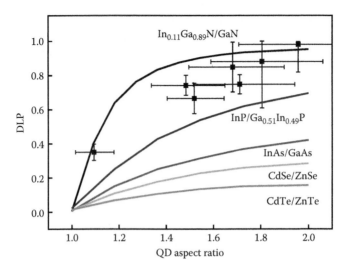

FIGURE 22.5 The degree of linear polarization (DLP) of elliptical QD emission for different QD materials and ellipse aspect ratio. The solid curves are from calculations. The data points are from experiments. All QD nanodisks have a thickness of 3 nm. The split-off energy for InGaN, InP, InAs, CdSe, and CdTe are 16, 108, 390, 420, and 930 meV, respectively. (From Teng, C.-H. et al., *Appl. Phys. Lett.*, 107, 191105, 2015. With permission.)

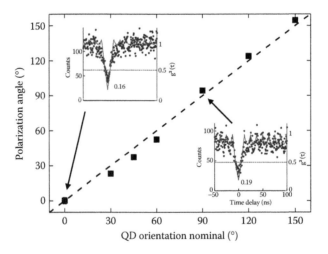

FIGURE 22.6 The direction of the linear polarization (polarization angle) from different elliptical InGaN QDs of different orientations. All QDs have nominal dimensions of 22 nm × 36 nm. The inset shows the second-order photon correlation measurements for two QDs oriented at 0° and 90°. (From Teng, C.-H. et al., *Appl. Phys. Lett.*, 107, 191105, 2015. With permission.)

The elliptical InGaN QDs made from the top-down approach can be applied to quantum key distribution (QKD). Because the polarization control is an intrinsic property of the elliptical QD and does not depend on any external structure [31], multiple QDs generating different polarizations can be placed in close proximity. Electrically driven single photon devices could also be readily made with proper planarization and metal contacts [25]. Figure 22.7 shows a potential design for an electrically driven QKD transmitter. It consists of four elliptical QDs generating 0°, 45°, 90°, and 135° polarizations (Figure 22.4) as two orthogonal bases for the polarization encoding following the BB84 protocol [32]. The emission wavelengths from these four QDs should be randomized (e.g., via temperature control using a random signal)

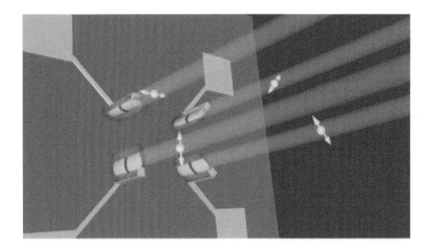

FIGURE 22.7 Schema of four elliptical QD single photon emitters generating four different polarizations as two basis sets for polarization encoding, for quantum key distribution. Quantum dots are embedded in nanopillars that are coated with an insulating layer. The top surface of the nanopillars is in contact with the top electrodes. The bottom electrodes are not shown in the graph. The polarized single photons are coupled by a common lens into the communication channel.

to avoid any association of the photon polarization to the wavelength. The proximity of these QDs can also greatly simplify the coupling of the QD output to a single communication channel.

22.4 Summary

In summary, group-III nitride semiconductor QDs have only been explored in recent years but have been shown to possess great potentials for quantum photonic applications. The large exciton binding energy helps to preserve quantum phenomena even at the room temperature [33,34]. Recently, InGaN nanostructures with a high indium composition have been successfully demonstrated, making III-nitride quantum light sources to potentially work directly at telecommunication wavelengths. These discoveries, in combination, with the mature manufacturing infrastructure already established for GaN LEDs and lasers, makes GaN a very attractive platform for quantum science and technology applications. In this chapter, a scalable GaN QD process was introduced based on the top-down approach which can produce elliptical QD structures that are otherwise difficult to achieve from a bottom-up process. Unique advantages of this approach in making elliptical QDs with a controllable and near-unity degree of linear polarization were overviewed. This work was financially supported by DARPA and NSF. We thank all students (in alphabetical order) who have contributed to the results presented in this chapter: Brandon Demory, Tyler Hill, Luke Lee, Chu-hsiang Teng, and Lei Zhang.

References

1. A. J. Shields, Semiconductor quantum light sources. *Nature Photonics*, 1(4): 215–223, 2007.
2. C. Santori, D. Fattal, and Y. Yamamoto, *Single-Photon Devices and Applications*. Wiley-VCH, 2010.
3. N. Somaschi et al., Near optimal single photon sources in the solid state. *Nature Photonics*, 10(2): 1–6, 2016.
4. Y.-M. He et al., On-demand semiconductor single-photon source with near-unity indistinguishability. *Nature Nanotechnology*, 8(3): 213–217, 2013.

5. M. Müller, S. Bounouar, K. D. Jöns, M. Glässl, and P. Michler, On-demand generation of indistinguishable polarization-entangled photon pairs. *Nature Photonics*, 8(3): 224–228, 2014.

6. P. Ramvall, S. Tanaka, S. Nomura, P. Riblet, and Y. Aoyagi, Observation of confinement-dependent exciton binding energy of GaN quantum dots. *Applied Physics Letters*, 73(8): 1104, 1998.

7. S. Deshpande, T. Frost, A. Hazari, and P. Bhattacharya, Electrically pumped single-photon emission at room temperature from a single InGaN/GaN quantum dot. *Applied Physics Letters*, 105(14): 141109, 2014.

8. M. J. Holmes, K. Choi, S. Kako, M. Arita, and Y. Arakawa, Room-temperature triggered single photon emission from a III-nitride site-controlled nanowire quantum dot. *Nano Letters*, 14(2): 982–986, 2014.

9. D. Bayerl and E. Kioupakis, Visible-wavelength polarized-light emission with small-diameter InN nanowires. *Nano Letters*, 14(7): 3709–3714, 2014.

10. Y. L. Chang, Z. Mi, and F. Li, Photoluminescence properties of a nearly intrinsic single InN nanowire. *Advanced Functional Materials*, 20: 4146, 2010.

11. R. J. Hughes et al., Practical free-space quantum key distribution over 10 km in daylight and at night. *New Journal of Physics*, 4: 1–14, 2002.

12. G. Vallone et al., Free-space quantum key distribution by rotation-invariant twisted photons. *Physical Review Letters*, 113(6): 1–5, 2014.

13. K. Takemoto et al., Quantum key distribution over 120 km using ultrahigh purity single-photon source and superconducting single-photon detectors. *Scientific Reports*, 5: 14383, 2015.

14. B. Korzh et al., Provably secure and practical quantum key distribution over 307 km of optical fibre. *Nature Photonics*, 9(3): 163–168, 2015.

15. H. Jayakumar, A. Predojević, T. Kauten, T. Huber, G. S. Solomon, and G. Weihs, Time-bin entangled photons from a quantum dot. *Nature Communications*, 5: 1, 2014.

16. H. P. Wagner, A. Schätz, W. Langbein, J. M. Hvam, and A. L. Smirl, Interaction-induced effects in the nonlinear coherent response of quantum-well excitons. *Physical Review B*, 60(7): 4454–4457, 1999.

17. X. Li et al., An all-optical quantum gate in a semiconductor quantum dot. *Science*, 301(5634): 809–11, 2003.

18. D. Bimberg, M. Grundmann, and N. N. Ledentsov, *Quantum Dot Heterostructures*. Chichester: Wiley, 1999.

19. L. Zhang, T. A. Hill, C.-H. Teng, B. Demory, P.-C. Ku, and H. Deng, Carrier dynamics in site- and structure-controlled InGaN/GaN quantum dots. *Physical Review B*, 90(24): 245311, 2014.

20. L. Zhang, L. K. Lee, C.-H. Teng, T. Hill, P.-C. Ku, and H. Deng, How much better are InGaN/GaN nanodisks than quantum wells—oscillator strength enhancement and changes in optical properties. *Applied Physics Letters*, 104(5): 51116, 2014.

21. C.-H. Teng, L. Zhang, H. Deng, and P.-C. Ku, Strain-induced red-green-blue wavelength tuning in InGaN quantum wells. *Applied Physics Letters*, 108: 71104, 2016..

22. L. K. Lee, L. Zhang, H. Deng, and P. C. Ku, Room-temperature quantum-dot-like luminescence from site-controlled InGaN quantum disks. *Applied Physics Letters*, 99(26): 263105, 2011.

23. L. Zhang et al., Single photon emission from site-controlled InGaN/GaN quantum dots. *Applied Physics Letters*, 103(19): 192114, 2013.

24. L. K. Lee and P. C. Ku, Fabrication of site-controlled InGaN quantum dots using reactive-ion etching. *Physica Status Solidi (c)*, 9: 609, 2012.

25. L. Zhang, C. H. Teng, P. C. Ku, and H. Deng, Site-controlled InGaN/GaN single-photon-emitting diode. *Applied Physics Letters*, 108(15): 2–6, 2016.

26. T. Jemsson, H. Machhadani, K. F. Karlsson, C.-W. Hsu, and P.-O. Holtz, Linearly polarized single photon antibunching from a site-controlled InGaN quantum dot. *Applied Physics Letters*, 105(8): 81901, 2014.

27. A. Lundskog et al., Direct generation of linearly polarized photon emission with designated orientations from site-controlled InGaN quantum dots. *Light: Science & Applications*, 3(1) e139, 2014.

28. J. B. Schlager et al., Steady-state and time-resolved photoluminescence from relaxed and strained GaN nanowires grown by catalyst-free molecular-beam epitaxy Steady-state and time-resolved photoluminescence from relaxed and strained GaN nanowires grown by catalyst-free molecula. *Journal of Applied Physics*, 103: 124309, 2008.

29. C.-H. Teng, L. Zhang, T. A. Hill, B. Demory, H. Deng, and P.-C. Ku, Elliptical quantum dots as on-demand single photons sources with deterministic polarization states. *Applied Physics Letters*, 107(19): 191105, 2015.

30. S. L. Chuang, *Physics of Photonic Devices*. New York: Wiley, 2009.

31. S. Strauf, N. G. Stoltz, M. T. Rakher, L. a. Coldren, P. M. Petroff, and D. Bouwmeester, High-frequency single-photon source with polarization control. *Nature Photonics*, 1(12): 704–708, 2007.

32. C. H. Bennett and G. Brassard, Quantum cryptography: Public key distribution and coin tossing. *Theoretical Computer Science*, 560(P1): 7–11, 2014.

33. P. Bhattacharya, T. Frost, S. Deshpande, M. Z. Baten, A. Hazari, and A. Das, Room temperature electrically injected polariton laser. *Physical Review Letters*, 112(23): 29–31, 2014.

34. A. Das, P. Bhattacharya, J. Heo, A. Banerjee, and W. Guo, Polariton Bose-Einstein condensate at room temperature in an Al(Ga)N nanowire-dielectric microcavity with a spatial potential trap. *Proceedings of the National Academy of Sciences of the United States of America*, 110(8): 2735–2740, 2013.

Index

Note: Page numbers followed by f and t refer to figures and tables, respectively.

Printed and bound by CPI Group (UK) Ltd, Croydon, CR0 4YY

01/11/2024

01782603-0016